Physiology for Nursing Practice

Physiology for Nursing Practice

Edited by

SUSAN M. HINCHLIFF BA, SRN, RNT

Visiting Lecturer in Nursing Studies and Consultant to Distance Learning Centre, Department of Nursing and Community Health Studies, South Bank Polytechnic, London; and Visiting Lecturer, Royal College of Nursing, London

SUSAN E. MONTAGUE BSc (Hons), SRN, HVDip, RNT

Visiting Lecturer in Nursing Studies, Department of Nursing and Community Health Studies, South Bank Polytechnic, London

Baillière Tindall
London Philadelphia Toronto Sydney Tokyo

Baillière Tindall   24–28 Oval Road
W. B. Saunders   London NW1 7DX

The Curtis Center
Independence Square West
Philadelphia, PA 19106–3399, USA

55 Homer Avenue
Toronto, Ontario M8Z 4X6, Canada

Harcourt Brace Jovanovich Group (Australia) Pty Ltd.
32–52 Smidmore Street
Marrickville, NSW 2204, Australia

Harcourt Brace Jovanovich (Japan) Inc.
Ichibancho Central Building, 22–1 Ichibancho
Chiyoda-ku, Tokyo 102, Japan

© 1988 Baillière Tindall

First published 1988
Reprinted 1989 and 1991

British Library Cataloguing in Publication Data

Physiology for Nursing Practice
   1. Man. Physiology – For nursing
   I. Hinchliff, S. II. Montague, S.
   612′.0024613

ISBN 0–7020–1194–0

Printed in Great Britain by The Bath Press, Avon

Contributors

Jennifer A. Alison MSc, DipPhty, Research Physiotherapist, Royal Prince Alfred Hospital, Sydney, NSW 2000, Australia

Douglas Allan RGN, RMN, RCNT, Clinical Teacher, Institute of Neurological Studies, Glasgow

Jennifer R. P. Boore PhD, BSc(Hons), SRN, SCM, RNT, Professor, Department of Nursing Studies, University of Ulster, Coleraine, Northern Ireland

Margaret Clarke MPhil, BSc(Hons), SRN, RNT, Director, Institute of Nursing Studies, University of Hull, Hull

Susan M. Goodinson BSc(Hons), SRN, Lecturer in Nursing Studies, Department of Biochemistry, University of Surrey, Guildford, Surrey

Rosamund A. Herbert MSc, BSc(Hons), SRN, Thorn Lecturer in Nursing Studies, Department of Nursing Studies, King's College, University of London, London

Susan M. Hinchliff BA, SRN, RNT, Visiting Lecturer in Nursing Studies and Consultant to Distance Learning Centre, Department of Nursing and Community Health Studies, South Bank Polytechnic, London; and Visiting Lecturer to Royal College of Nursing, London

Michael Hunter PhD, BSc(Hons), Cert Ed, Senior Lecturer, University of Newcastle, New South Wales, Australia; Former Lecturer, University of Aston, Birmingham

Diana G. Langelaan BA, SRN, RNT, Former Lecturer in Nursing Studies, Department of Nursing and Community Health Studies, South Bank Polytechnic, London

Susan E. Montague BSc(Hons), SRN, HVDip, RNT, Visiting Lecturer in Nursing Studies, Department of Nursing and Community Health Studies, South Bank Polytechnic, London

Valerie M. Nie BSc (Hons), MIBiol, Cert Ed., Former Lecturer, University of Aston, Birmingham

Janet Stocks PhD, BSc(Hons), SRN, HVDip, Lecturer in Respiratory Physiology, Respiratory and Anaesthetic Unit, Institute of Child Health, London

Jennifer Storer MSc, BSc(Hons), SRN, HVDip, Former Health Visitor, Harrow Area Health Authority

Carolynn Williams BA(Hons), MSc, SRN, RNT, Senior Tutor, Nightingale School, St Thomas' Hospital, London

Acknowledgements

We would like to thank all our students who provided the inspiration for this book, and our colleagues and friends who so readily and ably contributed to it.

Sean Duggan was instrumental in the initiation and early stages of the project and to him we are grateful.

We are indebted to Rosemary Morris formerly of Baillière Tindall for all her positive support and encouragement. Her enthusiasm for the project helped us to complete it. Thanks are also due to Carrie Bennett formerly of Baillière Tindall for her interest and help. We also thank those who were asked to comment on the text for their thorough reviews and helpful suggestions, which so far as possible have been incorporated in the text. We are grateful to Mary Benstead for her professional expertise in the preparation of the artwork and for her friendly efficiency.

Rosemary Clarke of the South Bank Polytechnic library has been most patient and helpful to us in our requests for bibliographical information.

Finally, we would like to thank all those who through their advice, encouragement and skill have helped to produce this book.

Preface

As a result of teaching nurses for a number of years, we became increasingly aware of the need for a British textbook which presents a comprehensive review of physiology and also indicates how physiological principles can provide the basis for understanding patient care.

In our experience, pre-registration nursing students thirst for applications and clinical nursing examples when learning physiology. On the other hand, qualified nurses on post-basic courses often have rewarding experiences in class – as they suddenly understand why a patient has a certain problem and thus the biological rationale for his medical and nursing care.

There are many physiology textbooks available to our students, but none has completely fulfilled their requirements. Some books have little or no nursing application; some present anatomy separately from physiology, failing to relate structure to function; and there is no consensus among authors about the level of physiology suitable for practising nurses.

It was for these reasons that *Physiology for Nursing Practice* was written. In preparing the book, we had four main aims:

1. to enable readers to understand the principles and mechanisms of normal body function, so that they can appreciate how these mechanisms alter in illness;
2. to provide readers with a rational basis for assessing a patient's health problems and the planning, delivery and evaluation of care;
3. to convey a sense of enthusiasm about how the body works and to motivate readers to further their studies;
4. to express all of the above as simply as possible, but not at the expense of scientific accuracy.

Knowledge of any scientific discipline is continually being modified and extended through research. Every effort has been made to ensure that the physiology presented in this book reflects the consensus of opinion at the time of writing.

In fulfilling the above aims, we have compiled a textbook which explains how the body functions. The structure of the body (human anatomy) is presented only in sufficient detail to enable the reader to understand how it subserves function. Clinical applications are integrated with the text and not isolated at the end of each chapter, but the reader will not find detailed descriptions of patient care since we have not set out to teach nursing.

The contributors to this book – the majority of whom are nurses – were chosen because of their knowledge of biological sciences in relation to nursing. In addition, most are experienced in teaching physiology to nurses.

It is likely that readers will study the individual chapters in this book as they become relevant (for example, during a taught course and when they are extending their knowledge of a specific approach to, or aspect of, patient care) rather than reading the book straight through. For this reason, we have indicated important cross-references between chapters.

Rather than following the classical systemic approach, we have divided the book into six sections which logically illustrate the functioning of the body as a complex homeostatic system. These sections are:

1. Characteristics of Living Matter
2. Control and Co-ordination
3. Mobility and Support
4. Internal Transport
5. The Acquisition of Nutrients and Removal of Waste
6. Protection and Survival.

Each section is then sub-divided into chapters, all of which have the following features:

(a) Learning objectives are given, both to enable the reader rapidly to assess the content of the chapter and to provide guidance on specific learning outcomes;

(b) Throughout the text, key terms appear in

bold type, in the place where their definition or explanation is provided. We feel that this is more helpful than the inclusion of a glossary, since the terms are explained in context.

(c) The text is consistently illustrated with clear diagrams and flow charts, drawn by an experienced medical artist. Occasionally, photographs have been included in order to demonstrate clinical features.

(d) At the end of each chapter, we have included review questions in multi-choice format, to allow readers to test their knowledge and understanding of the text. Answers to the multi-choice questions are given, but the readers are encouraged to return first to the appropriate part of the chapter to check their comprehension.

(e) Suggestions for practical work are also given, where this can helpfully illustrate theoretical points made in the text. These do not require elaborate laboratory equipment or facilities, and in most cases the work can be done at home or undertaken during clinical experience.

(f) In order to avoid frequent interruptions of the text, references have been kept to a minimum, but each author has included those which are of particular interest or relevance. These should be easily available in nursing and medical libraries, and the readers are encouraged to further their knowledge by consulting these sources. We appreciate, however, that the physiological theory presented in this book has its origins in the work of many researchers whom (for reasons of brevity) we are unable to acknowledge. Each author has also made suggestions for further reading.

For convenience, throughout the text, reference either to the 'person' or 'patient' has been made rather than to the 'person/client/patient'. Similarly, the feminine gender has been used in referring to the nurse, and the masculine when reference to the patient is made.

The book is written using the International System of Units (S.I.). The S.I. units given are, as far as possible, in accordance with current clinical usage in the United Kingdom. A full explanation of S.I. units, including notations of presentation and conversions to Imperial units, can be found in the Appendix.

The Appendix also presents other information that we consider necessary to support the text and aid conceptual understanding. There is a comprehensive index.

Physiology for Nursing Practice has been written primarily for nurses, but we consider that other health professionals will also find its content helpful.

The level of physiology presented here goes beyond that traditionally found in books written for nurses. This is because we feel strongly that nurses require a thorough understanding of body function in order to practise intelligently and effectively, and so gain satisfaction from their work and improve standards of care. We are confident that they will gain just such an understanding from our book.

We hope that the book will be seen as a major learning resource by anyone involved in teaching nurses about the care of patients. It does not underestimate the abilities of student nurses in general training, and will form a stimulating core text for undergraduate nurses. We anticipate that trained nurses in clinical practice will also find it useful in extending their knowledge. It will be particularly appropriate for students on post-basic clinical courses, and especially those studying the revised curriculum for the Diploma in Nursing of the University of London, where it will provide a foundation for Unit 1 – 'The Human Organism; Unit 3 – 'The Application of Care; and Unit 6 – 'Nursing'.

Finally, whilst we have made every attempt to ensure that the content of this book meets our reader's needs, we would value comments to help us in the preparation of future editions.

SUSAN M. HINCHLIFF
SUSAN E. MONTAGUE

Contents

We would like to dedicate this book to our husbands, Alan and Philip, for their love, patience and support during its preparation, and to our children, Sarah and James Montague and Katy Hinchliff. All our children arrived during the gestation of this book and have brought us great happiness.

Introduction

The human body is composed of billions of cells, most of which are not in direct contact with the outside world. In health the co-ordinated activity of all these cells maintains optimum conditions for survival. For example, in order to function efficiently, each cell must remain within a relatively narrow range of temperature and pH; receive an adequate supply of oxygen and other nutrients; and dispose of the waste products of its metabolism. Each cell achieves this through dynamic exchange with its immediate environment, the thin film of fluid between it and other cells. The immediate environment of the cell is called the 'internal environment'. Homeostasis is the process by which the composition of the internal environment is kept within certain limits which are compatible with optimum function and health.

In the human body, groups of cells are specialised to form tissues and organ systems which are in direct contact with the external environment – for example, the lungs, gastrointestinal tract and kidneys. These cells allow the exchange between the internal and external environments which is essential to homeostasis.

This book describes how the body maintains homeostasis. It is divided into six sections. The first contains introductory material describing the characteristics of living matter. The remaining sections each present a major aspect of homeostasis.

Physiology for Nursing Practice is primarily concerned with normal body function, but we have indicated throughout the book how homeostasis becomes deranged in illness and how knowledge of normal function provides a basis for understanding abnormality and a rationale for patient care.

Section 1
The Characteristics of Living Matter

Section 1
The Characteristics of Living Matter

Carolynn Williams
Susan E. Montague

Learning objectives

After studying this section the reader should be able to

1. Describe the components of a 'typical' cell, relating structure to function.
2. Describe mechanisms by which substances are transported across the cytoplasmic membrane.
3. Describe the process of mitosis and explain its functional significance.
4. Compare and contrast the processes of mitosis and meiosis, after also studying Chapter 6.3.
5. Explain how DNA functions as the genetic material in controlling cellular protein synthesis.
6. Describe the mechanism of protein synthesis, briefly stating how abnormalities may occur.
7. Define the term 'tissue', and classify the major types of tissue which compose the human body.
8. Describe the two basic arrangements of tissues forming organs.
9. Define the term 'internal environment', and state the functional significance of the internal environment in higher animals.
10. State the basic requirements and function of a homeostatic mechanism.
11. Discuss the concept of homeostasis and its relationship with health.

THE ORGANIZATION OF LIVING MATTER

The entire substance of each living thing is called **protoplasm**. Protoplasm is composed essentially of water, organic substances, such as proteins, fats and carbohydrates, and inorganic substances, such as mineral salts. Protoplasm is an unstable system and can be maintained only by a continuous expenditure of energy. During life, chemical processes release energy from absorbed nutrients. Protoplasm is thus an open system through which energy flows by the various chemical reactions which comprise metabolism. Although some energy is required to maintain protoplasmic organization, a greater amount is needed for other activities characteristic of living things, such as growth, movement and reproduction.

Protoplasm is organized into cells, and these cells are the structural and functional units of life. Each cell belonging to an organism of a particular species contains, at some stage of its development, all the genetic material required to produce that organism.

A cell comprises both its outer limiting membrane and its contents. Typically it is a semifluid mass of microscopic dimensions – most mammalian cells are 7–20 µm in diameter – completely enclosed within a thin, selectively permeable, cytoplasmic membrane. The cell usually contains two distinct regions – the nucleus and the cytoplasm. The nucleus is enclosed by a nuclear

3

membrane and contains the chromatin (genetic material) and one or more dense areas called nucleoli. Within the cytoplasm are many organelles, such as the mitochondria, the Golgi region, the endoplasmic reticulum, lysosomes and centrioles.

The variety of different shapes assumed by cells is related to their particular function. Although many cells, because of surface tension forces, will assume a spherical shape if freed from restraining influences, other cells retain their shape under most conditions because of their characteristic cytoskeleton or framework of microtubules and/or microfilaments.

At birth the human baby has about 2×10^{12} cells. This immense number has come from a single fertilized egg, which, by repeated cell divisions, develops into a baby usually weighing more than 2.5 kg. Cells do not divide at the same rate and some cells stop dividing altogether at various points in the life cycle. The growth of an organism is not merely due to an increase in the number of cells, but also involves increase in cell size. A human adult has approximately 6×10^{13} cells.

The lifespan of different human cells varies. For example, some nerve cells may persist throughout life whereas a red blood cell lives for about 120 days. Extensive cell death is constantly occurring within the organism because of the wear and tear of existence and as a part of development. Some parts of the body require a constant replacement of cells: friction wears away the outer cells of the skin, and the movement of food along the gastrointestinal tract removes lining cells. This loss is made up by rapid cell division. Cells also undergo a process of ageing. At some point in the life cycle of most cells there is a decrease in the chemical activity of their enzymes, a slowing down of metabolism, a breakdown of cell substance and the formation of inert material. In certain cases parts of the cell may persist after its death, for example, the keratin of hair and finger- and toe-nails.

CELL STRUCTURE (Fig. 1.1)

The structure of cells is characterized by both membranous and non-membranous components. The limiting boundary of the cell, its outer barrier, is known as the cytoplasmic membrane, cell-surface membrane or plasma membrane. Within the limits of the cytoplasmic membrane there are structurally and functionally distinct units.

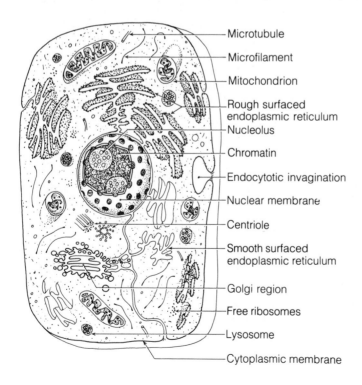

- Microtubule
- Microfilament
- Mitochondrion
- Rough surfaced endoplasmic reticulum
- Nucleolus
- Chromatin
- Endocytotic invagination
- Nuclear membrane
- Centriole
- Smooth surfaced endoplasmic reticulum
- Golgi region
- Free ribosomes
- Lysosome
- Cytoplasmic membrane

Figure 1.1 Diagram of the ultrastructure of a cell

Generally, cells have certain basic features in common, but there are exceptions. For example, the cell nucleus is a characteristic structure of most cells, but mature red blood cells do not contain nuclei.

In order to understand the general structure of the cellular subunits (organelles), it is useful to consider a brief definition of each. This is followed by a more detailed description.

The **cytoplasmic membrane** is the outer limiting boundary of each cell, separating the intracellular area from the extracellular environment.

The **nuclear membrane** is the outer limiting boundary of the nucleus, separating the nucleoplasm from the cytoplasm.

Cytoplasm is the cellular substance between the cytoplasmic membrane and the nuclear membrane.

The **nucleus**, defined by the nuclear membrane, is the cell structure that contains the genetic information.

Nucleoplasm is the cellular substance within the nuclear membrane.

Chromatin is the substance in the nucleus containing the chromosomes in an extended and diffuse state, permitting their maximum contact with the surrounding nucleoplasm.

Chromosomes are structures within the nucleus containing hereditary information in the form of the chemical substance deoxyribonucleic acid (DNA).

The **endoplasmic reticulum** is a network of membranous channels within the cytoplasm involved in the synthesis and transport of proteins.

Ribosomes, composed of ribonucleic acid (RNA) and protein, appear as free granules in the cytoplasm, giving endoplasmic reticulum a rough appearance.

Mitochondria are small cytoplasmic structures having outer and inner membranes which produce energy in the form of the chemical substance adenosine triphosphate (ATP).

The **Golgi region** is a collection of smooth membranes found in the cytoplasm, near the nucleus, which stores cell secretions.

Lysosomes are membranous sacs of varying size and shape which contain enzymes capable of breaking down protein, carbohydrate, DNA, RNA and other large molecules.

Microtubules are very small rods (20 nm in diameter) within the cytoplasm giving an organizational framework for co-ordinated cellular actions such as cell division, and supporting structures such as cilia.

Microfilaments are smaller rods (4–6 nm in diameter) having contractile properties facilitating cellular cohesion and events such as muscle contraction.

Centrioles are structures near the nucleus which are active in cell division.

The cytoplasmic membrane

The cytoplasmic membrane has a protective function and plays a vital role in the transfer of substances between the intracellular and extracellular fluids. The properties of the membrane also influence cell to cell recognition and immunological responses.

Biochemically, cell membranes are composed of protein and lipid with some carbohydrate. The fluid mosaic model of membrane structure was proposed by Singer and Nicolson in 1972. This model is well supported by experimental evidence and has largely superseded the earlier Davson–Danielli–Robertson model of unit membrane structure. The fluid mosaic model stresses the dynamic aspects of membrane structure and function, in that protein and lipoprotein complexes are mobile within the membrane. According to the fluid mosaic model, cell membranes consist of phospholipid molecules which are aligned in two layers (Fig. 1.2) with their hydrophobic (water-repellent) ends turned inwards and their hydrophilic (water-attracting) ends facing outwards. This structure is not rigid, the interior of the membrane being in an almost fluid state. Protein molecules are closely associated with the lipid bilayer. Some proteins extend right through the bilayer and others are located on only the outer or inner surface (Fig. 1.2). Thus the membrane proteins appear to float, like icebergs, in a sea of lipid and form a mosaic pattern on the surface of the membrane.

Some glycoproteins and glycolipids associated with the outer surface of cell membranes are probably involved in the cell's recognition of, and response to, factors in its environment.

No holes or pores in the membrane have been identified, but the proteins that extend all the way through the membrane may provide the channels which allow ions, for example, sodium (Na^+),

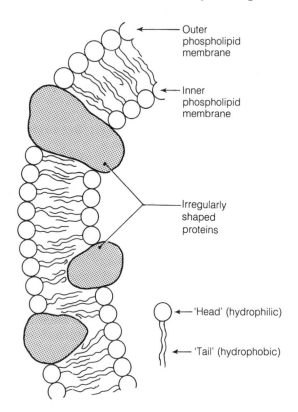

Outer phospholipid membrane

Inner phospholipid membrane

Irregularly shaped proteins

'Head' (hydrophilic)

'Tail' (hydrophobic)

Figure 1.2 Diagram of the fluid mosaic model of cell membrane structure

potassium (K^+) and chloride (Cl^-), and other lipid-insoluble substances, for example urea, to cross cell membranes. Lipid-soluble substances are able to pass through the lipid bilayer.

The cytoplasmic membrane's outer surface may be structurally modified in relation to adjoining cells. There may be variable distances between adhering surfaces, a tight junction or a gap junction, which physiologically provide degrees of cellular adhesion and communication. Gap junctions may allow small molecules to pass between cells but block the passage of large molecules.

Areas of the cytoplasmic membrane which form the surface of a tissue may be specifically modified to facilitate its function. For example, the fine microvilli (brush border) of intestinal epithelium and kidney tubule cells increase their surface area, thereby accelerating the process of absorption with which these cells are involved.

Cilia, which are motile processes, are present, extending from the surfaces of cells of certain epithelial linings. They occur in the respiratory tract and the uterine tubes and, in a somewhat more elaborate form, in the spermatozoan tail. A single cilium is covered by cytoplasmic membrane. Its internal structure is composed of microtubular filaments in a constant arrangement. Cilia of surface cells, orientated in the same direction, move rapidly forward and very slowly backward, effecting movement of fluid lying adjacent to the cells.

MECHANISMS BY WHICH SUBSTANCES ARE TRANSPORTED ACROSS THE CYTOPLASMIC MEMBRANE

Diffusion. This is the passive movement of a substance from an area of high concentration to an area of low concentration. It is due to the spontaneous movement of molecules. This process can cause the movement of substances through the extracellular fluid to the cytoplasmic membrane. Some of the substances are soluble in the lipid whereas others are soluble in the protein. By diffusion through the medium in which the substance is soluble, the required substances enter the cell. The same mechanism allows diffusion in the opposite direction when a concentration gradient is established by the accumulation of metabolic wastes. Diffusion also acts as a transport system within the cell.

Facilitated diffusion. This refers to transport involving carrier molecules. Although this type of diffusion is more rapid, there must be a concentration gradient similar to that required for simple diffusion. Glucose is transported into erythrocytes and liver cells in this manner. The carrier molecule is specific for the transported substance.

Osmosis. This is the flow of solvent (which in living things is water) across a semipermeable membrane, from a dilute to a more concentrated solution (or from an area of high water concentration to an area of low water concentration). This process occurs until the concentrations on either side of the membrane are equal. The cytoplasmic membrane functions as a semipermeable membrane, that is, *as if* it contained pores which permit only the passage of substances of, or below, a certain molecular weight. Although these 'pores' have not been visualized, biological membranes allow the passage of small solute molecules and ions, as well as water. A fuller explanation of osmosis is given in Chapter 4.2.

Active transport. Energy is necessary in order to transport substances when there is no concentration gradient or against a concentration gradient. The movement of substances using energy is called active transport. Substances actively transported through cytoplasmic membranes include the ions of sodium, potassium, calcium, iron, hydrogen, chlorine, iodine and urea, some simple sugars and amino acids. The mechanism of active transport requires energy, a carrier molecule with an affinity for the substance to be transported, and enzymes to catalyse the attachment and release of the transported substances from the carrier molecule. In primary active transport the energy requirement is thought to be met by a chemical source such as adenosine triphosphate (ATP). This occurs in the transport of sodium. In secondary active transport the energy source is the membrane potential and/or ion gradients. For example, monosaccharides and some amino acids are transported across intestinal epithelial cell membranes, coupled with the active transport of sodium ions providing an ion gradient (see Chapter 5.1)

Endocytosis. The process of endocytosis includes **phagocytosis**, which is the engulfing of particulate substances, and **pinocytosis**, which is the engulfing of water. Endocytosis enables larger molecules and substances to enter the cell. Phagocytosis involves the folding of the cytoplasmic membrane around macromolecules of microorganisms and occurs in some motile cells, for example polymorphonuclear leucocytes. Particles engulfed in this manner form intracellular vesicles. Small drops of fluid containing protein may be engulfed in a similar fashion during pinocytosis. The fate of the phagocytic and pinocytic vesicles formed involves the lysosomes – enzyme-containing vesicles within the cell. Each endocytotic vesicle fuses with lysosomes to form a digestive vesicle in which enzymatic breakdown of the engulfed substances occurs. The inner lining of the vesicles (which was the outer lining of the cytoplasmic membrane) alters physiologically, providing increased permeability which allows the products of digestion to be transferred more readily into the cytoplasm. The indigestible components of the vesicle are released into the extracellular fluid.

The nuclear membrane

The nucleus is surrounded and defined by a double membrane, each layer of which is similar in structure to the cytoplasmic membrane. The inner and outer membranes are separated from one another by a space of 10–15 nm. The outer membrane may be continuous with the endoplasmic reticulum of the cytoplasm and have ribosomes attached on its cytoplasmic surface. As a result it may appear wrinkled or corrugated. Fusion occurs between the outer and inner membrane components around 'nuclear pores' (60–90 nm in diameter). Transport mechanisms through the nuclear membrane are not as well understood as those of the cytoplasmic membrane. There is evidence that compounds of low molecular weight diffuse through the membrane readily, and that large molecules, such as proteins, pass through the nuclear pores. Binding sites for specific molecules inside the nucleus determine the concentration of certain materials within it. During the events of cell division the nuclear membrane disappears and later reforms by rearrangement of the endoplasmic reticulum.

Nuclear changes seen in cells infected with viruses or altered by carcinogens include increases in size, structural distortions of various kinds and invagination of the nuclear membrane which may produce inclusions containing cytoplasmic organelles. Ultraviolet and ionizing radiation, viruses and a variety of other toxic substances also bring about nuclear aberrations.

The nucleolus

The nucleolus has two major structural components. The core is a dense spherical collection of structural bars rich in DNA. The bars are called nucleonemas. The core is thought to be the structure concerned with organizing the functions of the nucleolus. In the peripheral area, filaments and granules are thought to be stages in the synthesis of ribosomal RNA. There is no definite limiting membrane. The number of nucleoli per cell varies, but generally there are two per diploid cell.

During cell division the nucleoli disperse and reform in the new cells. It has been suggested that the nucleolus is a focal point in cell division determining which parts of the genetic code are used. If this hypothesis is correct, the inheritance of disease might be explained in terms of initial nucleolar integrity. It has been noted that many human malignant tumour cells have an excess of abnormal nucleolar material. The possibility of early diagnosis based on this finding is being explored (Horrobin, 1981).

The endoplasmic reticulum

A system of interconnecting membranes (5 nm in diameter) is found throughout the cytoplasm of all nucleated cells. The membranes form a loose-meshed, irregular network of branching and anastomosing tubules with saccular expansions, called cisternae, and isolated vesicles. The exact configuration and number of each element (tubules, cisternae or vesicles) vary between cell types and during cellular activity.

If the outer surface of the endoplasmic reticulum (ER) bears large numbers of ribosomes, it is called **rough-surfaced** or **granular endoplasmic reticulum** because of its appearance. Glandular cells have a high proportion of rough-surfaced endoplasmic reticulum. Proteins assembled at the ribosomal sites are transported via the system of membranous pathways of the endoplasmic reticulum to the Golgi region of the cell where they are concentrated in the form of granules or droplets. Some carcinogenic chemicals cause prolonged detachment of ribosomes from membranes, leaving sites of unusual enzymatic activity exposed. This results in a change in the control of protein synthesis.

Smooth-surfaced or **agranular endoplasmic reticulum** lacks ribosomes and seldom forms cisternae. It is often associated with the Golgi region. Ascribed functions include metabolic detoxification and lipid and cholesterol metabolism in liver cells, and lipid transport in intestinal epithelium. The sarcoplasmic reticulum in skeletal and cardiac muscle cells is identical to smooth ER. The amount of smooth endoplasmic reticulum and the activity of related metabolic enzymes are increased with barbiturate administration. Some of these enzymes inactivate barbiturates which results in barbiturate tolerance and the need for increased dosage.

Mitochondria

Numerous small organelles, the mitochondria, show active movement within the cytoplasm. There is great variation in shape, size, mobility, orientation and number. Mitochondria have an inherent plasticity and appear as long slender or rounded structures with gradation of shapes between these forms. They all have the same basic structure (Fig. 1.3).

The outer boundary membrane (6–8 nm) has a smooth surface separated by a space (8 nm) from an inner membrane (6–8 nm) whose thin folds, called **cristae**, project into the cavity. Enzymes are present on the inner membrane surface. This structure provides a large area of membrane, spatially arranged to facilitate sequential chemical reactions. Mitochondria are the site of aerobic (in the presence of oxygen) cellular respiration. During cellular (or internal) respiration, energy is produced from the breakdown of glucose and is stored in the two high-energy phosphate bonds of ATP. The production of ATP takes place in mitochondria during a process called **oxidative phosphorylation**.

During the complete metabolic oxidation of 1 molecule of glucose, sufficient energy is produced to synthesize 38 molecules of adenosine triphosphate (ATP) from the same number of molecules of adenosine diphosphate (ADP), (that is, 38 high energy phosphate bonds are formed). Most of these (and hence most of the body's energy store)

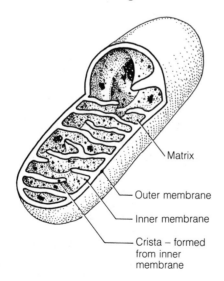

Matrix

Outer membrane

Inner membrane

Crista – formed from inner membrane

Figure 1.3   Diagram of a mitochondrion

are generated during the **Krebs'** or **citric acid cycle**, a series of chemical changes which also takes place in the mitochondria and couples directly with oxidative phosphorylation. The conversion of one mole (the amount equivalent to the molecular weight in grams) of ATP to ADP releases approximately 30 kJ (7 kcal) of energy for metabolic use.

The number and inner complexity of the mitochondria in a cell vary directly with that cell's metabolic activity.

The Golgi region

The Golgi region of the cytoplasm is a compact network of membranes near the nucleus. It is a system of canals and has connections with the smooth endoplasmic reticulum. The components are arranged in parallel layers of flat saccules (cisternae) whose continuous ends form the boundary of a closed cavity. Small vesicles cluster around the outer convex surface; and the inner cisternae may be partially distended and associated with spherical vacuoles containing secretory products. The Golgi region is larger in secretory cells.

Cell secretions move from the smooth endoplasmic reticulum to the Golgi region prior to leaving the cell. The functions of the organelle appear to be concerned with transport, concentration, temporary storage and channelling of these secretions. They may also be involved in synthesizing glycoproteins in some mucus-secreting cells. The Golgi region also has a role in the production of lysosomes at its surface.

Lysosomes

Lysosomes are sacs formed of membrane found in the cytoplasm. They contain enzymes and appear in diverse shapes of variable internal density. The enzymes are lytic, that is, they act to break down substances. It is essential that these enzymes remain separated from the cytoplasm within their membranous sacs otherwise their enzymatic action could digest the cell itself. They arise either from the Golgi region or from the endoplasmic reticulum.

The lysosomes act on pinocytotic and phagocytic vesicles, as described earlier. For example, thyroglobulin (from the thyroid follicles) is taken into the thyroid cells by pinocytosis and degraded by the action of lysosomes, resulting in the release of thyroid hormones.

Phagocytic cells discharge the contents of their lysosomes to digest engulfed bacteria. Lysosomes also function to clear damaged parts of cells or entire cells (**autolysis**) if required. The lysosomes of injured or dying cells are thought to rupture spontaneously.

Microtubules

Microtubules are fine, straight tubular structures, 20–24 nm in diameter and of indefinite length, found in the cytoplasm. They are arranged in bundles and give structural support, maintaining cell shape. Microtubules are found in the core of the flagella of spermatozoa, within cilia and in the axoplasm of neurones where they lie parallel to the long axis of each cell. Their hollow appearance is due to the lower density of the central protein material.

They form the spindle apparatus of the dividing cell, and are found randomly dispersed in the cytoplasm of many cells and at the periphery of red blood cells and platelets. The microtubules of the spindle apparatus which appears during cell division are dispersed by the cytotoxic drugs vincristine and vinblastine and hence these drugs inhibit cell division.

Microfilaments

A network of fine filaments, 4–6 nm in diameter and of indefinite length, is found in the cytoplasm of most cells. In the periphery of the cell these filaments function to prevent the nucleus from impinging on the cytoplasmic membrane. Microfilaments are of great significance in the physiology of striated muscle, in which they are composed of the proteins myosin and actin and called **myofibrils**.

Their function is linked with the potential contractility of the cell, the separation (cleavage) stage of cell division, cell motility and the movement of segments of cytoplasmic membrane which occurs during endocytosis. The extrusion of the nucleus during the maturation of red blood cells is associated with the disappearance of contractile microfilaments.

Cell organelles are surrounded by intracellular fluid within the cytoplasmic membrane. Outside the cell membrane, extracellular fluid surrounds the cell populations providing a homeostatically controlled physical and chemical environment. Living cells distinguish themselves from inanimate, non-living entities by their organized structure and the balance of physiological processes maintained in each cell and between cells.

CELL DIVISION – MITOSIS

Cells arise from the division of pre-existing cells. All the cells of the human body originate from the division of a single cell, the **zygote**, formed from the fusion of an ovum and a spermatozoon. The mechanism of cell division (mitosis) is one of the basic functions of living cells. It provides for growth, and for the transmission of inherited characteristics from one cell generation to the next.

There are two distinct phases of mitosis: the division of the nucleus and the division of the cytoplasm. These two phases usually occur at the same time, but there are occasions when the nucleus may divide a number of times without a corresponding division of the cytoplasm. In such a case the resulting mass of protoplasm containing many nuclei is referred to as a multinucleate cell. Skeletal muscle is an example. A many-celled mass formed by fusion of cells that have lost the typical structure of their cell membranes, such as cardiac muscle, is called a **syncytium**.

The nucleus contains **chromatin**. This substance is the form in which chromosomes are seen between cell divisions. Chromosomes carry the hereditary information in the form of DNA. During cell division the structure of each chromosome becomes well defined. At the beginning of mitosis each chromosome is seen as a double structure because the DNA content of chromatin has replicated during the period (**interphase**) prior to cell division. Replication of DNA occurs by the double-stranded helical molecule unwinding (Fig. 1.4). As each strand of DNA unwinds, enzymes catalyse the formation of a new strand precisely complementary to it. Two identical double helices are therefore formed at the end of the process.

During mitosis the shape and size of the chromosomes are constant, and each chromosome has

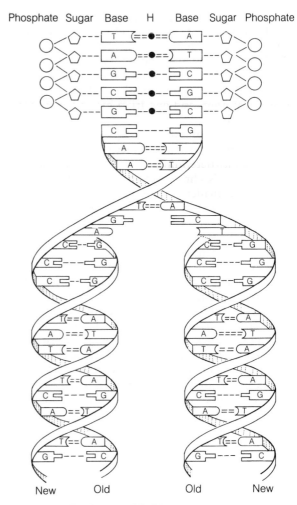

Phosphate   Sugar   Base   H   Base   Sugar   Phosphate

New   Old   Old   New

Figure 1.4   Replication of DNA

structural characteristics which enable it to be identified. For example, the shape of a double chromosome depends on the location of the **centromere** (see Fig. 1.5ii). This is the point at which the two **chromatids**, which constitute each double chromosome, are joined. Some chromosomes have their centromeres at the midpoint, and are classified as 'metacentric'. Others have centromeres close to one end and so have limbs of unequal length. These are called 'acrocentric'.

With the exception of the ovum and spermatozoon (the gametes), the cells of the human body each contain 46 chromosomes. The gametes each contain 23 – a half set or **haploid number**. This reduction in chromosome number is achieved during **meiosis**, a type of cell division which is described in Chapter 6.3.

When, using specialized techniques, the 46 chromosomes of human cells are studied, it is found that they can be matched into 23 pairs. Of each pair, one chromosome has come from the person's mother, and one from his or her father. Twenty-two of the pairs are called 'autosomes' and one pair constitutes the 'sex chromosomes', which are called X and Y. The presence of the sex chromosomes determines genetic sex: a female has two identical X chromosomes and a male has one X and one, smaller, Y chromosome.

The ovum and spermatozoon each provides a haploid set of chromosomes (23) and so, with their union at fertilization, a somatic or **diploid number** (46) is established.

Mitosis ensures that each daughter cell receives an identical chromosome complement to that of the parent cell. A cell usually functions abnormally if it fails to receive a normal chromosome complement.

It is not known what triggers cells to divide. In part, this may depend on the attainment of a critical mass and/or nuclear : cytoplasmic ratio, but some cells are able to divide without prior growth in size. Cells may be prevented from dividing by administration of the drug colchicine, and exposure to radiation may cause nuclear abnormalities and prevent division. Many other factors influence the process of cell division, including nutrition and the degree of cell specialization. Generally, the more specialized the cell, the less likely it is to divide and so reproduce itself.

Whatever other factors may be operating, a cell must complete certain preparations before it can undergo mitosis. All these occur during the period between cell divisions, which is called interphase. These preparations include:

(a) replication of the DNA complement of the chromosomes,
(b) replication of the centrioles,
(c) synthesis of the proteins from which the fibres of the mitotic spindle are formed,
(d) metabolic production of adequate energy to undergo cell division.

During interphase, a pair of highly organized structures is present in the cytoplasm, near the nuclear membrane and Golgi region. These are the **centrioles**. Each centriole is composed of nine groups of three protein rods, arranged in a

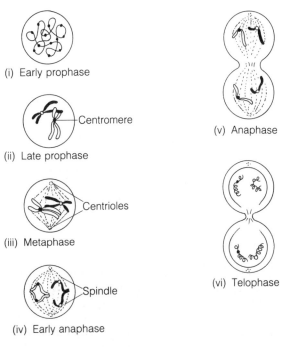

(i) Early prophase

Centromere

(ii) Late prophase

Centrioles

(iii) Metaphase

Spindle

(iv) Early anaphase

(v) Anaphase

(vi) Telophase

Figure 1.5 Stages of mitosis

circle, giving a cylindrical shape. The two centrioles lie with their axes at right-angles to each other. Just before cell division, each centriole replicates and, at the beginning of mitosis, one of each resulting pair migrates to opposite poles of the cell (see Fig. 1.5iii). Protein fibres, structures identical to microtubules, then form between the pairs of centrioles (see Fig. 1.5iii). The structure formed is called the **division spindle** or **mitotic apparatus**.

Mitosis is divided, for ease of description, into four stages, although, in reality, each stage merges into the next (Figs. 1.5i–vi).

Prophase (Figs. 1.5i and 1.5ii)

At the beginning of prophase the centrioles migrate towards diametrically opposite sides of the nucleus. At the same time the spindle fibres begin to aggregate, eventually resulting in the formation of the division spindle. The nuclear membrane disappears and the nucleolus disperses. Each chromosome condenses into a visible unit and can be seen to be made up of a double filament. Each longitudinal half-chromosome (chromatid) has the identical nucleic acids of the chromosome before duplication occurred. The paired chromatids of a chromosome are joined at a single point by a centromere.

Metaphase (Fig. 1.5iii)

The nuclear membrane has now disappeared and the spindle fibres make connections with the centromere of each double chromosome. Each chromosome moves to a central position in the cell, and the centromeres are evenly and precisely spaced along the equatorial plane.

Anaphase (Figs. 1.5iv and 1.5v)

Up to this time, each double chromosome has a single centromere holding the two chromatids together. Now the centromere splits so that two independent daughter chromosomes, each with its own centromere, are formed. The two chromatids of each double chromosome now separate. The same tension of the attached spindle fibres that pulled the double chromosomes into the equatorial plane now pulls the daughter chromosomes (former chromatids) toward opposite poles. One set of chromosomes can be seen moving towards one pole whilst the other set moves to the opposite pole along the spindle fibres.

Telophase (Fig. 1.5vi)

When the daughter chromosomes reach their respective poles, telophase has begun. A cleavage furrow encircles the surface of the cell, eventually constricting the cell into two daughter cells. The disappearance of the spindle fibres, gradual reappearance of the chromatin network from distinct chromosomes, reformation of the nuclear membrane, recondensation of the nucleolus, and the division of the cytoplasm all occur, in each daughter cell, during this final phase of mitosis.

The result of cell division is the formation of two daughter cells, each with an identical gene set and each potentially the same as the mother cell. Cell division is essential for growth and replacement of damaged tissue, as in wound healing. It involves the assimilation of materials from the cell's environment, their transformation through breakdown and synthesis into new cellular components, and the ultilization of energy.

The events which occur between one cell division and the next are termed the **cell cycle**. Cells may have a rapid cell cycle and proliferation rate, for example enterocytes, or a longer cell cycle with a slow rate of turnover, for example glial cells.

NUCLEIC ACIDS AND PROTEIN SYNTHESIS

The deoxyribonucleic acid (DNA) in the nucleus of a cell is the species-specific, inherited, genetic material. In order to understand how DNA functions as the genetic material, it is necessary to understand how it controls the synthesis and metabolism of the cell. This is a complex problem, which scientists have not yet fully explained, but the control of cellular protein synthesis by DNA is now largely understood. Proteins may be enzymes, catalysing cellular reactions, or structural proteins. Since the metabolic activity of a cell is dependent on the regulated synthesis of its enzymes, by controlling protein synthesis, the nuclear genes of DNA control cellular function.

The genetic code

The sequence of the nitrogenous bases (adenine [A], thymine [T], guanine [G] and cytosine [C]) in a DNA molecule makes up the letters of the genetic code for cellular protein construction, by specifying the amino acid sequence in a protein molecule. Each nitrogenous base is a 'letter' in the code and the 'words' formed consist of groups of three (or triplets of) bases.

As there are three bases in a 'word' and four bases in both DNA and RNA, there are $4^3 = 64$ different 'words' available to encode the 20 amino acids found in proteins. A two-letter codeword would be inadequate to specify 20 alternatives; it would give only $4^2 = 16$ different words. The triplet code, with its 64 alternatives, has more than one word specifying a given amino acid. There are also triplets of bases coding for 'initiate' and 'terminate', so defining the limits of a particular DNA message. Thus, if a protein contains 300 amino acids, the part of the DNA molecule that codes for its synthesis must have 300 codewords or base triplets, that is 900 nitrogenous bases in linear sequence.

Protein synthesis

Most protein synthesis takes place in the cytoplasm, on **ribosomes** – small granules composed of 60% RNA and 40% protein. Ribosomes are usually found attached to the endoplasmic reticulum (giving the latter a 'rough' appearance) but they may be free in the cytoplasm. The precise function of **ribosomal RNA** (r-RNA) is unknown, but it apparently has no direct involvement in coding for protein synthesis.

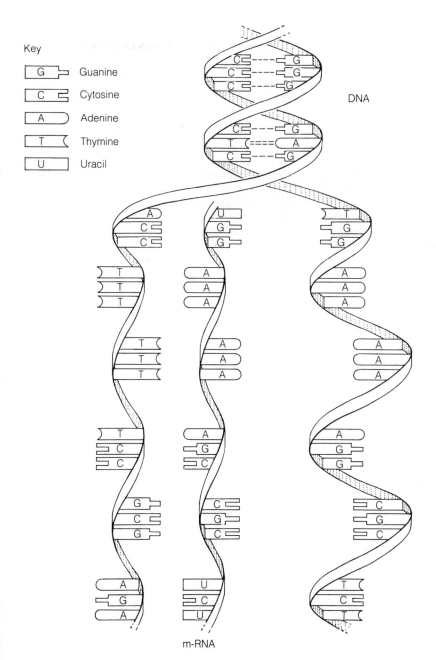

Key

G	Guanine
C	Cytosine
A	Adenine
T	Thymine
U	Uracil

DNA

m-RNA

Figure 1.6 Transcription of a strand of DNA by messenger RNA (m-RNA)

The precise replication of DNA, prior to cell division, was described in the preceding section. As well as acting as a template to replicate itself, DNA can act as a template to synthesize an RNA molecule, since the nitrogenous bases of RNA can pair with those of DNA. That is:

DNA base	pairs with	*RNA base*
Adenine		Uracil
Thymine		Adenine
Guanine		Cytosine
Cytosine		Guanine

This enzymatically catalysed process, which is called **transcription**, occurs in the nucleus. It produces a molecule of RNA which is a precise mirror image of the DNA template. Transcription is illustrated in Fig. 1.6. The information content of the RNA so formed is identical to that of the DNA on which it was transcribed, since each triplet codeword of the DNA is reproduced by a corresponding 'antiword' or **codon** of the RNA by the base pairing rule.

The RNA molecule so formed then leaves the nucleus and moves into the cytoplasm where it becomes attached to the surface of ribosomes. This type of RNA is called **messenger RNA (m-RNA)** because it acts as a messenger between nucleus and cytoplasm.

Just as a strand of DNA functions as a template for the synthesis of m-RNA, so does a m-RNA molecule function as a template for the synthesis of a particular protein. While a DNA molecule may be visualized as being divided up into many regions, each of which codes for the synthesis of a particular protein, the m-RNA molecule encodes a single protein. As described above, the code upon which this template is based has been discovered, and Fig. 1.7 shows the m-RNA codons now known to code for specific amino acids for the initiation and termination of a polypeptide chain.

Thus the alignment of amino acids in a polypeptide chain of a protein is dictated by the base sequence in the m-RNA, which is, in turn, a direct replica of the base sequence of the DNA.

Amino acids, in the cell's cytoplasm, become bound to amino acid specific activating enzymes, a process which requires energy in the form of ATP. They are then attached to a small molecule of RNA called **transfer RNA (t-RNA)**. There is a specific t-RNA for each type of amino acid. The function of t-RNA molecules is to carry their specific amino acid to the ribosomal site of protein

First base of codon	Second base of codon: U	C	A	G	Third base of codon
U	UUU Phenylalanine	UCU Serine	UAU Tyrosine	UGU Cysteine	U
	UUC	UCC	UAC	UGC	C
	UUA Leucine	UCA	UAA Terminate	UGA Terminate	A
	UUG	UCG	UAG	UGG Tryptophan	G
C	CUU Leucine	CCU Proline	CAU Histidine	CGU Arginine	U
	CUC	CCC	CAC	CGC	C
	CUA	CCA	CAA Glutamine	CGA	A
	CUG	CCG	CAG	CGG	G
A	AUU Isoleucine	ACU Threonine	AAU Asparagine	AGU Serine	U
	AUC	ACC	AAC	AGC	C
	AUA	ACA	AAA Lysine	AGA Arginine	A
	AUG Methionine + initiate	ACG	AAG	AGG	G
G	GUU Valine + initiate	GCU Alanine	GAU Aspartic acid	GGU Glycine	U
	GUC	GCC	GAC	GGC	C
	GUA	GCA	GAA Glutamic acid	GGA	A
	GUG	GCG	GAG	GGG	G

Figure 1.7 Messenger RNA (m-RNA) code words

synthesis and allow it to become aligned at the correct position on the m-RNA molecule. This occurs because a region of each t-RNA molecule contains a base triplet complementary to a specific m-RNA codon. The t-RNA triplet of bases which binds to a codon of m-RNA is called an **anti-codon**. The t-RNA lines up on the m-RNA template and so places its amino acid correctly in sequence with others also brought into position by their t-RNAs. As a result, the order of amino acids in the protein is established and it only remains for enzymes to catalyse the formation of the peptide bonds to connect them together. When the polypeptide chain is complete it is released from the ribosome and becomes folded in its active con-

figuration. The synthesis of part of the molecule of the protein hormone glucagon is illustrated in Fig. 1.8.

It is essential that the genetic code contained in DNA can transfer its 'blueprint' for protein construction into the product with both speed and accuracy. An error such as a change (mutation) in the code, altering the sequence of bases, can lead to a protein being produced which has one or more incorrect amino acids and resultant abnormal function. This is how genetic abnormalities are produced.

There are then two major steps in the synthesis of proteins. The first is the transcription of the DNA code by RNA, and the second the

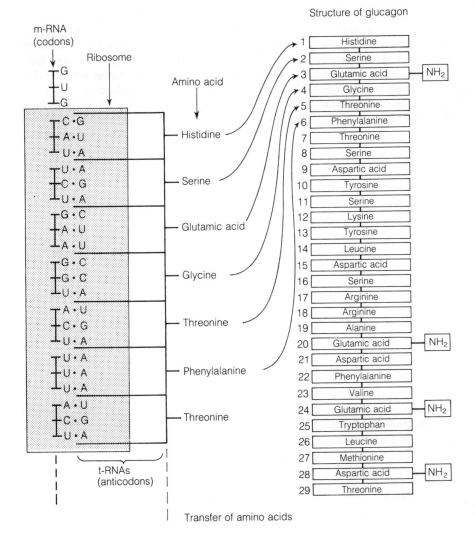

Figure 1.8 An example of protein synthesis: glucagon

transformation of that information into the correct amino acid sequence of a protein.

The same diploid set of genes (DNA) is found in the nucleus of all cells except the gametes, which contain a haploid set. A single gene may be considered to be the segment of DNA which encodes a particular protein. It is probable that, in addition to vital information, human genes also contain obsolete information which is in the process of being modified. The selection of genes to be actively expressed is crucial in determining cell function and is controlled by sophisticated mechanisms which select and activate parts of the DNA molecule, thereby limiting and determining functional gene sequences.

One model for such a regulatory mechanism – the operon model – was proposed by Jacob and Monod (1961), and strong evidence has since accumulated to support it. The operon model suggests that regulation of gene expression occurs at the level of transcription – the synthesis of m-RNA. Further description of this and other proposed control mechanisms, which remain the subject of extensive research, may be found in the texts listed at the end of this chapter.

TISSUES AND ORGANS

Tissues

It is impossible for each of the cells of a higher organism to perform all of the functions necessary to maintain the life of that organism. Instead, a division of labour has evolved in that certain cells have become specialized for specific functions and, as a result, usually have a specialized structure. The cells of the human body, for example, vary considerably in both structure and function, even though they each contain an identical set of genes. The process through which cells, with the same set of genetic information, come to differ from each other, even when they share a similar environment, is called **differentiation**. Differentiation involves the progressive development of specialized structure and function along with the gradual loss of ability to develop in other ways. The precise mechanisms through which differentiation occurs, during embryological development, are the subject of research, and accounts of current knowledge can be found in experimental embryology books and journals. The regulatory mech-

anisms mentioned at the end of the last section suggest ways in which a cell may selectively express its genetic information and so develop different characteristics from other cells.

Cells specialized for the performance of a common function are grouped together, with their associated intercellular material, to form **tissues**. For example, muscle cells have become specialized to contract and hence bring about movement. One result of cell specialization is that some cells have become so adapted to perform one particular function that they have lost the ability to perform others. An example of this is the mature human red blood cell, which has become so specialized in its ability to carry and exchange respiratory gases that it has lost its ability to reproduce.

The existence of different tissues was first recognized during anatomical dissection, but it was not understood that these were collections of specialized cells until use of the light microscope revealed this. The study of tissues is called **histology**, and the histology texts listed at the end of this chapter provide illustrated accounts of the features and functions of the various tissues. A brief classification of the major types of tissue which compose the human body is given below, and fuller descriptions of some of these tissues can be found in the relevant chapters of this book.

CLASSIFICATION OF THE BASIC TISSUES

Epithelial tissues
(a) *Surface epithelia* – these cells cover or line all surfaces of the body, except the joint cavities.
(b) *Glandular epithelia* – single cells, or groups of cells, specialized for secretion.
(c) *Specialized epithelia* – cells specialized in sensory perception (of taste, smell, vision and hearing) and in reproduction (germinal epithelium).

Connective tissues
(a) *Connective tissues 'proper'*:
   (i) loose connective tissues, for example areolar connective tissue, adipose (fatty) tissue;
   (ii) dense connective tissues, for example tendons and ligaments.
(b) *Blood, blood-forming tissues and lymph.*
(c) *Supportive connective tissues* – cartilage and bone.

Muscle tissue
(a) *Smooth (non-striated or 'involuntary') muscle.*
(b) *Skeletal (striated or 'voluntary') muscle.*
(c) *Cardiac muscle.*

Nervous tissue
(a) *Neurones.*
(b) *Non-neuronal cells* – for example, Schwann cells of the peripheral nervous system and glial cells of the central nervous system.

The arrangement of the specialized cells in each of the above tissues is as essential to body function as the biochemical characteristics of the molecules from which they are formed. The structure and function of the cells composing a tissue and controlled cooperation between tissues underlie and determine the form and function – in growth, maintenance and repair – of all parts of the body.

Organs

An organ is an orderly arrangement of tissues into a functional unit, for example the heart or the kidney. Generally, organs are composed of several different types of tissue and conform to two basic arrangements:

(a) tubular or hollow organs, and
(b) compact or parenchymal organs.

TUBULAR ORGANS

Through an understanding of embryological development, the body may be visualized as a large tube containing several internal tubes or tracts, such as the cardiovascular system, the respiratory system, the digestive system, the urinary system and the reproductive system. Each of these tracts, although modified for various functions, both as a whole and at different points along its length, is structurally similar to the others in that they are all formed of layers of tissue superimposed on one another in a certain way.

Each tubular organ has three basic layers.

(a) *An inner layer* composed of epithelium and its underlying connective tissue.
(b) *A middle layer* composed of alternating layers of muscle and connective tissue.
(c) *An external layer* composed of connective tissue and epithelium.

COMPACT ORGANS

Compact organs are characterized by their localized solid form, but vary in size and shape. They may be large, such as the liver, or small, such as the ovaries. Like tubular organs, compact organs all have a common basic pattern. They are usually enclosed by a dense, connective tissue **capsule**. If the organ is suspended in a body cavity, for example the abdominal cavity, it is covered by a serous membrane, which is a moist membrane bathed in serous fluid derived from blood plasma. On one side of the organ there is a thicker area of connective tissue which penetrates the organ, forming the **hilus**. Compact organs have an extensive connective tissue framework, or **stroma**. Strands of connective tissue, called **trabeculae** or **septae**, extend into the organ from the capsule and hilus, sometimes dividing the organ into complete sections called **lobules**. Delicate reticular fibres interlace through the rest of the organ forming a framework for the **parenchyma**, which is the predominant functional tissue of the organ. Parenchymal cells may occur in masses, cords, strands or tubules, depending on the organ. The parenchyma may be divided into two further functionally distinct regions, a subcapsular **cortex** and a deeper **medulla**.

It can be helpful to bear in mind the general patterns of tubular and compact organs while studying each body system, since this provides the basic plan to which the specialized structure and function of each component may be added.

HOMEOSTASIS

The principle of maintaining constant cell composition

The normal function of the cells of any animal or plant depends on their ability to maintain, within relatively narrow limits, the physical and chemical properties of their constituents, such as temperature, state of hydration, concentration of nutrients, waste products, electrolytes and hydrogen ions. This is essential for normal cellular metabolism and because the cell's structural proteins and enzymes can be destroyed or inactivated by abnormal temperature and acidity.

As all cells are dynamic, open systems, continually using and producing substances as part of their metabolism, they must be able continually to exchange these substances with their immediate environment if they are to maintain constant composition.

If life is to be maintained, not only must the cell's environment be capable of supplying its requirements to maintain constant composition, but also the cell must be able to protect itself from the effects of fluctuations in aspects of its environment. The external environment of free-living, unicellular animals, for example, is usually vast in comparison with their volume. Hence, such animals have wide access to a supply of nutrients, and the dissipation of their waste products negligibly changes the composition of their environment. On the other hand, such animals are directly exposed to the effects of adverse changes in their environment and, if they are to survive these, must develop mechanisms to protect themselves.

The consequence of increasing multicellularity

As larger, multicellular animals evolved, certain cells differentiated and became specialized in the acquisition of nutrients and removal of waste, forming the respiratory system, gastrointestinal system and renal system. In order to fulfil their functions, some cells of these systems communicate directly with the external environment. However, an inevitable consequence of increasing multicellularity and cell specialization was that some cells lost contact with the external environment and with each other. As exchange with their immediate environment became insufficient for these cells to maintain constant composition, so a biological need arose for a means of connecting each specialized group of cells with others and with the external environment – for an internal transport system. Hence, other cells became specialized to fulfil this function, forming the cardiovascular system and the fluid medium it contains, the blood. Through the function of these cells, indirect exchange between all cells and the external environment is made possible.

The internal environment

In a complex, multicellular animal then, few cells

are in direct contact with the external environment. Instead, the immediate environment of the majority of cells is an 'internal' environment composed of the small amount of fluid that surrounds them. The watery fluid medium inside cells is, logically, called **intracellular fluid**, and that outside cells **extracellular fluid**.

BODY WATER

Water constitutes between 50% and 70% of the total human body weight. The exact percentage depends on the amount of adipose tissue present, as this has a much lower water content relative to that of other tissues. As the amount of body fat is dependent on factors such as age, sex and dietary habits, total body water is also influenced by these variables. For example, it is lower in females and obese individuals, in whom body fat stores are greater, and is greatest in infants and children, in whom body fat stores are relatively less. After childhood, total body water gradually diminishes with age, forming, for example, about 60% of total body weight in a 70-kg adult man.

The volume of total body water is approximately 41 litres in a human adult and, together with the electrolytes and non-electrolytes that it contains in solution, this fluid is distributed between the **intracellular fluid** (25 litres) and the **extracellular fluid** (16 litres) (Fig. 1.9). The human extracellular fluid compartment is therefore

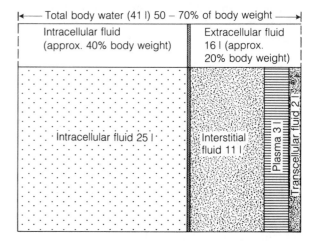

Figure 1.9 Size of the major body fluid compartments in a 70-kg adult man

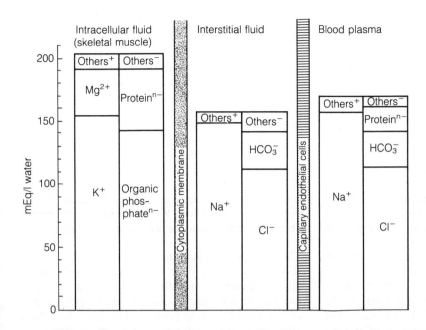

Figure 1.10 Electrolyte composition of body fluid compartments (electrolyte concentrations are given in milliequivalents (mEq) rather than millimoles (mmol), so that it is possible to illustrate that total positive charges are neutralized by total negative charges within any one compartment)

considerably smaller than the intracellular fluid compartment.

Blood plasma forms 3 litres of the extracellular fluid, and the **interstitial fluid** forms 11 litres. Interstitial fluid directly bathes the cells and is often called **tissue fluid**. It is the medium through which exchanges between cells and the external environment occur. A further minor subdivision of the extracellular fluid compartment is provided by **transcellular fluid** (2 litres), which is separated from blood not only by the capillary endothelium, but also by a layer of epithelium. It includes those fluids found in the body cavities such as intraocular, pleural, peritoneal, synovial and cerebrospinal fluids and the digestive secretions.

COMPOSITION OF FLUID COMPARTMENTS

Each major fluid compartment has a unique composition; but whereas plasma and interstitial fluid (both of which are extracellular) are similar, there are striking differences between these and intracellular fluid (Fig. 1.10).

In intracellular fluid the predominant positively charged ions (cations) are potassium (K^+) and magnesium (Mg^{2+}), with a very small amount of sodium (Na^+). The major negatively charged ions (anions) in intracellular fluid are organic phosphate (Org. PO_4^{n-}) and protein (Pr^{n-}), with

very small amounts of hydrogen carbonate (HCO_3^-) and chloride (Cl^-).

In contrast, the major cation in extracellular fluid is sodium, with very small amounts of potassium, magnesium and calcium (Ca^{2+}). The anions chloride and hydrogen carbonate predominate in interstitial fluids, and chloride, hydrogen carbonate *and* protein predominate in plasma. Small amounts of hydrogen phosphate (HPO_4^{2-}), sulphate (SO_4^{2-}) and organic ions are also found in extracellular fluid.

Thus, a major difference which exists between plasma and interstitial fluid is in the protein concentration, which is relatively high in plasma and low in interstitial fluid. The selective permeability of the capillary endothelium prevents large plasma protein molecules from entering interstitial fluid, although very small quantities do filter across and are returned to the vascular system via lymphatic drainage. While the selective permeability of the capillary endothelium accounts for the differences in protein concentration between plasma and interstitial fluid, the differences in electrolyte concentration are explained by the fact that since proteins are negatively charged anions (at the slightly alkaline pH of plasma – 7.4), they attract diffusible anions, e.g. Na^+ and K^+, and repel diffusible cations, e.g. Cl^-. Since the physicochemical requirement of electroneutrality demands that in any solution the sum of the positive charges should equal the sum of the negative charges, this leads to an uneven distribution of diffusible ions across the capillary wall, with a small excess inside the capillary. This uneven distribution is referred to as the **Gibbs–Donnan membrane distribution**, and the small excess of diffusible ions in plasma is known as the **Donnan excess**. The Gibbs–Donnan membrane distribution (or equilibrium) is further described in Chapter 2.1.

Differences in composition between intracellular and interstitial fluid are attributable to the selective permeability of the cytoplasmic membrane and the presence of an active transport, sodium–potassium ATPase pump which continually extrudes sodium ions in exchange for potassium ions. Bearing in mind the fact that the cell contains large, non-diffusible anions such as protein and phosphate, the distribution of ions between these two compartments is also influenced by the Gibbs–Donnan membrane equilibrium. In addition, the intracellular proteins also bind some ions, another factor which influences differences in composition between intracellular and interstitial fluid.

In the nineteenth century, the French physiologist Claude Bernard realized that, in higher animals, the interstitial fluid forms the true internal environment of the cells, and functions as a 'middleman' for all exchanges of matter between a cell and any other cell or with the external environment. It follows that it is essential that the composition of the interstitial fluid is prevented from fluctuating widely in order to provide the cells with the chemically stable, thermostatically controlled environment that they require. The process of preserving the constancy of the internal environment was called **homeostasis** by the American physiologist Walter Cannon (1932).

Homeostatic mechanisms

A large number of physiological mechanisms operate to preserve the necessary conditions for life in the internal environment. Such mechanisms are called homeostatic mechanisms. Homeostatic mechanisms operate via the systems which communicate with the external environment (respiratory, gastrointestinal and renal) and are controlled and coordinated by further specialized cells which comprise the nervous and endocrine systems.

A homeostatic mechanism is triggered by a change in a property of the extracellular fluid. It acts by negative feedback to restore or preserve the normal value by producing a change in the opposite direction. Some mechanisms are more complex than others. At its simplest, a homeostatic mechanism requires the following.

Detectors
These are usually nervous receptors which monitor the magnitude of the variables to be controlled.

Effectors
These are cells comprising muscles, glands, blood vessels, the heart and kidneys, which can bring about the necessary compensatory changes.

Coordinating mechanisms
These may be nervous or endocrine, and couple the receptors to the effectors, ensuring that the magnitude and timing of responses are appropriate.

For example, most people drink more fluid than they require in order to excrete waste products in the urine. When excess water is drunk and absorbed from the gastrointestinal tract into the circulation, the osmolality of plasma (normally about 285 mosmol/kg water) falls. This fall is detected by osmoreceptors in the hypothalamus in the brain which transmit impulses to a controlling centre in the hypothalamus. This centre regulates the synthesis and release of antidiuretic hormone (ADH) from the posterior pituitary gland into the blood. The level of ADH in the blood determines the rate at which the kidneys excrete water. In this case, ADH secretion is reduced and so the kidneys excrete more water, restoring water balance. This homeostatic mechanism controls the osmolality of plasma so that, in health, it fluctuates over only a narrow range.

Homeostasis and health

In health, each of the following physiological needs of a human being is kept in homeostatic balance.

(a) To maintain an adequate intake of oxygen and other nutrients.
(b) To eliminate waste products and toxic substances.
(c) To maintain normal water balance, electrolyte and hydrogen ion concentration.

(d) To maintain body temperature within the normal range.
(e) To maintain intact defence mechanisms.
(f) To move and maintain normal posture.
(g) To rest and to sleep.

A supplementary physiological requirement which, although not essential for an individual's survival, is essential for the survival of the human species, is the need to reproduce.

The fulfilment of all these needs is dependent upon the normal activity of the cells which perform these functions and also on the normal activity of the cells of the coordinating and integrating systems (nervous, endocrine and cardiovascular) which 'link up' homeostatic mechanisms. If any part of these systems is malfunctioning, this can have a widespread adverse effect on homeostasis and health.

The multitude of homeostatic mechanisms that maintain life and health make up a large part of the subject matter of physiology, and hence the content of this book. Most physical health problems may be understood as being the result of a breakdown in homeostasis. It follows that knowledge of homeostatic mechanisms, and the factors which can affect them, forms one of the major bases for the rational planning, delivery and evaluation of patient care.

Review questions

The answers to all these questions can be found in the text. In each case there is at least one correct and at least one incorrect answer.

1. The cytoplasmic membrane

   (a) is a double membrane containing pores
   (b) is mainly composed of protein and lipid
   (c) breaks down at the beginning of mitosis
   (d) may be structurally modified to increase the cell's surface area

2. Which of the following are features of mitochondria?

   (a) Their position in the cytoplasm is constant
   (b) They are the site of cellular aerobic respiration
   (c) They are storage sites for cellular secretions
   (d) They contain lytic enzymes

3. The DNA molecule

   (a) has a double-stranded helical configuration
   (b) contains the nitrogenous bases adenine, cytosine, guanine and uracil
   (c) is found chiefly in the cytoplasm
   (d) is the material basis of the genetic code

4. Mitotic cell division

   (a) is preceded by the duplication of the DNA of the chromatin
   (b) occurs with the same frequency in all types of cell
   (c) ensures that each daughter cell receives an identical chromosome complement to that of the parent cell
   (d) can be prevented by administration of the drug colchicine

5. Messenger RNA molecules

   (a) form anticodons complementary to DNA
   (b) carry the genetic code from nucleus to cytoplasm
   (c) function as templates for the synthesis of proteins
   (d) attach to amino acids in the cytoplasm

6. Compact organs

   (a) are composed of three basic layers of tissues
   (b) are usually enclosed by a connective tissue capsule
   (c) may be divided into lobules
   (d) are mainly composed of parenchymal cells

7. Interstitial fluid

   (a) forms the major part of the extracellular fluid
   (b) has a greater volume than the intracellular fluid
   (c) constitutes the immediate environment of most cells
   (d) has its composition homeostatically maintained within narrow limits

8. Homeostatic mechanisms

   (a) prevent large variations in the composition of the internal environment

(b) require receptors to monitor the variables to be controlled

(c) usually act by positive feedback

(d) are coordinated by cells of the nervous and endocrine systems

Answers to review questions

1. b and d
2. b
3. a and d
4. a, c and d
5. b and c
6. b, c and d
7. a, c and d
8. a, b and d

Short answer and essay topics

1. Relating structure to function, describe the features of a cell of a tissue type of your choice.
2. After reading Section 1 and Chapter 6.3, discuss the functional significance of mitotic and meiotic cell division.
3. With the aid of diagrams, explain how proteins are synthesized.
4. Define the term 'homeostasis' and explain the physiological significance of this process.

References

Cannon, W. B. (1932) *The Wisdom of the Body*. New York: Norton.

Horrobin, D. (1981) Nucleus or nucleolus: which runs the cell? *New Scientist* 89(1238): 266–269.

Jacob, F. & Monod, J. (1961) Genetic regulatory mechanisms in the synthesis of proteins. *Journal of Molecular Biology* 3: 318.

Singer, S. J. & Nicolson, G. L. (1972) The fluid mosaic model of the structure of cell membranes. *Science 175* (18 Feb.): 720–731.

Suggestions for further reading

Beck, F. & Lloyd, J. B. (Eds) (1974–1976) *The Cell in Medical Science, Volumes 1–4*. London: Academic Press.

Bernard, C. (1874) *Lectures on the Phenomena of Life Common to Animals and Plants*. Springfield, Illinois: Charles C Thomas.

Borysenko, M., Borysenko, J., Beringer, T. & Gustafson, A. (1984) *Functional Histology – A Core Text*. 2nd edn. Boston: Little, Brown.

Fawcett, D. W. (1981) *The Cell: Its Organelles and Inclusions*, 2nd edn. Philadelphia: W. B. Saunders.

Finean, J. B., Coleman, R. & Michell, R. H. (1984) *Membranes and their Cellular Functions*, 3rd edn. Oxford: Blackwell Scientific.

Giese, A. C. (1979) *Cell Physiology*, 5th edn. Philadelphia: W. B. Saunders.

King, B. (Ed) (1986) *Cell Biology*. London: Allen and Unwin.

Kranse, W. (1986) *A Concise Text of Histology*, 2nd edn. London: Williams and Wilkins.

Le Gros Clark, W. E. (1975) *The Tissues of the Body*, 6th edn. Oxford: Clarendon Press.

McElroy, W. D. (1971) *Cell Physiology and Biochemistry*, 3rd edn. New Jersey: Prentice-Hall.

Holmes, R. (Ed) (1974) *Physiology of Cells and Organisms; Cells and their Ultrastructure*, Units 0–2; *Membranes and Transport*, Units 3–4. Milton Keynes: Open University Press.

Robinson, J. R. (1975) *A Prelude to Physiology*. Oxford: Blackwell Scientific.

Rogers, A. W. (1983) *Cells and Tissues – an Introduction to Histology and Cell Biology*. London: Academic Press.

Sheeler, P. & Bianchi, D. E. (1983) *Cell Biology: Structure Biochemistry and Function*, 2nd edn. London: Wiley.

Swanson, C. P. & Webster, P. L. (1985) *The Cell*, 5th edn. New Jersey: Prentice-Hall.

Watson, J. D. (1968) *The Double Helix*. New York: Atheneum.

Watson, J. & Crick, F. H. C. (1953) A structure of deoxyribose nucleic acid. *Nature 177*: 737–738.

Woese, C. R. (1967) *The Genetic Code*. New York: Harper and Row.

Section 2
Control and Coordination

Chapter 2.1
Structure and Function of Nervous Tissue

Valerie M. Nie
Michael Hunter
with Douglas Allan

Learning objectives

After studying this chapter the reader should be able to

1 Identify the major structural features of a nerve cell.
2 Outline the putative functions of neuroglial cells.
3 Describe the structure of a mixed nerve.
4 Explain why a resting potential difference exists across a neuronal cell membrane and describe how the resting potential is maintained.
5 Outline the mechanism by which an action potential is generated and propagated along a nerve fibre.
6 Identify the major features of peripheral neuropathy.
7 Illustrate the major structural features of a chemical synapse.
8 Give an account of the mechanism of chemical transmission at a synapse.
9 Describe the electrochemical events constituting excitatory and inhibitory post-synaptic potentials.
10 Define the terms temporal summation and spatial summation.
11 Identify the principal structural components of a reflex arc.
12 Outline the characteristic properties of reflex responses.
13 Describe the structural elements of and functions of the flexion reflex, the crossed-extensor reflex and the stretch reflex.
14 Explain the importance of the assessment of reflexes in neurology.

Introduction

The maintenance of the human body in a healthy state depends on the body's ability to respond appropriately to environmental changes in a coordinated and organized fashion. A rapid means of communication is essential for this coordination.

There are two systems for communication within the body: the nervous system and the endocrine system. The nervous system is capable of transmitting electrical signals very rapidly – up to 120 m/s. The endocrine glands secrete hormones into the bloodstream which then modify the action of target organs which are often distant from the gland itself. This chapter is concerned with the structure and function of nervous tissue.

The nervous system is not only important as an internal system of communication, it also provides the individual with conscious awareness and the ability to communicate with others. Thus, the nervous system plays a vital role in establishing the relationship between nurse and patient and in directing the care that is provided.

ORGANIZATION OF THE NERVOUS SYSTEM

The human nervous system can be divided into the **central nervous system (CNS)** and the

peripheral nervous system (PNS). The CNS comprises the brain, which is protected by the bones of the skull, and the spinal cord, which lies within the vertebral column or backbone. The CNS integrates and interprets signals received from all parts of the body and controls signal output. The PNS includes all nerve tissue other than that of the brain and spinal cord, that is, all nerves which run between the CNS and other organs of the body.

Some of the peripheral nerve fibres terminate at specialized structures that are capable of detecting environmental changes. These structures are called **sensory receptors**. Receptors may respond to light, heat, mechanical energy or chemical energy. They may lie near the surface of the body (i.e. close to the external environment) or they may lie deeper within the body and detect internal environmental changes such as those of blood pressure. Each receptor responds preferentially to one kind of energy and converts it to tiny electrical signals that are transmitted to the CNS via the **sensory nerve fibres**. Sensory nerve fibres conveying information to the CNS are called **afferent nerve fibres**.

Some peripheral nerve fibres transmit signals in the opposite direction, that is, away from the CNS and towards distant body organs. These nerve fibres are called **efferents** or **motor nerve fibres**. Motor efferents terminate in structures called **effector organs** which are capable of making a response. Most effector organs are muscles that respond to contracting and developing force, although some motor nerve fibres terminate in glands which respond by increasing their secretions.

Part of the PNS controls the activities of the viscera (heart, gut etc.). The visceral nervous system is referred to as the **autonomic nervous system (ANS)**. The strictest definition of the ANS is that it is the motor nerve supply to visceral muscle, including cardiac muscle, and associated glands. Both visceral and cardiac muscle differ from skeletal muscle in structure and are not generally under conscious voluntary control. Motor signals are transmitted to the viscera in response to sensory signals. However, these sensory nerve fibres are included as part of the remainder of the PNS, the **somatic nervous system**. The ANS can be subdivided into two anatomically distinct parts, the **sympathetic** and the **parasympathetic nervous systems**. The

parasympathetic system maintains the body in a normal resting state, while the sympathetic system prepares it to cope with (or adapt to) stress. The ANS is considered in greater detail in Chapter 2.4.

COMPONENTS OF NERVE TISSUE

The brain, spinal cord and peripheral nerves are organs that are composed principally of nervous tissue. However, in addition, blood vessels and connective tissue permeate them. The blood supply provides nutrients and oxygen while connective tissue provides mechanical strength and support.

Nervous tissue, in common with other tissues, comprises cellular elements in a small amount of fluid matrix, the extracellular fluid. Two types of cell can be identified: **neurones**, which generate and propagate the electrical signals, and **neuroglia**, or satellite cells, which are thought to maintain the neurones in a healthy functioning state.

Neurones

The shape of the neurone is adapted for transmission of electrical signals over long distances but the ultrastructure is essentially the same as that of any animal cell (see Section 1).

The integrity of the neurone is maintained by a lipoprotein semipermeable unit membrane. This separates intracellular from extracellular fluid and permits differences in the composition of each.

The neurone has a cell body from which a number of fine processes arise. Cell body size is quite variable, ranging from 5 μm to 120 μm in diameter. The short processes are called **dendrites** – because they resemble the branches of a tree (*dendron* = Greek for tree) – and they increase the surface area available for connections with other neurones. Dendrites themselves bear tiny processes called dendritic spines, which are visible with high-powered electron microscopes. The soma bears one longer process, the **axon**, or nerve fibre, which may be branched. The axon conducts electrical signals from the neurone soma towards the axonal terminations.

The soma contains organelles typical of any animal cell: a nucleus, mitochondria, Golgi appar-

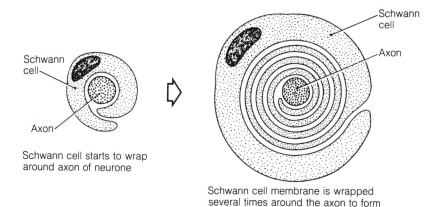

Schwann cell starts to wrap around axon of neurone

Schwann cell membrane is wrapped several times around the axon to form a segment of myelin

Figure 2.1.1 Myelination of a nerve axon

atus and lysosomes are all clearly visible. Light microscopy reveals fibres in the cytoplasm called **neurofibrillae**. Neurofibrillae and mitochondria are also found in axoplasm (cytoplasm of the axon). **Nissl granules**, which are ribosomal in nature, are visible in the cytoplasm of the soma but are not found in axoplasm.

Neurologlial cells are associated with the axons of neurones. In the CNS these cells are called **oligodendrocytes**, whereas in the PNS they are referred to as **Schwann cells.** Oligodendrocytes and Schwann cells associate with neuronal axons in one of two ways. In the first case, the neurone is said to be **myelinated**. The Schwann cells (or oligodendrocytes) wrap their cell membranes around an axon a number of times (Fig. 2.1.1). The Schwann cell (or oligodendrocyte) membranes form a segmented sheath around the axon, the **myelin sheath**. Myelin is white in colour and it accounts for the appearance of white matter in the CNS. **White matter** is largely composed of myelinated nerve fibres, as opposed to the **grey matter** which consists mainly of non-myelinated cell bodies. The gaps between segments of the myelin sheath represent junctions between adjacent neuroglial cells and are approximately 1 mm apart. The tiny gaps of naked membrane are called **nodes of Ranvier**. A myelin sheath forms a relatively good electrical insulator and in part determines the efficiency of transmission of electrical signals along a nerve fibre.

The second type of relationship between an axon and Schwann cells (or oligodendrocytes) is found in a **non-myelinated** neurone. In this case, the axon is enveloped by neuroglial cells but a myelin sheath is not formed. Each neuroglial cell may envelop a number of nerve fibres. In both myelinated and non-myelinated neurones the Schwann cell membrane surrounding the axon is termed the **neurilemma**.

Neurones can be classified on a functional or a structural basis. They may be classed as **sensory** or **motor neurones**, with interconnecting neurones in the CNS being termed **interneurones**. Neurones may also be categorized according to the physical dimensions which determine their functional efficiency in conducting nerve signals. This will be referred to again later in the chapter.

Neurones are easily classified according to the number of processes that arise from the soma. Three groups can be identified, namely **unipolar**, **bipolar** and **multipolar neurones**.

Unipolar neurones have only one true process arising from the cell body. Sensory neurones conveying information from the body surface (e.g. temperature, touch) are of this type. The cell bodies of these neurones are collected in groups called **ganglia** which lie on the dorsal nerve roots of the spinal cord. The axon of a unipolar cell divides into two branches; one enters the spinal cord and transmits information to the brain while the other terminates at a sensory receptor near the body surface.

Bipolar neurones have two processes arising from the soma, a central axon and a peripheral dendrite. Bipolar neurones are found in the eye and in the ear. The peripheral dendrite transmits signals from the light receptor in the eye or sound receptor in the ear through the cell body to the central axon. The central axon terminates on another neurone which conducts the signal to the brain.

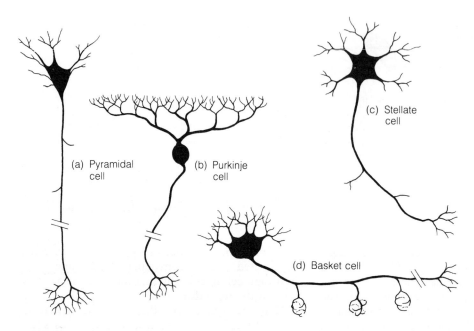

(a) Pyramidal cell

(b) Purkinje cell

(c) Stellate cell

(d) Basket cell

Figure 2.1.2 Variety of forms of multipolar neurones

Multipolar neurones are found throughout the peripheral and central nervous system. A multipolar neurone has three or more processes (two or more dendrites and an axon) arising from the neuronal cell body. It may be myelinated or non-myelinated. For example, all motor neurones supplying skeletal muscles are myelinated and multipolar. They vary in shape and size and in spread and number of dendritic processes. Axons may vary from a few micrometres up to a metre in length. The variability in appearance of multipolar neurones is illustrated in Fig. 2.1.2.

Neuroglia

Neuroglial cells are found in large numbers in the CNS. Indeed, they are said to outnumber neurones in a ratio of 10:1. Estimates of the number of neurones in the human brain vary considerably, but 10 000 million is probably a minimum for a young person. The numbers of neuroglial cells are therefore enormous, but despite this their functions are still not clearly understood.

Neuroglial cells are subdivided into three main categories on the basis of their size, shape and structural relationship with neurones. An **astrocyte,** as the name suggests, is a star-shaped cell with fine, long processes arising from the cell body. In brain tissue preparations, astrocytes are commonly observed to have one or more cell processes in close proximity to neurones, and other processes closely applied to blood capillary walls. That is, astrocytes appear to 'span' gaps between neurones and capillary endothelium.

Oligodendrocytes in the CNS and **Schwann** cells in the PNS are somewhat smaller than astrocytes and they have fewer, shorter cell processes. They are found in rows lying adjacent to neuronal axons throughout the nervous system. Their function in producing myelin has already been discussed.

Microglia are small neuroglial cells, found throughout the CNS, which are probably tissue macrophages and uncommitted stem cells.

Neuroglial cells have been implicated in a wide range of functions associated with the nervous system. Evidence indicates that neuroglial cells do not contribute to electrical signalling in the CNS. However, studies have shown that these cells may undergo electrical disturbances if surrounding neurones are active. Some of the functions which have been assigned to neuroglial cells are briefly considered below.

TRANSPORT CHANNEL FOR NEURONES

Doubtful evidence exists to suggest that neuroglial cells serve to distribute nutrients to neurones. This is now thought to be unlikely.

SECRETION AND UPTAKE OF NEUROTRANSMITTERS

Neurotransmitters are chemicals that influence the passage of electrical signals across the gaps between adjacent neurones in a nerve net. One such transmitter is gamma-aminobutyric acid (GABA). GABA is an inhibitory transmitter, that is, it reduces the likelihood of a signal passing from one neurone to the next. GABA, and other neurotransmitters, may be taken up and stored in neuroglial cells. There is also some evidence to suggest that neuroglial cells may secrete neurotransmitters or, at least, may be essential for the synthesis of transmitters in neurone terminations.

NEUROGLIAL CELLS IN THE DEVELOPMENT AND MAINTENANCE OF THE NERVOUS SYSTEM

Differentiated neurones are not capable of cell division and cannot therefore replace damaged or ageing tissue. However, neuroglial cells can divide and increase in number to occupy atrophied areas in the brain. Indeed, brain tumours generally occur in neuroglial cells rather than in neurones; the names of such tumours are related to their glial cell of origin, e.g. oligodendrocytoma, astrocytoma.

Damaged neurones may regenerate provided that the cell bodies remain intact. In this case, the course of axonal regeneration is directed along a pathway marked out by Schwann cells. Thus, neuroglial cells apparently aid regeneration of damaged axons as well as the establishment of precise neuronal connections during fetal development.

MECHANICAL SUPPORT AND ELECTRICAL INSULATION

Early histological investigations suggested that neuroglial cells formed a packing tissue around neurones, providing support and preventing electrical interference between neurones. No other firm evidence to support this has been forthcoming.

BLOOD–BRAIN BARRIER

The blood–brain barrier is a term used to describe a functional concept based on observations that certain chemicals transported in the blood, which gain access to soft tissues such as liver and kidney, are preferentially excluded from the brain. The normal functional importance of the blood–brain barrier may rest more with its role in preventing substances from getting out of the CNS rather than with its role in preventing the entry of neurotoxins. However, the latter cannot be ignored. The relative immaturity of the blood–brain barrier in young children permits the accumulation of lead in the brains of exposed individuals. This may cause serious brain damage and behavioural deficits. The blood–brain barrier may also be important for the exclusion of potentially neurotoxic therapeutic agents. However, it is almost certainly of importance in rigidly maintaining the composition of the extracellular environment of neurones and in maintaining neurotransmitters within the vicinity of neuronal connections.

Astrocytes have been proposed as structural correlates of the blood–brain barrier based on histological evidence demonstrating close association between astrocyte processes, capillary endothelium, and neurone membranes. Experiments have also identified the sites of the blood–brain barrier at specialized tight junctions between endothelial cells of brain capillaries. Such sites are associated with many astrocytic processes and it may be that astrocytes induce the tight endothelial junctions during early development.

STRUCTURE OF A NERVE

Nerve fibres (axons) are bound together by connective tissue to form whole nerves, which vary in length and diameter. The term **mixed nerve** is often employed. This refers to the fact that a nerve may contain both afferent and efferent nerve fibres innervating a number of different muscles or glands.

Each single nerve fibre is embedded in a delicate fibrous connective tissue, the **endoneurium**. Groups of nerve fibres are held together in bundles by a connective tissue sheath called **perineurium**. Finally, the bundles of nerve fibres are all enclosed in an outer fibrous sheath, the **epineurium** (Fig. 2.1.3). The connective tissue framework provides mechanical support for both nerve fibres and blood capillaries within the whole nerve.

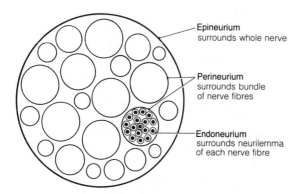

Epineurium
surrounds whole nerve

Perineurium
surrounds bundle
of nerve fibres

Endoneurium
surrounds neurilemma
of each nerve fibre

Figure 2.1.3 Transverse section through a whole nerve. For clarity, individual fibres are only shown in one bundle of nerve fibres

THE NERVE IMPULSE

Nerve fibres and muscle fibres are both capable of conducting electrochemical signals in the body. For this reason, nerve and muscle are both referred to as excitable tissues. The ability to conduct nerve signals is dependent upon the integrity of both the intra- and extracellular environments.

Intracellular environments differ from extracellular environments in a number of ways. Potassium ion (K^+) concentration is generally considerably greater inside cells. In neurones, the concentration of K^+ is approximately 30 times greater inside than outside the cell. In contrast, sodium (Na^+) is the most abundant extracellular cation (positively charged ion). The concentration of Na^+ is some ten times greater outside the nerve cell membrane. Intracellular anions (i.e. negatively charged ions) include protein and amino acids and organophosphates. The most abundant extracellular anion is chloride (Cl^-) which is some 10 to 15 times more concentrated outside the nerve cell membrane.

These differences in the distribution of ions between intracellular and extracellular fluids and the presence of a complex, selectively permeable membrane separating the two, result in the production of an electrical potential difference across the cell membrane. The membrane is said to be polarized. In excitable tissues this potential difference, called the **resting potential**, can be reversibly changed in a specific manner. The potential change is called an **action potential**, and it represents the nerve signal or impulse.

It is worth noting that all cells are surrounded by a semipermeable membrane and contain specialized intracellular fluid. They therefore also have the capacity to produce a resting potential across the membrane. However, only excitable tissues can repeatedly produce a specific reversible alteration in the potential difference. This property of excitable tissue is fundamental to the function of signal conduction.

The resting potential

The potential difference across the nerve cell membrane is very small. It is variable in size, but is generally within the range of 70–90 mV (thousandths of a volt), the inside of the cell being negative with respect to the extracellular fluid.

The resting potential is described as a negative value, approximately − 70 mV. The value relates to the electrical status of the inside of the cell compared with the outside. It should be noted that the term 'resting potential' implies that the neurone is not conducting impulses.

The resting potential of any one neurone is more or less constant throughout the life of the cell. The way in which the resting potential is generated and maintained will now be considered.

Essentially, the nerve cell membrane separates two fluids of differing ionic concentrations. If a simple example is taken in which two solutions of sodium chloride of differing concentrations are allowed to come into contact, a potential difference arises at the junction between the two. This is known as a **liquid junction potential**. It arises because the sodium ions (Na^+) and the chloride ions (CL^-) move at different rates through the fluid. That is, they have different ionic mobilities. Chloride ions move more easily through water and therefore tend to leave excess Na^+ behind. The Cl^- ions move from the stronger to the weaker solution. Sodium ions move in the same direction but lag behind the Cl^-. The strong solution is effectively left with an excess of positively charged Na^+ for a short period of time, and a potential difference is generated as a result. The liquid junction potential is very small and is transient since the Na^+ will also eventually redistribute themselves evenly throughout the two solutions.

The development of a liquid junction potential is a much simpler situation than that which exists

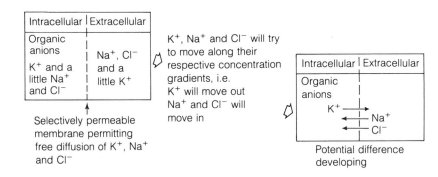

Figure 2.1.4 shows the following labelled panels:

Intracellular | Extracellular

| Organic anions | |
| K⁺ and a little Na⁺ and Cl⁻ | Na⁺, Cl⁻ and a little K⁺ |

Selectively permeable membrane permitting free diffusion of K⁺, Na⁺ and Cl⁻

K⁺, Na⁺ and Cl⁻ will try to move along their respective concentration gradients, i.e. K⁺ will move out Na⁺ and Cl⁻ will move in

Intracellular | Extracellular

Organic anions
$K^+ \longrightarrow$
$\longleftarrow Na^+$
$\longleftarrow Cl^-$

Potential difference developing

Figure 2.1.4 The development of a potential difference across a cell membrane – assuming organic anions are non-diffusible and K^+, Cl^- and Na^+ are freely diffusible

across the nerve membrane. Nevertheless, it demonstrates how solutions of differing ionic strengths, connected in such a way as to permit the movement of at least some charged ions, can generate an electrical potential difference.

The major sources of anions in axoplasm are amino acids and organophosphates. In general, these ions are relatively large, much larger than the inorganic K^+, Na^+ and Cl^-. They are too large to pass through pores in the neuronal membrane and therefore they cannot diffuse into extracellular spaces. Since K^+, Na^+ and Cl^- are all fairly small ions, even in the hydrated state, it can be assumed that they should be able to diffuse through the cell membrane. Indeed, as noted in the discussion of liquid junction potentials, some ions must be able to diffuse across the barrier if a potential difference is to be generated.

If non-diffusible organic anions and diffusible K^+, Na^+ and Cl^- are all present in differing concentrations intra- and extracellularly, then an equilibrium should be established with a potential difference across the membrane. Given the concentrations described for nerve cells, the development of the potential difference would then follow the pattern illustrated in Fig. 2.1.4.

It is useful to consider each ion separately. Figure 2.1.4 shows that as K^+ moves out of the cell, organic anions will be left in excess because they are too large to follow K^+ through the membrane. In this way the inside tends to become progressively negative with respect to the outside. In the absence of any other ionic movements, a net outward flow of K^+ would continue until a dynamic equilibrium was established whereby the potential difference would be sufficiently large to prevent further movement. That is, the concentration gradient would balance the electrical gradient. The electrical gradient reflects the fact

that the negative anionic charges attract positive cations and repel other negative anions.

The same principle would apply to the movement of the other diffusible ions: chloride ions would move into the cell along the Cl^- concentration gradient. However, the excess negativity resulting from the non-diffusible organic anions would tend to repel Cl^- movement into the cell. An equilibrium would be established when the inside was sufficiently negative to prevent further inward flux. Sodium ions would move into the cell along the Na^+ concentration gradient. The organic anions would also attract Na^+ into the cell and, in the absence of other ion movements, an equilibrium would only be reached if the inside of the cell became positive.

If the membrane were freely permeable to all of these ions, they would redistribute themselves so that the transmembrane potential difference was at a point between the values predicted for each ion separately. Under these conditions, the intracellular to extracellular concentration ratio for all diffusible cations (i.e. K^+ and Na^+) would be the same. The ratio for diffusible anions (i.e. Cl^-) would be the reciprocal of this. The equilibrium point so obtained is called the **Donnan membrane equilibrium**, whereby:

$$\frac{[K^+]e}{[K^+]i} = \frac{[Na^+]e}{[Na^+]i} = \frac{[Cl^-]i}{[Cl^-]e} = r$$

where $\begin{array}{l} [\ ]e = \text{extracellular concentration} \\ [\ ]i = \text{intracellular concentration} \end{array}$

The potential difference generated is dependent on this concentration ratio.

Measurements of the resting potential across nerve membranes have been experimentally obtained. In addition, the composition of intra-

Table 2.1.1 Characteristics of a resting nerve cell with a resting potential of $-70\,mV$

Ions	Large organic anions	Chloride (Cl^-)	Potassium (K^+)	Sodium (Na^+)
Distribution	Intracellular	More concentrated extracellularly	More concentrated intracellularly	More concentration extracellularly
Membrane permeability	Impermeable	Freely permeable	Freely permeable	Resting nerve cell membrane has a very low permeability for Na^+
Passive diffusion through membrane	Nil: ions cannot pass through membrane	Ions can pass through membrane and influx equals efflux	The resting potential is not quite large enough to prevent a small loss of K^+ from the cell (i.e. passive efflux is slightly greater than passive influx)	Passive influx of Na^+. Only a small amount of Na^+ diffuses into the cell because of the low sodium membrane permeability, despite the large electrochemical gradient
Active transport through membrane – ionic pumping	No pump for organic anions	No pump for Cl^-	Actively pumps K^+ into the cell against the concentration gradient, tending to restore the passive outward leak	Actively pumps Na^+ out of the cell against the concentration gradient, tending to restore the passive inward leak

cellular and extracellular fluid has been determined. Many of these measures have been taken from the giant axon of the squid. This axon has a diameter of approximately 1 mm and is sufficiently large to allow detailed electrophysiological and biochemical investigations of its structure and function. However, the underlying physiology of mammalian nerves appears to be the same.

Table 2.1.1 lists the approximate distribution of ions across a typical mammalian nerve cell membrane. It is clear that the distribution ratio for K^+ is very different from that for Na^+, and a little different from the reciprocal of the Cl^- ratio. In other words, a simple Donnan membrane equilibrium does not exist. This suggests that nerve membrane is not normally freely permeable to all of these ions.

If the concentration ratio of each ion is considered separately, it is possible to establish whether the distribution of any of the ions represents an equilibrium between electrical force and concentration force. The equilibrium potential for Cl^- can be calculated to be approximately $-68\,mV$. Since the measured resting potential difference is around $-70\,mV$, Cl^- can be assumed to be more or less in a state of equilibrium. It should be noted here that different literature sources cite slightly different values for intra- and extracellular ionic concentrations and for measured resting potential differences. Nevertheless, it is generally accepted that Cl^- ions are in

equilibrium when a nerve fibre is resting and not conducting signals.

The calculated equilibrium potential for K^+ is about $-91\,mV$ and is greater than the measured value for the resting potential. Potassium ions are not in a state of equilibrium and there is a tendency for a net loss of K^+ from the well. That is, the resting potential of $-70\,mV$ is not sufficiently large for efflux of K^+ to balance influx. Potassium ions diffuse passively out of the cell, tending to shift the potential differences towards $-91\,mV$, but this loss is replaced by an active, energy-consuming pump which can pump K^+ back into the cell. This allows a relatively constant potential difference of $-70\,mV$ to be maintained.

If the equilibrium potential for sodium is calculated, a very different situation exists. In this case, the inside of the cell should be approximately $62\,mV$ positive with respect to the outside in order to prevent any net movement of Na^+, if simple passive diffusion is assumed to occur. A resting potential difference of $-70\,mV$ (inside negative) is therefore very far removed from the equilibrium potential for Na^+. Indeed, Na^+ should be attracted into the cell. Thus, the membrane is more or less impermeable to Na^+, or there is a very powerful pump pumping Na^+ out of the cell or there is a combination of low permeability and an active sodium pump. Experimental evidence suggests that, under resting conditions, nerve membrane has a very low permeability to Na^+.

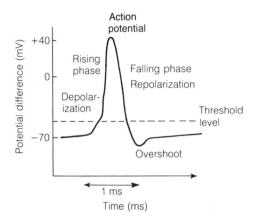

Figure 2.1.5 Time course of an action potential

The resting membrane is approximately 50 to 75 times more permeable to K^+. A passive flow of Na^+ into the cell does occur, but the flow rate in ions per unit time is small because of the low permeability. This small influx of Na^+ would be sufficient slowly to reduce the transmembrane potential difference, but the passive influx is countered by an active removal of Na^+ from the cell.

The characteristics determining the generation and maintenance of the resting potential are summarized in Table 2.1.1. Potassium is continually pumped into the cell whilst Na^+ are actively removed. Experimental evidence indicates that, overall, three Na^+ are exchanged for every two K^+. That is, a straight exchange of cations does not occur and there is a net transfer of positive charge out of the cell. This current flow would tend to increase the potential difference across the membrane. However, in practice, the effect is minimal because the absolute Na^+ efflux is small and tends to be matched by outward Cl^- movement.

The precise nature of the **ionic pump** referred to above has yet to be discovered. It incorporates an enzyme which is capable of splitting adenosine triphosphate (ATP), yielding the energy for the ionic transfer. There is also good evidence that the active transfer of K^+ and Na^+ is coupled in some way.

The action potential

Viable cells – nerve, muscle, blood or skin – generally exhibit resting potentials across their membranes. However, nerve and muscle fibres are able to alter the transmembrane potential difference reversibly. When this occurs, the membrane is **depolarized**, and the inside of the cell transiently becomes positive with respect to the outside. This shift in the pontential difference is referred to as the **action potential**. It has an amplitude of approximately 110 mV and lasts for about 1 millisecond (1 ms), after which time the resting potential is restored. The time course of events is illustrated in Fig. 2.1.5.

GENERATION OF A NERVE ACTION POTENTIAL

The initiating event in the generation of an action potential is the **stimulus**, which represents the delivery of energy to the nerve cell membrane. This energy can take the form of light, heat, mechanical force, chemical energy or electrical energy.

Essentially, the action potential reflects a change in the permeability characteristics of the nerve membrane (muscle membrane also behaves in such a way: see Chap. 3.1).

The characteristics of resting nerve membrane have already been considered. The membrane is polarized, with a potential difference across it of approximately $-70\,mV$ (inside negative). When stimulus energy is applied to nerve membrane it is depolarized, that is, the potential difference across it decreases. This depolarization appears to result in an increase in the sodium permeability of the membrane; Na^+ therefore enter the nerve fibre according to the electrochemical gradient. This transference of positive charge depolarizes the membrane further, which then results in a further increase in Na^+ permeability. In other words, there is a positive feedback between the depolarization of the membrane and the increase in Na^+ permeability. There is an explosive increase in the Na^+ current into the cell such that Na^+ movement dominates other ionic movements. The potential difference across the membrane approaches the equilibrium potential for Na^+.

The inside of the cell briefly becomes positive to about 40 mV. At the peak of the action potential, the membrane is in fact polarized in the opposite direction to the resting potential. That is, the depolarization is followed by a positive polarization. The changes in Na^+ permeability only last for a very short time, normally less than 1 ms.

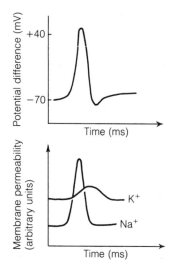

Figure 2.1.6   Time course of the changes in membrane permeability to sodium and potassium ions which occur during an action potential

At the peak of the action potential the inward Na^+ current rapidly reduces, and the increase in Na^+ permeability is said to be inactivated. Meanwhile, K^+ permeability increases. This rise in K^+ permeability, beginning just before the peak of the action potential, and the reduced Na^+ permeability explain why the sodium equilibrium potential is not quite reached. Since the inside of the cell is now positive with respect to the outside, and since K^+ has a concentration gradient favouring an outward K^+ current, K^+ rapidly move out of the cell. Movement of K^+ now becomes dominant, and the resting potential is rapidly restored by the transference of positive charge out of the cell. In fact, the repolarization commonly overshoots the resting potential, and the equilibrium potential for K^+ is approached (i.e. there is a transient overshoot hyperpolarization, see Fig. 2.1.5). The relationship between the action potential and Na^+ and K^+ membrane permeability changes is demonstrated in Figs. 2.1.6. The rising phase of the action potential is associated with a rise in Na^+ permeability, whereas the falling phase of repolarization is associated with a rise in K^+ permeability.

A millisecond or so after the delivery of a stimulus to nerve membrane, the resting potential is restored. However, there has been a very small change in the composition of both intra- and extracellular fluids. A small amount of Na^+ has been transferred into the cell, whilst a small amount of K^+ has been transferred to the extracellular field. After a single action potential, the resultant change in intra- and extracellular concentrations of these ions is negligible. Studies have shown that the intracellular K^+ concentration only changes by a factor of one thousandth to one millionth during the course of a single action potential. The smallest mammalian nerve fibres show changes of the order of one thousandth, but the concentration change is considerably smaller in larger fibres. Many action potentials can be transmitted before a noticeable alteration occurs in the composition of intra- and extracellular fluid. However, normally the Na^+/K^+ exchange pump restores the resting ionic concentrations.

Metabolic energy is not required for the generation of an action potential since the ions move through the membrane under the influence of electrochemical forces. However, metabolic energy is required for the exchange pump and the re-establishment of normal ionic concentrations. This is important if nerve cells are to respond continuously for long periods of time.

The reversible changes in ionic permeability associated with an action potential are peculiar to excitable tissue membrane. The membrane behaves as though it contains gated channels for Na^+ and K^+, such that the gates can be opened and closed in response to certain specified conditions. Thus, when the membrane is stimulated, gates across Na^+ channels in the membrane appear to open, permitting the entry of Na^+, as directed by the electrochemical gradient. This opening of the gates is only transient. At the peak of the action potential the gates apparently close again and Na^+ entry is inactivated. Gates across K^+ channels, which are normally closed under resting conditions, then appear to open, and K^+ flow out of the cell. These gates are also only open for a short period of time, and close when the resting potential is restored.

CHARACTERISTICS OF ACTION POTENTIALS

Action potentials are said to obey the **all-or-none law**. That is, they are not graded responses and are either full sized or absent, depending on stimulus strength.

The size of an action potential varies slightly from neurone to neurone, but generally has an amplitude of approximately 110 mV and a duration of 1 ms or less. However, action potentials produced by any one axon are of more or less constant size provided that the neurone remains healthy.

Neither the amplitude nor the duration of an action potential varies with the strength of the stimulus. Stimulus intensity will affect the frequency of action potential generation and may also influence the number of neurones that are activated, but it will not change the size of the action potentials produced. However, the stimulus must be sufficiently strong to permit a threshold level of membrane depolarization to be reached. If the stimulus is only very weak, the changes in membrane polarity and permeability may be too small to set in motion the positive feedback mechanism linking membrane depolarization with increased Na^+ permeability. The weak stimulus will only slightly depolarize the membrane, which will then return to the resting state without producing the explosive changes associated with an action potential. On the other hand, if the stimulus is sufficiently large to allow the threshold level of depolarization to be exceeded, then a full-sized action potential is always produced.

Threshold

The **threshold** refers to a critical level of depolarization of the membrane rather than to the strength of stimulus required to reach it, although these two factors are linked.

The ease with which nerve fibre membrane reaches threshold can vary. For example, if an action potential has just been generated, then a stronger than normal stimulus may be required to elicit a second one, i.e. the threshold is temporarily raised (see 'Refractory period' below).

In terms of ion movements, the threshold represents an unstable equilibrium between the Na^+ transfer into the cell, and K^+ loss from the cell, and Cl^- entry into the cell. At this point the potential difference across the membrane is of the order of 10 mV less than the resting potential. If extra Na^+ then diffuse into the cell, the positive feedback loop is set in motion whereby entry of Na^+ depolarizes the membrane, which then increases Na^+ permeability, permitting still further depolarization of the membrane and so producing the explosive changes of an action potential. If this occurs, threshold is exceeded and the neurone is said to fire an impulse. On the other hand, if extra K^+ diffuse out of the cell, the poten-

tial difference across the membrane increases again towards the resting potential, and resting conditions will be restored. Therefore, in order to fire on impulse, the stimulus must be of sufficient intensity to cause the unstable equilibrium of threshold to shift in the direction of increased influx of Na^+.

Refractory period

If a second stimulus is delivered to neurone membrane while it is undergoing the events of an action potential, a second impulse will not be produced. That is, a second action potential cannot be generated during the time course of the first spike. Action potentials therefore cannot summate and become superimposed on each other. During this period of unresponsiveness the nerve membrane is said to be in a **refractory state**. Whilst a spike is taking place, a second impulse will not be produced regardless of the intensity of the second stimulus. This time interval is referred to as the **absolute refractory period**. It generally lasts for no more than 1 ms, and may be considerably shorter (e.g. 0.5 ms) in some large mammalian myelinated nerve fibres.

Following the absolute refractory period there is an interval of some 3–5 ms when it is more difficult than normal to elicit a second action potential, i.e. a stronger stimulus than that normally required to reach threshold is necessary. This time interval is referred to as the **relative refractory period**. Action potentials produced during this interval are somewhat distorted in amplitude and duration. The rate of depolarization of the membrane and the final amplitude of the action potential are both reduced below the norm. The increase in stimulus intensity that is required to fire the neurone represents a rise in the threshold level of depolarization.

It may at first sight appear as though the 'all-or-none law' is not obeyed here, since action potentials of variable amplitude can be produced during the relative refractory period. However, for any specified degree of refractoriness, the action potential amplitude and duration will be constant. Under these specified conditions the stimulus is either large enough to produce all of the response or too small to elicit a propagated impulse.

Changes occurring in the membrane permeability to Na^+ and K^+, associated with an action potential, can explain the phenomenon of the refractory state of nerve membrane. During the rising phase of the action potential, Na^+ gates are open and there is a positive feedback operating between the degree of membrane depolarization and the increase in Na^+ permeability. If a second stimulus is delivered during the course of these events, any effects it might have on membrane permeability will be masked, since Na^+ channels are already open. The second stimulus cannot open up another quite separate set of Na^+ gates. Thus, a stimulus occurring during the rising phase of the spike (the first stage of the absolute refractory period) can have no detectable effect on the changes in membrane polarity and permeability. During the falling phase of the action potential, Na^+ permeability is inactivated and K^+ permeability is increased. Both of these changes tend to restore the resting potential and therefore oppose any attempt to depolarize the membrane. In fact, Na^+ inactivation is not simply the closure of gates in the membrane, but appears to be a separate physicochemical event analogous to locking the gates. If membrane depolarization in response to a stimulus is to exceed a threshold level, then Na^+ current into the cell must exceed K^+ current out of (plus Cl^- current into) the cell. Sodium inactivation associated with the falling phase of the spike coupled with the corresponding increase in K^+ permeability makes it impossible for such a threshold to be reached (the second stage of the absolute refractory period). During the ensuing interval of the relative refractory period, Na^+ inactivation wears off and K^+ permeability also reduces, thereby increasing the ease with which threshold can be reached. An increased stimulus intensity during this time can overcome the opposing factors and permit Na^+ entry to exceed the net K^+ and Cl^- flux. At the end of the refractory period, Na^+ and K^+ permeabilities return to normal and a normal action potential can be generated.

PROPAGATION OF ACTION POTENTIALS

Nerve impulses can be generated at one end of an axon and recorded at the other end. That is, impulses can be propagated along the axon. The mechanism by which action potentials apparently move along nerve fibres will now be considered.

Propagation in non-myelinated nerve fibres
When a stimulus is delivered to a nerve fibre it elicits local changes in membrane permeability which then give rise to an action potential. In other words, the stimulus opens Na^+ gates only at the point of stimulation and not over the whole surface area of the axon. A tiny segment of membrane is activated and this lies adjacent to the remaining resting membrane. The transmembrane potential is quite different in the active and resting segments. In the active region the inside of the axon is positive with respect to the outside by about $+40\,mV$, whilst the normal resting potential of around $-70\,mV$ exists elsewhere. This is illustrated in Fig. 2.1.7a. Similar changes occur in the extracellular fluid. These potential differences cause currents in the form of ions to flow. Current (denoted as positive charge) flows through the cytoplasm from the active to the resting segment of the axon and in the reverse direction through the extracellular fluid. (These local currents are shown in Fig. 2.1.7a.). In fact, recorded electrical changes associated with local current flow occurs much more rapidly than can be explained by the actual physical transfer of ions between active and resting segments of the axon. The local current flow really represents a 'knock-on' effect whereby the entry of a positive charge (Na^+) through the active segment membrane causes a realignment of ions already in the axoplasm, since like charges tend to attract and unlike charges to repel each other. Thus, the inside surface of adjacent membrane becomes more positive and the outside more negative, both of which tend to depolarize the membrane. This depolarization is normally sufficient to reach threshold, and the adjacent membrane then generates an action potential. In other words, the action potential itself is not propagated but is generated anew at each point on the membrane. The active segment then repolarizes, during which time it is refractory. These changes are illustrated in Fig. 2.1.7b.

Local currents, as the name implies, do not travel long distances down the axon. Local current flow decreases rapidly as the distance from the point of stimulation increases. This is principally because the membrane is not a perfect electrical insulator, indeed, if it were completely impervious to current flow, action potentials would not occur at all. The membrane is said to be leaky. Current flows out of the membrane and is dissipated in the relatively large volume of extracellular fluid. However, current is only required to travel very short distances in order to stimulate adjacent resting membrane.

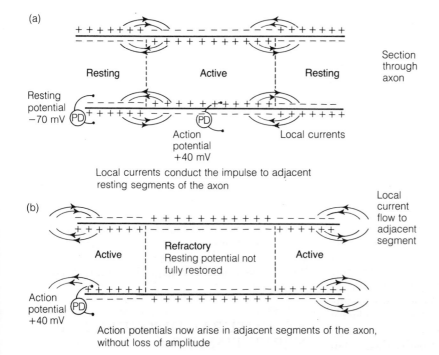

(a)

Section through axon

Resting   Active   Resting

Resting potential −70 mV

Action potential +40 mV

Local currents

Local currents conduct the impulse to adjacent resting segments of the axon

(b)

Local current flow to adjacent segment

Active   Refractory
Resting potential not fully restored   Active

Action potential +40 mV

Action potentials now arise in adjacent segments of the axon, without loss of amplitude

Figure 2.1.7 Propagation of an action potential along a nerve fibre

Direction of propagation

Action potentials are always propagated away from the point of stimulation. This is because each activated region subsequently becomes refractory and cannot be restimulated by local currents (see Fig. 2.1.7b). Within the nervous system, impulses are generally only conducted in one direction along a nerve fibre. This is not because of any inherent property of axons, indeed in isolated tissue, signals can be transmitted in either direction. However, the anatomy of the nervous system results in stimuli being delivered to one particular end of each fibre. Moreover, the junctions between neurones only allow transmission in one direction. These junctions, called synapses, are described later (see p. 39).

Propagation in myelinated nerve fibres

The mechanism of action potential propagation in myelinated nerve fibres is essentially the same as in non-myelinated fibres. However, the segmented myelin sheath alters the electrical properties of the fibre. The sheath increases the electrical resistance of the membrane, which is only appreciably leaky to charge at the nodes of Ranvier. An action potential sets up a local current flow but in mye-linated fibres the circuit is only completed when current leaks out through the membrane at an adjacent node of Ranvier. In other words, the impulse is transmitted by means of local current flow along the myelinated segments of the axon and action potentials are only generated at the nodes (Fig. 2.1.8). Action potentials therefore appear to jump from node to node, and impulse conduction in myelinated nerve fibres is referred to as **saltatory conduction** (from the Latin *saltare*, meaning to leap). Saltatory conduction is considerably faster than that occurring in non-myelinated neurones. In non-myelinated axons new action potentials have to be generated across each adjacent segment of membrane. That is, for a given length of axon the impulse has to be regenerated more times in the non-myelinated compared with the myelinated neurone. The process of impulse generation involves the physical transfer of ions across the membrane, which is a much slower process than the charge realignment consequent on local current flow. Thus, an important consequence of myelination in the nervous system is that it speeds up nerve signalling. Rates of up to 120 m/s can be recorded in large myelinated motor neurones.

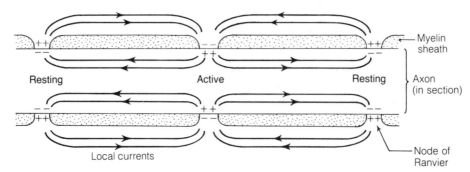

Figure 2.1.8 Saltatory conduction in a myelinated nerve fibre. Local currents conduct the impulse from node to node. The action potential is regenerated at each node of Ranvier

Velocity of conduction of nerve impulses
The velocity at which nerve impulses are conducted is variable and depends primarily on two factors, myelination and axon diameter. Myelinated nerve fibres conduct impulses faster than non-myelinated fibres of comparable diameter. In addition impulse conduction velocity depends on the diameter of the nerve fibre. Large fibres conduct impulses more rapidly than small fibres. This is because the longitudinal resistance to current flow decreases as axonal diameter increases and so current flows more readily and more rapidly in large fibres.

Classification of nerve fibres according to their conduction

VELOCITIES

Group A fibres are all myelinated and conduct impulses at speeds of up to 120 m/s. They are subdivided into α, β, γ and δ fibres. Group B fibres are also myelinated and are all preganglionic fibres

of the autonomic nervous system (see Chap. 2.4). Group C fibres are all non-myelinated and conduct impulses at speeds as low as 1 m/s. This classification system is summarized in Table 2.1.2.

STRUCTURAL DISORDERS OF NERVE TISSUES

Neurones and neuroglia may become damaged from a wide range of factors, including malignancy, dietary deficiencies, and heavy metal exposures. Damage may occur in the CNS or in the PNS and may affect sensory pathways and/or motor pathways and/or higher functions such as memory and intellect.

Malignancy of the nervous system is usually due to abnormal reproduction of neuroglia; hence the term **gliomas**. Gliomas can occur at any site and may compress and damage surrounding structures so that nervous dysfunction may be widespread.

Peripheral neuropathy results from degener-

Table 2.1.2 Classification of nerve fibres according to conduction velocity

Group	Divisions	Function	Diameter (μm)	Conduction velocity (m/s)
A All myelinated	Alpha (α) Beta (β) Gamma (γ) Delta (δ)	Motor and sensory nerve fibres. Largest fibres innervate muscle; smallest fibres convey pain and temperature	α and β 5–20 μ γ and δ 1–7 μ	α fastest and δ slowest α and β 45–120 γ and δ 2–45
B Myelinated	–	Preganglionic nerve fibres of ANS	Up to 3	3–15
C Non-myelinated	–	Some sensory nerve fibres; postganglionic fibres of ANS	Approx. 0.2–1.5	Up to 2.5

ation of the nerves, principally in the feet. It can be a complication of diabetes mellitus, alcoholism, vitamin B deficiency and other systemic disorders. Trauma can similarly cause neuropathy, e.g. a radial nerve may be trapped due to the prolonged use of crutches.

All of these patients will be in danger as a result of the loss of protective sensory input and therefore must avoid contact with extremes of temperature, e.g. hot-water bottles. The nurse should be aware of this vulnerability when planning nursing care and should pay particular attention to the skin over the affected area.

Multiple sclerosis (MS), a demyelination disorder of the nervous system, is manifested by the presence of hardened plaques in the myelin sheath. The cause is unknown, although research has indicated that MS sufferers have an increased susceptibility to viral infections. It is a disorder of early adulthood, often presenting between the ages of 20 and 30, and shows a slight female bias.

As the plaques occur at random, their presence will affect different functions of the nervous system to varying degrees in each individual, hence the illness is characterized by a series of remissions and relapses. It is this diversity of presentation which can make diagnosis difficult and outcome often impossible to predict. Some patients will quickly become wheelchair bound and fully dependent, while others remain in remission for a prolonged period of time. The potential nursing problems that may present are wide ranging. Psychological care is of the utmost importance; fears about the disease and for the future, based on limited knowledge, can result in the patient regressing. They will not communicate with other patients or staff and may adopt an attitude of complete helplessness despite the fact that they may be capable of performing many physical tasks. This can be very distressing for the family, and it is essential that the nurse ensures that there is always an opportunity to discuss the situation with both the relatives and the patient, and answers their questions as honestly as possible within the bounds of her responsibility.

Physical care will focus on managing urinary and bowel problems and those related to skin care. Incontinence with overflow often occurs, along with the attendant problem of infection. Management will include the use of simple external urinary collection devices or bladder catheterization. Female patients are more difficult to manage, as no satisfactory external devices yet exist. Some patients respond to bladder training programmes. Constipation frequently occurs, and management of this does not vary from that of any other patient. Loss of sensation in the skin will expose the patient to the danger of breakdown and infection. Education of the patient with regard to the dangers will help to obviate the situation, with the emphasis being placed on regular relief of pressure.

A patient and family teaching programme should be instituted immediately upon diagnosis. It should include advice on care of the bladder, bowels and skin, e.g. teaching the relatives how to transfer the patient safely from chair to bed and emphasizing the need for scrupulous personal hygiene. An explanation of the drug regimen and any side-effects should be provided, together with advice about returning to work and following as normal as possible a social life. The plan should also deal with problems relating to sexuality and personal relationships and availability of financial assistance.

SYNAPTIC TRANSMISSION

For a nerve impulse to travel from one nerve cell to another, a mechanism is required which will allow the nerve impulse to be transmitted across the small interstitial gap between the two cells. The specialized junctions between nerve cells or between nerve and muscle cells are called **synapses**, and the process by which a nerve impulse is transferred across a synapse is referred to as **synaptic transmission**.

Electrical transmission

Essentially what is required of synaptic transmission is for the action potential generated in one nerve cell to be transferred to the next excitable cell. The simplest scheme, conceptually, for such a process would be passive current flow. The process of passive current flow which is responsible for conduction along axons might also be capable of transferring a nerve impulse directly across a synapse. Such a mechanism would require the unit membranes of both cells to possess a very low resistance, and also the distance between the two cells, the synaptic gap or cleft, to be very

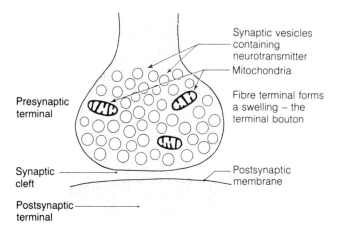

Figure 2.1.9 Ultrastructure of a neuronal synapse

narrow. The latter is necessary if the current flow is not going to be diffused away from the next cell by the extracellular fluid, which has a relatively low electrical resistance.

Electrical synaptic transmission of this kind has been described at synapses in which the outer layers of the unit membranes of the two cells are separated by as little as 2 nm. Indeed, the unit membranes may even be fused at these tight gap junctions. The process of electrical synaptic transmission assumes that the properties of axons which account for the passive transmission of signals along the nerve fibre will be sufficient to explain the transmission of signals between neurones. However, such assumptions do not seem to be acceptable for the vast majority of synapses. For example, the distance between cells at a synapse will normally be 20 nm or more, rather than the 2 nm of the tight gap junctions. Observations such as these suggest that some other process is involved in transmission at synapses.

Chemical transmission

The vast majority of synaptic connections seem to depend on a process involving chemical transmission of the nerve signal. The arrival of an impulse at the terminal of the presynaptic nerve fibre causes the release of a small amount of chemical agent called the **chemical transmitter**. The chemical transmitter diffuses across the synaptic cleft and, when it reaches the postsynaptic cell, it has the effect of altering the potential difference which exists across the postsynaptic cell membrane. This alteration in the transmembrane potential difference may then give rise to an action potential.

STRUCTURE OF A CHEMICAL SYNAPSE

The major features of a synapse are illustrated in Fig. 2.1.9. In addition to components of the synapse that have already been mentioned, such as the separation of the unit membranes of the cells by the synaptic cleft, other important features can be seen. For example, the spindly fibre of the presynaptic axon swells into a round or oval-shaped knob at its end. This is commonly referred to as the **terminal bouton** or **knob**. It is the flattened membrane of the terminal bouton that is juxtaposed to the next cell and which forms the presynaptic membrane. The area covered by the presynaptic membrane is something of the order of 1–2 μm. Within the terminal bouton, other intracellular structures have been identified. In particular, mitochondria are commonly observed and their presence suggests that the cellular activities of nerve terminals are energy-requiring ones.

An important and distinctive cellular organelle of the terminal bouton is the **synaptic vesicle**. Synaptic vesicles are tiny sacs which occur in great profusion at the nerve ending and are formed by pinocytosis, the pinching off of segments of the unit membrane of the cell. They show a tendency to move towards the presynaptic membrane, where they fuse with the membrane and extrude their contents into the synaptic cleft by the process of exocytosis. Synaptic vesicles range in size from 40 nm to 200 nm.

Stage 1: Release of chemical transmitter

Presynaptic
terminal

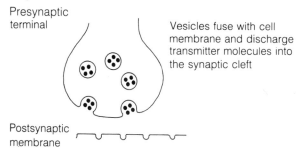

Vesicles fuse with cell
membrane and discharge
transmitter molecules into
the synaptic cleft

Postsynaptic
membrane

Stage 2: Diffusion of chemical transmitter

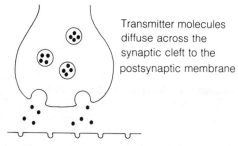

Transmitter molecules
diffuse across the
synaptic cleft to the
postsynaptic membrane

Stage 3: Attachment of transmitter
to postsynaptic membrane

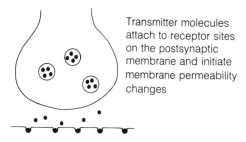

Transmitter molecules
attach to receptor sites
on the postsynaptic
membrane and initiate
membrane permeability
changes

Stage 4: Removal of transmitter; recovery

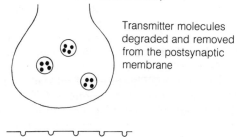

Transmitter molecules
degraded and removed
from the postsynaptic
membrane

Figure 2.1.10   Chemical transmission at a synapse

MECHANISM OF CHEMICAL TRANSMISSION

Electrical signals are passed from one cell to the
next across a synaptic cleft by means of an intermediate stage of chemical transmission. It is possible to identify four stages of the transmission and
these are summarized in Fig. 2.1.10. First, a
chemical is released from the presynaptic membrane in response to the nerve impulse. Then the
chemical moves from the presynaptic to the postsynaptic membrane, to which it attaches and where
it exerts its effects. Finally, a state of readiness to
transmit another signal is re-established.

In discussing these stages, reference will be
made to the synapse at the skeletal neuromuscular
junction, since much of the detail of synaptic
transmission has emerged from experiments
involving the neuromuscular synapse. However,
it must be noted that a nerve-to-muscle synapse
differs in certain ways from the typical nerve-to-
nerve synapse. For example, the morphology of
the neuromuscular synapse is distinctive. The
motor nerve terminals run in specialized channels
that invaginate the muscle cell membrane. Such
channels are not seen in the postsynaptic cell of a
nerve-to-nerve synapse. Nevertheless, it can be
assumed that the processes involved in neural
synaptic transmission follow a similar pattern to
those that have been found at the neuromuscular
synapse.

NEUROTRANSMITTERS

One aspect of the synapse at the skeletal neuromuscular junction that has been clearly demonstrated is the nature of the **neurotransmitter**
involved. It is known that this transmitter is
acetylcholine. Similarly, it is known that acetylcholine and noradrenaline are both neurotransmitters in the autonomic nervous system (see
Chap. 2.4).

Acetylcholine is considered to act as a central
transmitter in addition to its role at the neuromuscular junction. It belongs to a group of compounds referred to as monoamines since each
chemical only contains a single amine group.
Other monoamines that have been identified as
central excitatory transmitters (i.e. chemicals promoting impulse transmission) include serotonin
(5-hydroxytryptamine), dopamine and noradrena-

line. The last two of these (dopamine and nor-adrenaline) belong to a subgroup of monoamines called catecholamines.

Certain amino acids have also been identified as central neurotransmitters. Gamma-aminobutyric acid (GABA), an amino acid that is not incorporated into proteins, appears to be a common inhibitory transmitter (i.e. a chemical inhibiting impulse transmission) in the brain. Glycine has been shown to act as an inhibitory transmitter in the spinal cord. Other amino acids considered to act as neurotransmitters include glutamic acid and aspartic acid.

In addition to the above, an increasing number of **neuropeptides** have been identified and are regarded, with varying degrees of certainty, as central neurotransmitters. These substances, which consist of short chains of amino acids, have been located at neural synapses, although some are found in other body areas as well as within the nervous system. This group of substances includes adrenocorticotrophic hormone (ACTH), leucine enkephalin, methionine enkephalin, endorphin, angiotensin II, luteinizing hormone releasing hormone, somatostatin, and substance P. Reference will be made to some of these in later chapters.

There seems little doubt that the list of substances identified as centrally acting neurotransmitters will increase in the future.

STAGE 1 OF CHEMICAL TRANSMISSION: TRANSMITTER SYNTHESIS, STORAGE AND RELEASE

Biosynthesis of the transmitter substance within the terminal bouton has been described for several neurotransmitters, e.g. acetylcholine and the catecholamines.

Release of the chemical transmitter from the presynaptic membrane is achieved by the alteration of membrane permeability caused by the arrival of an action potential at the terminal bouton. Exactly how this occurs is not understood, although evidence points to the involvement of Ca ions. The arrival of a nerve impulse at the terminal bouton is followed by an influx of Ca ions. The sharp but brief rise in free Ca levels within the terminal bouton apparently leads to the fusion of synaptic vesicles with the presynaptic membrane. The fusion of the synaptic vesicles with the presynaptic membrane and the extrusion of their contents into the synaptic cleft are seen as the way

in which the presynaptic cell releases its transmitter substance.

The presence of vesicles at chemical synapses has led to the widely held hypothesis that they are responsible for transmitter storage and release. It has been demonstrated, for example, that transmitters are secreted in multimolecular packets – or **quanta** – rather than in a continuously graded quantity. In other words, the smallest amount of a transmitter that can be released would be one quantum, perhaps 10 000 molecules of transmitter, the next smallest amount would be two quanta, and so on. Quantal release of transmitter is compatible with a vesicular theory of transmitter release.

STAGE 2 OF CHEMICAL TRANSMISSION: DIFFUSION OF TRANSMITTER

The second stage of synaptic transmission is the passage of the transmitter across the synaptic cleft. This is assumed to occur by means of passive diffusion. The time lag of 0.5 ms or more between the arrival of an impulse at the terminal bouton and the observation of its effects on the postsynaptic cell membrane is congruent with the notion of passive diffusion.

STAGE 3 OF CHEMICAL TRANSMISSION: ACTION AT THE POSTSYNAPTIC CELL

The third stage of synaptic transmission is the action of the transmitter at the postsynaptic cell membrane. On arrival at the postsynaptic cell, the chemical transmitter initiates a change in the potential difference across the membrane. This change can either depolarize the cell, that is, reduce the potential difference and increase the likelihood of action potential generation, or it can hyperpolarize the membrane so that the resting potential is increased and the chance of an action potential occurring is reduced.

It seems that there are at least two processes whereby a chemical transmitter can cause an alteration to the postsynaptic membrane potential. First, the transmitter substance can act at specific receptor sites on the postsynaptic membrane, directly opening ionic gates. The alteration in the membrane potential is thus attributed to the passive diffusion of ions through these opened gates. For example, the excitatory effects of acetylcholine at the neuromuscular junction are attributed to simultaneous increases in permeability of the postsynaptic membrane to K^+ and Na^+. Since Na^+ are far from equilibrium at rest, Na movement will dominate, despite the increase in K^+ permeability, and depolarization occurs. Again, the hyperpolarizing effects of the inhibitory transmitter GABA (γ-aminobutyric acid) are attributed to the opening of Cl^- gates. A rise in Cl^- permeability will maintain the resting potential and resist any tendency towards depolarization, although an actual increase in the transmembrane potential difference may not be seen.

The second mode of action of the transmitter substance is less direct. Here, the transmitter is seen as triggering a chemical reaction, or a series of reactions, in the postsynaptic cell. Under this system the transmitter binds with a receptor which is linked to a 'second-messenger'. The second-messenger reacts within the cell to effect certain cellular changes. For example, the central neurotransmitter dopamine has been linked with the action of the postsynaptic membrane-bound molecule adenylcyclase. Stimulation by dopamine results in adenylcyclase catalysing the transformation of cellular ATP to another molecule called cyclic adenosine monophosphate (cAMP). This molecule then acts as a second-messenger and initiates the postsynaptic response. Details of how this process eventually leads to the alteration of the membrane potential of the postsynaptic cell are still not understood.

Central to the question of the manner in which transmitters affect postsynaptic cells is the concept of **receptor sites**. Receptors are assumed to be molecules on the postsynaptic membrane which are responsible for recognizing the chemical involved in synaptic transmission. Recognition of the transmitter will be the first event occurring at the postsynaptic cell, regardless of whether that cell is another nerve cell or a muscle cell. A receptor needs to be specific in its recognition of a particular chemical compound, yet any one cell may respond to a number of different kinds of receptor distributed throughout their membranes. To add to the complexity, it has been found that any one neurotransmitter can act on more than one type of receptor. For example, acetylcholine has been found to act on muscarinic and nicotonic receptors. **Muscarinic receptors** are so called because at these sites muscarine has actions similar to those of acetylcholine. These sites occur in smooth muscle, cardiac muscle and

exocrine glands. The receptors can be competitively blocked by atropine. On the other hand, **nicotinic receptors**, which occur in the autonomic ganglia and in skeletal muscle, are insensitive to muscarine but respond to nicotine as well as to acetylcholine. A further complication arises because the nicotinic receptors in the autonomic ganglia and in skeletal muscle are not identical. The effects of acetylcholine in the autonomic ganglia are blocked by hexamethonium, while the receptors at the skeletal nueromuscular junction are blocked by d-tubocurarine. d-Tubocurarine is used clinically as a muscle relaxant during surgery, or when a patient requires assisted ventilation.

Postsynaptic potentials
The binding of neurotransmitter with receptors on the postsynaptic membrane leads to a change in the potential difference across that membrane. These effects can be recorded in the postsynaptic cell and are called **postsynaptic potentials**. Postsynaptic potentials can be excitatory or inhibitory, and are consequently referred to as **excitatory postsynaptic potentials (epsps)** or **inhibitory postsynaptic potentials (ipsps)**. The excitatory postsynaptic potentials recorded at the neuromuscular junction are known as **end-plate potentials (epps)**. Certain features of a postsynaptic potential should be noted. First, it shows a continuously graded response up to a maximum; second, it spreads for only short distances, perhaps 1 mm or less, before becoming severely attenuated; and third, it has a relatively long duration, lasting 10–15 ms.

Excitatory postsynaptic potentials and end-plate potentials represent a simultaneous increase in Na^+ and K^+ permeability of the postsynaptic membrane, caused by the binding of transmitter molecules with receptors. The fact that membrane permeability to both Na^+ and K^+ occurs simultaneously distinguishes this process from the time-locked changes in membrane permeability to Na^+ and K^+ which lead to action potential generation. It might be thought that the simultaneous opening of Na^+ and K^+ gates would lead to no change in transmembrane potential, since both ions would flow down their concentration gradients and consequently cations would flow into and also out of the cell. However, under resting conditions, K^+ is much nearer its equilibrium than Na^+ and as a consequence of this there will be a tendency for Na^+ entry to dominate. This will lead to a net gain of positive charge by the cell and a corresponding tendency to depolarize the postsynaptic membrane. In contrast, an inhibitory postsynaptic potential results from changes in membrane permeability to K^+ and Cl^-. These changes will stabilize the transmembrane potential difference at its resting level and may even hyperpolarize the postsynaptic cell.

The continuous gradation of synaptic potential amplitude implies that the size of the change in potential difference across the postsynaptic membrane is a function of the number of receptors affected by the transmitter and hence the amount of transmitter release. In other words, the greater the number of postsynaptic membrane receptors that are affected, the greater the potential change.

At the neuromuscular junction, one impulse travelling along the motor neurone will be sufficient to depolarize the muscle fibre. This will not be the case at a neural synapse, where the situation is more complicated. Any one nerve cell may have many hundreds or even thousands of other nerve cells making contact with it. Moreover, some contacts will be inhibitory and some excitatory, and they will not all be active at the same time. The activity of the cell will therefore depend on the aggregate effects of all the excitatory and inhibitory influences operating on it at any one time. For example, if it is assumed that the cell is dominated by excitatory influences, then the spread of depolarization from the excitatory synapses will tend to spread along the dendrites towards the axon hillock (axonal bulge adjacent to neuronal soma). The cell membrane of the axon hillock has a much lower threshold than the rest of the cell membrane. Consequently action potentials are produced more readily and more rapidly here. The axon hillock is therefore the site of generation of action potentials in the postsynaptic cell. On the other hand, if inhibitory influences dominate, then any depolarization will be countered and will not be sufficient to reach threshold level.

Since the axon hillock is remote from the synaptic connections on the dendrites, it permits all epsps and ipsps occurring at any one time to be algebraicly added together. At the dendrites, the transmembrane potential changes reflect the activity of local synapses rather than all of the synapses acting on the cell. The much lower threshold of the axon hillock membrane ensures that action potentials are generated at the axon hillock, where all synaptic activity can be taken into account.

STAGE 4 OF CHEMICAL TRANSMISSION: RECOVERY

The final stage of chemical neurotransmission is that of recovery. Recovery is the process involved in preparing for another impulse to arrive at the synapse. In order that each impulse remains a discrete signal, the transmitter substance released as a result of each impulse must be inactivated or removed by the time the next one arrives at the synapse. Removal will partly be achieved by diffusion of the transmitter away from the area of the synapse, but diffusion is a relatively slow process compared with the ability of a nerve to generate action potentials every 3 or 4 ms. The major components of the process of re-establishing a state of readiness are biochemical deactivation of transmitter and reuptake of transmitter by the presynaptic neurone.

Deactivation of the transmitter substance by enzymatic action has been described for several neurotransmitters. For example, **acetylcholinesterase**, which is localized in postsynaptic neuronal membranes, hydrolyses and hence deactivates acetylcholine. This process transforms acetylcholine into choline plus acetate. The choline is then available for reabsorption by the presynaptic neurone, where it will be used once more for the synthesis of acetylcholine.

It has been suggested that an active reuptake mechanism, an amine pump, in the presynaptic neurone can account for the removal of most of the noradrenaline released at adrenergic synapses. The portion of noradrenaline that is not accounted for in this way will be deaminated by monoamine oxidase.

The fact that neurotransmitters are deactivated by enzymatic action means that nerve signalling can be influenced by drugs which affect these deactivating enzymes. In other words, in addition to affecting directly synaptic transmission by blocking or mimicking the neurotransmitter, pharmacological agents can affect transmission by means of their action on the enzymes which degrade the neurotransmitters. For example, monoamine oxidase inhibitors, which are used clinically for their antidepressant effects, cause an increase in the catecholamine content of adrenergic nerves.

SYNAPTIC INTEGRATION

The nervous system is responsible for selecting and integrating information, enabling the individual to make sense of the world and to respond appropriately. This process necessitates the careful channelling and filtering of information, as can be seen in many aspects of everyday behaviour. For example, people living in houses located by a busy road tend to be less aware of the traffic noise than their visitors. The noise is a constant background and is presumably filtered out of their conscious awareness. Again, when someone accidentally touches a hot saucepan the immediate response is a reflex withdrawal of the hand. In addition to this, several other responses occur: the individual orientates his gaze towards the saucepan, he alters his posture to compensate for the sharp withdrawal of the hand, and he may utter a scream or an appropriate verbalization. These separate components of the behavioural response are controlled by different parts of the nervous system, and for them to be integrated into a complete piece of behaviour the relevant information must be channelled appropriately.

The process of integrating information depends upon the synapse, since the synapse represents the basic switching mechanism of the nervous system. This basic mechanism yields great flexibility. For example, the human brain consists of something of the order of 10^{11} nerve cells, and any one of these cells may have many hundreds of synaptic contacts impinging on it. This demonstrates the immense capacity that the nervous system possesses for integrating information. The following discussion considers how this integration takes place.

Temporal and spatial summation

A single epsp at a single synapse will almost never cause an action potential in the postsynaptic cell. However, one epsp may not be completed before another action potential arrives at the synapse. Consequently, the change in the transmembrane potential caused by the second epsp will be added to what is left of the change caused by the first. The effects of the two epsps will summate over time, a process known as **temporal summation**. A single epsp at any single synapse may not be sufficient to generate an action potential in the postsynaptic cell, but a train of several epsps coming quickly one after the other may be able to do so. Therefore temporal summation is dependent upon the frequency of firing of the presynaptic cell.

In addition to temporal summation, **spatial summation** can also occur. A single nerve cell may have many hundreds of synaptic contacts impinging on it and hence it is likely that many excitatory synapses will be operating at the same time. Although a single epsp only alters the post-synaptic cell's transmembrane potential by a small amount in a localized manner, the effects of many epsps occurring at different synapses at the same time may summate to reach the threshold level.

It seems then that at any particular time a nerve cell membrane is being shifted either towards or away from its threshold potential by the effects of all its excitatory and inhibitory influences, and that these are combining by the processes of temporal and spatial summation. The generation of an action potential at the axon hillock will depend on these summated inhibitory and excitatory potentials.

Sensitization

When a neuronal pathway is consistently and regularly stimulated, the synapses along the circuit appear to transmit the nerve impulses with increasing ease. In other words, the repetition of firing of a pattern of neuronal connections leads to the synapses becoming sensitized to impulses travelling along that pathway. As a neurone becomes sensitized to a particular pattern of impulses, the influx of Ca^{++} into the presynaptic membrane apparently increases. This in turn leads to an increase in the release of the transmitter, which improves the chances of firing the postsynaptic cell. It is likely that sensitization of neuronal circuits takes place during learning.

REFLEXES

Some of the basic mechanisms whereby neurones influence the behaviour of other neurones and muscles have now been considered. In general, the nervous control of behaviour involves the transmission of nerve signals through complex networks of fibres. However, some aspects of behaviour can be broken down into components which can be investigated to illustrate the relationship between anatomical connections and function. A reflex is such an example of behaviour that can be examined to see how the events occurring at synapses are integrated to produce a response.

A reflex may be defined as an innate involuntary response to peripheral stimulation requiring the presence of part of the CNS. This definition immediately presents some difficulties in deciding whether a particular response is reflex or not. So-called conditioned reflexes would be excluded because the relationship between the stimulus and the response develops after a period of training or conditioning. That is, it develops after birth. Pavlov's famous experiments with dogs illustrate the difference. Dogs salivate when food is placed in the mouth, a reflex response which is innate. Dogs can be conditioned to salivate in response to a ringing bell if during training the bell rings at the time that food is presented. Salivation to the bell is a **conditioned reflex response** and the training process is known as **classical conditioning**. The response of salivation is the same in both cases, but the stimuli eliciting the responses differ. Distinguishing between the two may be useful when examining patterns of behaviour, but is less useful when considering the underlying physiology of the response.

The involuntary nature of a reflex response is also not always clear cut. The withdrawal reflex that occurs in response to pain stimulation is involuntary, but it can be modified, depending on the individual's perception of the circumstances in which stimulation occurs. If a hot object is accidentally picked up, the withdrawal reflex generally results in the object being dropped. However, if the object happens to be an expensive piece of pottery, the reflex may be overridden and conscious voluntary control may be exerted in order to replace the object without damaging it. Eye blinking will occur reflexly and involuntarily in response to a threatening gesture. However, the same response can occur entirely voluntarily or involuntarily without an obvious stimulus eliciting it.

The stimulus which elicits the reflex response may occur at the body surface or in an internal organ. Strong stimuli which gave rise to the sensation of pain can elicit withdrawal reflex responses when applied to the skin. Light directed through the pupil of the eye will cause a reflex constriction of the pupil, and visual stimuli which are subconsciously interpreted as a threatening gesture may cause reflex blinking. There are many examples of reflexes arising from internal stimulation. These include the reflex maintenance of muscular tone (which will be considered in more detail below), cardiovascular reflexes involved in the maintenance of arterial blood pressure, coughing, swallowing, vomiting and salivary secretion. Some of these responses are quite complex and involve the coordination of a number of muscles. In some cases the response is an increase in a glandular secretion rather than a change in the degree of muscle contraction.

Reflex pathways operate via the CNS. In some cases, the CNS component may be the brain. For example, coughing, swallowing, vomiting and blinking all involve the medulla, which is the part of the brain adjacent to the spinal cord. However, a number of reflexes operate via the spinal cord and do not involve the brain at all. The withdrawal reflex, which is of importance in protecting the body from harm, involves a pathway through the spinal cord. The maintenance of muscle tone (i.e. normal stage of contraction of muscle) is another spinal reflex which is of considerable importance in posture control.

When considering reflexes, it is important to bear in mind that they rarely operate as isolated systems. Nerve axons characteristically branch, and the sensory signal that initiates the reflex response may also be conveyed to other parts of the nervous system. Thus, the activation of a reflex pathway may lead to the activation of other reflexes and also to the activation of the cerebral cortex, the part of the brain which provides for conscious awareness. For example, when a withdrawal reflex operates via the spine to cause the removal of the foot from a sharp object, nerve impulses are also directed along other pathways to initiate maintenance of posture and to elicit conscious awareness of pain. Other actions may follow. The individual may vocalize, grimace, rub the affected area, and so on. These other actions, and the conscious perception of pain, are not part of the reflex as such, but they do form part of the total behaviour of the individual and represent a complex level of integration by the nervous system.

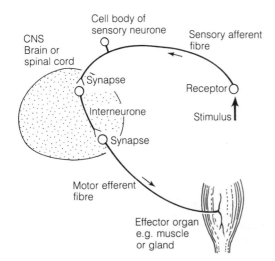

Figure 2.1.11 The component structures of a reflex arc

Component structures in a reflex pathway

The term **reflex arc** is often employed to describe the pathway from the sensory receptor, which detects the stimulus, to the CNS and, via synapses, to the effector organ which produces the response. Indeed, the term 'reflex' indicates that the sensory signal is *reflected* to the muscles.

The component structures in a reflex pathway obviously vary depending on the particular reflex concerned, but some general features can be identified.

SENSORY RECEPTOR

Since a reflex is a response to peripheral stimulation, it is clear that receptors must be present since these are the only structures that can transduce stimulus energy into nerve signals.

SENSORY AFFERENT NERVE FIBRE

Nerve impulses, once generated at receptors, must be conveyed to the CNS by sensory afferent fibres.

SYNAPSE IN THE CNS

Sensory nerve fibres cannot directly synapse onto muscle, therefore nerve impulses generated in receptors must be passed through at least one synapse in the brain or spinal cord before reaching the motor neurones. In general, reflex pathways involve more than one synapse and interneurones form part of the pathway through the CNS.

MOTOR EFFERENT FIBRE (MOTOR NEURONE)

Nerve signals pass out of the CNS via the motor neurones to the effector organs that will make the reflex response.

EFFECTOR ORGANS

The reflex response itself is a change in either muscular contraction or glandular secretion. Muscles or glands form the final structures in the reflex pathway.

The anatomical relationship between these structures are illustrated in Fig. 2.1.11. It is worth noting that a reflex response normally involves many receptors, sensory afferents and motor efferents. Summation therefore occurs at the CNS synapses. The diagrammatic representation of a reflex pathway in Fig. 2.1.11 is therefore an over-simplification.

Characteristics of reflex activity

The majority of reflexes are excitatory and result in increased muscle contraction or glandular secretion. However, some reflexes are inhibitory, and when stimulated cause a discharge of nerve impulses which pass from the sensory afferents to interneurones, which then inhibit motor neurones innervating the effector organs. Reflex inhibition of antagonist muscles generally accompanies reflex excitation of protagonist muscles. For example, when flexor muscles attached to a limb reflexly contract, the corresponding extensor muscles relax, thereby facilitating movement of the limb. This pairing of excitation and inhibition illustrates the principle of **reciprocal inhibition**.

Withdrawal (flexion) reflex

Noxious stimuli applied to receptors in the skin may give rise to the sensation of pain. Pain can be regarded as protective since it directs attention to

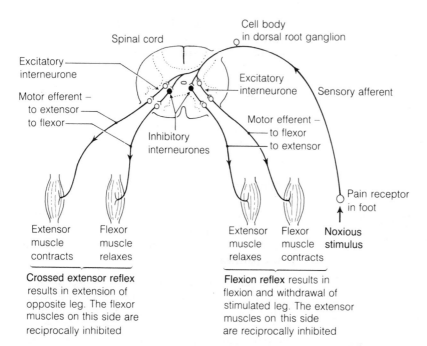

Figure 2.1.12 Withdrawal reflexes: the flexion and crossed extension reflex pathways

a potentially harmful situation which the affected individual may then be able to control. The perception of pain and the interpretation of the circumstances giving rise to it require the activation of a number of different systems in the brain. This activation must take place before the individual can consciously respond to the stimulus. A more immediate response is the subconscious reflex withdrawal of the part of the body affected. This is a very important protective reflex since it permits a more rapid removal from the source of harm than could occur via conscious control of the body.

The withdrawal reflex operates via a pathway through the spinal cord and it exhibits many of the general properties of reflex action which have already been described. The nature of the response depends in part on the site of stimulation, but generally involves a flexion movement, hence the name flexion reflex. It involves more than one synapse in the spinal cord and is termed a **polysynaptic reflex arc**. Temporal and spatial summation occur at these synapses. The number of muscles involved in the response and the tension developed by each muscle therefore depends on the area, intensity and duration of the stimulus.

The flexion reflex pathway is diagrammatically illustrated in Fig. 2.1.12. When the pain receptor is stimulated, nerve impulses are generated which pass through interneurones in the spinal cord and which then excite motor neurones supplying the flexor muscle. The muscle contracts, thereby removing the body from the source of stimulation.

A simple flexion reflex results in contraction and flexion of a muscle together with reciprocal inhibition and relaxation of antagonist muscles. If the intensity of the stimulus is increased, other muscle groups may also reflexly contract. For example, if a person stands on a drawing pin, flexion may occur at the ankle, knee and hips. Even the trunk of the body may flex under these conditions. Moreover, extensor muscles may reflexly contract, resulting in extension of other parts of the body. In the example of standing on a drawing pin, it is of obvious importance that posture is maintained at the same time as the foot is withdrawn, otherwise the person might fall back onto the drawing pin. The opposite leg therefore extends with contraction of the extensor muscles. This associated response is termed the **crossed-extensor reflex**. The withdrawal reflex therefore has the dual function of withdrawal from the noxious stimulus and maintenance of posture.

Crossed-extensor reflex

Nerve signals from the sensory afferents are transmitted, via interneurones, across the spinal cord to the other side of the body (Fig. 2.1.12). These interneurones synapse with motor neurones supplying extensor muscles, which then contract. At the same time, the associated flexor muscles are reciprocally inhibited.

The crossed-extensor reflex is most easily elicited if noxious stimuli are delivered to the foot. Contraction of extensor muscles in the opposite limb to that stimulated is very important for maintenance of posture, and indeed the whole weight of the body may be acting on the extended leg. The crossed-extensor muscles and the motor neurones innervating them are therefore recruited into the withdrawal response.

Plantar reflex and Babinski response

If the sole of the foot is firmly stroked, the toes reflexly curl downwards. This is a healthy plantar response and is said to be flexor despite the fact that the toes apparently show an extension movement downwards. It is more useful to consider this response as flexion since it is part of the normal reflex withdrawal response to noxious stimulation of the foot.

The plantar response is of considerable importance in neurological examinations. Disease of the motor pathways from the brain (the corticospinal tracts) may result in an extensor plantar response. Here, the big toe turns upwards and the other toes spread out when the sole of the foot is firmly stroked. This extensor plantar response was first described by Babinski in 1896 and is often referred to as a positive Babinski response.

The plantar response is also of interest in child development. Infants normally exhibit extensor plantar responses before they are able to stand and walk. The flexor response develops when myelination of the corticospinal tracts occurs, and this corresponds with the infant's ability to stand.

The stretch reflex

The stretch reflex is a spinal reflex in which the stretching of a muscle causes contraction of the same muscle. Sensory afferent fibres from stretch receptors synapse directly onto motor neurones in the spinal cord. Interneurones are not involved, and the reflex is said to be **monosynaptic**. The functional integrity of stretch reflex pathways is essential for normal maintenance of muscle tone.

The stretch receptors are spindle-shaped structures which lie embedded in skeletal muscle, parallel to the muscle fibres. The ends of these **muscle spindles** are attached to the connective tissue framework of the muscle. Thus, a change in the length of the muscle causes a corresponding change in the length of the spindles contained within it. Activation of muscle spindles causes nerve signals to be transmitted to lower centres of the brain which may then modify stretch reflex activity. This will not give rise to a conscious sensation.

The structure of a muscle spindle is illustrated in Fig. 2.1.13. It contains six to ten small specialized muscle fibres, termed intrafusal muscle fibres to distinguish them from the normal extrafusal fibres of the main body of the muscle. The polar ends of the intrafusal fibres are striated and contractile and in essence function like small-scale extrafusal fibres. The middle portions of the intrafusal fibres contain the nuclei and are non-contractile. Thus the centre portion of the whole spindle is also non-contractile. A large afferent nerve fibre is wrapped spiral fashion around the middle portion of the spindle. This is the primary annulospiral nerve ending. This afferent fibre belongs to the nerve fibre group A α- (sometimes called IA). It may approach $20\,\mu m$ in diameter, with a conduction velocity of over $100\,m\,s$. Two secondary sensory nerve endings are also wrapped around the spindle, one on each side of the primary afferent. These are the secondary flowerspray endings. These secondary afferents are smaller, less than $10\,\mu m$ in diameter, and they conduct impulses at much slower velocities. These belong to the A γ group (sometimes called group II). The contractile poles of the spindle have their own motor efferent supply, the gamma (γ) motor neurones, belonging to the A γ group of fibres.

The stretch reflex pathway is illustrated in Fig. 2.1.14. When stretch is rapidly applied to the muscle, the spindles are stretched and this causes discharges of nerve impulses in the spindle afferents.

The spindle afferents convey nerve signals to the spinal cord, entering via the dorsal roots. The

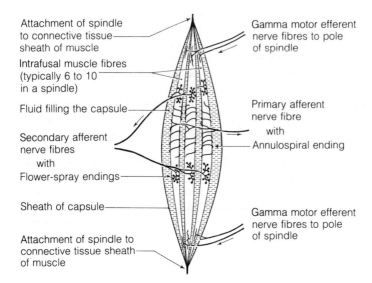

Attachment of spindle
to connective tissue
sheath of muscle

Intrafusal muscle fibres
(typically 6 to 10
in a spindle)

Fluid filling the capsule

Secondary afferent
nerve fibres
  with
Flower-spray endings

Sheath of capsule

Attachment of spindle to
connective tissue sheath
of muscle

Gamma motor efferent
nerve fibres to pole
of spindle

Primary afferent
nerve fibre
  with
Annulospiral ending

Gamma motor efferent
nerve fibres to pole
of spindle

Figure 2.1.13   A muscle spindle

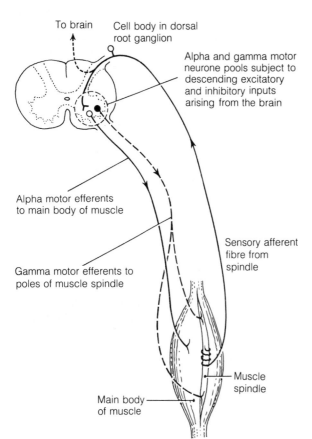

To brain

Cell body in dorsal
root ganglion

Alpha and gamma motor
neurone pools subject to
descending excitatory
and inhibitory inputs
arising from the brain

Alpha motor efferents
to main body of muscle

Gamma motor efferents to
poles of muscle spindle

Sensory afferent
fibre from
spindle

Muscle
spindle

Main body
of muscle

Figure 2.1.14   The stretch reflex pathway. When the
muscle spindle is stretched, the sensory afferent nerve
terminal is stimulated. The resulting sensory signal reflexly
stimulates the alpha-motor neurones and the main body of
the muscle contracts

afferents synapse directly onto the α- motor neurones which supply the extrafusal muscle fibres (α motor neurones belong to group A α). The α motor neurones conduct the signals via the ventral roots of the spinal cord to the muscle, which then contracts and shortens, thereby removing the stretch. As the muscle shortens, the spindles within it shorten and the stimulation of the afferents is reduced. Some afferents terminate on interneurones which simultaneously inhibit antagonist muscles.

This pattern of events occurs in tendon-jerk reflexes, which utilize the stretch reflex pathways. In the knee-jerk reflex, the patellar tendon is sharply tapped and this stretches the quadriceps muscle in the thigh. Muscle spindle afferents in the quadriceps are activated and impulses are conveyed to the α motor neurone pool in the spinal cord. These α motor neurones conduct impulses to the muscle causing it to contract, and the leg kicks forwards and upwards as the quadriceps muscle shortens. An ankle jerk can be elicited in the same way, by sharply tapping the Achilles tendon. In tendon-jerk reflexes, the rate of change of stretch is very important. Pressure slowly applied to the tendon does not cause a jerk response.

Tendon-jerk reflex responses are important in the neurological assessment of patients since they can provide information about the integrity of the stretch reflex pathways. However, the stretch reflex is not only elicited when a tendon is sharply tapped. Stretch can also be applied to muscle spindles by gravitational pull or by internal stimulation arising from the brain. It is by these means that the reflex operates in the control of muscle tone.

When an individual is standing, gravity causes some stretching of postural muscles. This activates the stretch reflex pathway and causes contraction of the muscles. The stretch reflex therefore plays a part in the maintenance of posture.

Golgi tendon organ reflex

The muscle spindles essentially monitor the length of the muscle, but normally when a muscle contracts it develops tension in addition to shortening. Embedded in the tendinous parts of the

muscle are specialized receptors which detect tension development. These receptors are called **Golgi tendon organs**. They consist of bundles of collagen fibres around which are wrapped afferent nerve terminals. When activated, these receptors generate nerve impulses which stimulate inhibitory interneurones in the spinal cord. These interneurones synapse onto the motor neurones and decrease their activity. Golgi tendon organ activity therefore inhibits muscle contraction.

Golgi tendon organs and the tendon organ reflex pathways undoubtedly play a part in the control of muscle contraction, although their precise role is not clearly understood. Signals will also be transmitted from the Golgi tendon organs to motor control centres in the brain, and these centres may then modify the output to the motor neurones. In addition, the Golgi tendon organ reflex appears to have a protective role. If the tension in a muscle increases to an extent which could be damaging and which could cause muscle tearing, then tendon organ reflex inhibition will cause a reduction in muscle contraction and a lengthening reaction. This can be demonstrated if tension is passively applied to a spastic limb in an attempt to bend it. At first the limb resists the movement and tension builds up. The limb then suddenly gives way and bends, 'the clasp-knife' response, and this is caused by reflex tendon organ inhibition of the motor neurones.

Reflexes in the control of posture

The maintenance of posture and balance is accomplished by a number of complex reflexes. The sensory afferent components include vision receptors in the eye, balance receptors in the ear, touch receptors in the skin, stretch and tension receptors in muscle, and various position sense receptors in joints. The CNS components may be in lower centres of the brain or in the spinal cord, whereas the final motor efferent component is the same in all cases, namely the motor neurones in the spinal cord.

In sighted persons, vision appears to be a dominant sensory input for balance control. A person standing on one leg will sway considerably more if the eyes are closed. However, blind persons adapt to the lack of visual input and maintain balance very well, with only a slight loss of precision. In general, it is the combination of all afferent inputs which reflexly determines muscle contraction in maintenance of posture.

Reflexes in neurological assessment

Tendon-jerk reflex testing forms part of routine neurological assessment. A tendon-jerk response will be diminished or even abolished if any part of the pathway from receptor to muscle is damaged. If a patient suffers from muscle disease, for example muscular dystrophy, then reflex responses involving affected muscles will be diminished. Similarly, if either sensory afferent nerve fibres or motor efferent nerve fibres are damaged, the reflex response will be reduced. For example, dorsal spinal nerve root damage which can occur in association with syphilis (tabes dorsalis) causes primary afferent nerve damage with a resultant diminution of affected reflexes. Motor neurone disorders, such as poliomyelitis, damage the efferent pathway and also result in diminution or abolition of affected reflexes. Certain disorders of the brain, such as coma or raised intracranial pressure, may also cause a reduction in reflex responses, presumably because of a reduction in the central control of spinal motor neurone activity. A knowledge of the parts of the body affected and the extent to which reflex responses are diminished contributes to diagnosis of the disorder.

Some individuals quite normally have brisker and more readily elicited tendon-jerk reflex responses than others. However, tendon reflexes may become exaggerated under certain conditions, especially in anxiety states associated with increased muscle tension. Clinically exaggerated tendon-jerk reflexes may indicate damage to the motor pathways from the brain to the spinal motor neurone pool. In this case, the damage has caused an imbalance between excitatory and inhibitory inputs, resulting in excessive stretch reflex activity. In spasticity the stretch reflex is so powerful that the muscles are in a permanent state of extreme tension.

If the spinal cord has been transected, as may occur in road traffic accidents, reflex responses are initially abolished below the level of transection and a state of 'spinal shock' exists. It is believed that the sudden cessation of nerve signals travelling to and from the brain causes disorder of function of even spinal reflexes. However, after a period of time, spinal reflex responses do not only return but are considerably exaggerated.

Review questions

The answers to all these questions can be found in the text. In each case there is at least one correct and at least one incorrect answer.

1 The blood–brain barrier

   (a) prevents access of blood to some parts of the brain
   (b) prevents the passage of some chemicals between the blood and the brain
   (c) is associated with special tight junctions between the cells of capillary endothelium in the brain
   (d) is a less effective barrier in children than in adults

2 The resting nerve cell membrane

   (a) is freely permeable to chloride ions
   (b) is freely permeable to sodium ions
   (c) has a potential difference across it which is negative inside with respect to the outside of the cell
   (d) has a potential difference across it of 40 mV

3 The initiation of an action potential is characterized by

   (a) a rise in sodium ion permeability of the nerve cell membrane
   (b) a rise in potassium ion permeability of the nerve cell membrane
   (c) entry of sodium ions into the nerve cell
   (d) entry of potassium ions into the nerve cell

4 Saltatory conduction

   (a) is the mode of nerve impulse conduction in myelinated nerve fibres
   (b) is generally slower than non-saltatory conduction of a nerve impulse
   (c) occurs when the nerve impulse 'leaps' from neurone to neurone
   (d) occurs when the nerve impulse is propagated as a local current flow between nodes of Ranvier

5 Nerve fibres classified as belonging to group A

(a) are all myelinated
(b) have lower conduction velocities when compared to nerve fibres in the C group
(c) are of larger diameter than fibres in the C group
(d) are all motor nerve fibres

6. At a neuronal synapse

(a) the nerve impulse can be transmitted in only one direction
(b) the synaptic cleft is 40 nm or more wide
(c) the enzyme for degrading acetylcholine is located in vesicles in the presynaptic terminal
(d) the neurotransmitter is actively transported from the pre- to the postsynaptic cell

7 An excitatory postsynpatic potential

(a) obeys the 'all-or-none law'
(b) lasts for about 1 ms
(c) is propagated along the postsynaptic nerve fibre
(d) is caused by simultaneous increase in membrane permeability to sodium and potassium ions

8 The flexion reflex

(a) may occur in response to pain stimulation
(b) operates via a nerve fibre pathway through the brainstem
(c) operates via interneurones in the central nervous system
(d) results in vigorous contraction of flexor and extensor muscles

9 The plantar reflex

(a) is elicited by tapping the knee
(b) is part of the normal withdrawal reflex response following stimulation of the foot

(c) develops when myelination of the corticospinal tracts occurs
(d) is also known as Babcock's response

10 The stretch reflex

(a) is a spinal reflex
(b) is polysynaptic
(c) is essential for maintenance of normal muscle tone
(d) gives rise to a conscious sensation

Answers to review questions

1 b, c and d
2 a and c
3 a and c
4 a and d
5 a and c
6 a and b
7 d
8 a and c
9 b and c
10 a and c

Suggestions for further reading

Eccles, J. C. (1977) *The Understanding of the Brain*, 2nd edn. London: McGraw-Hill.

Kandel, E. R. & Schwartz, J. H. (1981) *Principles of Neural Science*. Amsterdam: Elsevier North Holland.

Katz, B. (1966) *Nerve Muscle and Synapse*. London: McGraw-Hill.

Kuffler, S. W. & Nicholls, J. G. (1976) *From Neurone to Brain*. Sunderland, Mass. Sinauer.

Mitchell, P. H., Ozuna, J., Cammermeyer, M. & Woods, N. F. (1984) *Neurological Assessment for Nursing Practice*. Reston.

Purchese, G. & Allan, D. (1984) *Neuromedical and Neurosurgical Nursing*, 2nd edn. London: Baillière Tindall.

Stein, J. F. (1982) *An Introduction to Neurophysiology*. Oxford: Blackwell Scientific.

Chapter 2.2
Sensory Receptors and Sense Organs

Valerie M. Nie
Michael Hunter
with Douglas Allan

Learning objectives

After studying this chapter the reader should be able to

1 Describe the varieties of form of sensory receptors.
2 Describe the characteristics of a generator potential.
3 Explain how the intensity and locality of a stimulus are coded in nerve signals from sensory units.
4 Define the term 'lateral inhibition' and explain its importance to perception of a stimulus.
5 Describe in outline the structural and functional differences between tactile receptors, proprioceptors, thermal receptors and pain receptors.
6 Describe the mechanism of the production and drainage of tear fluid in the eye.
7 Draw a labelled diagram of a cross-section through a human eyeball and describe the functions of the labelled parts.
8 Account for the direction and extent of refraction of light as it passes through the eye to the retina.
9 Define the term 'power of accommodation' and explain its role in normal vision.
10 Distinguish the differences in the optics of normal vision, myopia, hypermetropia and astigmatism.
11 Describe the structure of the retina.
12 Describe in outline the photochemical reaction between light and rhodopsin.
13 Define the term 'visual acuity' and describe a method of its measurement.
14 Describe how changes in the electrical potential of photoreceptors are transmitted and processed by cells of the intermediate layer of the retina.
15 Identify and describe the functions of the major components of the auditory apparatus of the ear.
16 Describe the structure and innervation of the organ of Corti.
17 Define the term 'endolymphatic potential'.
18 Explain how perceived pitch and loudness relate to the frequency and intensity of sound vibration impinging on the ear.
19 Describe the structure of the vestibular apparatus of the ear.
20 Explain the mechanism by which the vestibular apparatus responds to rotational movements of the head and changes in position of the head relative to gravitational pull.
21 Outline the contribution of the vestibular apparatus to the control of posture and eye movement.
22 Describe the structure of a taste bud and illustrate the distribution of taste buds in the mouth.

23 Describe the structure of olfactory epithelium.
24 Describe in outline the mechanisms of taste and smell.

Introduction

Changes in both the external and internal environment of the body are detected and coded into nervous impulses by sensory receptors. The central nervous system is therefore continuously appraised of conditions within and outside the body. The CNS integrates the many sensory inputs, it 'compares' them with previous experience, and makes appropriate responses. Sensory signals may be integrated in parts of the brain or spinal cord which do not give rise to conscious awareness, for example in spinal and brainstem reflexes.

Sensory receptors are either adapted nerve fibre terminals or specialized cells associated with nerve terminals. They are diverse in structure, and together they convey the many qualities of sensation that can be experienced. These include touch, pressure, pain, temperature, light, sound, taste and smell. Sensory receptors also convey information which assists in the maintenance of balance and motor control, in regulating the chemical composition and hydrostatic pressure of arterial blood, and in regulating gut motility.

Sensory afferents arising from the body are grouped together as the **somatosensory system**. Somatosensation includes sensory information arising from the skin, the muscles and joints and the viscera. The **special senses** are associated with specialized sense organs in the head. They include vision, hearing, balance, taste and smell.

This chapter is principally concerned with the structure and function of sensory receptors and sense organs. The pathways conveying sensory signals to the CNS together with the CNS centres for integration of sensory inputs are described in Chapter 2.3.

CLASSIFICATION OF RECEPTORS

Sensory receptors can be classified on a structural or functional basis. Anatomical studies of receptors have revealed a bewildering variety of structural types. Certain common anatomical features permit a simplified structural classification consisting of only three groups: unspecialized free nerve endings; specialized or encapsulated nerve endings; and specialized non-neuronal receptor cells.

Unspecialized free nerve endings are found in considerable numbers in both the epidermis and the dermis of the skin and also in muscles and the viscera. Free nerve endings are certainly involved in detecting painful stimuli. However, in parts of the body which contain only this type of receptor, for example the cornea which covers the exposed surface of the eye, free nerve endings may also detect touch, pressure and temperature. It is unclear to what extent free nerve endings elsewhere convey these other sensations. Those in the epidermis are probably associated with touch and/or temperature since the superficial epidermis apparently does not contain pain receptors and it can be peeled off without eliciting pain.

Specialized nerve endings are found in muscle spindles, Golgi tendon organs and in olfactory epithelium in the nose. They are also found in abundance in the skin. Some are found in the deeper growing layer of the epidermis, notably the tactile (Merkel's) discs, which are believed to detect light touch. Various encapsulated nerve endings exist in the dermis. Meissner's corpuscles and basket nerve endings wrapped around hair bases are both thought to respond to touch, whereas the deeper lying Pacinian corpuscles are believed to respond to deeper pressure. Krause's end-bulbs and Ruffini's endings have been associated with cold and warmth detection respectively, but many studies have shown that free nerve endings can be temperature detectors, and Krause's end-bulbs and Ruffini's endings are also candidates for pressure receptors.

Specialized non-neuronal receptor cells are found in the ear, the eye and in the taste buds of the mouth. In response to appropriate stimuli, these cells produce electrical potential changes which influence the generation of action potentials in associated afferent nerve fibres.

Sensory receptors can be functionally classified according to the nature of the stimulus energy to which they respond. There are mechanoreceptors, chemoreceptors, thermoreceptors, and electromagnetic receptors.

Mechanoreceptors are sensitive to mechanical

energy and exist in many forms. Touch and pressure cause mechanical deformation and hence stimulation of mechanoreceptors found in the skin. Stretch receptors in muscle spindles, in the gut, in lung tissue and in blood vessels (baroreceptors) similarly respond to mechanical energy. Sound receptors and balance receptors in the ear respond to mechanical pressure waves in air, and gravity and head movements respectively. Golgi tendon organs respond to muscle tension, and joint receptors to changes of angle and position of joints.

Chemoreceptors are sensitive to chemical changes in the vicinity of the receptors. They include gustatory (taste) receptors, olfactory (smell) receptors, chemoreceptors which detect oxygen, carbon dioxide and acidity levels in arterial blood, and osmoreceptors detecting changes in salt concentration and hence the osmotic pressure of arterial blood.

Thermoreceptors respond to thermal energy. Cold and warmth receptors have been identified in the skin.

Electromagnetic receptors respond to light and are found in the visual receptor layer (the retina) of the eye.

Pain receptors may respond to more than one kind of stimulus energy, but characteristically pain is only elicited by intense and potentially damaging stimuli. Pain may be elicited after tissue damage has occurred when the tissue is inflamed and oedematous, thereby causing mechanical deformation of receptors contained within it. Inflamed tissue may release local chemicals such as histamine, bradykinin and prostaglandins. The stimulus for pain is therefore presumably either mechanical or chemical, although the tissue damage might have been caused by some other type of energy, for example ultraviolet irradiation.

Somatosensory receptors can also be classified according to their general location. Thus, there are cutaneous receptors in the skin, receptors in the viscera and receptors in muscles and joints. Receptors in muscles and joints contribute information about body position and body movement and are collectively called **proprioceptors**.

PROPERTIES OF RECEPTORS

Specificity

In general terms, each receptor elicits only one kind of sensory experience. The various types of sensation are termed modalities of sensation, for example touch, pain and temperature. Hence, receptors are referred to as being modality specific. The term adequate stimulus describes the nature of the stimulus to which a receptor is normally sensitive. Thus, light is the adequate stimulus for visual receptors in the eye, whereas sound pressure waves form the adequate stimulus for auditory receptors in the ear.

Regardless of the type of energy which stimulates a receptor, it will always elicit the same modality of sensation. Modality of sensation reflects the nature of the stimulated receptor together with its terminations in the brain.

Specific pain receptors are sensitive to several kinds of stimulus energy and it is therefore difficult to define the adequate stimulus. Adequate stimuli for pain receptors may include high intensity mechanical, electromagnetic, chemical and thermal energies. However, specific pain receptors and pathways can be identified. Free nerve endings may subserve touch, pressure and temperature in addition to pain, and there is probably some overlap of function between the many specialized nerve ending receptors. However, any single receptor, for example one free nerve ending, will only transmit one modality of sensation.

Thresholds

Sensory receptors transduce stimulus energy into nervous impulses which are conducted to the CNS via the sensory afferent neurones. Since a threshold level of stimulation must be reached in order to generate action potentials in a nerve fibre, it is clear that sensory receptors must also reach a threshold intensity of stimulation in order to excite nerve impulses in sensory afferents. Perception, however, is dependent on the transmission pathway and on brain organization as well as on the electrophysiological properties of the receptors.

Adaptation

All sensory receptors adapt to stimulation after a short period of time. That is, the frequency of impulse generation reduces to a lower level or even to zero despite the maintenance of stimulus input to the receptor.

Adaptation permits receptors to signal changes in environmental conditions.

Electrophysiological properties

When a receptor is appropriately stimulated, the transmembrane potential shifts and depolarizes. This potential change is a graded phenomenon, and the amplitude of the depolarization depends on stimulus intensity, duration of stimulation and receptor adaptation. This graded potential change is termed a **generator potential** if it occurs in a neuronal receptor, or a **receptor potential** if it occurs in a non-neuronal receptor cell. If the generator (or receptor) potential is large enough, action potentials are generated in the associated sensory afferent nerve fibres. In general terms, the larger the generator potential amplitude, the greater the frequency of impulse generation.

Generator (or receptor) potentials can be compared with the postsynaptic potentials already described in Chapter 2.1. Like postsynaptic potentials, they are not all-or-none events, and generator potentials can summate if stimulation is maintained or reapplied before the resting potential is re-established. The duration of the generator potential is variable, but for a single stimulus application it typically lasts for 5–10 ms.

As stimulus intensity (or rate of change of stimulus intensity) increases, so the generator potential increases to its maximum level. Increases in intensity beyond this point cannot be detected by the receptor. The relationship between stimulus intensity and generator potential amplitude is illustrated in Fig. 2.2.1.

Generator (or receptor) potentials have an ionic basis. The application of an adequate stimulus

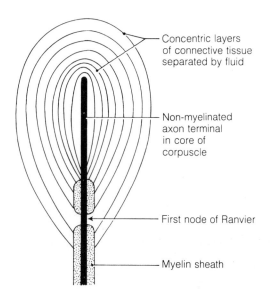

Figure 2.2.2 Section through a Pacinian corpuscle

causes a generalized increase in membrane permeability to small ions. Since at rest Na^+ are far removed from equilibrium, a net inward flow of these positive ions occurs on stimulation, resulting in membrane depolarization.

The adequate stimulus for a skin Pacinian corpuscle (Fig. 2.2.2) is pressure and mechanical deformation. When pressure is applied to one side of the capsule, the deformation results in a shifting of the connective tissue and enclosed fluid layers relative to one another, and this apparently results in membrane permeability changes in the unmyelinated nerve terminal, which in turn gives rise to the generator potential. The generator potential spreads passively from the site of stimulation and, if it is of sufficiently large amplitude, it excites the production of action potentials at the first node of Ranvier. The first node of Ranvier has a lower threshold for action potentials when compared with the unmyelinated portion of the nerve terminal. Action potential generation continues as long as the generator potential is maintained, and characteristically a burst of impulses occurs when a Pacinian corpuscle is stimulated.

SENSORY UNITS

A single afferent nerve fibre branches to produce a number of peripheral terminals, each of which is

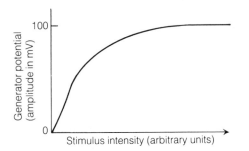

Figure 2.2.1 The relationship between the generator potential and stimulus intensity in a sensory receptor

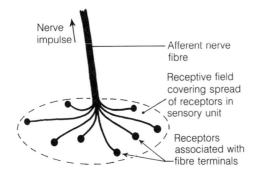

Figure 2.2.3   The receptive field of a sensory unit

associated with a receptor. The area innervated by these terminals is termed the **receptive field** of the neurone. The afferent neurone, together with its associated receptive field, is termed a **sensory unit** (Fig. 2.2.3).

INFORMATION CONTENT OF SENSORY SIGNALS

All sensory receptors transduce stimuli into action potentials, yet each burst of impulses provides unique information regarding intensity, modality and locality of the stimulus. The transmitted elements are simple, i.e. action potentials along sensory afferents. It is the *pattern* of transmitted elements together with the transmission pathway that is essential for the transfer of information in the signal.

Intensity code

The brain can determine stimulus intensity despite the fact that the action potential is of fixed amplitude. Information about intensity is provided in three ways.

Frequency of impulse generation at the receptor. Up to a maximum, the greater the stimulus intensity the greater the generator potential and the higher the frequency of action potential generation at each receptor.

Recruitment of other receptors in the sensory unit. As stimulus intensity increases, higher threshold receptors in the sensory unit will be activated and these will also produce bursts of impulses. These

will be conducted to the final common path of the receptive field – the sensory afferent fibre. Provided that the membrane is not refractory, the recruited receptors will increase the frequency of impulses in the sensory afferent fibre.

Recruitment of other sensory units. High intensity stimuli usually stimulate larger areas of tissue than low intensity stimuli. Thus, higher intensity stimuli may stimulate more than one sensory unit, and recruitment of sensory units can be indicative of an increase in intensity of stimulation.

Modality code

The modality code of the sensory input is a function of the receptor type together with its particular pathway to and termination point in the brain.

Location code

The transmission pathway from receptor to brain is also important for localizing a stimulus. In general terms, the greater the degree of overlap of receptive fields and the smaller the individual fields, the more precise the location of the stimulus. Thus, touch can be precisely located when applied to the skin of the fingers, where receptive fields are small and overlap considerably, but precision is lost if the skin of the back is so stimulated since

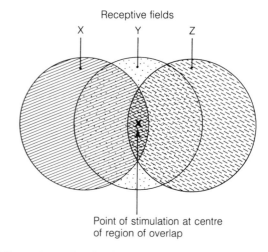

Figure 2.2.4   Overlapping receptive fields enable localization of the stimulus (see text)

receptive fields here are larger and overlap less. This is illustrated in Fig. 2.2.4.

Stimulation occurs at the centre of field Y and on the periphery of fields X and Z. Since low threshold receptors generally occur at the centres of receptive fields and high threshold receptors occur at the edges, it follows that action potentials will be elicited more readily in sensory afferent Y than in X and Z. The pattern of discharge in the three sensory units therefore permits a fairly precise location of the stimulus, and the higher rate of discharge in Y when compared with the other two indicates that the stimulus is nearer the centre of field Y and towards the periphery of fields X and Z.

Coding of boundaries

Characteristically, contrast is intensified in the transmission of sensory signals such that weaker signals tend to be filtered out and inhibited and stronger signals boosted, thereby providing sharp boundaries of stimulation. The term **lateral inhibition** describes the process of inhibition of weak signals arising from peripheral areas of stimulation. This can be readily demonstrated by pressing a pointed object firmly into the skin of a fingertip. The point is clearly felt and localized, despite the fact that surrounding tissue is also deformed and indented. Pressure receptors are stimulated in all deformed areas, but the weaker signals from the tissues surrounding the point are inhibited at synapses in the transmission pathway to the brain. The same phenomenon is also demonstrated in the visual system, where edges to shapes and patterns are clearly seen. Again, signals from visual receptors which are only weakly stimulated and which surround strongly stimulated areas are laterally inhibited.

SOMATOSENSATION

Somatosensory signals arise from the skin, muscles, joints and the viscera. Somatosensory receptors can be classed as tactile, proprioceptive, thermal and pain receptors. In addition, there are stretch receptors in the viscera. Stretch receptors in the gut are chiefly concerned with the initiation of *local* reflex contractions of the gut muscle, whereas stretch receptors in the arterial tree detect changes in blood pressure and send signals to the lower centres of the brain only. The major somatosensory input terminates in higher centres of the brain. Afferent fibres conveying somatosensory input from the trunk and limbs arise from unipolar cell bodies which are aggregated together in ganglia on the dorsal roots of the spinal cord. These unipolar neurones give rise to central afferents which convey the sensory signals to the higher centres of the brain.

TACTILE RECEPTORS

Tactile receptors are receptors which respond to touch, pressure or vibration. In each case the stimulus energy is mechanical. Pressure simply implies a greater intensity of stimulation than touch. Vibration is rapid, repetitive mechanical deformation.

Tactile receptors vary in structure. Meissner's corpuscles consist of spiral nerve endings embedded in corpuscles of epithelial tissue. They are found just beneath the epidermis of the skin, especially in hairless areas such as the fingertips. They respond to light touch and adapt readily to zero if stimulation is maintained. They therefore respond best to light movements across the skin. In contrast, Merkel's discs, which are also found in abundance in hairless areas of skin, adapt more slowly and respond to continuous touch stimuli. Basket nerve endings around hair bases are found in all other areas of the skin. They are sensitive to touch and respond when the hair is bent. Like Meissner's corpuscles, they adapt readily to stimulation.

Pressure receptors generally lie deeper within the tissues. For example, Pacinian corpuscles are found deep within the dermis, in joints and also in parts of the viscera. They adapt very rapidly to pressure stimuli and can therefore signal vibration. In contrast, Ruffini's endings, which are also found in deeper tissues, adapt more slowly and may signal continuous deformation.

Most of the tactile receptors send signals along larger diameter sensory fibres belonging to group Aβ. However, free nerve endings and most tactile nerve endings associated with hair bases transmit signals in Aδ and C fibres. These slower conducting pathways transmit rather poorly localized stimuli when compared with the Aβ fibre group.

PROPRIOCEPTORS

Proprioceptors transmit signals which provide information about body position and movement. Muscle spindles and Golgi tendon organs signal muscle length and muscle tension respectively.

Receptors in joints also signal important information about movement. Ruffini's endings are abundant in joints. They maintain a steady discharge in response to continuous stimulation, although the initial firing rate at the onset of stimulation is much greater. As the angle of a joint is changed, so particular Ruffini's endings are switched on or off. Each receptor responds maximally to a particular angle or position of joint. *Rate of change* of angle at the joint is probably detected, at least in part, by the initial strong rate of firing before the Ruffini's receptors adapt to the lower steady rate of activity. The *initial* receptor response to stimulation is certainly proportional to the rate of movement. In addition, rapidly adapting Pacinian corpuscles in the joints probably aid in the detection of rate of movement. Proprioceptor signals are transmitted in the larger diameter $A\alpha$ and $A\beta$ fibres.

THERMAL RECEPTORS

The many gradations of temperature which an individual can detect seem to depend on the relative activities of two populations of receptors, namely cold and warmth receptors. In addition, pain receptors may be stimulated in extremes of cold or heat. Cold and warmth receptors are probably free nerve endings, although specialized Krause's end-bulbs and Ruffini's endings have been associated with cold and warmth respectively. In any case, there are more cold receptors than warmth receptors.

The stimulus for temperature receptors is apparently thermal energy. However, it is probable that a change in temperature which causes a change in the metabolic rate of a receptor is only indirectly acting as a stimulus, and the receptor is in fact directly activated by chemical stimulation resulting from the changed metabolic rate.

Thermal signals are transmitted to the brain largely in $A\delta$ fibres but also in small C fibres.

PAIN RECEPTORS

Pain receptors are all free nerve endings. They are located in the skin, muscles, joints, arterial walls, the membranes surrounding the brain, and also in the hollow organs of the gut. Pain is variously described as pricking, cutting, burning, aching, stabbing or crushing. The quality of the pain experienced seems to depend on the part of the body affected and on the nature and extent of the injury which causes it. The perception of pain depends partly on the physiological response of the receptors and partly on a psychological reaction to the pain itself. This section is restricted to a consideration of the physiology of pain receptors.

The stimulus for pain may be physical or chemical. Physical stimuli may be thermal, mechanical or electrical energy of high intensity. For example, excessive mechanical stretching may elicit pain and this may contribute to the pain of inflammation. Excessive dilation of intracranial arteries may cause headache. It is possible that physical stimuli actually act through chemical mediators. Certainly it is known that a number of chemicals can elicit pain when applied to tissues. Severe physical blows or burns produce immediate pain, but pain may well continue after cessation of physical stimulation, and it is thought likely that chemical factors are responsible for this. Prostaglandins, histamine, bradykinin and 5-hydroxytryptamine (serotonin) are all chemicals which may be produced by the body when tissue damage occurs and these will all elicit pain if applied to a blister on the skin.

Pain receptors do not appear to adapt, and while stimulation is maintained, pain continues to be transmitted to the CNS. Pain is transmitted in unmyelinated C fibres or in slightly larger myelinated $A\delta$ fibres. The $A\delta$ fibres conduct signals more rapidly than the C fibres, and painful stimulation of skin elicits a double pain, a fast pricking sensation transmitted by $A\delta$ fibres and a slower burning sensation transmitted by C fibres.

SPECIAL SENSES: THE EYE AND VISION (see also Chap. 2.3)

Anatomy of the eye

Each eyeball is 1.5–2 cm in diameter and is located in a conical bony orbit of the skull. The exposed surface of the eye and the inner surfaces of the eyelids are covered with a delicate membrane, the **conjunctiva**, which helps to prevent drying out of the tissues. The eye is also protected by reflex blinking, by the eyelashes and eyebrows, and by the production of tear fluid which bathes the surface of the eye.

The more or less regular blinking, which occurs every few seconds throughout life, aids in washing tear fluid across the surface of the eye. If this process is impaired, for example as sometimes occurs with the wearing of contact lenses, then the transparent exposed surface of the eye, the **cornea**, may become inflamed and even ulcerated. An inability to blink effectively may also occur in patients who have had a stroke or in those in intensive care units who have been treated with muscle-relaxant drugs. If the problem is severe, the patient's eyelids may be sewn together, a procedure referred to as tarsorrhaphy. Blinking is normally initiated by stimulation of the conjunctiva, the cornea or of the eyelid itself. Blinking can also operate as a protective reflex response to a threatening gesture or in response to bright light.

The production of **tear fluid** is of considerable importance to the maintenance of a healthy cornea. Tear fluid contains salts and protein but very little glucose. It also contains oil secreted by sebaceous glands, which affects its physical properties and helps to prevent overflow of tears at the lid margins. Tears also contain an enzyme, lysozyme, which is responsible for the mild bacteriocidal action of tear fluid. Tears are produced by the **lacrimal apparatus**, which is located in the upper outer corner of each bony orbit, partly by the lacrimal gland itself and partly by microscopic accessory glands. The production of tears by the lacrimal gland is normally a reflex response to irritation or drying of the cornea and/or conjunctiva. It is controlled by the parasympathetic nervous system. However, there is also a sympathetic nervous system input to the lacrimal gland which is involved in emotional weeping. Tear fluid is produced continuously by the accessory glands, and the eyes will not normally suffer dryness even if the lacrimal gland itself is removed.

Adequate drainage of tear fluid is also important. Tears normally drain into small channels, **canaliculi**, which open at a number of small holes, the **puncta**, at the inner ends of the lid margins. The canaliculi drain tears into the **lacrimal sac**, which is located on the medial surface of the orbit. The lacrimal sac is continuous with the **lacrimal duct** which finally drains the tears into the nasal cavity. The tear drainage system may become blocked, causing a watery eye condition called

epiphora. Apart from the nuisance aspect of epiphora, if it is untreated, dermatitis and secondary infections may ensue. Blockage of tear ducts can occur in newborn babies, where it may be caused by maldevelopment or by blockage with mucus. Massage may clear a plug of mucus, but surgery may be necessary to clear the duct adequately. **Dacryocystitis** is an inflammation of the lacrimal sac which may also result in obstruction of the tear drainage system. This condition is more common in older women and may require surgery to open up a by-pass passage into the nose. A lot of elderly people also suffer from watery eyes. This commonly occurs because the lids are malaligned and turn outwards or inwards such that the puncta cannot adequately collect the tear fluid. Surgical repair of the lids may be necessary in severely affected individuals.

The eyeball is attached to the skull bones by three pairs of extraocular muscles which permit rotational, horizontal and vertical movements of the eye. The lateral and medial rectus muscles move the eye from side to side. They are controlled by the VIth and IIIrd cranial nerves respectively. Normal positioning of the eye in the horizontal plane depends on the relative activities of these muscles. For example, weakness or paralysis of the lateral rectus leads to excessive medial (nasal) movement of the eye and a squint develops. The superior and inferior rectus muscles cause upward and downward movement of the forward facing eye and are controlled by the IIIrd cranial nerve. The superior and inferior oblique muscles cause rotational movements of the forward facing eye and are controlled by the IVth and IIIrd cranial nerves respectively.

The structure of the eye is illustrated in Fig. 2.2.5. The outer layer of the wall of the eyeball is composed of mechanically strong fibrous tissue. It is white in colour over most of the eyeball and forms the **sclera**, the visible part of which is the 'white of the eye'. The most anterior part of the wall of the eye is transparent and forms the **cornea**. The cornea is a protective window through which light can pass into the globe of the eyeball. Since its surface is curved, it acts as a 'fixed lens', and considerable bending (refraction) of light rays occur at the cornea.

The sclera is lined by a very vascular layer of tissue called the **choroid**, which contains the pigment melanin and absorbs scattered light. Over the posterior two-thirds of the wall of the eye,

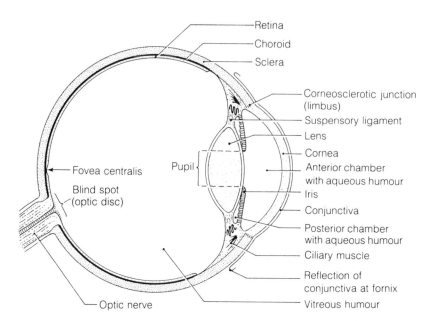

Retina
Choroid
Sclera
Corneosclerotic junction (limbus)
Suspensory ligament
Lens
Cornea
Anterior chamber with aqueous humour
Iris
Conjunctiva
Posterior chamber with aqueous humour
Ciliary muscle
Reflection of conjunctiva at fornix
Vitreous humour
Fovea centralis
Pupil
Blind spot (optic disc)
Optic nerve

Figure 2.2.5   Section through the human eye

there is an inner photosensitive layer of tissue called the **retina**. Lying immediately adjacent to the choroid is the retinal pigmented epithelium. This is heavily pigmented with melanin and it absorbs light that has passed through the photoreceptors of the retina. The receptor layer, which is attached to the pigmented epithelium, is stimulated by light energy. The receptors transmit visual signals through other cells in the retina.

The surface of the retina is covered with a 'mat' of nerve fibres which arise from retinal cells and which are collected together to form the **optic nerve**. The optic nerve leaves the bony orbit of the skull from a position which is slightly medial to the optical axis of the eye. The point where the optic nerve leaves the eyeball is called the **optic disc** or **blind spot**. The blind spot does not contain any receptors and it is therefore insensitive to light. If the intracranial pressure is raised, for example as a result of a brain tumour, the optic disc bulges and appears oedematous when observed with an ophthalmoscope. This condition is known as **papilloedema**. Optic neuritis, an inflammation of the optic nerve, may also cause papilloedema, but if the inflammation occurs central (brain side) to the optic disc, papilloedema is not seen and the condition is known as retrobulbar neuritis. In contrast to the papilloedema associated with raised intracranial pressure, papilloedema resulting from optic neuritis is associated with immediate rapid deterioration in vision.

The blind spot is not apparent in normal vision. This is largely because the image falling on the blind spot of each eye is not exactly the same. Thus, the left eye can provide the visual information that is missing from the right eye and vice versa. The brain also assists in 'filling in' missing detail.

There is a small depression in the retina lying lateral (temporal) to the blind spot. This depression is about 0.4 mm in diameter and is called the **fovea centralis**. In this part of the retina the receptor layer is exposed directly to the light, whereas elsewhere in the retina light is filtered through several layers of retinal cells. The fovea is important for detailed discrimination of visual images.

Anteriorly, the choroid layer bulges to form a ring of smooth muscle called **ciliary muscle** to which the lens is attached, and anterior to this the choroid projects into the eye as the **iris**. The iris is the coloured part of the eye, the colour of which is determined genetically. The iris forms a variable aperture, the **pupil**, through which light is directed into the eye. It contains circularly arranged smooth muscle fibres which are innervated by the parasympathetic nervous system. Contraction of these muscle fibres makes the iris aperture smaller and the pupil constricts. Parasympathetic stimulation can be blocked if eyedrops containing homatropine, an atropine derivative, are administered. The drug blocks the action of acetylcholine, the parasympathetic transmitter, and the pupils then dilate, facilitating clinical examination of the eye. The iris also contains radially arranged smooth muscle fibres which are innervated by the sympathetic nervous system. Activation of these muscle fibres causes pupillary dilation. Emotional arousal is associated with activation of the sympthetic nervous system and it results in dilated pupils.

Shining a light into the eye evokes a pupillary reflex response in which the pupil constricts. It is a bilateral response, that is, if light is directed through only one eye, the pupils of both eyes constrict. The pupillary reflex pathway is described in Chapter 2.3.

The **lens** of the eye is attached to the ciliary muscle by means of radially orientated suspensory ligaments. The lens has convex surfaces and causes refraction and convergence of light rays passing through it. The anterior surface is more curved than the posterior surface and therefore causes more bending of the light. The lens is composed of concentric layers of epithelial cells in an elastic capsule. It is not rigid and when it is detached from the ligaments it takes up a more or less spherical shape. The central core of the lens has a higher refractive index than other parts of the lens and therefore has the greatest potential for bending light passing through it. The lens is normally transparent, but old age or exposure to infra-red radiation may cause coagulation of lens protein leading to cloudiness. This condition is known as **cataract**. It can also present as a congenital disorder or as a complication of diabetes mellitus.

The narrow space between the iris and the lens forms the **posterior chamber**, and the space between the iris and the cornea forms the **anterior chamber**. Both chambers are filled with a fluid derived from plasma called **aqueous humour**. Aqueous humour supplies oxygen and nutrients to the lens, which does not have an independent blood supply. Aqueous humour is secreted in the posterior chamber by blood vessels in ciliary processes at the base of the ciliary muscle. It is formed at the rate of 2 or 3 mm^3 per min and it then passes through the pupil into the anterior chamber. It is reabsorbed into ocular veins at the **angle of Schlemm**, which is the angle formed between the iris and the cornea. The rate of secretion and reabsorption of aqueous humour determines the intraocular pressure. In glaucoma, the reabsorption of aqueous humour is impaired and intraocular

pressure increases. If untreated, the raised pressure is transmitted to the retinal artery and blindness may ensue because of impaired oxygen and nutrient supply to the retina.

The space between the lens and the retina does not communicate directly with the posterior or anterior chambers of the eye. It is filled with a gelatinous mass called **vitreous humour**.

The eye as an optical system

The primary function of the eye is to convert light energy into nervous impulses which can be 'interpreted' by the brain as visual images. In order to do this, the eye behaves like a camera such that light radiating from an object is brought into sharp focus on the photoreceptor layer of the retina. The retina is the equivalent of photosensitive film in the back of a camera. The image which is being viewed is reproduced, without blurring, on the retina and the pattern of stimulation of photoreceptors therefore accurately reflects the pattern of light radiating from the object. Thus, clear vision is dependent not only on the neurophysiological properties of the eye, but also on its optical properties.

LIGHT

The behaviour of light can be described in terms of straight lines or rays of energy which enter the eye through the pupil. Light rays from distant points appear to be parallel, whereas those from near points appear to diverge.

Refraction of light in the eye
Regardless of whether light arrives at the surface of the eye as divergent, parallel or convergent rays, it passes through the cornea, the aqueous humour, the lens, the vitreous humour and the surface layers of the retinal cells before impinging on the photoreceptors. As light passes through these layers it will be refracted to varying degrees, ultimately to be focused on the receptors.

Biconvex lenses cause refraction and convergence of light rays passing through them, and parallel rays of light are brought together to a point at a finite distance behind the lens. This is known as the **focal length** (Fig. 2.2.6a). Concave lenses cause refraction and divergence of light rays passing through them, such that parallel rays of

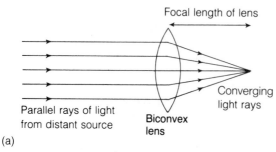

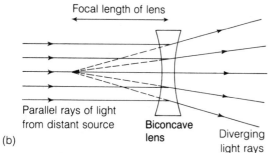

Figure 2.2.6 Refraction of light through a lens: (a) refraction through a biconvex lens, (b) refraction through a biconcave lens

light passing through a biconcave lens appear to originate from a point, a fixed distance in front of the lens (Fig. 2.2.6b).

Refraction of light occurs at several points in the transmission pathway through the eye. When the eye is focused on a near object, the curvature of the lens increases to cause greater refraction of the diverging rays of light so that they will be brought to a focus on the retina. The refractive power of the lens increases, thus reducing the focal length of the lens so that when it is focused on near objects, parallel rays of light from distant points are focused in front of the retina and images of distant objects are then blurred (Fig. 2.2.7).

Convex lenses cause an inversion of the image

of the object in both vertical and horizontal planes (Fig. 2.2.8). The visual world is therefore upside down and laterally reversed on the retina. This is of no importance to visual perception since the brain effectively 'learns' to organize the visual signals from the inverted retinal image into a view of the world which is the right way up.

ACCOMMODATION

The ability of the lens to change shape is termed the **power of accommodation**. When distant objects are viewed, the lens is flattened and the ciliary muscle to which it is attached is relaxed. When the eye focuses on a near object, the ring of ciliary muscle reflexly contracts in response to parasympathetic stimulation. The ciliary muscle is largely composed of circularly arranged fibres, so that when they contract the ring of muscle constricts and less tension is applied to the attached suspensory ligaments. The slackening of the ligaments slackens the pull on the lens. Since the lens is flexible and relatively elastic, it takes up a more curved shape in response to ciliary muscle contraction and it therefore increases its refractive power. The refractive power increases so as to focus divergent light rays from near objects onto the retina.

In addition to the circular muscle fibres, there are some ciliary muscle fibres with radial attachments to the corneoscleral junction. Contraction of these fibres tends to pull forwards on the ring of muscle, which facilitates increased curvature of the anterior surface of the lens.

When near objects are viewed, accommodation is accompanied by reflex pupillary constriction. Parasympathetic stimulation of circular muscle fibres in the iris accompanies contraction of ciliary muscle. The resultant pupillary constriction serves to shut down the periphery of the lens so that light rays entering the eye are directed through the most curved central part of the lens. The combi-

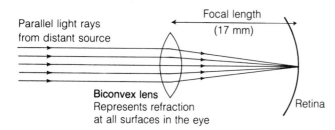

Figure 2.2.7 The reduced eye: adjusted for viewing distance objects. Parallel rays of light from a distant source are brought to a focus on the retina

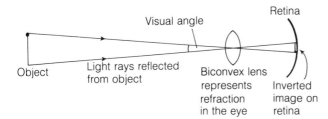

Figure 2.2.8 Inversion of the image on the retina

nation of contraction of ciliary muscle and pupillary constriction forms the **accommodation reflex**.

NEAR-POINT AND PRESBYOPIA

The nearest source of light which can be brought into focus on the retina is determined by the maximum increase in refractive power of the lens which can be achieved by accommodation. The power of accommodation reduces with increasing age, slowly at first and then rapidly from age 40 to 50 years onwards. This condition is called **presbyopia**. Thus, with increasing age, the nearest point which can clearly be focused on the retina recedes. Normally most elderly people require correcting spectacles with biconvex lenses for close work such as reading or sewing. Some examples of typical near-points at various ages are given in Table 2.2.1.

OPTICAL DEFECTS OF THE EYE

The most common are short-sightedness, long-sightedness and astigmatism.

Short sight (**myopia**) describes the condition in which only near objects can be focused on the retina. It usually occurs because the eyeball is 'too long' for the refracting power of the lens (Fig. 2.2.9a). Myopia can be corrected with an appropriate biconcave lens which will cause divergence of distant rays of light, enabling them

Table 2.2.1 Changes in the near-point with age

Near-point (cm from eye)	Age (years)
9	10
10	20
12.5	30
18	40
50	50
100	70

to be focused on the retina instead of in front of it (Fig. 2.2.9b).

In contrast to short sight, long sight (**hypermetropia**) normally occurs when the eyeball is 'too short' for the refractive power of the lens and divergent light rays from near objects are then focused behind the retina. A person with hypermetropia may even require some accommodation of the lens to focus on distant objects, and hence hypertrophy (enlargement) of ciliary muscle may be evident. Even with a maximally accommodated lens, the hypermetropic person cannot focus on near objects and requires correction with a biconvex lens to cause some convergence of light rays before they enter the eye (Figs. 2.2.10a and 2.2.10b).

Astigmatism is a defect of the eye in which the curvature of the cornea in the horizontal plane is not the same as that in the vertical plane. Less commonly, the astigmatism may affect the lens

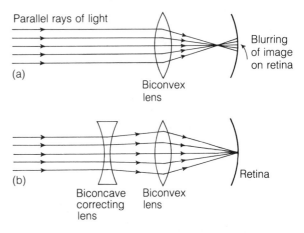

Figure 2.2.9 Myopia. (a) The eyeball is 'too long' relative to the refractive power of the lens, and parallel rays of light from distance sources are focused in front of the retina. (b) Correction with a biconcave lens: an appropriate biconcave lens causes divergence of parallel light rays before they impinge on the eye and the lens can then focus them onto the retina

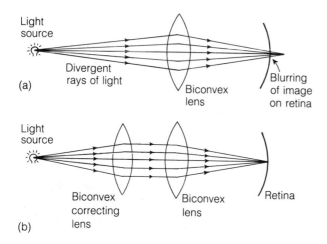

(a)

(b)

Figure 2.2.10   Hypermetropia. (a) The eyeball is 'too short' relative to the refractive power of the lens and divergent rays of light from a near object are focused behind the retina. (b) Correction with a biconvex lens. An appropriate convex lens causes convergence of light rays before they impinge on the eye and the lens can then focus them onto the retina

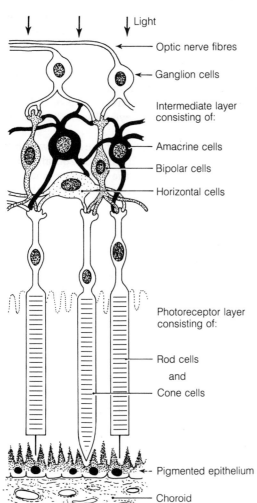

Figure 2.2.11   The cell layers of the retina

rather than the cornea. An astigmatism results in blurring of parts of a visual image.

Physiology of the retina

STRUCTURE

The retina is a complex layer of tissue lining the eye which is responsible for the conversion of light into visual nervous signals. The structure of the retina is schematically illustrated in Fig. 2.2.11.

Light passes through the mat of optic nerve fibres and two major layers of retinal cells before impinging on the photoreceptors. This may appear to be a rather 'back to front' arrangement. However, the photoreceptors lie adjacent to the pigment epithelium which is intimately associated with the photochemistry of the receptors, and this probably accounts for the positioning of the photoreceptor layer. Clearly, the retinal cell layers and the mat of optic nerves do not prevent the transmission of light.

The photoreceptor layer can separate from the pigmented epithelium, leading to the condition of **retinal detachment**. The retina, especially the photoreceptor layer, is dependent on the choroid and pigmented epithelium for much of its nutrient and oxygen supply. Consequently, retinal detachment, if prolonged, can lead to inadequate nutrition and to deterioration of the retina. However, in the short term the retina may remain viable because nutrients can diffuse across from the choroid.

The photoreceptor layer of the retina contains two types of receptors called **rods** and **cones**.

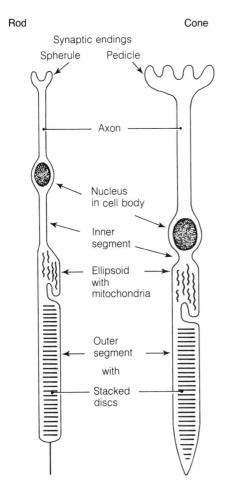

Rod   Cone

Synaptic endings

Spherule   Pedicle

Axon

Nucleus
in cell body

Inner
segment

Ellipsoid
with
mitochondria

Outer
segment

with

Stacked
discs

Figure 2.2.12   Structure of the photoreceptor cells

Their structures are illustrated in Fig. 2.2.12. Each receptor has an outer segment containing stacked piles of membranous discs on which are located the photosensitive chemical pigments. The outer segment of each receptor lies directly adjacent to the pigmented epithelium. Note that the outer segment of a cone is, as its name suggests, cone shaped, whereas that of a rod is rod shaped. The photochemical itself is manufactured in the inner segment before transfer to the outer segment, and hence the inner segment is characterized by the presence of mitochondria.

Rods and cones differ with respect to the organization of the outer segment. In rods, the membranous discs are continuously produced at the base of the outer segment and removed at the apex, where the discs are then absorbed into the pigmented epithelium. In cones, the membranous discs are not continuously replaced and the same complement of discs remains throughout life. The nature of the photosensitive chemical is also different in the two types of receptors. Rods all contain the same photopigment, **rhodopsin**, which is purple in colour and is often referred to as visual purple. Since the rods all contain the same photosensitive chemical, they can only respond in one way to light, that is, they can either be stimulated or not. Different rods cannot respond differentially to light of various wavelengths and hence cannot respond differentially to colour. Cones, on the other hand, do not all contain the same pigment. One group contains a pigment called **erythrolabe** which is most reactive to light at the red end of the spectrum. A second group contains **chlorolabe** which responds best to green light; and a third group contains **cyanolabe** and responds preferentially to blue light. Cones are essential for colour vision.

Adjacent to the inner segment of each receptor is the cell nucleus, which is attached to a synaptic ending. Rods and cones are not true neurones, but they do make synaptic contacts with other retinal cells and the synapses are characterized by the presence of vesicles. The synaptic ending of a cone is broad and flattened and is called a **pedicle**, whereas the synapse of a rod forms a smaller bulbous contact called a **spherule**.

Rods and cones are not randomly distributed across the retina. The fovea centralis contains only cones, whereas there are no cones at all in the most peripheral parts of the retina. Foveal cones are structurally distinct and appear to be longer and thinner than cones elsewhere. Altogether there are approximately 6 or 7 million cones in each human retina. Rods are found in greatest numbers in the periphery of the retina. There are many more rods than cones, with estimates varying over a range of 120–125 million rods per retina.

The intermediate layer of retinal cells consists of bipolar cells, horizontal cells and amacrine cells. Bipolar cells principally transmit signals from rods and cones to the third retinal layer of cells, which consists of ganglion cells. The ganglion cells give rise to optic nerve fibres, and all signals from other retinal cells are finally transmitted to the ganglion cell layer.

RECEPTOR PHOTOCHEMISTRY

The photosensitive chemicals in both rods and cones are stored on the membranous discs of the outer segments of the receptors. The stacked disc arrangement maximizes the exposure of the contained photopigments to incident light. This process involves a chemical change in the photopigment brought about by light energy which then induces a receptor potential in the rod or cone.

Rods contain the photopigment rhodopsin which consists of vitamin A aldehyde, called **retinene**, combined with a protein referred to as **opsin**. When light is directed onto rhodopsin, the retinene portion of the molecule *isomerizes*, that is it undergoes a change in shape from a curly to a straight form, without a change in chemical composition. The curly form of retinene, termed the all cis form, fits snugly onto its attachment site on the protein opsin. However, the straight form, the all trans form of retinene, is the wrong shape for the attachment, and so the isomerization induced by light makes the photopigment unstable until finally the opsin splits completely from the retinene. The final products of the conversion are lighter in colour when compared to the purple of rhodopsin, and light is therefore said to bleach the photopigment. Bleaching actually takes several seconds, whereas receptor potentials are produced within a fraction of a second of light exposure. Clearly, the chemical change inducing the receptor potential must occur during the decay process and before complete bleaching of the pigment occurs.

In order for rods to continue to function, rhodopsin must be reformed so that it can once again 'capture' light and induce a receptor potential. The first stage in its reformation involves the reconversion of all trans (straight) retinene into all cis (curly) retinene. This reaction requires an isomerase enzyme and energy. Once the all cis retinene is formed, it readily recombines with opsin, and the reconstituted rhodopsin then remains stable until light energy is again absorbed.

Retinene is formed from vitamin A by a dehydrogenation reaction. Normally both cis and trans forms of vitamin A are present as well as cis and trans forms of retinene. Large stores of vitamin A are located in the pigment epithelium. In conditions of darkness, virtually all the retinene is converted to rhodopsin and much of the stored vitamin A is then converted first to retinene and then to rhodopsin.

Cone photopigments are similar to rhodopsin. They all contain retinene but it is bound to different opsins. Thus, erythrolabe, chlorolabe and cyanolabe represent three different opsins bound to retinene. Essentially the same process of decomposition of the pigment is thought to occur, and vitamin A is essential for cone as well as rod function.

FUNCTIONAL DIFFERENCES BETWEEN RODS AND CONES

Rods and cones are functionally as well as structurally distinct. Their relative contributions to colour vision have already been referred to, and colour vision depends on the integrity of cones rather than of rods. Rods and cones also differ in their relative sensitivities to light. Rods have low thresholds to light and are easily stimulated, even by quite low light intensities. However, in bright light rods appear to adapt to stimulation and they no longer respond. On the other hand, cones have higher thresholds to light. They require a greater intensity of light in order to be stimulated and they therefore operate under conditions of bright light when rod function is reduced.

Since rods function best under conditions of relatively low light intensity when cones are not stimulated, and since rods do not contribute to colour vision, it is to be expected that vision in semidarkness will depend on rods and will provide information on relative brightness but not colour of the objects viewed. In moonlight, everything appears in shades of silvery grey, with those objects that appear lighter in shade reflecting more light than apparently dark grey or black objects.

Rods and cones also respond differentially to different wavelengths and hence colours of light. This was first described by Purkinje following observations that the relative brightness of colours changes as daylight fades to dusk. Reds look less bright whereas blue-greens increases in brightness in dusk conditions. This shift in spectral sensitivity is now known as the Purkinje shift, and it reflects a change from predominantly cone vision in daylight to predominantly rod vision at dusk. In daylight, all wavelengths of light are absorbed by cone pigments but peak absorption and peak sensitivity occur with yellow-green light. Cone vision in daylight is termed **photopic vision**. As dusk falls, rods take over from cones. Rhodopsin maximally absorbs blue-green light and therefore rod sensitivity is maximal for light of this wavelength. Thus, blue-green objects appear brighter with rod vision when compared to cone vision. Rod vision at dusk is termed **scotopic vision**.

Night blindness is a condition associated with a severe deficiency of vitamin A. Vitamin A is essential for the formation of photopigments, and a severe deficiency might therefore be expected to impair both rod and cone function. However, the 'blindness' is most apparent in conditions of low light intensity when vision is dependent on rod function. If vitamin A deficiency is very prolonged and severe, cone function also deteriorates and blindness progresses. It is thought that cones are less susceptible than rods because they operate under daylight conditions of relatively high light intensity, when sufficient light is available to elicit a response despite the reduction in available photopigment.

VISUAL ACUITY

The ability to discriminate visually between two separate points of light is called **visual acuity**. Visual acuity permits detail to be appreciated.

An emmetropic eye (one with normal accommodation and refraction) can generally distinguish between two separate light sources provided that the angle that they subtend at the eye, the visual angle, is at least 1 minute (1′). However, some individuals have much better visual acuity than this. Visual acuity is most acute when the visual image falls on the fovea. It has been suggested that visual acuity is at its limit when two stimulated foveal cones are separated by one unstimulated cone.

Visual acuity is measured using **Snellen's type**, an example of which is shown in Fig. 2.2.13. Snellen's type usually consists of rows of letters with each row being of a different size. The letters subtend an angle of 1 minute at the eye when viewed from a specified distance. Typically, there are 60, 36, 24, 18, 12, 9, 6 and 5 metre lines. The type is viewed from a distance of 6 m by each eye separately and the visual acuity for each eye is expressed as 6 over the number of the line of smallest letters which is accurately recognized.

COLOUR VISION

Colour vision is dependent on the integrity of the

Clement Clarke International Ltd

H
AL
TNC
OLHA
ECTNO
CLOHNA
AENLOHCT
HTNELACO
AECONHTL

Figure 2.2.13 An example of Snellen's type (photograph supplied by Clement Clarke International Ltd)

cones. The human eye is stimulated by light of wavelengths 400–750 nm and this represents an ability to recognize 150–200 separate hues. The short wavelengths of light are from the violet end of the spectrum and the long wavelengths are from the red end.

The human eye is not equally sensitive to the three primary light colours, and if blue light, green light and red light are matched for intensity, the red light is perceived as being the brightest and the blue light appears to be the dimmest. The difference in sensitivity to red light and green light is not very great, but the eye is considerably less sensitive to blue light. This difference in sensitivity may reflect differences in photopigment absorption properties, differences in amounts of photopigment contained in the three cone types, or differences in the numbers and distribution of the three cone types. It is known that blue cones are not found in the fovea and therefore colour vision is dichromatic here. Furthermore, the pattern of distribution and connection of blue cones is more diffuse and shows greater convergence onto ganglion cells than that shown by the red and green cones.

DEFECTS OF COLOUR VISION

Although colour defects can occur if the retina is damaged, they generally result from genetic factors. Colour defects can be classified as a reduction of, or absence of, the ability to detect red, green or blue light. Total **colour blindness**, that is, monochromatic vision, is very rare.

The most common colour defect is so-called red–green colour blindness. Red–green colour blindness may arise from a number of inherited conditions and results in an inadequacy or inability to distinguish between reds and greens.

Colour defects are most commonly assessed using Ishihara's colour charts, which consist of arrays of coloured dots with a figure or pattern picked out in different colours from the background. For example, a number may be picked out in red dots on a green dot background. The number can be recognized by a person with normal colour vision but not by someone with red–green colour blindness.

ADAPTATION

All sensory receptors adapt to stimulation to some extent, and cones and rods adapt to light energy. The phenomenon of dark adaptation is readily appreciated when one enters a dark room, such as a cinema auditorium, from a well-lit area. Initially, considerable difficulty is experienced in seeing anything, but as the eyes adapt to the darkness vision improves. Complete dark adaptation may take over half an hour, by which time the sensitivity of the eye to light may be increased by up to a million times.

Electrophysiology of retinal cells

When light falls on retinal receptors, an electrical change is induced which is transmitted by cells in the intermediate layer of the retina. The signal may undergo modification here before transmission to a ganglion cell and its associated optic nerve fibre which finally transmits the signal to the brain.

ELECTRICAL CHANGES IN RECEPTORS

Neither rods nor cones can generate action potentials, and the predominant electrical response to light stimulation in a visual receptor is a slow, graded membrane **hyperpolarization**. Retinal receptors in the dark are partially depolarized to around $-30\,\text{mV}$ because there is a steady leak of Na^+ into the cells. When a receptor is stimulated by light, the Na permeability is *reduced*, and fewer Na ions leak into the cell. The membrane potential then increases to a more negative value and is hyperpolarized. Thus, retinal receptors are partially excited in the dark and are inhibited by light. However, in the dark the receptors spontaneously release transmitters which can inhibit intermediate retinal cells. Thus, in the dark, partial excitation of receptors can result in inhibition of bipolar cells. Stimulation by light then reduces this inhibition because the receptor hyperpolarization reduces the release of inhibitory transmitter. The disinhibition so caused then results in increased activity of the bipolar cells.

PROCESSING OF VISUAL SIGNALS IN THE INTERMEDIATE LAYER

In common with visual receptors, the cells of the intermediate layer of the retina do not appear to transmit action potentials. Instead, they produce slow, graded membrane potential changes, some of which are depolarizing and some of which are hyperpolarizing.

ELECTRICAL CHANGES IN GANGLION CELLS

Ganglion cells constitute the final stage in retinal processing of visual signals. Ganglion cell axons carry the signals in the optic nerves to further processing centres in the brain.

SPECIAL SENSES: THE EAR AND AUDITION (see also Chap. 2.3)

Structure of the auditory apparatus

The human ear is sensitive to an enormous range of sounds. The auditory apparatus of the ear is able to code the various characteristics of sound into patterns of nervous signals which the brain can interpret.

The ear is subdivided into three main compartments, the outer ear, the middle ear and the inner ear (Fig. 2.2.14). The outer ear funnels the sound energy into the middle ear which serves to transmit the sound to the auditory apparatus of the inner ear, which transduces the mechanical energy of sound vibration into electrical energy. An electrical potential change is produced in the inner ear which gives rise to patterns of action potentials in the auditory branch of the VIIIth cranial nerve. The auditory nerve transmits the signals to the brain.

THE OUTER EAR

The two ears are cartilaginous outgrowths covered by skin. They are called the **pinnae** or **auricles**. In man the pinna may play a minor role in protection since it surrounds the opening to the **external auditory meatus** (the auditory canal), which transmits sounds to the middle ear, and so the pinna helps to prevent hair and skin debris from entering the ear.

The external auditory meatus is 'S' shaped and about 25 mm in length. It terminates in the **tympanic membrane** (the ear-drum). The shape of the canal does not impair the transmission of sound; indeed, it aids sound amplification. It probably also plays a defensive role. The 'S' shape will help to prevent the ingress of particulate matter and it will also reduce the effects of air currents which could otherwise damage the ear-drum. The external auditory meatus is lined by skin which contains the **ceruminous glands**. These glands secrete a waxy substance which prevents drying out of the tissue. Ear wax is sticky, and sloughed off skin cells and dust will tend to stick to it and hence be removed from the ear. However, excess production of ear wax may impair sound transmission through the outer ear, especially if a plug of wax attaches to the ear-drum.

The tympanic membrane is composed of connective tissue and it stretches across the aperture

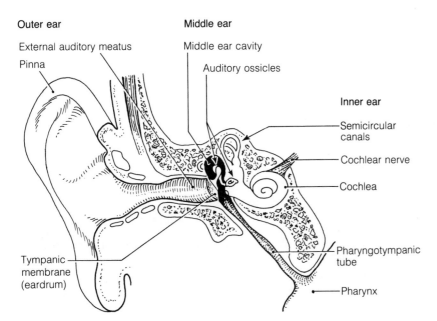

Outer ear

External auditory meatus

Pinna

Middle ear

Middle ear cavity

Auditory ossicles

Inner ear

Semicircular canals

Cochlear nerve

Cochlea

Tympanic membrane (eardrum)

Pharyngotympanic tube

Pharynx

Figure 2.2.14 Structure of the ear

of the inner end of the auditory canal. It separates the outer ear from the middle ear. It is covered by skin on the outer ear side and by mucous membrane on the middle ear side. It can easily be observed with an otoscope. The tympanic membrane is slightly conical in shape, with the apex of the cone pointing into the middle ear. The shape aids funnelling of sound. The tympanic membrane vibrates and moves in and out in response to sound.

MIDDLE EAR

The middle ear is an air-filled cavity containing three tiny bones, the **ossicles**, which are called individually the **malleus**, the **incus** and the **stapes**. The ossicles transmit sound energy from the tympanic membrane to a much smaller membrane, the **oval window**. The oval window separates the middle ear from the inner ear.

The malleus is hammer shaped and the hammer handle is attached by ligaments to the tympanic membrane. The head of the hammer is firmly attached to the incus (the anvil) which articulates with the stapes (the stirrup). The face-plate of the stapes is attached by an annular ligament to the oval window (Fig. 2.2.15a). The ligaments between the malleus and the incus effectively prevent relative movement between the two bones and they act as a single unit. When the tympanic membrane vibrates, the malleus and incus are displaced. The displacement is such that the hammer head acts as a pivot and the long process of the incus alternately pushes and pulls against the stapes which in turn acts as a plunger on the oval window. The oval window, like the tympanic membrane, is flexible. It vibrates in phase with the plunging action of the stapes and transmits the sound vibration to the contents of the inner ear. The plunging action of the ossicles can be damped by two small muscles: the **tensor tympani** muscle pulls on the tympanic membrane and the **stapedius** muscle pulls on the neck of the stapes (Fig. 2.2.15b). Both muscles reflexly contract in response to loud sound, although the tensor tympani is now also thought to be important in controlling the relative positions of the ossicles in relation to head and body movement. The stapedius reflex, that is the reflex contraction of the stapedius muscles, occurs with a latency of about

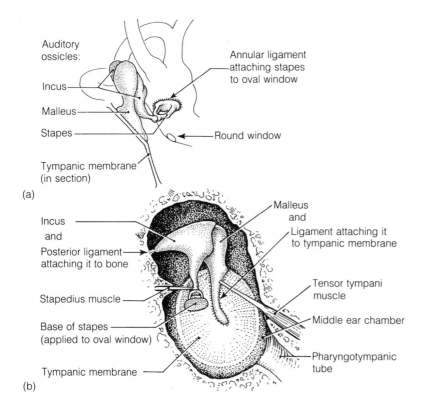

(a)

Auditory ossicles:

Incus

Malleus

Stapes

Tympanic membrane (in section)

Annular ligament attaching stapes to oval window

Round window

(b)

Incus and Posterior ligament attaching it to bone

Stapedius muscle

Base of stapes (applied to oval window)

Tympanic membrane

Malleus and Ligament attaching it to tympanic membrane

Tensor tympani muscle

Middle ear chamber

Pharyngotympanic tube

Figure 2.2.15 The middle ear ossicles. (a) Anterior aspect of ossicles (based on a figure in Elias, H. and Pauly, J. E. (1966) *Human Microanatomy*. Chichester: Wiley. (b) Medial aspect of ossicles (based on a figure in Garven, H. S. D. (1965) *A Student's Histology*. Edinburgh: Livingstone)

30 ms in response to high intensity sound. It stiffens the ossicles and damps down transmission of sound vibration to the inner ear. It therefore has a protective function in attenuating high intensity sound vibration, which is potentially damaging. High intensity sound can cause destruction of auditory receptors in the inner ear, resulting in a hearing loss. However, the latency of the reflex is such that it does not afford any protection against loud impulsive sound, such as that produced by hammering.

It has already been noted that the middle ear is air filled. In order that the tympanic membrane can 'follow' the vibrations of sound, the air pressures on the two sides of the membrane need to be equalized. The air on the outer ear side is continuously open to the atmosphere. However, the middle ear chamber is only open to the atmosphere periodically. A narrow tube, the pharyngotympanic tube (formerly called the Eustachian tube), passes between the middle ear cavity and the pharynx but the pharyngeal opening is normally sealed (see Fig. 2.2.14). It opens during yawning and swallowing and at these times the air in the middle ear equilibrates with the atmosphere. This explains why yawning or swallowing relieves the feelings of pressure in the ear which occurs during rapid changes in altitude. The change in atmospheric pressure which occurs with a change in altitude causes an air pressure difference across the tympanic membrane and deliberate swallowing or yawning then allows the pressure in the middle ear to equilibrate with the atmosphere. Catarrh may block the pharyngotympanic tube and 'popping' of the ears will not then

be possible. Furthermore, the air in the middle ear may be absorbed and the tympanic membrane will then bulge inwards causing pain and loss of hearing.

INNER EAR

The inner ear is composed of several fluid-filled chambers encased in a bony labyrinth in the temporal bone. The semicircular canals are important for balance. The inner ear auditory apparatus is the **cochlea**, which is a closed-ended tube, coiled like a snail's shell. There are about two and a half turns in the spiral of the cochlea, with the basal turn connecting to the middle ear chamber. The cochlea is coiled around a central bony core, the **modiolus**.

For simplicity, the cochlea can be illustrated as an uncoiled straight tube (Fig. 2.2.16). It is divided into two main chambers along most of its length by a structure called the **basilar membrane**. The upper chamber is the **scala vestibuli**, which is sealed at its basal end by the oval window. The lower chamber is the **scala tympani**, which is sealed at its basal end by another flexible membrane, the **round window**. Like the oval window, the round window separates the inner ear from the middle ear. The scala vestibuli and scala tympani are continuous with each other at the **helicotrema**, which is at the apex of the cochlea. Both chambers therefore contain fluid of the same composition. They are both filled with **perilymph**, which is similar in ionic composition to other extracellular fluids. Perilymph is secreted by arterioles in the periosteum and it drains into

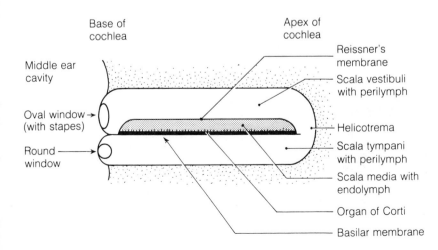

Base of cochlea
Apex of cochlea
Middle ear cavity
Oval window → (with stapes)
Round window
Reissner's membrane
Scala vestibuli with perilymph
Helicotrema
Scala tympani with perilymph
Scala media with endolymph
Organ of Corti
Basilar membrane

Figure 2.2.16 Schematic diagram of the cochlea

the subarachnoid space via the perilymphatic duct.

The scala vestibuli is actually separated from the basilar membrane by another very thin membrane, **Reissner's membrane**. Reissner's membrane and the basilar membrane enclose a third chamber, the **scala media** or **cochlear duct**. The scala media is roughly triangular in cross-section, the sides of the triangle being formed by Reissner's membrane, the basilar membrane and the **stria vascularis**. The stria vascularis is a vascular tissue located in the outer wall of the cochlear spiral (Fig. 2.2.17). The scala media is quite separate from the scala vestibuli and scala tympani and contains a fluid called **endolymph** which is very different in composition from perilymph. Endolymph composition is much more like that of intracellular fluid. Endolymph is believed to be formed by the stria vascularis. It also fills the vestibular apparatus (which is described later). Endolymph drains into the venous sinuses of the brain via the endolymphatic duct (see Fig. 2.2.19).

The scala media contains the receptor mechanism for audition. Auditory receptor cells are situated on the basilar membrane in a structure called the **organ of Corti**, which extends the length of the cochlea as far as the helicotrema. The organ of Corti is innervated by terminals of bipolar cells, the cell bodies of which are located in the **spiral ganglion** which extends through the bony core of the modiolus. The central axons of the bipolar cells give rise to the auditory branch of the VIIIth cranial nerve which transmits to the brain-stem. The organ of Corti also has an *efferent* nerve supply, and these descending nerve fibres arise from structures in the brain-stem called the **olives**. These olivocochlear fibres modify the afferent output from the organ of Corti.

When the oval window vibrates with the plunging action of the stapes, the column of perilymph in the scala vestibuli is displaced. As the oval window moves inwards, so the fluid in the scala vestibuli is displaced towards the apex of the cochlea. If movements of the oval window are slow, fluid is pushed through the helicotrema into the scala tympani, and the result of fluid displacement here is that the round window bulges towards the middle ear cavity. However, sound vibrations generally occur too fast to allow fluid to be pushed all the way round from base to apex and back to the base again. Thus, as the oval window vibrates, the vibration is transmitted to the basilar membrane which is flexible and moves up and down in phase with the displacement of perilymph in the scala vestibuli. Reissner's membrane is so thin

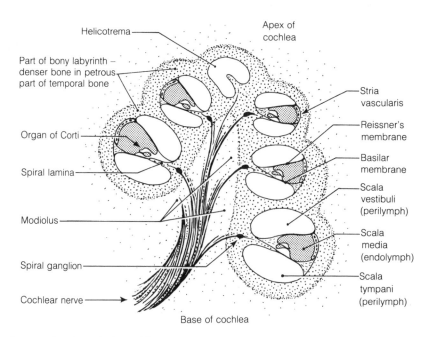

Figure 2.2.17 Cross-section through the cochlea

that it does not impede transmission of vibration. As the basilar membrane vibrates, it stimulates auditory receptors which signal sound to the brain. Movement of the basilar membrane causes displacement of fluid in the scala tympani and consequent movement of the round window. Thus, under normal conditions, the sound vibration effectively takes a 'short-cut' from the scala vestibuli through the basilar membrane to the scala tympani and little flow of perilymph occurs through the helicotrema.

Transmission of sound through the ear

The primary function of the outer and middle ear chambers is the transmission of sound to the auditory receptors of the inner ear.

CHARACTERISTICS OF SOUND AND SOUND PERCEPTION

Sound is produced by mechanical vibration and the audibility of a sound is determined by how strongly and how rapidly the vibration occurs. The sound is transmitted away from the source because of a 'knock-on' effect whereby the disturbed molecules of air adjacent to the sound source in turn cause displacement of air molecules next to them.

The rate at which a sound source vibrates determines the spacing of bands of compression (and rarefraction). The distance between two adjacent compressions is a measure of the wavelength of the sound and it is directly related to the frequency of vibration. Conventionally, sound waves are described in terms of frequency of vibration rather than wavelength. The unit of frequency is the hertz, where 1 hertz (Hz) equals one complete cycle of vibration which occurs between two compression (or rarefraction) bands.

Speech is a complex sound composed of many frequencies, but particularly those in the 400–500 Hz range. Noise is a mixture of a wide range of frequencies, and the term 'white noise' is employed to describe the sound produced by a mixture of all frequencies detectable by the human ear.

The term **pitch** describes the perception of the frequency of a sound. High frequency vibration produces a high pitch, and low frequency vibration a low pitch. The human ear is able to hear sounds covering an enormous frequency range, approximately 20 Hz to 20 000 Hz. The audible range shows individual variability and is particularly affected by age. In old age, there is a loss of sensitivity to high frequencies, a phenomenon referred to as **presbyacusis**. In the elderly, the audible range may be foreshortened at the high frequency end to 5000 Hz, depending on the intensity of the sound.

Although the range of audible frequencies is large, the human ear is not equally sensitive across the whole range. The greatest sensitivity occurs for frequencies in the range 1000–4000 Hz, and the audibility threshold intensity of sound is lowest for this frequency band.

The intensity of a sound reflects the degree to which the molecules of the medium are compressed. When sound is transmitted through air, the compression bands represent bands of increased air pressure, the change in air pressure being related to the intensity (strength) of vibration of the sound source. A large difference in pressure, which is associated with a high intensity sound, causes more displacement of the ear-drum than a small difference in pressure. The intensity of a sound can therefore be described in terms of pressure.

The intensity of a sound is actually a measure of the magnitude of the energy flow from a sound source to the surrounding medium.

Changes in intensity of sound are perceived as changes in loudness. As intensity increases, so the difference threshold for a detectable change in loudness increases. Thus, a change in apparent loudness appears to be dependent on a relative change in intensity, and a logarithmic scale of relative intensity better reflects perception of loudness.

RESONANCE IN THE EAR

When a solid object or column of air or liquid is struck, it vibrates with its own natural frequency. In other words, the material can become a sound source emitting a sound of a characteristic frequency. This phenomenon of 'natural vibration' causes structures to amplify or resonate sound waves of the same frequency. The air-filled chambers of the ear have their own characteristic resonance frequencies. The shape and size of the chambers are such that specific frequencies are amplified. Resonance in the outer and middle ears probably plays a large part in determining the frequency characteristics of the ear whereby sensitivity is greatest for sounds of 1000–4000Hz.

IMPEDANCE MATCHING IN THE MIDDLE EAR

The middle ear serves to conduct sound travelling in air from the outer ear to the fluid-filled cochlea of the inner ear. The fluid in the cochlea has much greater inertia than the air in the outer and middle ear chambers. Owing to its greater mass, more force is required to displace the fluid. Consequently, a mechanism is needed to overcome the inertia of cochlear fluid in order to effect an equivalent displacement of the oval window to that of the tympanic membrane.

The middle ear ossicles are vitally important for the transmission of sound to the auditory receptor mechanism of the inner ear. First, the lever system operating round a pivot on the malleus head has some mechanical advantage because the stapes is attached nearer to the pivot than is the vibrating tympanic membrane. Consequently, when a force is applied to the tympanic membrane, the effect of the ossicle lever system is to amplify the vibration at the oval window. The second important advantage of the ossicle system is that it permits all of the force applied to the tympanic membrane to be directed onto the much smaller oval window membrane.

The advantages of the lever system together cause an approximate twentyfold increase in pressure on the oval window and this is sufficient for virtually perfect impedance matching. Sound vibration is transmitted to the basilar membrane of the cochlea with little loss.

FREQUENCY SELECTIVITY IN THE MIDDLE EAR

The presence of one sound reduces the ability of the ear to respond to other less intense sounds and this phenomenon is referred to as **masking**. The extent to which a sound is masked depends not only on its intensity but also on its frequency, and particularly on the degree to which its frequency differs from the masking sound.

BONE CONDUCTION OF SOUND

Since the cochlea is situated in a bony labyrinth,

vibration of the temporal bone can induce cochlear fluid displacement and stimulation of auditory receptors. Thus, sound vibration can be transmitted through bone, and bone conduction is used diagnostically to assess whether a hearing impairment is the result of primary damage to the auditory receptor mechanism of the inner ear or secondary to impaired transmission of sound through the outer and middle ears. Under normal circumstances, bone conduction only plays a very small part in the transmission of sound since sound waves in air will only cause slight vibration of the skull bones.

Neurophysiology of audition

THE ORGAN OF CORTI

The organ of Corti is situated in the scala media of the cochlea. Its structure is illustrated in Fig. 2.2.18.

The 'floor' of the organ of Corti is formed by the basilar membrane, which is composed of laterally orientated fibres which run between the bony modiolus and the spiral ligament of the cochlea. The basilar membrane is narrow and stiff at the basal turn of the cochlea but progressively widens and becomes more elastic towards the apex. This is despite the fact that the chambers of the cochlea actually get smaller towards the apex.

The auditory receptors are called **hair cells**. They are arranged in orderly rows and are attached to the basilar membrane by supporting **Deiter's cells**. There is a single row of inner hair cells, which lies nearest to the central core of the cochlea spiral, and three to five rows of outer hair cells. The inner and outer hair cells are separated by a triangular-shaped tunnel, the tunnel of Corti, which is bounded by the rods of Corti. The tunnel of Corti contains perilymph, whereas the main chamber of the scala media contains endolymph.

The hair cells are so called because prominent hairs, or cilia, project from their upper surfaces. The hairs project through the dense granular layer of the **recticular lamina** and into the **tectorial membrane**. The tectorial membrane is gelatinous in texture and is composed of glycoprotein. On the inner side of the cochlear spiral the tectorial membrane is attached to the modiolus and on the outer side to the upper surfaces of supporting cells situated on the basilar membrane (see

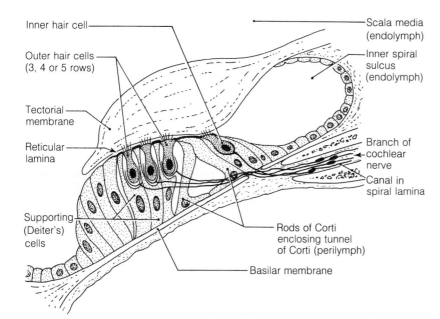

Inner hair cell

Outer hair cells
(3, 4 or 5 rows)

Tectorial
membrane

Reticular
lamina

Supporting
(Deiter's)
cells

Scala media
(endolymph)

Inner spiral
sulcus
(endolymph)

Branch of
cochlear
nerve

Canal in
spiral lamina

Rods of Corti
enclosing tunnel
of Corti (perilymph)

Basilar membrane

Figure 2.2.18 Structure of the organ of Corti (based on a figure in Garven, H. S. D. (1965) *A Student's Histology*. Edinburgh: Livingstone)

Fig. 2.2.18). Thus the tectorial membrane effectively forms the 'roof' of the organ of Corti.

The projecting hairs of the hair cells are quite stiff and bristle like. Mechanical bending of the hairs occurs when the basilar membrane vibrates in response to sound. This movement is the mechanical event which induces receptor potentials in the hair cells and hence action potentials in the cochlear nerve fibres.

The afferent nerve supply to the organ of Corti is the cochlear nerve branch of the VIIIth cranial nerve. Intimately associated with the base of each hair cell is one or more afferent nerve terminals derived from the dendrites of bipolar cells. The bipolar cell bodies are located in the spiral ganglion. The central axons of these cells form the cochlear nerve which ascends to the brain-stem and then to the auditory cortex. The pathway to the auditory cortex is described in Chapter 2.3. The afferent 'wiring' of the hair cells differs for inner and outer hair cells. There are over 3000 inner hair cells, each of which is innervated by approximately 10 afferent nerve fibres, which do not overlap or converge up to the level of the brain-stem. Inner hair cells receive fibres from approximately 90% of the total afferent supply to the cochlea. In contrast, the outer hair cells converge onto the bipolar cells. Typically, 10 outer hair cells receive terminals from one single afferent nerve fibre and there is considerable overlap of

innervation. The outer hair cells only receive about 10% of the total afferent supply to the cochlea, despite the fact that they outnumber inner hair cells in a ratio approaching 4:1.

The efferent nerve supply to the cochlea, provided by the olivocochlear fibres, also shows a differential pattern of input to inner and outer hair cells. Most efferent fibres appear to terminate on the outer hair cells, although some efferents make close contacts with the afferent dendritic supply of the inner hair cells. The precise role of the efferent nerve supply is not understood at present. However, increased activity in efferent fibres appears to result in an inhibition of afferent activity in the cochlear nerve. The efferents therefore apparently influence the sensitivity of auditory detection of sound.

THE ENDOLYMPHATIC POTENTIAL

The scala media contains endolymph, which itself contains Na^+ and K^+ ions in concentrations similar to those found in intracellular fluid. In contrast, the scalae vestibuli and tympani contain perilymph, which has a higher Na^+ ion and lower K^+ ion concentration than endolymph. The composition of perilymph is similar to that of extracellular fluid.

The difference in composition between perilymph and endolymph is maintained because Reissner's membrane and the reticular lamina are impermeable to Na^+ and K^+ ions. These ions can readily penetrate the basilar membrane and so it is assumed that endolymph surrounds the organ of Corti but only penetrates it between the tectorial membrane and the reticular lamina. This means that the nerve terminals associated with the hair cell bases are surrounded by perilymph.

When no sound is impinging on the ear, a steady potential difference exists between endolymph and perilymph. This potential difference is referred to as the endolymphatic potential. It is normally of 50–80 mV amplitude, with endolymph being positive with respect to perilymph. The sign of the endolymphatic potential is somewhat surprising: typically, intracellular fluid is negative with respect to extracellular fluid, and hence the endolymphatic potential cannot result from the differences in distribution of Na and K ions between endolymph and periplymph. The endolymphatic potential is maintained metabolically by the stria vascularis.

TRANSDUCTION OF THE MECHANICAL VIBRATION OF SOUND INTO ELECTRICAL SIGNALS

When sound waves impinge on the ear, the energy is transmitted to the cochlea and the basilar membrane vibrates. As the basilar membrane moves up and down, so the hairs of the cells bend. In effect, the basilar membrane, which is attached on both sides, alternately bows upwards and dips downwards, and the change in curvature of its surface causes changes in the orientation of the hairs on the hair cells. Since the tips of the hair cells are embedded in the relatively firm, gelatinous mass of the tectorial membrane, the vibration of the hair cell bases results in a shearing or bending of the projecting hairs. This shearing movement then results in the generation of receptor potentials.

FREQUENCY CODING IN THE EAR

The human ear can discriminate small changes in the frequency of sound, an ability which appears to depend primarily on the nature of the mechanical displacement of the basilar membrane.

The most widely accepted theory to explain frequency coding of a sound is the place theory, first proposed by Helmholtz in 1863 and later refined by Georg von Békésy in the 1920s (Békésy, 1960). In essence, the place theory states that different parts of the basilar membrane resonate at different frequencies. The basal end of the basilar membrane is narrow, stiff and composed of short lateral fibres. This favours fast frequency vibration. In contrast, the apical end of the basilar membrane is wider and more elastic and it vibrates maximally in response to low frequency sounds.

The place theory proposes that the hair cells in the region of the basilar membrane which is maximally displaced will produce the greatest response in the afferent fibres associated with them. The pattern of afferent input to the brain is then interpreted as a sound of a specific frequency.

INTENSITY CODING IN THE EAR

The intensity of a sound is primarily coded by the frequency of impulse discharge in stimulated afferents. High intensity (loud) sounds produce a higher frequency of discharge than low intensity sounds. In addition, high intensity sounds produce a greater displacement over a wider area of the basilar membrane, although the region of maximal displacement remains a function of the frequency of the sound source.

LOCALIZATION OF A SOUND SOURCE

The majority of people can locate the source of a sound reasonably accurately. This ability depends on two factors. First, the timing of the arrival of a sound wave at the two ear-drums will differ, depending on the position of the sound source. This will in turn influence the timing of the impulse discharge in the afferents from the two ears, and this discrepancy can be interpreted by the brain to enable localization of the sound. Second, the head casts an 'acoustic shadow' such that sounds originating from the left side produce marginally higher intensity vibrations in the left ear when compared with the right ear. Again, the discrepancy in intensity will result in a small difference in the frequency of impulse discharge arising from the two ears.

Hearing loss: implications for nursing

Impaired hearing may result from impaired conduction of sound through the outer and middle ear; from damage to the cochlea; or from damage to the auditory pathway beyond the ear. It is important to distinguish between deafness which is conductive (the most common) and sensorineural, which may indicate important neurological disease.

Complete deafness is uncommon; the adult 'labelled' as deaf is much more likely to have a markedly reduced sense of hearing. This often develops gradually and can be well advanced before the person realizes. Some children are born deaf, the most common cause is contraction of rubella by the mother during the first trimester of pregnancy.

Intact hearing is essential for proper communication and maintaining a safe environment. When communicating with the deaf patient, the nurse should not speak until she has the person's attention. Having achieved this, the nurse should speak slowly and clearly, ensuring that the deaf person can see the movement of her lips. Exaggerated lip movements are not necessary. If the patient has a 'good' ear, this should be identified and utilized. When the patient has difficulty understanding, it is better to rephrase your statement rather than to repeat continually the same misunderstood phrase.

As a general rule, vowels are more readily heard than consonants (Stalker, 1984).

It is imperative to take your time and display patience and tolerance while the deaf person communicates with you. Any outward displays of irritability or impatience on the part of the nurse will serve only to discourage contact with the deaf person.

Some patients will be trained to lip read and others will benefit from the use of an appropriate and properly fitted hearing aid. People with impaired hearing benefit from receiving advice and guidance from the Royal National Institute for the Deaf, an organization which specializes in the welfare of the deaf person.

SPECIAL SENSES: THE EAR AND BALANCE

The inner era contains the **vestibular apparatus** in addition to the cochlea. The vestibular apparatus signals rotational movements of the head and changes in its position relative to the pull

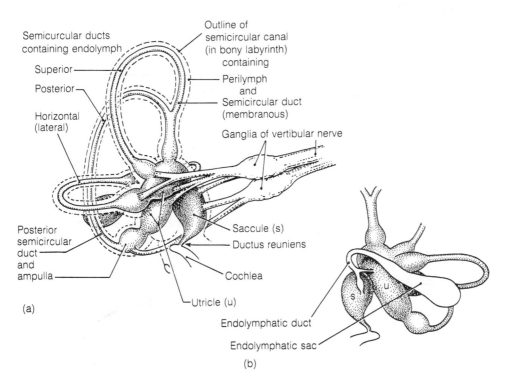

Figure 2.2.19 Vestibular apparatus: (a) anterior aspect, (b) posterior aspect (based on figures in Elias, H. and Pauly, J. E. (1966) *Human Microanatomy*, Chichester: Wiley)

of gravity. These sensory signals are involved in the reflex control of posture and in the control of eye movements.

Anatomy of the vestibular apparatus

The vestibular apparatus consists of a **membranous labyrinth** contained within the **bony labyrinth** of the temporal bone. The major features of the vestibular apparatus are illustrated in Fig. 2.2.19.

There are three **semicircular canals** orientated at right-angles to each other. The membranous labyrinth of the canals is filled with endolymph and is surrounded by perilymph contained within a bony cavity in the temporal bone. At the base of each canal is a swelling, the **ampulla**, which contains a ridge called the **crista**. Hair cells are situated on the crista and the hairs project into a gelatinous mass, the **copula**, which divides the semicircular canal into two parts (Fig. 2.2.20). The bases of the hair cells are innervated by afferent nerve terminals of bipolar vestibular nerve cells. The cell bodies of these afferents are located in the vestibular ganglion, which is illustrated in Fig. 2.2.19a. The central axons of the fibres project to the brain-stem as part of the VIIIth cranial nerve. The pathway is described in Chapter 2.3.

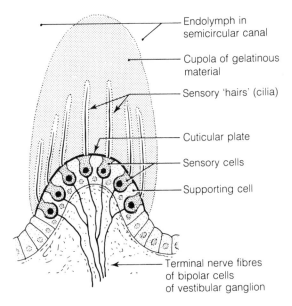

Figure 2.2.20 Structure of the ampulla of a semicircular canal

The semicircular canals communicate with the **utricle**, which in turn communicates with the **saccule**. The utricle and saccule are filled with endolymph and they are directly connected to the scala media of the cochlea by the ductus reuniens. The utricle and saccule each contain a ridge, the **macula**, from which hair cells project into a gelatinous copula in much the same way as in the ampullae of the semicircular canals. However, the cupulae in the utricle and saccule contain calcium carbonate particles called **otoliths** (Fig. 2.2.21). When posture is erect the macula of the utricle projects horizontally, whilst in the saccule the macula projects vertically. As in the semicircular canals, the hair cells of the utricle and saccule are associated with afferent terminals of the vestibular nerve.

The hair cells of the vestibular apparatus are the receptors which detect head movement. Bending of the projecting hairs initiates an impulse discharge in the vestibular nerve. The mechanism is assumed to be similar to the receptor mechanism in the cochlea.

THE SEMICIRCULAR CANALS

Receptors in the three **semicircular canals** in each ear are stimulated by rotational movements of the head. Since the canals are orientated at right-angles to each other, they can respond to rotation in any plane. Thus, the receptors in the horizontal canal are primarily stimulated when the head is rotated from side to side, as when gesturing 'no'; those in the superior canal are stimulated when the head is rotated up and down, as when gesturing 'yes'; and those in the posterior canal are stimulated when the head is rotated in a transverse plane tilting the head towards one shoulder.

When the head is rotated laterally, the hair cells of the two horizontal canals are stimulated. As the head moves, the semicircular canals move with it. However, the endolymph contained within each canal is unattached and, because of its own inertia, its movement lags behind movement of the head. Thus, the relatively stationary columns of endolymph displace the copulae in the two ampullae and this in turn causes bending of the projecting hairs. In each copula, bending of the hairs in one direction causes an increase in the impulse discharge in the associated afferent nerve terminal, whereas bending in the opposite direction reduces

Labels in figure: Endolymph in semicircular canal; Cupola of gelatinous material; Sensory 'hairs' (cilia); Cuticular plate; Sensory cells; Supporting cell; Terminal nerve fibres of bipolar cells of vestibular ganglion

the discharge. As the head is rotated to the left, the impulse discharge in the vestibular nerve from the left ear decreases but that from the right ear increases, and vice versa. This imbalance in input to the brain is presumed to form the basis for interpretation of direction of movement.

The bending of the hairs only occurs when the rotational movement accelerates or decelerates. During acceleration the endolymph rapidly overcomes its inertia and it then moves in unison with the semicircular canals. The cupolae then resume their normal orientations and the bending force on the hair cells is removed. The impulse discharge then returns to its resting level. During deceleration the events occur in reverse. The endolymph continues to move under its own momentum after cessation of movement of the head. Thus, the hair cells are again bent until the endolymph also becomes stationary. The nervous discharge to the CNS from the two ears also reverses since deceleration is equivalent to movement in the opposite direction. Thus, as a rotational movement to the left decelerates, the left ear impulse discharge increases and the right ear discharge decreases.

Since the semicircular canals are only stimulated during acceleration or deceleration, they can perform a predictive function. The impulse discharge from the semicircular canals signals a rate of change of position of the head and the brain can therefore predict the position of the head in advance of the movement. This is thought to be an important afferent input enabling the brain to modify the output to postural muscles to effect coordinated movement without loss of balance.

The events described can occur in the three pairs of semicircular canals, depending on the plane of rotation of the head.

THE UTRICLE AND SACCULE

The utricle and saccule are stimulated by changes in position of the head relative to the pull of gravity. They are collectively referred to as the **otolith organ**. When the head is tilted away from its normal upright position, the otoliths in the cupolae tend to move under the influence of the gravitational pull and this causes bending of the projecting hairs of the hair cells. The projecting hairs differ in their orientations and therefore movement of the otoliths causes maximal stimulation of only some of the hair cells, depending on the direction of tilt of the head. Stimulation of hair cells gives rise to an impulse discharge in the associated afferent terminals. Bending of the hairs in one direction causes an increased impulse discharge, whilst bending in the opposite direction results in a decrease in impulse discharge. The pattern of impulse discharge arising from the otolith organ codes the direction of movement of the head. The frequency of the impulse discharge is primarily dependent on the distance moved by the head.

The nervous discharge shows little adaptation, in contrast to the rapidly adapting signal transmitted from the semicircular canals. The otolith organ therefore continually appraises the CNS of the position of the head in space.

The vestibular apparatus in the control of posture and eye movements

The vestibular apparatus signals changes in position of the head in space and, together with the visual input and proprioceptor input, contributes to the reflex maintenance of posture and balance during movement.

The otolith organ continuously signals the position of the head even when the body is not moving, and the signal is involved in the reflex control of postural muscles. When linear acceleration occurs, the otolith organ is again stimulated.

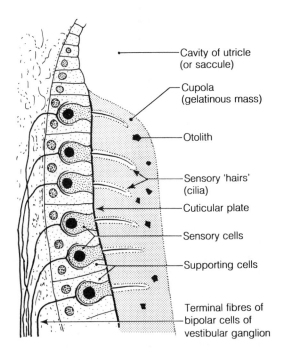

Cavity of utricle (or saccule)

Cupola (gelatinous mass)

Otolith

Sensory 'hairs' (cilia)

Cuticular plate

Sensory cells

Supporting cells

Terminal fibres of bipolar cells of vestibular ganglion

Figure 2.2.21   Structure of the macula of an otolith organ

The otoliths in the utricles and/or saccules experience a thrust in the opposite direction to that of the movement, and this can also initiate stimulation of hair cells. The utricle and saccule provide important afferent inputs for posture control during the linear acceleration which occurs when an individual starts to walk forwards.

The otolith organ also contributes to the control of eye movements which occur when the head position is changed. Eye movements occur in order to enable visual fixation of an image despite head movement. Thus, as the head is tilted upwards, the eyes are lowered, and vice versa. (Visual input from the eyes also contributes to the reflex control of the extrinsic eye muscles.)

During rapid rotational movements, reflex compensatory eye movements also occur in an attempt to maintain a fixed visual reference point. The eyes move back and forth, a rotational nystagmus. The nystagmus results from the input to the brain from stimulated semicircular canals. When rotation occurs to the left the slow component of the nystagmus is towards the right in an attempt to keep a fixed visual reference point. The fast component occurs towards the left, the direction of rotation, in order to pick up a new visual reference point. When rotation ceases, the semicircular canals will at first be stimulated as though rotation were occurring in the opposite direction, and a postrotational nystagmus occurs.

In addition to the reflex control of eye movements, the semicircular canals also provide an input for posture and balance control. Thus, when rotating to the left, extensor muscles in the left limbs contract to maintain balance. When the rotation stops, the initial effect on the semicircular canals is as though the body is rotating to the right, and the right limb extensors therefore reflexly contract. However, since the body has in fact stopped rotating, extension of the right limbs causes the body to fall towards the left and there is a temporary loss of balance. Dancers can overcome the effects of postrotational nystagmus and loss of balance by making short sharp movements of the head in the opposite direction to that of rotation, thereby preventing movement of endolymph in the semicircular canals.

Disturbances of the vestibular apparatus

Ear infections can cause abnormalities of vestibular function, resulting in disturbances of balance and dizziness. **Menière's disease** is a condition in which there is excessive production of endolymph. This can cause auditory disturbances but it also commonly causes dizziness and nausea as a result of impaired function of the vestibular apparatus. Patients with these distressing symptoms are advised to rest in bed in a quiet, darkened room. This is a frightening time for the sufferer, and the nurse will need to stay with the patient, to provide reassurance during the worst stages of the attack.

A relatively common disturbance associated with vestibular function is that of **motion sickness**. This appears to result when movement is erratic and there is poor correlation between movement detected by the vestibular apparatus and movement detected by the eyes. People vary in their susceptibility to motion sickness.

SPECIAL SENSES: THE CHEMICAL SENSES OF GUSTATION AND OLFACTION

Taste (gustation) and smell (olfaction) are both chemical senses which depend on the detection of a local change in the chemical environment of appropriate receptors.

Gustation (see also Chaps. 2.3 and 5.1)

Taste results from stimulation of chemical receptors which are located primarily on the tongue (see Fig. 5.1.3). There appear to be four primary tastes which can be detected by these receptors: namely, sour, sweet, salt and bitter. All the flavours which a human being can experience depend on the relative activation of these four primary taste receptor types and on associated olfactory sensations. The involvement of olfaction in taste is demonstrated by the reduction in ability to taste food when a heavy cold interferes with smell. Olfaction is an important component for the identification of particular tastes. It is also possible that texture and chewiness may contribute to the recognition of foods. Taste identification therefore may involve several sensory areas in the brain.

TASTE RECEPTORS

Taste receptors are grouped together in oval-

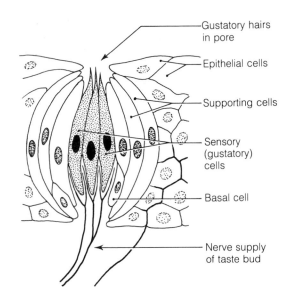

Gustatory hairs in pore

Epithelial cells

Supporting cells

Sensory (gustatory) cells

Basal cell

Nerve supply of taste bud

Figure 2.2.22 Structure of a taste bud

shaped structures called **taste buds** which are located primarily on the tongue. A few are found on the soft palate and at the back of the throat. There may be up to 10 000 taste buds in the mouth. The structure of a taste bud is illustrated in Fig. 2.2.22. Each bud is about 70 μm long and up to 40 μm wide and contains up to 20 functioning taste receptor cells. The taste receptor cells are derived from epithelial cells. At the exposed tip of the bud, microscopic taste hairs, which arise from the taste cells, project through a pore. The hairs (or microvilli) bear receptor sites for the attachment of chemicals which stimulate the cells. The taste pore is the route of access for the receptors. All substances which are tasted are dissolved in saliva which bathes the surface of the taste buds and penetrates the taste pores.

Taste buds are located in the sides of small connective tissue elevations called **papillae**, which give the surface of the tongue its characteristic rough appearance. The taste buds are found in the fungiform knob-shaped papillae, which are located primarily at the tip and the sides of the tongue, and in circumvallate (circular) papillae, which form a prominent inverted V-shape at the back of the tongue (see Fig. 5.1.3).

Each taste bud is associated with two or three large afferent nerve fibres which branch to form a network of fine unmyelinated fibres innervating the receptor cells. The terminals of the fibres lie in folds in the taste receptor cell membranes so there is intimate contact between the receptors and the sensory afferent fibres. There is some overlap of innervation of taste buds. One large afferent fibre may have branches supplying more than one taste bud. The afferent fibres from taste buds join cranial nerves which convey the nerve signals to the brain. Fibres from taste buds in the anterior two-thirds of the tongue, which are predominantly located in the fungiform papillae, travel in the VIIth cranial nerve, the facial nerve. Fibres from taste buds located in the posterior third of the tongue travel in the IXth, the glossopharyngeal nerve, whilst fibres from the throat area travel in the Xth, the vagus nerve.

THE MECHANISM OF TASTE

Electrophysiological studies of taste afferents have shown that any single taste bud may respond to several, even up to four, of the primary tastes of sourness, saltiness, sweetness and bitterness. However, each taste bud shows differential sensitivity to the four tastes such that a maximum sensitivity is shown to only one taste type. In addition, single taste receptor cells within a bud may respond to more than one taste. Each cell may therefore bear receptor sites on microvilli for more than one of the primary tastes.

DISTRIBUTION OF SENSITIVITY TO THE PRIMARY TASTES

There is a distinct pattern in the distribution of taste buds in terms of the predominant taste to which they are sensitive. Thus, taste buds which are maximally sensitive to sour chemicals are located on the sides of the tongue, those which primarily respond to sweetness are found at the tip, and those for bitterness at the back of the tongue (see Fig. 5.1.3). Buds which predominantly respond to saltiness are found in all of these areas, but particularly at the tip of the tongue.

THRESHOLD AND SENSITIVITY TO TASTE

The threshold concentration required to elicit a taste sensation is an individual characteristic of a chemical. However, there do appear to be general differences in threshold between the four primary tastes. Very much weaker concentrations of bitter substances are required to elicit a bitter sensation when compared with threshold concentrations for sweet or salt chemicals. Receptor cells for sourness tend to have thresholds which are intermediate.

Sensitivity to the four primary tastes is influenced by a number of factors, chiefly individual variability, age and temperature. Individual variability amongst people is clearly to be expected, but one particular aspect is of interest genetically. About 3% or 4% of a Caucasian population has a very high threshold for the taste of a bitter chemical, called phenylthiourea. Their measured thresholds are of the order of 400 times that expected for the rest of the population. The ability to taste phenylthiourea is determined by a single dominant gene. It is of particular interest since it is so specific to one chemical and there is no apparent shift in threshold to other bitter substances, nor to the other primary tastes.

Age has an effect on sensitivity to taste. Newborn infants are apparently relatively insensitive to taste, whereas young children are often very sensitive to, and indeed intolerant of, bitter or salty substances. Adult sensitivity is less than that found in children and it decreases further in old age. These changes in sensitivity may contribute to the differences in food preferences which are often exhibited by children and adults.

Temperature also appears to alter sensitivity to tastes, but the effect is quite complex. Salt and bitter tastes are generally stronger at lower temperatures, and an adequately salted hot casserole may seem oversalted as it cools a little. In contrast, sweetness sensitivity appears to increase to a maximum at a temperature of about 40°C. Sensitivity to sourness is relatively unaffected by temperature changes.

ADAPTATION

Adaptation to sapid (taste-producing) substances occurs rapidly, although the reduction in sensation is minimized if the food is moved continuously around the mouth. The adaptation appears partly to reflect receptor adaptation to stimulation but also to depend on central brain mechanisms.

TASTE AND HEALTH

Taste is probably of greatest clinical significance in terms of changes that occur secondarily to other disorders or conditions. Poor oral hygiene can be a direct cause of an impaired taste sensation and loss of appetite. This serves to emphasize the importance of assessment of the patient's mouth during illness. Those most at risk of developing a dry mouth are patients receiving oxygen therapy or those constantly breathing through their mouths. The patient with a swallowing difficulty, who may be leaving food debris in the mouth, and the infected or dehydrated patient are all at risk. Effective oral hygiene should be performed as each individual patient's needs demand. The patient with an impaired sense of taste will need to be tempted to eat, and this can be enhanced by proper presentation and service of meals.

Olfaction (see also Chap. 2.3)

The sense of smell is even less well understood than taste. So far, primary odour types have not been identified. Smell may play a role in survival since many, though by no means all, noxious volatile materials have characteristic odours.

Smell may be involved in sexual attraction. Certainly, humans, insects and animals secrete attracting chemicals as a forerunner to mating. There is currently debate as to whether humans are subconsciously attracted by body odours (see Chap. 6.3), but sexual secretions do have characteristic odours.

Olfactory receptors

The olfactory receptors are found in olfactory epithelial tissue which is located in the roof of the nasal cavity. It extends down to cover the superior turbinate bone in each nostril and also the upper part of the walls of the dividing nasal septum.

The olfactory receptors are derived from dendritic terminals of bipolar cells, the central axons of which unite to form the olfactory nerves. The olfactory nerves transmit to the two olfactory bulbs which lie under the frontal poles of the brain. The dendrites of the olfactory bipolar cells extend as olfactory rods which bear microscopic hairs, or cilia. The olfactory hairs bear receptor sites for chemical attachment and they are embedded in the mucous lining of the nasal cavity (Fig. 2.2.23).

MECHANISM OF OLFACTION

The olfactory cilia bear receptor sites for attach-

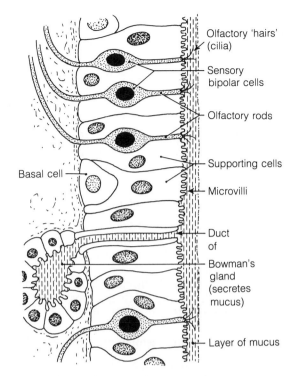

Olfactory 'hairs' (cilia)

Sensory bipolar cells

Olfactory rods

Supporting cells

Basal cell

Microvilli

Duct of

Bowman's gland (secretes mucus)

Layer of mucus

Figure 2.2.23 Olfactory receptors in olfactory epithelium

ment of odoriferous chemicals, which are brought into contact with the olfactory epithelium by the passage of air through the nose. Since the epithelium is situated in the roof of the nasal cavity, sniffing helps to draw air up across the surface of the sensory cells. Any odoriferous chemical must clearly be sufficiently volatile to produce a vapour which can be carried in the airstream through the nasal cavity. In order to gain access to receptor sites on the olfactory cilia, a chemical must be able, at least partially, to dissolve in the aqueous mucous lining of the nasal cavity. Lipid solubility is probably also important since the olfactory cilia membranes largely consist of lipoprotein.

The olfactory epithelium appears to show regional variations in terms of the type of odoriferous chemicals to which the receptors respond (c.f. the distribution of the primary tastes on the tongue). However, since no primary odour types can be identified, the pattern of distribution of receptors in relation to odour is not precise.

THRESHOLD AND SENSITIVITY TO SMELL

Different chemicals will require different concen-

trations in air in order to achieve the threshold level for odour detection. Moreover, the pattern of breathing will influence the quantity of chemical which is brought into contact with the olfactory epithelium.

As with taste, odour sensitivity is not a constant, and individuals vary widely in their sensitivities to different odoriferous chemicals.

ADAPTATION

Adaptation to odour occurs very rapidly and an odour may not be detectable seconds or minutes after the beginning of exposure, the time to extinction being dependent on the particular chemical. Most people have experienced how even a very unpleasant smell rapidly fades and apparently disappears. As with taste adaptation, olfactory adaptation apparently partly depends on central brain mechanisms.

OLFACTION AND HEALTH

As with taste, changes in olfactory sensitivity are chiefly of clinical importance as secondary signs of disorder. The main problem encountered is decreased stimulation of appetite, as taste is enhanced by the sense of smell. The person who lives alone and whose sense of smell is not acute, runs the risk of being exposed to noxious gases and needs to be warned of this potential danger.

Review questions

The answers to all these questions can be found in the text. In each case there is at least one correct and at least one incorrect answer.

1 The 'adequate stimulus' for a sensory receptor

(a) is the stimulus intensity required to elicit an action potential in the afferent nerve fibre
(b) is the stimulus intensity required in order to be perceived by a subject
(c) is the type of stimulus energy to which the receptor is most sensitive
(d) is the only type of stimulus energy to which the receptor will respond

2 Generator potentials

(a) last for about 1 ms
(b) can summate to produce a larger response
(c) result from the movement of small ions across the receptor membrane
(d) increase in amplitude in proportion to increase in stimulus intensity

3 Aqueous humour

(a) fills the space between the lens and the retina of the eye
(b) is secreted by blood vessels in the posterior chamber of the eye
(c) through its rate of secretion and absorption, determines the intraocular pressure
(d) supplies oxygen and nutrients to the lens

4 Visual acuity

(a) is a measure of the sensitivity of the eye to light
(b) is the ability to discriminate between two separate points of light
(c) is measured using Snellen's type
(d) is greatest when the visual image falls on the fovea of the retina

5 The cones of the photoreceptor layer

(a) are distributed throughout the retina
(b) are essential for colour vision
(c) are more sensitive to light when compared to the rods
(d) are responsible for photopic vision in daylight

6 The middle ear

(a) contains air which is normally at atmospheric pressure
(b) contains three small ossicles which transmit sound vibration from the tympanic membrane to the round window membrane
(c) cannot conduct sound vibration if the ossicles are removed
(d) ossicles transmit sound energy more efficiently when the tensor tympani and stapedius muscles are contracted

7 The basilar membrane

(a) vibrates at the same frequency as an applied sound source
(b) at the apex of the cochlear is maximally displaced by high frequency sound vibration
(c) is displayed to an extent which is dependent on the frequency but not the intensity of sound vibration
(d) separates the scala vestibuli from the scala tympani

8 The endolymphatic potential

(a) is a potential difference between the scala vestibuli and the scala tympani
(b) is maintained metabolically by the stria vascularis
(c) results from a differential distribution of sodium and potassium ions
(d) is about 110 mV in amplitude

9 The membranous semicircular canals

(a) contain endolymph
(b) are only stimulated during acceleration of rotation of the head
(c) contain hair cells embedded in a gelatinous cupula containing calcium carbonate particles
(d) provide a sensory input to the brain which contributes to the control of eye movements

10 The saccule

(a) communicates with the semicircular canals of the vestibular apparatus
(b) responds to changes in position of the head relative to the pull of gravity
(c) contains a ridge, the macula, which projects vertically when posture is erect
(d) contains receptors which rapidly adapt to stimulation

11 Taste buds

(a) are selective such that any single taste bud is stimulated by only one of the primary taste types
(b) which are most sensitive to a sour taste are located predominantly on the sides of the tongue

(c) each contain up to 20 taste receptor cells
(d) located at the tip of the tongue are innervated by the glossopharyngeal nerve

12 Pain receptors

(a) are all free nerve endings
(b) only send nerve impulses along the smallest, unmyelinated, C group of nerve fibres
(c) adapt rapidly to stimulation
(d) can be stimulated by histamine and bradykinin

Answers to review questions

1 c
2 b, c and d
3 b, c and d
4 b, c and d
5 b and d
6 a
7 a and d
8 b
9 a and d
10 a, b and c
11 b and c
12 a and d

References

Békésy, G. von (1960) *Experiments in Hearing*. London: McGraw-Hill.

Stalker, A. E. (1984) *ENT Nursing*, 6th edn. London: Baillière Tindall.

Suggestions for further reading

Coren, S., Porac, C. & Ward, L. M. (1979) *Sensation and Perception*. London: Academic Press.
Kandel, E. R. & Schwartz, J. H. (1981) *Principles of Neural Science*. Amsterdam: Elsevier North Holland.
Ludel, J. (1978) *Introduction to Sensory Processes*. New York: W. H. Freeman.
Pick, G. F. (1978) The cochlea – normal and pathological. *Nursing Mirror* (Feb. 9): 21–23.
Purchese, G. & Allan, D. (1984) *Neuromedical and Neurosurgical Nursing*, 2nd edn. London: Baillière Tindall.
Rosenberg, M. E. (1982) *Sound and Hearing*. Studies in Biology No. 145. London: Edward Arnold.
Stein, J. F. (1982) *An Introduction to Neurophysiology*. Oxford: Blackwell Scientific.

Useful addresses

Royal National Institute for the Blind
224 Great Portland Street
London
W1N 6AA
Tel: 01 388 1266

Royal National Institute for the Deaf
105 Gower Street
London
WC1E 6AH
Tel: 01 387 8033

Chapter 2.3
The Central Nervous System

Valerie M. Nie
Michael Hunter
with Douglas Allan

Learning objectives

After studying this chapter the reader should be able to

1 Identify the main features of the embryological development of the central nervous system (CNS).
2 Locate and describe the appearance of the four ventricles and the aqueduct in the human brain.
3 Identify the separate layers of the meninges.
4 Describe the site of production, path of flow and site of reabsorption of cerebrospinal fluid (CSF).
5 Describe the function and composition of CSF and demonstrate the importance of CSF analysis to the identification of disease.
6 Identify the major blood vessels supplying the brain.
7 Describe the relationship between the spinal cord, spinal nerves and the vertebral column.
8 Describe the structural features of grey and white matter in the spinal cord.
9 Name the 12 pairs of cranial nerves and describe their separate functions.
10 Identify the major structural features of the brain and describe in outline their functions.
11 Identify the major pathways for somatosensory input to the brain and describe the cortical representation of somatosensation.
12 Describe the underlying physiology of pain perception.
13 Identify the visual pathway from the retinae to the brain and explain how the visual input is processed by the cerebral cortex.
14 Describe the pathways involved in visual reflexes and eye movements.
15 Describe the pathway and cortical representation of audition.
16 Describe the nervous connections and functions of the vestibular nuclei.
17 Identify the gustatory and olfactory pathways to the cerebral cortex.
18 Describe the major features of the pyramidal and extrapyramidal pathways for motor control.
19 Explain the role of the cerebellum in motor coordination and control of balance.
20 Explain the role of the brain-stem in the control of states of consciousness.
21 Describe the EEG and behavioural characteristics of the major types of epilepsy.
22 Relate structure to function in the CNS in the control of motivation and emotion.

EMBRYOLOGICAL DEVELOPMENT OF THE CENTRAL NERVOUS SYSTEM

The central nervous system (CNS) comprises the brain and spinal cord. In the human embryo the

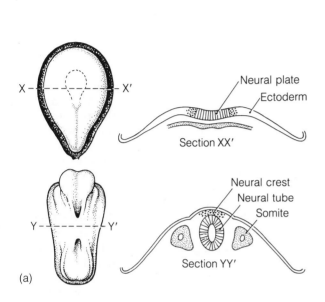

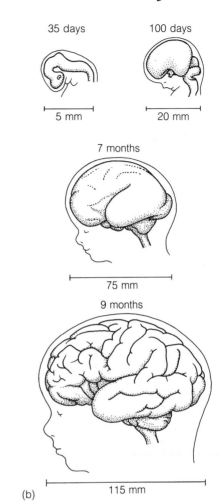

Figure 2.3.1 Embryological development of the human nervous system: (a) development of the neural tube, (b) development of the fetal brain (based on illustrations in Cowan, W. M. (1979) The development of the brain. In: *The Brain*. San Francisco: W. H. Freeman)

CNS develops from a plate of ectodermal cells along the midline of the back of the embryo. This neural plate is at first exposed to the amniotic fluid, but as it elongates the lateral edges of the plate rise and grow medially until they meet and unite to form a **neural tube** (Fig. 2.3.1). An arrest in this process leading to incomplete closure of the neural tube (and an associated malformation of the vertebral column) results in the condition of **spina bifida**. This condition varies in severity depending on the site and extent of the abnormality, and it occurs in approximately 1 in every 200 pregnancies.

The neural tube is normally formed by about the 20th day of gestation, after which it becomes detached from the ectoderm and sinks beneath the surface layers of the developing embryo. As the neural tube elongates it also changes shape. The tube bulges at the head end to form three enlarged areas which represent the developing brain. From the furthest forward to the one which is contiguous with the rest of the neural tube, these three enlargements are called the **prosencephalon**, or forebrain; the **mesencephalon**, or midbrain; and the **rhombencephalon**, or hindbrain. Early in the second month of gestation these three enlargements increase in number to five: the prosencephalon subdivides into the **telencephalon** (endbrain) and the **diencephalon** (between-brain) and the rhombencephalon subdivides into the **metencephalon** (after-brain) and the **myelencephalon** (spinal brain).

Further alterations to the structure of the brain are made by the twists, or 'flexures', of the neural tube and by the differential enlargements of parts of the developing brain. Nevertheless, the tube arrangement is retained throughout, with the cavity of the tube becoming the central canal of the spinal cord, which expands at the head end into the ventricular system of the brain.

In the third month of gestation the lateral cerebral fissure appears, and by the end of the third month of fetal life the main outlines of the brain are recognizable. Indeed, the integrity of the system can be demonstrated at this stage, since by the 14th week of gestation tactile stimulation of the face of the fetus will evoke primitive reflex responses such as rotation of the head and pelvis, and contraction of the trunk musculature.

Further cell division continues throughout the rest of gestation, and at birth the brain weighs about 350 g. By the end of the first year of postnatal life, the brain weight has increased to about 1000 g, but this is not caused by any large increase in the number of nerve cells. It appears that cell division of neurones is more or less complete at birth. However, nerve tissue which is already present at birth continues to grow postnatally, especially in the first 3 years of life. The process of myelination of nerve fibres continues during infancy and adolescence. Myelination is more or less completed by about the age of 15 years. By puberty, the brain has almost reached its full development and weighs about 1300 g. An average adult human brain weighs about 1400 g.

THE VENTRICULAR SYSTEM OF THE BRAIN

It has been noted that a cavity within the neural tube is retained throughout embryological development, so that in a mature brain a central canal exists within the spinal cord and this expands into larger cavities, or 'ventricles', in the brain.

There are four **ventricles** within the brain. The most posterior, lying in the midline of the hindbrain, is called the **fourth ventricle**. This is joined by the **cerebral aqueduct**, the midbrain equivalent of the neural tube, to the **third ventricle**, which also lies in the midline. A pair of openings called the **foramina of Monro** then joins the third ventricle with the two **lateral ventricles** which extend in a horseshoe shape either side of the midline within each cerebral hemisphere. The third ventricle and the two lateral ventricles are the cavities of the forebrain (Fig. 2.3.2).

Each ventricle contains a spongy network of capillary blood vessels called a **choroid plexus**. The choroid plexuses are believed to be responsible for producing **cerebrospinal fluid** (CSF), which circulates around the brain and within the ventricular system (Fig. 2.3.3).

The ventricular system can be viewed using computerized axial tomography (CAT scan), a radiographic technique whereby serial sections through the brain can be viewed as X-ray plates,

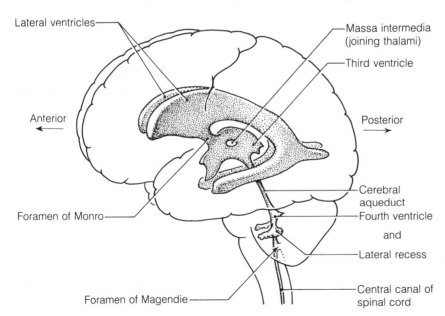

Figure 2.3.2 The ventricular system of the brain, lateral view

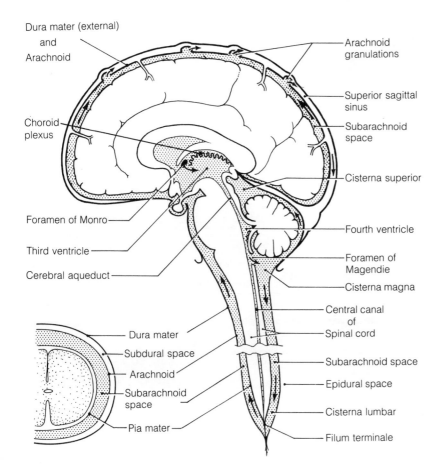

Dura mater (external) and Arachnoid

Choroid plexus

Foramen of Monro

Third ventricle

Cerebral aqueduct

Dura mater

Subdural space

Arachnoid

Subarachnoid space

Pia mater

Arachnoid granulations

Superior sagittal sinus

Subarachnoid space

Cisterna superior

Fourth ventricle

Foramen of Magendie

Cisterna magna

Central canal of Spinal cord

Subarachnoid space

Epidural space

Cisterna lumbar

Filum terminale

Figure 2.3.3 Path of flow of cerebrospinal fluid (Based on an illustration by R. J. Demarest in Noback, C. R. (1967) *The Human Nervous System*. New York: McGraw-Hill)

with the aid of computer integration (Fig. 2.3.4). When this is done, any distortions of shape and size of the ventricles which may be of clinical importance can be seen. For example, hydrocephalus, which results from an abnormal build up of the pressure of CSF, can lead to enlarged and distorted ventricles. Also, a space-occupying lesion in the brain might lead to the ventricles being pushed away from their normal position with respect to the midline.

THE MENINGES OF THE BRAIN AND THE CEREBROSPINAL FLUID

The brain and spinal cord are surrounded by three layers of connective tissue which are collectively referred to as the brain **meninges**. The first of these is a thin, filmy layer of connective tissue called the **pia mater**, which closely adheres to the surface of the brain and spinal cord and follows every indentation and fissure. Between the pia mater and the next layer of tissue is a space which contains circulating cerebrospinal fluid. The space is referred to as the subarachnoid space and the next meninge is called the **arachnoid layer**. The arachnoid layer surrounds the brain and spinal cord and follows the contours of the bony casing of the skull and backbone rather than the neural surface itself. The outermost meninge is a thick membrane principally composed of tough fibrous tissue and is called the **dura mater**. The dura lies in close association with the arachnoid layer and follows the contours of the inside of the cranium and spinal column. It also extends as a double layer into the midsagittal fissure between the cerebral hemispheres as the **falx cerebri**, and into the area between the occipital lobe and cerebellum as the **tentorium**. The dura encloses the major

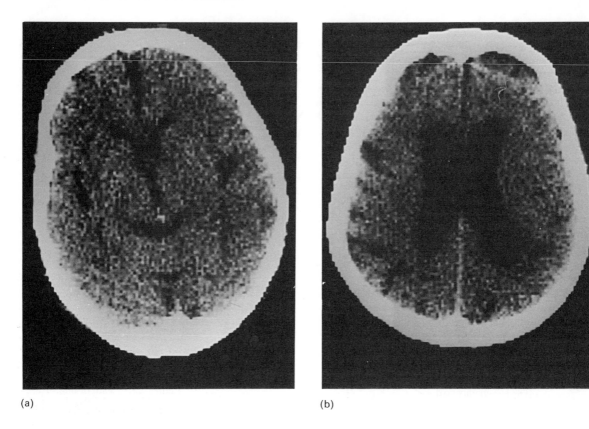

(a)   (b)

Figure 2.3.4 Computerized axial tomography (CT scans): (a) showing a normal adult brain, (b) showing an adult brain with enlarged ventricles

sinuses of the venous drainage system such as the superior sagittal sinus which runs in the midline above the midsagittal fissure (see Fig. 2.3.3). The area between the thick dural membrane and the periosteum of the vertebrae of the spinal column forms the **epidural space**. This space is used clinically at sacral levels of the vertebral column for the injection of anaesthetics in order to block pain sensations arising from the lower part of the body. The procedure of epidural anaesthesia is particularly associated with the relief of pain during childbirth.

Cerebrospinal fluid is a clear and colourless liquid, similar in composition to blood plasma, from which it is derived. However, there are differences between CSF and blood plasma. The CSF contains considerably less protein and reduced amounts of glucose when compared with a plasma filtrate. In addition, there are differences in the electrolyte content of the two fluids: CSF contains less K^+ and Ca^{++} but more Na^+ and Cl^-

than plasma. The CSF typically contains a few leucocytes, but generally the cell count is less than 5×10^6 leucocytes/l of fluid.

An adult brain may have about 130–150 ml of circulating CSF at any one time, of which 80–100 ml will be in the brain ventricles and the remainder in the subarachnoid space. Production of CSF has been estimated to be about 300 ml per day. It is produced by the **choroid plexuses** of the brain ventricles. These plexuses are complex capillary networks covered by a specialized epithelium, the cells of which contain mitochondria and other organelles essential for active transport. Formation of CSF is a selective secretory process with active transport of certain constituents, and this accounts for the differences in composition between CSF and plasma.

The pressure of CSF varies, but is of the order of 1.33 kPa (10 mmHg) when the subject is supine. The specific gravity of CSF is approximately the same as that of brain tissue, so that brain and spinal cord float in a bath of CSF. The support provided by the CSF and meninges protects the brain from damage caused by movements of the head. In this way the CSF acts as a shock absorber between the rigid bones of the cranium and the soft malleable tissue of the brain. The CSF probably fulfils other functions as well, some of which may be related to metabolic activity and others to additional protective processes such as the blood–brain barrier. However, these latter imputed functions are not clearly understood at present.

The CSF, once secreted, flows from the lateral ventricles into the third and fourth ventricles and then through three small apertures in the roof of the hindbrain into the subarachnoid space. The apertures are the two lateral **foramina of Luschka** and the medial **foramen of Magendie**, and they open into an enlargement of the subarachnoid space known as the **cisterna magna** (see Fig. 2.3.3). The CSF then circulates within the subarachnoid space before diffusing into the venous blood supply. This diffusion largely occurs through **arachnoid granulations** which are projections of the arachnoid layer into the major venous sinuses of the brain.

Towards the tail-end of the spinal column, at the level of the second lumbar vertebra, the spinal cord terminates and the associated pia mater forms the fine, hairlike **filum terminale** which anchors the cord to the base of the vertebral column. The space between the spinal cord and the arachnoid layer forms the lumbar cisterna, which contains the roots of the lower spinal nerves (called the **cauda equina**) and CSF. The space of the lumbar cisterna is sufficient to allow the drawing off of samples of CSF without any serious threat to the neural tissue. This may be done for several reasons, for example in order to extract samples of CSF for diagnostic analysis; for the estimation of CSF pressure; and for the introduction of drugs. The procedure is referred to as **lumbar puncture** and it involves puncturing the dura with a hollow needle between lumbar vertebrae L3 and L4 or between L4 and L5. One direct result of drawing off any large sample of CSF will be the lowering of CSF pressure. Performance of this procedure in the patient with undiagnosed raised intracranial pressure is contraindicated. The release of pressure at the lumbar region encourages herniation of the brain-stem with subsequent loss of consciousness. If this pressure falls much below normal, the effectiveness of CSF as a shock absorber between brain and skull will be lessened, and as a consequence any head movement made by the patient may result in headache. It should be noted, however, that the pain is not being mediated by the brain itself since the brain has no sensory receptors. Rather, it is the meninges and blood vessels which will mediate the pain since they are extensively innervated by pain afferents. The standard patient care following lumbar puncture includes 24 hours bedrest with one pillow and administration of extra oral fluids to enhance the replacement of CSF. The puncture site is checked frequently for leakage, and routine neurological observations should be performed.

The circulation of CSF, its hydrostatic pressure and its composition can be altered in certain pathological states of the CNS. For example, intracranial pressure may be raised when a cerebral tumour develops. The raised pressure may distort and partially block the apertures in the roof of the fourth ventricle and in addition may squash the cerebellum (part of the hindbrain) into the foramen magnum, which is the opening at the base of the skull through which the spinal cord passes. The resulting compression of the hindbrain and spinal cord may occlude the spinal canal, from which about a fifth of total CSF is normally absorbed. As a consequence of compression, CSF reabsorption will be greatly impaired.

Hydrocephaly is a condition in which there is a discrepancy between CSF production and reabsorption, resulting in an excess of fluid. This excess causes an increase in intracranial pressure and can lead to enlargement and distortion of the brain ventricles. The condition is particularly apparent in affected infants since the suture lines of their skull bones can become widely separated, so that the whole head becomes abnormally enlarged. There are many causes of hydrocephaly, including cerebral tumour, infections, and congenital and developmental abnormalities. The cerebral aqueduct is particularly vulnerable to blockage owing to its narrowness. Certain congenital disorders and infections can cause obstruction in the flow of CSF through the aqueduct. A disorder such as this, which causes a blockage of the CSF circulation through the brain ventricles, is referred to as 'non-communicating hydrocephalus'. Hydrocephalus caused by abnormal reabsorption of CSF by the arachnoid granulations is called communicating hydrocephalus.

Treatment of hydrocephalus essentially involves removing or bypassing the block in the system. Congenital hydrocephalus and some forms of communicating hydrocephalus can be treated by shunting the excess CSF through a valve into the peritoneum or right atrium of the heart.

The composition and appearance of CSF may also become abnormal in certain diseases of the CNS. The CSF is normally clear and colourless, but turbidity develops when there is an excess of leucocytes, which is typically associated with meningeal irritation. The excess of cells in CSF may result from a raised polymorphonuclear cell count, which is usually associated with acute infections, or a raised mononuclear cell count, which is usually associated with chronic infections. However, acute viral infections predominantly raise the mononuclear cell count, in addition to the lymphocyte count, and this typically occurs in viral meningitis. **Meningitis** may be caused by a range of micro-organisms including bacteria, spirochaetes, fungi and viruses, but in all cases inflammation of the meninges, particularly of the pia mater and arachnoid, is the major feature. Blood is not normally present in CSF unless there is a pre-existing subarachnoid haemorrhage. A **subarachnoid haemorrhage** may result from rupture of an intracranial aneurysm, from a cerebral haemorrhage breaking into the ventricular system, or, rarely, from severe head injury.

The protein content of CSF is normally 0.2–0.4 g/l, with albumin and globulin being present in a ratio of about 8 : 1. Inflammatory diseases such as meningitis, encephalitis, poliomyelitis, multiple sclerosis and syphilis may all result in a moderate rise in the protein content. A large rise in protein content to 5 g/l or greater is usually indicative of either obstruction of the spinal subarachnoid space or acute infective polyneuritis.

Glucose levels in CSF may also show abnormalities, and typically the glucose content of CSF is reduced in bacterial or fungal meningitis.

THE BLOOD SUPPLY TO THE BRAIN

Blood flow through the adult brain is of the order of 750 ml per min, which represents about 15% of total cardiac output at rest. The supply of blood to the brain is closely associated with metabolic

activity of brain tissue, so that factors affecting metabolic activity quickly affect blood flow. For example, within limits, an increase in carbon dioxide concentration, or an increase in hydrogen ion concentration, or a decrease in oxygen concentration, will lead to a rapid increase in cerebral blood flow. The arterial blood supply to the brain passes through two pairs of major arteries, the vertebral arteries and the internal carotid arteries. The vertebral arteries enter the base of the cranial cavity and then unite to form the basilar artery. The basilar artery then runs up on the underside of the brain-stem structures until it reaches the level of the midbrain, where it again bifurcates into the two posterior cerebral arteries. The vertebral arteries supply the brain-stem, the cerebellum and posterior portions of the forebrain. The internal carotid arteries enter the cranial cavity and take up a position on either side of the pituitary gland. The internal carotid arteries divide into several branches including the anterior and middle cerebral arteries. The blood flowing through these arteries supplies the remainder of the forebrain.

Venous drainage from the brain passes through superficial venous plexuses and dural sinuses into the internal jugular veins of the neck. The dural sinuses are valveless channels formed by the dura mater. One such sinus is the midsagittal sinus which lies in the dura above the midline fissure of the cerebrum. It drains blood from the upper lateral and medial aspects of the cerebral hemispheres.

The brain requires a constant supply of blood in order to maintain metabolic activity. Indeed, consciousness is lost if the blood supply is cut off for only a few seconds. Nevertheless, neuronal activity will not be constant over all the brain at any particular time. In other words, particular areas of the brain will be relatively more or less active depending upon the behaviour of the individual at the time. One consequence of this will be that cerebral blood flow will be directed more intensively towards those areas of brain that are relatively more active. The metabolic demands of the neurones in these areas will be greater. Following this line of argument, a measure of localized cerebral blood flow would indicate whether an area of brain is relatively more or less active at any one time. Recently, radiographic techniques have been developed to record local cerebral blood flow in conscious human subjects. These techniques have enhanced our knowledge of head injury and its effects.

THE SPINAL CORD

The spinal cord lies in the flexible bony column of the vertebrae and stretches from the base of the brain-stem to the level of the second lumbar vertebra. In cross-section, the spinal cord comprises a central canal surrounded by a butterfly-shaped area of grey matter, which is in turn surrounded by white matter.

A series of 31 pairs of **spinal nerves** leave the spinal cord and pass through successive intervertebral foramina. These pairs of nerves are identified according to the segment of the spinal column from which they emerge. There are

 8 cervical pairs
12 thoracic pairs
 5 lumbar pairs
 5 sacral pairs
 1 coccygeal pair

of spinal nerves. These pairs of nerves correspond to the number of vertebrae in each region of the spine, except in the case of the cervical nerves where there are only 7 vertebrae and the coccygeal nerve where 4 coccygeal vertebrae are fused to form the coccyx. The relationship between the spinal nerves, segments of the spinal column and the vertebral column is illustrated in Fig. 2.3.5.

Each spinal nerve, which is surrounded by a meningeal sleeve of dura, has a dorsal root and a ventral root. The **dorsal root** consists of sensory afferent fibres which have their cell bodies gathered together in the dorsal root ganglion located within the intervertebral foramen. The **ventral root** consists of motor efferents which have cell bodies lying within the grey matter of the ventral horn of the spinal cord. The roots of the spinal nerves below the level of the second lumbar vertebra, that is below the caudal end of the spinal cord, gather together as the **cauda equina** (mare's tail) before they exit through lower lumbar and sacral intervertebral foramina.

Each spinal nerve is distributed to a particular part of the body. Consequently, the dorsal root of each spinal nerve conducts sensory information from a particular body area. These body areas – **dermatomes** (Fig. 2.3.6) – show a considerable degree of overlap. Hence, damage to a single

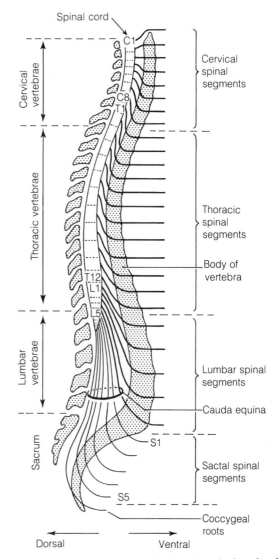

Figure 2.3.5 The relationship between the spinal cord and the vertebral column

spinal nerve need not lead necessarily to any great loss of sensation. The ventral root similarly projects to localized muscle groups.

Peripheral nerves may become damaged and may degenerate, resulting in **neuropathy**. Compression, such as that resulting from a slipped intervertebral disc, is a common physical cause of neuropathy, but some virus infections, such as herpes zoster (shingles), can also cause neuropathy as a result of acute inflammation of a peripheral nerve. The term **polyneuropathy** refers to a condition in which many peripheral

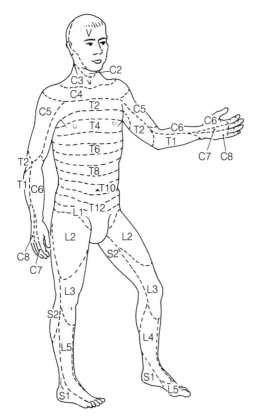

Figure 2.3.6   Sensory dermatomes: the distribution of somatosensory innervation

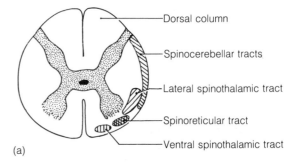

(a)

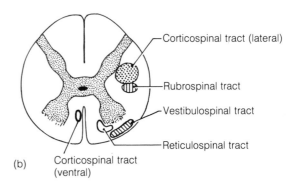

(b)

Figure 2.3.7   Major nerve tracts of the spinal cord: (a) ascending nerve tracts, (b) descending nerve tracts

nerves are simultaneously impaired and which usually results in flaccid muscular weakness and sensory disturbances. This presents a number of problems for the patient. He needs to be aware of the dangers of contact with extremes of temperature, e.g. sitting too close to a fire. The loss of sensory information may result in damage without the patient's knowledge. Regular skin inspection is necessary as these patients will be more prone to skin breakdown.

The use of aids may be indicated to support weakened muscles, e.g. light-weight splints may be applied to the affected arm or leg, and cutlery with padded handles may be useful for the patient with a sensory disturbance affecting the hands.

The grey matter of the spinal cord is organized into left and right dorsal (sensory) horns, intermediate zones, and left and right ventral (motor) horns. Further subdivisions can be made. The dorsal horn can be divided into three main groups of cells which receive the major somatic sensory input. The intermediate zone includes columns of cells associated with sympathetic (thoracolumbar region, lateral horns) and parasympathetic (sacral region) autonomic control, in addition to groups of cells involved in somatic sensory and motor functions. The ventral horn can be subdivided according to the groups of muscles innervated. These will vary at different levels of the spinal cord.

The white matter of the spinal cord can also be differentiated since it is made up of bundles (or tracts) of nerve fibres travelling along the longitudinal axis of the cord. Some of these tracts are illustrated in Fig. 2.3.7.

The shape of the spinal cord varies at different levels. In particular, two enlargements of the grey matter at cervical and lumbar levels reflect the greater number of cells required for innervation of the arms, legs and extremities. At higher levels, more white matter is present since ascending sensory fibres are added by progressively more rostral dermatomes and at the same time descending motor fibres are still coursing caudally to their target muscles and glands.

CRANIAL NERVES

Just as spinal nerves are peripheral nerves of the spinal cord, so the 12 cranial nerves are peripheral nerves of the brain. The particular significance of the cranial nerves is that they conduct sensory information from the important special sense organs such as the eye and the ear and they also control the voluntary muscles involved in facial expression and speech. In addition, they are involved in sensory and motor functions of visceral organs. The sensory branches of the cranial nerves are arranged in a similar manner to the dorsal roots of the spinal cord, with the cell bodies of the afferent fibres grouped in ganglia which lie outside the CNS. Thus, the cell bodies of the optic nerve are the ganglion cells of the retina, and the spiral ganglion of the cochlea contains the cell bodies of the auditory nerve.

The cranial nerves can be identified with little

Table 2.3.1 Functions of the cranial nerves

Cranial nerve		Function
I	Olfactory	Afferent – smell
II	Optic	Afferent – vision
III	Oculomotor	Efferent – eye movements (to inferior and superior rectus, internal rectus and inferior oblique muscles, also to iris and ciliary muscles)
IV	Trochlear	Efferent – eye movements (to superior oblique muscles)
V	Trigeminal	Afferent – somatic sense (from anterior half of head, including face, nose, mouth and teeth) Efferent – to muscles of mastication
VI	Abducens	Efferent – eye movements (to external rectus muscle)
VII	Facial	Afferent – taste and somatic sense (from tongue and soft palate) Efferent – to muscles of face (controlling facial expression), plus parasympathetic outflow to salivary glands (submaxilliary and sublingual glands)
VIII	Vestibulocochlear (auditory)	Afferent – hearing and balance
IX	Glossopharyngeal	Afferent – taste and somatic sense from posterior third of tongue and from pharynx Efferent – to pharyngeal muscles (controlling swallowing), plus parasympathetic outflow to salivary glands (parotid gland)
X	Vagus	Efferent – taste (from epiglotis) and sensory nerves from heart, lungs, bronchi, trachea, pharynx, digestive tract and external ear Efferent – parasympathetic outflow to heart, lungs, bronchi and digestrive tract
XI	Accessory	Efferent – to larynx and pharynx, plus to muscles of neck and shoulder (controlling head and shoulder movement)
XII	Hypoglossal	Efferent – to muscles of tongue and neck

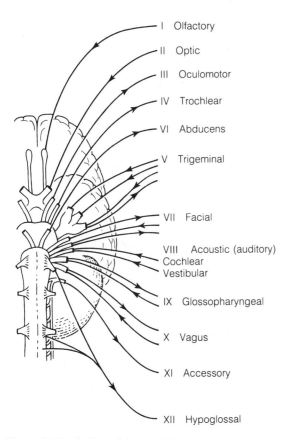

I   Olfactory
II   Optic
III   Oculomotor
IV   Trochlear
VI   Abducens
V   Trigeminal
VII   Facial
VIII   Acoustic (auditory)
Cochlear
Vestibular
IX   Glossopharyngeal
X   Vagus
XI   Accessory
XII   Hypoglossal

Figure 2.3.8 Origins of the cranial nerves

difficulty when the brain is viewed from the inferior (ventral) aspect. Figure 2.3.8 illustrates the points of entry of the cranial nerves and Table 2.3.1 summarizes their individual functions.

OUTLINE OF BRAIN ANATOMY

Before a brief outline of the various parts of the brain can be considered, a number of issues regarding nomenclature need to be addressed.

The term **cortex** is used in association with several brain structures, as in 'cerebral cortex' and 'cerebellar cortex'. The term cortex refers to the mantle of cell bodies (grey matter) which surrounds the agglomerated nerve axons (white matter). This arrangement of grey matter outside and white matter inside is the reverse of that found in the spinal cord where the nerve fibres course up and down the cord in bundles which surround the central area of grey matter. This reversal of the relative positions of grey and white matter allows ascending fibres in the spinal cord to fan out in the brain to reach the cells of higher brain structures and descending fibres from the brain to gather together as they descend to the cord.

The term nucleus is also used in a specific way when referring to parts of the nervous system. A nucleus refers to a mass of cell bodies which are grouped together because of their proximity and their structural and functional relatedness. Nuclei appear as discrete masses of grey matter embedded in the core of the brain. This use of the term nucleus must be distinguished from the conventional use of the word to indicate the nucleus of an individual cell.

The cerebrum

The human brain, looked at from the side, is dominated by the cerebrum and is shaped rather like an oversized boxing glove, with the lateral sulcus (commonly also called the Sylvian sulcus) representing the separation between the thumb and the fist of the glove (Fig. 2.3.9). Instead of having the smooth contours of a boxing glove, the brain shows many convolutions. The raised ridges of these convolutions are called **gyri** (singular: gyrus) and the slit-like indentations are called **sulci** (singular: sulcus). The outer surface of the cerebrum consists entirely of cortex which follows the contours of the gyri and sulci. The corrugations of the cerebral cortex enable approximately $0.2\,\text{m}^2$ of cortex to fit inside the cranium.

If the brain is viewed from above, it can be seen that the cerebrum is composed of two halves, or **hemispheres,** with a clear line of separation running along the midline. This deep midline sulcus is called the **midsagittal sulcus,** or midsagittal fissure. In each hemisphere of the brain, another

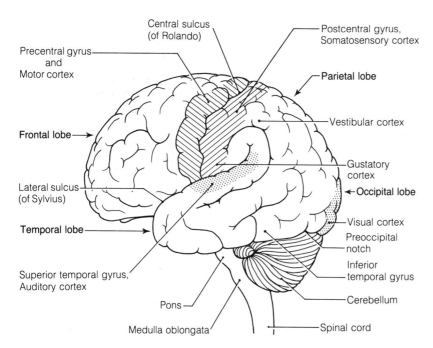

Figure 2.3.9 Lateral aspect of the human brain

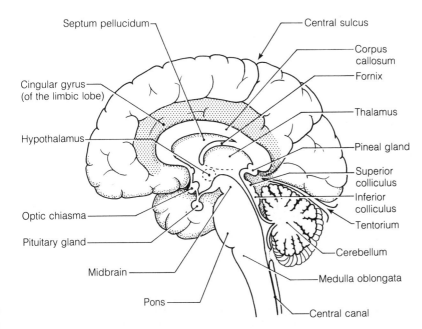

Septum pellucidum

Central sulcus

Corpus callosum

Cingular gyrus (of the limbic lobe)

Fornix

Thalamus

Hypothalamus

Pineal gland

Superior colliculus

Inferior colliculus

Optic chiasma

Tentorium

Pituitary gland

Cerebellum

Midbrain

Medulla oblongata

Pons

Central canal

Figure 2.3.10 Medial aspect of the human brain

clearly seen sulcus runs at right-angles to the mid-sagittal fissure down the convexity of the lateral surface of the brain, roughly in the midcoronal plane. This is called the **central sulcus** and is also sometimes referred to as the sulcus of Rolando (Fig. 2.3.9).

Each cerebral hemisphere can be subdivided into six lobes. The large area anterior to the central sulcus and superior to the lateral sulcus is referred to as the **frontal lobe**. The area inferior to the lateral sulcus (which comprises the thumb of the boxing glove) is called the **temporal lobe**. The area posterior to the central sulcus is called the **parietal lobe**, and the area at the very back of the brain, which is posterior to a line taken roughly vertically from the preoccipital notch (Fig. 2.3.9), is called the **occipital lobe**. The major portion of the occipital lobe lies within the midsagittal fissure and can only be seen when the two halves of the brain are separated. The **insula**, or central lobe of cerebral cortex, is revealed if the frontal and temporal lobes are eased apart along the line of the lateral sulcus, thus separating the thumb and the fist of the boxing glove. The insula cortex lies in the depths of the lateral sulcus. Finally, the **limbic lobe** is a ring of cortex on the medial surface of the brain which can be seen if the brain is divided along the line of the midsagittal fissure (Fig. 2.3.10). In evolutionary terms, the limbic

lobe contains cortex which is much older than that found in other areas of the cerebrum.

Another gross subdivision of the cerebral cortex is the distinction made between **primary projection areas** and **association areas**. The projection areas comprise the primary sensory and primary motor areas of cerebral cortex. The primary sensory areas of the cerebral cortex receive sensory afferent inputs arising from the periphery. The strip of cortex of the postcentral gyrus, which follows the line of the central sulcus and lies posterior to it, is the projection area for somatosensation (principally touch and position sense). The area at the posterior extremity of the occipital lobe is the primary projection area for the sense of vision. The primary projection area for the sense of hearing lies in the depths of the lateral sulcus. Primary projection areas for taste and smell can also be identified (see Figs. 2.3.20 and 2.3.21). The primary motor area of cortex is that part of the cerebral cortex which projects descending efferent nerve fibres to the motor neurones of the spinal cord. The primary motor cortex lies in the precentral gyrus, the strip of cortex which follows the line of the central sulcus and is just anterior to it.

Six horizontal cell layers can be distinguished over most of the cerebral cortex but there are only two major subtypes of nerve cells within the cortex. The two subtypes are **pyramidal cells,** which vary in size and tend to have long axons, and **stellate** (or granule) **cells,** which tend to have much shorter axons and make fewer synaptic contacts.

Midline structures

If the brain is bisected along the line of the mid-sagittal fissure, a view of midline structures is revealed. This is represented in Fig. 2.3.10.

FOREBRAIN STRUCTURES

It can be seen from Fig. 2.3.10 that the cerebral cortex descends into the midsagittal fissure. In particular, the cingulate gyrus forms part of the ring of the limbic lobe of the cerebral cortex. The **corpus callosum** is a thick band of nerve fibres which connect the two cerebral hemispheres and which are hence seen in cross-section. The corpus callosum maintains a constant flow of information between the two halves of the cerebral cortex.

The **thalamus** is a major centre of neural integration. It can be thought of as a relay-station interpolated between the cerebral cortex and lower brain structures. The two thalami (one thalamus in each half brain) are located on either side of the thin third ventricle.

The **hypothalamus** is situated inferior to the thalamus and is functionally related to the autonomic nervous system (see Chap. 2.4). Cells of the hypothalamus are involved in the control of food and water intake, in body temperature regulation, and in emotional expression. The hypothalamus is also directly involved in the control of the activity of the **pituitary gland**. The pituitary gland, as the 'master gland', in turn controls and integrates the activity of other endocrine glands (see Chap. 2.5).

The **pineal body** is attached to the roof of the third ventricle and lies directly in the midline. It is a secretory organ associated with the endocrine system and contains melatonin, serotonin and other biologically active amines. As yet, its functions remain unclear, although it is implicated in sexual development at puberty.

MIDBRAIN STRUCTURES

The **superior colliculi** and the **inferior col-**

liculi form the dorsal surface of the midbrain. The superior colliculi are concerned with visual functions and the inferior colliculi with audition.

Ventral to the cerebral aqueduct, the major visible midbrain structures are the **cerebral peduncles** which are large trunks of ascending and descending nerve fibres lying either side of the midline.

HINDBRAIN STRUCTURES

The major hindbrain structures include the pons, the medulla oblongata, and the cerebellum. The **pons** and the **medulla oblongata** lie ventral to the fourth ventricle, and are the level at which most cranial nerves enter the brain. The pons has many clearly defined transverse fibres crossing the midline, and both pons and medulla contain ascending and descending nerve fibres.

The **reticular formation** is a loose network of neuronal connections situated throughout the length of the pons and medulla and extending into the midbrain. Its major functions include motor control and cerebral arousal.

The pons and medulla also contain nuclei responsible for the reflex control of essential functions such as respiration, heart rate and vasomotor activity. These functions are termed 'brain-stem reflexes'.

The **cerebellum** is a large outgrowth forming the roof of the fourth ventricle. It consists of an outer mantle of cortex, which is highly convoluted, together with underlying associated white matter and deeper-lying nuclei. The cerebellum is essential for normal motor coordination.

Structures lateral to the midline

FOREBRAIN STRUCTURES

The **basal ganglia** are made up of three large nuclei which lie lateral to the thalamus (see Fig. 2.3.22). The three structures are the **globus pallidus**, the **putamen** and the **caudate nucleus**. The putamen and the globus pallidus together form the **lentiform nucleus**. The caudate nucleus and putamen are together referred to as the **corpus striatum**. The basal ganglia exercise important influences on motor activity.

The caudate nucleus has a head and a long tail which sweeps around underneath the lateral ventricle (see Fig. 2.3.24). The tail of the caudate merges into the amygdaloid nucleus. The **amygdaloid nucleus**, deep within the temporal lobe, is associated with the olfactory system and with emotional behaviour.

The **hippocampus** lies on the medial wall of the temporal lobe. It follows a path similar to that of the tail of the caudate nucleus, curving around beneath the lateral ventricle. The hippocampus represents that portion of the limbic lobe which is composed of evolutionarily older cortex. The hippocampus has been associated with several different behavioural functions such as memory, spatial orientation and emotional responses.

The **internal capsule** is the large bundle of fibres passing to and from the cerebral cortex and linking the cerebrum with lower brain structures. The fibres of the internal capsule pass to and from the cerebral peduncles, between the thalamus and the lentiform nucleus. They then fan out to the cortex.

MIDBRAIN STRUCTURES

The **subthalamus**, the **red nucleus** and the **substantia nigra** are all midbrain nuclei lying lateral to the midline (see Fig. 2.3.22). They are all concerned with the control of movement and they have extensive connections with the basal ganglia.

THE SOMATOSENSORY SYSTEM

The pathways of the somatosensory system within the CNS connect the primary afferents from receptors to the sensory projection areas of the cerebral cortex. The pathways involve synapses at several levels including the spinal cord, the nuclei of the lower brain-stem, and the sensory relay nuclei of the thalamus. Axons from the thalamic nuclei then project to the primary sensory cortex, which lies along the postcentral gyrus, and to secondary sensory cortex. There are also descending pathways from the cerebral cortex which terminate in the main relay nuclei of the sensory projection pathways (see Fig. 2.3.9).

The ascending sensory pathways can be divided into the **dorsal column–medial lemniscal pathway** and the **spinothalamic tract**. This division has functional relevance since the two pathways transmit information which has qualitative differences. The spinothalamic tract transmits important information concerning pain and temperature but provides rather poor localization and poor discrimination of the stimulus. Sensory information of this somewhat diffuse, all-or-none kind is referred to as **protopathic**. The dorsal columns, on the other hand, are a more recent evolutionary development and mediate tactile and **kinaesthetic** information which is highly discriminatory and can be precisely localized. This more discriminative type of information is referred to as **epicritic**.

The dorsal column – medial lemniscal pathway (Fig. 2.3.11)

The dorsal columns are large bundles of nerve fibres lying in the dorsal (posterior) quadrant of the spinal cord. The columns consist of ascending branches of dorsal root fibres. As the dorsal root fibres enter the dorsal horn, they immediately enter the dorsal columns without synapsing, although collateral branches synapse with cells of the dorsal horn. The dorsal root fibres of the dorsal columns tend to be large-diameter myelinated A fibres, which can transmit impulses at relatively high velocities (around 70 m/s). The nerve fibres are arranged somatotopically, such that nerve fibres from sacral and lumbar roots take up a medial position and those from higher levels of the spinal cord assume a more lateral position. As the cord is ascended, the nerve fibres from higher levels are added to the lateral aspect of the dorsal columns so that the higher the level of origin, the more lateral will be their position within the dorsal columns.

The first synapse of the dorsal column nerve fibres is at the level of the lower medulla. Here, the axons terminate on cells of the two dorsal column nuclei, called the **nucleus gracilis** and the **nucleus cuneatus**. The cells of these brain-stem sensory nuclei then project **second-order nerve fibres** to the thalamus.

The second-order fibres cross the midline and join a large fibre bundle called the **medial lemniscus**. Crossing the midline results in the projections from one half of the body travelling to the contralateral half of the brain. Consequently, at the level of the cerebral cortex, the left side of the body is represented in the right sensory projection

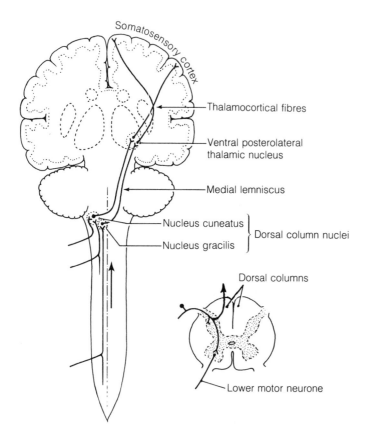

Somatosensory cortex

Thalamocortical fibres

Ventral posterolateral
thalamic nucleus

Medial lemniscus

Nucleus cuneatus ⎫
⎬ Dorsal column nuclei
Nucleus gracilis ⎭

Dorsal columns

Lower motor neurone

Figure 2.3.11 The dorsal column:
medial lemniscal sensory pathway to
the brain

area of the cortex and the right side of the body is represented in the left sensory cortex. The second-order nerve fibres ascend in the medial lemniscus to terminate in the **sensory relay nucleus** of the thalamus – the ventroposterolateral nucleus (VPL). Cells of this nucleus then project to the sensory projection areas of cerebral cortex, principally to the primary sensory area.

The spinothalamic tract (Fig. 2.3.12)

The dorsal root fibres of the protopathic pathway tend to be of the smaller diameter myelinated Aδ or unmyelinated C fibre type. The conduction velocities of these nerve fibres are consequently slower than those typical of the dorsal column fibres.

The dorsal root fibres enter the spinal cord and may either ascend or descend for a few segments (usually from one to six segments) before terminating on cells of the dorsal horn. In the dorsal horn, the cells of the intermediate grey matter, the

cells of the **substantia gelatinosa**, which caps the dorsal horn and the marginal cells lying outside the substantia gelatinosa all receive input from dorsal root fibres either directly or via interneurones lying within the dorsal horn itself. Dorsal horn cells then project fibres which usually ascend for a few segments before crossing the midline in the ventral (anterior) commissure of the spinal cord. These fibres then ascend the spinal cord as the spinothalamic tract. The nerve fibres of the spinothalamic tract travel in the ventrolateral (anterolateral) quadrant of the cord, maintaining a roughly somatotopic organization.

Most of the fibres terminate on cells of the ventroposterolateral nucleus of the thalamus, although some synapse with cells of the posterior group of thalamic nuclei. Cells of the thalamic relay nuclei then project fibres to both the primary and secondary sensory projection areas of the cerebral cortex, but mainly to the primary sensory area.

It is worth noting here that there is another pathway originating from groups of cells in the

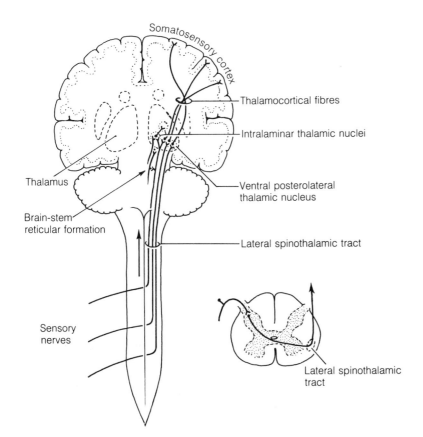

Figure 2.3.12 The spinothalamic sensory pathway to the brain

dorsal horn similar to those giving rise to the spinothalamic tract. This pathway arises mainly from cells of the intermediate grey matter and also ascends the spinal cord in the ventrolateral quadrant. This is the **spinoreticular tract** which projects fibres to the reticular formation of the brain-stem. The ascending influences of the reticular formation have widespread effects on the arousal level of the brain and upon conscious awareness. The closeness of the association between the spinothalamic tract and the spinoreticular tract presumably rests upon the necessity for a protopathic input, such as a noxious stimulus, to be responded to immediately. Control and direction of such a rapid response may require a generalized increase in arousal level, probably associated with a shift in the direction of attention.

The trigeminal nerve

The somatosensory fibres of the trigeminal nerve,

the Vth cranial nerve, transmit sensory information from the face. The peripheral fibres synapse with cells of two brain-stem nuclei. Second-order fibres then join the spinothalamic tract and the medial lemniscus. The fibres joining the spinothalamic tract mostly cross to the opposite side of the brain, but a portion of the fibres joining the medial lemniscus remain uncrossed.

Spinal cord lesions

Lesions of the spinal cord can be caused by physical insult or disease. Motor car, industrial and other accidents account for the majority of cases of spinal cord lesion, but clinical conditions such as tumours, cervical spondylosis (associated with degeneration of intervertebral discs), syringomyelia (which involves the development of elongated cavities close to the central canal of the cord), and myelitis (inflammation of the cord) may all cause compression or lesions of the spinal cord.

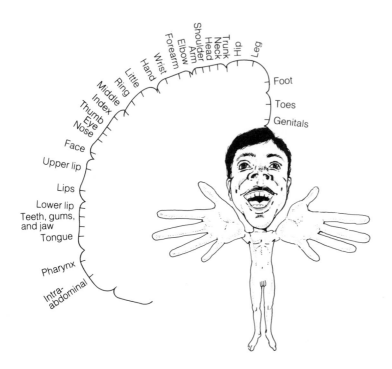

Figure 2.3.13 Sensory homunculus (Reprinted with permission from Penfield, W. & Rasmussen, T. (1950) *The Cerebral Cortex of Man*. London: Macmillan)

Spinal cord lesions will cause motor as well as sensory disturbances, but only the sensory disturbances are considered here.

Destruction of the dorsal columns causes a loss of fine tactile discrimination, a loss of appreciation of vibration, of passive movement of joints and of position sense below the level of and on the same side as the lesion. Damage to the spinothalamic tract causes a loss of appreciation of pain (analgesia) and a loss of temperature sensitivity (thermoanaesthesia) on the opposite side of the body. Moreover, since the nerve fibres of the spinothalamic tract may ascend for a few segments before crossing the midline, the actual zone of anaesthesia is likely to be slightly below the level of the lesion.

Cortical representation

Nerve fibres from thalamic nuclei, in particular the ventroposterolateral nucleus, project to the primary and secondary sensory areas of the cerebral cortex. Projections to the primary sensory area maintain a strict somatotopic organization such that the representation of body parts in the postcentral gyrus forms a **sensory homunculus** of the opposite half of the body (Fig. 2.3.13). The size of the area of cortex receiving sensory input from any particular part of the body is a function of the degree of innervation rather than of the physical size of that body part. Consequently, the sensory homunculus appears distorted because of the greater innervation of the lips, tongue and thumb compared with that of the trunk and legs.

Extensive parietal lobe lesions are associated with sensory inattention and disturbances of the body schema. That is, a patient may show a disordered appreciation of his own body image. For example, a patient may ignore parts of his own body, feeling that the arm or leg contralateral to the lesion is not his. This may even extend to the patient washing and shaving only half the face and combing hair on only one side of the head. Other extensive parietal lobe lesions, especially those involving the lower portions of the parietal lobe, can lead to fairly generalized disturbances of spatial relationships. These can be demonstrated by the inability of patients to learn finger mazes or to comprehend maps and follow directions, or to assemble and compare objects and patterns in two or three dimensions.

PAIN

The function of pain

The sensation of pain provides an important defence mechanism against tissue damage and is therefore essential for survival, even though the experience of pain is distressing. In the very rare cases of congenital analgesia or the more common cases of analgesia caused by lesions of nervous pathways, patients may sustain extensive tissue damage without being aware of any noxious stimulus. Such patients constantly have to be wary of potential harm, and nursing such patients also requires special procedures. For example, it is necessary to move regularly and alter the position of paraplegic patients in order to avoid constant pressure on parts of the body, such as the back and the buttocks, which could result in pressure sores.

Pain as a protective mechanism has some shortcomings. Irradiation of the body with either ionizing or non-ionizing radiation frequently does not elicit pain at the time of stimulation, although extensive tissue damage may occur. Nearly everyone has experienced painful sunburn following an apparently pleasant and comfortable spell of sunbathing; pain in this instance does not serve any warning function. Similarly, pain is not generally associated with many cancers until a tumour is quite extensive. Indeed, the tumour may have metastasized and produced 'secondaries' elsewhere before pain is appreciated. Chronic pain associated with degenerative disorders, such as arthritis, can also hardly be regarded as protective.

Perception of pain

The perception of pain is dependent on both physiological and psychological factors. The location and intensity of the stimulus will affect the quality and severity of the perceived pain. Cutaneous pain may be described as pricking or burning pain and can readily be distinguished from pain arising from joints or muscles and from that arising from the gut. In general, the greater the intensity of the stimulus, the more severe the perceived pain. However, pain perception is also determined by psychological factors such as personality and mood and by psychic factors which are influenced by culture. In other words, the severity of pain experienced depends on a physiological response to stimulation and on a psychological reaction to it. The psychological reaction to painful input probably largely accounts for the great variability in individual perception and tolerance of pain.

In general, emotionally anxious people are less able to withstand pain than those with a more phlegmatic personality, although the degree of emotional arousal is important, and pain perception is apparently reduced in states of high arousal.

Culture also appears to play a part in the tolerance of pain. This aspect of the psychological reaction may be so powerful as to cause pain where there is no obvious physiological stimulation. In certain primitive cultures, child birth is associated with apparent 'labour pains' in the father whilst the mother quietly, and apparently without undue discomfort, produces the infant.

Transmission of pain signals

Pain signals appear to be transmitted in both myelinated Aδ and unmyelinated C fibres. It has been suggested that the division of painful stimuli into those causing sharper pricking sensations and those causing the duller burning or aching sensations may be associated with these two nerve fibre types. Sharp pricking sensations are thought to be transmitted by the faster conducting Aδ fibres, whereas aching sensations are conducted by the slower C fibres. Certainly, experiments in which nerve fibres are subjected to ischaemia induce a block of fast, pricking pain first. Ischaemia blocks larger myelinated fibres before unmyelinated ones. On the other hand, experiments utilizing local anaesthetics, such as procaine, block the burning or aching pain preferentially, and local anaesthetics are known to have a predilection for small unmyelinated C fibres. Nevertheless, it is still not clear whether different receptors as well as different types of nerve axon are responsible

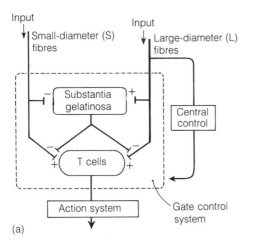

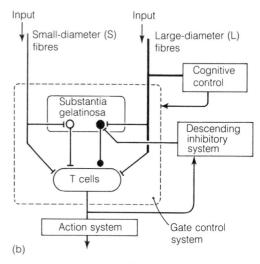

Figure 2.3.14   The gate control theory of pain. (a) Original formulation of the theory. Large diameter (L) and small diameter (S) peripheral nerve fibres input to the substantia gelatinosa (SG) and to the first central transmission (T) cells of the spinal cord. The inhibitory effect exerted by the SG on the T cells is increased by activity of the S fibres (pain fibres). The central control mechanisms are represented as running from the L fibre system and feeding back to the gate control. (b) Updated model. On the basis of subsequent evidence, Melzack and Wall formulated the gate control theory to include excitatory (white circle) links from the SG to the T cells, as well as descending inhibitory control from the brain-stem. All synaptic connections are excitatory except the inhibitory link from SG to T. The round knob at this inhibitory synapse implies that its action may be presynaptic, postsynaptic, or both (Reproduced with permission from Melzack, R. & Wall, P. D. (1982). *The Challenge of Pain*. Harmondsworth: Penguin)

for the differentation of pain into these two qualitatively separable experiences.

The **gate control theory** of pain proposed by Melzack and Wall in 1965 (Melzack and Wall, 1965) suggests that there is a relationship between the inputs from touch receptors and pain receptors at the level of the spinal cord. In particular, it suggests that interneurones in the substantia gelatinosa of the dorsal horn can regulate the conduction of ascending afferent input. The theory proposes that large-diameter fibres from touch receptors normally presynaptically inhibit small-diameter fibres from pain receptors. This constitutes a 'gate' which inhibits the transmission of input from pain receptors unless a great deal of small-diameter fibre activity forces open the gate to allow transmission along pain pathways in the spinothalamic and spinoreticular tracts. Melzack and Wall also propose that descending fibres from the brain synapse in the same area of the dorsal horn of the spinal cord and can modify the transmission of pain signals (Fig. 2.3.14).

The gate control theory helps to explain a number of observable pain phenomena. For example, the destruction of large-diameter afferents following amputation of a limb and the consequent release of inhibitory influences over small-diameter nerve fibres may account for the experience of **phantom limb pain**. Similarly, the destruction of, or damage to, touch fibres in inflamed tissue may in part account for the phenomenon of hyperalgesia whereby inflamed tissue is more sensitive to pain. The presence of descending influences from the brain also helps to explain the relationship between psychological factors and pain perception.

Since their original paper, research relating to Melzack and Wall's gate control theory has indeed established the existence of inhibitory interactions between large- and small-diameter afferent nerve fibres in the dorsal horn, although these interactions now appear to be rather more complicated than was at first suggested. Despite this, there is no doubt that the theory has led to many important developments. For example, the use of **transcutaneous electrical stimulation** as a means of relieving pain is now a well established and widely used technique (Allan, 1981). This development owes much to Melzack and Wall's ideas, even though the effectiveness of this treatment may not lie entirely in the action of stimulated large-diameter afferents inhibiting the activity of small-diameter pain fibres. In addition to the inter-actions of dorsal horn cells, the presence of a powerful descending inhibitory pathway from nuclei in the brain-stem is now firmly established, and stimulation of these nuclei causes marked analgesia.

Representation of pain in the brain

Many parts of the brain contribute to the pain experience. The thalamic nuclei seem to influence the conscious appreciation of pain, and damage to the thalamus can cause severe and intractable pain. This pain is experienced as arising from those parts of the body which normally transmit afferent input to the damaged areas of the thalamus. Extensive damage to the somatosensory cortex on the postcentral gyrus does not abolish pain, and it appears that the adjacent area of cortex is more important for the integration and interpretation of pain signals.

Referred pain

The phenomenon of 'referred pain' has attracted considerable attention in pain research. The term referred pain describes the way in which pain arising from damage in one part of the body is actually experienced as though it were arising in a different part of the body. It is thought to result from the fact that the pain fibres from the dam-aged visceral organs enter the spinal cord at the same level as afferents from the referred area of the body.

Pain transmitters

A major recent development in the study of pain has been the discovery of various peptides which have been implicated in pain transmission and pain relief.

Substance P is a neurotransmitter or neuro-modulator which has been identified in various parts of the nervous system, including the sub-stantia gelatinosa of the dorsal horn. Noxious stimulation causes the release of substance P from dorsal root afferents and consequently it has been suggested that substance P acts, at least partly, as a transmitter for pain.

While substance P is implicated in the trans-mission of pain, other peptides have been dis-covered which seem to possess analgesic prop-erties. The discovery of these peptides, the **enkephalins** and **endorphins**, stems from research into the powerful analgesic action of opiate derivative drugs such as morphine (Hughes *et al.*, 1975). In performing its analgesic action, morphine binds to receptors on the nerve cell membrane. These receptors are, at least partly, located presynaptically. The existence of recep-tors to which morphine can attach implied the existence of some endogenous substance, similar to morphine, to which receptors would bind under normal conditions (i.e. without the presence of morphine). The search for this endo-genous opiate-like substance led to the discovery of the enkephalins. Hughes and colleagues were the first to isolate and identify these endogenous opiates in 1975 (Hughes *et al.*, 1975). They iden-tified two peptide chains, comprising just five amino acids each, which they called methionine and leucine enkephalin. These peptides are very unstable and have a half-life of 1 min or less in brain tissue, but they are fragments of a much longer peptide chain, β-endorphin. β-Endorphin comprises 30 amino acids and is a much more stable molecule, and it too binds with opiate receptors. In turn, β-endorphin is a fragment of the pituitary hormone β-lipotrophin, which is itself derived from the same precursor as adrenocorticotrophic hormone (ACTH).

The endogenous opiates have been shown to inhibit prostaglandin formation and **prosta-glandins** are putative chemical stimuli for pain. In addition, endogenous opiates have been shown experimentally to inhibit the actions of a number of transmitters, including substance P, the suggested transmitter for pain.

The enkephalins and endorphins have an anal-gesic action similar to that of morphine, and it is thought that the analgesia of acupuncture may be attributable to enkephalin activity. In spite of the many questions that still remain about the pre-cise mode of action and range of influence of enkephalins and endorphins in the central and peripheral nervous systems, their existence does go some way to explain phenomena such as the **placebo response**, where an individual per-ceives pain relief even though no analgesic agent is administered. It may be that in such cases the mere expectation of pain relief is sufficient to release psychogenically the endogenous opiates, which would then cause genuine analgesia even without the administration of an analgesic drug.

Pain treatment

Since pain perception depends in part on the psychological state of the patient, pain tolerance may be improved if anxiety is reduced by providing the patient with information about his condition and the likely extent and severity of his pain. This requires careful management, but studies have shown that requirements for post-operative analgesic drugs may be reduced in patients who are provided with specific information concerning the extent and duration of postoperative pain (Hayward, 1975).

Clearly, analgesic drugs, ranging from aspirin to morphine and other opiate derivatives, play an important part in pain control. In some situations, for example terminal illness, psychoactive drugs may be required in addition to narcotic analgesics in order to alleviate depression or reduce anxiety, both of which may contribute to the pain experience.

In some types of severe intractable pain, cutting pain pathways may offer relief. However, it should be noted that severing pain pathways does not necessarily result in the permanent relief of pain, which may return in whole or in part after a number of years.

Some less orthodox methods of pain control are also in use. These include hypnosis, which apparently reduces attention to pain and may even reduce reflex responses, transcutaneous electrical stimulation, which is thought to work by inhibiting pain through stimulating larger touch afferents in the skin, and acupuncture, which may also depend, at least partly, on stimulation of large afferents which inhibit pain transmission pathways.

THE VISUAL SYSTEM (see also Chap. 2.2)

The visual pathways

The ganglion cells of the retina project their axons towards the posterior pole of the eye when they converge and pass out of the eye as the **optic**

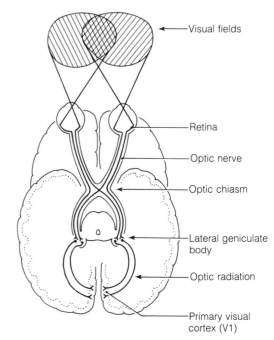

Figure 2.3.15   Visual pathways to the brain

nerve. Upon leaving the eye, the cells of the optic nerve acquire a myelin sheath.

The optic nerves pass to the **optic chiasma** which is situated at the base of the brain, anterior to the pituitary stalk. At the optic chiasma, a proportion of the nerve fibres cross the midline in a partial decussation. The fibres from the nasal halves of the retinae cross to the opposite side of the brain, whereas fibres from the temporal halves of the retinae maintain their ipsilateral (same-sided) projection (Fig. 2.3.15). Consequently, fibres from the temporal half of one retina are joined by fibres from the nasal half of the other, and in this way fibres leaving the right side of each eye project to the right half of the brain and fibres from the left side of each eye project to the left side of the brain. As a result, the right visual hemifield is projected to the left cerebral hemisphere, and the left visual hemifield is projected to the right cerebral hemisphere.

After the partial decussation of the optic chiasma, the axons of the retinal ganglion cells continue as two diverging optic tracts before terminating on cells of the **lateral geniculate nuclei** (LGN). These thalamic relay nuclei contain six layers of cells receiving ipsilateral and contralateral input from the eyes.

The nerve fibres from the LGN constitute the optic radiations which arc around the lateral ventricles and terminate on cells of the striate cortex of the occipital lobe, the primary visual cortex. The strict organization of the projection from the retina is maintained at the level of the primary visual cortex, with the lower halves of the retinae (representing the upper halves of the visual field) projecting to the visual cortex below the calcarine sulcus and the upper halves of the retinae projecting to the visual cortex above the calcarine sulcus. Retinal projections from the fovea, which represent vision in the centre of the visual field, project to the posterior part of the primary visual cortex.

Visual processing

The way in which the nervous system interprets the visual world is still far from being clearly understood. Nevertheless, the use of single-cell electrophysiological recording techniques has led to considerable advances over the last three decades, revealing details of the function of nerve cells at different levels of the visual system.

In humans, the inability to recognize objects visually in the absence of any obvious peripheral sensory loss is called **visual agnosia**. This is a dysfunction associated with lesions in posterior parts of the brain. The patient is unable to recognize and name an object when it is presented in the visual modality, even though he may be able to recognize the same object if he is allowed to touch it or experience it in some other non-visual way. Visual agnosia can vary greatly in severity, but one particularly severe, although rare, form is **prosopagnosia**. In this condition the patient is unable to recognize faces, and, consequently, is disturbingly unable to recognize his family or friends by sight.

Visual perception

It is necessary to relate the way in which the visual system processes visual signals to the way in which the visual world is perceived.

PATTERN RECOGNITION

The basis of pattern recognition is laid down in the sensitivity of the cells of the visual cortex. These cells are able to detect the barriers to areas of light and dark which lie in particular orientations and in particular parts of the visual field. Separating the visual world into blocks of light and shade provides a first step to identifying objects. The visual system also seems to enhance the distinction between areas of light and dark at the boundaries. For example, if you observe a large sheet of white paper against a blackboard, the paper appears to be whiter at the edges than it does in the centre. In other words, the visual system responds vigorously to contrast.

MOVEMENT

At the level of the retina, the perception of movement in the visual scene is probably linked with the activity of the y-type ganglion cells. These cells produce fast, transient responses and the axons project both to the LGN and also to the superior colliculus, where they presumably have some effect on the control of eye movements.

STEREOSCOPY AND DEPTH PERCEPTION

There are several factors which contribute to the perception of depth. The relative sizes of objects may provide one clue. For example, a person standing 50 m distant will provide an optical image which is much smaller than that of a person standing only 5 m away. However, we do not interpret such a visual image as meaning that one person is only a fraction of the size of the other, since memory suggests very strongly that objects show a constancy of size. Consequently, we assume that the two people are of roughly the same height but are at different distances from us. Shadows may also play a part in depth perception. The casting of a shadow of one object onto another provides information about the relative positions of the objects in space. Movement may also play a role in depth perception. As we travel in a car or railway train, stationary objects outside, such as trees or fences, which are close to the vehicle, pass across the visual field very quickly, whereas distant objects pass across the visual field much more slowly. This is referred to as **motion parallax**. Another very important factor in depth perception is that we observe the visual world binocularly and are thus able to decipher information about the relative positions of objects from the disparity

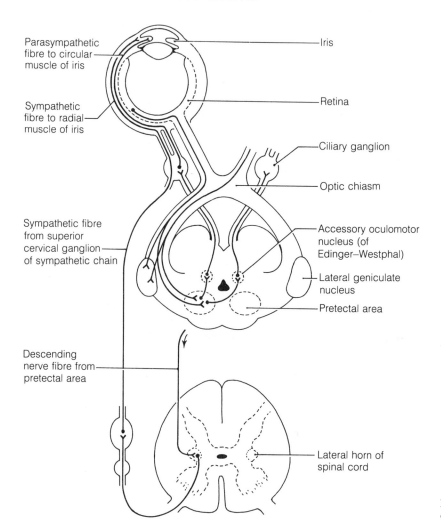

Parasympathetic fibre to circular muscle of iris

Sympathetic fibre to radial muscle of iris

Sympathetic fibre from superior cervical ganglion of sympathetic chain

Descending nerve fibre from pretectal area

Iris

Retina

Ciliary ganglion

Optic chiasm

Accessory oculomotor nucleus (of Edinger–Westphal)

Lateral geniculate nucleus

Pretectal area

Lateral horn of spinal cord

Figure 2.3.16 Nerve pathway of the pupillary reflex

of the images presented to the two eyes. It has been reported in some animal experiments that some binocular cortical cells do respond preferentially to the disparity between the images of an object presented to the two eyes.

COLOUR

A variety of colour-sensitive cells has been described at various levels of the visual system. The opponent colour coding found in the retina is also present at the LGN. The colour-sensitive cells are concentrated in the area of cortex receiving information from the fovea (see Chap. 2.2).

Visual reflexes and eye movements

The **pupillary light reflex** refers to the constriction of the pupil in response to the illumination of the eye. The reflex pathway is multisynaptic and bilateral (Fig. 2.3.16). The reflex arc begins with retinal ganglion cells which project via the optic nerve to the pretectal area of the midbrain. Axons of pretectal neurones project bilaterally to the accessory occulomotor nuclei and in turn axons from these nuclei join the oculomotor nerves (IIIrd cranial nerves) to the ciliary ganglia lying just behind the eyes. Here, they make contact with postganglionic parasympathetic neurones which innervate the circular muscle fibres of the iris of the eye and cause the pupillary constriction.

Dilation of the pupil acts through descending sympathetic pathways which stimulate the preganglionic sympathetic neurones of the lateral horns of the upper thoracic spinal cord. These preganglionic fibres then ascend in the sympathetic chain and synapse in the superior cervical ganglion. From here, postganglionic fibres project anteriorly along the branches of the internal carotid artery before terminating on the radial muscle fibres (dilator muscle) of the iris.

The **accommodation reflex** alters the curvature of the lens of the eye in order to focus the eye on objects at different distances. Movements of the eyes within the orbits are effected by the six extraocular muscles and these are innervated by the IIIrd, IVth and VIth cranial nerves. The frontal eye fields and the occipital eye fields of the cerebral cortex influence the cranial nerve nuclei via the superior colliculus and the pretectal area. The frontal eye fields are concerned with voluntary scanning of the visual scene and the larger occipital eye fields are associated with pursuit movements which occur in the voluntary pursuit of a moving visual target.

Clinical testing

OPHTHALMOSCOPY

Papilloedema (swelling of the optic disc) may occur in many medical conditions, and consequently examination of the fundus of the eye with an ophthalmoscope is of great importance. Papilloedema may be caused by the raised intracranial pressure associated with brain abscesses, intracranial tumours or meningitis. It also occurs in optic neuritis (inflammation of the optic nerve), which may be associated with demyelinating diseases such as multiple sclerosis.

VISUAL FIELD DEFECTS

Reduction of the normal size of the visual fields or the development of areas of blindness (scotomas) within the visual field can be assessed by **perimetry**. The patient is asked to fixate each eye successively on a central target, and then to indicate when a stimulus enters his peripheral vision. The stimulus, usually a small white or coloured disc, is then moved inwards from various points around the periphery of the visual field. Damage to the visual pathway at various stages will result in the visual field defects illustrated in Fig. 2.3.17.

Lesions of the optic chiasma which result in a bitemporal hemianopia (half visual field loss) may occur in association with a pituitary tumour. Damage to one optic tract posterior to the chiasma causes a crossed homonymous hemianopia. That is, damage to the left optic tract will cause a loss of vision in the right half of the visual field, and damage to the right optic tract will cause a visual defect in the left half of the visual field.

EYE MOVEMENT DEFECTS

Defects in the control of eye movements may result from damage to the nerve pathways involved in controlling the ocular muscles or from damage to the muscles themselves. For example, paralysis of the ocular muscles, **ophthalmoplegia**, may be caused by subcortical damage, by brain-stem lesions associated with the IIIrd, IVth and VIth cranial nerves, or by muscular disorders such as myasthenia gravis. However, cases of ophthalmoplegia caused by muscle disorders are rare.

Disorders of eye movement control may result in conjugate ophthalmoplegia, strabismus or diplopia. **Conjugate ophthalmoplegia** refers to an impairment of conjugate eye movements. In other words, the patient loses the ability to move both eyes harmoniously together. **Strabismus**, or **squint**, refers to a condition in which one eye moves normally but, owing to some weakness in the ocular muscles, the other eye sometimes does not. **Diplopia**, or double vision, occurs when a dysfunction of ocular muscle control results in the image of a visual object falling on noncorresponding parts of the retinae of the two eyes. For example, a target in the centre of the visual field will fall on the fovea in the normal eye but may be displaced to the side in the weaker eye. Consequently, a double image is perceived. Diplopia may be a presenting feature in the early stages of multiple sclerosis.

Visual defects: implications for nursing

Hemianopia is the term used to describe a visual loss in any section of the visual field, and homonymous hemianopia refers to a complete or partial

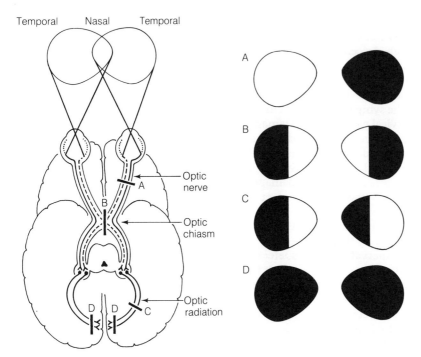

Temporal   Nasal   Temporal

—Optic nerve

—Optic chiasm

—Optic radiation

Figure 2.3.17 Visual field defects related to damage of the visual pathway. In A, the right optic nerve is cut, resulting in blindness in the right eye. In B, the optic chiasma is cut, causing blindness in the temporal half of the visual fields of each eye. In C, the right optic radiations are cut, causing blindness in the left half of the visual field of each eye (left homonymous hemianopia). In D, the visual input to the cortex is cut bilaterally, leading to complete blindness

loss of the nasal half of vision in one eye and of the temporal half in the other. This type of defect can arise in several disorders affecting the nervous system, including stroke and intracranial tumour. In the latter, encroachment on the visual pathways by the tumour will produce a defect, the extent of which will depend on the location and size of the offending tumour. The stroke victim's visual field defect is a consequence of the deprivation of the blood supply to the affected area of the brain.

To assist the patient with a defect such as hemianopia, the nurse should always approach the patient from his unaffected side and remind him to turn his head to compensate for his visual defects. Impairment of vision can pose a safety threat, and the nurse needs to be aware of dangers within the ward and home environments. Establishing and maintaining communication is essential, and to this end the position of the patient's bed is important. The visually impaired patient should be allocated a bed which permits maximum use of his unaffected field of vision; to do otherwise would essentially render the patient 'blind' and reduce contact between other patients and members of staff to a minimum.

Double vision can often be abolished with the use of an eye patch, and blurring of vision, which will result in decreased visual acuity, is dealt with as it arises.

Complete blindness, when it occurs, rarely happens in both eyes at the same time, but it is common for blindness in one eye to be followed by that in the other. Individuals are dependent upon visual images for sensory input and stimulation, and a loss of these will pose considerable problems for the affected person. The blind in-patient requires to be orientated to his new surroundings; this may be by verbal description or, if the patient's condition permits, providing a conducted tour. Once the patient is familiar with his new layout, alterations should not be made, e.g. moving furniture to a new location. When escorting a blind person, allow him to take your arm, and he can follow your body movements more easily if you gently squeeze your arm against your body. The nurse should enquire of the patient how he normally copes with the activities of living at home, and attempt as far as possible to accommodate the patient's normal prehospital routine. It may be that this is not always possible, and when a new arrangement is proposed, adequate explanation must be provided for the patient beforehand.

The blind person should be addressed on approach and the nurse should identify herself

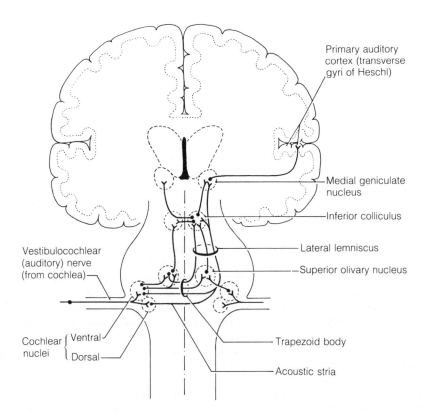

Primary auditory cortex (transverse gyri of Heschl)

Medial geniculate nucleus

Inferior colliculus

Lateral lemniscus

Superior olivary nucleus

Vestibulocochlear (auditory) nerve (from cochlea)

Trapezoid body

Cochlear nuclei { Ventral / Dorsal

Acoustic stria

Figure 2.3.18 Auditory pathways to the brain

and state what she is doing, even though this may not directly affect the blind person. When performing nursing care directly concerned with the blind patient, adequate explanation is absolutely crucial. The recently blinded person will require a lot of support, reassurance and explanation. A return to independence is encouraged as soon as possible, with guidance being provided on all aspects of the activities of daily living.

The Royal National Institute for the Blind provides a comprehensive welfare service, supplying information and facilities to the blind person. Being registered as blind or partially sighted will allow the person to become eligible for local and national services and for additional tax relief and supplementary benefit.

THE AUDITORY SYSTEM
(see also Chap. 2.2)

Axons from cells of the spiral ganglion of the cochlea leave the inner ear as the VIIIth (auditory) cranial nerve and terminate on cells of the cochlear nucleus in the medulla. Afferent fibres from the base of the cochlea, signalling high frequencies, penetrate deep into the nucleus before dividing. Afferent fibres from the apex of the cochlea, signalling low frequencies, divide at more superficial levels. In this way the strict organization of the frequencies of sound, which exists in the cochlea, is maintained at the level of the cochlear nucleus.

The cells of the cochlear nucleus give rise to axons which cross the midline in either the acoustic stria or the trapezoid body (Fig. 2.3.18). The nerve fibres of the acoustic stria arise from the dorsal part of the cochlear nucleus, cross the midline and ascend in the lateral lemniscus to terminate on cells of the inferior colliculus. The fibres of the trapezoid body arise from the cells of the ventral part of the cochlear nucleus, cross the midline and synapse in the superior olive. This nucleus in turn projects axons which join the lateral lemniscus and terminate in the inferior colliculus.

Although the majority of nerve fibres cross the midline, a sizeable minority synapse in the ipsilateral superior olive. Some cells of the superior olive receive input from ipsilateral and contralateral axons, and these 'binaural' cells are sensitive to the differences in the time of arrival of auditory signals from the two ears. Such time differences, although very small, are an important cue for the localizing of a sound source in the environment.

The inferior colliculus retains the tonotopic organization of the sensory input. The probable function of the inferior colliculus is the orienting response when a sound is being attended to. The connections between the inferior colliculus and the superior colliculus are presumably important in performing this role. The cells of the inferior colliculus project axons to the thalamic sensory relay nucleus of the auditory pathway, the medial geniculate nucleus (MGN).

The cells of the MGN project to the primary auditory cortex in the superior temporal gyrus, also known as the transverse gyri of Heschl. Here, too, an ordered tonotopic map of the sensory input is retained.

There is asymmetry in function between the two cerebral hemispheres. The left hemisphere contains an area referred to as **Wernicke's area**, which is concerned with the analysis of speech. Damage to this area causes a 'sensory (or receptive) aphasia' such that the patient is able to hear sounds but cannot interpret speech. Speech sounds may become as unintelligible as listening to the gabble of an unfamiliar foreign language. Damage to the corresponding area in the right hemisphere will leave the perception of language intact but will cause dysfunction in the recall, recognition and discrimination of non-verbal sounds. Such dysfunctions, called amusias, lead to the inability to discriminate between melodies.

Speech problems will lead to impairment of the patient's ability to communicate. The speech therapist should be involved at an early stage, in order that the best possible recovery is achieved. A skilled assessment is performed to establish the extent of the disorder, and an individualized therapy programme is then devised. This may range from the use of exercises and massage to the use of sophisticated microcomputer systems. It is imperative that the nurse attempts to communicate with the dysphasic patient on a day-to-day basis in order to restore the patient's self-esteem and avoid any feelings of isolation.

THE VESTIBULAR SYSTEM
(see also Chap. 2.2)

The central representation of the vestibular system begins with the bipolar cells of the vestibular ganglion. These cells are peripherally connected with the hair cells of the vestibular apparatus of the inner ear and project centrally along the VIIIth cranial nerve to the four vestibular nuclei of the medulla (Fig. 2.3.19). However, some fibres ascend directly to terminate in the cerebellum. The four brain-stem nuclei – the superior, inferior, lateral and medial vestibular nuclei – comprise the vestibular nuclear complex.

The lateral vestibular nucleus receives relatively few primary vestibular fibres but has input from the cerebellum and also from the spinal cord. The cells of this nucleus then project axons which descend as the vestibulospinal tract in the ventral columns of the spinal cord and terminate in the ipsilateral ventral horn of the cord at all levels from cervical to lumbar regions. This pathway has a facilitatory influence on muscle tone and spinal reflex activity.

The medial and superior vestibular nuclei send axons which ascend to terminate bilaterally on the nuclei controlling the extraocular muscles, that is, the nuclei of the IIIrd, IVth and VIth cranial nerves. The connections between the vestibular fibres and the oculomotor centres are concerned with the reflex pathway which produces eye movements compensating for head movements.

The medial and inferior vestibular nuclei project descending fibres to the cervical segments of the spinal cord. These axons are concerned with the control of head and neck muscles. In addition, these nuclei project ascending axons to the cerebellum.

As yet, there is no clear understanding of how vestibular sensation reaches conscious experience. A small thalamic nucleus, which projects to the parietal lobe, may be involved in this but there is little clear evidence at present.

Disturbances of the central vestibular system can cause dysfunctions in the control of eye movements and the control of balance. For example, a disturbance of ocular muscle control leading to nystagmus may be caused by peripheral or central damage to the vestibular system. **Vertigo**, which implies a disorientation of the body in space and

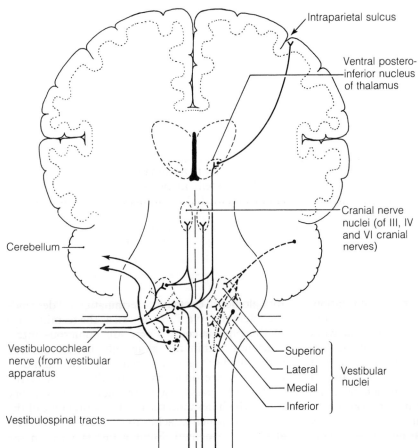

Figure 2.3.19 Vestibular pathways: note that, for clarity, some inputs are indicated on the left and some on the right side of the brain only

which may be accompanied by nausea, sweating, pallor and changes in heart rate, may also be a sign of vestibular damage.

THE GUSTATORY AND OLFACTORY SYSTEMS

Gustation

The peripheral afferent nerve fibres from the taste buds travel from the tongue via the VIIth (facial), IXth (glossopharyngeal) and Xth (vagus) cranial nerves to the solitary nucleus of the medulla. From here, cells project axons which ascend to the ventral posteromedial nucleus of the thalamus, and cells from this nucleus in turn project to the gustatory area of cortex at the bottom of the postcentral gyrus (Fig. 2.3.20). Little is known about

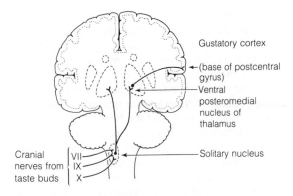

Figure 2.3.20 Gustatory pathways of the brain

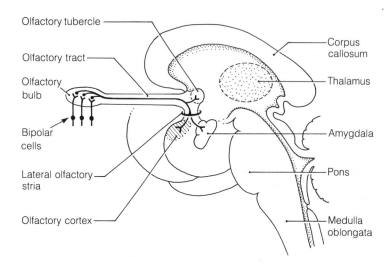

Olfactory tubercle

Olfactory tract

Olfactory bulb

Bipolar cells

Lateral olfactory stria

Olfactory cortex

Corpus callosum

Thalamus

Amygdala

Pons

Medulla oblongata

Figure 2.3.21 Olfactory pathways of the brain

the central integration of gustatory information, and the status of taste as a relatively minor sensory experience has meant that little experimental work has been undertaken in this area. However, dysfunctions of the gustatory system do occur. For example, some epileptic patients report an aura of a distinctive taste in the mouth which precedes an epileptic seizure. The aura may be caused by abnormal firing of cells in the area of the gustatory cortex or the closely neighbouring temporal cortex. In particular, these hallucinations of taste are associated with irritative lesions of part of the limbic lobe of the cortex.

Olfaction

Central representation of olfaction occurs in the olfactory bulb and in several areas of the limbic lobe. The pathway begins with the olfactory bipolar cells, the peripheral terminations of which serve as olfactory sense receptors. Centrally, the thin unmyelinated axons of the bipolar cells pass through the cribriform plate into the skull cavity and synapse in the olfactory bulb (Fig. 2.3.21).

The **olfactory bulb** is the primary integrating centre for olfactory impulses. It is a small, oval-shaped structure which contains a multiplicity of cells and synaptic contacts grouped together in bundles called **olfactory glomeruli**. The olfac-

tory glomeruli contain the terminations of descending efferent nerve fibres as well as the terminations of the bipolar cells. There are numerous interneurones within the glomeruli, including large cells called mitral cells which form the main outflow from the olfactory bulb. The axons of the mitral cells pass posteriorly along the olfactory tract, which lies underneath the frontal lobe of the brain and which attaches to the base of the brain at its posterior end. Some axons synapse in the anterior olfactory nucleus. This nucleus is in fact made up of small groups of neurones which are strung out along the olfactory tract. Other olfactory afferents terminate in the olfactory tubercle which lies at the posterior end of the tract. However, the main branch of the olfactory tract passes posteriorly and laterally as the lateral olfactory stria, to terminate in the amygdala and in the olfactory cortex of the limbic lobe, which lies adjacent to the anterior hippocampus.

The olfactory system is unique as a sensory system in that it does not have a specific relay nucleus within the thalamus. However, both the olfactory cortex of the limbic lobe and the olfactory tubercle send large numbers of fibres to the medial nucleus of the thalamus. In turn this projects to a wide area of the frontal lobes.

The sense of smell may be implicated in temporal lobe epilepsy since patients may hallucinate smells as part of the aura in the same way as tastes are sometimes hallucinated.

THE CENTRAL NERVOUS SYSTEM AND MOTOR CONTROL

Introduction

Motor activity ultimately involves the effects of peripheral motor neurones upon muscle cells. For example, the activity involved in a skilled act, such as catching a cricket ball, eventually rests upon the integration of motor units so that tension and force in various muscles are altered appropriately. The complexity of this integration can be seen when the muscle coordination necessary for a fielder to catch a cricket ball is considered in detail. The muscles directly involved in the catch, that is those of the fingers, thumb and hand, contract in a coordinate fashion at precisely the right moment. Other muscle groups which are not directly involved in the catch act synergistically. For example, the muscles of the shoulder and arm act to position the hand correctly in space. Postural muscles of the trunk and legs position the body to receive the ball and adapt to the rapid deceleration of the ball when it reaches the hand. The flight of the approaching ball is followed by smooth pursuit movements of the eyes. Moreover, if the ball is temporarily obscured from view, the fielder is able to predict its likely position when it comes back into view, and he can adjust his position in relation to this predicted flightpath. In summary, the execution of this type of skilled act involves the coordination of many muscle groups in a precisely controlled manner and in such a way as to allow for the alteration and adjustment of activity in response to moment-to-moment feedback from sensory receptors.

Many aspects of the central nervous control of motor activity have been elucidated and these will be discuseed in this section. There are many brain structures that are implicated in the control of motor activity. Traditionally, motor control has been subdivided into three neural systems. First, the **pyramidal system**, which consists of a fast and direct descending pathway from motor cortex; second, the **extrapyramidal system**, which consists of a multisynaptic pathway involving many brain structures, of which the most important are the basal ganglia; and third, the **cerebellum**, which interacts with both the pryamidal and extrapyramidal motor pathways. In addition to this division of the motor system, the traditional conception of motor control has implied a hier-

archical arrangement of these various subunits. Thus, the pyramidal system has been regarded as the most important subunit since its influences upon the musculature are fast, direct and concentrated particularly upon distal muscles, such as those of the fingers and hand, which carry out skilled acts. The extrapyramidal system has been seen as performing a supporting role since it consists of more diffuse and slower pathways and its influences are generally more strongly concentrated upon the proximal muscle groups of the arms, legs and trunk. However, it has been argued more recently that this traditional view of central motor control underrepresents the role of the extrapyramidal system.

The problem of ascribing a subservient role to the extrapyramidal system is clear when the effects of lesions to the pyramidal system are considered. For example, damage to the hand and arm areas of the motor cortex leads to severe motor impairment of the contralateral forelimb musculature. However, this impairment is, for the most part, only temporary. After a period of time, motor control is usually regained over all but the most distal muscle groups. An inability to clasp the fingers in a grip is the only permanent disability found in experimental primates. It is assumed that the extrapyramidal motor pathways are responsible for re-establishing this high degree of motor control.

The traditional view of central motor control has stressed the overriding importance of the pyramidal tract and the motor cortex. It has even been suggested that the conscious intent, or 'will', to move develops in the motor cortex. However, clinical investigations involving the electrical stimulation of the motor cortex in conscious patients have disproved this hypothesis. Patients stimulated in this way develop twitches and movements in peripheral muscle groups but they report no feeling of any intention to make these movements. Indeed, they report that the movements seem to be outside their control. In addition, more recent evidence from experimental studies in animals has shown that a great deal of neural activity in many different areas of the brain precedes an intentional motor act. It seems reasonable to assume that this widespread preparatory neural activity is responsible for converting a conscious intention into a planned and precisely controlled coordination of muscular activity. Thus, the motor cortex rather than being the location of the intention to act, in fact appears to be involved only towards the end of the process of neural integration.

The spinal cord

UPPER AND LOWER MOTOR NEURONES

At the level of the spinal cord, motor nerve cells are divided, by convention, into 'upper' and 'lower' motor neurones. Strictly speaking, the term upper motor neurone is a misnomer since only nerve fibres innervating muscle cells can truly be described as motor neurones. Nevertheless, the division of the nerve pathways into upper and lower motor neurones is valuable clinically and hence the terms continue to be used.

The term **lower motor neurone** is applied to cells of the ventral horn of the spinal cord and their associated nerve fibres which innervate muscle fibres. The axon of a single lower motor neurone branches at its distal end so that it innervates a number of muscle fibres – anything from a few to a few hundred. The muscle fibres innervated by a single motor neurone are referred to collectively as a **motor unit**. Damage to the lower motor neurone effectively disconnects the muscle from its nerve supply. This results in a **flaccid paralysis**, in which muscle tone is lowered (hypotonia); a marked reduction in resistance to passive movement occurs, and a loss of spinal reflexes ensues. In addition, denervation of the muscle may result in twitches, called fasciculations, and the muscle may also show signs of atrophy.

The term **upper motor neurone** is applied to the nerve fibres of the spinal cord which descend to synapse on lower motor neurones or upon spinal cord interneurones. Upper motor neurone nerve fibres make up the long, descending spinal tracts of the central motor system. Damage to upper motor neurones tends to cause dysfunction in muscle groups rather than in individual muscle units, and consequently the motor deficit tends to involve disruption of whole movements. This contrasts with lower motor neurone damage which affects single muscles. In addition, upper motor neurone damage does not cause the disconnection of the muscle from its nerve supply since upper motor neurone damage is not necessarily accompanied by lower motor neurone damage.

Consequently, upper motor neurone damage is not associated with the muscle atrophy seen in lower motor neurone injury, although prolonged lack of use of affected muscles will cause wastage in the longer term.

Upper motor neurone damage is also normally associated with **spastic paralysis**. Spastic paralysis involves an increase in muscle tone (hypertonus) which is assumed to be caused by a reduction or removal of descending inhibitory influences following an upper motor neurone lesion. This same removal of descending inhibitory influences would also account for the exaggeration of spinal reflexes which occurs below the level of the lesion following upper motor neurone injury. An increased resistance to passive movement also occurs after upper motor neurone injury, resulting in the 'clasp-knife' response. The Babinski response is also associated with upper motor neurone lesions in adults.

The cell bodies of lower motor neurones lie within the ventral horn of the spinal cord. It can be seen at the segmental level that the motor neurones are arranged within the ventral horn in an organized manner such that those cells projecting to proximal muscles are located medially and those projecting to more distal muscles are located progressively more laterally.

Upper motor neurones may synapse directly on α and γ lower motor neurones, or they may terminate on spinal interneurones. Direct termination on lower motor neurones will allow for immediate changes in muscle tension following alterations in the descending upper motor neurone discharge. On the other hand, spinal interneurones can act as gating mechanisms, so that peripheral sensory input via spinal reflexes such as the stretch reflex can be switched on or off, thereby enhancing or inhibiting the reflex activity. In fact, the major proportion of descending influences are integrated with ongoing sensory events via spinal interneurones. These spinal interneurones are referred to as propriospinal neurones.

DESCENDING PATHWAYS (Fig. 2.3.22)

The corticospinal, rubrospinal, vestibulospinal and reticulospinal tracts constitute the major descending nerve pathways of the spinal cord.

The **corticospinal tracts**, which are also referred to as the cerebrospinal or pyramidal

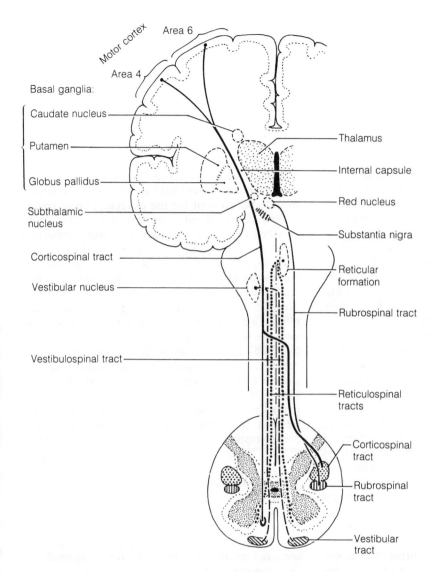

Basal ganglia:
- Caudate nucleus
- Putamen
- Globus pallidus

Motor cortex
Area 6
Area 4

Subthalamic nucleus

Corticospinal tract

Vestibular nucleus

Vestibulospinal tract

Thalamus

Internal capsule

Red nucleus

Substantia nigra

Reticular formation

Rubrospinal tract

Reticulospinal tracts

Corticospinal tract

Rubrospinal tract

Vestibular tract

Figure 2.3.22 Descending motor pathways from the brain

tracts, originate from widespread areas of the cerebral cortex. The nerve fibres descend in the internal capsule to the brain-stem, where they form easily identifiable lumps (or pyramids) on the ventral surface of the medulla. The term pyramidal tract derives from this anatomical feature. Approximately 90% of the nerve fibres cross the midline and descend further as the lateral corticospinal tract of the spinal cord. This tract extends throughout the length of the spinal cord. The other 10% of nerve fibres descend ipsilaterally as the ventral (anterior) corticospinal tract. Although the nerve fibres of the ventral corticospinal tract descend ipsilaterally, most of them

cross the midline before synapsing in the spinal grey matter.

The **rubrospinal tract** originates from cells of the red nucleus of the midbrain. This nucleus forms one of the final integrating centres in the extrapyramidal motor pathway. It is topographically organized and receives major inputs from the cortex, via collaterals of pyramidal tract axons, and also from the cerebellum. Additional inputs come from the reticular formation and from the spinal cord. The red nucleus projects descending nerve fibres which comprise the rubrospinal tract. Rubrospinal fibres decussate and then descend the spinal cord alongside lateral corticospinal fibres to terminate in the grey matter of the cord. Functionally, the rubrospinal pathway closely mirrors the pyramidal tract system. For example, it has been found in primates that removal of the red nucleus and the closely surrounding reticular formation results in motor dysfunction which is similar to that seen following pyramidotomy. The only major difference seems to be in the control of the most distal musculature, in particular the ability to grip with the fingers.

The **vestibulospinal tracts** originate from cells of the vestibular nuclei of the brain-stem. The lateral vestibulospinal tract projects from the lateral vestibular nucleus and descends, uncrossed, throughout the length of the spinal cord. The medial vestibular tract originates from the medial vestibular nucleus and comprises both crossed and uncrossed nerve fibres which terminate at cervical levels of the spinal cord. Stimulation of vestibulospinal fibres has a facilitatory effect and enhances muscle tone.

The **reticulospinal tracts** comprise the descending projections of cells of the brain-stem reticular formation. The medial reticulospinal tract projects from cells of the pons and descends as an uncrossed tract throughout the length of the spinal cord. The lateral reticulospinal tract projects from cells of the medulla and descends, both as a crossed and uncrossed pathway, through the entire length of the cord. These two tracts appear to function in a reciprocal manner through their indirect influences on α and γ motor neurones. The medial reticulospinal tract facilitates extensor and inhibits flexor muscle reflexes, whereas the lateral reticulospinal tract inhibits extensor and facilitates flexor reflexes. However, the precise nature of reticular control over motor activity may be more complex than this description implies, since the reticular formation receives descending input from many brain structures, including the cerebral cortex (particularly the pre-motor area) and limbic forebrain structures as well as from extrapyramidal motor fibres. It also receives ascending sensory input.

Complete or partial spinal transection has marked effects on motor activity. A complete section through the spinal cord results in **spinal shock,** in which there is no activity below the level of transection. Subsequently, spinal reflex activity recovers. Loss of voluntary control of muscle activity persists, owing to the sectioning of all upper motor neurones. Paraplegia describes a

paralysis of the lower half of the body, whereas quadriplegia is a paralysis from the neck downwards. The degree of motor deficit following partial spinal transection depends on the location and extent of the lesion. For example, a hemisection results in paralysis of the muscles on the side of the lesion and below the level of the lesion.

A patient suffering from paraplegia or quadriplegia relies, in the acute phase, on skilled nursing care to exploit residual ability to the full. Meticulous care must be taken when positioning the patient in bed, to ensure that paralysed limbs are maintained in good alignment and properly supported. The patient's position also needs to be changed frequently in order to prevent ischaemic damage to the tissues from unrelieved pressure.

Bladder and bowel dysfunction can also be common features. A high-fibre diet and the use of stool softeners usually suffice to produce a formed bowel movement every other day. Urinary catheterization is often used in the early stages, however the aim is to remove the catheter as soon as possible and retrain the patient's bladder to empty reflexly. This requires perseverance on the patient's part and encouragement from the nurse. However, the achievement of continence is a major morale booster. Every opportunity should be taken to educate the patient, as he may need to accept responsibility for a lot of his future care.

The pyramidal system

PYRAMIDAL TRACT

The terms pyramidal tract and corticospinal tract are used interchangeably. The term pyramidal in this context refers to the appearance of the descending tracts at the level of the medulla, rather than to the pyramid-shaped neurones, the axons of which make up the tracts.

Primary motor cortex lies along the precentral gyrus and contains very large pyramid-shaped cells, called **Betz cells,** which are exclusive to this area of cortex and which give rise to a small proportion of the nerve fibres of the pyramidal tracts. However, by far the greatest proportion (approximately 50%) of nerve fibres in the corticospinal tracts project from smaller, pyramidal-shaped cells in the larger anterior portion of the motor cortex.

The pyramidal tract is commonly represented

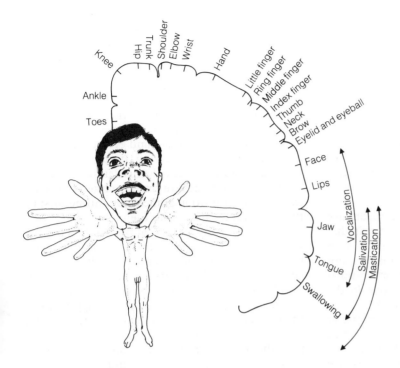

Figure 2.3.23 Motor homunculus (Reprinted with permission from Penfield, W. & Rasmussen, T. (1950) *The Cerebral Cortex of Man.* New York: Macmillan)

as consisting of nerve fibres which project directly from motor cortex to motor neurones or spinal cord interneurones. In fact, most axons leaving the motor cortex do not reach the spinal cord. Moreover, those fibres that do enter the pyramidal tracts project axon collaterals to many other brain structures. For example, axon collaterals of pyramidal fibres project to the red nucleus, the basal ganglia, the thalamus and the brain-stem reticular formation.

MOTOR CORTEX

The motor cortex occupies the precentral gyrus of the cerebral cortex, extending medially into the midsagittal fissure. It is topographically organized so that the muscle groups to which its cells project can be represented as a homunculus. This homunculus (Fig. 2.3.23) extends from the toes and feet, represented in the depths of the midsagittal fissure, through the areas projecting to the trunk, arm and hand, and finally to the cells projecting to the face, lips and tongue, which lie at the lateral end of the precentral gyrus. The various body parts of the homunculus appear distorted in size and shape according to the relative extent of the innervation of the musculature.

The motor cortex receives inputs from the supplementary motor cortex, from somatosensory cortex, from the contralateral motor cortex of the opposite cerebral hemisphere, and from the thalamus. The thalamic nuclei concerned are those of the ventroposterolateral nucleus, the ventrolateral nucleus and the ventral anterior nucleus. The ventrolateral nucleus is of particular importance since it relays signals from the cerebellum and from the basal ganglia and thus allows for the interplay of neural signals between the various subunits of the motor system.

The efferent fibres of the motor cortex project in a topographically organized manner to the basal ganglia, the thalamus, the red nucleus, the lateral reticular formation and the spinal cord.

PYRAMIDAL TRACT NEURONES

Since the motor cortex contributes to extra-pyramidal pathways as well as to the corticospinal tracts, it is not possible to identify the motor cortex as being exclusively part of the pyramidal system. Nevertheless, the medullary pyramids consist almost entirely of corticospinal axons and, therefore, it is possible to manipulate pyramidal tract fibres exclusively at this level.

Damage to the corticospinal tract at the medullary pyramids reveals an uncontaminated picture of the results of pyramidal tract injury. The major difference between a section of the medullary pyramid and removal of the motor cortex is that the spasticity and exaggerated reflexes seen following damage to the motor cortex are not present with a pure section of the corticospinal tract. Instead, a decrease in muscle tone in the limbs is observed, together with a loss of reflexes. From this it can be assumed that the spasticity seen after damage to the motor cortex is the result of damage to extrapyramidal pathways. One response which is of particular clinical importance as being diagnostic of pyramidal tract damage is the Babinski response.

PYRAMIDAL DISORDERS

Stroke (cerebrovascular accident)

Cerebrovascular accident (stroke) is the third most common cause of death in the Western world and occurs most frequently in the elderly population. It is characterized by a variable degree of neurological deficit produced either as a consequence of cerebral ischaemia caused by thrombus, embolism and/or atherosclerosis, or as a result of cerebral haemorrhage.

It frequently occurs in the region of the internal capsule (see Fig. 2.3.22), causing disruption of major inhibitory and facilitatory descending motor influences. Depending on the severity of the disturbance, a weakness or paralysis of one-half of the body musculature (hemiparesis/hemiplegia) occurs contralateral to the site of the damage. If the stroke occurs in the left hemisphere, speech may also be lost because of damage to the speech motor pathways.

Initially, the paralysis is flaccid, presumably a shock response to a massive loss of descending influences, but usually hypertonus (spastic paralysis) of the affected muscles follows. Considerable recovery of motor activity may occur following a stroke, particularly in the lower limbs.

The motor deficit associated with stroke may be accompanied by sensory loss, depending upon the extent to which ascending sensory pathways in the internal capsule are also damaged.

A characteristic feature of the disruption of descending motor pathways which may be caused by stroke is the persistence and even exaggeration of emotional facial expression, despite the loss of voluntary control of facial muscles. This occurs because the pathways controlling emotional facial expression arise from parts of the limbic lobe of the brain rather than from the motor cortex.

The physical problems that present in a patient who has suffered a major stroke, then, will include some or all of the following: alteration in conscious level, hemiparesis/hemiplegia, speech problems and visual field defects. Some stroke victims are managed at home, with support from their general practitioner and community nursing services, whereas others merit admission to hospital due to the seriousness of their condition or because of their home background.

A small number of patients are treated by surgical means; carotid endarterectomy involves the removal of atheroma from the internal carotid artery, thereby re-establishing circulation to the brain. A second procedure involves anastomosing the superior temporal artery to the middle cerebral artery, creating a bypass collateral to the brain. However, for the vast majority of patients, treatment is conservative and is aimed towards rehabilitating the individual to his or her fullest potential. Nursing management is individually tailored according to the main presenting problems.

Alteration in conscious level. The care appropriate for the unconscious patient is applied. Emphasis must be placed on the maintenance of adequate pulmonary function, necessitating a patent airway and efficient gaseous exchange; this is a priority.

Hemiparesis/hemiplegia. This is a weakness or paralysis down one side of the body. Proper positioning and support of the affected limbs in good alignment will assist in better mobility. Frequent alteration of body position is also necessary, to avoid the development of skin breakdown.

Speech problems. These commonly involve dysphasia, which may be expressive (an inability to express oneself) or receptive (an inability to comprehend the spoken word). Specialist advice and help are obtained from the speech therapist, who will assess the patient's deficit and create a treatment programme. It is beneficial for the nurse to be aware of the patient's therapy so that it may continue in the therapist's absence.

Visual problems. These may include field defects, diplopia (double vision) or blurring of vision. Diplopia can be abolished with the use of an eye patch, and the nurse should be aware of the decreased visual acuity and its attendant problems for the patient with blurred vision. Identification of the visual defect and compensating for this by approaching the patient from his unaffected side are the appropriate management. (This is elaborated upon in Chapter 2.2.) Many other problems can arise, involving intellectual and emotional deficits, e.g. lability or confusion, and other physical aspects, e.g. swallowing difficulties.

A multidisciplinary approach is advocated in most centres. The overriding aim of care is to make the patient as independent as possible in order that he may return home. This demands an enthusiastic rehabilitation programme, individually planned to meet each of the patient's needs and expectations. A key member of the team is the nurse, whose own attitude and standard of care will determine, in many instances, the quality of future life for the patient.

Extrapyramidal motor control

THE BASAL GANGLIA

Strictly speaking, extrapyramidal motor control refers to all neural integration of motor activity occurring outside the influence of the corticospinal tracts. Nevertheless, the cerebellum is a clearly distinguishable brain structure and hence its function tends to be assessed separately. Again, the red nucleus forms a major extrapyramidal motor pathway, but it is perhaps better considered alongside the pyramidal system because of the close parallels between the functions of the pyramidal tract and the rubrospinal tract. Consequently, the term extrapyramidal motor control has come to refer to the influences upon motor activity of a group of brain structures, the most important of which are the basal ganglia.

The basal ganglia comprise the **caudate nucleus, putamen** and the **globus pallidus** (see Figs. 2.3.22 and 2.3.24). The caudate nucleus and putamen are together referred to as the **corpus striatum** and make up by far the largest subcortical cell mass in the human brain. Caudate and putamen both consist of numerous small neurones interspersed with a scattering of a few large cells. The globus pallidus, on the other hand, contains large, widely spaced neurones and is divisible into a medial and a lateral part. The putamen lies immediately lateral to the globus pallidus, and together these two nuclei are referred to collectively as the **lentiform** (or **lenticular**) **nucleus**.

EXTRAPYRAMIDAL PATHWAYS

In each hemisphere, all areas of neocortex send fibres to both the caudate nucleus and the putamen of the same side. From the corpus striatum (caudate plus putamen), nerve axons pass through the globus pallidus and descend to the **substantia nigra,** a nucleus located in the brain-stem. Axon collaterals of this pathway synapse with cells of the globus pallidus. Axons from the medial part of the globus pallidus project to the thalamus, in particular to the ventral anterior and ventrolateral nuclei. The ventrolateral nucleus of the thalamus in turn projects axons to the cortex. Together, all these synaptic connections make up a feedback loop from cortex to striatum to globus pallidus to thalamus and back to the cortex (Fig. 2.3.25). The implication of this is that information from the entire neocortex can be processed and fed back to the motor cortex via the basal ganglia and thalamus in preparation for motor activity.

In addition to their contribution to the corticocortical feedback loop, the basal ganglia also make other important neural connections. For example, the basal ganglia receive input from the intralaminar nuclei of the thalamus, to which they also project. This connects the basal ganglia with the activity of the reticular formation since the reticular formation provides a major source of input to the intralaminar nuclei of the thalamus. The cerebellum also has indirect access to the extrapyramidal circuits of the basal ganglia since cerebellar fibres terminate in areas of the thalamus (the ventral anterior and ventrolateral nuclei) which overlap with the areas of termination of fibres from the globus pallidus. The basal ganglia also have reciprocal connections with the **sub-**

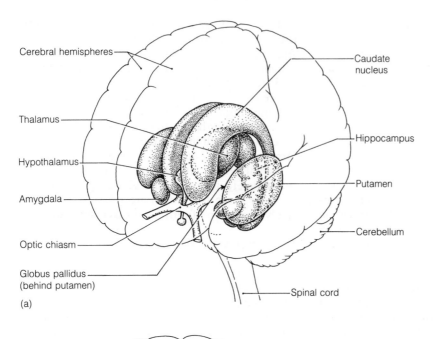

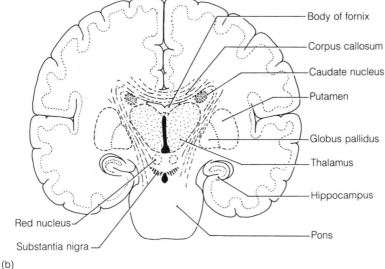

Figure 2.3.24 Basal ganglia of the brain: (a) position of basal ganglia in the brain, (b) basal ganglia viewed in a coronal cross-section of the brain

thalamic nucleus and the substantia nigra (see Figs. 2.3.22 and 2.3.25). Indirect pathways via the substantia nigra also project to the superior colliculus and the reticular formation.

EXTRAPYRAMIDAL FUNCTION

The complexities of the extrapyramidal neural connections suggest that the basal ganglia have diffuse and widespread influences on motor con-trol. Unfortunately, the detailed functions of these neural connections remain largely unknown. In terms of their physiological characteristics, it is known that the caudate nucleus and putamen are generally excitatory, whereas the outflow from the globus pallidus tends to be inhibitory. It has also been found that neurones of the basal ganglia tend to fire before those of the motor cortex and cerebellum. Thus, the latencies between extra-pyramidal activity and an associated muscle con-

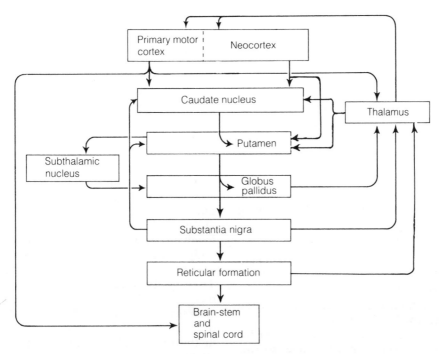

Figure 2.3.25 Schematic representation of the neural interconnections of the extrapyramidal motor system (Adapted from DeLong, M. R. (1974) Motor functions of the basal ganglia: single unit activity during movement. In *The Neurosciences Third Study Programme*, Schmitt, F. O. & Worden, F. G. (eds.). Cambridge, Mass.: MTP Press)

traction tend to be longer than those seen in the motor cortex or cerebellum.

EXTRAPYRAMIDAL DISORDERS

Parkinson's disease
The neural connections between the striatum and the substantia nigra have been implicated in the neurological condition known as Parkinson's disease or Parkinsonism.

Parkinson's disease (paralysis agitans) is characterized by three features. First, patients display a rhythmical tremor at a frequency of approximately six times a second, which is usually seen most clearly in the head and hands. The tremor disappears during sleep and when an intentional act, such as reaching and grasping, is carried out. For this reason it is referred to as a non-intention tremor. A second feature is a unique muscular rigidity which is clearly seen in the facial muscles and gives the patient an expressionless 'mask-like' appearance. The third feature of Parkinson's disease is a slowness in the initiation and execution of movements (bradykinesia). Parkinson's

disease is progressive and is generally associated with the elderly.

The tremor and ridigity of Parkinson's disease are principally attributable to the loss of the inhibitory influences of the basal ganglia, leading to an exaggeration of excitatory descending cortical output. The mechanism involved in this disinhibition appears to begin with the degeneration of cells in the substantia nigra. The basal ganglia and substantia nigra have a reciprocal neural connection (see Fig. 2.3.25). The neurotransmitter used in the striatonigral circuit appears to be **gamma amino butyric acid** (GABA). The reciprocal nigrostriatal pathway uses **dopamine** as the neurotransmitter. In Parkinsonism, the cells of the substantia nigra, which synthesize dopamine, degenerate. The degeneration of these cells closely parallels a reduction in dopamine levels in the corpus striatum. Pharmacological methods of treatment for Parkinsonism can raise dopamine levels in the striatum and can significantly ameliorate patients' symptoms. Administration of the dopamine precursor, **L-dopa**, especially in association with the administration of carbidopa, which blocks the decarboxylation and transamination of L-dopa, has been found to result in a marked remission of the symptoms of Parkinsonism.

In spite of this clinical success, it should be noted that L-dopa therapy only provides short-term relief from Parkinson's disease. Unfortunately, this therapy does not halt the degeneration of cells of the substantia nigra, and the cause and mechanism of this degeneration remain unknown.

The tremor, rigidity and bradykinesia experienced by these patients pose the main nursing problems. These can be overcome in the following ways

1 The provision of a low bed with a firm mattress, located near to toilet facilities. This will help to overcome the problem of rising out of bed and poor bladder control.

2 Adapted clothing, e.g. replacing buttons and zips with Velcro, will help the patient.

3 Rising from a high-backed chair is easier for the patient rather than from a low one, so this should be provided.

4 Initiating walking may pose a problem and to overcome this the attendant should gently rock the patient back and forward.

5 Occasionally a patient will freeze up and be unable to move; this may be remedied by suggesting the patient imagine that there is a step to get over and this may achieve recommencement of movement.

Chorea
The term chorea refers to involuntary, rapid, jumpy movements of the limbs and facial muscles. These abnormal movements are associated with a number of clinical conditions including the severe and fatal Huntington's disease. Huntington's chorea is a hereditary disease which usually only shows itself in the fourth or fifth decade of life. It is a progressive disorder involving chorea, dementia and eventual death.

Recently, it has been demonstrated that Huntington's disease is associated with the degeneration of cells of the striatum. In particular, small cholinergic intrastriatal neurones are destroyed along with others which project to the substantia nigra in the GABA-ergic striatonigral pathway. It has been suggested that the destruction of these cells disinhibits the substantia nigra. The sub-

stantia nigra, by way of its reciprocal pathway, then exerts excessive inhibitory influences upon the outflow from the basal ganglia to the thalamus. In this way the motor outflow to the cortex is affected and the abnormal movements result. At present, however, these suggestions for the mechanism of the disorder remain speculative.

Athetosis, ballism and tardive dyskinesia
Other signs of extrapyramidal disorder include athetosis and ballisms. **Athetosis** refers to slow writhing movements of the fingers, hands and arms and it has been associated with damage to the putamen and globus pallidus. **Ballisms** are violent and unexpected flailing movements, often involving the proximal muscles of one side of the body. These abnormal movements have been associated with damage to the subthalamic nucleus. **Tardive dyskinesia** is an abnormality of movement caused by exposure to antipsychotic drugs, such as chlorpromazine (Largactil). Antipsychotic drugs have an effect upon dopaminergic pathways in the brain and consequently these drugs can affect normal extrapyramidal activity. Therefore, tardive dyskinesia is a side-effect of antipsychotic drug therapy.

The cerebellum

ANATOMY

The cerebellum is a very large brain structure which lies underneath the occipital lobes and is separated from them by a fold in the dura mater called the **tentorium** (see Fig. 2.3.10). The surface of the cerebellum appears as a highly convoluted mantle of cells, the **cerebellar cortex**. Beneath the cerebellar cortex lies white matter and in the depths of the white matter are three pairs of deep-lying cerebellar nuceli – the **fastigial**, **interpositus** and **dentate nuclei** (Fig. 2.3.26). These nuclei provide nearly all the efferent outflow of the cerebellum.

The cerebellum appears as a bilaterally symmetrical structure with a central portion, the **vermis**, separating the two halves. Unlike the neocortex, the cerebellum does not possess any interhemispheric nerve fibres, and so does not transfer information from one side to the other.

The cerebellum can be separated into three different sections based upon the phylogenetic origin

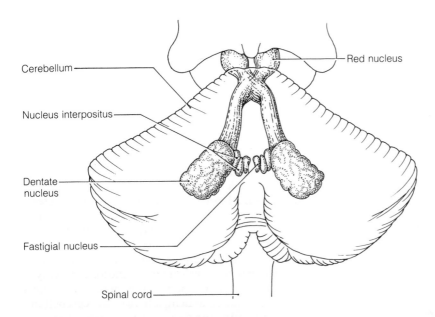

Cerebellum

Nucleus interpositus

Dentate nucleus

Fastigial nucleus

Spinal cord

Red nucleus

Figure 2.3.26 The cerebellum and cerebellar nuclei

of the various parts (Fig. 2.3.27). Such a division is also of functional significance. The oldest part of the cerebellum, the **archicerebellum**, chiefly consists of the flocculonodular lobe, which has neural connections with the vestibular system and is involved with the control of posture. The **paleocerebellum** comprises the anterior lobe of the cerebellum together with the caudal parts of the vermis. The anterior lobe can be experimentally mapped into sensory and motor representations of the body surface. These sensory and motor maps are superimposed exactly on one another. Finally, the **neocerebellum** consists of the posterior lobe of the cerebellum and the remainder of the vermis. The posterior lobe also shows sensory and motor maps of the body surface, which are again superimposed on one another. Both the paleocerebellum and the neocerebellum are involved in motor coordination.

Another way of subdividing the cerebellum, which is again of functional significance, is in terms of the deep-lying cerebellar nuclei through which the efferent signals of the cerebellum are passed. In general, this divides the cerebellum along the sagittal plane. The vermis projects to the fastigial nuclei; longitudinal strips lateral to the vermis project to the interpositus nuclei, and the lateral parts of the cerebellar hemispheres project to the dentate nuclei. The flocculonodular lobe projects to the vestibular nuclei and to the fastigial nuclei.

The structure and organization of the cerebellar cortex are uniform and relatively simple. The cortical mantle contains three layers of cells. The outer layer, the molecular layer, contains small interneurones, basket cells and stellate cells. The middle layer consists of large Purkinje cells. Dendrites of the cells of the Purkinje layer extend into the molecular layer, whilst the axons of the Purkinje cells provide the only efferent path from the cerebellar cortex. The inner layer of cortex, the granular layer, is composed of a huge number of densely packed granule (small stellate) cells.

CEREBELLAR FUNCTION

The cerebellum is involved in motor coordination. Essentially, it monitors both signal output from the motor cortex to muscles and also the execution of a motor act via sensory inputs arising from proprioceptive and touch receptors in the periphery. It is speculated that the cerebellum 'assesses' the force of contraction, rate of contraction and distance moved and then feeds back signals to the motor cortex, resulting in adjustments to the motor signal output so that the movement is executed in a smooth and coordinated fashion. For example, consider the execution of a relatively simple motor act such as drinking a cup of tea. The cerebellum monitors the muscle contraction and provides continuous feedback to the motor cortex so that, at a subconscious level, the

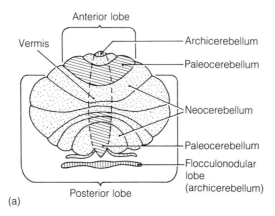

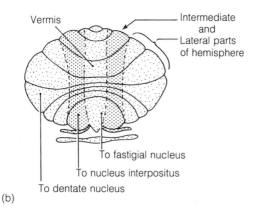

Figure 2.3.27 Functional and anatomical divisions of the cerebellum: (a) showing archicerebellum (phylogenetically oldest part of the cerebellum), paleocerebellum and neocerebellum (phylogenetically most recent part of the cerebellum), (b) showing the relationship between parts of the cerebellum and the cerebellar nuclei to which their efferent nerve fibres project

cortex is receiving signals which might be expressed as 'commands', such as 'move to the right', 'move to the left', 'move up', 'move down', 'move more quickly', 'move more slowly'. These commands occur rapidly on a moment-to-moment basis and the cup is brought to the mouth directly and smoothly, even without the aid of visual cues. It should be noted that visual and auditory inputs to the cerebellum do occur via the superior and inferior colliculi.

In addition to its role in motor coordination, the cerebellum is essential for normal balance and posture control and is connected to the vestibular nuclei of the brain-stem.

The **afferent inputs** to the cerebellum arise from the whole of the body as well as from higher brain centres, most importantly from the cerebral cortex (Fig. 2.3.28). The input from peripheral receptors carries proprioceptive and tactile information and reaches the cerebellum either directly via the spinocerebellar pathways, or indirectly via brain-stem nuclei, particularly the inferior olives. The spinocerebellar pathways arise from proprioceptive and skin afferents and project to the ipsilateral hemisphere of the cerebellum. The input from the inferior olives is contralateral and again arises from widespread sources, since the olivary nuclei receive input from all quadrants of the spinal cord as well as receiving collateral branches of descending motor axons. The input from the cerebral cortex to the cerebellum arrives via the corticopontocerebellar pathway. Collaterals from descending pyramidal tract neurones, particularly those arising in the motor cortex, synapse in nuclei of the pons. From here, cells project to the contralateral hemisphere of the cerebellum.

All the **efferent fibres** of the cerebellum arise from the Purkinje cells (Fig. 2.3.28). Almost all these axons synapse in the deep-lying nuclei of the cerebellum. The only exception to this is the axons of those cells of the flocculonodular lobe, which pass directly and ipsilaterally to the vestibular nuclei of the brain stem. The cells of the fastigial nuclei project bilaterally to the vestibular nuclei, thereby complementing the cells of the flocculonodular lobe in exerting their influence upon vestibular control. Each interpositus nucleus projects mainly to the contralateral red nucleus, but also sends axons to the ventrolateral nucleus of the contralateral thalamus. Cells from this thalamic nucleus project to the motor cortex. The dentate nucleus projects mainly to the ventrolateral and ventral anterior nuclei of the contralateral thalamus, and from here cells project to the motor cortex. It should be noted that although the interpositus and dentate nuclei project contralaterally, their influences are exerted over the ipsilateral musculature, since the descending axons of the red nucleus and motor cortex also decussate in the brain-stem. This effectively constitutes a crossing and then a recrossing of nerve fibres across the midline, such that damage to one side of the cerebellum will result in dysfunction of the musculature on the same side of the body.

The cerebellar output is entirely inhibitory, with the Purkinje cells using the inhibitory transmitter gamma-aminobutyric acid (GABA). The effects of this inhibitory influence are clearly seen in the activity of vestibular nuclei, following damage to the cerebellum. Such damage removes the cerebellum's inhibitory influence upon the vestibular nuclei and, as a consequence, the increase in vestibular excitation of motor neurones leads to an exaggeration of muscle tone.

EFFECTS OF DAMAGE

Damage to the cerebellum results in a number of disturbances of posture and movement which reflect the various neural connections of the cerebellum. Injury to the cerebellum may disturb the maintenance of balance, such that standing and walking are performed unsteadily. The patient will walk with a wide and unsteady gait, tending to deviate towards the side of the injury. Disturbance of eye movement control, such as nystagmus, may also be present, as may abnormalities of reflex movements. Disturbances of voluntary movement include an 'intention tremor'. Here, a tremor develops when the patient carries out an intentional act. Voluntary movements are also performed in a jerky and uncoor-

dinated fashion, and misreaching may be easily demonstrated by asking the patient to touch his nose with his hand. All these disturbances of fine coordination of movement can be interpreted as illustrating the cerebellum's activity as an adjuster of ongoing motor actions. For example, the neural connections between the cerebral cortex and the cerebellum relay information concerning an intended action. The information from peripheral proprioceptors to the cerebellum will in turn relay information relating the status of that action in terms of position of a limb, speed of movement and intensity of muscular contraction. The cerebellum is then able to affect the execution of the act, both via the feedback loop to the motor cortex and also by means of its outputs to rubrospinal and vestibulospinal motor neurones. It has been speculated that this intervention acts as a damping mechanism in order to achieve a smooth, coordinated action. Damage to the cerebellum eliminates this damping action so that an intentional movement becomes jerky. The initiation of an intentional act causes a muscle contraction, and if this is not damped by the cerebellum, it tends

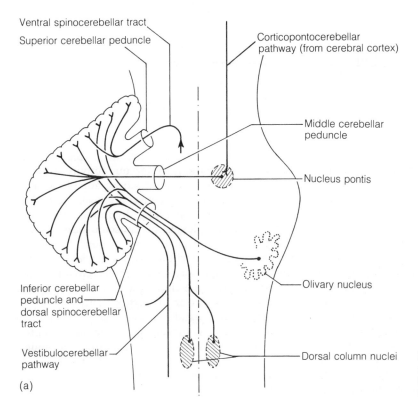

Ventral spinocerebellar tract

Superior cerebellar peduncle

Corticopontocerebellar pathway (from cerebral cortex)

Middle cerebellar peduncle

Nucleus pontis

Olivary nucleus

Inferior cerebellar peduncle and dorsal spinocerebellar tract

Vestibulocerebellar pathway

Dorsal column nuclei

(a)

Figure 2.3.28 Neural connections of the cerebellum: (a) inputs to the cerebellum, (b) outputs from the cerebellum

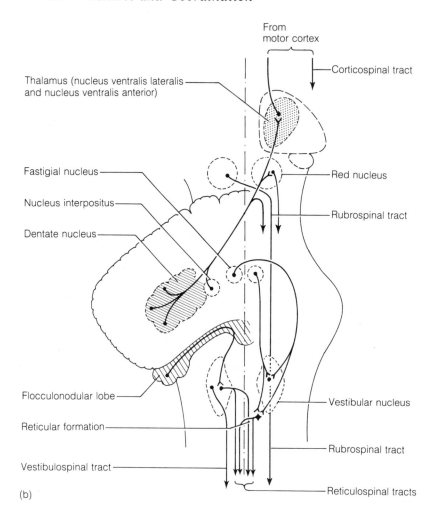

Figure 2.3.28 Continued

to overshoot and, as a consequence, the cerebral cortex compensates by initiating opposing muscular contractions. However, these again overshoot and the resultant effect is that of the tremor associated with intentional acts.

STATES OF CONSCIOUSNESS

A state of conscious awareness implies that the nervous system is operating at a level of functioning which allows deliberate volitional interactions with the environment. Such a broad definition covers a very wide range of states that occur in normal everyday experience – from euphoric excitement, through placid relaxation, to deep

depression. All these different states will be reflected in variations in nervous activity. Beyond these, other states exist which lie outside what is normally accepted as conscious awareness. For example, sleep, coma and hypnotic trance all reflect different levels of consciousness. Not a great deal is known at present about the neural processes that underlie states of consciousness and the concept of consciousness itself remains a baffling biological question. For example, we spend almost a third of our lives asleep and yet a comprehensive biological reason for sleep has still to be established. At present it is generally accepted that sleep serves restorative functions, although why these need to be accompanied by the dramatic changes in behaviour that are seen in sleep is not clear.

The sleep–wakefulness cycle

During the first weeks of life the newborn baby demonstrates a polyphasic sleep–waking cycle (i.e. with many alternating periods of sleep and waking throughout the day). This changes within weeks to a pattern of long periods of sleep at night and shorter periods of sleep during the day. By the second or third year of life the infant has adopted the typical circadian biorhythm of sleeping at night and waking during the day: the child has adapted to the rhythm of the light–dark cycle. That the light–dark cycle plays a role in controlling the sleep–wakefulness cycle can be seen when subjects are allowed to 'free-run'. In this type of study, the subjects are kept in conditions of continuous light or continuous darkness with no knowledge of the time of day. In such circumstances, subjects tend to adopt a pattern of activity and rest which either extends beyond 24 hours or is less than 24 hours. In other words, left to themselves, subjects behave as if the 'day' is either longer or shorter than 24 hours. Much the most common finding is for subjects to behave as if the day is slightly longer than 24 hours, perhaps 25 or 26 hours. However, subjects have been reported to adopt cycles of as long as 48 hours and as little as 16 hours. From this evidence it seems that the normal cycle of activity and rest is at least partly controlled by the light–dark cycle. However, even in conditions of 'free-running', all subjects maintain a typical biphasic pattern of sleep and activity. In other words, sleep is followed by activity and then sleep again and so on, in a regular diurnal pattern.

Neurophysiology of sleep and waking

At present, there is little clear understanding of the changes in nervous activity associated with changes in states of consciousness. The complex behaviour patterns that are nightly carried out in preparation for sleep, such as drinking cocoa, brushing teeth, donning pyjamas and getting into bed, suggest that the higher nerve centres of the cerebral cortex are involved in controlling the smooth transition from being awake to falling asleep. Nevertheless, there is no clear understanding of the relationship between cerebral cortical activity and the activity of the brain-stem

nuclei which have been found to influence consciousness.

The part of the brain that has received most study in respect to its influence on levels of consciousness is the brain-stem **reticular formation**. The reticular formation is a diffusely organized network of cells and fibres that constitutes the central core of the brain-stem. It is surrounded by the cranial nerves and their relay centres as well as the long ascending and descending fibre systems of the brain-stem. At its caudal end, the reticular formation is contiguous with the substantia intermedia of the spinal cord, and at its rostral end, it passes into the intralaminar nuclei of the thalamus. Thus, it stretches throughout the entire length of the brain-stem. Several nuclei can be differentiated within the reticular formation, but the simplest way of distinguishing between these separate cell clusters is to subdivide them into three parallel nuclear groups, stretching along the neuraxis. The medial strip, adjacent to the central canal, consists of the raphe nuclei, an intermediate strip contains many large cells, and the lateral zone contains smaller cells. The lateral group of nuclei project to brain-stem motor nuclei and are probably concerned with brain-stem reflex activity. Of importance for sleep are the large cells of the intermediate zone and the cells of the raphe nuclei. The large cells of the intermediate zone project axons which bifurcate and send long ascending branches to forebrain areas and long descending branches to the spinal cord. These cells receive projections from a wide area, including the spinal cord, the cranial nerve nuclei, the cerebellum and the forebrain. These cells are implicated in both sensory and motor pathways. The medial raphe nuclei also show a complexity of projection pathways, both descending to the spinal cord and ascending to forebrain structures.

The normal pattern of sleep and waking undoubtedly involves a complex interaction between lower brain structures and forebrain areas.

Other states of consciousness

Temporary loss of consciousness may occur because of **fainting** (syncope). This is most often caused by a reduction in cerebral blood flow and can be quickly rectified by ensuring the head is at, or below, the level of the heart. Heat syncope is caused by peripheral pooling of blood as the body tries to lose heat in a hot environment. If the patient is removed to a cooler environment, consciousness may be quickly restored. On the other hand, coma implies a relatively permanent state of unresponsiveness which may end in death.

Coma may be the result of large lesions of the midbrain and forebrain areas (usually distinguished as supratentorial, i.e. above the level of the tentorial membrane of dura mater), lesions of the upper brain-stem (designated subtentorial lesions), or metabolic disorders.

Supratentorial lesions cause coma because of the compression that they exert on brain-stem areas. For example, a large space-occupying lesion, such as a tumour in the cerebrum, may cause distortion of the temporal lobe over the medial edge of the tentorial membrane and this will cause compression of upper brain-stem structures. As the pressure is exerted more caudally, a progression of dysfunction occurs. Initially, the patient may be roused by verbal command or shaking. As unconsciousness deepens, the patient may only be roused by painful stimuli but the pupils of the eyes are reactive and Cheyne–Stokes respiration is present (see Chap. 5.3). As the midbrain is affected, signs of decerebrate rigidity, together with disconjugate eye movements and fixed pupils, become apparent and the patient becomes unresponsive. As the brain-stem is affected, respiration becomes grossly irregular, the pupils dilate and no eye movements can be elicited. The patient is totally unresponsive and, as the respiratory centres of the medulla are affected, terminal gasping eventually ceases and death ensues.

Coma caused by subtentorial lesions is principally associated with damage to the upper brain-stem and the associated reticular formation. The role of lower brain-stem areas in causing coma is less clear because damage to these areas will affect the respiratory and cardiovascular centres and any such damage is likely to prove rapidly fatal.

Metabolic coma may be a symptom of many conditions. For example, coma may be the result of brain anoxia, hypoglycaemia, uraemia, liver failure and meningitis, as well as a consequence of the ingestion of drugs such as opiates and barbiturates. In most, but not all, cases of metabolic coma the pupillary light reflex is retained and eye movements are not affected.

More efficient resuscitation techniques have led to the need for an acceptable criterion of **brain death** when dealing with patients in deep coma (Harrison, 1980; Allan, 1984). In practice, the criteria used rely on tests of brain-stem reflex activity to assess the likelihood of recovery from coma. Testing is normally carried out if the patient is believed to have suffered damage to the brain which is not amenable to treatment. The damage may be the result of cardiac arrest, cerebral haemorrhage or other causes, but coma resulting from hypothermia or ingestion of drugs is not sufficient cause. The patient is tested for responsiveness by applying intense and painful stimuli and looking for signs of reaction. Any head, face or eye movements is looked for in response to an appropriately applied painful stimulus, including supra-orbital pressure lest the patient has an undetected spinal injury. In addition to these tests, specific brain-stem reflexes are tested. Presence or absence of the pupillary light reflex is tested for in both eyes, together with the presence or absence of any eye-blink responses. Eye movements are tested by vigorous rolling movements of the head, the doll's head manoeuvre. In drowsy patients, this head movement will cause rolling eye movements in the opposite direction, but in patients in deep coma, these counter-rolling movements are absent. Eye movements are also tested by irrigating the external auditory canal with ice-cold water. This will cause a shift of gaze towards the affected side in a drowsy patient but such a response will be absent in deep coma. The presence or absence of spontaneous respiration is tested by allowing a slow rise in blood carbon dioxide level and observing for signs of respiratory movements. Tests for respiratory reflexes, such as the cough reflex, are also carried out. If all these tests show negative results, it is customary to repeat and confirm them before confirming death.

Epilepsy

Epilepsy represents an abnormality of cerebral function which is frequently, though not invariably, associated with partial or complete loss of consciousness. Epilepsy may be associated with structural damage to many different parts of the brain but it can also occur as a functional disturbance, which is not accompanied by an identifiable pathology. The latter type is termed idiopathic epilepsy. The behavioural manifesta-

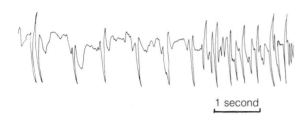

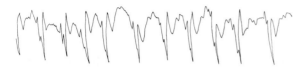

1 second

Figure 2.3.29 Epileptic EEG: upper trace shows paroxysmal spikes at onset of a seizure; lower trace shows spike and wave epileptic discharge

tions of epilepsy also vary, depending upon the site and extent of the epileptic activity in the brain. Therefore, epilepsy is probably better described as a group of disorders, the epilepsies, which have one characteristic feature in common, namely the presence of abnormal paroxysmal electrical discharges in the brain, which can be recorded on an EEG (Fig. 2.3.29). The epileptic EEG shows the presence of paroxysmal spike discharges and/or slow waves, and the EEG is an important aid to classifying the type of epilepsy. In 1980, the World Health Organization issued a new, simplified classification system for epilepsy.

GENERALIZED SEIZURE DISORDERS

Petit mal. Petit mal attacks are characterized by recurrent brief episodes of partial or total loss of conscious awareness, which last only a few seconds. The frequency of the attacks varies, but they may occur several times a day. During an attack the patient usually stops what he is doing and stares in a dazed fashion but he does not fall and no convulsions are seen. At the end of the attack he resumes his normal activity. The EEG of petit mal shows a characteristic 3-Hz spike and slow wave discharge over both frontal lobes. This pattern of discharge may also be seen in the absence of overt behavioural change – subclinical attacks. Petit mal is most common in young children and is thought to result from abnormal electrical discharges arising from the diencephalon (thalamus and hypothalamus). However, there is no identifi-

able lesion in the brain and petit mal is a form of idopathic epilepsy.

Tonic–clonic seizures (grand mal). These epileptic attacks are characterized by loss of consciousness and muscular convulsions. They are frequently preceded by a warning aura, the nature of which is believed to be related to the site of origin of epileptic activity in the brain. Auras may take the form of peculiar smells, visual hallucinations, apparent noises, emotions such as fear, or feelings of physical discomfort, especially in the abdomen. The attack proper begins with the patient falling to the ground as consciousness is lost. This is followed by a phase of tonic spasm of the muscles, usually lasting for just a few seconds, during which respiration may cease. The tonic phase is followed by the clonic phase in which the muscles show frequent short, sharp jerking contractions. At the cessation of convulsions the patient remains unconscious for some minutes, up to half an hour.

PARTIAL SEIZURE DISORDERS

Temporal lobe epilepsy. This is commonly associated with a lesion in the medial part of the temporal lobe. It is characterized by disturbances in sensory perception accompanied by changes in motor behaviour. The sensory abnormality may take the form of quite elaborate visual or auditory hallucinations, for example complete pictures may be 'seen' or a piece of music 'heard'. Size and distance may be distorted and the patient may experience déjà-vu. Characteristic smells or tastes are commonly reported and unpleasant emotional states, such as fear, often occur. The wide range of experiences that can occur in temporal lobe epilepsy reflect the range of functions associated with the temporal lobe – learning, memory, sensation and emotion. The motor component of the attack also varies. The patient may show no response at all or he may respond to commands, he may carry out a complex but routine activity or he may become aggressive. In all cases the dominant EEG feature is a spike discharge arising from the temporal lobe.

FOCAL SEIZURES

These arise in a specific part of one primary motor cortex and begin with clonic jerking contractions of affected muscles, usually the digits, of the contralateral side of the body. If the fit progresses, the convulsions increase in strength and spread to involve other parts of the body, frequently becoming bilateral. This is then termed a **Jacksonian seizure**. The EEG shows a clear focus of abnormal paroxysmal spike discharges.

Other, less frequently occurring, seizures include **myoclonic jerks**, a sudden shock-like jerk of the limb, **akinetic seizures**, in which there is a sudden loss of movement, and **infantile convulsions**, which are usually caused by a pyrexia (when they are sometimes termed febrile convulsions).

Status epilepticus is said to occur when one seizure, usually of the tonic–clonic variety, is rapidly followed by another. This can continue for a prolonged period of time and requires urgent intervention to stop it.

TREATMENT

The treatment of epilepsy commonly involves the use of anticonvulsant drugs such as phenytoin sodium (Epanutin) and sodium valproate (Epilim). These drugs may control the frequency and severity of attacks but do not act as a cure. Occasionally, surgery may be appropriate, especially for the treatment of temporal lobe epilepsy.

Epilepsy is perhaps the most misunderstood disorder of the nervous system; this is unfortunate as it only serves to make the resocialization of the epilepsy sufferer more difficult. Social rehabilitation is vital. The patient should be given clear advice on all aspects, including the following.

Drug regimens. The patient should be reminded to maintain his anticonvulsant therapy. It should be stressed that a seizure-free period indicates that the medication is working and is not an indication to stop. Abstention from alcohol is necessary as it inhibits the effect of many anticonvulsants.

Precautions. If the patient is taught to adopt a few simple common sense precautions, e.g. having a shower rather than a bath, this will avoid some of the problems of seizure activity. Education of the patient's family and friends prior to them witnessing a seizure will lessen their anxiety. Similarly, teaching simple first-aid measures will prepare them to cope with potential problems. Patients can be advised about wearing Medic-Alert jewellery

or an identification card. The patient should be warned of restrictions on driving.

Leisure and work. Some changes in the patient's usual work pattern may be necessary, e.g. he should not be in contact with moving machinery. Most leisure activities can be resumed and indeed this should be encouraged. Dangerous sports such as mountain climbing should be avoided.

Every member of the health care team should know what to do in the event of someone having a seizure. The following is a list of actions to be initiated: clear the airway, get help, prevent the patient from harming himself, maintain a record of the seizure, and reassure the patient and any onlookers. The patient must be allowed to recover at his own speed and should not be hurried.

MOTIVATION AND EMOTION

The **hypothalamus** has been identified as a brain structure of major importance in the control of motivation. It lies beneath the thalamus and forms the floor and part of the walls of the third ventricle. Several nuclei can be distinguished within the hypothalamus, although large areas consist of a diffuse neural matrix.

The hypothalamus can be seen as occupying the nodal position in a series of complex, interconnecting ascending and descending pathways (Fig. 2.3.30). For example, it is connected to forebrain structures such as the preoptic area, the septum, the thalamus, limbic lobe cortex and frontal lobe cortex, as well as to midbrain and hindbrain centres, such as the reticular formation, the visceral and somatic motor centres of the brain-stem and the spinal cord. The major nerve fibre pathway passing through the hypothalamus and involved in these ascending and descending connections is the **medial forebrain bundle**.

In addition to these neural connections, the hypothalamus has a controlling influence over the activity of the pituitary gland, the major endocrine gland. The pituitary gland lies directly beneath the hypothalamus, separated from it by the infundibular stalk (see Chap. 2.5). The posterior pituitary comprises the axon terminals of two distinct hypothalamic nuclei, the paraventricular nucleus and the supraoptic nucleus. These two nuclei synthesize the hormones oxytocin and antidiuretic

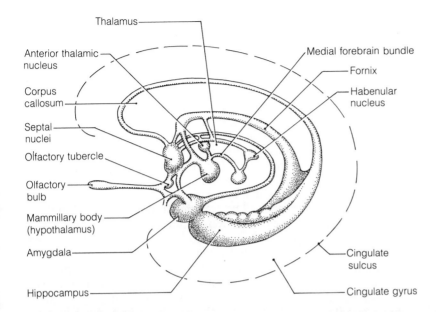

Thalamus

Anterior thalamic nucleus

Corpus callosum

Septal nuclei

Olfactory tubercle

Olfactory bulb

Mammillary body (hypothalamus)

Amygdala

Hippocampus

Medial forebrain bundle

Fornix

Habenular nucleus

Cingulate sulcus

Cingulate gyrus

Figure 2.3.30 Limbic lobe connections: the diagram illustrates the major cortical and subcortical brain structures implicated in the so-called limbic system

hormone (vasopressin) respectively. The cells of these nuclei transport the hormones along their axons down the infundibular stalk and into the posterior pituitary for release directly into the circulatory system. Smaller cells from a much wider area of the hypothalamus indirectly control the activity of the anterior pituitary by extruding releasing hormones (or releasing factors) and inhibiting hormones. These releasing and inhibiting hormones pass from axon terminals at a position midway down the infundibular stalk (i.e. at the median eminence) into the pituitary portal blood vessels. This specialized portal system conducts the hormones to the anterior pituitary, where they affect the release of anterior pituitary hormones.

An example of the role of the hypothalamus in the control of homeostatic mechanisms can be seen in its influence over body temperature control. Stimulation of anterior regions of the hypothalamus in unanaesthetized animals results in vasodilation and a drop in body temperature. Damage to this area causes hyperthermia. Contrastingly, stimulation of posterior regions of the hypothalamus causes shivering and vasoconstriction consistent with conservation of body heat. Damage to the posterior regions results in chronic hypothermia. Central receptors for both sets of responses appear to lie in the anterior hypothalamus and to be sensitive to local temperature changes, presumably caused by locally circulating

blood (see Chap. 6.1). In addition, signals from peripheral temperature receptors in the skin are involved in this control.

The hypothalamus is also implicated in the control of eating and drinking. Destruction of the ventromedial nuclei of the hypothalamus in experimental animals leads to hyperphagia (overeating) and severe obesity, whereas electrical stimulation of this area suppresses feeding. Contrastingly, destruction of the lateral nuclei causes aphagia (not eating) and adipsia (not drinking) and the animals die unless they are force fed and hydrated. Again, electrical stimulation of this area has the opposite effect, causing feeding to occur. Unfortunately, recent experimental evidence has suggested that this apparently straightforward formulation of a lateral hypothalamic 'feeding centre' and a ventromedial hypothalamic 'satiety centre' is incomplete. It has been suggested that damage to some of the nerve fibres passing through the hypothalamus may contribute to the effects of these lesions. For example, lateral hypothalamic lesions may damage the trigeminal nerve pathways and the resultant sensory and motor loss may contribute to the reluctance to eat. Lesions of the hypothalamus can also affect the release of hormones and this may indirectly affect feeding behaviour. For example, damage to the medial hypothalamus can cause an increase in insulin release, which in turn might be partially responsible for the hyperphagia and weight gain seen in ventromedial hypothalamic lesions. It appears that the role of the hypothalamus in eating and drinking is far from clear. Nevertheless, this can hardly be regarded as surprising in view of the complex interaction of visceral cues, environmental stimuli and learned behaviour patterns that contribute to eating and drinking.

In addition to motivation related to specific behaviour patterns, the hypothalamus also appears to be involved with general mechanisms of reward and punishment. These more general mechanisms of motivation have become known as **pleasure centres** and **pain centres**. In laboratory animals, experiments using electrical stimulation as a reward have mapped pleasure centres in a large number of brain structures, from the reticular formation to the hippocampus and limbic cortex. However, the areas of brain in which stimulation is most likely to elicit these responses are the septum, the hypothalamus and their associated nerve fibre pathway, the medial forebrain bundle.

In addition to pleasure centres, pain centres have been identified. These are more accurately described as **aversion centres**, since it is not known whether stimulation actually causes pain. Aversion centres tend to be fewer in number and are associated with more medial and midline areas of the brain. In these areas, stimulation leads to an immediate reduction and cessation of responding.

Stimulation of pleasure and aversion centres in humans has occurred as a consequence of some clinical procedures. Patients report that stimulation is either generally pleasurable or somewhat unpleasant, depending on whether a pleasure or an aversion centre is stimulated. However, the human experience would appear to be much less extreme, and consequently much less highly motivating, than that implied by the behaviour of laboratory rats.

Emotion

In addition to its role in motivated behaviour, the hypothalamus, along with other brain structures, also seems to be involved in emotional responses, although its precise role is not completely understood.

The importance of the hypothalamus in emotional behaviour would appear to lie in its position at the nodal point of a complex of ascending and descending nerve pathways. The descending pathways influence both the voluntary and involuntary motor responses. For example, the voluntary motor responses to a threatening situation will involve fight or flight. The involuntary responses will involve the associated sympathetic activity of the autonomic nervous system (see Chap. 2.4). The complexities of the ascending nerve pathways are much less clearly understood, but are often said to involve the limbic system.

The **limbic system** is not really an identifiable system at all. It comprises a large number of forebrain structures, including the limbic lobe of cortex, which are interconnected in a very complex manner. The functional significance of most of these connections is not understood at present. It will suffice here to say that the limbic system involves links between the hypothalamus, the septal area, the thalamus, the hippocampus, the amygdala, the limbic lobe of cortex, and inferior (orbital) areas of the frontal cortex. All these structures, which have been implicated in emotional responses, presumably operate through a complex pattern of nerve connections and feedback loops, although suggestions of how these mechanisms work are only speculative at the moment.

HIGHER MENTAL FUNCTIONS

Higher mental functions, such as learning, memory, language and thought, have been studied extensively by psychologists. Although insights have been gained into the nature of these important human abilities, it is impossible at the present time to relate these functions closely to physiological mechanisms. It is reasonable to assume that the nervous system subserves all these forms of mental functioning, but the mechanisms involved are not understood.

Learning and memory

Learning implies a modification of an organism's behaviour dependent upon experience. If any such modification takes to a degree of relative permanence, then we can impute the existence of a memory system. At present little is known of the changes in nervous activity that underlie learning.

Attempts to understand where memories lie within the nervous system have so far achieved no more success than those investigations concerned with the question of how learning and memory occur. Investigations of patients with neurological problems have suggested that the hippocampus may play a role in the establishment or recall of short-term memories. However, the evidence has been obtained from patients with large bilateral temporal lobe lesions, and the precise role of the hippocampus in short-term memory has yet to be elucidated.

The cerebral cortex

The cerebral cortex constitutes a much larger proportion of the brain in humans and other primates when compared with lower animals. It is therefore assumed that the cerebral cortex is responsible for our highest mental activities – activities such as thinking, communicating with others, problem solving and so on. Indeed, studies of patients with neurological problems have shown that damage to the cerebral cortex can result in serious disruption of higher mental functions.

Disorders

Certain clinical conditions lead to severe disruption of higher mental functioning. For example, disturbances occurring in psychosis and dementia may involve disordered thought processes, perceptions and memories. **Schizophrenia** is typically associated with thought disorder, hallucinations (particularly auditory hallucinations of voices) and disturbed emotional status, while the development of **senile** or **presenile dementia** may extend from minor lapses of memory to a state of utter confusion and debility in which the whole personality of the individual changes. At present, knowledge of the causes of these psychopathological states is limited. Nevertheless, some insights into physiological mechanisms have emerged recently. For example, in the case of schizophrenia, it has been discovered that certain drugs, which can alleviate the symptoms of psychosis, affect dopaminergic pathways in the brain. These drugs, such as chlorpromazine, block dopamine receptors and consequently lower the activity of central dopaminergic nerves. The effectiveness of antipsychotic drugs has led to their wide use. Even so,

it must be remembered that they are not effective in all cases, and are not equally effective in the patients that are helped by them. Moreover, it has not been possible, so far, to extrapolate from treatment with antipsychotic drugs to any full understanding of the causes of the behavioural and intellectual impairments seen in schizophrenic patients.

Review questions

The answers to all these questions can be found in the text. In each case there is at least one correct and at least one incorrect answer.

1 Cerebrospinal fluid
   (a) contains more sodium and chloride ions than plasma
   (b) provides most of the nutritional requirements of brain cells
   (c) samples are usually obtained by a lumbar puncture between the first and second lumbar vertebrae
   (d) is normally reabsorbed, via arachnoid granulations, into the major venous sinuses of the brain

2 Cerebrospinal fluid is produced
   (a) at the rate of about 30 ml/day in a healthy person
   (b) by the choroid plexuses in the ventricles of the brain
   (c) more rapidly than it is reabsorbed in the condition of hydrocephaly
   (d) by passive diffusion from plasma

3 The blood supply to the human brain
   (a) is provided by about a quarter of the resting cardiac output
   (b) is decreased if there is a moderate fall in the pH reaction of the blood
   (c) is largely provided by the internal carotid and vertebral arteries
   (d) is drained, via venous sinuses, into the external jugular vein

4 The spinal cord
   (a) is organized with white matter surrounding a roughly H-shaped zone of great matter

(b) gives rise to seven pairs of cervical spinal nerves
(c) ventral roots receive the sensory input whilst the dorsal roots supply the motor output
(d) is covered by the three layers of meninges

5 Human cerebral cortex

(a) of the occipital lobe lies at the posterior end of the cerebrum
(b) consists largely of white matter
(c) is composed of six horizontal cell layers over most of its area
(d) contains stellate cells which have longer axons and more extensive synaptic connections with the pyramidal cells

6 The dorsal column–medial lemniscal pathway for somatosensation

(a) contains nerve fibres which arise from cell bodies in the dorsal horns of the spinal cord
(b) contains nerve fibres which synapse in brain-stem nuclei
(c) contains nerve fibres arising from proprioceptors
(d) terminates in the precentral gyrus of the cerebral cortex

7 The spinothalamic tracts

(a) convey pain and temperature sensations
(b) travel in the ventrolateral quadrants of the spinal cord
(c) cross over the midline at the level of the brain-stem
(d) consist of smaller diameter nerve fibres when compared with the dorsal column tracts

8 Enkephalins

(a) possess analgesic properties
(b) are high molecular weight proteins
(c) are derived from the pituitary hormone β-lipotrophin
(d) have been shown experimentally to stimulate the production of prostaglandins

9 The vestibular nuclear complex

(a) is located in the midbrain
(b) comprises the superior, inferior, medial and lateral vestibular nuclei
(c) receives inputs from the VIIIth cranial nerve
(d) gives rise to the vestibulospinal tract, which has an inhibitory influence on muscle tone

10 The olfactory tracts

(a) arise from mitral cells in the olfactory bulbs
(b) project directly to the thalamus
(c) contain nerve fibres which terminate in limbic cortex
(d) contain nerve fibres which project to the olfactory tubercle

11 Primary motor cortex

(a) is located on the precentral gyrus
(b) is mapped out with the feet represented at the inferior end of the central sulcus
(c) gives rise to 90% of the fibres of the pyramidal tract
(d) has projections to the basal ganglia

12 Parkinson's disease

(a) results from degeneration of the pyramidal tracts
(b) is characterized by tremor when an intentional act is carried out
(c) is characterized by muscular rigidity
(d) is associated with a reduction in dopamine levels in the corpus striatum

13 The left cerebellar hemisphere

(a) is connected to the right cerebellar hemisphere via interhemispheric fibre tracts running through the vermis
(b) receives inputs from the right motor cortex
(c) is important for coordination of the musculature of the right side of the body
(d) projects, via the interpositus nucleus, to the contralateral red nucleus

14 The hypothalamus

(a) is part of the limbic system
(b) forms the floor and part of the walls of the fourth ventricle
(c) has connections with the frontal lobe and the thalamus
(d) is believed to be involved in the control of emotional behaviour

Answers to review questions

1 a and d
2 b and c
3 c
4 a and d
5 a and c
6 b and c
7 a, b and d
8 a and c
9 b and c
10 a, c and d
11 a and d
12 c and d
13 b and d
14 a, c and d

References

Allan, D. (1981) The use of transcutaneous nerve stimulation in patients with severe pain. *Nursing Times* 77: 40, 1721.

Allan, D. (1984) Brain death. *Nursing* 2: 23, 671.

Harrison, M. J. G. (1980) The diagnosis of brain death. *Medicine* 32: 1652.

Hayward, J. (1975) *Information, a Prescription Against Pain*. London: Royal College of Nursing.

Hughes, J., Smith, T. W., Kosterlitz, H. W., Fothergill, L. A., Morgan, B. A. & Morris, H. R. (1975) Identification of two related pentapeptides from the brain with opiate agonist activity. *Nature* 258: 577.

Melzack, R. & Wall, P. D. (1965) Pain mechanisms: a new theory. *Science* 150: 971.

Suggestions for further reading

Bannister, R. (1985) *Brain's Clinical Neurology*. Oxford: Oxford University Press.

Blakemore, C. & Cooper, G. F. (1970) Development of the brain depends on the visual environment. *Nature* 228: 477.

Bowsher, D. (1975) *Introduction to the Anatomy and Physiology of the Nervous System*. Oxford: Blackwell Scientific.

Bowsher, D. (1978) *Mechanisms of Nervous Disorder: An Introduction*. Oxford: Blackwell Scientific.

Fallon, B. (1975) *So You're Paralysed*. London: Spinal Injuries Association.

Gresh, C. (1980) Helpful tips you can give your patients with Parkinson's disease. *Nursing (USA)* 10: 1, 26.

Hickey, J. V. (1986) *The Clinical Practice of Neurological and Neurosurgical Nursing*, 2nd edn. Philadelphia: J. B. Lippincott.

Jennet, W. B. (1983) Brain death. *The Practitioner 277*: 1377, 451.

Kandel, E. R. & Schwartz, J. H., (1981) *Principles of Neural Science*. Amsterdam: Elsevier North Holland.

Lindsay, M. (1982) Living with epilepsy – 1. *Nursing Times 78*: 26, 1115.

Lindsay, M. (1982) Epilepsy – 2. People with epilepsy in the job market. *Nurising Times 78*: 27, 1155.

Noback, C. R. (1977) *The Nervous System: Introduction and Review*. New York: McGraw-Hill.

Norman, S. E., Browne, T. R. & Tucker, C. A. (1981) Seizure disorders, *American Journal of Nursing 81*: 5, 983.

Purchese, G. & Allan, D. (1984) *Neuromedical and Neurosurgical Nursing*. London: Baillière Tindall. pp. 100–109, 182–184, 215–217, 218–229 & 262.

Rogers, E. C. (1979) Paralysed patients and their nursing care. *Nursing 1*: 5, 207.

Rogers, M. A. (1979) Paralysis – how it affects movement and daily life. *Nursing 1*: 5, 203.

Ross Russell, R. W. & Wiles, C. M. (1985) *Neurology*. London: Heinemann medical.

Smith, C. (1980) Peripheral nerve lesions. *Nursing Times 76*: 47, 2057.

Smith, C. (1980) Peripheral nerve lesions – 2. The upper limbs. *Nursing Times 76*: 48, 2116.

Smith, C. (1980) Peripheral nerve lesions – 3. The lower limbs. *Nursing Times 76*: 49, 2159.

Walsh, K. W. (1978) *Neuropsychology a Clinical Approach*. Edinburgh: Churchill Livingstone.

The Autonomic Nervous System

Margaret Clarke

Learning objectives

After studying this chapter the reader should be able to

1 Classify the autonomic nervous system (ANS) into its divisions.
2 Relate the autonomic nervous system to the central nervous system and the peripheral nervous system.
3 Enumerate the anatomical differences between the two divisions of the ANS.
4 Name the neurotransmitters involved in the ANS and the way in which they are inactivated.
5 Explain the function of the sympathetic nervous system.
6 Explain the different actions of the parasympathetic nervous system.
7 Describe the structural reasons for the diffuse nature of sympathetic activity.
8 Recognize the complexity of interactions between the two divisions of the ANS.
9 Recognize the importance of the ANS to the maintenance of homeostasis.
10 Explain the signs of shock in terms of autonomic activity.
11 Discuss the relationship of autonomic activity to the overall experience of stress.
12 State how drugs acting upon the ANS exert their effect.

Introduction

The autonomic nervous system (ANS) has a crucial function in the maintenance of homeostais. Exactly how crucial is underlined by the knowledge that it is responsible, for example, for the fine adjustments of the heart, vascular system and respiratory system. The ANS is vital to the physiological aspects of coping during stress and forms one of the links between the nervous and endocrine control of behaviour. Anatomically and pharmacologically, it is important since it is the only example of nervous tissue in which synapses occur outside the protection of the skull and spinal column.

The **autonomic nervous system** may be considered as part of the peripheral nervous system and defined as the **motor efferent system** distributed to visceral muscle, cardiac muscle and associated glandular tissue. In terms of the anatomy of the system, this definition will be adhered to in this chapter. However, the most characteristic function of the control systems of the body is their interrelatedness and interdependence. To isolate the function of an efferent system without a consideration of the information input or control of that system leads to an arbitrary and incomplete description of its operation. Thus, wherever appropriate, the function of visceral sensory nerves will be considered as well as the integration and controlling aspects of the ANS.

STRUCTURE

In common with the rest of the nervous system, the ANS is comprised of neurones, neuroglia and other connective tissue. Structurally, however, it displays many distinctive features. For descriptive purposes, it can conveniently be dealt with in the traditional two divisions, namely the **sympathetic division** and the **parasympathetic division**. This anatomical classification coincides with a functional differentiation between the two divisions.

The efferent nerves of the sympathetic division arise from the thoracic spinal cord and the first two lumbar spinal segments, whereas those of the parasympathetic division arise from the brainstem and the sacral spinal segments. Consequently, the sympathetic division is sometimes called the **thoracicolumbar division**, and the parasympathetic division may occasionally be called the **craniosacral division**.

The fact that the neurones of this system synapse after leaving the CNS has been referred to above, but the manner in which this occurs is another feature which distinguishes the sympathetic division from the parasympathetic division. The sympathetic division will be described first.

Sympathetic division (Fig. 2.4.1)

From the fact that there is a synapse outside the CNS, it can be deduced that essentially the system is formed from the unit structure of two neurones in series, the **presynaptic neurone** and the **postsynaptic neurone**. The cell bodies of the presynaptic neurones lie in the intermediolateral horn of grey matter of their respective spinal segments. The fibres emerge and travel within the spinal nerves on leaving the spinal canal, but almost immediately leave the spinal nerves to run into the paired chain of **sympathetic ganglia** which lie on either side of the spinal column. This presynaptic nerve fibre is myelinated, and the short branches of the spinal nerve running to the sympathetic chain are called the **white rami communicans**, reflecting the appearance of the myelination. Within the sympathetic chain of ganglia, most of the presynaptic fibres synapse, although a few run through the chain intact, to synapse within collateral ganglia lying nearer to the visceral organs, which the postsynaptic fibre will ultimately supply.

Postsynaptic fibres are mainly unmyelinated and, on leaving the paravertebral sympathetic chain, many run a short distance as the **grey rami communicans** before re-entering spinal nerves. They are distributed to effector tissues (blood vessels, sweat glands, and piloerector muscles), with the appropriate spinal and peripheral nerves to each dermatome. Each preganglionic fibre synapses on average with 8–9 postganglionic neurones. This arrangement ensures that sympathetic activity tends to be diffuse.

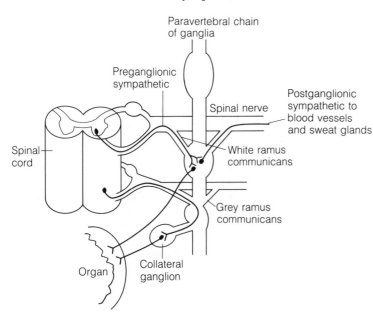

Figure 2.4.1 Arrangement of pre- and postganglionic fibres in the sympathetic division of the autonomic nervous system

The **paravertebral chain of ganglia** extends superiorly to the level of the cervical vertebrae, and some of the post-ganglionic fibres which arise from this cervical extension are distributed to the head. Within this cervical extension there are three ganglia: the superior, middle and inferior cervical ganglia. Postganglionic fibres arising from the superior cervical ganglion are distributed to the blood vessels and skin of the head, the pupil of the eye, and the salivary glands. Postganglionic fibres arising from all three cervical ganglia are distributed to the lungs and heart.

There is also an inferior extension of the sympathetic ganglionic chain. Four ganglia extend it to the level of the lower lumbar segments of the vertebral column. Fibres from these segments run to the inferior mesenteric ganglia. From there, the postganglionic fibres are distributed to the anal and bladder sphincters, and to the sexual organs.

As a general rule, the characteristic anatomy of the sympathetic division is the relative shortness of the preganglionic fibres and the relative length of the postganglionic fibres. There are exceptions to this characteristic pattern, the most important and striking being the case of the adrenal medulla. Here, an exceptionally long preganglionic set of

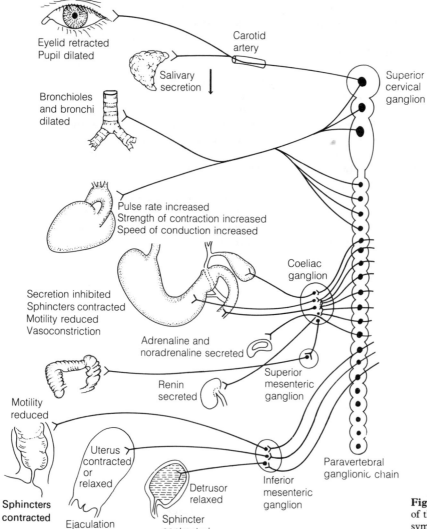

Figure 2.4.2a Diagrammatic view of the sympathetic ganglia chain and sympathetic innervation of some of the key organs

fibres runs to the organ itself. The **adrenal medulla** is considered to be a sympathetic ganglion in which the postganglionic cells have lost their axons and become highly specialized, secreting the neurotransmitters adrenaline and noradrenaline directly into the bloodstream. The stimulus for this secretion is the acetylcholine released by the preganglionic fibres arriving at the adrenal medulla.

Another exception to the normal pattern is the uterus, where it is believed that apparently long preganglionic fibres synapse with the postganglionic neurones within the organ itself.

Collateral sympathetic ganglia also form the synaptic origin of postsynaptic sympathetic fibres which are distributed diffusely to the thoracic and abdominal visceral organs. These collateral ganglia are the coeliac and mesenteric plexuses which distribute the coeliac and mesenteric nerves to the visceral organs (Fig. 2.4.2a).

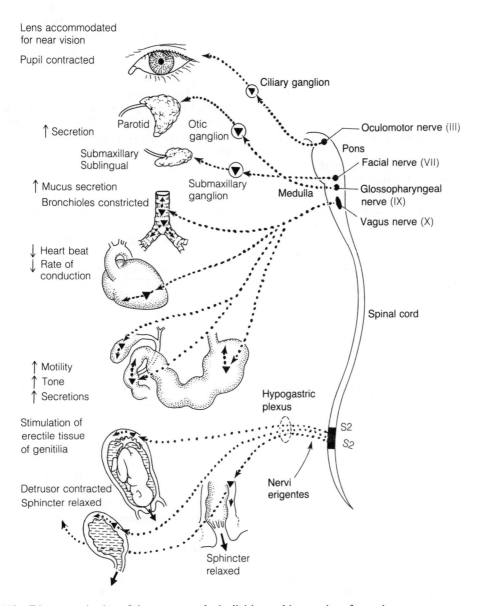

Figure 2.4.2b Diagrammatic view of the parasympathetic division and innervation of some key organs

It can be seen from this that the sympathetic division is by no means simple in its structure. This complexity is inevitable when one realizes that effector organs are found in the most peripheral areas of skin as well as core visceral organs, whilst the originating preganglionic cells occur within the limits of the thoracic spinal cord segments and two lumbar spinal cord segments.

The anatomical distinctiveness of the sympathetic chain of ganglia lying on either side of the vertebral column has led to the possibility of the operation **sympathectomy** being performed. This is very occasionally carried out for the congenital idiopathic condition of hyperhidrosis (excessive sweating), since the sympathetic nerves innervate the sweat glands. Other reasons for carrying out this operation will be discussed later.

The parasympathetic division (Fig. 2.4.2b)

The basic unit of the structure of the parasympathetic division is also that of two neurones in series. As described above, the preganglionic nerve cells are situated either in the brain-stem (as one component of some cranial nerve nuclei) or in the medial lateral grey matter of sacral segments.

Preganglionic fibres of the parasympathetic division are long; ganglia lie close to, or within the effector organ. Again, the system is complex as it is distributed to the eye and salivary glands in the head region and to the viscera of the thoracic, abdominal and pelvic cavities. Unlike the sympathetic division, it is not distributed to blood vessels, sweat glands and piloerector muscles of the skin.

From this we can see that most of the organs of the thoracic, abdominal and pelvic cavities have both a sympathetic and a parasympathetic nerve supply and they act as effector organs to *both* divisions.

Due to the length of the preganglionic fibres of the parasympathetic division, the parasympathetic component of peripheral nerves tends to be preganglionic for the most part.

PARASYMPATHETIC DIVISION AND CRANIAL NERVES

Parasympathetic fibres are found as a component of the IIIrd cranial nerve (oculomotor). They run to the ciliary ganglia in the orbit, where they

synapse. The postganglionic fibres innervate the ciliary muscle of the pupil, causing pupillary constriction (miosis).

The facial (VIIth cranial) nerve also contains parasympathetic fibres. Some innervate the lachrymal glands and others supply the submaxillary and sublingual salivary glands. The parotid salivary gland, however, derives its parasympathetic innervation from a branch of the glossopharyngeal (IXth cranial) nerve.

The most complex and important parasympathetic outflow is that contained in the **vagus (Xth cranial) nerve**. Seventy-five per cent of all parasympathetic fibres run in the vagus nerve and these form a substantial proportion of its substance.

Ganglionic synapses include the cardiac plexus, from whence postganglionic fibres proceed to the heart. From the pulmonary plexus, postganglionic fibres run to the bronchi. Other branches of the vagus nerve carry preganglionic fibres to intramural plexuses within the oesophagus, stomach and intestines. Short postganglionic fibres innervate these organs.

A lower motor neurone lesion of the facial nerve can cause loss of lachrymation (tear formation) on the affected side, although salivation still occurs since salivary glands have two sources of innervation.

A lesion affecting the IIIrd cranial nerve, as in early tentorial herniation, causes the dilatory action of the sympathetic supply to act unopposed, giving a dilated pupil on the affected side. Tentorial herniation is a potential outcome when there is dangerously increased intracranial pressure, as may occur, for example, in extradural haematoma complicating a head injury. The dilating pupil (together with an increasingly sluggish response to light) should be reported immediately it is observed so that emergency action may be taken.

The operation **vagotomy** does not, as its name implies, involve the whole of the vagus nerve, but refers instead to a division of the branch which supplies the pyloric muscle of the stomach and the secretory glands of the stomach (thus the term **selective vagotomy** is sometimes used). This prevents the vagus nerve from stimulating the secretion of digestive juice; consequently, acidity is reduced. The operation also allows more rapid emptying of the contents of the stomach into the duodenum. This operation may be carried out to treat peptic ulceration.

Parasympathetic sacral outflow

The preganglionic fibres arising from spinal cord sacral segments 2, 3 and 4 run with the pelvic splanchnic nerve and then to the vesicular plexus, from which they continue to synapse in ganglia within those organs which are innervated from the sacral part of the parasympathetic division – namely, the rectum and anus, the bladder and its sphincter, the blood vessels of the sexual organs. It is particularly important to note that these organs also have a sympathetic nerve supply, and in addition the sphincter muscles have a voluntary nerve supply as well. Normal function of the emptying of the bladder and bowel relies not only upon an intact nervous supply but also upon coordinated control of the input to the organs from the various types of nervous supply. This is also true of sexual activity in the male, since both parasympathetic and sympathetic activities are involved in erection and ejaculation. It is probably also true of sexual activity in the female.

NEUROTRANSMITTERS

The anatomical distinctiveness of the ANS in having synapses within ganglia outside the CNS has been mentioned already. This has allowed the study of the neurotransmitters involved in the ANS and the development of drugs which can block or mimic the effect of those neurotransmitters.

In this chapter, general characteristics of the neurotransmitters and synapses within the ANS only will be described (see Chap. 2.1 for further information on neurotransmitters). Since the system comprises two sets of neurones in series with one another, it follows that there are two sets of neurotransmitters involved: one set at the presynaptic nerve ending, affecting the postsynaptic neurone and situated within the ganglia, and the other set released from the postganglionic nerve ending at the effector organ.

To take the preganglionic neurone first, the neurotransmitter released from this nerve ending in *both* divisions of the autonomic nervous system, is acetylcholine. The transmission at these ganglionic synapses is characterized as **nicotinic** because the postganglionic membrane responds to nicotine in a way which is similar to its response to acetylcholine.

It is at the postganglionic nerve ending that the two divisions differ from one another. Within the effector organs, the neurotransmitter at the postganglionic nerve ending of the parasympathetic division is again acetylcholine (they are said to be **cholinergic** nerve endings). However, there is a difference between the action of acetylcholine at this site compared with its action at the preganglionic site, since it has been found that it is muscarine which mimics the action of acetylcholine release at the effector organs. Thus acetylcholine is said to be **muscarinic** at the parasympathetic postganglionic nerve ending. Atropine, the drug which is used as a premedication, will block the muscarinic effects of acetylcholine without affecting the nicotinic effects. Thus it blocks the parasympathetic stimulation of the effector organs without affecting the sympathetic effects of the nicotinic action of acetylcholine at the preganglionic nerve ending.

To turn now to the sympathetic division of the ANS, the neurotransmitter within the effector organ is noradrenaline at the majority of postganglionic nerve endings. Here, at least two different types of membrane receptors have been identified which respond differently to drugs. These have been called **α and β (adrenergic) receptors**. α-Adrenergic receptors are stimulated by noradrenaline to a greater extent than by adrenaline, whereas β-adrenergic receptors are stimulated by adrenaline to a greater extent than by noradrenaline. Both adrenaline and noradrenaline can reach the peripheral receptor sites of the sympathetic division from the bloodstream, and both substances are secreted into the blood by the adrenal gland. This reinforces the widespread diffuse action of the sympathetic nervous system. It will be recalled that preganglionic sympathetic nerve endings act directly upon cells within the adrenal medulla. These nerve endings release acetylcholine which stimulates the release of adrenaline and noradrenaline into the blood flowing through the gland.

Noradrenaline which is released at postganglionic sympathetic nerve endings is inactivated mainly by being taken up again into the nerve ending, although some is inactivated by monoamine oxidase.

A small number of sympathetic nerve endings are not adrenergic but **cholinergic**. These are the nerve endings which act on the sweat glands of the skin, although the sweat glands in the palm of the

Table 2.4.1 Structure, transmitter and action of the autonomic nervous system in relation to effector structures

Effector organ	Autonomic division	Receptor type	Action
Eye			
Pupil	Sympathetic	α	Pupillary dilation
Ciliary muscle		β	Accommodation for far vision
	Parasympathetic	Cholinergic	Pupillary constriction
			Accommodation for near vision
Lachrymal gland	Sympathetic	?	Vasoconstriction
	Parasympathetic	Cholinergic	Tear secretion
Salivary glands			
Parotid	Sympathetic	α	Vasoconstriction
Submandibular			Secretion of mucus, low enzyme
Sublingual			content
Parotid	Parasympathetic	Cholinergic	Secretion of watery saliva with high
			enzyme content
			Vasodilation
Submandibular	Parasympathetic	Cholinergic	Watery saliva with high enzyme
Sublingual			content
			Vasodilation
Heart	Sympathetic	β	Coronary artery dilation
			Acceleration of heart rate
			Increased force of contraction
			Increased rate of conduction
		α	Coronary artery constriction
	Parasympathetic	Cholinergic	Deceleration of heart rate
			Reduced force of contraction
			Reduced rate of conduction
			Coronary artery constriction
Bronchi	Sympathetic	β	Bronchial dilation
	Parasympathetic	Cholinergic	Bronchial constriction
			Secretion
Oesophagus	Sympathetic	α	Vasoconstriction
	Parasympathetic	Cholinergic	Peristalsis
			Secretion
Stomach ⎱	Sympathetic	β	Inhibition of peristalsis and secretion
Intestine ⎰		α	Vasoconstriction
			Sphincter contraction
	Parasympathetic	Cholinergic	Peristalsis and secretion
Spleen	Sympathetic	α	Contraction
Adrenal medulla	Sympathetic	–	Adrenaline ⎱ Secreted into blood
		–	Noradrenaline ⎰
Liver	Sympathetic	β	Glycogenolysis
Gall bladder	Sympathetic	β	Relaxation
	Parasympathetic	Cholinergic	Contraction
Pancreatic islet cells	Sympathetic	α	Inhibition of insulin secretion
		β	Insulin secretion
Descending colon	Sympathetic	β	Inhibition of peristalsis and secretion
		α	Vasoconstriction
	Parasympathetic	Cholinergic	Peristalsis and secretion

Table 2.4.1 Continued

Sigmoid colon ⎫ Rectum ⎬ Anus ⎭	Sympathetic	β α	Inhibition of peristalsis and secretion Constriction of sphincter
	Parasympathetic	Cholinergic	Peristalsis and secretion Relaxation of sphincter
Bladder	Sympathetic	β α	Relaxation of detrusor muscle Contraction of sphincter
	Parasympathetic	Cholinergic	Contraction of detrusor Relaxation of sphincter

Sex organs
  Penis
  Uterus
	Sympathetic	
		
		α
		β
			Ejaculation
			Contraction
			Relaxation

Penis Clitoris	Parasympathetic	Cholinergic	Erection

Blood vessels
  Skin
  Mucosa
	Sympathetic	α	Constriction

Muscle	Sympathetic	Cholinergic	Dilation
Renal	Sympathetic	α	Constriction
Pulmonary	Sympathetic	α	Constriction
Intracranial	Sympathetic	α	Slight constriction

Sweat glands
  General
  Hands
	Sympathetic
	Sympathetic
		Cholinergic
		α
			Sweating
			Sweating

Pilomotor muscles	Sympathetic	α	Contraction
Adipose tissue	Sympathetic	β	Lipolysis
Juxtaglomerular cells	Sympathetic	β	Renin secretion

hand are an exception to this, since the sympathetic nerve endings here are adrenergic. Another cholinergic sympathetic nerve ending site is in the muscles where stimulation by these sympathetic postganglionic fibres brings about vasodilation. These sympathetic cholinergic nerve endings are muscarinic in type.

FUNCTIONS (Table 2.4.1)

The importance of the functioning of the ANS in the maintenance of homeostasis has already been mentioned. It provides the automatic neural control within vital organs, which is the background to the voluntary and purposeful activity of daily life. To illustrate this point: unless the blood supply were maintained to the brain on standing and an increased blood supply delivered to the muscles in exercise, voluntary actions would be impossible even though the somatic nervous system were intact.

It was once believed that no voluntary control could be exerted over the functions carried out by the ANS. However, biofeedback techniques have shown that a certain degree of control is possible over the heart rate and blood pressure, for example through conditioning techniques. Implicit in the description of the anatomy of the ANS is one of its unique features, that of the dual innervation (involving both sympathetic and parasympathetic fibres) to many of the effector organs. It follows from this that the interaction of the two divisions is extremely complex.

The sympathetic division

As an oversimplification, the sympathetic division can be characterized as underlying many of the physiological coping mechanisms during threat – those of the **fight and flight response** described by Cannon (1932). On the other hand, the para-sympathetic division can be characterized as underlying relaxed, restful, sleepy circumstances. It is also concerned in the digestion of food and in anabolic metabolism.

The classical description of someone in a state of sympathetic arousal illustrates the action of this division of the ANS. It should be noted, however, that this is an 'ideal' model of the function of the system and that real people in real situations may fail in several respects to portray the classical picture.

The presence in the bloodstream of adrenaline and noradrenaline derived from the adrenal gland ensures that stimulation by the sympathetic nervous system, is widely diffused, since these hormones attach to effector organ membrane sites as neurotransmitters.

In sympathetic arousal, classically, the person has dilated pupils due to the stimulation of the radial muscle of the iris from the cervical ganglion. The ciliary muscle of the lens is relaxed, bringing about adjustment for distance vision. Skin arterioles are in a state of constriction, brought about by the contraction of circular muscle in their walls. This can be seen as pallor. Sympathetic nerves to the cardiovascular system, enhanced by the secretion of adrenaline and noradrenaline from the adrenal gland, ensure increased heart rate, increased conduction velocity of the specialized conducting tissue of the heart, and increased contractility of cardiac muscle – all contributing to increased cardiac output. Coronary blood vessels are dilated and so are the blood vessels within skeletal muscle. On the other hand, blood vessels supplying the gastrointestinal organs are constricted. The spleen contracts, ejecting blood into the system. Renin secretion from the juxtaglomerular apparatus in the kidney is enhanced, activating angiotensinogen and stimulating the production of aldosterone which aids the direct conservation of sodium chloride and the indirect conservation of water. Together the net effect of these adjustments increases the blood pressure.

In general, respirations will be quiet, since bronchial muscles are relaxed. Increase in rate and depth of respirations is less a direct effect of sympathetic stimulation than an indirect effect of the other changes, in particular the cardiovascular changes.

Motility and tone of the muscles of the oesophagus, stomach and intestine will be reduced, and secretion of digestive juice is inhibited. The anal sphincter is (usually) contracted.

The detrusor muscle of the bladder will be in a state of relaxation and the bladder sphincter will (usually) be contracted.

Stimulation of the adrenal gland to produce increased amounts of adrenaline and noradrenaline has already been mentioned, and all the effects of the sympathetic system enumerated above are thereby enhanced and reinforced.

Adrenaline and noradrenaline increase the arousal effect of the ascending reticular system. Insulin secretion is depressed and blood glucose levels are increased through glycogenolysis. Lipolysis increases the free fatty acid levels in the blood. The tendency of blood to clot is enhanced and this is aided by the constriction of peripheral blood vessels mentioned above.

Other effects of sympathetic arousal include contraction of the piloerector muscles. This may be insignificant in humans, but in furry animals it leads to an apparent increase in size. There is a slight localized secretion of the sweat glands in the palm of the hand (adrenogenic or adrenergic sweating). In addition, the normal insensible fluid loss through the skin may stand out as 'sweat' in the absence of heat from the blood to evaporate this fluid during the 'shut down' of skin blood vessels.

These physiological adjustments have great survival value when an individual is under physical threat. However, many of all of these adjustments can occur in humans when there is psychological threat but no physical threat. Such 'useless' action of the sympathetic division, if it occurs frequently or is prolonged, far from having survival value, can be detrimental. This will be considered in discussing stress.

Not included in the account given above of sympathetic arousal is the action of sympathetic stimulation during sexual function of men and women. This is incompletely understood, especially in the case of women, but it is known that in both sexes parasympathetic stimulation is important as well, and that unless the two types of innervation are acting in a coordinated way, sexual dysfunction occurs. However, psychogenic and other CNS factors are also crucially important to sexual functioning in humans. The case of sexual activity underlines and reinforces the point made already, that in reality extremely complex interactions take place between the two autonomic divisions and that to isolate the function of one system and describe it can lead to a false picture unless the complexity is fully appreciated.

The parasympathetic division

Unlike the sympathetic division, during activity of the parasympathetic division the postsynaptic neurotransmitter (acetylcholine) is not found in the bloodstream so no widespread autonomic stimulation occurs. Therefore, in general, the effects of the parasympathetic division of the ANS are discrete and of short duration. This is not only due to the more discrete release of acetylcholine, but also to the highly effective action of cholinesterase in splitting acetyl choline and thus deactivating it. This can be contrasted with the case of the sympathetic system in which the destruction of catecholamines is less prominent than their reuptake by postsynaptic nerve endings. This mechanism of reuptake not only prolongs the effect of sympathetic activity, but contributes to a significant extent to the level of noradrenaline in the circulation. It is not surprising, then, to note the more diffuse effect of sympathetic activity.

However, since parasympathetic activity is discrete, it is less accurate to portray a global abstracted picture of an individual during parasympathetic activity and instead more helpful to consider its effects upon the different effector tissues.

Parasympathetic stimulation of the gastrointestinal tract aids the digestion and absorption of food through an increase in the motility of the muscle, an increase in digestive secretions, and relaxation of the pyloric sphincter. The gall bladder contracts, releasing bile, and insulin secretion from the pancreas is favoured. Salivary glands are stimulated to produce profuse watery secretions, and the blood vessels supplying them are dilated.

It will be recalled that the vagus nerve has an important role in carrying parasympathetic fibres and distributing them widely to the thoracic and abdominal organs. Its role in the functions related to digestion and absorption has already been dealt with. In relation to the respiratory system, parasympathetic stimulation increases the secretion from nasopharyngeal and bronchial glands and favours the contraction of bronchial muscle.

Parasympathetic stimulation via the vagus nerve decreases the heart rate and also decreases the contractility of muscle and the conduction velocity through the specialized conducting tissue of the cardiac muscle. In extremely rare circumstances (fortunately), the vagus nerve can bring about complete cessation of the heart beat (vagal arrest). This is incompletely understood.

In the eye, parasympathetic stimulation acts on the sphincter muscle of the iris to constrict the pupil. It also contracts the ciliary muscle of the lens, aiding near vision. It is activity of the parasympathetic division which leads to weeping, through stimulation of the lachrymal glands.

Turning to the function of parasympathetic fibres having their origin in the sacral segments of the spinal cord, those supplying the bladder stimulate contraction of the detrusor muscle and relaxation of the sphincter, bringing about micturition, although it will be realized that, after infancy, voluntary control also plays a very important part in this act. Similarly, relaxation of the anal sphincter at an involuntary level is a function of the parasympathetic supply to the muscle.

Sexual activity has been mentioned before. Erection in the male is a function of the parasympathetic supply, whilst ejaculation requires sympathetic stimulation.

In practice, it may be difficult in some circumstances to distinguish the role of stimulatory effects of the parasympathetic division from the absence of stimulation by the sympathetic division.

AFFERENT NERVES AND HIGHER LEVEL CONTROL

Clearly, the ANS could not carry out its functions in the absence of information and control. It has already been stated that, strictly speaking, the ANS is a purely efferent system and in the description of its anatomy this definition was adhered to. However, in discussing function it would be misleading to omit all mention of sensory information and those controlling centres which play a crucial role in ensuring that the function of the ANS is adaptive.

Whether or not visceral sensory fibres are considered part of the ANS is controversial (Walton, 1983). Sensory information from a proportion of the visceral organs is carried by sensory nerve fibres which run with autonomic nerves, enters the spinal cord via posterior nerve roots, and is then conducted centrally with somatic sensation. Visceral sensory information from remaining organs is conveyed to the brain in cranial nerves. This aspect of function is considered in other chapters when reflex or central control of cardiovascular and respiratory function is described and as part of the consideration of the control of other visceral organs.

Although the control of the ANS is complex, one can state confidently that the hypothalamus is concerned in the control of many of the ANS functions.

The hypothalamus is situated at the base of the brain. It lies immediately above the pituitary gland and it has both a nervous and a portal blood connection with the pituitary gland. Nervous centres controlling visceral function occur in a hierarchy of levels. For example, simple reflexes controlling the contraction of the full bladder are integrated in the spinal cord. More complex reflexes controlling heart rate and respiration are integrated in the brain-stem. The homeostatic functions of controlling core body temperature and blood glucose levels occur in the hypothalamus. The hypothalamus is also a constituent of the limbic system which underlies emotional and instinctual behaviour. The final control of all voluntary behaviour is exerted by the cerebral cortex. (See Chap. 2.3 for full details.)

DISORDERS

Since many organs have a dual autonomic innervation, it is possible that an interruption of the nerve supply from one ANS division leaves the supply from the other division acting unopposed. An example of this in relation to the pupil of the eye has already been mentioned. When the IIIrd cranial nerve nucleus is subjected to pressure during very early tentorial herniation, it leaves the sympathetic innervation to the pupil acting unopposed (giving dilation).

Unopposed action of the parasympathetic supply to the pupil can occur in **Horner's syndrome**. This syndrome arises when there is a lesion of the cervical sympathetic ganglion supplying the eye, head and neck. The features of Horner's syndrome are: constriction of the pupil (miosis); drooping of the eyelid (ptosis); slight retraction of the globe of the eye within the orbit (enophthalmos); and loss of sweating on the affected side of the head and neck (anhidrosis). An aneurysm or

thrombosis of the internal carotid artery affecting the sympathetic nerve may occasionally cause all these signs with the exception of anhidrosis, since the nerve fibres affecting sweating are carried with the external carotid artery.

Lesions of the cauda equina and spinal cord

A severe lesion of the cauda equina affects bladder and bowel function as a result of damage to the sacral parasympathetic outflow. The detrusor muscle of the bladder is affected and reflex contraction no longer occurs in response to distension. The bladder wall itself has a certain amount of elasticity, and as pressure mounts the elasticity forces some urine into the urethra. However, the unopposed sympathetic supply to the sphincter muscle keeps it contracted and closed, and dribbling incontinence occurs. A similar situation arises with regard to the bowel and anal sphincter.

A lesion of the spinal cord itself, in which voluntary control of the bowel and bladder is affected, but which does not affect the parasympathetic outflow, leads to an automatic bladder (once spinal shock has worn off). Such lesions can affect the sympathetic innervation to the skin. Initially during spinal shock, the skin below the level of the lesion is dry, pale and cool, but at a later stage, profuse sweating in the affected area is common and can be provoked by many cutaneous or other stimuli.

SYMPATHECTOMY

This operation was referred to earlier in relation to the anatomy of the sympathetic division of the ANS, since the paravertebral chain of ganglia are rendered relatively accessible for surgical intervention.

There are several reasons for considering such an operation, in addition to the condition of hyperhidrosis already referred to.

Vascular
(a) To improve the blood flow to peripheral areas whose physiological integrity is threatened by vasospasm.
(b) To reduce blood pressure more generally in hypertension.

In chronic pain

(a) For causalgia (burning pain in the cutaneous distribution of an injured peripheral nerve).

(b) For phantom limb pain.

It is important to note, however, that with the increased knowledge available today, this operation is carried out only as a last resort, since less drastic and non-invasive methods of opposing the action of the sympathetic nervous system are available.

Vascular effects of sympathetic stimulation can be opposed with drugs having very specific effects upon the blood vessels (see below).

In the case of chronic pain, behavioural methods have been developed which are potent in helping patients (McCaffery, 1979).

In any case, before sympathectomy is carried out, the effect of an injection of local anaesthetic into the paravertebral sympathetic ganglionic chain at the required level is checked to ensure that the operation is likely to be successful.

A brief explanation of the mechanisms by which this operation is believed to be effective will be given, in spite of its rarity, since the rationale helps to underline the function of the sympathetic division of the ANS.

Understanding the reasons why sympathectomy can alter vasomotor tone is simple, since the smooth muscle in blood vessel walls is entirely innervated by sympathetic fibres. These control the extent of contraction of muscle and hence the calibre of blood vessels. Removal of the impulses stimulating constriction leads to vasodilation. Post-operatively, hypotension is a potential problem.

The rationale for sympathectomy in the treatment of causalgia and phantom limb pain is much more complex and not entirely understood. However, the afferent peripheral pathway for visceral sensation runs with the efferent autonomic nerves, the sympathetic ones in particular. Visceral sensory fibres, having entered the posterior nerve roots, then travel centrally within the spinal cord with somatic afferents. It is believed that the phenomenon of referred pain is related to this close anatomical relationship between sensory nerve fibres in the spinal cord. An example with which nurses will be familiar is the referral of cardiac pain (visceral) to the somatic areas of the left arm and substernal region.

It is believed that afferent visceral fibres may provide an alternative path for pain from areas deprived of somatic sensory nerves. Certainly pain may be carried by afferents accompanying the autonomic supply to blood vessels. This occurs in migraine, for example.

Following an incomplete lesion of a peripheral nerve, a syndrome of causalgia may arise in which there is excessive sweating, excessive sensitivity to touch, and pain in the area normally supplied by the nerve. Afferent stimuli responsible for this are carried with the autonomic nerve, and removal of the sympathetic supply to the area relieves the pain.

Phantom limb pain appears to be a special example of causalgia and it can be relieved by sympathectomy in the same way. Prior to operation the stump may be cold, cyanotic and covered with perspiration, whilst exposure to cold and excessive emotional stress increase the pain. Post-sympathectomy, the stump becomes warm, dry and non-tender.

DRUGS AND THE AUTONOMIC NERVOUS SYSTEM

A large number of drugs have an effect upon the ANS and not all can be mentioned here. One reason why so many such drugs have been discovered or developed is that effector organs supplied by autonomic nerves are relatively easy to isolate and to use to test the pharmacological action of both naturally occurring and synthetic substances. The frog heart and ileum are examples of such organs.

Drugs affecting the sympathetic division

SYMPATHOMIMETIC AGENTS

These are drugs which stimulate the sympathetic nervous system. They can be classified according to the receptor sites on which they have their major effect. This may be the α or β receptor site. Beta receptor sites in turn can be further subdivided into β_1 and β_2, the former being present in the heart.

Two naturally occurring and powerful sympathomimetic agents are adrenaline and noradrenaline. These, of course, are the hormones/neurotransmitters which are released during sympathetic nervous activity and both have α and β receptor activity to different degrees. Adrenaline, with comparatively more β receptor activity, is of value in the relief of bronchospasm. It is also used in acute anaphylaxis, since it has the effect of counteracting histamine. It is occasionally used for its vasoconstrictor effects as an additive to some local anaesthetics, thus prolonging their effectiveness.

Noradrenaline, with its comparatively greater α receptor effects, is a very powerful vasoconstrictor agent; so powerful, indeed, that it must be administered by slow intravenous infusion. If it comes into contact with tissue other than the endothelial lining of blood vessels, it causes tissue necrosis. There are other agents which are effective in raising blood pressure and these may be preferred, e.g. metaraminol.

Dopamine is the naturally occurring precursor to noradrenaline (see Chap. 2.1). It is a very potent agent in increasing the force of the heart beat, thereby improving cardiac output. It also appears to act on dopaminergic receptors in the kidney to increase renal blood flow.

Other drugs have been developed for their more selective effects upon the different adrenergic receptors. Drugs having mainly α receptor effects raise blood pressure through their vasoconstrictor action. They are also effective in slowing the heart rate, so may be useful in tachycardia. Examples of such drugs are methoxamine and phenylephrine.

Beta-1 activity improves the force of the heart without affecting its rate. An example here is prenalterol. Beta-2 activity is effective in relaxing bronchial muscles in asthma. Drugs with β_2 action are ephedrine and salbutamol.

CENTRALLY ACTING DRUGS AFFECTING THE ADRENERGIC SYSTEM

These include monoamine oxidase inhibitors (MAOIs). Since the reuptake of noradrenaline at nerve endings is a more important method of removing its activity than enzymic degradation, drugs which interfere with the metabolism of amines have a greater effect centrally than peripherally since dopa, dopamine, and serotonin appear to be central neurotransmitters, and are all dependent upon monoamine oxidase for their degradation. The MAOIs (for example, phenelzine) are psychoactive drugs, having the effect of lifting depression. Patients who take them should avoid food and drink containing large amounts of amines, such as cheese and yeast products, since these cannot be properly metabolized.

SYMPATHETIC BLOCKING DRUGS

The action of adrenaline and noradrenaline can be blocked by drugs which prevent their access to receptor sites. Again the effect can be selective for particular types of receptor sites.

Alpha blocking agents

Drugs which block the α receptor effects of sympathetic stimulation have a vasodilator action. Tolazoline and phenoxybenzamine are examples of α blockers which are used to relieve peripheral vasospasm, as in Raynaud's disease. They may produce some tachycardia as a side-effect. Other α blocking drugs are used in hypertension to reduce blood pressure. Indoramin is an example; it has little or no effect upon the heart rate. Labetalol blocks both α and β receptors. It is a good hypotensive agent since the peripheral vascular dilation is not counteracted by increased cardiac output.

Beta blocking agents

These may be effective against both types of β receptor, or may be selective. Propranolol is an example of an agent which is effective against both β_1 and β_2 activity. It inhibits the normal response to cardiac stimulation and reduces the contractile force of heart muscle. It is used in angina, myocardial infarction and cardiac arrhythmia. In very low doses, it can act to prevent migraine.

Drugs which selectively block β_1 receptors are very useful in angina. Examples include acebutolol and atenolol.

Drugs which oppose the stimulative action of sympathetic nerves in other ways are frequently used as hypotensive drugs. One example is guanethidine. This both reduces the store of noradrenaline at nerve endings and also prevents its reuptake after release, to a degree.

Clonidine is a drug which reduces the release of noradrenaline from nerve endings and also has the central effect of reducing the flow of impulses to sympathetic nerves. The drug methyldopa interferes with the metabolic pathway through which noradrenaline is produced and this is the way in which it exerts its hypotensive effects.

Drugs affecting the parasympathetic division

In contrast to the sympathetic division, the naturally occurring parasympathetic postsynaptic neurotransmitter acetylcholine is broken down so rapidly by enzyme action (cholinesterase) that it is of no therapeutic value itself, and so similar substances less easily degraded are used as drugs.

Carbachol has most of the properties of acetylcholine but it is stable and effective orally. It has the effect of causing contraction of both bladder and intestinal muscle. It can be used to treat retention of urine and intestinal atony.

Natural cholinergic activity can be enhanced by the use of drugs which interfere with the activity of cholinesterases. Neostigmine is an example of such a drug, but it is used chiefly for its stimulating effect upon the voluntary motor end-plate, another site where acetylcholine is active. It can also be used occasionally for postoperative paralytic ileus and urinary retention.

ANTICHOLINERGIC DRUGS

Atropine is used widely for its effect in blocking the muscarinic activities of acetylcholine in the parasympathetic nervous system. It blocks the access of acetylcholine to the receptors and so reduces the tone of smooth muscle and gastrointestinal activity in general. It is particularly useful in the relief of renal, biliary and intestinal colic. Used preoperatively, it decreases bronchial and salivary secretions. As eyedrops it is used in ophthalmology to dilate the pupil. Atropine has a stimulant effect upon the CNS. Hyoscine is similar in its effects upon secretions and smooth muscles, but has a depressant effect upon the CNS. For this reason it may be preferred over atropine as a premedication.

THE AUTONOMIC NERVOUS SYSTEM AND STRESS

Stress can be defined in terms of physiology or in terms of psychology, although both of these aspects are necessary for a complete understanding of the concept. However, physiological aspects of stress have been studied over a longer period of time and this has resulted in a greater knowledge of their effects.

Although the word stress has come to have a generally negative connotation, it should be emphasized that stress is usually adaptive. We should particularly note Selye's view (1956) that complete freedom from stress equates with death.

The adaptive aspect of the physiological stress response can be seen by nurses in relation to **shock** following trauma, either of an accidental or a surgical nature. This response includes arousal of the sympathetic nervous system and the accompanying hormonal effects.

The sympathetic activity brings about peripheral vasoconstriction giving pallor; there is increased cardiac output due to an increased blood return to the heart; increased force of the cardiac muscle contraction and increased heart rate. Together these responses maintain the blood flow to vital organs and help to maintain the blood pressure. It is only if homeostasis is not restored and intense vasoconstriction continues that the response becomes non-adaptive and irreversible shock ensues. The sympathetic response forms a part (and an important part) of the **alarm** reaction, which is the initial phase of the **triphasic** stress response described by Selye (1956; see Chap. 2.5).

Hypoglycaemia is another condition in which the nurse can identify the effects of sympathetic nervous system activity. Indeed, these effects usually precede signs of CNS involvement. Therefore it is useful to be able to recognize signs of sympathetic activity. These are: pallor, piloerection, tachycardia, trembling and perspiration. Subjectively, the sufferer experiences nervousness, weakness, hunger, palpitations and irritability. Three of the four symptoms most commonly reported by patients are sympathetic in origin; these are nervousness, weakness and perspiration. The fourth symptom, mental confusion, is due to deprivation of glucose in the CNS.

Clearly, in clinical practice it is useful to identify the physiological components of stress. However, stress in a broader sense is also important in nursing since:

(a) nurses may themselves experience stress;
(b) patients frequently experience avoidable stress, from negative aspects of hospitalization or treatment;
(c) stress is implicated in the aetiology and maintenance of some disease states.

In relation to these more general aspects of stress, a psychological definition will be used

(Clarke, 1984). **Stress** can be said to arise out of an individual's appraisal of a mismatch between demands made on him or her and his or her ability to cope, where **demand** is an internal or external stimulus, perceived by the individual as requiring an adaptive response; and **coping** is a response carried out by the individual and appraised by him or her as either satisfactorily or unsatisfactorily affecting the demand in the desired direction.

The difficulty for humans is that whilst stress arises most commonly from psychological demands, nonetheless a physiological response occurs, even though it is maladaptive in conditions where energetic physical activity is inappropriate. A response which occurs alongside the action of the sympathetic nervous system which has not been mentioned so far, is tensing of the voluntary muscles – clearly, a crucial aspect of a stereotyped 'fight or flight' response.

Thus 'maladaptive' aspects of stress include:

(a) pain due to tensed muscles, e.g. headache due to tensed neck muscles, tooth-grinding, tensed jaw, etc;
(b) increased blood pressure due to vasoconstriction and increased cardiac output;
(c) cardiac arrhythmias due to adrenaline/noradrenaline activity;
(d) increased levels of blood glucose due to gly-coneogenesis;
(e) increased lipolysis leading to increased blood free fatty acid levels.

If such responses occur frequently or are prolonged, it is believed that they may contribute to:

hypertension
myocardial infarction
diabetes mellitus of mature onset.

Since the short- and long-term effects of severe stress can be so negative, methods of helping people to cope have been developed. These include the development of improved problem-solving techniques; giving information which allows anticipatory coping; and counselling. Two techniques which are used frequently and appear to have effects upon the autonomic nervous system are relaxation techniques and biofeedback techniques.

Relaxation techniques, of which there are many, include two elements: peripheral muscular relaxation and more central relaxation. Some authorities believe that the techniques used to 'switch' into central relaxation trigger the para-sympathetic nervous system into activity. There is little or no evidence for this, however, and since the system is less diffuse in its activity, it is more likely, logically, that such techniques act to damp the sympathetic nervous system.

Biofeedback techniques rely upon the observation that it is possible, by feeding back information to the individual, to train, by conditioning, such activities as the ability to lower one's own heart rate or blood pressure or to increase α wave activity on the EEG, for example. Clearly these are useful techniques, but it is possible that they do not occur beyond the training environment.

Stress and the parasympathetic nervous system

At least one stress disease (peptic ulceration) is related to activity of the parasympathetic nervous system. In patients with a gastrostomy, hyperaemia and hypermotility of the stomach have been observed in emotional states.

Excessive secretion of highly acid gastric juice and increased gastric motility are factors contributing to peptic ulceration in individuals with stressful occupations. Stress ulcers can occur in patients who have been severely burned or who have severe neurosurgical conditions. Although vagotomy may be performed to relieve excessive gastric secretion, anticholinergic drugs with specific actions are available for treatment which avoids the drastic effects of major surgery.

The fact that both the sympathetic and para-sympathetic divisions of the ANS are implicated in stress diseases underlines the complexity of the interactions of the two divisions. Our understanding of the ANS is still far from complete, and more research is needed, both of a basic and of an applied nature.

Review questions

The answers to all these questions can be found in the text. In each case there is at least one correct and at least one incorrect answer.

1 The autonomic nervous system is defined as

(a) part of the central nervous system
(b) an efferent system

(c) an afferent system
(d) part of the peripheral nervous system

2 Postganglionic nerve endings in the sympathetic nervous system are

(a) entirely adrenergic
(b) entirely cholinergic
(c) a mixture of cholinergic and adrenergic
(d) nicotinic in type

3 The vagus nerve is of crucial importance because

(a) it controls blood pressure
(b) it dilates bronchial muscle
(c) it stimulates the action of the heart
(d) it participates in the control of the heart rate

4 Noradrenaline

(a) is only found in nerve endings
(b) is rapidly destroyed by enzymes
(c) has only β receptor activity
(d) is a neurotransmitter

5 Action of the sympathetic nervous system

(a) increases intestinal motility
(b) contracts sphincter muscles
(c) contracts smooth muscle in blood vessels
(d) constricts the pupil

6 Action of the parasympathetic nervous system

(a) relaxes the ciliary muscle of the eye
(b) stimulates secretion from the salivary glands
(c) causes relaxation of the gall bladder
(d) causes glycogenolysis

7 Action of the sympathetic nervous system in stress

(a) is adaptive
(b) is non-adaptive
(c) is part of the alarm reaction
(d) is inappropriate to life today

8 The sympathetic and parasympathetic nervous systems

(a) are antagonistic to one another

(b) have one division always active when the other is inactive
(c) have actions which reinforce one another
(d) interact together in a complex way

9 Study of the autonomic nervous system is useful

(a) because it helps our understanding of stress
(b) because its action underlies homeostasis
(c) because it helps our understanding of the way in which drugs work
(d) because it helps us to understand somatic control

Answers to review questions

1 b and d
2 c
3 d
4 d
5 b and c
6 b
7 c
8 d
9 a, b and c

Short answer and essay topics

1 Using knowledge of the anatomy and neurotransmitters of the ANS, discuss three reasons why the action of the sympathetic division is so diffuse in nature.
2 Discuss the action of the parasympathetic fibres contained in the vagus nerve.
3 Explain the different ways in which drugs could in theory have an effect on the ANS in terms of their interference with normal neurotransmitter function.
4 Discuss the role of the ANS in the act of micturition.

Suggestions for practical work

1 Interview a patient who has had a premedication of atropine or hyoscine. (Delay the interview until the patient has recovered from the procedure.) Ask what the patient's subjective

feelings were whilst the drug was working. List these.

2 Think of the way you yourself felt prior to an experience which you classified as stressful. Possible examples are:

before a crucial A level examination

before your interview for a place on a nursing course

the morning of your first day on the ward.

List the feelings you experienced. If possible, also identify how you might have appeared to an onlooker. For each item on your list, identify the probable controlling system involved.

3 Carry out a relaxation exercise as follows. Sit in a comfortable chair. Consciously relax all your muscles, including facial muscles. Take deep, slow breaths. Close your eyes and think of a pleasant, relaxing experience. Possible examples are:

listening to soft music

relaxing after a pleasant meal

lying in bed on a morning when you don't have to go to work.

Continue this relaxation for at least 10 minutes. List the way you felt during this exercise and, if possible, how you might have looked to an onlooker. For each item on your list, state the probable controlling system involved. NB. Instead of sitting in a chair for this exercise, you may prefer to lie on your back on the floor with a small cushion under your head. You should flex your knees to ensure that the small of your back is supported on the floor.

4 Identify the signs of anxiety (the subjective experience frequently associated with stress) during the admission of a patient to the ward. List these signs.

5 Interview a patient who is anxious. In the interview, explore his/her feelings and what is the cause of those feelings, i.e. what makes him/her feel anxious. At the end of the interview, assess signs of anxiety. List the subjective feelings of the patient. State whether, from the objective signs, you believe that talking has helped the patient or not.

6 Identify the signs and symptoms displayed by a patient in shock. Write down and identify the underlying physiological mechanism for each.

References

Cannon, W. B. (1932) *The Wisdom of the Body*. New York: Appleton.

Clarke, M. (1984) Stress and coping: constructs for nursing. *Journal of Advanced Nursing* 9 (1): 3–14.

McCaffery, M. (1979) *Nursing Management of the Patient with Pain*, 2nd edn. Philadelphia: Lippincott.

Selye, H. (1956) *The Stress of Life*. New York: McGraw-Hill.

Sherrington, C. S. (1947) *The Integrative Action of the Nervous System*, 2nd edn. New Haven, Conn.: Yale University Press.

Walton, J. (1987) *Introduction to Clinical Neurosciences*, 2nd edn. London: Baillière Tindall.

Suggestions for further reading

Burgen, A. S. V. & Mitchell, J. F. (1978) *Gaddum's Pharmacology*, 8th edn. London: Oxford University Press.

Burt, W. L., Clements, M. L., Hilyard, D., Lower, J. S., Van Riper, J., Van Riper, S. G. & Wilson, R. F. (1984) *Shock. Nursing New Series*.Springhouse, Pa.: Springhouse.

Clarke, M. & Montague, S. E. (eds.) (1980) Stress. *Nursing Series 1, 10*: Oxford: Medical Education Int.

Cox, T. (1978) *Stress*. London: Macmillan.

Ganong, W. (1979) *Review of Medical Physiology*, 9th edn. Los Altos, Calif: Lange.

Chapter 2.5
Endocrine Function

Jennifer R. P. Boore

Learning objectives

After studying this chapter the reader should be able to

1 Describe the general organization of the endocrine system and the relationship between the nervous and endocrine systems.
2 Describe the two main groups of hormones and discuss the methods of action of hormones on target cells.
3 Discuss the secretion, transport, metabolism and excretion of hormones and indicate how these factors affect the timescale of the activity of different hormones.
4 Discuss the role of the thyroid hormones in the control of metabolic rate and contrast the effects of secretion of inadequate and excessive amounts of these hormones.
5 Describe the metabolic changes which take place to maintain the blood glucose level within normal limits and the role of hormones in controlling these changes.
6 Describe the endocrine regulation of fluid and electrolyte balance.
7 Indicate the roles played by calcium in the body and discuss the way in which the blood calcium level is controlled.
8 Discuss the involvement of the endocrine system in coordinated growth and development.
9 Describe the body's hormonal response to environmental stressors and the importance of this in adaptation.

Introduction

The endocrine and nervous systems are the two controlling systems of the body. They function in different ways, but there is some interaction between them and their activities complement each other. The nervous system provides a rapid response to the information it receives, and the fast transmission of nervous impulses modulates the activity of the muscle and secretory cells innervated. In contrast, the endocrine system functions more slowly as cell function is modified and body metabolism controlled.

The endocrine, or hormonal, system consists of a number of endocrine (ductless) glands situated in different parts of the body (Table 2.5.1). These glands secrete hormones directly into the bloodstream, in contrast to the exocrine glands which pass their secretions directly into a body cavity or onto the surface of the skin. The hormones released are carried in the bloodstream until they reach their target organ, where they are active.

ORGANIZATION OF THE ENDOCRINE SYSTEM

There are a considerable number of organs which have an endocrine function, and not all of these will be discussed in detail in this chapter as some are dealt with elsewhere. For example, the hormones released from and regulating the activity of the gastrointestinal tract are discussed in

Table 2.5.1 Major endocrine glands of the body and positions

Hypothalamus	Base of brain
Pituitary gland (hypophysis)	
Posterior pituitary gland (neurohypophysis)	Pituitary fossa of cranium
Anterior pituitary gland (adenohypophysis)	
Thyroid gland	Neck, below larynx
Parathyroid glands (4)	Posterior surface of thyroid gland
Adrenal glands (2)	Superior pole of kidneys
Adrenal cortex	Outer layers of adrenal gland
Adrenal medulla	Inner core of adrenal gland
Islets of Langerhans	Embedded in pancreas
Ovaries (2) (female)	Pelvic cavity
Testes (2) (male)	Scrotum
Trophoblast (followed by placenta)	Pregnant uterus
Pineal gland	Cranial cavity, behind midbrain
Thymus	Mediastinum
Kidneys	Retroperitoneal position
Gastric and intestinal mucosa	Gut wall
Skin	Areas exposed to sunlight

Chap. 5.1. The activity of a number of endocrine glands is regulated through the activity of the hypothalamus in the brain, while others respond to the concentration of particular constituents of the extracellular fluid.

The hypothalamus

The hypothalamus plays a major role in the maintenance of the internal environment in several ways. These are summarized in Fig. 2.5.1. The hypothalamus has nervous connections with many parts of the brain, particularly the limbic system. The limbic system consists of a number of deep structures in the brain and a rim of cortical tissue around the hilus (point of attachment to the midbrain) of the cerebral hemisphere. It is particularily involved in emotions, both the development and the expression of them, and while there are few nervous connections with the cerebral cortex, cerebral activity influences the limbic system, and vice versa. Through the connections

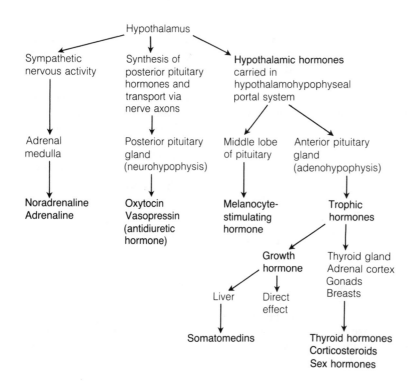

Figure 2.5.1 The role of the hypothalamus in endocrine regulation

Table 2.5.2 Hypothalamic hormones influencing pituitary secretion, and organs affected

Hypothalamus	Anterior pituitary	Tissues affected	Action
Corticotrophin-releasing hormone or factor (CRH or CRF)	Adrenocorticotrophic hormone (ACTH) ↑	Adrenal cortex	Glucocorticoid secretion ↑ Aldosterone secretion ↑ temporarily
Thyrotrophin-releasing hormone (TRH)	Thyroid-stimulating hormone (TSH) ↑	Thyroid gland	Thyroid hormones secretion ↑
Growth hormone-releasing hormone (GRH)	Growth hormone (GH) ↑	Most body tissues, particularly liver	Somatomedin secretion ↑ Growth increased
Growth hormone release-inhibiting hormone (GI)	Growth hormone ↓		
Prolactin-releasing hormone (PRH)	Prolactin (luteotrophic hormone, LH) ↑	Breasts	Development of breasts
Prolactin release-inhibiting hormone (PIH)	Prolactin ↓		Secretion of milk
Luteinizing hormone/follicle-stimulating hormone-releasing hormone (LHRH/FRH) *or* Gonadotrophin-releasing hormone (GnRH) } *	Luteinizing hormone (LH)/interstitial cell-stimulating hormone (ICSH) ↑	Ovaries or testes	Ovulation, development of corpus luteum Testosterone secretion (males) ↑
	Follicle stimulating hormone (FSH) ↑	Ovaries Testes	Ovarian follicle growth ↑ Spermatogenesis ↑
	Middle lobe of pituitary		
Melanocyte-stimulating hormone releasing hormone (MSHRH)	Melanocyte-stimulating hormone (MSH) ↑	Skin	Deposition of melanin by melanocytes
Melanocyte-stimulating hormone release-inhibiting hormone (MSHRIH)	Melanocyte-stimulating hormone (MSH) ↓		

*It is uncertain whether there are one or two releasing hormones for the luteinizing and follicle-stimulating hormones

between the limbic system and the hypothalamus, the activity of the endocrine system is also modulated.

As Fig. 2.5.1 indicates, the hypothalamic hormones play a major role in the control of endocrine activity. These hormones are formed in the median eminence of the hypothalamus (the area connected to the pituitary gland by the pituitary stalk), absorbed into the capillary loops in the area, and carried in the venous circulation (in the hypothalamohypophyseal portal system) to the anterior pituitary gland and middle lobe of the pituitary. Nine different substances have been definitely identified as hypothalamic hormones, which either stimulate or inhibit the release of anterior and middle pituitary hormones (Table 2.5.2). Table 2.5.2 also shows the hormones secreted in response to these hypothalamic hormones, and the influence of the pituitary hormones on the tissues or organs affected.

The pituitary gland

The pituitary gland also plays a major and diverse role in endocrine regulation. In embryonic development this gland develops from the merging of different tissues. The anterior part originates from an upgrowth of the primitive mouth and is known as the **adenohypophysis**, while the posterior pituitary is derived from the base of the brain and is known as the **neurohypophysis** (Fig. 2.5.2). The posterior pituitary still has nervous connections with the hypothalamus of the brain, down which antidiuretic hormone and oxytocin pass after they are synthesized in the hypothalamus. The anterior pituitary, on the other hand, is only linked to the brain via the venous blood in the hypothalamohypophyseal portal system. Disorders of this gland can have extremely varied results. Destruction of the anterior pituitary gland will result in hypopituitarism, when changes due to the lack of trophic hormones

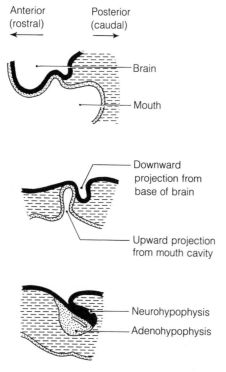

Anterior
(rostral)

Posterior
(caudal)

— Brain

— Mouth

— Downward
projection from
base of brain

— Upward projection
from mouth cavity

— Neurohypophysis
— Adenohypophysis

Figure 2.5.2   Development of the pituitary gland (adapted from Passmore, R. & Robson, J. S. (1976) *Companion to Medical Studies* 2nd edn. London: Blackwell Scientific.

develop (Macleod, 1984). This condition may sometimes be due to failure of the hypothalamus to secrete normally the hypothalamic releasing hormones. Tumours of the pituitary gland may cause increased secretion of hormones, but more commonly cause reduced secretion, or sometimes a combination of an increase of some hormones and a decrease in secretion of others.

Hormone secretion in response to changes in extracellular fluid

The hormones thus secreted are shown in Table 2.5.3, with some indication of their role in the control of body function.

Endocrine regulation of metabolism

This is summarized in Table 2.5.4.

Hormone structure

Hormones fall into two main groups: one group comprises the steroid hormones which are formed from cholesterol. Figure 2.5.3 shows the structure of cholesterol and some of the steroid hormones secreted by the body. As can be seen, they are very similar in structure, with only small chemical differences leading to the varied actions of the different hormones. The similarities of structure in these hormones explain the overlap in activity that occurs between the different steroid hormones. This group includes all the hormones of the adrenal cortex, the glucocorticoids, mineralocorticoids and sex hormones, and the sex hormones secreted by the ovaries and testes.

The second group of hormones is that derived from amino acids and includes single amino acid derivatives, polypeptides (short chains of amino acids), and proteins formed from a relatively long chain of amino acids.

Methods of action of hormones

Hormones vary in the way they act on their target

Table 2.5.3   Endocrine function directly controlled by the condition of the internal environment (i.e. the extracellular fluid)

Endocrine organ	Hormone secreted	Regulation of release	Activity
Pancreas B-cells A-cells	Insulin Glucagon	↑ in blood glucose Inhibited by ↑ in blood glucose }	Regulation of blood glucose level
Parathyroid glands (4) Kidneys Thyroid parafollicular cells	Parathormone 1,25-dihydroxycholecalciferol Calcitonin	↑ when blood calcium low ↑ by parathormone ↑ when blood calcium high }	Regulation of plasma concentration of calcium ions (Ca^{2+})
Adrenal cortex	Aldosterone	↑ when low serum sodium ↑ when high serum potassium ↑ by action of renin from kidney }	Regulation of sodium and potassium concentration in plasma
Kidneys	Erythropoetin	↑ by hypoxia	Production and release of erythrocytes

Table 2.5.4 Endocrine glands involved in aspects of body function

Control of metabolic rate	Thyroid gland, adrenal medulla
Control of glucose metabolism	Pancreas, adrenal cortex, anterior pituitary, adrenal medulla
Growth and development	Anterior pituitary, thyroid, gonads, adrenal cortex, pancreas, skin, kidneys
Fluid and electrolyte balance	Posterior pituitary, adrenal cortex, kidneys
Calcium and phosphorus balance	Parathyroid glands, thyroid, skin, kidneys
Response to environment	Adrenal cortex, adrenal medulla, thymus
Reproduction and nurturing (See Chap. 6.3)	Anterior pituitary, ovaries, testes, placenta, ?pineal gland

(For the hormones involved see relevant sections)

organs, with some having a very rapid action and others taking a considerable time to have any noticeable effect. A number of methods have been identified whereby hormones cause an alteration in the activity of specific cells. The metabolism of the cell is modified largely through regulation of enzyme activity, which can occur in two main ways: either by increasing the amount of enzyme present, or by regulating the level of activity of already formed enzyme. In addition, cell activity can be altered by control of entry of substances into the cell.

Two of the methods of hormone action identified act by increasing the amount of enzyme present. Some hormones, such as steroids and thyroxine, act by increasing the synthesis in the nucleus of messenger ribonucleic acid (m-RNA), which then passes into the cytoplasm of the cell and acts as a template for the formation of a specific enzyme. The synthesis of that enzyme is thus increased. A considerable period of time (hours or even days) of exposure to the hormone may be needed before any effect can be detected. The second method of increasing enzyme synthesis is

Cholesterol
Showing numbering of the carbon atoms forming the molecule, to allow the accurate identification of the chemical differences between molecules

Oestrone
One of the oestrogens

Progesterone

The female sex hormones

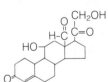

Cortisol
One of the glucocorticoid hormones

Aldosterone
The main mineralocorticoid hormone

Testosterone
One of the androgens – a male sex hormone

Figure 2.5.3 Cholesterol and some of the steroid hormones

by stimulation at the ribosome in the cell cyto- plasm, where the enzyme is formed against the m-RNA template.

The other two methods of hormone action appear to occur through the hormone binding to specific receptors in the cell membrane. Some hormones, e.g. insulin and the catecholamines (adrenaline and noradrenaline), alter the transport of substances across the cell membrane into the cell by direct binding to these receptors. Finally, also through receptor binding, hormones may act by regulating the level of **cyclic-adenosine monophosphate (cAMP)** in the cell. Cyclic AMP is a fairly small molecule, synthesized from adenosine triphosphate which activates a large variety of enzymes which regulate a number of metabolic pathways. The level of cAMP may be increased or decreased by the action of hormones, depending on the tissue involved.

An individual hormone may alter cell activity by more than one of the above methods. For example, in muscle and adipose tissue, insulin functions by increasing the transport of glucose across the cell membrane. However, in the liver some of the actions of insulin are mediated through a change in cAMP levels, while others seem to be due to an increase in synthesis of enzymes involved in glucose metabolism.

THE ROLE OF CALCIUM IN HORMONE SECRETION AND FUNCTION

Calcium plays an important role in hormone secretion and function. It is needed for the secre- tion of almost all hormones stored in granules within the cell of formation. Most of the protein hormone actions are inhibited in the absence of calcium, even though regulation of cAMP levels is unaffected. It appears that cAMP may act through its effect on the level of calcium in the cytoplasm of the cell. Some hormones appear to increase the entry of calcium into the cell, while cAMP seems to cause the release of calcium from a binding protein, **calmodulin**, within the cell.

PROSTAGLANDINS

Prostaglandins may also influence cell activity through modification of cAMP levels. These are substances formed from the essential fatty acids, particularly arachidonic acid. They have been found in virtually all tissues and are formed from

their precursors in the cell membrane. They are synthesized and released as necessary, and are rapidly broken down in many tissues, particularly the lungs, spleen and kidneys. There are a number of different groups of prostaglandins (PGs) with varying actions on the different types of cell. The most stable are PGD_2, PGE_2, and $PGF_{2\alpha}$, but a number of less stable prostaglandins are also active in modifying cell function. A number of factors, such as hypoxia, trauma, catecholamines, can alter the pattern of prostaglandin formation, which may be unique in a particular tissue.

The way in which PGs function is still not entirely clear. It has been suggested that the ratio of cAMP to cGMP (cyclic guanosine mono- phosphate – a molecule similar to cAMP, but with the chemical grouping of adenine replaced by guanine) controls cell function. In several tissues one PG raises the level of cAMP, while a different PG increases the level of cGMP; thus variation in PG synthesis will adjust the ratio between the two substances. Further work is being carried out on prostaglandins in an attempt to elucidate their modes of action. It seems unlikely that they act as hormones and are carried to distant organs, but they may act as messengers between adjacent cells, or modulate intracellular activity.

Regulation of hormone activity

Hormone action on a particular target organ or tissue depends on the level of free hormone in the blood, which is determined by the balance between three main factors.

1 The rate of synthesis and secretion of the hor- mone from the endocrine gland involved.
2 Transport of the hormone and the degree of binding to carrier proteins in the plasma.
3 Metabolism and excretion of the hormone from the body.

REGULATION OF SYNTHESIS AND SECRETION

Regulation of hormone release, involving both synthesis and secretion of the substance, is mainly through **negative feedback** or **feedback inhi- bition** of the endocrine gland involved. This sys- tem of control endeavours to keep the level of the substance being regulated within close limits,

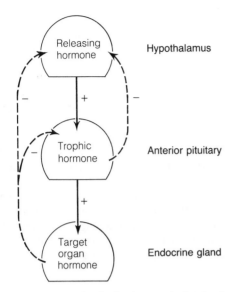

Figure 2.5.4 Negative feedback control of endocrine systems involving the hypothalamus, anterior pituitary and target endocrine gland

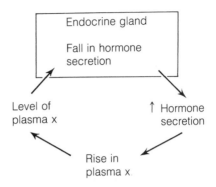

Figure 2.5.5 Endocrine regulation of plasma constituents

although there will be some vacillation within those limits. However, through the action of the hypothalamus and its links with other parts of the nervous system, the endocrine system can also respond to environmental and emotional change.

Those hormones released through the action of the hypothalamus and anterior pituitary gland act on one or both of these organs to reduce the secretion of the appropriate hypothalamic and pituitary hormones. The pituitary hormone may also act on the hypothalamus to regulate the secretion of the hypothalamic-releasing hormone (Fig. 2.5.4). Where a disorder of an endocrine gland prevents secretion of the hormone, the levels of the trophic hormone involved in its release are not inhibited and may become very high. In Addison's disease, due to atrophy of the adrenal cortex, corticosteroids are not secreted and cannot, therefore, inhibit the secretion of adrenocorticotrophic hormone (ACTH). Progressive pigmentation of the skin develops, due to an increase in secretion of melanocyte-stimulating hormone which occurs simultaneously with the rise in ACTH secretion, and due to the action of ACTH itself (Macleod, 1984). When hormones are released in direct response to a constituent of the extracellular fluid, the level of that substance controls secretion of the hormone through negative feedback, as shown in Fig. 2.5.5.

TRANSPORT AND PROTEIN BINDING

Hormones are secreted into the capillaries passing through an endocrine gland and transported through the circulation. The circulation time taken for a bitter substance injected into an arm vein to reach the tongue is about 15 seconds, which gives an indication of the speed with which hormones can reach the various organs of the body. The free hormone in the blood is available to act on the target organ, but is also exposed to metabolizing enzymes which degrade it into an inactive form.

The **half-life** of a hormone is the time taken for half of it to be removed from the plasma (for details see below). A number of hormones, such as the catecholamines and insulin, have a very short half-life, and these tend to be the hormones which are transported through the circulation in the free state. On the other hand, some hormones, e.g. thyroxine and the corticosteroids, are transported in the plasma bound to specific proteins and these hormones have a longer half-life. The bound hormone is in a state of equilibrium with the free hormone:

$$\text{free hormone} + \text{protein} \rightleftharpoons \text{hormone–protein}$$

As the free hormone is used up, more is released from the protein-bound state.

METABOLISM AND EXCRETION

The major organs involved in metabolism and excretion of hormones are the liver and kidneys. Some hormones are inactivated by the tissues on which they act, but the liver plays the major role in hormone degradation. Protein-bound hor mones are protected from destruction by liver enzymes.

Enzymes in the endoplasmic reticulum of the liver cells catalyse a number of chemical reactions which reduce the endocrine activity of the hormones, and, by making them water soluble, enable them to be more readily excreted from the body. Some hormones are conjugated (combined) with other chemicals (mainly glucuronic acid or sulphuric acid) before excretion and this also increases the water solubility of the hormone molecules. Excretion takes place in the bile from the liver and in the urine through the kidneys. Patients with liver disease may be unable to metabolize hormones normally and will show signs of high levels of particular hormones. For example, a male patient with cirrhosis of the liver may develop enlargement of the breasts (gynaecomastia) due to high levels of oestrogens (female sex hormones) in his blood (Macleod, 1984; see also Chap. 5.2).

CONTROL OF METABOLIC RATE

Body metabolism consists of all the chemical reactions taking place in the body cells. It includes catabolism and anabolism. In **catabolic reactions**, carbohydrates, proteins and fats are broken down through oxidation into smaller molecules, mainly to water and carbon dioxide, and energy is produced. The energy is released in small amounts which is stored within the body in high-energy phosphate compounds, the major one of which is adenosine triphosphate (ATP). Some of the energy, however, is liberated in the form of heat. In **anabolic reactions**, energy is required for the formation of more complex molecules.

During catabolism the amount of energy liberated is the same as the amount liberated if food is burnt outside the body. The energy is used for work, is stored in energy-rich compounds, and is released as heat. The metabolic rate is the amount of energy released in a given time.

The SI units of energy now used are the joule (J), the kilojoule (kJ = 1000 J, or 10^3 J), or the megajoule (mJ = 1 000 000 J, or 10^6 J). The older unit, still sometimes used clinically and almost invariably by the general public, is the calorie (the heat required to raise 1 g of water through 1 °C), or the kilocalorie (kcal), usually known in nutrition as the Calorie (1 Cal = 1000 calories). One Calorie equals 4.2 kJ.

Energy is released during the breakdown of nutrients, and oxygen is required for these reactions to occur with maximum efficiency. While the energy released from the different types of nutrients varies somewhat, the amount of energy released per litre of oxygen utilized is approximately 20 kJ. Therefore, by measuring the quantity of oxygen used in a given time, it is possible to calculate with an acceptable degree of accuracy the amount of energy released, i.e. the **metabolic rate**. The **basal metabolic rate** is the rate at which energy is used in the body while the person is awake, but is at complete rest. In disease conditions when the metabolic rate is altered, such as in thyrotoxicosis or myxoedema, it is expressed as plus or minus a percentage from normal for an individual of that age, sex and size.

Factors influencing metabolic rate

A number of factors influence the metabolic rate and these are shown in Table 2.5.5. Hormones are also involved, particularly the thyroid hormones, thyroxine and tri-iodothyronine, but also the hormones released from the adrenal medulla with sympathetic nervous system stimulation (see Chap. 2.4 and p. 192), which cause a rise in the metabolic rate.

Thyroid hormones markedly affect the metabolic rate of the body, as well as being essential for normal growth and development. The metabolic rate may be as low as -40% to -50% when there is a complete lack of thyroid secretion, or as

Table 2.5.5 Factors influencing metabolic rate

Gender	Females lower than males
Age	Declines with increasing age
Emotional state	Anxiety → ↑ adrenaline secretion → ↑ metabolic rate
	Depression → ↓ metabolic rate
Pregnancy or menstruation	↑ Metabolic rate
Muscular exertion	↑ Metabolic rate
Ingestion of food	Specific dynamic action of food → ↑ metabolic rate (protein greater effect than carbohydrates and fat) Starvation → ↓ metabolic rate
Environmental temperature	Cold → ↑ metabolic rate Heat → ↓ metabolic rate until so hot that body temperature rises
Fever	Rise of 1°C → ↑ metabolic rate of 14%
Height, weight, surface area	Allowed for in calculation of metabolic rate

high as $+60\%$ or $+100\%$ when excessive thyroid hormones are secreted. Metabolic activity of the tissues of the body is maintained at the optimal level by regulated secretion of the thyroid hormones.

Function of thyroid hormones

As already indicated, the thyroid hormones, thyroxine and tri-iodothyronine, play a major role in the regulation of the metabolic rate of all body tissues. They increase the resting metabolic rate and thus increase oxygen consumption and heat production.

The role of thyroid hormones in growth and development is discussed later. The thyroid hormones act within the cell by binding to receptors within the nucleus. This is followed by the increased synthesis of m-RNA, leading to increased production of cell enzymes. The rise in oxygen consumption is due to the increase in respiratory enzymes within the mitochondria of the cell. This mode of action explains the considerable lag period before any observable effect when administering thyroid hormones to hypothyroid patients. Some effects will probably be seen within a week, but the full benefit of treatment may take several months to develop. At very high concentrations of thyroid hormones, protein synthesis may be increased, but so is protein breakdown for use as an energy source; negative nitrogen balance occurs and the nitrogen released is excreted from the body.

The effects of thyroid hormones were mainly identified by studying conditions in which secretion of these hormones is disturbed. Myxoedema (hypothyroidism) develops when there is a reduction in thyroid secretion and is five times more common among elderly women than men (Hodkinson and Irvine, 1985). Because it develops insidiously, it may not be recognized by those in frequent contact with the person involved. Hyperthyroidism (thyrotoxicosis) implies an increase in thyroid secretion and is also more common in women than in men.

HYPERTHYROIDISM

Metabolic changes
As virtually all body tissues are affected by thyroid hormones, an alteration in the level of secretion, whatever the cause, influences the activity of virtually all body systems. In hyperthyroidism, the metabolic rate is increased and much of the additional energy liberated is released in the form of heat, rather than stored in ATP. The sufferer, therefore, feels warm even in cold conditions and the body temperature may be raised. The increased metabolism causes increased utilization of oxygen and an increase in the amount of carbon dioxide formed. The rise in carbon dioxide produced stimulates an increase in the rate and depth of respiration. There is also an increased demand for nutrients, with an increased appetite and rate of absorption of these nutrients, but the patient still tends to lose weight. The nurse should encourage the patient to eat a high protein, high carbohydrate diet whilst remembering that an increase in the motility and secretion of the gastrointestinal tract may result in diarrhoea. Activity may be more vigorous, but eventually excessive protein breakdown causes muscle weakness and fatigue, and muscle tremor is commonly seen when the patient holds out his or her hand.

Cardiovascular disturbances
The changes described result in an increased demand on the cardiovascular system to supply the nutrients and oxygen required and to remove the increased waste products of metabolism. The rise in pulse and blood pressure which occurs indicates the resultant increase in cardiac output. Tachycardia persisting during sleep is one of the earliest signs of thyrotoxicosis, but the nurse may have difficulty in measuring a sleeping pulse as these patients wake from sleep very easily. Cardiac arrhythmias may develop, particularly in the elderly, and thyrotoxicosis is one of the three commonest causes of atrial fibrillation (Macleod, 1984). The nurse should plan care in order to encourage rest to reduce the demand on the cardiovascular system; but as these patients have considerable difficulty in resting, nursing measures alone may be insufficient to ensure relaxation and so sedative drugs may be necessary.

Nervous and reproductive disturbances
The effect of excessive thyroid secretion on the nervous system is particularly marked. Patients often demonstrate increased rapidity of thought, tend to be irritable, nervous and agitated, and may show exaggerated and purposeless movements (hyperkinesis). They may also show symptoms of

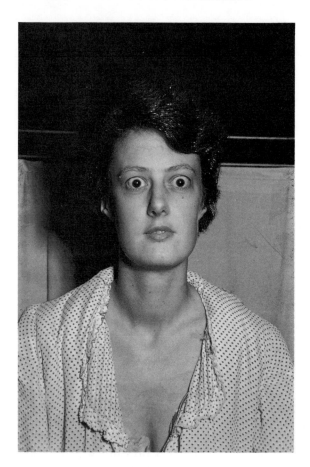

Figure 2.5.6 A person with Graves' disease

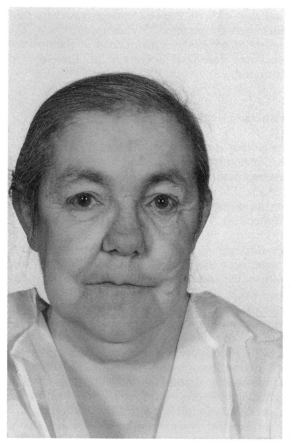

Figure 2.5.7 A person with myxoedema

an anxiety state and suffer from insomnia. Menstrual irregularities sometimes occur and the patient may become infertile.

Graves' disease (Fig. 2.5.6)
Graves' disease is one form of hyperthyroidism and is an autoimmune disease in which the patient forms antibodies against the thyroid-stimulating hormone (TSH) receptors in the thyroid cells. These antibodies also stimulate activity of the thyroid cells and are known as **thyroid-stimulating immunoglobulins**. In this condition, **exophthalmos** (protrusion of the eyeballs) also develops.

Treatment
Treatment of hyperthyroidism can be by the use of antithyroid drugs, by surgery or by the use of radioactive iodine.

MYXOEDEMA (Fig. 2.5.7)

The patient with myxoedema (reduced thyroid activity) demonstrates the opposite signs. The individual may feel cold and may even develop hypothermia as the body metabolism is no longer producing enough heat to maintain the temperature within normal limits. The patient's mental processes become slowed and most other body functions become sluggish. This is seen in slow, lethargic movements, slow pulse and respiratory rates, and a tendency to become constipated (a diet high in fibre should be encouraged). The skin becomes thickened and coarse, hair becomes thin and the patient puts on weight. However, in the adult, changes due to lack of thyroid hormones are readily reversed with the administration of exogenous hormones.

Formation and release of thyroid hormones

The thyroid gland is in the form of two lobes lying either side of the trachea just below the larynx; the lobes are connected by a bridge, called the **isthmus**, passing in front of the trachea. The gland is composed of many follicles lined with epithelial cells and filled with colloid, a substance secreted by the lining epithelial cells. The amount of colloid present varies with the degree of activity of the gland, being reduced when large amounts of thyroid hormones are released.

The epithelial cells of the thyroid secrete a complex protein molecule, **thyroglobulin**, into the colloid within the follicles of the gland. The amino acid tyrosine is one of the component parts of this molecule and it is this amino acid which is the basis for the formation of **thyroxine (T_4)** and **tri-iodothyronine (T_3)**, the thyroid hormones. Each molecule of hormone is formed from two molecules of tyrosine, with added iodine. Thyroxine contains four atoms of iodine, whereas tri-iodothyronine only contains three atoms of iodine per molecule of hormone. About 80 μg of T_4 and up to 40 μg of T_3 are secreted from the thyroid gland daily. It is now suggested that thyroxine is a form of prohormone as much of it is converted to tri-iodothyronine in the tissues, and it is this which is available to the tissues and is active at the cellular level. The thyroid hormones are stored in the form of thyroglobulin in the follicles of the gland for several weeks before release into the bloodstream.

The iodine required for the formation of the thyroid hormones is obtained in the diet, approximately 1.2 mmol being required daily, and is carried in the bloodstream. The cells of the thyroid gland have a particular affinity for iodine and are able to transfer iodine from the blood into the cells, by means of the iodide pump, and concentrate it in the follicles of the gland in combination with thyroglobulin. This ability to concentrate iodine is used in treatment of Carcinoma of the thyroid gland or hyperthyroidism with radioactive iodine, which is similarly concentrated in the thyroid gland, and thus the effect is localized.

Transport mechanisms within the thyroid cells engulf thyroglobulin from the colloid. Enzymes split the thyroid hormones off the large protein molecule and they are released into the blood capillary for transport around the body.

Transport and breakdown of thyroid hormones

On entering the blood, both hormones combine with plasma proteins, particularly thyroxine-binding globulin, and can be measured as protein-bound iodine. Thyroxine has a greater affinity for plasma proteins than has tri-iodothyronine. Only about 0.05% of thyroxine and 0.3% of tri-iodothyronine are in the free form in the plasma. However, the amount of T_4 secreted daily is greater than that of T_3, and the concentrations free in the plasma are 75–150 mmol/l of T_4 and 1.1–2.2 mmol/l of T_3. T_3 has a half-life of 24 hours and T_4 a half-life of 6 days. It is only the free hormone which is available for its action on the tissues, and for metabolism and excretion.

The thyroid hormones are broken down in many tissues and the iodine returned to the bloodstream. Some of this is then taken up again by the thyroid gland and re-used, while some is excreted in the urine. Thyroid hormones are conjugated in the liver and excreted in the bile, although some may be reabsorbed from the gut and then excreted by the kidneys.

Regulation of thyroid secretion (Fig. 2.5.8)

In order to maintain the metabolic rate within normal limits, it is essential that the level of thyroid hormones available to the tissues is precisely controlled, and this is achieved mainly through negative feedback from T_3.

Thyrotrophin-releasing hormone (TRH) from the hypothalamus influences the amount of **thyroid-stimulating hormone** (TSH) secreted from the anterior pituitary gland. The TSH stimulates the secretion of thyroid hormones by increasing the production of cAMP in the thyroid cells, thus influencing the activities of the thyroid gland. High levels of TSH secretion lead to the following changes in thyroid function.

(a) Extraction of iodine from the blood is increased.
(b) Thyroid hormone release from thyroglobulin is increased.

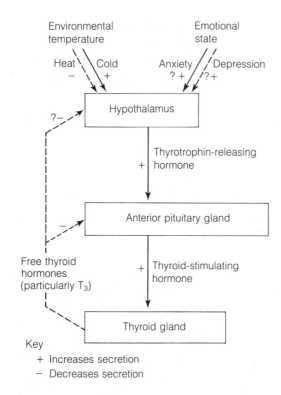

Key

+ Increases secretion

− Decreases secretion

Figure 2.5.8   Regulation of thyroid secretion

(c) Thyroid cells increase in size and secrete more hormone.

(d) The number of thyroid cells and development of follicles are increased.

Through negative feedback, the increase in level of T_3 in the blood causes a decrease in the secretion of TSH by the pituitary. When the thyroid gland is unable for some reason, e.g. inadequate supply of iodine, to produce enough thyroid hormones, lack of feedback inhibition results in a continued high level of TSH. This leads to an increase in both size and number of cells shown in an enlarged thyroid gland, or simple **goitre**. Even without TRH, some TSH is still secreted. It is known that environmental temperature, and it appears emotional states, can modify secretion of thyroid hormones. Exposure to cold produces a surge in TSH secretion, entirely due to an increase in TRH release from the hypothalamus. Noradrenaline also causes a rise in TRH secretion.

The metabolic rate is maintained within normal limits by negative feedback regulating the amount of thyroid hormones released, but can be modified

to respond to environmental and emotional changes.

CONTROL OF GLUCOSE METABOLISM

Glucose is the major nutrient used as a source of energy for body function. All cells use glucose, but most can also utilize fatty acids or ketones for energy, and may indeed do so in preference to glucose. However, under normal circumstances, certain tissues, and most importantly the brain, can only use glucose for energy. It is, therefore, essential that there is a continuous supply of glucose available. Brain cells can survive only a very short time without glucose or oxygen and a period of hypoglycaemia can result in permanent brain damage. After a period of some days, fasting brain cells begin to form the enzymes to utilize ketones as an energy source. The normal fasting blood glucose level is between 3.3 mmol/l and 5.5 mmol/l,

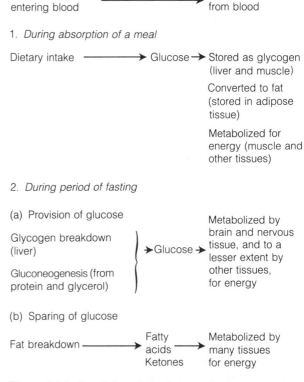

Figure 2.5.9   Regulation of blood glucose level

rising to around 7–9 mmol/l after a meal and falling to the fasting level again over a period of about 2 hours. Figure 2.5.9 summarizes the events involved in the maintenance of the blood glucose concentration within normal limits. The processes involved are mainly regulated by hormonal activity.

Glucose metabolism

AFTER A MEAL

Glucose is obtained from the digestion of carbohydrates in the diet. The simple sugars (monosaccharides) released are absorbed through the intestinal wall and transported in the hepatic portal system to the liver. Here, fructose and galactose are converted to glucose, although some fructose can be metabolised directly. After a meal, the blood glucose level rises and then begins to fall as it is either utilized by the tissues or stored.

Glucose in the blood is taken up by the cells and goes through a series of chemical reactions in

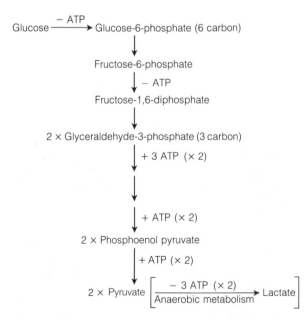

Glycolysis to pyruvate
+ 8 ATP

Pyruvate to lactate
− 6 ATP

Balance = 2 ATP from one glucose molecule

Figure 2.5.10 A summary of glycolysis

order to extract the energy within its molecular structure. The energy made available is stored in the high-energy molecule, **adenosine triphosphate** (**ATP**), and is then available for use by the cell. The first stage in the catabolism of glucose is the **glycolysis pathway** (Fig. 2.5.10) which takes place in the cytoplasm of the cell. During the series of reactions making up this pathway, the 6-carbon molecule (glucose) is split into two 3-carbon molecules of glyceraldehyde, and then converted to pyruvate. Under anaerobic conditions, when adequate oxygen is not available, pyruvate is converted to lactate and a net total of only two molecules of ATP are formed from the one molecule of glucose.

Under aerobic conditions, however, the pyruvate is transported into the mitochondria of the cell where it is converted to acetate (a 2-carbon molecule) in combination with coenzyme-A, to acetyl-CoA. The mitochondria have been described as the 'power-houses' of the cell, and metabolic pathways concerned mainly with the extraction of energy from substances and the storage of that energy in the form of ATP are located here.

Acetyl-CoA is at the junction of the major metabolic pathways. Particularly important amongst these is the **Krebs' cycle** (or **citric acid** or tricarboxylic acid (TCA) cycle) which is the major route for the formation of ATP. The Krebs' cycle is shown in Fig. 2.5.11. Acetyl-CoA (2 carbons) combines with oxaloacetate (4 carbons) to form the 6-carbon molecule, citrate. This is followed by a number of chemical reactions during which the 2 carbon atoms fed in as acetyl-CoA are released as carbon dioxide molecules, and the energy made available is stored in ATP. The constituents of the Kreb's cycle are regenerated. The breakdown of 1 glucose molecule, with adequate oxygen available, produces 38 molecules of ATP. The difference between this and the 2 molecules formed under anaerobic conditions helps to explain the lethargy and rapid tiring of patients with tissue hypoxia, whether due to respiratory, cardiac or blood disease.

If excessive amounts of acetyl-CoA are being formed, it is built up to form fatty acids. These combine with glycerol, composed from glyceraldehyde, to form neutral fat (or triglyceride) which is stored in adipose tissue. When necessary, the fatty acids can be released again and broken down to acetyl-CoA which can feed into the Krebs' cycle. Under certain conditions, excess acetyl-

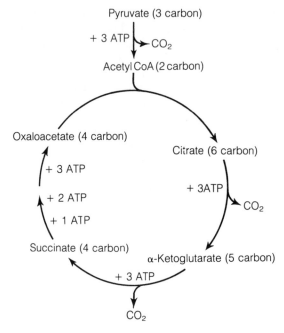

Pyruvate (3 carbon)

+ 3 ATP $\rightarrow$ CO_2

Acetyl CoA (2 carbon)

Oxaloacetate (4 carbon)

Citrate (6 carbon)

+ 3 ATP

+ 2 ATP

+ 1 ATP

+ 3ATP $\rightarrow$ CO_2

Succinate (4 carbon)

α-Ketoglutarate (5 carbon)

+ 3 ATP

CO_2

Energy released:
from pyruvate   15 ATP
from 2 × pyruvate formed from 1 glucose molecule   30 ATP
from glycolysis of 1 glucose molecule to 2 × pyruvate   8 ATP
Total from aerobic metabolism of 1 glucose molecule   38 ATP

Figure 2.5.11 A summary of the Krebs' cycle

CoA is converted to acetoacetyl-CoA, from which ketones are formed.

After a meal, in addition to the immediate use of glucose for energy and its conversion to fatty acids, glucose can be stored in the form of **glycogen** (Fig. 2.5.12). Glucose is converted to glycogen in the liver and in the muscle cells. Liver glycogen is used when required to maintain the blood glucose level. However, muscle glycogen is only used within the muscle cells for the provision of energy. During moderate levels of activity, with an adequate blood supply providing oxygen, the glycogen is broken down completely through the glycolysis pathway and the Krebs' cycle. On the other hand, during severe exercise, when the oxygen supply is not adequate, the glycogen is broken down as far as pyruvate and then to lactate. The lactate is transferred through the circulation to the liver where it is reconverted to pyruvate and utilized.

FASTING

During fasting there are two main considerations

in metabolism. First, it is essential to maintain the blood glucose level for brain function to continue. Second, it is necessary to supply an adequate energy source for all tissues of the body.

The blood glucose level is maintained in two ways. Liver glycogen is broken down (**glycogenolysis**) to glucose, which is released into the bloodstream. However, the hepatic glycogen store is relatively small; Cahill (1976) has reported a store of about 0.15 kg of glycogen in the liver, representing about 600 kcal (2.5 MJ). This will become depleted rapidly, possibly even overnight but certainly within 24–36 hours. The second source of blood glucose, which becomes important in prolonged fasting, is through **gluconeogenesis** (the formation of new glucose) from glycerol (released from triglycerides) and amino acids (from protein breakdown). Figure 2.5.13 illustrates the gluconeogenetic pathways, which are mainly the reverse of the glycolysis pathway. The pathways of glucose metabolism are summarized in Fig. 2.5.14.

The release of fatty acids from adipose tissue is the major source of energy for most other tissues

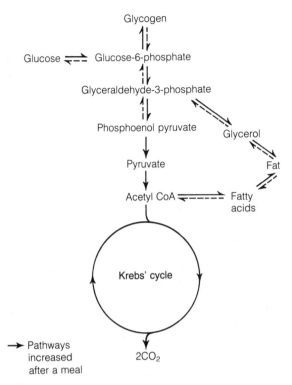

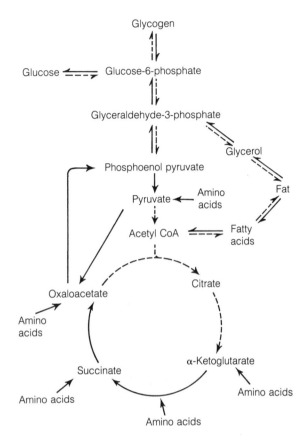

Figure 2.5.12   A summary of glucose metabolism after a meal

Figure 2.5.13   Gluconeogenesis

of the body. A non-obese man of about 70 kg weight will have fat stores equivalent to about 141 000 kcal (about 592 MJ) (Cahill, 1976). Fatty acids can be used directly, being broken down to acetyl-CoA and used through the Krebs' cycle. In addition, the liver can form ketones from the acetyl-CoA from the fatty acids. These are released into the bloodstream and utilized by the tissues through reconversion into acetyl-CoA, which again feeds into the Krebs' cycle.

Endocrine regulation of blood glucose level

Two hormones play a major role in the control of blood glucose level: **insulin**, produced by the β cells, and **glucagon**, produced in the α cells of the islets of Langerhans. These two hormones function in opposite directions; insulin acts to lower the blood glucose level and glucagon to cause an increase. A number of other hormones also influence blood glucose levels – somatostatin, growth hormone, glucocorticoids, adrenaline and thyroxine in particular.

INSULIN

The effects of insulin are anabolic in nature and it has been described as 'the hormone of nutrient storage'. The major metabolic effects of insulin are in muscle, adipose tissue and the liver, although it also regulates the metabolism of glucose in the eye (the disturbed metabolism possibly explaining why diabetic patients are more likely to develop cataracts).

Insulin binds firmly to a receptor site on the cell membrane and it appears that it carries out most of its functions by modifying cellular activity without entering the cell. The effects of insulin binding seem to be dependent on the presence of calcium, which becomes involved in the modification of cell enzymes. A high carbohydrate diet leads to an increased sensitivity of the tissues to insulin, and it is suggested that this may be due to a rise in the number of insulin receptors (Turner and Williamson, 1982).

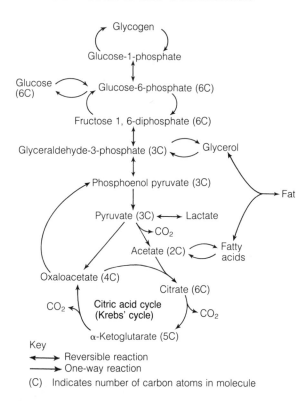

Figure 2.5.14   Summary of pathways of glucose metabolism

The effects of insulin on muscle and adipose tissue vary from its effect on the liver (Fig. 2.5.15). One of the main results of the action of insulin on muscle and adipose tissue is the stimulation of the transport of glucose, and a number of other substances including amino acids, potassium, phosphate and calcium, across the cell membrane into the cell. The increased glucose entry into the cell leads to an increase in all glucose metabolism within the cell, so that glycogen is laid down in muscle cells and the catabolism of glucose to acetyl-CoA is increased. Insulin also stimulates the synthesis and deposition of triglycerides in adipose tissue and inhibits the release of fatty acids. It also appears to stimulate protein formation from the amino acids entering the cell, and has been shown to reduce protein breakdown.

In the liver, on the other hand, the cell membrane is not a barrier to the passage of glucose, but it appears that part of the action of insulin in liver cells may still result from its binding to the cell membrane and its effect on calcium entry and binding within the cell. Insulin also increases the synthesis of the enzymes controlling the glycolysis pathway by affecting DNA transcription (i.e. synthesis of RNA from the DNA code) in the cell nucleus. This increase in enzyme synthesis increases the amount of glucose that can be dealt with by glycolysis, but takes some time to occur – 30–80 minutes (Turner and Williamson, 1982). More rapid control is achieved by rapid (within seconds) activation or inhibition of key enzymes. Thus, glycolysis, the Krebs' cycle and glycogenesis are all enhanced, so that utilization of glucose for energy and storage as glycogen are both increased. At the same time gluconeogenesis from protein breakdown is decreased.

Biosynthesis and metabolism

Insulin is a peptide hormone made up of two chains of 21 and 30 amino acids (51 in all) linked alongside each other by chemical bonds. It is synthesized in the endoplasmic reticulum of the β cells of the pancreas in the form of proinsulin, which has no, or very little, endocrine activity. Proinsulin is converted to insulin in the Golgi body of the cell, and is stored in secretory granules until required.

The pancreas stores about five times the amount of insulin secreted daily (daily secretion is about 50 International units (i.u.). The secretory granules containing the insulin move to the cell membrane, fuse with it, and liberate the free hormone into the extracellular fluid. It moves into the capillaries and is transported round the body in the unbound state. Insulin is secreted in a series of pulses at intervals of about 13 minutes in the basal state; the size of the pulses rising in response to a rise in blood glucose (Lang *et al.*, 1979).

Insulin has a very short half-life (about 5 minutes). After secretion, it is transported directly to the liver via the hepatic portal system and about half of it is removed immediately. Most of the rest is removed by the kidneys.

Regulation of insulin secretion

Insulin secretion can be influenced at two points: at the point of its release from the cell where secretion can be affected within 30 to 60 seconds, and at the point of formation of insulin within the cell where regulation may take up to two hours.

The major stimulus to insulin release is glucose, although other sugars which can be metabolized by the body, such as fructose, will also stimulate insulin release, but to a lesser extent. Other sub-

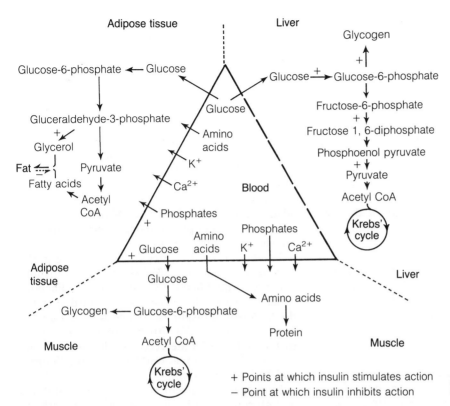

Figure 2.5.15 The major effects of insulin on metabolism

+ Points at which insulin stimulates action
– Point at which insulin inhibits action

stances, such as amino acids or gastrointestinal hormones, as well as stimulation of the vagus nerve all increase insulin secretion. Some of these factors cause a fall in the amount of glucose released from the liver, in anticipation of the absorption of nutrients from the gut (Turner and Williamson, 1982). On the other hand, adrenaline inhibits insulin secretion while at the same time liberating glucose from glycogen. This increases the blood glucose level but preserves it for utilization by the brain by preventing its entry into other body cells. Somatostatin also inhibits insulin secretion.

The amount of insulin secreted in response to a rise in blood glucose level is determined not only by the concentration of glucose in the blood, but also by the rate of change in that concentration. If the blood glucose level rises slowly, the amount of insulin required to ensure satisfactory utilization of a specific amount of glucose is less than if the blood glucose level rises rapidly. This is part of the explanation of the value of the high fibre, high carbohydrate diet now being recommended for diabetics, as the glucose from such diets enters the circulation slowly and thus reduces the amount of insulin required for diabetic control.

GLUCAGON

Glucagon released from the α cells of the pancreas gives rise to a rapid increase in the blood glucose level. It acts mainly on the liver and causes a rapid rise in cAMP within the cells. This leads to activation of the enzyme system, causing glycogenolysis, and, similarly to inhibition of glycogenesis. Thus, there is a rise in the amount of glucose released from the breakdown of glycogen. At the same time glucose breakdown is inhibited and the gluconeogenesis pathway is enhanced. Within adipose tissue, glucagon stimulates fat breakdown and the liberation of fatty acids. The formation of ketones in the liver is also increased. Glucagon and adrenaline have many of the same effects in stimulating both lipid and glycogen breakdown, but glucagon mainly affects the liver, while adrenaline mainly affects muscle and adipose tissue.

Biosynthesis and metabolism

Glucagon is a single polypeptide chain of 29 amino acids It has a short half-life (about 10 minutes) and is metabolized in the liver and kidneys.

Regulation of secretion

Glucagon secretion is stimulated by a low blood glucose level, and is inhibited by glucose, but only when insulin is present. Glucagon release is increased by many amino acids and inhibited by fatty acids. As amino acids stimulate glucagon and glucose stimulates insulin release, both hormones can be secreted after a meal and the overall effect will depend on the relative amounts of the two hormones present. Adrenaline also stimulates glucagon secretion.

Elevated levels of glucagon are found in severe diabetes mellitus and may contribute to the biochemical changes of the disease. These high levels are probably due to the lack of insulin present to inhibit the secretion of glucagon.

SOMATOSTATIN

This substance is secreted from several parts of the body. It acts as a hypothalamic hormone in the regulation of growth hormone secretion, but is also secreted from the δ cells of the islets of Langerhans of the pancreas. It is possible that it acts as a general intra-islet regulator as it inhibits secretion of both insulin and glucagon.

COORDINATED REGULATION OF GLUCOSE METABOLISM

As already stated, a number of hormones are involved in regulating or modifying glucose metabolism. Insulin alone acts to reduce the blood glucose concentration, while glucagon, catecholamines, glucococorticoids and growth hormone all contribute to a rise in blood glucose level. The influence of adrenaline and glucocorticoids is discussed later, but thyroxine also alters glucose metabolism. Hyperthyroid patients exhibit a rise in fasting blood sugar and hypothyroid patients a lower blood glucose level. Insulin secretion varies rapidly in response to a change in blood glucose levels. On the other hand, glucagon secretion remains relatively unchanged when subjects are eating a normal mixed diet. The other hormones mentioned are secreted mainly in response to stimuli unconnected to glucose metabolism, or in

response to clear hypoglycaemia. It thus appears that insulin plays the major role in the regulation of plasma glucose concentration.

Diabetes mellitus

This is a condition in which there is an absolute or relative lack of insulin for the requirements of the tissues. The insulin lack is often associated with higher than normal levels of glucagon in the blood, although these are returned to normal with adequate insulin replacement. Many of the metabolic changes and physiological effects in diabetes are shown in Fig. 2.5.16.

Juvenile-onset diabetics have a decrease in the number of β cells and an absolute lack of insulin. On the other hand, maturity-onset diabetics usually have a relative lack of insulin due to the development of insulin resistance, most commonly due to obesity. In this situation there is a decrease in the cell membrane receptors for insulin, but this returns to normal on losing weight (Turner and Williamson, 1982).

In diabetes the glycolytic enzymes are inhibited and gluconeogenetic enzymes activated, with the result that additional glucose is liberated into the bloodstream. However, this is not utilized normally for energy because, without insulin, it cannot cross the cell membrane into the muscle and adipose tissue, and the blood glucose thus rises further. Breakdown of protein is increased as amino acids are required for gluconeogenesis, and the nitrogen excreted in the urine as urea is increased. As the blood glucose level rises, it exceeds the renal threshold and glucose is excreted in the urine. As glucose is an osmotically active particle, additional water is also excreted, giving rise to polyuria. The passage of glucose-laden urine may lead to pruritus; it is important, therefore, that, for patients who have glycosuria, the nurse encourages scrupulous personal hygiene. As the patient becomes dehydrated he will also become very thirsty. The dehydration gives rise to many of the signs found in a patient with uncontrolled diabetes, and if severe can even lead to coma. This hyperglycaemic, hyperosmolar, non-ketotic coma is sometimes the presenting state with middle-aged or elderly patients with maturity-onset diabetes, who often do not require insulin replacement. They can be managed

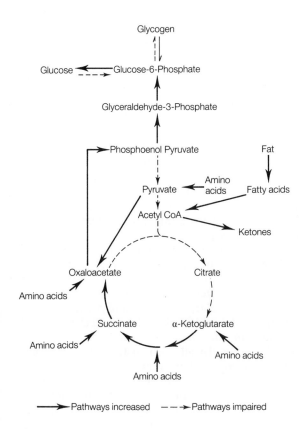

Pathways increased →   – – → Pathways impaired

(a)

through the control of diet alone or diet and oral hypoglycaemic drugs.

In addition to the rise in blood glucose there is a rise in the liberation of fatty acids from adipose tissue. In the liver acetyl-CoA is produced from the breakdown of fatty acids. However, the high concentration of fatty acids present inhibits the formation of further fatty acids and eventually inhibits the Krebs' cycle. The excess acetyl-CoA present is then used to form cholesterol and ketones. **Ketones** in moderate amounts can be utilized by many tissues of the body, but in uncontrolled diabetes they may be produced in excess. They will then be excreted in the urine and through the lungs, causing the sickly sweet smell of acetone. More importantly, ketones are acidic substances and the lowering in pH as a result of excessive ketone formation can lead to impaired brain function and coma. These patients require treatment with insulin and their diet has to be strictly controlled to balance the insulin. Insulin is a protein and digested by the gastrointestinal enzymes, and, therefore, must be given by injection. It is available in a number of forms with different durations of activity.

A number of hormones antagonize insulin, and patients secreting large amounts of these are more likely to develop diabetes mellitus. Growth hormone decreases glucose uptake in some tissues and in excess increases the risk of developing diabetes. Patients with Cushing's syndrome are also at increased risk, and people who are

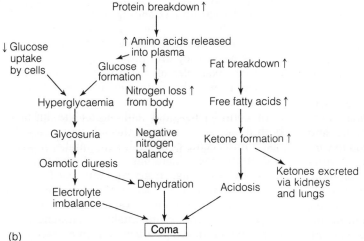

(b)

Figure 2.5.16 Diabetes mellitus:
(a) metabolic changes,
(b) physiological effects

prediposed to do so often develop the first signs of diabetes after a period of stress when there has been increased secretion of glucocorticosteroid hormones. Insulin requirements may increase in times of stress, in pregnancy, after surgery, and with infections.

A major nursing role in the care of the diabetic patient is to educate him or her (a) to adapt to and cope with the problems which the disease may present, and (b) to be aware of, and where possible prevent, potential complications.

Atherosclerosis, possibly due to the increased cholesterol produced from excess acetyl CoA and affecting small blood vessels in particular, may result in ischaemic ulceration of the feet. This may be further complicated by the neuropathies sometimes experienced by older diabetics. Neuropathy is thought to be caused by demyelination due to damage to Schwann cells. Foot care is therefore essential and regular chiropody desirable. Ideally, toenails should only be cut by one who is skilled in this.

Diabetics tend to be vulnerable to infection of the urinary tract and skin – probably due to the higher than normal glucose levels in both blood and urine. The nurse should assist the patient to recognize signs of impending infections and to take appropriate action. Both nurse and patient should feel confident in recognizing the clinical features of hypoglycaemic and hyperglycaemic comas. Education towards self-care must be a priority in the nursing care of both the juvenile and the elderly person with this chronic condition.

FLUID AND ELECTROLYTE BALANCE

The role of the kidneys in fluid and electrolyte balance is considered in Chap. 5.4; here, the role of the endocrine system in the regulation of water, sodium and potassium balance will be discussed.

As sodium (with chloride) is the major electrolyte in the extracellular fluid, it is one of the major determinants of body water content, and sodium and water content of the body tend to vary together. Although it is possible to have a disturbance of one without the other, this is uncommon. As a change in body sodium content is usually associated with an alteration in total body water, the concentration of sodium in the serum normally gives little indication of the total body sodium

content. If the patient develops depletion or accumulation of water alone, the increased concentration or dilution of the blood is reflected in a rise or fall in the serum sodium concentration. However, a change in serum sodium concentration is normally corrected rapidly by the action of hormones on the kidney. The normal serum sodium level is 135–146 mmol/l.

As the majority of potassium is within the cells of the body, the serum potassium concentration may give little indication of body potassium content. Normal serum potassium concentration is 3.5–5.2 mmol/l and a deviation outside these limits may indicate a much larger disturbance in the total body potassium.

Disturbances in osmolality and specific electrolyte content of the extracellular fluid will affect the intracellular fluid and the distribution of water through the different fluid compartments of the body (intracellular, interstitial and plasma). Alterations in total body fluid volume affect the blood pressure and distribution of substances throughout the body. In all such disturbances cell and body function will be impaired. In the last 20 years a much clearer understanding of the mechanisms of fluid and electrolyte balance has developed, along with recognition of its clinical importance. Medical skills in monitoring and regulating water and electrolyte levels in the body have improved considerably. Normal cellular, and therefore organ and body, function depends on the maintenance of a normal fluid and electrolyte balance.

The signs and symptoms of disturbances of fluid and electrolyte balance are rather nonspecific, and diagnosis depends partly on suspicion. A clear understanding by the nurse of the normal physiology, the possible disturbances which can occur and knowledge of the circumstances when they are most likely to develop is essential. Examples of disturbances in hormonal regulation will be indicated in the following discussion.

In considering fluid and electrolyte balance, both intake and output of the substance must be taken into account. In health the intake of water and electrolytes (including water formed within the body) will balance the amounts lost from the body by all routes. In theory, the body content of a substance can be regulated at both the intake and the output point. Fluid intake is influenced by thirst, which develops after a fall in extracellular fluid volume, or when the osmoreceptors of the hypothalamus are stimulated by a rise in the plasma osmotic pressure. However, many other factors, such as the social situation, also influences water intake. Sodium intake also is not carefully regulated physiologically, and many people ingest far more than is required. Therefore, the amounts of these substances within the body are regulated at the point of output, largely by the action of hormones on renal tubules. The major hormones involved are antidiuretic hormone and aldosterone.

Antidiuretic hormone (ADH, vasopressin)

Antidiuretic hormone is one of the hormones synthesized in the hypothalamus and released from the posterior pituitary gland. As its name suggests, it increases the reabsorption of water from the renal tubules, causing a reduction in the volume of urine excreted. Water is retained within the body and the plasma osmolality is maintained at about 290 mmol/kg of water (normal range 285–295 mmol/kg). Antidiuretic hormone also has a direct effect on the smooth muscle of the blood vessels, leading to a rise in blood pressure, but only when given in doses much larger than required for fluid balance regulation. Thus the name vasopressin is less appropriate than antidiuretic hormone.

Antidiuretic hormone regulates the amount of water retained in the body by increasing the permeability to water of the distal convoluted tubule and collecting duct of each renal nephron. Thus, water in the tubule passes passively into the interstitial spaces of the kidney medulla. As the countercurrent exchange mechanism (see Chap. 5.4) creates a highly concentrated interstitial fluid in the renal medulla, and particularly in the pyramids, a great deal of water is passively drawn out of the renal tubule in the presence of ADH. Thus, a small volume of highly concentrated urine is formed. In the complete absence of ADH, water is not reabsorbed in this part of the nephron and a large volume of hypotonic urine is excreted. **Diabetes insipidus** is a condition in which ADH is not secreted at all, and 5–20 litres or more of very dilute urine may be passed in a 24-hour period. If the patient does not continue to drink large amounts of fluid for replacement, he will become extremely dehydrated very quickly (Macleod, 1984). Such dehydrated patients may require 2-hourly pressure area care to prevent damage to their skin, and frequent mouth care to refresh their dry mouths.

Diabetes insipidus develops in a number of patients who have a hypophysectomy (removal of the pituitary gland) but disappears in some as the axons passing from the neurosecretory cells in the hypothalamus recover and recommence ADH secretion. This condition is treated by administration of the hormone, or by the use of a nasal spray a few times a day.

The ADH acts through its effect on cAMP levels in the tubule cells. It binds to receptors on the blood side of the cells and activates the formation of cAMP, which causes a considerable increase in the permeability of the cells to water (Handler and Orloff, 1981). The effect of ADH on the kidney develops very rapidly, but only lasts for a short period.

SECRETION AND METABOLISM OF ANTIDIURETIC HORMONE

This is a peptide hormone composed of nine amino acids. It has a very similar structure to oxytocin, the other hormone secreted from the posterior pituitary gland, differing only in two amino acids. Both ADH and oxytocin are synthesized in nerve cell bodies in the hypothalamus and pass down the axons of the nerves to the posterior pituitary gland where they are stored in neurosecretory granules in combination with a protein, neurophysin. The release of either hormone from the axon of the specific neurosecretory cell is triggered by nerve impulses passing down the axon.

Antidiuretic hormone has a short half-life of about 9 minutes as it is not bound to plasma proteins and hence is rapidly inactivated in the liver and kidneys.

Regulation of secretion (Fig. 2.5.17)

Osmotic pressure. Nervous activity of the neurosecretory cells is increased, resulting in the release of ADH into the circulation, when the osmotic pressure of the plasma is increased. **Osmoreceptors** in the anterior hypothalamus are sensitive to a rise in osmotic pressure and stimulate the activity of the nerve cells containing the hormone. If the osmotic pressure falls, nervous secretion of the

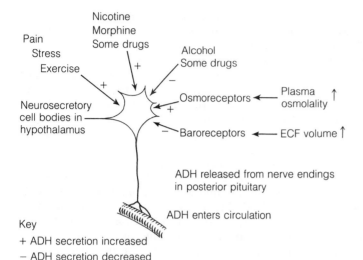

Pain
Stress
Exercise

Nicotine
Morphine
Some drugs

Alcohol
Some drugs

+

+

−

Osmoreceptors ←── Plasma osmolality ↑

Neurosecretory cell bodies in hypothalamus

+

− Baroreceptors ←── ECF volume ↑

ADH released from nerve endings in posterior pituitary

ADH enters circulation

Key
+ ADH secretion increased
− ADH secretion decreased

Figure 2.5.17 The regulation of antidiuretic hormone secretion

neurosecretory cells is inhibited. Secretion of ADH is stimulated when plasma osmolality is above about 280 mosmol/kg, and at about 290 mosmol/kg the amount of ADH secreted varies directly with the osmolality. Secretion of the hormone is markedly altered by even a 1% change in osmolality, so that the concentration of the plasma is maintained at, or very close to, 290 mosmol/kg.

Volume of extracellular fluid. The volume of the extracellular fluid (ECF) is the other major controlling factor of ADH secretion. If the ECF volume falls, ADH secretion rises, and vice versa. This is mediated by the baroreceptors in the vascular system. If the blood volume falls, without altering the blood pressure, impulses from the baroreceptors (the low pressure 'volume' receptors) in the venous side of the circulation, the great veins and atria, diminish. This allows a rise in the secretion of ADH which increases water reabsorption in the kidneys and thus corrects the hypovolaemia. If there is a blood loss sufficient to cause a fall in blood pressure, the activity from the arterial baroreceptors also drops, and ADH secretion rises further.

Other factors. Other factors can also influence ADH secretion. Pain, stress caused by an operation or by anxiety, exercise, and certain drugs, including nicotine and morphine, can all increase ADH secretion. This is of particular relevance when nursing surgical patients who may well be

producing large amounts of ADH after operation, and will, therefore, form small amounts of urine. If the patient is receiving an intravenous infusion, he is at some risk of developing water intoxication as he is unable to eliminate excess fluid from the body. Alcohol causes a fall in ADH secretion and, therefore, some degree of dehydration.

Aldosterone

Aldosterone is the major **mineralocorticoid hormone** secreted from the adrenal cortex. It causes a rise in serum sodium concentration by increasing reabsorption of sodium from renal tubules in exchange for potassium or hydrogen ions, and reducing excretion of sodium from the skin, salivary glands and gastrointestinal tract.

Aldosterone increases messenger-RNA formation, and then protein synthesis. The end result is that the active transport of sodium from the renal tubules to the bloodstream is increased. In exchange, the excretion of potassium and/or hydrogen ions is increased.

The effect of excessive aldosterone secretion is seen in Conn's syndrome – primary hyperaldosteronism (Conn, 1977). The patient is hypertensive as a result of the increase in extracellular fluid volume. However, oedema does not normally develop because, when the extracellular fluid expands beyond a certain point, sodium excretion occurs in spite of the continued action of aldosterone on the kidney tubules. These patients are deficient in potassium and will, therefore, suffer from muscle weakness, and many also have alkalosis resulting in tetany. However, the severity of the signs and symptoms of the condition is variable. It is treated by removal of the affected adrenal gland.

SECRETION, METABOLISM AND EXCRETION

Aldosterone is formed from cholesterol (see Fig. 2.5.3) and has a half-life of about 20 minutes. It is metabolized in the liver and kidneys and mainly excreted in the urine.

Regulation of secretion

While ACTH plays a permissive role in the secretion of aldosterone, it is relatively unimportant at physiological levels in controlling the amount of aldosterone secreted. However, in pharmacological doses, ACTH first results in a rise in aldosterone secretion, but with continued administration levels fall below normal (Drury *et al.*, 1982). A fall in serum sodium or a rise in potassium increases the release of aldosterone from the adrenal cortex. In addition, a fall in the extracellular fluid volume increases the secretion of aldosterone through its effect on the kidney and the **renin–angiotensin–aldosterone system** (Fig. 2.5.18; see also Chap. 5.4).

There is a baroreceptor system within the afferent arteriole of the renal glomerulus which stimulates renin release from the juxtaglomerular apparatus when there is a fall in the arteriolar pressure, and vice versa. In renal artery stenosis, the blood flow to the kidney is reduced and renin secretion stimulated by this mechanism. The macula densa of the juxtaglomerular apparatus is thought to stimulate renin secretion when the delivery of sodium and chloride to that point is low. Increased sympathetic activity, either through the sympathetic nervous system or from circulating catecholamines, also increases renin secretion. On the other hand, both angiotensin II and antidiuretic hormone reduce renin secretion.

Renin is a proteolytic enzyme which acts on angiotensinogen (a plasma protein formed in the liver) to release angiotensin I (a decapeptide). Angiotensin I is inactive, but is converted to the active angiotensin II (an octapeptide) by a converting enzyme found in endothelial cells widely distributed through the body, although the highest concentration is found in the lungs. Angiotensin II disappears very rapidly from the circulation, in about 30 seconds. Angiotensin II is changed to angiotensin III by aminopeptidase enzyme (Drury *et al.*, 1982).

Angiotensin II stimulates the secretion of aldosterone from the adrenal cortex. In addition, it is a potent vasoconstrictor and, formed in excess, leads to hypertension; this is the cause of hyper-

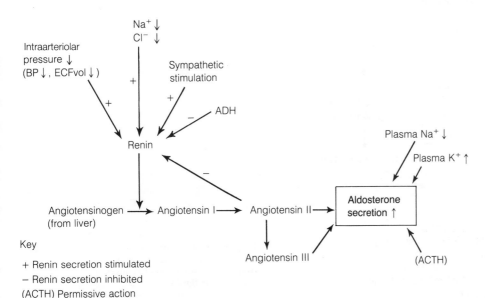

Figure 2.5.18 Control of aldosterone secretion

tension in patients with renal artery stenosis. Angiotensin III is thought to stimulate aldosterone to a greater extent than angiotensin II, but to have less effect on blood pressure.

Coordinated regulation of water and electrolyte balance

The two hormones, antidiuretic hormone and aldosterone, interact to maintain the balance of water and sodium in the body.

DISTURBANCES IN EXTRACELLULAR FLUID VOLUME

In Fig. 2.5.19, disturbances in ECF volume are considered. A fall in ECF volume (Fig. 2.5.19a) due to loss of body fluids causes an increase in renin secretion which leads to a rise in aldosterone

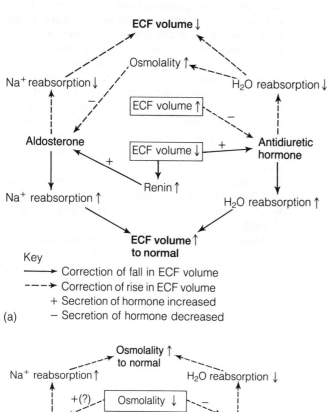

Key
→ Correction of fall in ECF volume
- - -→ Correction of rise in ECF volume
+ Secretion of hormone increased
− Secretion of hormone decreased

(a)

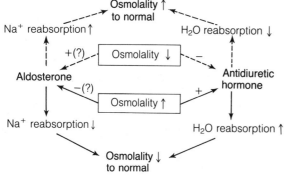

Key
——→ Correction of a rise in plasma osmolality
- - -→ Correction of a fall in plasma osmolality
+ Secretion of hormone increased
− Secretion of hormone decreased
? Uncertainty about effectiveness of stimulus on hormone secretion

(b)

Figure 2.5.19 Coordinated regulation of water and electrolyte balance: (a) correction of changes in extracellular fluid volume, (b) correction of changes in osmolality

release. This increases the reabsorption of sodium in renal tubules, thus raising the osmolality of the blood plasma. This stimulates ADH secretion (also stimulated directly by the fall in ECF volume) which causes reabsorption of water. Thus the ECF volume and osmolality are both restored to normal. The effect of a rise in volume is also shown in the figure.

CHANGES IN OSMOLALITY

Figure 2.5.19b illustrates the effect of a change in osmolality on the regulatory system. For example, a fall in osmolality accompanied by a rise in ECF volume due to drinking a large volume of water is rapidly corrected by a fall in ADH secretion. However, a rise in ECF volume without an alteration in osmolality, as may occur in a patient receiving excessive intravenous 0.9% sodium chloride solution (which is isotonic with plasma), is corrected more slowly. The change in osmolality is a more potent regulator of ADH secretion than is a change in ECF volume. A rise in osmolality is rapidly corrected both by release of ADH, increasing water reabsorption, and by thirst, increasing water consumption.

In congestive cardiac failure this regulation of water and sodium balance becomes disturbed. Because the heart is pumping less strongly, the pressure of blood within the arterioles of the kidney falls. As a result, renin secretion rises, leading to an increase in aldosterone release. The resulting rise in sodium reabsorption causes ADH secretion. The increase in the volume of the ECF eventually results in oedema.

CONTROL OF CALCIUM AND PHOSPHORUS METABOLISM

Calcium is the commonest mineral in the body (as much as 1200 g in an adult), with at least 99% of the total in the bones and teeth. However, that remaining in the body fluids and cells plays an important role in metabolism. Phosphorus also is of crucial importance in body function, and again most of it (85–90%) is found in the skeleton.

In the skeleton, most of the calcium and phosphorus is present as crystals of hydroxyapatite $(Ca_{10}(PO_4)_6(OH)_2)$ which are attached to the collagen fibres forming the matrix of the bone, and impart hardness to the structure (see also

Chap. 3.2). The deposition of these crystals, and their orientation, is partly controlled by the mechanical stresses on the bone, thus allowing adaptation to varied forces. However, in people who have little force acting on their bones for a period of time, for example astronauts in space and patients on bedrest, a reduction in the mineral content of the skeleton develops, leading to a fall in bone density and reduced strength of the bone. This is a condition known as osteoporosis, which can also occur as a result of hormonal deficiency or, possibly, a deficiency in calcium absorption. It is a condition which increases in incidence with advancing age and is most common in postmenopausal women.

The skeletal minerals are in a continual state of flux as the bone is being remodelled all the time. The minerals in bone are derived from and return to the calcium and phosphorus in the extracellular fluid. Normally there is an equilibrium between calcium released from bone by osteoclast activity and calcium deposited following osteoblast activity. Part of the calcium in bone is readily exchanged with the mineral in the ECF, but the larger part is stable and exchanges slowly.

The normal plasma concentration of calcium is 2.10–2.70 mmol/l, and of inorganic phosphate 0.70–1.40 mmol/l. The concentration in the ECF of both these minerals is determined by the balance between the amounts absorbed from the gut, excreted by the kidneys, and deposited in and released from the bone (Fig. 2.5.20).

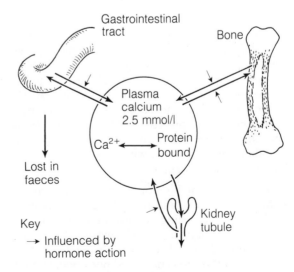

Figure 2.5.20 Calcium metabolism

Calcium

Calcium is present in the ECF about equally bound to protein (to albumin and globulin) and ionized, with a small amount combined with other substances such as citrate. In terms of physiological activity, it is the ionized calcium (Ca^{2+}) which is important. Calcium ions play an important role in the regulation of many vital cellular activities. They are principally involved in nerve and muscle function, hormonal actions, blood clotting, and cellular motility. A fall in the plasma concentration of Ca^{2+} leads to increased activity of motor units (motor nerves and the associated muscle fibres), resulting in **tetany** when the skeletal muscles go into spasm. This is occasionally seen if the parathyroid glands are inadvertently removed during a thyroidectomy.

Calcium has been described as a 'second messenger' as it is involved in transmitting the effect of many stimuli acting on the cell to the part of the cell to be affected. This appears to be mediated through **calmodulin**, a protein found within the cell that binds calcium ions when their concentration rises in response to a stimulus. When calcium is bound to calmodulin the activity of a great variety of cell enzymes is modulated. The importance of Ca^{2+} is indicated by the narrow limits within which plasma calcium levels are maintained.

Phosphorus

Phosphorus also plays a major role in cell function, largely in the form of phosphates – phosphate (PO_4^{3-}), hydrogen phosphate (HPO_4^{2-}), dihydrogen phosphate ($H_2PO_4^-$). These are found as components of nucleic acids, ATP, cAMP, other nucleotides, phospholipids, and some proteins, and are also present in the ECF and in small amounts intracellularly, as free ions. Phosphorus metabolism is regulated with less precision than that of calcium.

Three different hormones are involved in the control of calcium and phosphorus metabolism, the main effect being to maintain the concentration of calcium in the ECF within the narrow range normally found. The three hormones mainly involved in this are parathyroid hormone, vitamin D and calcitonin.

Parathyroid hormone (parathormone)

Parathyroid hormone is secreted from the four parathyroid glands which lie embedded within the thyroid gland. The main effects of parathyroid secretion are to elevate the concentration of Ca^{2+} in the ECF and to depress the plasma phosphate concentration.

Parathyroid hormone achieves these effects mainly by acting on the bone and kidneys. This hormone increases bone reabsorption by osteocytes, thus causing a rapid release of calcium and phosphorus into the ECF. In the kidney, calcium reabsorption from the tubules is increased, but the excretion of phosphate is greater. The net result is a rise in the level of Ca^{2+}, but a fall in phosphate levels in the plasma.

In addition to the direct effects on Ca^{2+} and phosphate levels, this hormone also modifies calcium absorption indirectly by increasing the activation of vitamin D by the kidneys.

BIOSYNTHESIS AND METABOLISM

Parathyroid hormone is a polypeptide of 84 amino acid residues. It is synthesized in the chief cells of the parathyroid glands in the form of a precursor which is then quickly converted to proparathyroid hormone. This is altered to parathyroid hormone in the Golgi body. Only small amounts of the hormone are stored in the cell, thus it is continually being synthesized and secreted.

Parathyroid hormone has a half-life of about 18 minutes and is degraded in the liver. Some of the products of degradation still have some parathyroid activity; they are further metabolized more slowly and excreted.

REGULATION OF SECRETION

Parathyroid hormone secretion is regulated directly by the concentration of Ca^{2+} in the ECF. When the level of Ca^{2+} rises, parathyroid hormone secretion is reduced and calcium is deposited in the bones. If the level of Ca^{2+} falls, hormone secretion rises. In conditions such as chronic renal failure when the level of plasma Ca^{2+} is chronically low, feedback to the parathyroid glands leads to hypertrophy and secondary hyperparathyroidism (Macleod, 1984). Chronic renal failure causes a low Ca^{2+} concentration because vitamin D is not activated by the diseased kidney.

Vitamin D

Vitamin D is a term which refers to a group of closely related steroid substances, the active form of which (**1,25-dihydroxycholecalciferol**) is the second hormone involved in the regulation of calcium and phosphorus metabolism. As the active form is produced in the kidney and it is transported in the circulation to its target organ, it fulfils the criteria of a hormone.

1,25-dihydroxycholecalciferol acts on the intestine, bone and kidneys with the overall effect of increasing the concentration of Ca^{2+} in the ECF.

The hormone enters the intestinal cell and stimulates calcium absorption from the gut. Phosphate absorption is also enhanced. In the kidneys, the reabsorption of phosphate from renal tubules is increased by 1,25-dihydroxycholecalciferol, but this is masked by the effect of parathyroid hormone on phosphate reabsorption. Calcium reabsorption in the kidney is also increased. Bone resorption is enhanced, releasing calcium into the ECF.

SOURCES OF VITAMIN D

The D vitamins are formed from provitamins – ergosterol (in plants) and 7-dehydrocholesterol (in animals). These are chemically modified by ultraviolet radiation to ergocalciferol (vitamin D_2) in plants, or to **cholecalciferol** (**vitamin D_3**) in the skin of animals. The two substances are very similar in structure and have equal biological activity.

Human beings have two sources of vitamin D – ingestion or the action of sunlight in forming cholecalciferol in the skin (Fig. 2.5.21). Whichever is the source, the vitamin D_2 or D_3 is protein bound and transported to the liver, where it is converted to **25-hydroxycholecalciferol**. This has no physiological action, and is the main storage form of vitamin D. This substance is converted in the kidney to 1,25-dihydroxycholecalciferol – the active hormone.

ACTIVATION OF VITAMIN D

The formation of the active hormone is regulated indirectly by the plasma Ca^{2+} level and directly by parathyroid hormone. Hypocalcaemia increases the secretion of parathyroid hormone and this increases the renal production of 1,25-dihydroxycholecalciferol. The hormone increases the plasma concentration of Ca^{2+}, and the secretion of parathyroid hormone and activation of vitamin D then fall.

Rickets occurs in children where there is an inadequate amount of vitamin D in the diet and insufficient exposure to sunlight to form chole-

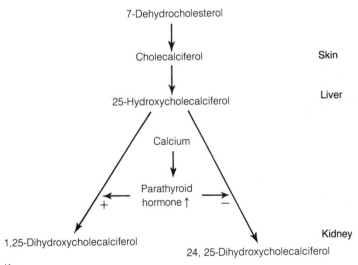

Key
+ Synthesis stimulated
− Synthesis inhibited

Figure 2.5.21 Formation and activation of vitamin D

calciferol. Inadequate amounts of calcium are absorbed from the gut and the bones are poorly calcified and so are soft and become deformed. These deformities become marked when the child begins to sit, crawl and walk, when knock knees or bow legs and deformities of the spine develop. Distortion of the pelvis may occur and lead to serious difficulties during childbirth in women who had rickets as children. When the condition is severe there may be a fall in concentration of calcium ions (Ca^{2+}) in the serum, and in infants tetany sometimes occurs. The equivalent condition in adults is **osteomalacia**. Both these conditions are more common in the Asian population of Britain, but not among those of West Indian descent. It is suggested that skin pigmentation is one factor amongst many in determining the amount of vitamin D synthesized in the skin (Dunnigan *et al.*, 1982).

Calcitonin

Calcitonin, the final hormone involved in the control of calcium and phosphorus metabolism, is secreted from the parafollicular cells (clear or C cells) of the thyroid gland, although it is possibly also secreted from other tissues. Its main effect is the reverse of parathyroid hormone; that is, it causes a fall in the plasma concentration of Ca^{2+} and of phosphate.

Calcitonin appears to play an important role in calcium metabolism during pregnancy and in childhood. During pregnancy it is possible that the hormone protects the mother from excessive loss of calcium to the baby, and it is suggested that it plays a role in the development of the child's skeleton during growth. However, it seems to be of little importance in adults, although it may play a part in the fine regulation of Ca^{2+} levels.

This hormone, like parathyroid hormone, also affects mainly bone and kidneys. Inhibition of bone resorption leads to a fall in the calcium released into the ECF and, combined with an increase in the urinary excretion of calcium, leads to a fall in the plasma concentration of Ca^{2+}. It also inhibits the activation of vitamin D.

SECRETION

Calcitonin is a small polypeptide containing 32 amino acid residues. The half-life is very short, only 4–12 minutes, supporting the suggestion that this hormone is involved in the fine control of Ca^{2+} levels.

The secretion of calcitonin occurs as a direct response to a rise in the circulating Ca^{2+} level. In addition, it appears that some of the gastrointestinal hormones, CCK-PZ, secretin, glucagon, and particularly gastrin, can stimulate calcitonin secretion. Thus, before calcium is absorbed from the gut, the hormone can be secreted to limit that

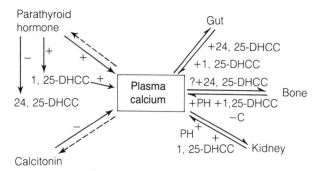

Key
- - -→ Negative feedback control of hormone secretion
by plasma calcium level
+ Increase in
− Decrease in
PH Parathyroid hormone
C Calcitonin
1, 25-DHCC 1, 25-Dihydroxycholecalciferol
24, 25-DHCC 24, 25-Dihydroxycholecalciferol

Figure 2.5.22 Regulation of plasma calcium concentration

absorption and the consequent rise in blood calcium levels.

Coordinated regulation of calcium ion concentration

As already indicated, in health the concentration of Ca^{2+} in the plasma is very precisely regulated (Fig. 2.5.22). Calcium levels are raised by the action of parathyroid hormone, which increases calcium released from bone and that reabsorbed from the renal tubules. 1,25-dihydroxycholecalciferol increases Ca^{2+} concentration by enhancing absorption from the gut and reabsorption in the kidneys, and by mobilizing calcium from bone. As parathyroid hormone stimulates the activation of 1,25-dihydroxycholecalciferol, the levels of active hormones in the blood are raised together. As the half-life of parathyroid hormone is short, the negative feedback regulation of hormone secretion through a change in Ca^{2+} concentration modulates hormone levels in the blood rapidly. This probably accounts for the major part of Ca^{2+} concentration control in adults. Calcitonin causes a fall in plasma Ca^{2+} by inhibiting bone resorption, but plays a small role in adults, probably in the fine control of Ca^{2+} concentration.

Some other hormones may also play a small part in calcium metabolism. Corticosteroids appear to lower Ca^{2+} concentration, while growth hormone causes a rise. Thyroid hormone in excess can also lead to hypercalcaemia and osteoporosis as calcium is lost in the urine.

Some disturbances of endocrine control of calcium metabolism

Both hyperparathyroidism and hypoparathyroidism can occur due to a number of different causes. As would be expected, **hyperparathyroidism** results in an increase in the concentration of Ca^{2+} in the ECF, due to mobilization of calcium and phosphorus from bone. Both calcium and phosphate are excreted in large amounts in the urine, which can lead to the development of renal calculi, and these can cause impairment of renal function. It is for this reason that nurses should encourage patients to increase their fluid intake to 2–2.5 litres per day. Measurement of serum Ca^{2+} concentration allows straightforward diagnosis.

Hypoparathyroidism causes a fall in plasma Ca^{2+} and a rise in phosphate levels, which results in **tetany**. This involves increased excitability of the nerves, often with paraesthesia and muscle spasm. **Carpopedal spasm** (spasm of the hands and feet) is the commonest finding in adults, with laryngeal spasm occurring in children but rarely in adults (Macleod, 1984). The development of tetany is a possible complication of thyroidectomy as the parathyroid glands may be inadvertently removed or damaged. Another cause of tetany is alkalosis. A rise in plasma pH, from whatever cause, leads to an increase in the amount of calcium bound to protein and a fall in the concentration of the ionized form of calcium in plasma, and thus tetany. In an emergency, tetany is treated with intravenous calcium gluconate.

GROWTH AND DEVELOPMENT

Growth involves an increase in length and size, not just in weight, and the deposition of additional protein. In development this is associated with a coordinated pattern of bodily changes culminating in the normally developed adult man or woman.

The normal pattern of rate and extent of growth and development is complex and is influenced by a number of different factors. Genetic factors lay down the basic guidelines, as indicated by the correlation of adult height between parents and children. The major influence superimposed on this is probably nutritional, although illness, trauma or other circumstances can also modify the processes involved in growth. A childhood diet inadequate in content or amount will result in an adult who does not reach his or her genetic potential in terms of height, but someone who is 'overnourished' in childhood will not exceed his or her genetically predetermined stature. This modification by the amount of food ingested during growth often occurs in areas of the world where the food supply is inadequate and is usually an adaptive response to the situation. The smaller adult requires less food than the larger individual.

The normal rate and pattern of growth and development are regulated through the endocrine system, and vary at different times of life. The rate of growth is greatest before birth, at about 4 months gestation. However, while the endocrine determination of sexual differentiation is under-

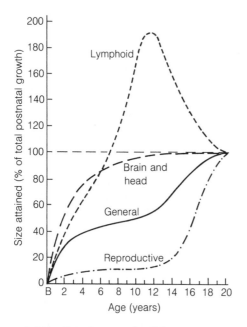

Figure 2.5.23   Growth curves for different body tissues (reprinted from Tanner, J. M. (1962) *Growth at Adolescence*, 2nd edn. Oxford: Blackwell)
Lymphoid – thymus, lymph nodes, intestinal lymph masses
Brain and head – brain and parts, dura, spinal cord, eyes, head size
General – body as whole, external dimensions (except head), respiratory and digestive organs, kidneys, aortic and pulmonary blood vessels, musculature, blood volume
Reproductive – male or female reproductive organs

stood quite well (see Chap. 6.3), the regulation of growth in utero is still not clearly comprehended. Infants with thyroid deficiencies may be of about normal size at birth, but show other abnormalities of growth and development, while the role of growth hormone is not yet elucidated.

The rapid rate of growth during fetal life continues into infancy, but significantly decelerates through early childhood. There is a further deceleration before puberty which is followed by the pubertal growth spurt. The age at which this growth spurt occurs varies considerably between, on average, 10.5 and 13 years in girls, and 12.5 and 15 years in boys. In general, the younger this growth spurt occurs the shorter the final height achieved. During this period there is very considerable variation in stature and development between individuals of the same chronological age. Different tissues also vary in the rate and timing of growth, and Fig. 2.5.23 demonstrates

the variation between tissues at different ages as a percentage of the size at age 20.

It is clear that a number of different hormones are involved in the coordinated regulation of growth and development, and this is discussed later. Many of these have already been mentioned in other contexts, and the role of the gonadotrophins and sex hormones in the development of the gonads and secondary sexual characteristics is discussed in Chapter 6.3. One hormone which plays a major part in the control of body growth is growth hormone.

Growth hormone (somatotrophin)

Growth hormone is a single polypeptide chain of 191 amino acids formed in the anterior pituitary gland. The structure of this hormone is very similar to that of prolactin (also formed in the anterior pituitary gland) and human placental lactogen, and it is probable that they have all evolved from the same hormone. As expected, there is some overlap in activity between growth hormone and prolactin.

Growth requires a modification of metabolism in a number of different ways and growth hormone has a general anabolic action through a number of different effects on the cell metabolism of various tissues (Table 2.5.6).

SOMATOMEDINS

Growth hormone affects cell activity through binding to specific receptors on the cell membranes of the target organs. However, it is now clear that a number of the effects of growth hormone are mediated through a group of substances, the somatomedins. These are synthesized in the liver, and possibly in other tissues, under the influence of growth hormone. One of the somatomedins causes most of the effects of growth hormone on cartilage, and it now appears that there are a number of other growth factors secreted into the circulation which stimulate the growth of specific tissues and organs. Two of these are similar in structure to proinsulin and are known as **insulin-like growth factors (IGF) I** and **II**.

The somatomedins travel in the blood bound to plasma protein, and therefore have a half-life of

Table 2.5.6 Effects of growth hormone

1 Increase in stature by stimulation of epiphyseal cartilages
 Increased growth of: connective tissue, skin, muscle viscera – heart, lungs, liver, kidneys, intestines, pancreas
2 Increased protein synthesis ⎫
 Decreased protein catabolism ⎭ → Positive nitrogen balance
3 Increased fat breakdown → ↑ Free fatty acids in plasma
 ↑ Fatty acid oxidation (long-term effect)
4 ↑ Gluconeogensis
 ↓ Glucose utilization
 Antagonism of insulin (can cause ↑ blood glucose in diabetics who cannot respond by ↑ in insulin secretion)
5 Positive calcium and phosphorus balance
 Retention of sodium → positive balance
 Retention of potassium → positive balance
6 Kidneys – increased glomerular filtration → ↓ blood urea
 Liver – increased ability to conjugate many substances
7 Some prolactin-like effects: growth and differentiation of breast tissue, ? effect on lactation

about 3 hours. Cyclic-AMP may be involved in the mechanism of action of somatomedins, and the IGFs appear to affect cell function by interacting with the insulin receptor on the cell membrane (Preece and Holder, 1982).

Some of the metabolic actions of growth hormone are, nevertheless, thought to be due to the direct effect of the hormone on the tissues involved.

EFFECTS OF GROWTH HORMONE

The increase in height achieved occurs through the effect of growth hormone, mediated by a somatomedin, on the epiphyseal cartilages. When growth is completed these cartilaginous plates between the shaft and the ends of the bone are finally ossified. Up to this time, growth hormone stimulates the activity of the cartilaginous tissue, the epiphyses become wider, and more bone is laid down, increasing the overall length of the bone (see Chap. 3.2). In children, excessive secretion of growth hormone leads to the condition known as **gigantism**. In adults, after full height is achieved and the bone fully ossified, an increased secretion of growth hormone, possibly as the result of a pituitary tumour, causes **acromegaly** (Fig. 2.5.24). In this condition the bones and soft tissues of the hands, feet, face and lower jaw become enlarged and the skin becomes coarse. Many of the internal organs increase in size, and metabolic disturbances may occur. The person with acromegaly may be very embarrassed by these changed features and will therefore require very sensitive support from all involved in his or her care. In the child, absence of growth hormone prevents the attain-

ment of normal stature and these patients remain as dwarfs with trunk and limbs usually in proportion. A similar clinical condition results from the inadequate formation of somatomedins. As dwarf-

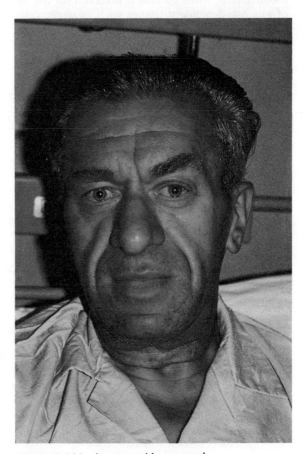

Figure 2.5.24 A person with acromegaly

ism due to lack of growth hormone, as opposed to inadequate formation of somatomedins, can be treated with injections of the hormone, careful diagnosis is essential.

Some of the metabolic effects of growth hormone shown in Table 2.5.6 are concerned with the formation of the new protein necessary for the growth of bone, of connective tissue and skin, and of the viscera. Potassium is retained within the body and is incorporated into the new cells formed. Growth hormone also increases the reabsorption of both calcium and phosphate from the renal tubules, thus facilitating the mineralization of the additional bone created. With an increase in body size there must be a concomitant enlargement of the extracellular fluid volume, and thus there is also increased reabsorption of sodium from the renal tubules. The energy source required for the anabolic activities is provided by the increased liberation of free fatty acids into the bloodstream, while glucose is reserved for the use of glucose-dependent tissues such as nerve cells. The effect of growth hormone on glucose metabolism explains the high incidence (30%) of diabetes mellitus among patients with acromegaly (Macleod, 1984).

SECRETION, METABOLISM AND EXCRETION

Human growth hormone is stored in the pituitary gland in large amounts. It is secreted in short bursts, with the largest episodes of secretion taking place during the early part of sleep at night. In the adult, other peaks of secretion also occur when awake after meals. The nocturnal secretion appears to be linked to the sleep–wakefulness pattern, rather than to an intrinsic circadian rhythm. The pattern of growth hormone secretion appears to alter with age; children only secrete the hormone while sleeping and the change to the adult pattern of secretion occurs at adolescence. In the elderly, the amount of growth hormone secretion falls to very low levels indeed. It has been estimated that 0.2–1.0 mg of growth hormone is secreted daily in adults.

The half-life of growth hormone is short, 20–30 minutes, as it is rapidly metabolized and excreted. Because it is secreted from the pituitary gland in bursts and is rapidly removed from the circulation, the levels in the blood vary greatly within a short period of time. Some of the actions of growth hormone are spread over a longer period and the effects of fluctuations in level are smoothed

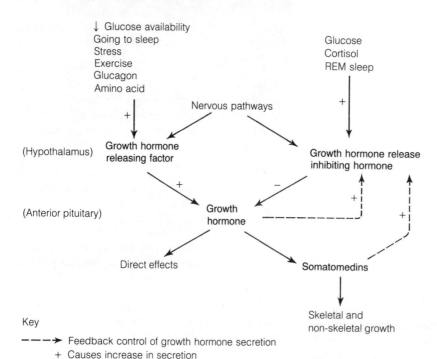

Figure 2.5.25 Control of growth hormone secretion

out because they are mediated through soma-tomedins which have a long half-life.

Regulation of secretion

The secretion of growth hormone is modified by a number of different factors (Fig. 2.5.25), but the main control of secretion is via the hypothalamus. The stimuli which influence growth hormone secretion appear to fall into three main groups; the first group includes conditions in which there is an actual or potential fall in the supply of an energy source for metabolism, second, an increase in amino acid levels in the blood, and finally, a variety of physical and psychological stimuli increase the secretion of growth hormone. Most of these stimuli are integrated in the hypothalamus or higher levels of the brain and are mediated through control of the release of hypothalamic hormones.

Two hypothalamic hormones involved in the regulation of growth hormone secretion have been identified. These are **growth hormone releasing factor (GRF)**, the structure of which is not yet clear, and **growth hormone release-inhibiting hormone (GIH, somatostatin)**. Somatostatin is a peptide of 14 amino acids and is found in the δ cells of the pancreas and the stomach as well as in the hypothalamus. Somatostatin inhibits the secretion of growth hormone, but also reduces the secretion of insulin, glucagon, thyroid-stimulating hormone and follicle-stimulating hormone.

Growth hormone itself appears to act through feedback control on the hypothalamus and the pituitary gland to reduce its own secretion. However, it also appears that there is a somatomedin feedback system (Fig. 2.5.25).

Coordinated control of growth and development

As indicated previously, growth is a complex process, basically determined by genetic makeup, modified by such factors as diet and illness, and requiring the coordinated activity of a number of hormones. Reduction in growth during illness is often followed by a period of 'catch-up' growth when the rate is considerably above normal to allow the child to return to the previous growth curve (Prader, 1978).

The number of hormones involved in the nor-mal growth and development of an individual is indicated by the range of abnormalities of hormone secretion which can result in disturbed growth and abnormal development (Brook, 1978). Obviously, growth hormone is involved, but so also are thyroid hormones, parathyroid hormone, vitamin D and calcitonin, glucocorticoids, insulin and the sex hormones, as well as the pituitary hormones regulating the release of any of these.

The lack of knowledge about the control of intrauterine growth has already been mentioned. However, it is clear that after birth adequate, controlled endocrine function is essential. Thyroid hormones appear to play a permissive role, in that growth hormone without thyroid hormones does not result in normal growth, but thyroid hormones alone do not stimulate growth. Thyroid hormones are necessary for the normal formation and storage of growth hormone in the pituitary gland. Lack of thyroid hormones during early childhood leads to **cretinism**, when the child is dwarfed with infantile bodily proportions and coarse features. During later childhood, hypothyroidism leads to growth retardation and deterioration in mental and physical ability. Thyroid hormones are essential for normal brain development and a delay in necessary treatment early in life leads to permanent mental impairment, the degree of which will vary according to the length of time before beginning treatment (Macleod, 1984).

Parathyroid hormone and vitamin D (and to a lesser extent calcitonin) are essential for normal bone formation and growth (see Chap. 3.2).

The glucocorticoid hormones may have a regulatory effect on tissue growth, and in excess they inhibit growth in height. In addition, the distribution of tissue (particularly adipose tissue) is altered. Children requiring steroid therapy need careful monitoring of growth and their therapy should be adjusted as necessary to ensure the attainment of adequate, if reduced, stature.

Insulin stimulates protein formation, and also inhibits protein catabolism for use as an energy source. Therefore, it has a role in normal growth in the formation of additional body protein.

During childhood the mononuclear phagocytic and lymphoid systems, including the thymus gland, grow fairly rapidly, and development of the immune system occurs (see Fig. 2.5.23). This involves the endocrine activity of the thymus gland.

The effect of the gonadotrophic and sex hormones on the gonads and the development of secondary sexual characteristics are discussed in Chap. 6.3. However, during puberty a number of other bodily changes also occur. There is a considerable increase in growth rate, resulting in a gain in height of about 28 cm in boys and 25 cm in girls, and changes in body size and shape. This growth spurt comes early in puberty in girls, but late in boys, and the reason for this difference is not known. Most of the height increase is in the trunk, in the length of the vertebral column, rather than in the legs. In addition, increased width of shoulders in boys and size of pelvis in girls occurs. The endocrine control for some of these changes is not entirely clear.

An increase in androgen secretion from the adrenal cortex occurs between the ages of 5 and 8 years in both boys and girls, and it is thought that this may be a preliminary to changes during puberty. The endocrine control of puberty is not fully understood. It has been suggested that as maturity approaches, an increase in secretion of gonadotrophic-releasing hormones occurs. On the other hand, it has been hypothesized that puberty results from the withdrawal of inhibition of endocrine events. It has been suggested that the pineal gland may be involved. Whatever the initiator, the growth spurt during puberty is due to the activity of both growth hormone and sex hormones, with the involvement of some other hormones such as insulin and parathyroid hormone. The sex hormones are thought to be mainly responsible for the changes in the vertebral column and shoulders and hips, while growth of the legs is due largely to the activity of growth hormone.

In childhood, distribution of different types of tissue within the body is similar in girls and boys. However, men have about one and a half times the lean body mass and skeletal mass of women, while women have approximately twice as much body fat as men. The differences in fat distribution and amount between the sexes are thought to be due to the secretion of testosterone preventing the female pattern of fat deposition. Presumably the increased body mass in men is due to androgen activity.

As can be seen, growth and development involve alterations in many parts of the body occurring over a period of about 20 years. The interrelated sequence of events can be affected by many factors, and requires coordinated activity of many hormones. Some of these, e.g. thyroid hormone, are essential in early life, but their importance in this area wanes, while others become prominent at puberty in controlling the dramatic changes in the body taking place at that time. With so many hormones involved in the control of growth, a number of different hormone deficiencies can result in disturbed (usually reduced) growth and/or development. As some causes of these disturbances can be readily treated, early and accurate diagnosis is essential.

RESPONSE TO THE EXTERNAL ENVIRONMENT

In order to survive in varying environmental conditions, an individual must be able to adapt to change, and the nervous system is of major importance in responses requiring thought or planned activity. However, the endocrine system is involved in some of the innate responses to environmental conditions.

The immune response

People are in continual contact with microorganisms and require a normally functioning immune system to prevent morbidity due to infection. The endocrine system is important in the normal development of the immune system (discussed in Chap. 6.2) through the activity of the thymus gland. In humans it has been established that three **thymic hormones** are involved, thymosin fraction V, thymic humoral factor and thymopoeitin. These probably act in sequence on precursor cells (Wara, 1981).

The stress response

Probably the major way in which the endocrine system is involved in reacting to environmental change is through the stress response. The body's ability to respond to stressful situations developed through evolution in conditions very different from those under which we live today. The stress response is important to aid adaptation, but under some circumstances the physiological and biochemical changes which occur are not beneficial, and may indeed be harmful (Selye, 1976).

It is clear that most of the endocrine glands of

the body are affected in stress, mediated through the hypothalamus. Thus, secretion of growth hormone is increased and the gonadotrophic hormone balance may be altered, resulting in disturbances in the menstrual cycle frequently seen in young women when they first move away from home. A rise in antidiuretic hormone secretion leads to fluid retention, which may be dangerous in surgical patients receiving intravenous fluids. However, the major effects which will be discussed here involve the secretions of the adrenal cortex (the corticosteroid hormones) and the adrenal medulla (the catecholamines). The adrenal medullary response occurs very rapidly and lasts a short time and is seen in the 'fright, flight, fight response' described by Cannon and de la Paz (1911). The response of the adrenal cortex is much slower to occur, and lasts for a considerable length of time.

THE ADRENAL MEDULLA

Functionally the adrenal medulla is a part of the sympathetic nervous system. During embryonic life it develops from an outgrowth of nervous tissue which becomes surrounded by what is, in terms of function, a completely separate gland, the adrenal cortex. (The sympathetic nervous system is discussed in Chap. 2.4.)

The neurotransmitter of the postganglionic sympathetic nerve fibres is **noradrenaline** (norepinephrine) and this is also one of the hormones secreted by the adrenal medulla. However, **adrenaline** (epinephrine) is the major hormone of this gland. The two hormones, which are very similar in structure, are formed from the amino acid tyrosine. The catecholamine hormones are stored in vesicles in the gland in the proportion of 80% adrenaline and 20% noradrenaline (Fig. 2.5.26).

Actions of adrenal medullary hormones
While the structures of the two hormones are very similar, their actions are not always the same. In most situations, stimulation of the sympathetic nervous system and secretion of hormones from the adrenal medulla are coordinated, with sympathetic stimulation being followed by hormone secretion as the stimulus continues.

In general, these hormones stimulate activity of the nervous system, act on the cardiovascular and respiratory systems, and have various metabolic effects. These actions prepare the body for imme-

diate activity in response to environmental challenge, and occur rapidly – within the time taken for the blood to circulate round the body. The activity of these hormones, as of the sympathetic nervous system, can be explained as being mediated through α and β receptors on the cell membranes of the target organs.

The **α receptors** are concerned with excitatory functions such as vasoconstriction. The **β receptors** are mainly concerned with inhibitory functions such as vasodilation and bronchodilation, as well as with excitation of the heart leading to an increased heart rate. This may be an oversimplification of the situation. It is suggested that there is only one type of adrenergic receptor, but that its properties will vary under different conditions (Young and Landsberg, 1979). However, the original concept of α and β receptors is still of value in medical treatment. Drugs have been developed that can selectively block these receptors; for example β-blocking drugs, such as propranolol, block the effect of the catecholamines on the heart, thus preventing additional cardiac work caused by sympathetic stimulation. The effect of these hormones seems to be produced through an increase of cAMP in the cell.

A number of the other hormones have been found to influence the responsiveness of body tissues to the catecholamines, and certain other factors also influence the effect of catecholamines on the tissues. Acidosis, hypoxia, sepsis and endotoxaemia have all been reported to reduce the effect of the hormones on the tissues (discussed in Young and Landsberg, 1979).

Effect on the nervous system. These hormones increase arousal of the nervous system, leading to increased wakefulness and emotion. However, the emotion experienced is defined by the individual's interpretation of the changes occurring. In addition, the degree of arousal affects the ability to carry out some activities, including learning. A U-shaped relationship has been described, with poor learning at both low and high levels of arousal. Thus, severe anxiety, which may be experienced by a patient on admission to hospital, will reduce his or her ability to understand and retain information, necessitating several repetitions of information.

Cardiovascular and respiratory effects. The effects of the two hormones, adrenaline and noradrena-

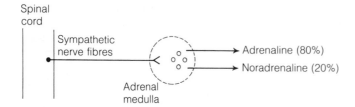

Spinal cord

Sympathetic nerve fibres → Adrenal medulla → Adrenaline (80%) / Noradrenaline (20%)

Figure 2.5.26 Secretions of the adrenal medulla

line, vary somewhat in this area. Adrenaline results in an increase in cardiac output (due to tachycardia) and a fall in peripheral resistance, actions which lead to a rise in systolic and a fall in diastolic blood pressure. The reduction in peripheral resistance results from the dilation of blood vessels in the skeletal muscle (and perhaps in cardiac muscle), even though the blood vessels supplying the gut and skin are constricted. The overall effect is to redistribute the blood to the areas of major activity in the 'fright, flight and fight response', i.e. to the muscles, heart and brain.

On the other hand, noradrenaline (normally released in small amounts from the adrenal medulla) has less effect on the heart, but causes a rise in blood pressure (systolic and diastolic) as peripheral resistance is increased. In **phaeochromocytoma** (a tumour of the adrenal medulla or sympathetic ganglia) the hormone usually secreted in greatest amounts is noradrenaline. Hypertension is the commonest sign to occur, along with other signs and symptoms of excessive catecholamine release such as pallor, sweating, palpitations and apprehension (Macleod, 1984).

The rate of respiration, rather than the depth, is increased by both hormones. However, adrenaline has a greater effect on bronchodilation than does noradrenaline.

Metabolic effects. The metabolic changes which occur as a result of secretion of catecholamines are of two types. There is an increase in metabolic activity leading to a rise in energy utilization and, secondly, there is a rise in the supply of nutrients made available.

The increase in metabolic rate results in a rise in heat production and oxygen consumption. Heat production is increased partly as a result of increased muscular or other activity, but partly because of a rise in thermogenesis by the brown adipose tissue. This is clearly a valuable response to cold (Young and Landsberg, 1979).

The supply of nutrients available to the cells for the increased activity is increased by mobilization of stored nutrients. This occurs both at a local and at a systemic level. Glycogen breakdown in skeletal and cardiac muscle, as well as fat breakdown in brown adipose tissue, only supply fuel for use within the cell. On the other hand, glycogenolysis in the liver and free fatty acid release from white adipose tissue supply nutrients for use in distant tissues.

It appears that both adrenaline secreted from the adrenal medulla and noradrenaline from the medulla and from the sympathetic nervous system play a part in these changes.

Secretion, metabolism and excretion
The effects of sympathetic activity, including those of the adrenal medullary hormones, are rapid both in onset and in disappearance, thus allowing speedy adaptation to environmental changes. This activity is regulated by the central nervous system from the nuclei in the brain-stem, which are the origins of the nerve supply to the sympathetic nervous system. Adrenal medullary secretion is activated by nerve impulses from the sympathetic preganglionic nerve supply to that gland (Fig. 2.5.26).

Activity of the brain-stem nuclei is affected by many internal and external factors. Nervous connections from the cerebral cortex, limbic system and hypothalamus allow anticipatory secretion, secretion as a result of emotions, and secretion as a response to changes in the temperature or composition of the internal environment (the extracellular fluid) which are monitored in the hypothalamus.

Secretion of the catecholamines usually occurs in bursts of activity, being low under conditions of rest and sleep. When there is a rise in secretion, adrenaline and noradrenaline are generally released in the proportions found within the gland (i.e. 80% adrenaline, 20% noradrenaline). However, hypoglycaemia appears to cause the secretion of adrenaline rather than noradrenaline (Young and Landsberg, 1979). It has also been suggested that noradrenaline secretion is increased by emotional situations which are familiar, while adrenaline secretion rises in unfamiliar conditions.

The catecholamines have a very short half-life, being rapidly metabolized by enzymes which are widely distributed throughout the body. They are excreted, mainly in the urine, in the form of a number of metabolites, some of them conjugated, others in the free form.

While the actions of these hormones were valuable for man living as a hunter–gatherer when immediate action was important, they may occur now on admission to hospital, prior to ward rounds, or in clashes with one's superior etc., when physical activity is not of benefit and may be impossible. In such circumstances these changes may be maladaptive, as the energy requirements of the body are increased with a rise in catabolism and cardiac activity. It has been suggested that these changes may result in hypertension in susceptible individuals (Raab, 1966).

THE ADRENAL CORTEX

The other main part of the stress response involves activity of the adrenal cortex, and the effects begin more slowly and last longer than those mediated through the adrenal medulla. The adrenal cortex has three layers which secrete the different hormones: the outer **zona glomerulosa** produces mineralocorticoids, i.e. aldosterone (discussed earlier), the middle **zona fasciculata** secretes glucocorticoids, and the inner **zona reticularis** secretes small amounts of glucocorticoids and sex hormones. Although the hormones produced by the adrenal cortex are classified into the three types, there is some overlap in the activity of the different groups of hormones. For example, the glucocorticoids such as cortisol also have some degree of mineralocorticoid activity, although much less than aldosterone. The structures of some of the hormones formed in the adrenal cortex are illustrated in Fig. 2.5.3.

In the stress response it is the activity of the **glucocorticoids** which is of most importance. Hans Selye (1976) described the **general adaptation syndrome** (Fig. 2.5.27), with its triad of signs and its three stages, which occurs when an animal is exposed to a **stressor** (a factor which initiates the stress response). The first stage is the **alarm reaction**, which develops on initial exposure to the stressor and, in the rats studied, caused:

> enlargement of the adrenal glands,
> shrinkage of the thymus and lymph glands,
> peptic ulceration.

With continued exposure to the stressor, the second stage of *adaptation* or *resistance* develops when all these changes are reversed and the organism has apparently adapted to the changed situation. However, if the stressor is severe and exposure continued, the animal may enter the final stage, the stage of **exhaustion**: all the signs found during the alarm reaction reoccur and the animal dies.

It is now known that the changes described by Selye are due to an increase in secretion of the glucocorticoid hormones. These hormones are essential for life; without the secretions of the adrenal cortex exposure to even relatively minor stressors leads to collapse and death.

Actions of glucocorticoid hormones
Steroid hormones, including the glucocorticoids, act on the nucleus of the cell to stimulate the formation of messenger-RNA, and thus of enzymes which modify cell function. In addition to the alterations in function caused by glucocorticoids,

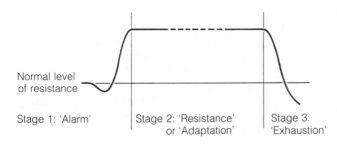

Normal level
of resistance

Stage 1: 'Alarm'   Stage 2: 'Resistance'   Stage 3:
                    or 'Adaptation'   'Exhaustion'

Figure 2.5.27 The three phases of the general adaptation syndrome (from Selye, 1976)

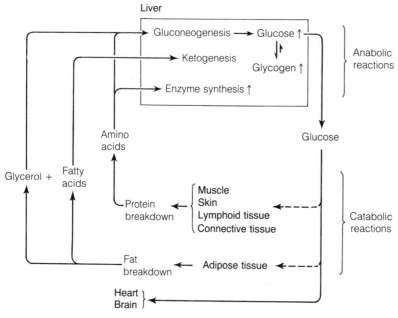

Liver

Key

———→ Pathways increased

- - - -→ Pathways decreased

Figure 2.5.28 Action of glucocorticoid hormones on carbohydrate, fat and protein metabolism

these hormones also have some permissive actions. They must be present to allow some other hormones to modify certain metabolic reactions; for instance, the catecholamines only influence a number of metabolic pathways in the presence of glucocorticoids.

The main actions of the glucocorticoids are concerned with modification of the metabolism of carbohydrates, fats and proteins. In peripheral tissues many of the actions are antagonistic to those of insulin (discussed earlier in the section on maintenance of blood glucose), leading to a rise in blood glucose level, but the glucocorticoid hormones are relatively inactive in heart and brain, and thus the extra glucose is available for use in these organs.

Figure 2.5.28 illustrates the effects of glucocorticoid action on carbohydrate, fat and protein metabolism. In the peripheral tissues, most of the changes are catabolic in nature, i.e. are concerned with breaking down larger molecules such as proteins and fats (in adipose tissue), and are sparing of glucose. Thus, the utilization of glucose in these tissues (muscle, lymphoid, adipose, etc.) is reduced. The changes in the liver, however, are mainly anabolic in nature. Gluconeogenesis is

stimulated and glucose is formed from the amino acids and glycerol reaching the liver from the peripheral tissues. Some of the additional glucose formed is deposited in the form of glycogen. In the liver, reactions which remove amino acids are increased. Thus protein synthesis, breakdown of amino acids (the residue used in gluconeogenesis) and formation of urea are all increased. The fatty acids released may be converted into ketones in the liver and used as an energy source by other tissues. Normally the formation of ketones is constrained by the insulin released in response to the rise in blood glucose.

At high levels of secretion glucocorticoid hormones have a number of other effects on body function.

The anti-inflammatory effect. This effect of glucocorticoids is used in the treatment of certain diseases. However, inflammation is an essential stage in the process of wound healing (see Chap. 6.1) and if this inflammation is prevented or reduced by corticosteroid therapy, wound healing will be slow as the initial fibrous tissue is not laid down normally.

Table 2.5.7 Causes of Cushing's syndrome

Cushing's disease due to excess secretion of ACTH from the pituitary gland

Adrenocortical tumours – adenoma or carcinoma

Ectopic ACTH syndrome – secretion of ACTH by malignant or benign tumours of non-endocrine tissue

Iatrogenic: (a) corticosteroid administration
    (b) ACTH administration
    (c) alcohol

Reduction in the immune response. The reduction in the immune response caused by these hormones is also of value in medical treatment as it is used to reduce the risk of transplant rejection, and in the treatment of autoimmune diseases in which the body is making antibodies against some of its own tissues. However, high levels of these hormones will also reduce antibody formation and white cell count in a patient invaded by bacteria or viruses, and will allow the rapid spread of infection (see Chap. 6.2).

Increase in gastric secretion of hydrochloric acid and pepsinogen. This effect of glucocorticoids leads to an increased risk of peptic ulceration.

Effect on the nervous system. This is indicated by personality changes with either below or above normal levels of these hormones.

Effect on fluid and electrolyte balance. These hormones have some degree of mineralocorticoid activity, increasing sodium and water retention and potassium excretion through the kidneys. However, glucocorticoids increase the glomerular filtration rate and may have some effect on the distal tubules and collecting ducts of the kidneys, allowing adequate secretion of water.

Many of the effects of glucocorticoid hormones described can have deleterious effects on an individual if secretion is maintained at a high level, although all the effects can also be of value, in moderation, under normal circumstances. **Cushing's syndrome**, which results from high levels of these hormones, can be due to a number of factors (Table 2.5.7), although it appears that changes in the pituitary gland are the commonest cause. The changes which occur in Cushing's syndrome are shown in Table 2.5.8. Similar changes are also found in individuals in a state of stress for a period of time, although they will usually be less severe.

Biosynthesis and transport of glucocorticoids
Glucocorticoids and the other corticosteroid hormones are formed in the adrenal cortex from free cholesterol. As this process is stimulated, more cholesterol is released (Brown *et al.*, 1979). Ascorbic acid is also present in large amounts in the adrenal cortex and the quantity drops when the gland is stimulated. A number of different

Table 2.5.8 Disturbances in Cushing's syndrome

Protein catabolism ↑ → protein depletion	Thinning of skin and subcutaneous tissue Muscles poorly developed Reduced protein replacement	Skin easily damaged → bruises, lacerations Lacking in strength Poor wound healing
Changed fat distribution	Moves from periphery Collects in face, upper back, abdominal wall	Thin limbs 'Hump' between shoulders Stretching → striae
Altered glucose metabolism	Hyperglycaemia, ↓ in glucose utilization	Diabetes mellitus in susceptible people
Altered fluid and electrolyte balance	Sodium and water retained Potassium depletion	Moon-face, hypertension (85% of patients) Muscle weakness
Effects on bone metabolism	Loss of collagen matrix Anti-vitamin D effect Increased calcium excretion	Osteoporosis – softening and demineralization of bone
Response to infection	↓ Fibrosis and inflammation ↓ Systemic effects of infection ↓ Antibody formation ↓ Eosinophil count	Infection spreads readily No obvious effects of spread Reduced ability to combat infection
Effect on nervous system	Acceleration in basic EEG rhythms → mental aberrations	Increased appetite, insomnia, euphoria Toxic psychoses

glucocorticoid hormones are synthesized, the commonest being cortisol (see Fig. 2.5.3) and corticosterone.

About 20 mg/day of cortisol and about 3 mg/day of corticosterone are produced from the adrenal cortex. There is considerable variation in the rate of secretion at different times in the 24 hours, with a minimum at about midnight and the maximum in the early morning in those working during the day.

The glucocorticoids are transported in the plasma bound to a globulin protein, **transcortin** or **corticosteroid-binding globulin**, and to a lesser extent bound to albumin. The different glucocorticosteroids have varying degrees of affinity for these proteins: cortisol is bound strongly, while corticosterone is less strongly bound, and the half-lives of the hormones reflect these differences. The half-life of cortisol is about 60–90 minutes, while that for corticosterone is about 50 minutes. It is only the free hormone which is physiologically active, and the amount of this is very small, about 5% of the total. The bound hormone acts as a reservoir to maintain the supply of free hormone to the tissues.

Metabolism and excretion of the glucocorticoids
The glucocorticoids are not altered in structure while exerting their physiological effects on the body.

The hormones secreted from the adrenal cortex are metabolized in the liver to a number of other steroid products – including **cortisone** which is widely used therapeutically but not secreted in appreciable amounts in the body. Most of these steroid products are conjugated with glucuronic acid, in which form they are freely water soluble. On re-entering the circulation they do not become bound to plasma proteins and are rapidly excreted in the urine. A relatively small amount (15–20%) of the conjugated steroids is excreted through the biliary system in the faeces. In disease of the liver, degradation of glucocorticoids is depressed.

Regulation of glucocorticoid secretion (Fig. 2.5.29)
The release of glucocorticoids from the adrenal cortex is controlled by the secretion of **adrenocorticotrophic hormone (ACTH)** from the anterior pituitary gland. Regulation of release of ACTH appears to take place at three levels: in the

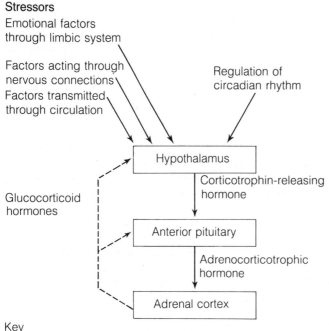

Stressors

Emotional factors through limbic system

Factors acting through nervous connections

Factors transmitted through circulation

Regulation of circadian rhythm

Glucocorticoid hormones

Hypothalamus

Corticotrophin-releasing hormone

Anterior pituitary

Adrenocorticotrophic hormone

Adrenal cortex

Key
----→ Feedback inhibition

Figure 2.5.29 Regulation of glucocorticoid secretion

higher centres of the nervous system, the hypothalamus, and the pituitary gland. Regulation from the higher centres appears to be mediated through an effect on secretion of **corticotrophin-releasing factor (CRF)** from the hypothalamus. It has been suggested that there may be more than one substance with CRF activity, and that ADH may have a role in the regulation of ACTH secretion (Mialhe *et al.*, 1979).

Three major factors are involved in the regulation of ACTH secretion: circadian rhythm, stressors, and negative feedback from glucocorticoids. The circadian rhythm in ACTH secretion seems to be mediated through release of CRF as the peak in ACTH levels follows the CRF peak by about 2 hours, and it is followed by the peak in corticosterone levels about 2 hours later still (Ixart *et al.*, 1977).

Stressors also act by modifying the release of CRF, either through the effect on the hypothalamus of substances carried in the blood, or through the nervous links with other parts of the brain, including the limbic system. Factors (several of which may be experienced by patients) which may act as stressors include environmental factors such as heat or cold, physiological factors such as hypoglycaemia, surgery or pyrogens, or psychological factors such as anxiety. At least some of these factors will only act as stressors if the situation is perceived as stressful by the individual (Lazarus, 1971).

Circulating glucocorticoids have an important role in the regulation of ACTH secretion. There appears to be both a rapid and a long-term inhibition of ACTH release. The rapid inhibition of secretion caused by glucocorticoids results from effects of the hormones at the hypothalamus, while the long-term effect occurs at both the hypothalamus and the pituitary gland. It is only the unbound hormone in the blood which acts on the pituitary, and this level is normally very low, with a fairly small inhibitory effect. However, if this level drops (as in chronic adrenal insufficiency – **Addison's disease**), ACTH release is not inhibited and the levels rise. **Melanocyte-stimulating hormone** release is also increased concurrently with ACTH, and causes the increased pigmentation found in this condition.

Secretion of ACTH occurs in irregular bursts lasting for a few minutes, and secretion of glucocorticoids rises and falls in response. The circadian rhythm in secretion is shown by the greatest release occurring during the early morning and the lowest in the evening.

ACTH is a polypeptide containing 39 amino acids but is synthesized as a part of a large precursor molecule which is also the precursor for lipotrophin, melanocyte-stimulating hormone and the endorphins and enkephalins. These last two are peptides with analgesic properties (the endogenous opiates).

The half-life of ACTH in the blood is about 10 minutes. It acts by binding to membrane receptors on adrenal cortical cells, where it stimulates the conversion of cholesterol to the steroid hormones. Uptake and the endogenous formation of cholesterol are increased, as is synthesis of adrenal gland protein. The ACTH also stimulates the release of glucocorticoids from the cell, but has only a small effect on the output of mineralocorticoid hormones.

The level of ACTH secretion, therefore, is the result of a balance between the stimulation caused by stressors and other factors mediated through the hypothalamus, and the inhibition due to glucocorticoids in the blood. The inhibition following the rise in glucocorticoids in the blood moderates the rise in ACTH secretion due to stressors.

Clinical implications

The rise in ACTH and glucocorticoid secretion in response to stressors is essential for survival, although the reason for this is not yet clear. Patients who have been treated with steroids for a period of time and whose adrenal cortical glands have lost the ability to secrete steroids adequately, require the administration of higher doses if exposed to a stressor such as illness or surgery. It is suggested that part of the reason for the requirement for glucocorticoids may be due to their effect on catecholamine action. Catecholamines are often released under the same conditions as glucocorticoids, and the latter have a permissive action for some of the actions of catecholamines, including the mobilization of free fatty acids as an energy source, and their actions on the vascular system. However, the adrenal medulla is not essential for life. Following bilateral adrenalectomy, a patient will remain well if given adequate corticosteroid replacement.

Most patients admitted to hospital are exposed to a number of stressors which cannot be modified, for example a surgical operation. While it is clear that the stress response is essential for adaptation and that patients with inadequate adrenal function recover badly from surgery (Hubay *et al.*, 1975), the actions of glucocorticoids in excess and if prolonged can be harmful. Therefore it is of value to patients to minimize exposure to those stressors which can be modified, for example fear of the unknown. Nursing intervention which aims to relieve anxiety through various types of communication has been shown to minimize the biochemical indicators of stress (Boore, 1978).

Review questions

The answers to all these questions can be found in the text. In each case there is at least one correct and at least one incorrect answer.

1 All endocrine glands

  (a) do not have ducts
  (b) produce hormones which are proteins
  (c) secrete hormones directly into a body cavity
  (d) are regulated through the activity of the hypothalamus

2 The anterior pituitary gland

  (a) is also known as the neurohypophysis
  (b) in embryonic development originates from the base of the brain
  (c) produces the trophic hormones
  (d) is stimulated by hypothalamic hormones

3 Thyroxine

  (a) is produced from the amino acid tyrosine
  (b) is converted to tri-iodothyronine in the tissues
  (c) has a shorter half-life than tri-iodothyronine
  (d) increases the resting metabolic rate

4 Insulin

  (a) is a steroid hormone
  (b) is synthesized in the α cells of the islets of Langerhans in the pancreas
  (c) increases the synthesis of glycolytic enzymes by liver cells
  (d) secretion is inhibited by adrenaline

5 In uncontrolled diabetes mellitus

  (a) there is an absolute or relative lack of insulin
  (b) muscle cells increase their uptake of glucose
  (c) urea production and excretion are increased
  (d) the pH of body fluids rises

6 Antidiuretic hormone

  (a) is produced by the hypothalamus
  (b) secretion is stimulated when the osmotic pressure of plasma falls
  (c) increases the reabsorption of water from the proximal convoluted tubules of renal nephrons
  (d) secretion is increased in stress

7 When aldosterone secretion is excessive

  (a) the condition is called Addison's disease
  (b) blood pressure is increased
  (c) oedema normally develops
  (d) potassium deficiency occurs

8 In hyperparathyroidism

  (a) calcitonin secretion is increased
  (b) serum calcium ion concentration is increased
  (c) renal calculi may develop
  (d) tetany may occur

9 Human growth hormone

  (a) secretion is controlled via the hypothalamus
  (b) has a longer half-life than the somatomedins
  (c) produces a negative nitrogen balance
  (d) in excess, may produce diabetes mellitus

10 Noradrenaline

  (a) is stored in vesicles in the adrenal cortex
  (b) is the neurotransmitter of the postganglionic sympathetic nerve fibre
  (c) increases peripheral resistance and hence blood pressure
  (d) is responsible for long-term responses to environmental changes

11 The changes which occur in Cushing's syndrome

(a) are frequently due to malfunction of the pituitary gland
(b) enhance wound healing
(c) increase susceptibility to infection
(d) are also found in severely stressed individuals

Answers to review questions

1 a
2 c and d
3 a, b and d
4 c and d
5 a and c
6 a and d
7 b and d
8 b and c
9 a and d
10 b and c
11 a, c and d

Short answer and essay topics

1 Draw a diagram illustrating the general adaptation syndrome and discuss the importance of the concept of stress to the nurse.
2 Discuss the disturbances of glucose metabolism that develop in a patient with diabetes mellitus and the physiological effects of these disturbances.
3 Explain the physiological reasons for the disturbances which occur in a patient with thyrotoxicosis, and the concomitant *nursing* measures which might be taken to relieve such effects.
4 Describe the effects of cortisol on metabolism and the disturbances which develop in a patient with Cushing's syndrome.
5 Discuss the role of the hypothalamus in the regulation of endocrine activity.
6 How is the secretion of aldosterone from the adrenal cortex regulated? What effect does this hormone have on electrolyte and fluid balance?
7 Discuss the actions of growth hormone and the control of its secretion from the pituitary gland.
8 Why is it important that the concentration of calcium ions in the blood is kept relatively constant? How is this achieved?

Suggestions for practical work

1 *Glucose tolerance test* (should be performed on normal subjects under supervision). Equipment required:

> 50 g glucose dissolved in tumbler of water (may be flavoured)
> Dextrostix
> Lancets and spirit swabs
> Washbottle
> Watch with second hand
> Spectrophotometer for reading Dextrostix (not essential)
> Clinistix
> Jug for collecting urine specimens
> Graph paper

Work in pairs, one person acting as subject. Subject should have fasted for at least 4 hours (preferably overnight) but may drink water as desired. During the test the subject may not smoke or eat. The glucose dissolved in water is taken by mouth. Urine samples are obtained just before the subject takes the glucose and at hourly intervals for the duration of the test, and are tested with Clinistix for glucose.

Capillary blood samples are tested for glucose with Dextrostix just before the glucose is ingested and at half-hourly intervals for $2\frac{1}{2}$ hours. The thumb, finger or ear-lobe is cleaned with a spirit swab, pierced smartly and one drop of blood applied to the test portion of the Dextrostix without smearing. After exactly 60 seconds, the blood is washed off by a steady stream of water. The colour on the Dextrostix is matched against the colour chart or read with the spectrophotometer and the quantity of glucose present is recorded. Results should be presented as a graph of glucose content plotted against time. Explain your results.

2 *Diuresis* (should be performed on normal subjects under supervision). Equipment required:

> 1 litre of water (may be flavoured)
> 1 litre of Normal saline (may be flavoured)
> 2 jugs to collect urine samples
> Measuring cylinder
> Urinometer or Multistix-SG

Work in groups of three or four, two people acting as subjects, who should have nothing to drink for at least 3 hours before beginning the experiment. The two subjects empty their bladders, and the volume and specific gravity of each sample are measured and recorded. One subject drinks the litre of water, the other subject the litre of Normal saline.

Each subject passes urine at 15-minute intervals for 2 hours. The volume and specific gravity of each sample are measured and recorded.

Draw a graph of volume against time and another of specific gravity against time for each subject. Explain the different findings from the two subjects.

3 *Surgical nursing.*
(a) Observe a patient undergoing a relatively minor operation. For what period of time is he deprived of food (including pre- and postoperative periods)? Test the urine for acetone postoperatively. Explain why acetone may be present.
(b) Observe a patient after a relatively major operation who has an intravenous infusion running. Note the fluid intake and the urine passed in the 24 hours immediately following operation. Why may there be an imbalance?

4 *Admission to hospital.*
Take the pulse and blood pressure of a patient as soon as he is admitted to your ward. Then spend some time with the patient. Explain the ward routine, show him around the ward and introduce him to other patients; give the patient the opportunity to talk about his anxieties and to get to know you. Take the pulse and blood pressure again. Is there any difference? If so, why?

5 *Thyrotoxicosis.*
Talk to a patient with thyrotoxicosis. Ask him about weight changes, sleep, appetite, feeling hot or cold; record the pulse rate and note the condition of the skin. Explain your findings.

6 *Cushing's syndrome*
Talk to a patient with Cushing's syndrome, or one who has been receiving long-term steroid therapy. Ask him about any changes in weight and in distribution of that weight, and whether he has noticed any changes in wound healing or in infection. Examine the skin. Test the urine for sugar. Explain your findings.

References

Boore, J. R. P. (1978) *Prescription for Recovery.* London: Royal College of Nursing.
Brook, C. G. D. (1978) Problems of growth and development in endocrinology. In: *Recent Advances in Endocrinology and Metabolism I*, O'Riordan, J. L. H. (ed.), Edinburgh: Churchill Livingstone.
Brown, M. S., Kovanen, P. T. & Goldstein, J. L. (1979) Receptor-mediated uptake of lipoprotein-cholesterol and its utilization for steroid synthesis in the adrenal cortex. *Recent Progress in Hormone Research* 35: 215–257.
Cahill, G. F. (1976) Insulin and glucagon. In: *Peptide Hormones*, Parsons, J. A. (ed.) London: Macmillan.
Cannon, W. B. & de la Paz, D. (1911) Emotional stimulation of adrenal secretion. *American Journal of Physiology* 28: 64–70.
Conn, J. W. (1977) Primary aldosteronism. In: *Hypertension*, Genest, J., Koiw, E. & Kuchel, O. (eds.), pp. 768–780. New York: McGraw-Hill.
Drury, P. L., Al-Dujaili, E. A. S. & Edwards, C. R. W. (1982) The renin–angiotensin–aldosterone system. In: *Recent Advances in Endocrinology and Metabolism II*, O'Riordan, J. L. H. (ed), pp. 157–186. Edinburgh: Churchill Livingstone.
Dunnigan, M. G., McIntosh, W. B., Ford, J. A. & Robertson, I. (1982) Acquired disorders of vitamin D metabolism. In: *Calcium Disorders*, Heath, D. & Marx, S. J. (eds.). London: Butterworths.
Handler, J. S. & Orloff, J. (1981) Antidiuretic hormone. *Annual Review of Physiology* 43: 611–624.
Hodkinson, H. M. & Irvine, R. E. (1985) Thyroid disease in old age. In: *Textbook of Geriatric Medicine and Gerontology*, 3rd edn. Edinburgh: Churchill Livingstone.
Hubay, C. A., Weckesser, E. C. & Levy, R. P. (1975) Occult adrenal insufficiency in surgical patients. *Annals of Surgery* 81: 325–332.
Ixart, G., Szafarczyk, A., Belugou, J. L. & Assenmacher, I. (1977) Temporal relationships between the diurnal rhythm of hypothalamic corticotrophin releasing factor, pituitary corticotrophin and plasma corticosterone in the rat. *Journal of Endocrinology* 72: 113–120.
Lang, D. A., Matthews, D. R., Peto, J. & Turner, R. C. (1979) Cyclical oscillations of basal plasma glucose and insulin concentrations in man. *New England Journal of Medicine 301*: 1023–1027.
Lazarus, R. S. (1971) The concepts of stress and disease. In: *Society, Stress and Disease, I. The Psychosocial Environment and Psychosomatic Disease*, Levi, L. (ed.). Oxford: Oxford University Press.
Macleod, J. (1984) *Davidson's Principles and Practice of Medicine*, 13th edn. Edinburgh: Churchill Livingstone.

Mialhe, C., Lutz-Bucher, B., Briaud, B., Schleiffer, R. & Koch, B. (1979) Corticotropin-releasing factor (CRF) and vasopressin in the regulation of corticotropin (ACTH) secretion. In: *Interaction within the Brain–Pituitary–Adrenocortical System*, Mortyn, T. J., Gillham, B., Dallman, M. & Chattopadhyay, S. (eds.). London: Academic Press.

Prader, A. (1978) Catch-up growth. *Postgraduate Medical Journal 54* (Suppl. 1): 133–143.

Preece, M. A. & Holder, A. T. (1982). The somatomedins: a family of serum growth factors. In *Recent Advances in Endocrinology and Metabolism II*, O'Riordan, J. L. H. (ed.). Edinburgh: Churchill Livingstone.

Raab, W. (ed.) (1966) *Prevention of Ischaemic Heart Disease*. Springfield, Illinois: Charles C Thomas.

Selye, H. (1976) *The Stress of Life*, 2nd edn. New York: McGraw-Hill.

Tanner, J. M. (1962) *Growth at Adolescence*, 2nd edn. Oxford: Blackwell.

Turner, R. C. & Williamson, D. H. (1982) Control of metabolism and alterations in diabetes. In: *Recent Advances in Endocrinology and Metabolism II*, O'Riorden, J. L. H. (ed.). Edinburgh: Churchill Livingstone.

Wara, D. W. (1981) Thymic hormones and the immune system. In *Advances in Paediatrics*, Barness, L. A. (ed.), Vol. 28, pp. 229–270. Chicago: Year Book Medical.

Young, J. B. & Landsberg, L. (1979) Catecholamines and the sympathoadrenal system: the regulation of metabolism. In: *Contemporary Endocrinology I*, Ingbar, S. H. (ed.), pp. 245–303. New York: Plenum Press.

Suggestions for further reading

Brook, C. G. D. (1981) Endocrinological control of growth at puberty. *Medical Bulletin 37* (3): 281–285.

Elattar, T. H. A. (1978) Prostaglandins: physiology, biochemistry and clinical applications. *Journal of Oral Pathology 7*: 175–207, 253–282.

Hintz, R. L. (1981). The somatomedins. In *Advances in Paediatrics*, Barness, L. (ed.), Vol. 28, pp. 293–317. Chicago: New York Medical.

Lee, J. & Laycock, J. (1983) *Essential Endocrinology*, 2nd edn. Oxford: Oxford University Press.

Martin, D. W., Mayes, P. A. & Rodwell, V. W. (eds.) (1985) *Harper's Review of Biochemistry*, 20th edn. Los Altos, Calif: Lange Medical.

Olley, P. M. & Coceani, F. (1980) The prostaglandins. *American Journal of Diseases of Childhood 134*: 688–693.

Pestana, C. (1981) *Fluids and Electrolytes in the Surgical Patient*. Baltimore: Williams & Wilkins.

Phillips, L. S. & Vassilopoulou-Sellin, R. (1980) Somatomedins. *New England Journal of Medicine 302*, 371–380, 438–446.

Tanner, J. M. (1978) *Foetus into Man*. Shepton Mallet: Open Book Publishers.

Section 3
Mobility and Support

Chapter 3.1
Skeletal Muscles

Susan M. Goodinson

Learning objectives

After studying this chapter, the reader should be able to
1 List the functions of skeletal muscle.
2 Describe the connective tissue support, nerve and blood supply to skeletal muscle.
3 Describe the different levels of organization of skeletal muscle in fasciculi, individual fibres and at the molecular level.
4 Discuss the events which lead to muscle contraction, i.e. stimulation of motor nerves, neuromuscular transmission, and excitation–contraction coupling.
5 Explain how each of the events mentioned in (4) may be allayed or prevented by pathological disorders, electrolyte disturbances and drugs, citing relevant clinical examples.
6 Demonstrate an understanding of the 'sliding filament hypothesis' of muscle contraction.
7 Describe the common disorders which affect skeletal muscle and their pathological effects on the organization of muscle fibres.
8 Review the major problems which arise in individuals afflicted with primary and secondary myopathies and discuss their nursing implications.

Introduction

The primary function of skeletal muscle is to allow movement, not only of the whole body, but of parts of the body relative to each other, so that an individual can move and explore the physical environment. Muscle is a tissue which is structurally specialized for contraction, a property which enables it to transform chemical energy into the mechanical work which causes movements of bones at a joint. Movements are not only generated at joints, but also in soft tissues, for contractions of skeletal muscle also bring about movements of the eyeball, palate, tongue and voluntary sphincters. In addition, muscles play an important role in maintaining body posture, for it is the sustained partial contractions of muscles which allow us to remain in one body position for long periods of time, as when sitting or standing.

Another important function of skeletal muscle is to assist body heat production. It has been estimated that only 25% of the chemical energy generated in skeletal muscle is actually used to do mechanical work, the rest is dissipated as heat. Heat production in muscle may be adjusted when a person is exposed to extremes of environmental temperature. In cold conditions, for example, shivering occurs. This process of involuntary muscle tremor increases heat production in muscle, in order to maintain body temperature when the environmental temperature falls (see Chap. 6.1).

GROSS ANATOMY

The skeletal muscle mass forms 40–50% of the body weight in an adult. It is divided into more

205

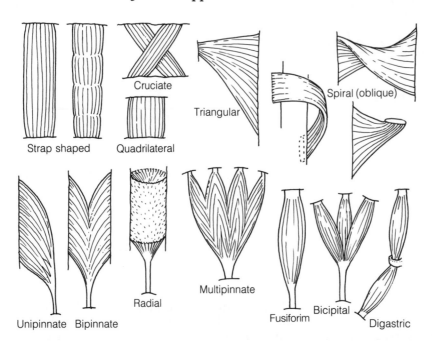

Strap shaped   Quadrilateral   Cruciate   Triangular   Spiral (oblique)

Unipinnate   Bipinnate   Radial   Multipinnate   Fusiforim   Bicipital   Digastric

Figure 3.1.1   Muscle fasciculi arrangements

than 600 individual muscles which vary in size, shape and arrangement of cells. Elongated muscle cells, which are also known as **fibres**, are bound together by connective tissue in bundles, or **fasciculi**. Some of the most common arrangements of fasciculi found within muscles are shown in Fig. 3.1.1.

Fibre length is crucial in determining the range of movements that may be performed by a muscle, whereas the strength of contraction is dependent on fibre numbers and size. If the mean length of fibres is increased, then so too is the range of movement, and vice versa. However, in a muscle with an oblique arrangement of fibres (Fig. 3.1.1), although the fibres are short, giving a limited range of movement, the fibre numbers are increased, giving the muscle greater strength. In health, a factor which influences fibre length and mass is the stress generated when a muscle is put through its full range of movement. If the workload of muscle is increased during athletic training, the fibres increase in size (hypertrophy) and muscle strength is greatly increased. However, consider for a moment the effect on fibre length and mass when muscles are not exercised due to limb pain, prolonged bedrest, joint immobilization or the flaccid paralysis which follows a cerebrovascular accident. In these circumstances muscles are not exercised through their full range of movement

and, as a consequence, the fibres atrophy and shorten and the joint may become stiff and immobile. Inevitably, limited range of movement occurs and the strength of muscle contraction is diminished, leading to weakness. Unless physiotherapy is instituted at an early stage, muscles may become permanently fixed, shortened and resist stretching, causing a **contracture deformity** which is difficult to treat. Atrophy is a common pathological change found in diseased muscles.

In Chapter 3.3, the range and type of movements which are produced when muscles contract at a joint are described. It is important to appreciate that in addition to muscle action, the shape of bone surfaces at the joint and the type of lever system present are of equal importance in determining the range and type of movement possible at a joint.

Prime movers, antagonists and synergists

Muscles do not act in isolation; more than one muscle supplies a joint and their action is integrated. Furthermore, muscles may cross more than one joint so that when a muscle contracts, movement occurs at all joints crossed by it. Unfortunately, this can present problems, for, unless action at some of the joints is prevented, a purposeful movement cannot be made. This difficulty is overcome by utilizing the combined actions of muscles as prime movers, antagonists and synergists. Muscles which initiate and maintain a desired movement are known as **prime movers**, whilst those which resist the actions of prime movers and are capable of producing the opposite movement are known as **antagonists**. When the prime movers shorten and contract, the antagonists relax and lengthen but their structural and physical properties offer resistance to the prime movers. Towards the end of the movement, a brief contraction of the antagonists opposes the prime movers in order to bring about 'braking' deceleration. In certain circumstances, prime movers and antagonists contract in unison in order to stabilize movements at a joint, most commonly when some part of the body must be held rigid and immobile, for example the knee joint in the standing position.

If a muscle crosses more than one joint, or if it produces more than one type of movement at a single joint, it is essential that unwanted movements which are generated during contraction are removed. **Synergistic muscles** are those which cancel out such unwanted movements, helping the prime movers to produce the desired effect.

Muscle attachment

Muscles are attached at the periosteum, to bones by tendons. Two types of attachment, known as **the origin** and **the insertion**, are made here. During muscle contraction the site of origin remains relatively fixed whilst the insertion moves. Therefore, contraction of a muscle brings the sites of origin and insertion closer together.

In addition to providing muscle with firm attachments, **tendons** must transmit the forces generated during muscle contraction to bone. In order to perform these functions effectively, tendons consist of white fibrous connective tissue containing parallel bundles of collagen fibres. This enables them to retain flexibility but resist stretching when a pulling force is applied. Fibrocartilage is present and gives added strength to the areas where tendons merge with bone. If tendons are severed by trauma, the longitudinal arrangement of collagen fibres makes resuturing difficult. Healing of tendons is also affected by a sparse arteriolar blood supply which limits their regeneration.

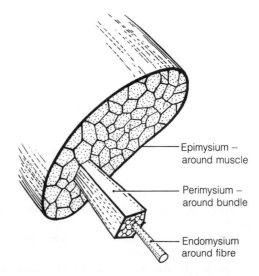

Figure 3.1.2 Connective tissue surrounding muscle

Epimysium – around muscle

Perimysium – around bundle

Endomysium around fibre

Connective tissue

Muscles are enclosed by a very extensive network of fibrous connective tissue which is penetrated by blood vessels, nerves and lymphatics. Surrounding the entire muscle is a smooth layer of **epimysium** which allows the muscles to slide over adjacent structures during movement (Fig. 3.1.2). A finer septum of connective tissue, the **perimysium**, subdivides the muscle into bundles of fibres, and finally each fibre or muscle cell is surrounded by a very fine layer of connective tissue, the **endomysium**.

In addition to providing muscle with a supporting framework, the connective tissue matrix condenses into the tendons which provide muscles with their origin and insertion on bone. Furthermore, the physical properties (viscosity, elasticity) of the connective tissue matrix allow relaxed antagonists to offer resistance to prime movers during movement. Elasticity is an important property of both the connective tissue matrix and the tendons. It is these structures which act like a spring during muscle contraction, transmitting the forces generated in muscle to bone.

Nerve supply

Each skeletal muscle is supplied by one or more nerves containing both sensory and motor fibres. Three types of motor nerve are present: **mye-**

linated **alpha efferent nerves**, which relay from the anterior horn cells of the spinal cord to muscle fibres; **myelinated gamma efferents**, which supply the muscle spindles; and **non-myelinated autonomic efferents**, which supply smooth muscle in the walls of arterioles.

A number of **sensory afferent fibres** are present, some of which terminate in proprioceptors in the muscle spindles, tendons and connective tissue. **Proprioceptors** are concerned with the detection and signalling of information on movement and body position to the central nervous system; their action is essential to coordinate and grade muscle contraction. They are stimulated by the mechanical deformation or vibration created when muscles and joints move and also by changes in body position. Muscle spindles relay information on length and on speed of contraction in muscle fibres. A fuller discussion on sensory receptors is given in Chap. 2.2, and on muscle spindles in Chap. 2.1.

Some of the sensory afferent nerves terminate as free-nerve endings in the endomysium and tendons, responding to the stimuli created by compression, inflammation, ischaemia and necrosis of muscle fibres. These nerves are concerned with the signalling of painful stimuli, although it is perhaps misleading to call one fibre a 'pain fibre'

since pain appears to be detected by an imbalance in fibre activities (see gate control mechanism, Chap. 2.3).

Contraction of skeletal muscle is voluntarily controlled. The areas of the brain which control voluntary movement are the cerebral cortex, cerebellum, basal ganglia and brain-stem nuclei. Motor tracts descend from these areas to synapse in the anterior horn of the spinal cord with the lower motor neurones which supply skeletal muscles (Fig. 3.1.3). Motor nerves may originate from one or all of the cervical, thoracic, lumbar or sacral segments of the spinal cord, ultimately terminating at the **motor end-plate** or **neuromuscular junction**.

On entering a skeletal muscle, the axon of a motor neurone divides into a number of unmyelinated branches which supply groups of muscle fibres. Each muscle fibre is innervated by one of these branches. A **motor unit** consists of one motor neurone and the muscle fibres supplied by its branches, which may range in number from as few as ten to several hundred. Very precise movements of the fingers or the eye are brought about by stimulation of the motor units containing smaller numbers of fibres.

Blood supply

Skeletal muscles have a very extensive blood supply, receiving in total 1 litre of blood each minute, which is 20% of the resting cardiac output. The capillary network is vast, consisting of 300–400 capillaries/ $=$ mm^3 of muscle tissue.

In healthy subjects performing maximum exercise at a steady level, the total muscle blood flow increases up to 15–20 litres/min to meet the additional oxygen requirements of muscle cells. However, in athletes, muscle blood flow during exercise may reach an astounding value of 30 litres/min! This increase in the blood supply to muscle during exercise is known as **exercise hyperaemia** and is brought about by the relaxation of precapillary sphincters.

The blood supply to skeletal muscle may be impaired in a number of disorders which include arterial thrombosis, polyarteritis, trauma and bone fractures. Prolonged ischaemia of muscle can lead to fibre necrosis and contracture deformities such as Volkmann's ischaemic contracture.

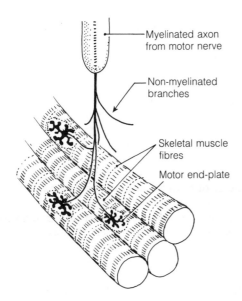

Myelinated axon
from motor nerve

Non-myelinated
branches

Skeletal muscle
fibres

Motor end-plate

Figure 3.1.3 Nerve supply to muscle

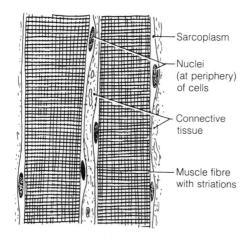

Figure 3.1.4 The skeletal muscle fibre

MICROANATOMY

Approximately 10–100 μm in diameter, skeletal muscle cells or fibres are elongated, cylindrical and taper at the ends (Fig. 3.1.4). Fibre length is variable, ranging from a few millimetres in short muscles to values of 30 cm which may be found in the long sartorius muscle of the upper leg.

A different cell terminology is used to describe some of the structural components of muscle cells. The **sarcolemma** forms the plasma membrane, the **sarcoplasm** the cytoplasm and the **sarcoplasmic reticulum** is similar to the endoplasmic reticulum of other cells. Whereas all body cells have one nucleus, skeletal muscle cells contain several nuclei distributed around the periphery of the sarcoplasm (Fig. 3.1.4). The term used to describe this feature, peculiar to the skeletal muscle cell, is **syncytium**. Fusion of uninucleate myoblast cells during embryological development gives rise to the syncytium.

Inside the cell, the sarcoplasm is filled with longitudinally running fibres 1–2 μm in diameter, known as **myofibrils**. Myofibrils possess a banded structure which gives skeletal muscle its characteristic 'striped' appearance under the microscope (Fig. 3.1.5a). Two different types of filament containing contractile proteins are found within each myofibril. Thick filaments are formed of the protein **myosin**, and thin filaments consist of three different proteins, **actin**, **troponin** and **tropomyosin**. Thick and thin filaments are alternately arranged in the myofibril, as shown in Fig. 3.1.5b.

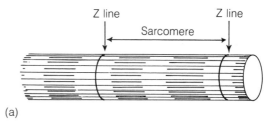

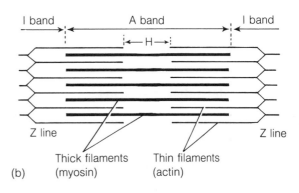

(a)

(b)

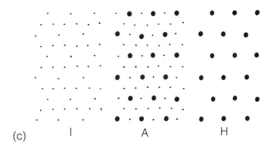

(c)

Figure 3.1.5 Microscopic structure of myofibrils: (a) myofibrils within a single muscle fibre; (b) myofilaments; (c) cross-sections of different bands

Due to the overlap of thick and thin filaments, the banded structure showing A, H, M, I and Z zones is visible on electron microscopy. The area between two Z-lines is known as a **sarcomere**, (as is clearly shown in Fig. 3.1.6). Several sarcomeres are arranged along each myofibril.

During contraction, the thick and thin filaments slide against one another, shortening the myofibrils. This reduces the width of the I and H zones, whilst no change occurs in the A band width.

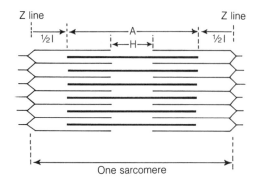

Figure 3.1.6   Zonation in the sarcomere

Intracellular tubular systems

Two tubular systems penetrate each muscle cell (Fig. 3.1.7). One of these is known as the **transverse tubular system** or **T-system**, which consists of tubules lined by sarcolemma, extending from the cell exterior into the sarcoplasm where they branch and terminate near the A–I band of the sarcomere.

The other, wholly internal, system is formed by fine tubules – **sarcotubules** – of the sarcoplasmic reticulum. These end in closed sacs, the terminal cisterns, near the A–I band of each sarcomere. Both tubular systems play an important role in the events which precede and follow muscle contraction.

Myoglobin

Myoglobin is a haem-protein, abundantly distributed throughout the sarcoplasm. Like haemoglobin, it combines reversibly with oxygen and is capable of providing a temporary store for oxygen. It performs this function when the capillary supply to muscle is transiently decreased during contraction, and when the oxygen demands of cells are increased during exercise.

An interesting property of myoglobin is that it combines with oxygen far more effectively than does haemoglobin at lower oxygen tensions. For example, at an arterial PO_2 of only $4\,kPa$ ($30\,mmHg$), myoglobin is 95% saturated with

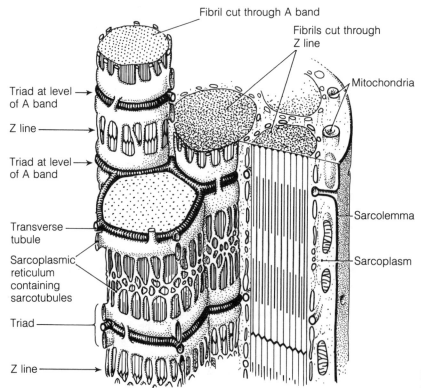

Figure 3.1.7   Intracellular tubular systems

oxygen whereas haemoglobin is only 50% saturated.

Myoglobin may escape into the circulation following crush injuries to muscle, or any other insult which may bring about severe necrosis of muscle, e.g. pressure sore necrosis, drug toxicity and in some of the disorders known collectively as the muscular myopathies. In these circumstances, myoglobin is excreted in the urine, but may form casts in the nephron and lead to acute renal failure.

Molecular structure of the thick and thin filaments

In order to understand how contraction is brought about in skeletal muscle, it is essential to visualize the arrangement of actin and myosin within the thin and thick filaments.

THICK FILAMENTS

More than 100 myosin molecules are found in a single thick filament. If it were possible to examine one thick filament in isolation, we would see a structure very much like that shown in Fig. 3.1.8. Each myosin molecule possesses a double, club-shaped head, followed by a double-stranded 'tail' which is coiled in a helix. All myosin molecules are assembled together in the thick filament so that the heads protrude at the surface and the tails form a cylindrical rod.

According to the **sliding filament hypothesis**, during contraction the globular myosin heads form cross-bridges which attach to binding sites on the thin actin filaments. Repeated attachments of the myosin heads slide the thin filaments towards the centre of the thick filaments, shortening the myofibril. It is this concerted shortening of myo-

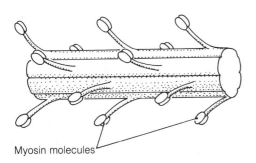

Figure 3.1.8   Structure of a thick filament

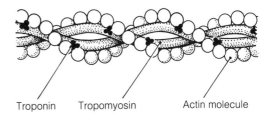

Figure 3.1.9   Structure of thin filament

fibrils which brings about contraction of the muscle cells.

THIN FILAMENTS

Actin, troponin and tropomyosin are the proteins which form the thin filaments. The actin molecules are arranged in two helical strands which are intertwined with a third strand of tropomyosin, as shown in Fig. 3.1.9. Another protein, troponin, is interspersed at regular intervals along the thin filament.

In resting muscle, the orientation of tropomyosin blocks the binding sites on actin, preventing the formation of cross-bridges. However, during muscle contraction, this inhibition is removed by calcium ions, allowing cross-bridges to form between actin and myosin.

MUSCLE CONTRACTION

Before muscle contraction can occur, the following sequence of events must take place.

1 Stimulation of the motor nerves. This leads to the arrival of an electrical impulse, the action potential, at a motor nerve terminal.

2 Neuromuscular transmission. The transfer of a chemical signal from the nerve terminal to the sarcolemma brings about a change in electrical potential, known as 'depolarization', of the membrane. Following this, an action potential is fired in the muscle cell.

3 Excitation–contraction coupling. The generation of an action potential in the muscle cell is followed by activation of the contractile filaments. Shortening of the myofilaments inside the muscle cell, mediated by calcium ions, brings about muscle contraction.

Disruption of these events may be caused by a number of disorders affecting nerve or muscle and may lead to problems of muscle weakness, impaired mobility or even paralysis. In the following pages, stimulation of motor nerves, neuromuscular transmission and excitation–contraction coupling are discussed individually, together with some of the disorders which may impair them.

Stimulation of motor nerves

Muscle contraction is brought about by stimulation of motor nerves controlled by areas of the brain such as the cerebral cortex, cerebellum, brainstem nuclei and basal ganglia. Upper motor neurones descend from these higher centres to synapse in the anterior horn of the spinal cord with the lower motor neurones which supply skeletal muscles. An electrical signal, the action potential, is transmitted along myelinated motor nerves by saltatory conduction, a process which is described in Chapter 2.1.

Inside the muscle, a motor neurone branches to form a number of unmyelinated terminals, each of which supplies an individual muscle cell at the motor end-plate or neuromuscular junction. The action potential finally arrives at this specialized junction between nerve and muscle, also known as the neuromuscular synapse (Fig. 3.1.10). Here, the nerve terminal is separated from the muscle cell sarcolemma by a cleft approximately 30–50 μm wide. Transmission of the signal from nerve to muscle is accomplished by release into the synapse of a chemical transmitter, acetylcholine.

Disorders affecting the relay of impulses by motor neurones are broadly classified as either upper or lower motor neurone lesions. If the nerves are damaged, then, clearly, muscle contraction will be impaired in some way. **Upper motor neurone lesions** are those which affect the centres in the brain which are concerned with the control of voluntary movement. Such lesions include Parkinson's disease, multiple sclerosis and cerebrovascular accidents. Typically, upper motor neurone disorders create problems of muscle weakness and paralysis. In addition, abnormalities in muscle tone may be present, such as spasticity or rigidity. Abnormalities of movement including ataxia, intention tremor, poor muscle coordination, jerky movements (chorea) or writhing movements (athetosis) may also feature in some disorders depending on the nature and site of the lesion.

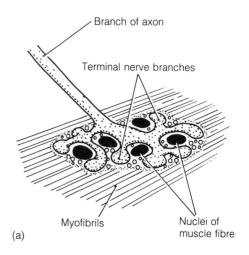

(a)

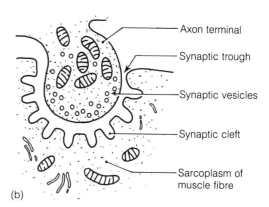

(b)

Figure 3.1.10   The neuromuscular junction

In contrast, **lower motor neurone disorders**, which include the peripheral neuropathies, lead to muscle weakness accompanied by atrophy of muscle fibres, and sometimes sensation is impaired.

Neuromuscular transmission

Acetylcholine is stored in a number of vesicles inside the axon terminal. Arrival of an action potential depolarizes the nerve terminal, triggering an influx of calcium ions from the extracellular fluid. Following this, the vesicles move towards the terminal membrane, fuse with it and discharge at least 1 million molecules of acetylcholine into the synaptic cleft. Acetylcholine diffuses across the cleft, binding to specific protein molecules in the sarcolemma which are known as receptors. Following this, an electrical signal (the action potential) is fired which spreads longitudinally along the muscle cell and is conducted down the T-system. Contraction of the muscle cell follows a few milliseconds later.

Acetylcholine then diffuses away from the receptors in the sarcolemma and is broken down (hydrolysis) by the enzyme acetylcholinesterase which is also present in the synaptic cleft. One of the products of this hydrolysis is choline, which is transported back into the nerve axon and used to resynthesize acetylcholine.

Electrical events

The muscle cell has a negative resting potential which is created by a separation of electrical charge across the plasma membrane. In effect, the inside of the membrane is negatively charged (− 90 mV) with respect to the outside. The reason for this negative potential, which is caused by an unequal distribution of ions across the cell membrane, is discussed in Chap. 2.1.

DEPOLARIZATION

Both muscle and nerve are 'excitable' cells, which means that it is possible to change their negative resting potential. A decrease in the membrane potential towards 0 mV (i.e. *less* negative) which increases excitability is known as **depolarization**, whereas an increase in the negative potential which decreases excitability is termed **hyperpolarization**. These changes are brought about by a phased selective increase in the permeability of the cell membrane to positively charged sodium and potassium ions, a process which generates the action potential in nerve and muscle cells.

THE ACTION POTENTIAL

The combination of acetylcholine with receptors in the sarcolemma of the muscle cell generates a small depolarization known as the end-plate potential. Channels open in the membrane allowing positively charged sodium ions to enter the cell, which decreases the negative membrane potential towards 0 mV (Fig. 3.1.11). Within 0.5 ms, this small depolarization moves the membrane towards a **critical threshold**, at which point a massive

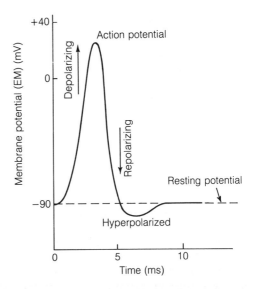

Figure 3.1.11 Action potential in muscle

depolarization occurs and an action potential is fired. Other channels open in the membrane, increasing its permeability to both sodium and potassium ions, although the permeability change initially favours the sodium ions. Positively charged sodium ions enter the cell, reversing the membrane potential which at the peak of the action potential is + 30 mV. At this point **inactivation** occurs, the sodium channels close, and potassium channels which have been opening slowly reach a maximum. Potassium ions move out of the cell, **repolarizing** the membrane back towards the negative resting potential. In fact, it is repolarized to a value slightly below the resting potential so that, technically speaking, at this point the cell is hyperpolarized. The sequence of electrical events from the initial depolarization of the end-plate to the conclusion of the action potential only takes 5–8 ms to accomplish. From the generation of the action potential, the muscle cell has gained a small quantity of sodium ions and lost potassium ions. The intracellular resting concentrations of these two ions are restored by the action of the Na^+/K^+ ATPase pump, returning the membrane potential to its resting value of − 90 mV.

Impaired neuromuscular transmission

There are several circumstances in which the

depolarization of the sarcolemma may be impaired or prevented. Some of the most common are

1 A decrease in the release of acetylcholine due to toxins or electrolyte imbalances.
2 A blockade of acetylcholine receptors by drugs, or by antibodies in the case of myasthenia gravis.
3 A change in excitability of the sarcolemma caused by drugs or electrolyte imbalances.

DECREASED ACETYLCHOLINE RELEASE

Acetylcholine release may be abolished by a number of toxins, which are fortunately rare! Perhaps one of the best known examples of this is the botulinum exotoxin which is produced following food poisoning with *Clostridium botulinum* bacteria. The exotoxin produced by the bacteria abolishes acetylcholine release at the skeletal neuromuscular junction and in the parasympathetic nervous system. This leads to skeletal muscle paralysis with involvement of the respiratory muscles. Effects of parasympathetic blockade include decreased gut motility and retention of urine. Death may be caused by respiratory failure unless the individual is admitted to an intensive care unit and mechanically ventilated.

Calcium ions play a vital role in mediating the release of acetylcholine from vesicles inside the nerve terminal – actions which are opposed by magnesium ions. If the extracellular concentration of magnesium ions is raised, then calcium-mediated acetylcholine release is blocked, leading to weakness or paralysis of skeletal muscle with respiratory embarrassment. Magnesium excess occurs if excretion is impaired by severe dehydration or by renal failure and in situations in which magnesium replacement therapy is excessive. The treatment consists of promoting magnesium excretion in patients with renal failure by dialysis and, in the case of severe dehydration, promoting urinary excretion by fluid replacement therapy. The effects of excess magnesium ions on calcium-mediated acetylcholine release can be reversed by administering intravenous calcium gluconate solution.

If the concentration of free calcium ions in the extracellular fluid is reduced (normal 2.00–2.55 mmol/l), the excitability of skeletal muscle is increased, thus leading to tetany. This is not what we would expect from a knowledge of the role of calcium ions on acetylcholine release! The explanation is that initial reductions in the calcium concentration are directed at muscle excitability; and it is only when the calcium concentration in extracellular fluid is reduced well below 1 mmol/l that neuromuscular blockade is observed.

ACETYLCHOLINE RECEPTOR BLOCKADE

Drugs such as tubocurarine and pancuronium are used therapeutically to block neuromuscular transmission. They act by competing with acetylcholine for receptor sites on the muscle cell sarcolemma. These drugs can be displaced from the receptors and their effects reversed by increasing the local concentration of acetylcholine at the synapse. This is achieved by the administration of anticholinesterase drugs such as neostigmine and pyridostigmine. Tubocurarine and pancuronium are both used to achieve muscle relaxation during surgery; to facilitate intermittent positive pressure ventilation (IPPV) and to control the severe muscle spasms caused by tetanus infection. As these drugs paralyse the respiratory muscles, IPPV is always employed whenever they are used.

MYASTHENIA GRAVIS

Myasthenia gravis is an uncommon disorder, affecting 1 in 18 000 individuals. It may occur at any age, but is most common after 30 years (Drachman, 1978).

Although the cause of this disorder is not fully understood, 75% of myasthenic patients have an abnormality of the thymus gland. More than 85% have raised anti-acetylcholine receptor antibodies present in their serum. These antibodies, which are produced by the thymus, bind to acetylcholine receptors reducing the numbers available to combine with released acetylcholine. As a result, when end-plate potentials are generated, they may be inadequate to trigger an action potential in the skeletal muscle cell. Muscle weakness and fatigue result (see Table 3.1.2).

It is a characteristic feature of this disorder that the muscle weakness becomes progressively worse as muscles are repeatedly used, probably because it takes time to resynthesize acetylcholine, and repeated use decreases the supply available to produce an effective depolarization. Rest improves this progressive weakness in some muscles. The severity of the disorder may fluctuate through periods of mild and severe exacerbation, with remissions in some individuals, or pursue an extreme course which terminates in death due to respiratory paralysis.

Treatment of myasthenia gravis

Drugs. Neuromuscular transmission may be improved by increasing the concentration of acetylcholine at the motor end-plate. Anticholinesterase drugs such as neostigmine (Prostigmin) and pyridostigmine bromide (Mestinon) achieve this by inhibiting the enzyme acetylcholinesterase which hydrolyses acetylcholine.

Immunosuppressive drugs such as prednisone, azathioprine and cyclophosphamide reduce the production of anti-acetylcholine receptor antibodies. These drugs are the treatment of choice for elderly patients or those with a poor response to removal of the thymus. In severe cases, immunosuppressive drugs may be used in conjunction with plasma exchange (see below).

Thymectomy. Although surgical removal of the thymus is effective in some cases, it is not usually successful in patients aged over 50 years.

Plasma exchange. Plasma exchange involves the removal of large volumes of plasma from the circulation and its replacement by fresh plasma, or a plasma equivalent, so that hypovolaemia is prevented. In the management of myasthenia gravis, the aim of plasma exchange is to deplete the patient's serum of anti-acetylcholine receptor antibodies. At present, plasma exchange is restricted to the management of myasthenia gravis in patients with a severe form of the disease. Great improvements in the condition of these patients have been reported following its use (Glassman, 1979).

Planning nursing care
As the degree of muscle weakness and immobility caused by this disorder varies, individual assessments should be made by the nurse of the patient's capacity to perform the activities of daily living, and care should be planned accordingly. Bearing in mind that the muscle weakness of myasthenia gravis is increased by repeated activity and alleviated by rest, nursing care is aimed at planning the activity of the patient's day to avoid muscle fatigue by supervising periods of rest and activity and

setting small goals which are easily achieved. Maximum use should be made where necessary of physiotherapy aids, such as frames and tripods to assist walking, and the maintenance of good posture. The nursing management of other problems associated with this disorder are discussed at the end of this chapter.

INCREASED EXCITABILITY OF THE MUSCLE CELL

As described earlier in this chapter, calcium ions play a vital role in mediating both acetylcholine release and excitation–contraction coupling.

If the concentration of calcium ions is decreased in plasma, the permeability of the muscle cell sarcolemma to sodium ions is increased. As a result, the excitability of the membrane is increased, bringing it nearer the critical threshold at which an action potential is fired. As the plasma calcium concentration falls, this increased excitability may precipitate a condition known as **hypocalcaemic tetany**. Features of this include numbness and tingling of the extremities, muscle cramps, carpopedal spasms, dysphasia and, in severe cases, laryngeal spasms and convulsions. A positive Chvostek's sign may be present, i.e. contraction of the facial muscles occurs when the facial nerve is tapped.

Hypocalcaemia may be caused by hypoparathyroidism, excessive urinary losses of calcium ions in chronic renal diseases and through the gut due to fistulae, diarrhoea, or prolonged nasogastric aspiration. Hypocalcaemic tetany is treated by administering either oral or intravenous calcium gluconate replacements.

Disturbances in potassium balance may also have repercussions on the depolarization of the sarcolemma. Remember that the resting potential of muscle cells is approximately $-90\,mV$. If the plasma potassium concentration falls, more potassium ions are lost from the cell. As a result, the negative resting potential increases in value above $-90\,mV$, hyperpolarizing the membrane and moving it further away from the threshold at which an action potential is fired. The result is muscle weakness and, in severe depletion states, paralysis. Potassium depletion may occur when excessive losses from the gut of potassium-rich fluid occur in prolonged vomiting, diarrhoea and in patients receiving nasogastric aspiration for intestinal obstruction. Hyperaldosteronism, Cushing's disease, respiratory and metabolic alkalosis and prolonged therapy with diuretics such as frusemide may also lead to potassium depletion. The treatment consists of careful oral or intravenous potassium replacement therapy.

Hyperkalaemia, which may occur in acute renal failure and in adrenocortical deficiency states, has complex effects on the membranes of nerve and muscle. Excitability is increased in slight to moderate hyperkalaemia, but it is reduced in severe hyperkalaemia where it leads to weakness and paralysis of skeletal muscle. However, the effects on the heart usually lead to cardiac arrest before any signs of paralysis have set in.

Alterations in the potassium concentrations in plasma and muscle cells are features of the hypo- and hyperkalaemic metabolic myopathies. Altered excitability of the cell membrane leads to intermittent attacks of muscle weakness and paralysis.

A number of drugs prevent cholinergic receptors responding to acetylcholine by maintaining the sarcolemma in a constant state of depolarization. This **depolarizing blockade** causes an immediate flaccid paralysis of skeletal muscle. Suxamethonium, which is used in the same circumstances as pancuronium, is an example of this type of drug.

Excitation–Contraction Coupling

The generation of an action potential in the muscle cell must be followed by activation of the contractile machinery in the myofibrils, in order that contraction may occur. This process is known as excitation–contraction coupling. As described earlier in this chapter, contraction is brought about by shortening of the myofibrils inside the muscle cell when the thin actin filaments slide inwards against the thick myosin filaments, propelled by the myosin cross-bridges.

In the resting cell, sliding of the filaments is prevented by the configuration of tropomyosin on the thin filament, which conceals crucial binding sites for the cross-bridges to attach. Therefore, an essential step in activating the contractile machinery of the muscle cell is to expose these actin-binding sites.

When an action potential is fired, it is transmitted longitudinally along the cell and down into the T-tubular system which is lined by sarcolemma. Depolarization of the sarcolemma lining the T-tubules triggers the release of calcium ions into the sarcoplasm from terminal sacs of the sarcotubular system.

Calcium ions then bind to troponin on the thin filaments, and allow the tropomyosin strand to alter position, exposing the binding sites on actin which are essential for cross-bridge formation. Myosin heads on the thick filaments then attach to the actin-binding sites and form a series of cross-bridges.

In order that the cross-bridges may propel the thin filaments inwards, energy must be supplied by the conversion of the energy-rich compound adenosine triphosphate (ATP) to adenosine diphosphate (ADP) + inorganic phosphate (P_i). This chemical reaction is accomplished by the enzyme adenosine triphosphatase which is present in the myosin head. Release of energy from ATP allows the myosin heads to move to a different angle (Fig. 3.1.12), pushing the thin filaments inwards towards the centre of the thick filaments.

As action potentials continue to arrive, this process is repeated many times. Cross-bridges form, break and reattach along the thin filaments, sliding them inwards. In this way, concerted shortening of the myofibrils inside the muscle cells brings about concentration of the muscle as calcium ions are repeatedly released.

In order to enable relaxation to take place, calcium ions dissociate from troponin and are pumped back into the sarcoplasmic reticulum by an ATP-powered pump. Removal of the calcium ions allows tropomyosin to block the active sites on the actin filaments again, breaking the cross-bridges. Table 3.1.1 summarizes the events occurring before, during and after muscle contraction. For a more detailed consideration of muscle contraction, see Murphy (1979) and Tregear and Marston (1979).

IMPAIRED EXCITATION–CONTRACTION COUPLING

Delayed relaxation of skeletal muscle – **myotonia** – is a feature of myotonia dystrophica, a muscular dystrophy of genetic origin. It has been suggested that this may be caused by a delay in returning calcium ions to the sarcoplasmic reticulum after muscle contraction.

In Duchenne muscular dystrophy, which is a genetically determined *non*-myotonic dystrophy, microscopic loss of areas of the sarcolemma has been reported. It is thought that this may increase

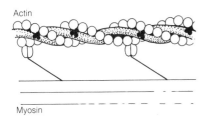

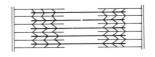

(a)   Sarcomere relaxed

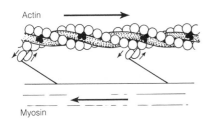

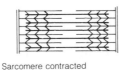

(b)   Sarcomere contracted

Figure 3.1.12   Cross-bridge formation

Table 3.1.1   A summary of events before, during and after muscle contraction

1   An action potential initiated in a motor neurone arrives at the axon terminal on the motor end-plate
2   Calcium ions move into the axon terminal, following which vesicles containing the neurotransmitter acetylcholine move to the presynaptic membrane and release their contents into the synaptic cleft
3   Acetylcholine molecules diffuse across the synaptic cleft and bind to receptors on the motor end-plate sarcolemma. An end-plate potential is generated
4   The end-plate potential depolarizes the sarcolemma to the critical threshold at which an action potential is fired. Acetylcholine diffuses away from the receptors and is hydrolysed by cholinesterases in the synaptic cleft
5   The muscle fibre action potential is conducted over the surface of the sarcolemma and down into the transverse tubular system
6   Calcium ions are released from terminal sacs of the sarcoplasmic reticulum and bind to troponin. Binding sites on actin are exposed
7   Myosin heads on the thick filaments attach to actin binding sites forming cross-bridges. ATPase in the myosin head is activated
8   The release of energy from ATP allows the myosin heads to move to a different angle, sliding the thin filaments inwards towards the thick filaments
9   ATP binds to myosin, breaking the cross-bridges. As action potentials continue to arrive, maintaining a high local calcium concentration, cycles of cross-bridge formation, dissociation and reattachment occur, moving the thin filaments inwards, and the muscle fibre contracts
10 Calcium ions are sequestered by an ATPase pump back into the terminal sacs of the sarcoplasmic reticulum. Actin binding sites are no longer exposed. Muscle fibre relaxation occurs

the entry of calcium ions into the muscle cell, causing hypercontraction of the myofibrils. This appears to be followed by degeneration of the fibres. Another possibility is that a defect in the sarcoplasmic reticulum increases the intracellular calcium concentration causing the structural changes which have been observed.

A number of drugs can inhibit excitation–contraction coupling. Dantrolene sulphate is used to prevent the spastic movements which often follow spinal cord injury, cerebrovascular accidents, cerebral palsy and multiple sclerosis. In many of the disorders which affect skeletal muscle, including muscular dystrophies, degenerative changes are seen in the structure of the muscle cells. In many of the congenital disorders, abnormal myofibrils are present. If the contractile machinery of the cell is damaged, muscle cells cannot contract effectively. It is not surprising, then, to find that muscle weakness is one of the major problems associated with these disorders.

Types of contraction

Contraction is the process of generating force in a muscle; it does not necessarily lead to shortening of the muscle. In the preceding section, the process whereby force (tension) is developed in the contractile filaments of the muscle fibres was described. Tension is transmitted by the muscle tendon to bone at the point of insertion.

Tension developed by the muscle is applied to a load, which is the object to be moved or supported. The load consists of the bone, the part of the body to be moved and any external object which is supported or lifted. Resisting the tension exerted by the muscle is the force which is produced by the weight of the load. In order to lift a load, the tension generated by the muscle must be greater than that exerted by the load (see also Lever systems, Chap. 3.3).

During an **isotonic contraction**, the tension developed in the muscle remains constant and the muscle shortens, lifting the load. Mechanical work is done, and movement is produced by the contraction. In contrast, during an **isometric contraction**, tension is increased in the muscle but it is insufficient to lift the load and shortening does not occur. In this case, mechanical work is not done, and movement is not produced. Isometric contractions are performed when supporting a load in a fixed position or when pushing against a flat immovable surface. Isometric exercises, which develop tension in muscles without bringing about movement, are frequently used as part of a physiotherapy programme to prevent weakness, atrophy and contractures in patients who are immobilized. They are particularly useful in situations where, for a variety of reasons, isotonic movements cannot be performed, for example, in patients with joint disease when isotonic, active or passive movements can precipitate painful spasms.

MUSCLE TONUS

Muscle tone is defined as the resistance offered to passive stretch of the muscle. It is determined by the muscle nerve supply and its control and also by the elastic and contractile properties of muscle.

Tonic contractions are the sustained partial contractions of muscle which are so important for maintaining posture; they are brought about by the asynchronous contraction of small numbers of muscle fibres producing tautness rather than movement.

Flaccidity (flabbiness) is the term used to describe a decrease in muscle tone (decreased resistance to passive stretch). It is a feature of disorders which directly affect muscle (myopathies) or its lower motor neurone supply (peripheral neuropathies). A flaccid paralysis is an early feature of cerebrovascular accidents or spinal cord injuries.

In contrast, **spasticity** is the increased muscle tone which accompanies higher neural lesions such as Parkinson's disease. It is caused by a decreased transmission of nerve impulses down inhibitory pyramidal paths (see Chap. 2.3). Excitatory impulses relayed to the anterior horn cells then predominate to a greater extent than normal, increasing muscle tone and causing either spasticity or rigidity. A rigid muscle has a very great resistance to passive stretch due to its sustained contraction.

TWITCH CONTRACTION

A twitch contraction is brought about in response to a single nerve stimulus. Figure 3.1.13 shows the characteristic response during an isometric and an isotonic twitch. In both, three stages can be observed: the latent, contraction and relaxation periods.

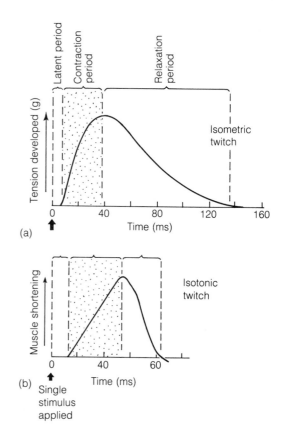

Figure 3.1.13   (a) Isometric and (b) isotonic twitch contraction

In fact, muscles usually contract in response to a series of action potentials, not just one. If the frequency of nerve stimulation is increased so that one stimulus arrives before the preceding contraction phase is over, the tension produced in the muscle will be greater than that created by a single twitch, and the contractile response will increase. In other words, **summation** or addition of contractions occurs. One reason for the summation in tension is that, as a series of action potentials arrives, more and more calcium ions are released from the sarcoplasmic reticulum, increasing the active sites where cross-bridges can form.

TETANUS

It is possible to increase the frequency of muscle stimulation to a maximum point where further summation of tension cannot occur. A sustained contraction known as a tetanus results from this. Muscle contractions are rarely single twitches; summations and incomplete tetanic contractions play the greater part.

RECRUITMENT

One way of increasing the tension (force) developed in the muscle is to increase the number of fibres which contract by a process known as recruitment. This is accomplished when more motor neurones are activated by higher control centres in order to increase the number of motor units which are active. This is one means by which muscle contractions are graded to match the demands of a particular task. As described earlier, muscle tension must exceed the force generated by a load if that load is to be lifted.

The force of muscle contraction is also increased if the resting length of the muscle fibres is increased.

ENERGY SUPPLY

The energy for contractions is provided by adenosine triphosphate (ATP). Energy is released when ATP is converted to $ADP + P_i$ by the action of the enzyme adenosine triphosphatase which is present in myosin. It is vital that ATP is rapidly reconstituted to replenish the energy supply. A way in which this can be accomplished is by using another high energy phosphate compound as a 'back-up' energy source; this is provided by **creatine phosphate**. The following chemical reaction proceeds in order to replenish the ATP supply, catalysed by the enzyme creatine phosphokinase (CPK).

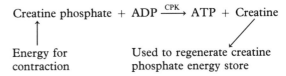

$$\text{Creatine phosphate} + \text{ADP} \xrightarrow{\text{CPK}} \text{ATP} + \text{Creatine}$$

Energy for contraction

Used to regenerate creatine phosphate energy store

Although creatine phosphate provides an immediate reserve energy supply at the start of contraction, additional ATP sources must be mobilized. Oxidative phosphorylation in the muscle cell supplies most of the ATP required during moderate activity. However, during intense exercise the oxygen delivery to muscle is transiently impaired by the contractions, which impede arterial blood flow. This reduces the ATP supply from oxidative phosphorylation. The ATP is then provided by glycolysis, a process which can proceed in the absence of oxygen, using glycogen stores in the muscle cell as an energy substrate. Lactic acid is the end-product of this process.

DISORDERS AFFECTING SKELETAL MUSCLE

All those disorders which primarily impair skeletal muscle are collectively known as **myopathies**. **Primary myopathies** are caused by intrinsic defects within the muscle fibre, for example the genetically determined myotonic and non-myotonic muscular dystrophies. In contrast, **secondary myopathies** arise as a sequel to endocrine, metabolic or vascular disorders; they are not caused by intrinsic defects in the muscle fibre. **Myositis** is the term used to describe an inflammatory myopathy, for example polymyositis and dermatomyositis. Here, inflammation and fibrosis of the muscle connective tissue are major features.

For many disorders, but particularly the muscular dystrophies and congenital myopathies, there is as yet no effective medical treatment which will allay the progression of muscle degeneration. However, the physiotherapy and nursing measures outlined on the following pages may do much to attenuate or delay disabilities. In order to provide information on the risks of transmitting hereditary disorders to offspring, genetic counselling is essential for individuals suffering from many of the muscular dystrophies and the periodic paralyses.

Degenerative changes in muscle

Muscle disorders are characterized by a number of degenerative changes which decrease the size of fibres and impair their intracellular structure and the integrity of surrounding connective tissue. If fibres are damaged, they cannot contract effectively, if at all. It is not surprising to find that in severely affected muscles these degenerative changes lead to pain, with decreased muscle strength, tone and mass, impairing voluntary movement.

ATROPHY

Atrophy is a decrease in muscle fibre mass, which may follow damage to the nerve supply or prolonged immobility. Individual muscle fibres shorten and decrease in diameter. Atrophy is a common feature of many muscular dystrophies. Exactly what the cellular basis for atrophy is, remains to be fully explained, but enhanced breakdown and decreased synthesis of proteins in the myofibrils may account for it.

HYPERTROPHY

Hypertrophy, an increase in muscle mass, is found in the form of dystrophy known as myotonia congenita. Both muscle fibre size and numbers are increased here.

As the increased muscle mass in Duchenne dystrophy is mainly due to fatty infiltration with only minor increases in fibre numbers, it is more correctly described as **pseudohypertrophy**; it predominantly affects the calf muscles.

NECROSIS (FIBRE DEATH)

Fibre necrosis is a common feature of many muscle disorders. It is present in the muscular dystrophies and metabolic myopathies and may follow trauma, ischaemia, inflammation and infection of muscle. Necrosis may also be caused by toxic injury when drugs such as pentazocine are administered, via the intramuscular route, for prolonged periods.

DEFECTS IN THE SARCOLEMMA AND SARCOPLASMIC RETICULUM

In muscle fibres, the sarcolemma (or cell membrane) consists of two areas: a basement lamina of glycoproteins, inside which is the lipoprotein plasma membrane. If the basement lamina and the endomysium of connective tissue outside it are extensively damaged, no 'scaffolding structure' exists to assist the regeneration of a new fibre. Traumatic and ischaemic injuries most commonly cause this type of 'non-selective' damage and this limits regeneration.

MYOFIBRILLAR ABNORMALITIES

Alterations in the structure of myofibrils are seen in many muscle diseases in which fibres are undergoing necrotic changes. Myofibrils may become hypercontracted, 'clumped' and disassembled due to the action of phagocytes.

CONGENITAL ABNORMALITIES

Abnormal structures in the myofibrils are found in the congenital myopathies. In central core disease, for example, cores of disorganized myofibrils are found in the centre of the fibre.

OTHER CHANGES

In a healthy muscle fibre, several nuclei are distributed around the periphery of the sarcoplasm. In a diseased muscle fibre, the nuclei increase in numbers and move towards the centre of the sarcoplasm. This may be a feature of regeneration.

Fatty infiltration and fibrosis of muscle are found in the muscular dystrophies. Inflammation and fibrosis of muscle connective tissue occurs in inflammatory myopathies such as polymyositis.

Regeneration of skeletal muscle

The full complement of skeletal muscle fibres appears to be established by birth or in the early postnatal period. Subsequent increases in muscle size are achieved by increasing the length and diameter of existing fibres.

Exactly how regeneration of skeletal muscle takes place following injury has not been fully explained. It has been suggested that primitive satellite cells which retain their powers of division persist in skeletal muscle. Following injury, they move towards areas of damage in the fibre. The satellite cells then divide, giving rise to immediate myoblast cells which fuse together to form a multinucleate muscle fibre. A healthy muscle will regenerate after necrosis if the blood and nerve supplies are intact and the cause of injury is removed.

To what extent regeneration takes place depends on the degree of damage to the fibre and adjacent tissues, and whether or not any pathological activity decreases.

Non-selective necrosis of the entire fibre with its basement lamina and the surrounding endomysium may lead to less effective regeneration because vital 'scaffolding' structures are damaged. Trauma and ischaemia are causes of this type of injury.

Regeneration is more effective if the basement lamina and endomysium are left intact, despite necrosis of the remainder of the fibre.

Investigations of muscle disease

ELECTROMYOGRAPHY

Electromyography is a technique which is invaluable in the diagnosis of muscle and nerve disorders. A needle electrode is inserted through the skin into the muscle in order to make an extracellular recording of motor unit action potentials. In healthy muscle, motor unit action potentials are usually 1–2 mV in amplitude, last for 2–8 ms and have two or three distinct phases. In muscle disease, degeneration of fibres leads to low amplitude, short duration motor unit action potentials which may also be polyphasic.

ENZYME TESTS

Enzymes are released into the circulation from muscle fibres which degenerate. Enzyme measurements are therefore useful in the diagnosis of skeletal muscle disease. Those most commonly used include the following.

Creatine phosphokinase (CPK). This is the most sensitive, and is increased during mild and severe degeneration.

Aldolase, aspartate aminotransferase (AST), and lactate dehydrogenase (LDH). These are all increased in serum during rapid degeneration of muscle.

BLOOD AND URINARY CREATINE

Creatine is a compound which is vital to muscle energy metabolism. Muscle diseases or trauma which cause severe fibre necrosis will result in an increase in the concentration of blood and urinary creatine.

MUSCLE BIOPSY

This is another useful technique in the diagnosis of muscle disease. Samples are taken for histopathological examination from an affected muscle.

Signs of primary muscle disorders

Some of the most common signs of muscle disorders are

1 Decreased muscle strength.
2 Decreased muscle mass (atrophy).
3 Decreased muscle tone (flaccidity).

Clinical manifestations resulting from muscle disorders include

1 Muscle weakness.
2 Fatigue.
3 Delayed relaxation (myotonia).
4 Intermittent paralysis.
5 Pain.

Medical assessment

A full medical assessment of the patient is carried out by the physician. This involves taking a medical history in which the family history and onset and duration of symptoms are noted. Body muscle groups are individually assessed by carrying out range of motion exercises and noting alterations in tone and muscle strength. Patterns of weakness are noted, together with the extent of atrophy. Tendon reflexes are also tested; they are reduced in muscle disease. A full examination of the nervous system is necessary if coordination is impaired and tremor or abnormal movements such as chorea, athetosis and dystonia are present. These are the signs of upper motor neurone lesions.

Table 3.1.2 Effects of muscle disorders on activities of daily living

Muscle group affected	Disorder	Activities affected problems
Neck muscles	Duchenne dystrophy Myotonia dystrophica Polymyositis	Inability to support head leading to difficulties during eating, restriction of vision, communication, reading
Respiratory muscles	Myasthenia gravis Duchenne dystrophy Polymyositis Hypokalaemic paralysis	Breathing impaired, leading to dyspnoea
Extraocular muscles	Myasthenia gravis Hypokalaemic paralysis Hyperthyroid myopathy	Impaired vision Diplopia Paresis of eye muscles
Jaw muscles	Myasthenia gravis Polymyositis Myotubular myopathy	Difficulty with chewing food
Facial muscles	Myasthenia gravis Myotonia dystrophica	Altered facial expression Ptosis impairing visual field
Muscles of palate, pharynx, larynx, tongue	Myasthenia gravis Polymyositis	Dysphagia, dysarthria, difficulty in coughing
Proximal muscles of the arms and legs	All primary myopathies	Restricted movement Difficulties meeting hygiene needs, dressing, walking on the flat and up stairs
Distal muscles of the arms and legs	Myotonia dystrophica Rarely, in the later stages of some other primary myopathies	Effects on hands and fingers lead to impaired fine movements, sewing, fastening buttons, writing, etc. Effects on leg muscles cause tripping, high stepping gait
Quadriceps muscles at the knee	Many primary myopathies	Difficulties in standing, walking

Planning care

A full nursing history should be taken and an assessment made in order to identify the problems which exist for the patient. Observations should be made on muscle strength, range of movement, mobility and posture in order to establish to what extent the patient is able to carry out the activities of daily living. Some of the most common problems which arise for patients with muscle disorders are discussed below, and the effects of muscle disorders on activities of daily living are summarized in Table 3.1.2.

WEAKNESS AND FATIGUE

Muscle weakness, wasting and fatigue are common problems resulting from muscle disorders. Weakness may be restricted to one group of muscles, for example, it is mainly confined to the pelvic girdle in corticosteroid myopathy, or it may become progressively more widespread involving the limbs, pelvic and shoulder girdles, as is the case in many muscular dystrophies, congenital and secondary myopathies. (Proximal muscles, nearer the trunk, are usually affected to a greater extent.) Depending on the extent and severity of weakness, mobility may be impaired. Weakness may affect the respiratory and pharyngeal muscles, leading to difficulties in breathing and swallowing for those afflicted with myasthenia gravis, polymyositis and hypokalaemic periodic paralysis.

Fatigue occurs in healthy muscle after intense exercise, but in diseased muscle it may occur readily following any activity. Affected individuals complain of tiring easily and may be exhausted by tasks which they finished with ease before the onset of illness. This is most noticeable in patients suffering from myasthenia gravis, where muscle weakness and fatigue increase excessively with repeated use and are improved by rest.

Bearing this in mind, nursing care is aimed at planning the activities of the patient's day to avoid exacerbating muscle weakness or precipitating fatigue and exhaustion, by supervising periods of rest and activity. Initially, small goals should be set which are easily achieved, to foster independence and boost morale. Activities can be gradually increased as muscle strength is regained. Rapid regression of muscle weakness occurs when the underlying adrenal or thyroid disorders are treated in the endocrine myopathies. Anticholinesterase drugs such as neostigmine and physostigmine are successfully used to alleviate muscle weakness in patients suffering from myasthenia gravis.

STIFFNESS

Delayed relaxation of muscles (myotonia) may lead to slow, stiff movements in patients suffering from the myotonic muscular dystrophies. In myotonia congenita, delayed relaxation of muscles is most severe after rest and is exacerbated by cold; warmth and exercise improve the condition. In this disorder, muscles are hypertrophied so the atrophy and weakness do not cause problems. Administration of drugs such as procainamide, phenytoin or quinine sulphate may improve delayed relaxation of voluntary muscle in the myotonic muscular dystrophies.

INTERMITTENT PARALYSIS

Attacks of intermittent muscle weakness and paralysis are features of the metabolic myopathies. In the hypokalaemic form of periodic paralysis, the attacks of muscle weakness tend to occur in the evening, often following a heavy, carbohydrate-rich meal or after heavy exercise. A fall in the plasma potassium ion concentration and a rise in the concentration inside the muscle cell occur, altering its excitability. Potassium chloride is given orally to treat the attacks and as a preventive measure in the evenings; acetazolamide is also used to treat the disorder. Patients should restrict their dietary intake in the evenings and reduce the carbohydrate content, as well as avoiding heavy exercise.

IMMOBILITY

The extent of immobility experienced by individuals with muscle weakness and wasting depends on the progressive nature of the disorder. Duchenne muscular dystrophy is so relentlessly progressive that walking may be impossible by the age of 10 years and many of those afflicted die in young adulthood. In contrast, the non-progressive nature of some of the congenital myopathies, e.g. central core disease, allows some of these individuals to remain active throughout life. Most patients afflicted with myotonia dystrophica are unable to walk by 35 years of age.

As described earlier in this chapter, muscle fibre length is crucial in deciding the range of movement possible, whilst fibre numbers and size influence muscle strength. Longer fibres give a greater range of movement, and increased size and numbers impart greater strength to muscle. A major influence on fibre size is the stress and strain imposed by daily use. If the workload of a muscle is increased, then healthy fibres increase in size (hypertrophy), giving the muscle greater strength. If muscles are not used, they undergo atrophy; the fibres shorten and decrease in diameter. Atrophy, leading to muscle weakness, wasting and impaired mobility, is a pathological feature in muscle disorders. It can also occur if the nerve supply to a muscle is impaired, or due to disuse if mobility is restricted.

A major problem is that muscles may atrophy and shorten, leading to a **contracture** at the joint which severely limits movement. A permanent deformity may result. Therefore, it is vital that any individual afflicted with a muscle disorder receives regular physiotherapy exercises during the course of the day, which will attenuate atrophy and build up muscle bulk in healthy fibres, thereby promoting strength and preventing disabling contractures. Active and passive exercises must be instituted to maintain range of movement, and maximum use must be made of all aids to ambulation – tripods, walking frames, etc. Immobile limbs must not be left in flexed positions which can cause flexion contractures, and other deformities such as foot or wrist drop must be prevented by the use of bed footboards and splints.

Contractures may also develop when prolonged ischaemia results in necrosis of muscle. This usually occurs in association with bony factures involving the humerus, femur or tibia, although it may follow thrombosis of the arterial supply to the muscle as a result of atherosclerotic disease.

Following fracture reduction, the formation of a haematoma and oedema may lead to obstruction of the muscle arterial supply. If the ischaemia is not relieved within 4 hours, serious muscle necrosis will take place. Subsequent fibrosis and scar tissue formation then lead to shortening of the muscle, causing a contracture deformity. A classic example of this is seen in **Volkmann's ischaemic contracture** which may follow a supracondylar fracture of the humerus at the elbow. Ischaemia of the arterial supply to the flexor muscles of the forearm leads to a flexion deformity of the forearm, wrist and fingers. In order to prevent this, nurses must be able to recognize and report the signs of 'muscle' ischaemia, which in this case include the absence of a radial pulse, severe pain with pallor and coolness of the extremities. Postoperative elevation of the arm will reduce oedema formation. Immediate treatment of this condition includes removal of tight bandages or splints to improve the circulation, and surgical relief of the cause of ischaemia, if necessary.

PAIN

Muscle pain may occur in healthy muscles as a result of severe exercise. We are all familiar with **cramps**, the short, painful, involuntary tetanic contractions of muscle which may follow sudden exercise, especially in those unaccustomed to taking it! Exercise-related pain in healthy muscle usually disappears within a few days.

However, pain may also follow traumatic injury, infection, inflammation and ischaemia of muscle. It may be diffuse or localized, depending on the nature of the disorder. In patients with peripheral vascular disease, an ischaemic muscle pain develops in the calf muscles on walking – **intermittent claudication**. Muscle spasms are sustained muscle contractions which may be reflexly induced by pain, ischaemia and mechanical damage. Persistent muscle pain which is not related to exercise or injury and is associated with decreased muscle tone, strength and atrophy suggests that an underlying muscle disease is present. Nurses should attempt to locate the source of the pain and assess its quality; any precipitating causes, such as exercise, should be identified. Analgesia and rest should then be provided together with immobiliz-

ation and elevation of a painful limb. Pain associated with severe inflammatory disorders such as polymyositis is alleviated when corticosteroid drugs are administered to reduce the inflammation. Ischaemic pain is treated by alleviating the underlying causes. Drugs such as Praxilene (naftidofuryl oxalate) may be useful in relieving muscle pain in patients suffering from peripheral vascular disease.

DYSPNOEA

Dyspnoea may sometimes occur in disorders which directly affect the muscles of respiration, e.g. myasthenia gravis, polymyositis, and in those which impair the nerve supply to the respiratory muscles, e.g. Guillain–Barré syndrome.

Resulting weakness of the respiratory muscles decreases the vital capacity of the lungs, and the patient may be unable to expectorate pulmonary secretions effectively; pooling of secretions then leads to dyspnoea. If dysphagia is also present, saliva or food debris may be inhaled into the trachea leading to airway obstruction. Chest physiotherapy, pharyngotracheal suction and nursing the patient in an upright position which allows maximal lung expansion are some essential preventive measures here.

Respiratory weakness may be so severe in some patients that respiratory failure develops. Intermittent positive pressure ventilation (IPPV) is then necessary.

DYSPHAGIA

Weakness may affect the muscles of the pharynx and jaw in patients suffering from myasthenia gravis or polymyositis, leading to difficulties in swallowing and chewing food. A softer diet with smaller, more frequent meals is recommended instead of large meals and bulky foods which require extensive chewing and swallowing. Prostigmin 1 mg may be given subcutaneously before meals to myasthenic patients, to assist chewing and swallowing. It may be necessary to meet the patient's nutritional needs either by nasogastric feeding or by parenteral nutrition.

Many other problems arise for patients suffering from muscle disorders. Weak, immobile patients are vulnerable to all the complications of immobility, including pressure sores, deep vein thrombosis, chest infections and disuse osteoporosis. Implementing all the physiotherapy and nursing measures which keep the patient as mobile as possible will do much to prevent these complications and other disabilities, as will the prophylactic use of sheepskins, air mattresses and at least 2-hourly changes of position for immobile patients. Depression and disturbances of body image may result from the physical changes brought about by muscle disease.

Review questions

The answers to all these questions can be found in the text. In each case there is at least one correct and one incorrect answer.

1 Prime movers are muscles which perform the following actions

   (a) initiate and maintain a desired movement
   (b) cancel out unwanted movements
   (c) oppose a desired movement to achieve a 'braking' effect
   (d) contract in unison with antagonists in order to stabilize movements at a joint when a limb must be held rigid

2 At rest, the blood supply to skeletal muscle comprises

   (a) 20% of the cardiac output
   (b) 5 litres per minute
   (c) 1 litre per minute
   (d) 3% of the cardiac output

3 Myofibrils contain the following contractile proteins

   (a) myoglobin
   (b) myosin
   (c) troponin
   (d) actin

4 At the skeletal neuromuscular junction, transmission of the signal from nerve to muscle is accomplished by release into the synapse of

   (a) adrenaline
   (b) histamine
   (c) acetylcholine
   (d) noradrenaline

5 In the skeletal muscle cell, depolarization of the sarcolemma may be impaired by

   (a) pancuronium
   (b) botulinum exotoxin
   (c) propranolol
   (d) phentolamine

6 Which of the following enzyme measurements is/are used in the diagnosis of skeletal muscle disease?

   (a) lactate dehydrogenase
   (b) carbonic anhydrase
   (c) creatine phosphokinase
   (d) aldolase

7 During an isotonic contraction

   (a) tension gradually increases but is insufficient to lift the load and shortening does not occur
   (b) tension developed in the muscle remains constant and it shortens, lifting the load
   (c) mechanical work is done and movement is produced by the contraction
   (d) mechanical work is not done and movement is not produced by the contraction

8 If the concentration of calcium ions is decreased in extracellular fluid, which of the following occur(s) in the muscle cell?

   (a) the permeability of the sarcolemma to sodium ions is increased
   (b) the excitability of the membrane is increased, bringing it nearer the 'critical threshold'
   (c) the permeability of the sarcolemma to sodium ions is decreased
   (d) hyperpolarization occurs

Answers to review questions

1 a and d
2 a and c
3 b, c and d
4 c
5 a and b
6 a, c and d
7 b and c
8 a and b

Short answer and essay topics

1 List the functions of skeletal muscle.
2 Define the terms 'prime mover', 'antagonist', and 'synergist'. How are the actions of these muscle groups coordinated during movement?
3 Describe the microstructure of a single skeletal muscle fibre. What specialist terminology is used to describe the intracellular structures present?
4 Describe the molecular structure of the thick and thin filaments within a myofibril. What is a sarcomere?
5 Give an account of the sliding filament hypothesis of muscle contraction.
6 Describe the process of neuromuscular transmission. Briefly explain how this process may be impaired or prevented.
7 Discuss the aetiology of myasthenia gravis; describe the medical treatment and nursing care which may be given to individuals afflicted with this disorder.
8 Explain the ionic events which form the basis of the action potential in a skeletal muscle cell. How may electrolyte imbalances (Ca^{2+}, K^+) affect muscle cell excitability?
9 Define the terms 'resting potential', 'depolarization', 'repolarization' and 'hyperpolarization'.
10 What is muscle tone? How is muscle tone altered in primary myopathies and upper motor neurone lesions?
11 Describe the common signs of muscle disorders. What clinical manifestations may be present?
12 Review the major problems which may arise in individuals afflicted with myotonia dystrophica, Duchenne dystrophy and polymyositis. What nursing care is indicated?
13 What is a contracture deformity? How may this be allayed or prevented?
14 Differentiate between an isotonic and an isometric muscle contraction.

References

Drachman, D. (1978) Myasthenia gravis. *New England Journal of Medicine 298* (3): 136–186.
Glassman, A. (1979) Immune responses: the rationale for plasmaphoresis. *Plasma Therapy 1* (1): 13.
Murphy, R. (1979) Filament organisation and contractile function in vertebrate skeletal muscle. *Annual Review of Physiology 41*: 737.
Tregear, R. & Marston, S. (1979). The crossbridge theory. *Annual Review of Physiology 41*: 723.

Suggestions for further reading

Appel, S. H. & Roses, A. D. (1977) Membranes and myotonia. In *Pathogenesis of Human Muscular Dystrophies*, Rowland, L. P. (ed.). Amsterdam: Excerpta Medica.
Bacon, P. A. A. (1971) Familial muscular dystrophy. *Journal of Neurology 34*, 93.
Eisenberg, E. & Greene, L. (1980) The relation of muscle biochemistry to muscle physiology. *Annual Review of Physiology 42*, 293.
Evarts, E. (1979) Brain mechanisms of movement. *Scientific American 241* (3): 146.
Gardner-Medwin, D. (1979) Controversies about Duchenne muscular dystrophy. *Journal of Developmental Medicine and Child Neurology 21*: 390.
Harper, P. S. (1979) Myotonic Dystrophy. Philadelphia: Saunders.
Hubel, D. (1979). The brain. *Scientific American 241* (3): 39.
Jolesz, F. & Streter, F. (1981). Development, innervation and activity – pattern indiced changes in skeletal muscle. *Annual Review of Physiology 43*: 531.
Katz, B. (1966) *Nerve, Muscle and Synapse*. New York: McGraw-Hill.
Keynes, R. (1979) Ion channels in the nerve cell membrane. *Scientific American 240*: 98–107.
Miles, F. A. (1969) *Excitable Cells*. London: Heinemann Medical.
Mokri, B. (1975) Duchenne dystrophy – electron microscopy findings. *Neurology 25*: 111.
Rosse, C. & Clawson, K. (1980) *The Musculoskeletal System in Health and Disease*. London: Harper and Row.
Stevens, C. (1979) The neuron. *Scientific American 241* (3): 49.
Walton, J. N. (1981) *Disorders of Voluntary Muscle*, 4th edn. Edinburgh: Churchill Livingstone.

Chapter 3.2
Bones

Susan M. Goodinson

Learning objectives

After studying this chapter, the reader should be able to
1 List the functions of bone.
2 Describe the architecture, composition and vascular supply of bone.
3 Discuss the origin and function of osteoblasts, osteocytes and osteoclasts.
4 Describe the processes of endochondral and intramembranous bone ossification.
5 Explain how growth in length and diameter of bone is achieved.
6 Describe the common metabolic disorders affecting bone, explaining how pathological changes give rise to symptoms.
7 Classify fractures of bone.
8 Discuss the stages in the healing of fractures.
9 List the major complications which may occur following the fracture of bone.
10 Describe the factors which promote healing of bone, and the means by which complications may be treated.

Introduction

Our present-day knowledge of the microstructure and physiological functions of bones is far from complete and in many instances there is controversy. Outward appearances can be misleading, for bone is far removed from the structurally inert, reinforced concrete so frequently envisaged. Instead, it is a dynamic tissue with a high metabolic activity, which is continuously undergoing complex, structural alterations under the influences of mechanical stressors and hormones.

Bone is considered to have a dual role, providing mechanical support and mineral homeostasis. Four major functions are ascribed to it.

1 Provision of a structural support for body tissues and an attachment for muscles, tendons and ligaments.
2 Formation of the lever systems which permit body movement (see Chap. 3.3).
3 Fulfillment of a major role in mineral homeostasis by providing a reservoir of body calcium, phosphorus and magnesium salts (see Chap. 2.5).
4 Formation and protection of haemopoietic tissue in the bone marrow (see Chap. 4.1).

Bone is a specialized form of connective tissue which is made durable by the deposition of minerals such as calcium and phosphate within its infrastructure. It is well fitted to withstand the deforming forces generated during body movement, for its ultimate tensile stress and compressive strength compares favourably with other structural materials such as granite and cast iron (Table 3.2.1). In addition, its elastic properties confer flexibility and a diminished likelihood of fracturing under stress, while protection against the impact of a deforming force is afforded by its soft tissue covering. In an adult, skeletal bone forms one of

Table 3.2.1 Physical characteristics of bone (adapted from Passmore and Robson, 1976)

	Ultimate tensile strength (kg/mm^2)	Ultimate compressive strength (kg/mm^2)
Cortical bone	7–10	14–21
Cast iron	7–21	42–100
Granite	10–20	9–26

the largest tissue masses in the body, weighing 10–12 kg.

COMPOSITION OF BONE

Three major components are found in bone.

1 An organic matrix of collagen which imparts its tensile strength.
2 A mineral matrix of calcium and phosphate which imparts its compressive strength and rigidity.
3 Bone cells, including osteoblasts, osteoclasts, osteocytes and fibroblasts.

The organic matrix of bone comprises 25% of its total weight and is known as **osteoid**. By far its greatest component is collagen (95%), with hyaluronic acid and chondroitin sulphate constituting the remainder of the matrix. In contrast, the mineral matrix of bone consists of amorphous calcium phosphate and a crystalline structure similar to hydroxyapatite. The mechanisms underlying the process of bone mineralization are poorly understood. Amorphous calcium phosphate is transformed into apatite crystals which become oriented along the collagen molecules of the organic matrix. The physical properties of the collagen fibres and, in particular, their banded structure, favour formation of the hydroxyapatite crystals $3Ca_3(PO_4)_2 Ca(OH)_2$. Surface ions of the hydroxyapatite crystals are hydrated, forming a shell through which exchange of ions with the body fluids occurs; this provides a basis for the modelling and remodelling of bone throughout an individual's lifetime. The process of mineralization is initially very rapid, with 70% mineralization occuring within a few days, but completion may take up to 6 weeks.

ARCHITECTURE OF BONE

Two forms of architecture are distinguished,

woven and lamellar bone. **Woven bone** is a transitional, relatively fragile form which is most commonly seen during phases of rapid bone formation in embryonic life or at zones of **endochondral** ossification (p. 233). It is also found during fracture repair, and some persists in the adult near sites of tendon insertion and ligament attachment.

In contrast, lamellar bone is of greater durability, manifest in its structure, which consists of layers in each of which the collagen fibres have a different orientation. Two forms of lamellar bone are distinguished – compact, cortical bone, and spongy, cancellous, trabecular bone.

Compact bone

Compact, cortical bone forms the outer area of all bones. It is found in the shafts of long bones where it encloses the marrow cavity, and in the outer and inner parts of flat bones. The functional units of compact bone are the **haversian systems** or **osteons**, which are 0.2–0.5 mm in diameter. These consist of a central canal oriented parallel to the long axis of the bone, through which pass an arteriole, capillary, venule and nerve. Each haversian canal is surrounded by concentric lamellae (as shown in Fig. 3.2.1) which contain **osteocytes** within grooves known as **canaliculi**. Adjacent osteocytes communicate via a system of microcanaliculi which permeate the bone, facilitating the exchange of nutrients and movement of waste products. This is particularly important since osteocytes are not in direct contact with blood vessels. Around each haversian system is a 'cement line' which the canaliculi do not cross. Due to the constant remodelling of bone, great variation in the mineralization of osteons and the diameter of haversian canals occurs. Some canals appear irregular and enlarged due to the removal of mineral and matrix which may be occurring. Remains of earlier systems are also found, forming the interstitial lamellae shown in Figure 3.2.1b. Finally, circumferential lamellae are found beneath the periosteum and endosteum.

Trabecular bone

In contrast to the regular structure found in compact bone, this form contains fewer haversian systems and is organized into a lattice of trabeculae in which red or fatty marrow fills the cavities. Bone

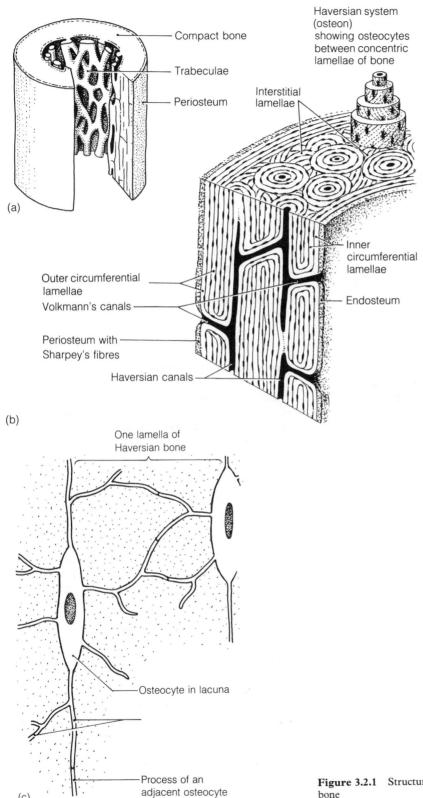

(a)

Compact bone

Trabeculae

Periosteum

Haversian system
(osteon)
showing osteocytes
between concentric
lamellae of bone

Interstitial
lamellae

Inner
circumferential
lamellae

Endosteum

Outer circumferential
lamellae

Volkmann's canals

Periosteum with
Sharpey's fibres

Haversian canals

(b)

One lamella of
Haversian bone

Osteocyte in lacuna

Process of an
adjacent osteocyte

(c)

Figure 3.2.1 Structure of compact
bone

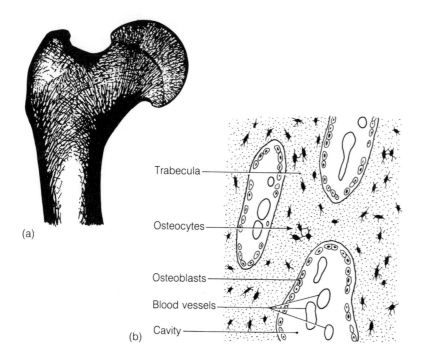

(a)

(b)

Trabecula

Osteocytes

Osteoblasts

Blood vessels

Cavity

Figure 3.2.2 Structure of trabecular bone

remodelling occurs on the surface of the trabeculae, and the lacunae containing osteocytes are scattered throughout the matrix. Spongy, cancellous, trabecular bone is found in vertebrae, flat bones and at the end of long bones (Fig. 3.2.2).

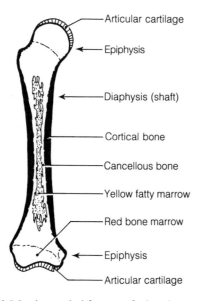

Articular cartilage

Epiphysis

Diaphysis (shaft)

Cortical bone

Cancellous bone

Yellow fatty marrow

Red bone marrow

Epiphysis

Articular cartilage

Figure 3.2.3 Anatomical features of a long bone

Epiphysis and diaphysis

Figure 3.2.3 shows the typical anatomical features of a long bone. The central shaft, or **diaphysis**, is formed of compact bone, while at each end is the **epiphysis**. The latter extends from beneath the articular cartilage inwards towards the epiphyseal plate formed of cancellous bone. The joint capsule is attached around the synovial joint formed at the epiphysis, as shown in Chap. 3.3. Adjacent to the growth plate is the growing end of the diaphysis, termed the **metaphysis**. During growth, the epiphyses are separated from the diaphyses by this growth plate of cartilage. However, once growth is completed in adult life, the plate becomes fully calcified and remains as the epiphyseal line.

Periosteum and endosteum

Contiguous over the surface of most bones is a tough layer of fibrous connective tissue known as the **periosteum**. It does not cover the articular cartilage surfaces of the synovial joint. The outer, fibrous layer of the periosteum transmits blood vessels and provides a site of attachment for ligaments and muscles. Beneath this is found a layer of **osteoblasts**, cells which contribute to the increased growth in diameter of bone in this region. Until growth is complete, the periosteum is not firmly attached to the underlying bone except at the epiphyseal plate. The periosteum is abundantly supplied with myelinated and unmyelinated nerve fibres. These comprise sensory fibres which subserve pain, pressure changes and vibration and some autonomic vasomotor fibres. Stimulation of the sensory fibres due to traumatic injury accompanied by tearing of the periosteum or to the increased pressure generated by a tumour can result in bone pain which is always severe and continuous in nature.

A fine, inner layer of tissue, known as the **endosteum**, lines the marrow cavity of the bone. This contains osteoblasts, osteoclasts and the cells which are their precursors.

Bone cells

The origin of bone cells remains controversial. One major theory proposes that precursor cells originate in bone marrow and that osteoblasts and osteocytes differentiate from haemopoietic stem cells, whereas osteoclasts differentiate separately, from mononuclear phagocytic cells (Vaughan, 1975).

OSTEOBLASTS

Osteoblasts are present on all bone surfaces, in single layers adjacent to the unmineralized osteoid of newly forming bone. They are uniform in size and are linked to adjacent osteoblasts by fine cytoplasmic processes. A characteristic feature is their high intracellular concentration of the enzyme alkaline phosphatase. This appears to play a key role in the mineralization of bone by destroying pyrophosphate which is an inhibitor of this process.

Functions of the osteoblast include the synthesis and secretion of the constituents of the organic matrix, including collagen and protein–polysaccharides (glycoproteins and proteoglycans). In addition, osteoblasts promote mineralization (calcification) of the matrix, controlling the rapid phases of this process in the initial stages (first 4 days).

OSTEOCYTES

Osteocytes are derived from osteoblasts which have become trapped in lacunae by the matrix which they have secreted. Fine cytoplasmic processes extend through the microcanaliculi of lacunae, linking adjacent osteocytes. At present, the precise role of these cells is uncertain. Bone adjacent to the wall of the osteocyte (perilacunar bone) contains fewer collagen fibres and is more soluble in terms of mineral content. Furthermore, the size of the lacunae appears to increase following either vitamin D or parathyroid hormone administration. It has been suggested (Arnold, 1971) that the osteocyte may act as a 'bone pump', releasing calcium from perilacunar bone in response to these hormones (**osteocytic osteolysis**). Other functions ascribed to the osteocytes include synthesis of some collagen and control of matrix mineralization, long-term calcium exchange and plasma protein uptake (Smith, 1979).

OSTEOCLASTS

These are responsible for the resorption of bone and are abundant on or near surfaces undergoing erosion. They show a great variation in size and are highly mobile cells, some are small and contain one nucleus, others are multinucleate giant cells. At their site of contact with bone is a brush border of microvilli which infiltrates the disintegrating bone surface. To facilitate bone resorption, osteoclasts contain a large number of enzymes for lysis of the mineral and organic matrix. They contain a high concentration of the enzyme acid phosphatase. A number of hormones are known to control the activities of the osteoclasts, notably parathyroid hormone, thyroxine, calcitonin, oestrogens and metabolites of vitamins A and D. In order to make calcium available for physiological requirements, osteoclast activity is enhanced by parathyroid hormone and thyroxine. This action is opposed by calcitonin when the opposite conditions apply (see Chap. 2.5).

Bone tissue fluid

Bone surfaces are covered by a layer of osteoblasts and osteocytes which separate the bone tissue fluid from the extracellular fluid and the capillary supply. Metabolites must cross this barrier in order to reach bone. Movement of ions and other metabolites may then occur via the bone fluid in the lacunae and canaliculi. The distribution of bone fluid is shown in Fig. 3.2.4. It differs from

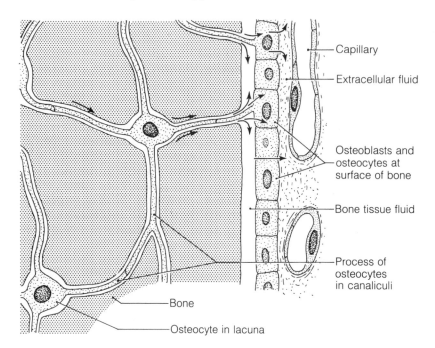

Capillary

Extracellular fluid

Osteoblasts and osteocytes at surface of bone

Bone tissue fluid

Process of osteocytes in canaliculi

Bone

Osteocyte in lacuna

Figure 3.2.4 Bone tissue fluid (based on Smith, 1979)

extracellular fluid in both mineral and protein content: the potassium concentration is higher than that of plasma, but the calcium, magnesium, sodium and albumin concentrations are all markedly lower.

Blood supply

Long bones are supplied by three major vessel types: the nutrient artery, periosteal arteries and the metaphyseal and epiphyseal arteries (as shown in Fig. 3.2.5). The nutrient artery arises from a systemic artery, pierces the diaphysis through a foramen and gives rise to ascending and descending medullary arteries within the marrow cavity. In turn, these give rise to arteries supplying the endosteum and diaphysis.

The periosteal blood supply passes from the periosteal network via Volkmann's canals to the lamellae of compact bone. Approximately one-third of cortical bone and a large proportion of epiphyseal and metaphyseal bone are supplied by periosteal vessels. The periosteal blood supply is increased when growth is complete and the periosteum becomes firmly attached to the underlying bone.

The metaphyseal and epiphyseal supply of long

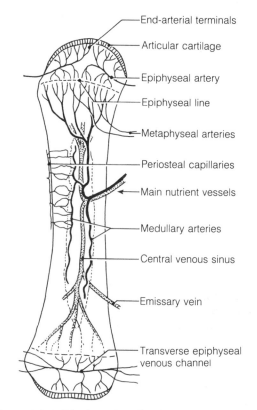

End-arterial terminals

Articular cartilage

Epiphyseal artery

Epiphyseal line

Metaphyseal arteries

Periosteal capillaries

Main nutrient vessels

Medullary arteries

Central venous sinus

Emissary vein

Transverse epiphyseal venous channel

Figure 3.2.5 Blood supply to long bones

bones changes with age. During growth, a special circulation to each epiphyseal growth plate exists, but once growth is complete, this changes. Metaphyseal and epiphyseal arteries arise from systemic arteries, entering the bone foramina as shown in Fig. 3.2.5. Note that they are separated by the growth plate. Nutrients are supplied to the epiphyseal ossification centres by the epiphyseal arteries and to newly formed bone by the metaphyseal arteries. Once growth is complete, the two groups of arteries anastomose freely.

At rest, the arterial flow rate to the skeleton is approximately 12% of the cardiac output, or $2-3\,\text{ml} \cdot 100\,\text{g}^{-1} \cdot \text{min}^{-1}$ (Brookes, 1971). At present, little is known of the mechanisms controlling the osseous circulation, but blood flow is greatly increased in the presence of inflammation, infection and following fracture. The flow rate to the marrow is increased in hypoxaemia, when red cell production is increased to raise the oxygen transport capacity.

DEVELOPMENT OF BONE

The earliest skeletal framework on the developing embryo arises from mesenchyme cells which lay down membranes of fibrous connective tissue. During fetal growth, the fibrous tissue is almost completely replaced by hyaline cartilage templates. Subsequently, bone may develop by ossification of cartilage (endochondral) or replacement of fibrous connective tissue (membranous) ossification. Ossification is begun in early fetal life but is not completed until the third decade of adult life.

Endochondral ossification

Here, pre-existing cartilage templates are replaced by bone. A signal that ossification is about to occur is provided by the appearance of centres of ossification within the shaft of the bone. Chondrocytes in this region hypertrophy and the lacunae surrounding them expand to form cavities within the matrix. The result is that the cartilage matrix degenerates rapidly to form irregular spicules. Subsequent calcification of the cartilage matrix cuts off the nutrient supply to the chondrocytes, which eventually die. While this is occurring, perichondral osteogenic cells are stimulated to

form a circumferential periosteal collar of bone around the mid-region of the shaft. As blood vessels grow inwards from the periosteum, they carry with them periosteal cells which give rise to marrow cells, osteoblasts and osteoclasts. The osteoblasts deposit bone matrix around the scaffolding of calcified cartilage spicules, forming trabecular bars. Within the shaft, spaces invaded by marrow cells become confluent, eventually forming a true marrow cavity. From the centre of the shaft, ossification extends towards the bone ends by the same sequence of events, and the central endochondral bone rapidly converges with the periosteal collar. Initial sites of bone formation in the shaft and periosteal collar are known as **primary centres of ossification**. They first appear in long bones during the first 7–8 months of intrauterine life. At birth, the ends of most long bones still consist of cartilage, but **secondary ossification centres** in the epiphyses appear during the first 5 years of childhood.

Intramembranous bone formation

In this form of ossification, direct replacement of fibrous connective tissue by bone occurs. Intramembranous bone formation occurs in bones of the cranium, lower jaw and clavicle. The imminent ossification of the tissue is signalled by its increase in vascularity. Following this, mesenchyme cells proliferate, hypertrophy and differentiate into osteoblasts which synthesize the collagen fibres and proteoglycans of the osteoid matrix. Under the influence of the osteoblasts, mineralization follows rapidly. Spicules of developing bone gradually enlarge, forming trabeculae in a closely interwoven network. Osteoblasts are eventually trapped within lacunae in the matrix which they synthesize. Here, they gradually lose the potential for proliferation and differentiate into osteocytes. In the final stages of intramembranous ossification, the trabeculae are remodelled and replaced by cancellous or compact bone. However, if the bone remains spongy, the labyrinth is rapidly filled by mesenchyme cells which develop into marrow.

GROWTH OF BONE

Growth in length

Growth in the length of a bone is achieved by

endochondral ossification, and growth in diameter by apposition on existing bone surfaces.

Longitudinal growth of bone takes place by endochondral ossification at the bone ends, a process which is dependent on the production of cartilage at the epiphyseal plate. At the epiphyseal growth plate, cartilage cells known as **chondrocytes** are arrayed in columns extending from the epiphysis inwards towards the shaft. Four zones are distinguished within the columns of chondrocytes. The outer zone is one of proliferation, where chondrocytes are actively dividing, and this is located farthest from the shaft. Beneath this lie zones of maturation and hypertrophy where chondrocytes are gradually enlarging, becoming vacuolated and degenerating. The innermost zone is one of calcification, where osteogenic cells encroach on the lacunae left by the chondrocytes, differentiate into osteoblasts and lay down endochondral bone, as described earlier. Thus, at one end of the epiphyseal plate, cartilage is produced, while at the other end it is degenerating, so growth in the length of bone is dependent on the proliferation of new cartilage cells. As this process continues, bone trabeculae in the diaphysis are eroded by osteoclasts to ensure that the marrow cavity lengthens as growth proceeds.

At the end of the growth period, under the influence of growth hormone and gonadal hormones, the epiphyseal plate is entirely replaced by bone, unifying the epiphysis and diaphysis, a process known as **synostosis**. In most individuals, the epiphyseal plates are fused and no further lengthwise growth is possible by the late teens/ early twenties. Premature arrest of bone growth can occur if the epiphyseal plate is fractured. Although growth in length of most bones is completed by 20 years of age, the clavicle is the last bone to ossify completely in the third decade of life.

Growth in diameter

Growth in diameter of the bone shaft is achieved by deposition of new membranous bone under the periosteum. This is rapidly ossified to keep pace with the growth in length attained by endochondral ossification.

One vestige of the original cartilaginous template of the long bone persists throughout adult life: this is the articular cartilage covering the sur-

face of the epiphysis, which remains as an integral structure in the synovial joint (Chap. 3.3).

REMODELLING OF BONE

The process of bone remodelling continues throughout life, long after growth has ceased. Remodelling comprises phases of bone formation in which new matrix is secreted and calcified, followed by resorption of mineralized bone. Younger bone has a more rapid turnover, for there is an increase in skeletal mass until the age of 50 years, followed by a slow decline. With the exception of growing bones, rates of deposition and resorption are equal, so the total bone mass remains the same in young adults. At any one time, 3–5% of the adult skeletal mass is being actively remodelled by osteoblasts, which continually deposit bone on the outer surface and cavities, and osteoclasts which resorb bone. Remodelling of bone occurs in cycles of activity in which resorption precedes formation. The process is initiated by activation of a group of osteoclasts over a period of a few hours or days. This is followed by a period of resorption extending from 1 to 3 weeks in which osteoclasts erode tunnels 1 mm in diameter through the bone. Resorption is succeeded by bone formation for a period of 3 months: osteoblasts deposit bone inside the tunnels around blood vessels. In this manner, new osteons are formed and the haversian canal is all that remains of the original cavity created by the osteoclasts. In the new mineralized bone, the osteoblasts form osteocytes.

The value of remodelling is that it allows bone to adapt to external stressors, adjusting its formation to increase strength when necessary. In addition, the shape of bones can be rearranged to support mechanical forces more effectively. A number of internal factors also control bone remodelling, for example vitamin D, calcitonin and parathyroid hormone, all of which are concerned with calcium and phosphate homeostasis.

Mechanical stress and pressure

Bone is deposited in proportion to the load it must carry; the converse also applies, for an immobilized bone rapidly decalcifies, a phenomenon known as **disuse osteoporosis**. Very rapid losses of calcium from weight-bearing bones have been observed in healthy individuals confined to bed for a period of 6 weeks (Hulley, 1971). This was reflected by an increase in urinary hydoxyproline excretion, hypercalciuria and an increase in urinary nitrogen losses due to muscle wasting. These effects were reversed once normal activity was resumed. Similar effects have been observed in astronauts, who are subjected to recumbency and weightlessness for prolonged periods. In the Gemini series of flights between 1965 and 1970, some individuals lost up to 20% of their bone mass. The problem was alleviated on later flights by instituting a planned exercise programme (Pace, 1977).

Since the work of Bassett (1971), the possible role that biophysical events may play in controlling bone remodelling has been acknowledged, but remains enigmatic. It has been suggested that the stimulus for bone formation and destruction in remodelling appears to be mainly electrical. A deforming force may generate mechanical stress in bone producing an electric signal. In turn, this could alert the bone mesenchymal cells to adjust the bone's mechanical properties to meet the need that has arisen by activating the formation of osteoblasts or osteoclasts (see also Fracture healing).

HORMONES AND MINERAL HOMEOSTASIS

The maintenance of blood calcium and phosphate concentrations is to a significant degree dependent on the resorption and mineralization of bone. These processes are regulated by vitamin D, parathyroid hormone and calcitonin. If the plasma calcium concentration falls, release of parathyroid hormone is evoked and this in turn stimulates the formation of $1,25 (OH)_2$ vitamin D in the kidney. These act to increase the resorption of bone, and the intestinal and renal reabsorption of calcium, raising the plasma calcium concentration. In contrast, if the plasma calcium concentration rises, calcitonin is released which inhibits bone resorption, achieving a reduction in the plasma calcium concentration. Controversy has surrounded the extent to which bone contributes to mineral homeostasis, but its contribution to short-term regulation cannot be denied. It also provides a reservoir for the long-term buffering of plasma calcium over a period of months and years. The endocrine control of calcium homeostasis is described in Chap. 2.5.

Other hormones

A number of other hormones and vitamins also exert effects on calcium and phosphate homeostasis, but they are primarily directed towards skeletal homeostasis and so they will be considered here.

Growth hormone (GH). This is essential for healthy bone growth, particularly in the epiphysis. A deficiency of the hormone causes dwarfism, while an excess produces gigantism and, after fusion of the epiphyses, **acromegaly**. In acromegaly, the size of the skeleton is increased but this is accompanied by reduced bone density with osteoporotic changes.

Thyroxine. In excess this produces marked skeletal changes. An increased bone turnover occurs, in which resorption predominates, resulting in increased plasma calcium and phosphate concentrations with hypercalciuria. In contrast, a deficiency of the hormone in childhood (**cretinism**) causes growth retardation with a delay in the appearance of the epiphyseal centres; in adults, bone turnover is decreased.

Oestrogen and testosterone. Deficiency of these hormones contributes to osteoporosis. In females, oestrogen deficiency contributes to **postmenopausal osteoporosis**, probably by enhancing the sensitivity of bone cells to parathyroid hormone. Together with growth hormone, the sex hormones bring about the increased growth of bone which leads to fusion of the epiphyses in young adults.

Insulin. Insulin promotes the uptake of amino acids into bone and enhances their incorporation in components of the organic matrix.

Glucagon. This can inhibit bone resorption and possibly bone formation in certain circumstances, but its role in bone metabolism is still unclear.

Cortisol. In excess, as occurs in individuals with Cushing's syndrome and during steroid therapy, this causes a decrease in bone formation accompanied by increased resorption. High doses suppress bone resorption and low doses stimulate it. The changes produced by corticosteriods on bone may lead to osteoporosis and pathological fracture.

Prostaglandins. These are 20-carbon, short chain fatty acids with potent systemic effects. The E series comprises stimulators of bone resorption in vitro and these may be involved in mediating the effects of neoplasms on bone structure.

VITAMINS

A number of vitamins are required for the healthy growth of bone, notably vitamins A and C.

Vitamin A. This is essential for the synthesis of matrix components, such as collagen and glycosaminoglycans, and also promotes the activity of osteoclasts and their production from progenitor cells. Deficiency of vitamin A causes remodelling failure associated with bony overgrowth, while an excess increases demineralization and vulnerability to fracture.

Vitamin C. This is also essential for the synthesis of collagen. In deficiency states, bone formation is impaired and healing is delayed.

DISORDERS AFFECTING BONE STRUCTURE

A comprehensive description of these disorders is beyond the scope of this text and so the enthusiast is directed to the additional reading recommended in the bibliography. The texts by Vaughan (1975) and Rasmussen and Bordier (1975) are classics amongst those written on bone physiology. However, the major features of four common disorders are described below.

Paget's disease: osteitis deformans

This is a disorder of bone architecture, of unknown aetiology. An uncontrolled, intense increase in the activity of both osteoblasts and osteoclasts is present and the cortical and cancellous bone is replaced by coarse trabeculae. This new bone is soft, inadequately mineralized and, as it invades the medullary cavity, the dimensions of the bones increase. Eventually the bone marrow is displaced

by a highly vascular, fibrous tissue. Bones of the axial skeleton are most commonly affected (the skull, vertebral column, pelvis) and the disorder has an overall incidence of 3% in adults under 40 years of age. The problems which result can be formidable, including severe disability due to deformity, bone pain related to arthritic changes, microfissuring and fractures and pressure generated by encroachment of nerves. Nerve compression may result in deafness, and other neurological complications may set in due to bone distortion and overgrowth. Cardiac failure may occur, related to the increased blood flow through the affected bones. The treatment of Paget's disease is aimed at reducing symptoms, preventing complications and using drug therapy to suppress the excessive bone turnover. Mithramycin, calcitonin, glucagon and the diphosphonate drugs have all been used with some success in the suppression of bone turnover. The most effective combination appears to be mithramycin, which decreases bone formation (an antimitotic), with calcitonin, which reduces osteoclastic bone resorption.

Osteitis fibrosa: hyperparathyroidism

The primary cause of hyperparathyroidism is an adenoma of the parathyroid which secretes excessive amounts of PTH. However, the disorder is also found in chronic renal failure accompanied by hypocalcaemia, which stimulates excessive PTH secretion. The pathological features include an increase in the activity of the osteoclasts, leading to excessive bone resorption. As a compensatory response, an increase in the production of woven bone occurs in which mineralization is slowed. The major problems which result for the affected individual include skeletal pain, tenderness and fractures. As calcium ions are lost from bone, the plasma calcium level rises, leading to weakness and lethargy, polyuria and thirst. Renal stones may form as a consequence of hypercalciuria, and an acute arthritis may be precipitated. The treatment comprises surgical removal of the adenoma, together with short-term replacement of calcium reserves by an increased dietary intake. Postoperative hypocalcaemia may be a problem where the bone disease has been extensive; this can be prevented by the administration of high-dose vitamin D.

Osteomalacia

This disorder is the adult form of rickets, a disorder of bone in which defective mineralization of the organic matrix is present. Excess osteoid is the characteristic feature, leading to problems of bone pain, persistent tenderness, deformity, fracture and delayed healing. Muscle weakness may be triggered by a related myopathy, causing immobility and abnormal gait or posture. The disorder is caused by vitamin D deficiency, which may be due to a poor intake in the diet, defective synthesis due to lack of exposure to sunlight or to malabsorption syndromes. It may also be associated with chronic hepatic or renal failure, in which vitamin D cannot be metabolically activated. The treatment of osteomalacia is focused on providing oral vitamin D supplements (25–125 mg daily) and correcting the dietary intake if malnutrition is a contributing cause. In children, rickets may result from inadequate exposure to ultraviolet light or from dietary deficiency of vitamin D.

Osteoporosis

Osteoporotic bone is normal in composition but reduced in quantity, due to an excess of bone resorption over formation. In many cases the aetiology is obscure, particularly when it occurs in childhood or young adults. Deficient oestrogen secretion during the menopause may result in osteoporosis or it may occur as a result of the predominance of bone resorption over secretion which is a feature of old age. In both sexes the latter may be related to hormonal deficiencies, or to a poor dietary intake of calcium and phosphate, or failure of renal calcium excretion to adjust to a low calcium intake. Disuse, immobilization and excessive cortisol secretion can also cause osteoporosis. Whatever the aetiology, the pathological features include thinning of cortical bone by resorption, enlargement of the haversian canals, and loss of trabeculae from cancellous bone. The bone appears fragile and porous and may fracture spontaneously. The problems experienced by the affected individual include pain related to fractures, or deformity, disability and shortening of limbs if fractures become compressed. Remedial measures include the maintenance of regular physical activity in the elderly, ensuring an adequate dietary intake of calcium, and the use of oestrogen replacement therapy in menopausal women who are affected.

Any immobile patient is at risk of developing disuse osteoporosis. It is an alarming fact that the plasma calcium rises, as does urinary excretion, after only 1–2 days bedrest. Over a period of 4–6 weeks, this can amount to losses of 10–14 g calcium, serious enough to complicate recovery following bone fracture. In addition, hypercalciuria may lead to the deposition of stones in the renal tract; a complication which may be prevented by increasing fluid intake sufficiently to raise the flow rate of urine. Thus, early ambulation and physiotherapy are essential if disuse osteoporosis is to be prevented in patients confined to bed. In limbs immobilized by a plaster cast, isometric exercises may be sufficient to prevent the problem.

FRACTURES

Bone fractures or breaks when its capacity to absorb energy is exceeded; most commonly following traumatic injury. Fractures are classified on the basis of location or complexity, as shown in Fig. 3.2.6.

Fracture healing

A number of stages are distinguished in fracture healing: the inflammatory, reparative and remodelling stages. **The inflammatory stage** follows the fracture, when extensive disruption of the blood vessels, endosteum, periosteum and muscles may have occurred. Haemorrhage is followed by clot formation, during which the fibrin network enmeshes bone and muscle debris, erythrocytes, leucocytes and marrow cells. Within 12 hours, blood vessels dilate, allowing neutrophils to infiltrate the site. After 24 hours has elapsed, monocytes infiltrate, phagocytosing the tissue debris and fibrin, assisted by the neutrophils which lyse fibrin and tissue fragments. The extensive invasion of blood capillaries at 72 hours concludes the inflammatory stage. Fibroblasts originating from the periosteal connective tissue, marrow and endosteum also infiltrate the site at this stage, proliferating to form chondroblasts or osteoblasts. In the subsequent **reparative stage**, the chondroblasts synthesize collagen and proteoglycans, uniting the bone ends in a fibrous connective tissue

(a) *On the basis of severity*

Type of fracture	Features
Closed (simple)	Skin intact No communication with surface
Open (compound)	Skin broken Communication from bone to surface Open wound
Complicated	Fractured bone penetrates adjacent organs, blood vessels, nerves, etc.
Complete	Fracture extends completely through the bone
Incomplete 'Greenstick'	Fracture extends only partially through bone; occurs only in pliable bones of child
Comminuted	Bone broken in two or more places Fragmented
Displaced	Bone fragments separated

(b) *Direction of fracture*

Type of fracture	Features
Transverse	Across the bone
Oblique	At an oblique angle to the longitudinal axis of the bone
Spiral	Fracture forms spiral 'twist' encircling bone; produced by rotatory force
Linear	Parallel to the longitudinal axis of the bone

(c) *According to deforming force*

Type of fracture	Features
Compression	Adjacent cancellous bones compacted; usually heals rapidly due to minimal soft tissue injury caused by deforming force
Avulsion	Bone pulled apart; ligaments remain intact
Stress	Undisplaced microfracture caused by repetitive stress, e.g. athletic training may summate to macrofracture

(d) *According to anatomical location*

Type of fracture	Features
Osteochondrial	Involves articular cartilage at a joint and underlying bone
Extra- or intracapsular	Without or with capsular involvement at the joint
According to the major bone and area involved	i.e. Supracondylar fractures of the humerus

This classification is not comprehensive. Pathological fractures occur in bone which is weakened by disease, e.g. in Paget's disease and in osteoporosis.

Some fractures are still characteristically named after the individual who first described them, e.g. Colles' fracture of the distal radius and Pott's fracture of the distal fibula.

Figure 3.2.6 Classification and diagrams of fractures

Table 3.2.2 Complications following fractures: prevention and treatment

Complication	Cause	Prevention/treatment
Avascular necrosis of bone which delays union	Bone and soft tissue damage resulting in diminished accessibility of fracture site to blood vessels May be caused by initial trauma or surgical intervention, e.g. pin and plates, insertion of metal medullary rods Extensive necrosis may occur, e.g.head of femur following subcapital fracture of neck	Avoidance of weight bearing until healing adequate to provide support without risk Prosthetic replacement of necrotic head of femur Surgical arthrodesis (fixation of a joint) advised in certain cases
Joint stiffness and contractures	Prolonged immobility following fracture Formation of scar tissues, fibrosis, ischaemic contracture of muscles Local oedema following fracture (e.g. hand injury)	Maintain maximum activity during immobilization with appropriate active/passive physiotherapy exercises Isometric exercises may be sufficient to prevent muscle wasting in an immobilized limb Volkmann's ischaemic contracture may require surgical release (see 'Muscle') Elevate limb following injury, to decrease oedema
Fat embolism	Embolism caused by flocculation of fat chylomicrons in plasma Triggered by release of kinins and thromboplastins from damaged tissue May lead to cerebral, pulmonary, or renal infarction	Early recognition of signs vital: cyanosis, haemoptysis, hypoxaemia, tachycardia, pyrexia, convulsions, confusion, coma Full supportive measures including IPPV in ITU Low dose heparin therapy to clear lipaemia Intravenous dextran to improve circulation to ischaemic tissue
Deep vein thrombosis	Venous stasis due to immobility Clotting cascade triggered by release of kinins and thromboplastins released from damaged tissue	Early detection of signs vital: calf pain on dorsiflexion of the foot, pyrexia, local swelling and tenderness Prophylactic or therapeutic heparin administration Intravenous dextrans to prevent circulatory sludging Deep breathing exercises and limb exercises to prevent venous stasis Application of Tubigrip stockings Early remobilization and resumption of physical activities
Local oedema	Following removal of splints from limbs Cause uncertain ? Temporary disequilibration of Starling forces	Apply firm bandage Elevate limb intermittently Initiate exercises as advised by physiotherapist
Infection	Following common compound fracture or open reduction	Early recognition of signs – pyrexia, pain, swelling, discharge of pus through sinus – not always visible if splint in situ Antibiotics following identification of organism Strict asepsis during wound excision and dressing procedures High protein/energy diet with adequate vitamin C to promote healing

Table 3.2.2   Continued

Complication	Cause	Prevention/treatment
Pressure sores	Immobility: prolonged pressure on bony prominences May arise under splints and plaster due to local friction Prolonged pressure due to incorrect splinting can lead to nerve palsy	Alleviation of local pressure by 2-hourly or more frequent turning, use of ripple mattresses, sheepskins Scrupulous attention to skin hygiene Early remobilization Careful padding of splints and plasters over bony prominences Maintain nutritional status: ensure diet contains adequate protein, energy, vitamins to meet individual needs Avoid heavy night sedation which decreases spontaneous movements during sleep Established sores: implement above, with other specific treatments, e.g. aseptic technique during application of sterile dressings, antibiotics if infection occurs, application of 'opsite' dressing, treatment with ultraviolet light

known as **callus**. Within 14–17 days, the callus has calcified and, as hydroxyapatite, becomes deposited in the connective tissue matrix; ossification is gradually completed. Osteoblasts lay down trabeculae of cancellous bone at the fracture ends and osteoclasts destroy dead bone, keeping the marrow cavity patent. **Remodelling** of bone by osteoblasts and osteoclasts concludes the healing, a process which takes several months to complete and involves replacement of cancellous by compact bone.

FACTORS AFFECTING FRACTURE HEALING

A number of factors may impair the healing of a fracture. Some of the most important are loss of the blood supply, displacement of bone fragments, loss of the blood clot in open fractures and infection. Less commonly, the administration of corticosteriods, lack of vitamin D and protein-energy malnutrition may all hinder healing. The principles of fracture treatment are therefore to minimize all the factors listed above, to correct any displacement that has occurred and to facilitate healing in a position which will ensure the maximum retention of function. Correction of displacement is achieved by performing either an open or closed reduction, depending on the nature of the fracture, under local or general anaesthetic. Immobilization may be achieved in three ways: splinting in a

plaster of Paris cast, applying traction via a metal pin inserted in the bone, or using internal fixation with a pin and plate. Traction may be initially applied to correct displacement, disengage the bone ends and overcome excessive muscle spasm which enhances displacement. Healing may be promoted by the application of local electrical currents. Electrodes are attached on either side of the fracture and a small current is passed to increase the activity of the osteoblasts.

The complications which may follow bone fractures and the nursing care which may prevent them are summarized in Table 3.2.2.

Ensuring an adequate nutritional intake of energy, protein, vitamins A, B, C and D, calcium and phosphate is essential to promote the healing of fractures. In a recent nutritional survey by Dickerson *et al.* (1986), the postoperative food intake of elderly female patients who had undergone surgery for fractured neck of femur was measured. In many, their nutritional intake of protein, energy, calcium, thiamin, vitamins C, A and D and iron was found to be inadequate despite the adequate provision of food. It is vital that nurses supervise patients at mealtimes and alert the nutritional support team if, for any reason, food intake is impaired. Failure to provide nutritional support in appropriate circumstances could lead to delayed healing, slower remobilization and an increased risk of postoperative complications.

Review questions

The answers to all these questions can be found in the text. In each case there is at least one correct and at least one incorrect answer.

1 The ultimate tensile strength of bone is

    (a) $30 \, kg/mm^2$
    (b) $14–21 \, kg/mm^2$
    (c) $7–10 \, kg/mm^2$
    (d) $50 \, kg/mm^2$

2 The greatest component of the organic matrix of bone is constituted by

    (a) hyaluronic acid
    (b) collagen
    (c) calcium phosphate
    (d) chondroitin sulphate

3 Haversian systems are the functional units of

    (a) compact bone
    (b) woven bone
    (c) cartilage
    (d) transitional bone

4 The metaphysis of a long bone is defined as

    (a) the growing end of the diaphysis adjacent to the growth plate
    (b) the area directly beneath the articular cartilage
    (c) the entire central shaft of the long bone
    (d) an area extending over the entire bone surface below the periosteum

5 Osteoblasts function to

    (a) control osteocytic osteolysis
    (b) promote demineralization of the bone matrix
    (c) synthesize and secrete the organic matrix of bone
    (d) enhance bone resorption on or near surfaces undergoing erosion

6 In order to make calcium available for physiological requirements, the activity of osteoclasts is enhanced by

    (a) parathyroid hormone
    (b) pyrophosphate
    (c) calcitonin
    (d) alkaline phosphatase

7 During endochondral ossification, which of the following occur(s)?

    (a) cartilage templates are replaced by bone.
    (b) fibrous tissue is directly replaced by bone.
    (c) bones of the cranium, lower jaw and clavicle become calcified.
    (d) fibrous connective tissue templates are replaced by cartilage.

8 Growth in length of bone is achieved by

    (a) deposition of new bone under the periosteum
    (b) decreased production of cartilage at the epiphyseal plate
    (c) intramembranous ossification
    (d) endochondral ossification

Answers to review questions

1 c
2 b
3 a
4 a
5 c
6 a
7 a
8 d

Short answer and essay topics

1 List the major functions ascribed to bone.
2 Describe the three major components of bone.
3 Write short notes on each of the following: (a) woven bone, (b) compact bone, (c) trabecular bone, (d) haversian systems.
4 Draw a diagrammatic section through a named long bone, labelling the major anatomical features including the vascular supply.
5 Discuss the origin and functions of osteoblasts, osteoclasts and osteocytes.
6 Briefly explain how bone tissue fluid differs in composition from extracellular fluid.
7 Compare and contrast the processes of endochondral and intramembranous ossification.
8 Explain how growth in length and diameter of bone is achieved.

9 'Bone is deposited in proportion to the load it must carry'. Discuss this statement with reference to the effects of prolonged bedrest and weightlessness on the skeleton.

10 Describe the pathological features and problems resulting from the following disorders of bone: (a) osteoporosis, (b) osteomalacia, (c) osteitis fibrosa, (d) Paget's disease.

11 Describe the stages which occur as a fracture heals. What factors may accelerate the healing process?

References

Arnold, J. S. (1971) The osteocyte as a bone pump. *Clinical Orthopaedics 78*: 47–55.

Bassett, C. A. L. (1971) Biophysical principles affecting bone structure. In *The Biochemistry and Physiology of Bone*, 2nd edn. Bourne, G. H. (ed.), pp. 1–76. London: Academic Press.

Brookes, M. (1971) *The Blood Supply of Bone: An Approach to Bone Biology*. London: Butterworths.

Dickerson, J., Fekkes, J., Goodinson, S. M. & Older, M. (1986) Post-operative food intake of elderly fracture patients. *Proceedings of the Nutrition Society 45*: 7a.

Hulley, S. (1971) The effect of supplemental oral phosphate on the bone mineral changes during prolonged bed-rest. *Journal of Clinical Investigation 50*: 2506–2518.

Pace, N. (1977) Weightlessness: a matter of gravity. *New England Journal of Medicine 297*: 32–37.

Passmore, R. & Robson, J. S. (1976) *Companion to Medical Studies*, 2nd edn. London: Blackwell Scientific.

Rasmussen, H. & Bordier, P. (1974) *The Physiological and Cellular Basis of Metabolic Bone Disease*. Baltimore: Williams and Wilkins.

Smith, R. (1979) *Biochemical Disorders of the Skeleton*. London: Butterworths.

Vaughan, J. M. (1975) *The Physiology of Bone*, 2nd edn. Oxford: Claremont Press.

Chapter 3.3
Joints

Susan M. Goodinson

Learning objectives

After studying this chapter, the reader should be able to

1 Describe the general functions of joints in the musculoskeletal system.
2 Explain how movements in humans is brought about by the concerted actions of muscles, bones and joints within lever systems.
3 Describe how lever systems may be manipulated to gain a mechanical advantage.
4 Classify joints according to the structure and degree of movement they permit.
5 List the characteristic features of fibrous, cartilaginous and synovial joints.
6 Describe the structure, function and anatomical sites of synovial, cartilaginous and fibrous joints.
7 Provide a classification of the common disorders affecting joints.
8 Describe the pathophysiological features which occur in inflammatory and degenerative joint disease.
9 Relate pathophysiological changes to limitations in movement and other problems which result from common disorders affecting synovial and cartilaginous joints.
10 Discuss the rationale for planning care for the patient with problems arising from joint disease.

Introduction

Joints, or articulations, are the specialized structures which are found where two or more bones meet. They have the following functions.

1 To facilitate movement of different parts of the body relative to each other. Immovable fibrous joints provide the exception to this.
2 To confer stability on movement.
3 To assist in the maintenance of body posture.

Preceding chapters have considered aspects of mobility and support in relation to the physiology of muscle and bone. Movement in humans is brought about by lever systems in which the integral components are provided by bones, muscles and joints acting in concert. Accordingly, a fuller discussion on movement is given here, and this chapter begins with a necessary though simple presentation of lever principles. Understanding the properties of lever systems is both useful and relevant, for, as will be shown, they can be manipulated to advantage in clinical nursing practice. The physiology of movement, however, which includes lever mechanics, is a complex subject, so for the enthusiast suggestions for further reading can be found in the bibliography.

LEVER SYSTEMS

A lever is best visualized as a rigid bar which rotates about a pivot, or fulcrum (plural: fulcra),

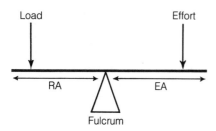

Figure 3.3.1 First class lever system. In the body, bones are the levers and joints are the fulcra. The position of muscle insertion on bone is the application point of effort, and the weight of the bone plus the weight of that part of the body to be moved form the load (RA, resistance arm; EA, effort arm)

when a force is applied to the lever. The distance between the fulcrum and the point of application of effort, or applied force, is known as the effort arm (EA; Fig. 3.3.1). A resistance arm (RA) is represented by the distance between the fulcrum and point of application of the load, or resisting force. This resisting force tends to rotate the lever in a direction opposite to that produced by the effort.

In the body, the components of a lever system are provided by bones which act as levers, and joints which act as fulcra. The position of muscle insertion on bone is the application point of effort,

and the weight of the bone itself, plus the weight of that part of the body to be moved, forms the load.

Classification of lever systems

Lever systems are classified according to three arrangements along the lever of its components: the fulcrum, and the application points of effort and load.

THE FIRST CLASS LEVER

Here, the fulcrum occupies an intermediary position between the points of application of effort and load (Fig. 3.3.1). An example of a first class lever system is found in elbow extension. The elbow joints provide the fulcrum, the triceps muscle insertion is the point of effort application, and the weight of the arm is the load (Fig. 3.3.2).

THE SECOND CLASS LEVER

In this system, the point of load application is located between the fulcrum and the point of effort application (Fig. 3.3.3). Rising on tiptoe utilizes a second class lever. Here, the joint at the ball of the foot provides a fulcrum. Effort is applied at the

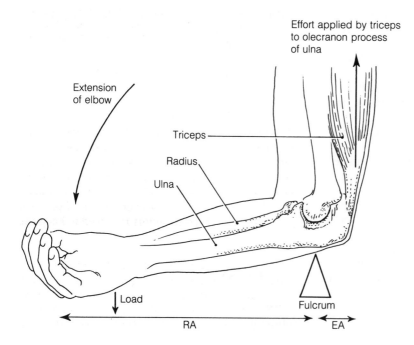

Figure 3.3.2 First class lever at elbow joint

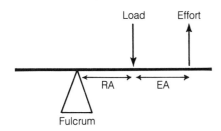

Figure 3.3.3 Second class lever system

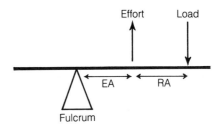

Figure 3.3.5 Third class lever system

insertion point of the flexor muscles of the lower leg, and body weight is the load (Fig. 3.3.4).

THE THIRD CLASS LEVER

In this class, the point of effort application is situated between the fulcrum and the load (Fig.

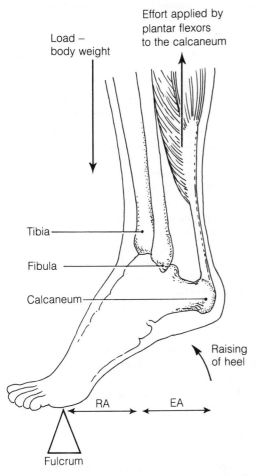

Figure 3.3.4 Second class lever at the tarsometatarsal joint

3.3.5). Third class levers are the most commonly found in the body. Unfortunately, this type of lever requires the application of a large effort to produce movement of comparatively small loads. A classic example is when the biceps muscle flexes the forearm. Here, the elbow joint forms the fulcrum, the biceps muscle insertion is the point of effort application, and the weight of the arm is the load (Fig. 3.3.6).

Manipulation of lever systems

Traumatic injuries to bones, muscles, ligaments and joints may arise where lever systems are overloaded. To explain how this may come about, let us consider to begin with a lever of the first class (Fig. 3.3.7).

In this system, a 5 kg weight 20 cm from the fulcrum is just balanced by a 10 kg weight 10 cm from the fulcrum. The mathematical principle involved here is:

$$\text{Effort} \times \text{effort arm (length)}$$

$$= \text{resistance} \times \text{resistance arm (length)}$$

The effort needed to keep the lever in balance can be lessened by lengthening the effort arm. If the effort arm was 40 cm in length, the effort required would be only 2.5 kg. If the effort arm was shorter, say 5 cm, 20 kg effort would be required to keep the lever in balance. However, if the effort arm was very *short*, then the effort required to keep the lever in balance might be so great as to bend or break the lever.

Most body lever systems are of the third class, where the effort arm and fulcrum are anatomically fixed, and the effort arm is indeed very small (Figs. 3.3.5 and 3.3.6). To gain maximum advantage in these systems, the length of the resistance arm can

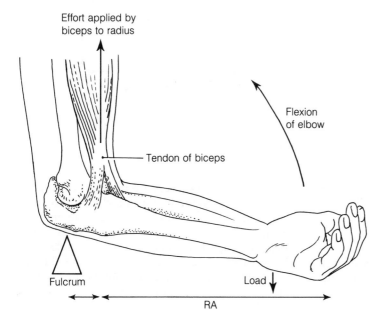

Figure 3.3.6 Third class lever system at elbow joint

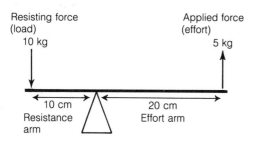

Figure 3.3.7 Manipulation of lever systems: (1) a first class lever in balance.

Principle.
Applied force (effort) (kg) × effort arm length (cm) = resisting force (load) (kg) × resistance arm length (cm)

Insertion of the values shown in the diagram into this equation shows that the forces on either side are balanced.

However, if the effort arm was lengthened to 40 cm:
$X = 10\,kg \times 10\,cm$
$X = \frac{100}{40}$
$X = 2.5\,kg$
i.e. 2.5 kg effort must now be applied to balance the lever.

But, if the effort arm was shortened to 5 cm:
$X = 10\,kg \times 10\,cm$
$X = \frac{100}{5}$
$X = 20\,kg$
i.e. 20 kg effort must be applied to balance the lever.

Conclusion. The shorter the effort arm, the greater is the applied force required to balance or lift the load

be altered since effort arm length cannot be altered. Transposing the measurements from Fig. 3.3.7 to a third class lever (Fig. 3.3.8), we find that if the resistance arm is shortened to 10 cm, then only 5 kg effort is required to keep the lever in balance; whereas if the resistance arm is longer, say 40 cm, a 20 kg effort will be required.

The application of such principles is clearly of great importance to the nurse when lifting or turning patients, or assisting patients with lifting exercises. By carrying loads as close to the body as possible, the resistance arm is decreased and thus less force needs to be exerted by muscles, and less strain is exerted on the body lever systems. For example, lifting a straight limb requires the development of considerably more force by muscle than when the limb is bent at the knee joint. Any action with a short resistance arm is less tiring, and is also more economical in terms of energy expenditure. However, there are situations when moving straight limbs and developing greater forces in muscle are used to advantage. Following meniscectomy (removal of meniscus cartilages in the knee joint), straight-leg raising exercises both strengthen and increase muscle mass, giving the joint greater stability, an essential postoperative requirement.

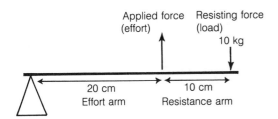

Figure 3.3.8 Manipulation of lever systems: (2) a third class lever. Most body levers are of the third class.

Question. What effort must be applied to the lever system above in order to balance the lever?

Transposing the measurements from Fig. 3.3.7, the effort (kg) required to balance the lever is:

$X \times 20\,cm = 10\,kg \times 10\,cm$

$X = \frac{100}{20}$

$X = 5\,kg$

However, if the resistance arm is lengthened to 40 cm, the effort required to balance the lever is:

$X \times 20\,cm = 10\,kg \times 10\,cm$

$X = \frac{400}{20}$

$X = 20\,kg$

Conclusion. The shorter the resistance arm, the smaller is the applied force required to balance or lift the load

CLASSIFICATION OF JOINTS

Joints are classified according to structure and the type of movement they permit – as fibrous, cartilaginous or synovial. A number of subclasses are distinguished within each of these categories, and a few exceptions are found in which features of more than one class merge.

Fibrous joints – synarthroses

Two features are characteristic of this class.

1 Either no movement is permitted, or it is severely limited.
2 Fibrous connective tissue is present between the articulating surfaces, merging into the periosteum on either side.

Bones of the skull are united by short, compact fibrous strands, the sutural ligament, to form an immovable fibrous joint (Fig. 3.3.9). This articulation is reduced in adult life as ossification progresses. An equally simple fibrous joint is formed between the tooth and its socket. Although serving to anchor the tooth firmly in the jaw, this type of joint does allow slight movements to take place,

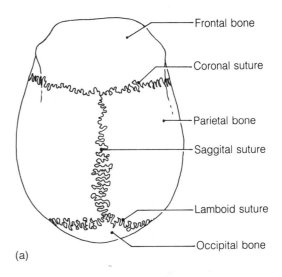

(a)

(b)

Figure 3.3.9 Fibrous joint (skull): (a) superior aspect of skull; (b) transverse section of fibrous joint

when food is chewed. Another fibrous joint is formed where the distal end of the fibula articulates with the tibia. At this joint, some slight movement is possible as the fibrous strands uniting the bones are long, forming what is known as the interosseus ligament.

Cartilaginous joints – amphiarthroses

Here, a limited amount of movement is permitted by the flexible fibrocartilage present between opposing bone ends. Two subclasses are distinguished, the symphyses and the synchondroses; the latter usually ossify in adult life and are thus immovable.

SYMPHYSES

At symphyses, the fibrocartilage takes the form of a pad or disc inserted between the hyaline plates of the articulating surfaces. Typical examples are

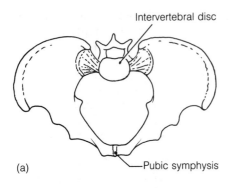

(a)

—Intervertebral disc

—Pubic symphysis

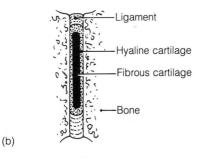

—Ligament

—Hyaline cartilage

—Fibrous cartilage

—Bone

(b)

Figure 3.3.10   Cartilaginous joint (pelvis): (a) superior aspect of female pelvis; (b) transverse section of pubic symphysis

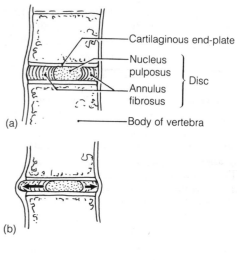

—Cartilaginous end-plate

—Nucleus pulposus

—Annulus fibrosus

Disc

(a)   —Body of vertebra

(b)

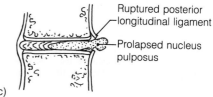

—Ruptured posterior longitudinal ligament

—Prolapsed nucleus pulposus

(c)

Figure 3.3.11   Compression of vertebral disc

found at the pelvic symphysis pubis and in the intervertebral joints of the spinal column (Fig. 3.3.10). Compression of the intervertebral disc permits movements which are limited to small degrees of flexion, side flexion and extension.

Extensive and complex arrangements of ligaments and muscles act as important stabilizing structures on the joint, and together with the thickness of the disc they act to limit movement. Resistance to movement is provided by the elastic structure of the disc, which allows bones to recoil to their original position once a compressing force is removed. The elasticity of the intervertebral disc also enables it to act very much like a shock absorber, uniformly redistributing pressures which are generated during movement and lifting heavy loads. It is a formidable thought that intervertebral joints may be subjected to stresses equivalent to several hundred kilogrammes weight when the spine is flexed and loads are lifted. Rupture of the disc has occurred when loads of 500–600 kg have been applied, the frac-

ture usually occurring at the weakest spot, the hyaline end-plates.

Prolapse of intervertebral discs
Intervertebral discs may slip out of place (prolapse) as a result of degenerative changes or flexion injuries which are caused by heavy lifting. Degenerative changes are seen in the structure of the intervertebral disc, usually after the third decade of life, and are probably caused by mechanical stress. Once degeneration has occurred, the disc can no longer function as an effective shock absorber and this increases the load borne by the joints of the vertebral arch, leading to back pain. Degenerative changes may alone be sufficient to lead to prolapse of the disc, or it may be triggered by minor trauma caused by, for example, coughing or sneezing, associated with degeneration.

A prolapsed disc (Fig. 3.3.11) may compress the spinal nerve or its roots and stretch or tear adjacent tissues. In some cases, the prolapse may be severe enough to compress the spinal cord. It can occur in any region of the vertebral column, but the lumbar vertebrae are most commonly affected by prolapse.

Major problems for the patient are pain and limited movement. Back pain may radiate to the arm or leg, depending on the site of prolapse, and is aggravated by straight-leg raising, bending or lifting. Numbness and tingling may be present in affected limbs and movement is limited by local muscle spasm around the affected vertebrae. If the spinal cord is compressed, the effects on motor nerve tracts may lead to weakness, uncoordinated movements and abnormal gait. Bladder function may also be impaired.

Treatment includes bedrest on a firm mattress and the provision of anti-inflammatory analgesics, to allow the pain and inflammation to subside. A cervical collar may be worn to support and immobilize the neck following a cervical disc prolapse. In some circumstances, cervical or pelvic traction is applied to increase the distance between adjacent vertebrae, allowing the prolapse to subside, and to relieve painful muscle spasms. Surgical excision of the disc may be necessary if these measures fail or if control of bladder function is impaired.

SYNCHONDROSES

Synchondroses are cartilaginous joints which ossify in adult life and thus permit little, if any, movement. Typical synchondroses are found where the epiphysial plate of cartilage is present between the epiphysis and diaphysis of long bones, and where the upper ribs articulate with the sternum at the sternocostal joints.

Synovial joints – diarthroses

Most of the body joints are synovial, that is to say, they are freely movable in a number of planes. At present they are classified by anatomical structure and the range and axes of movement they allow.

A classification of synovial joints together with the range of movement they permit are presented in Table 3.3.1. Definitions of the types of movement permitted are listed and illustrated in Table 3.3.2. Clearly, the range and type of movements which are characteristic for a synovial joint may be modified by diseases, but when making a nursing assessment of the patient with limited mobility, the nurse should bear the following points in mind.

1 Range of movement at a joint decreases with age. The presence of stiffness or even increased mobility may not be a feature of an underlying disease.
2 If impaired movement is present at any joint, the effects on coordinated movements, posture and balance as a whole must be established.

STRUCTURE OF SYNOVIAL JOINTS

A number of structural features are common to all synovial joints, irrespective of classification (Fig. 3.3.12). A capsule of fibrous connective tissue surrounds the articulating bones like a sleeve. Lining the capsule is a highly specialized synovial membrane which secretes a lubricant fluid into the cavity of the joint. Hyaline cartilage covers the opposing bone ends, which are not in direct contact. All of these structural components possess properties which allow movement in a number of planes.

The joint capsule

A capsule of fibrous connective tissue containing a high percentage of collagen fibres and abundantly supplied with blood capillaries encases the joint. The capsule is attached on either side to the periosteum of the articulating bones. Stability of the joint is increased if intra- or extracapsular ligaments are present (Fig. 3.3.13), and by the mass and tension in surrounding muscles.

Innervation of the capsule is complex: the main supply branches from the nerve supply to the adjacent muscles which produce movement at that joint. Additionally, a number of afferent sensory nerves are found within the capsule and its surrounding ligaments. Myelinated afferent nerves terminate in a number of sensory receptors, known collectively as proprioceptors, and convey information on joint movement to the somatosensory cortex and cerebellum. Together with visual, auditory and vestibular nerve inputs, this information subserves our sense of position, and facilitates balanced and coordinated movements.

The synovial membrane

Beneath the fibrous capsule lies a thin, highly vascular membrane which covers all surfaces inside the joint with the exception of the articular cartilages. Its surface is highly folded in some ares to form microscopic projections known as **villi**.

Table 3.3.1 Classification and range of movement in synovial joints

Synovial joint	Site	Articulating surfaces	Range of movements
1 Hinge joint	(a) Elbow	Distal humerus with proximal ulna (humeroulnar joint)	Movement is restricted about a single transverse axis
	(b) Toes Fingers	Between phalanges (interphalangeal joint)	Flexion, extension
	(c) Ankle	Distal tibia and fibula with talus (talocrural joint)	Flexion (dorsiflexion) Extension (plantar flexion)
	(d) Knee	Femur and tibia	Flexion, extension, slight rotation when leg flexed
2 Pivot joint	(a) Elbow	Radius with ulna (proximal radioulnar joint) The head of the radius forms a pivot which rotates within a ring formed by an ulnar notch	Rotation (pronation, supination)
	(b) Vertebral column	First cervical vertebra (atlas) with second cervical vertebra (axis). Here, the atlas (ring) rotates about the odontoid process of the axis (pivot)	Rotation
3 Gliding joint	(a) Shoulder girdle	Apposed flat bone surfaces glide together Movement is limited by extensive ligature 1 Clavicle with sternum and cartilage of 1st rib (sternoclavicular joint) 2 Clavicle with scapula (acromioclavicular joint)	Gliding, limited motion in several directions / Gliding, rotation of scapula on clavicle
	(b) Hand	Some articulations between carpals	In concert with radiocarpal joints, describe varying degrees of flexion, extension, adduction, abduction, circumduction
	Foot	Articulations between bases of metatarsals (tarsometatarsal joint)	Limited to slight gliding movements
	(c) Vertebral column	Facets of articular processes of adjacent vertebrae	Range of movements between adjacent vertebrae is small but extensive in vertebral column as a whole
		Cervical vertebrae 2–7 Thoracic vertebrae	Flexion and extension, side flexion and rotation are limited cervically
		Lumbar vertebrae	Flexion, extension, side flexion, no rotation

(1) Hinge joint

(2) Pivot joint

(3) Gliding joint

Table 3.3.1 Continued

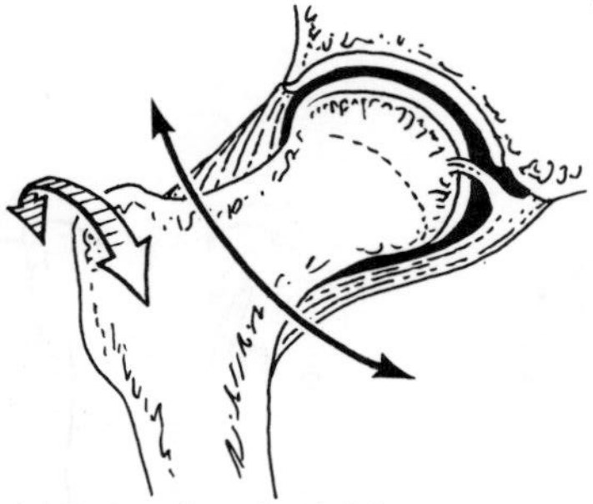

(4) Ball and socket joint

(5) Saddle joint

(6) Ellipsoid joint

4 Ball and socket joint		One articulating surface is shaped to form a spherical head, which rotates in a cuplike depression in the reciprocal surface	Very extensive range of movements
	(a) Hip	Head of the femur with the acetabulum of the pelvis	Flexion, extension, abduction, adduction, internal and external rotation, circumduction
	(b) Shoulder	Head of the humerus with the glenoid cavity of the scapula	
5 Saddle (sellar) joint		Articulating surfaces both saddle shaped, the convex surface of one bone is inserted into the concave surface of the other	Movements permitted in two directions at 90° to each other
	Hand Base of thumb	First metacarpal with carpal (trapezium bone)	Flexion, extension, abduction, adduction, opposition
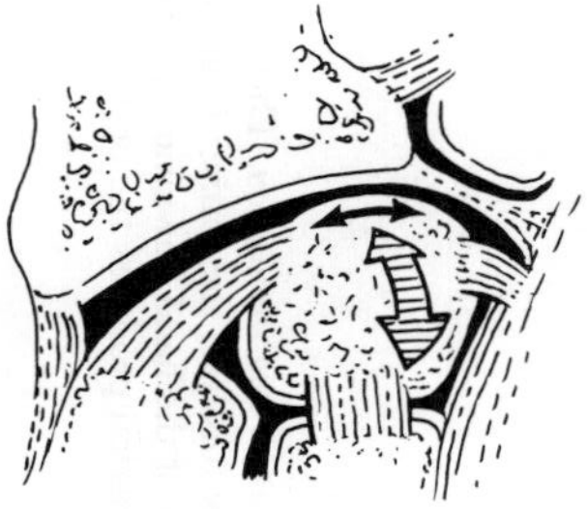 6 Ellipsoid joint		Both surfaces ellipsoidal, one with a longer radius of curvature	Movements permitted in two directions at 90° to each other
	Wrist	Distal radius with 3 carpals; the scaphoid lunate and triquetral bones (radiocarpal joint)	Flexion, extension, radial flexion, moving hand towards thumb Ulnar flexion moving hand towards little finger True rotation is impossible but circling movements may be produced by a combination of wrist and forearm movements
	Hand	Metacarpals with phalanges	Flexion, extension, adduction, abduction
	Foot	Metatarsals with phalanges	

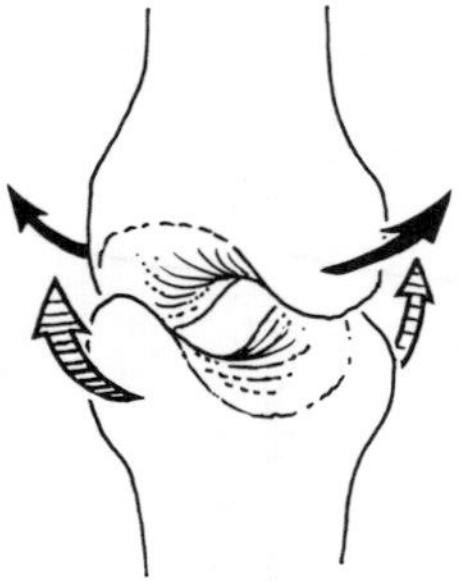

Table 3.3.2   Range and types of movement at synovial joints

1   *Flexion.* A movement which decreases the angle
between adjacent bones, e.g. movement of the hand
towards the shoulder, decreasing the angle at the elbow
joint

2   *Extension.* Increases the angle between adjacent bones,
e.g. when the hand is moved away from the shoulder
and the arm is straightened

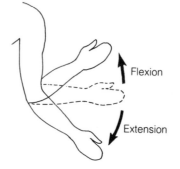

3   *Circumduction.* As the bone moves, it describes a cone
while its end describes a circle, e.g. stretching the arms
out straight and making a circular movement from the
shoulder joint

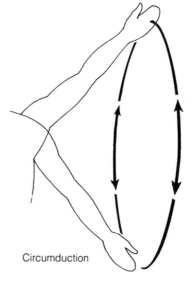

4   *Eversion.* A special movement at the ankle, turning the
sole of the foot outwards; *inversion* is the converse
movement

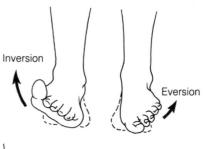

5   *Pronation and supination.* These movements are peculiar
to the forearm, the former is produced by turning the
palm of the hand downwards, rotating the radius;
supination produces the opposite effect

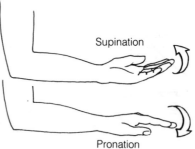

Table 3.3.2 Continued

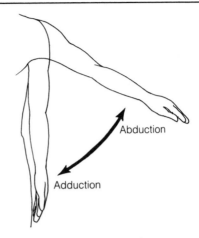

6 *Abduction.* A movement of the bone away from the body midline; *adduction* produces the converse movement towards the body midline, e.g. moving the arms straight out to the sides and back again

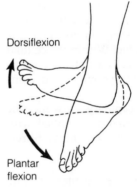

7 *Dorsiflexion and plantar flexion.* Describe movements of the sole of the foot; the former upwards towards the shin and the latter downwards away from the shin

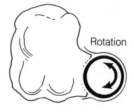

8 *Rotation.* Rotational movement of a bone on its own axis, e.g. moving the head from side to side where the atlas rotates around the odontoid process or turning the palm of the hand up and down which rotates the radius

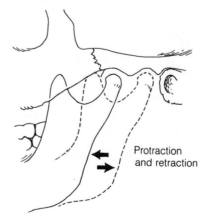

9 *Protraction and retraction.* Movement of a bone forwards and backwards, e.g. when the mandible of the lower jaw is thrust forwards and backwards

Table 3.3.2   Continued

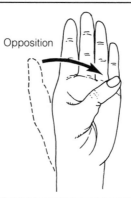

10   *Opposition*. Movement of the thumb across the palm of the hand

Two regions are found within the membrane: an outer cellular layer next to the joint cavity approximately three cells thick, resting on an inner fine meshwork of connective tissue. Within the cellular layer are phagocytic type B cells which are responsible for the production of synovial fluid components such as hyaluronic acid.

In addition to its rich capillary network, the synovial membrane is abundantly supplied with lymphatic vessels and some sensory nerve fibres.

Synovial fluid
In composition, synovial fluid resembles a filtrate of plasma, and as such is similar to the fluid found in the pleural and peritoneal cavities, where body surfaces are also in close apposition. The protein content is low ($<20\,g\,l$, cf. plasma 60–$80\,g\,l$), and most of it is combined with hyaluronic acid. In health the volume of synovial fluid is small, approximately 0.2–$0.4\,ml$ in the knee joint. It is a highly viscous fluid, but when subjected to the shearing forces produced by movement, its viscosity is lowered. It is, therefore, an ideal joint lubricant. The infinitely small film of fluid, dispersed between the articulating cartilages when they move relative to one another, effectively reduces friction.

It is the presence of the hyaluronic acid–protein complex which gives synovial fluid its characteristically high viscosity and lubricant properties.

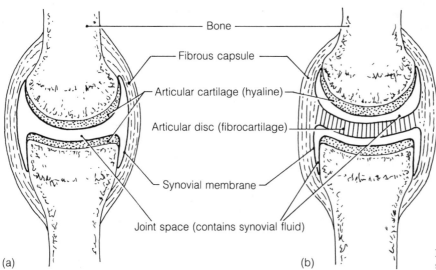

Bone

Fibrous capsule

Articular cartilage (hyaline)

Articular disc (fibrocartilage)

Synovial membrane

Joint space (contains synovial fluid)

(a)   (b)

Figure 3.3.12   Diagrammatic section through a synovial joint

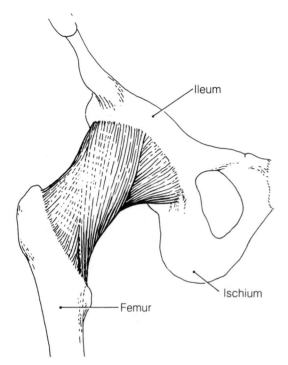

Figure 3.3.13 Arrangement of extracapsular ligaments at the synovial hip joint.

Some of the hyaluronic acid–protein complex appears to be directly adsorbed onto the cartilage, giving added lubrication and protection from erosion. However, the exact mode of action of the synovial fluid as a lubricant is as yet incompletely understood.

A number of cellular components are also present in synovial fluid, mainly leucocytes and synovial cells. In health their number is very small, and most cells are phagocytic, functioning mainly in the removal of the debris which appears with normal use.

In addition to its function as a lubricant, synovial fluid also fulfils a nutritive role in permitting the free diffusion of nutrients to articular cartilage and other intracapsular structures.

Articular cartilage
Opposing bone ends are covered by closely moulded hyaline cartilage, which may reach a depth of 2–3 mm in the hip joint. In composition, the cartilage consists of cells known as chondrocytes embedded in a matrix which they both maintain and secrete. Collagen fibres are present in the matrix together with proteoglycans (protein–polysaccharide complexes), mainly chondroitin and keratin sulphate. Water forms a considerable proportion of the matrix – at least 70% of the wet weight – and is osmotically held in place by the proteoglycans. Deeper zones of the cartilage may be calcified in the adult.

It has been estimated that the articular cartilage on the head of the femur may be subjected to loads of two and a half to five times body weight during walking. How then does cartilage withstand these continual stresses without mechanical failure? Present evidence suggests that when cartilage is subjected to forces generated during joint movement, due to its high proteoglycan/water content, it deforms elastically. Under conditions of sustained pressure, fluid may be squeezed out of the matrix. The ability to deform elastically reduces the forces acting on the cartilage, and protects the underlying bone surfaces. In this way, articular cartilage is able to withstand considerable mechanical loading without failure.

THE PATHOPHYSIOLOGY OF JOINT DYSFUNCTION

A number of disorders may affect the integral components of synovial joints and their related structures – muscles, tendons, ligaments and nerves. Whilst a detailed consideration of the aetiology and pathology of disease processes is not appropriate here, the pathophysiology of inflammatory, metabolic and degenerative diseases affecting joints will be related to the specific joint components. These pathophysiological effects will in turn be considered in relation to the clinical nursing problems which subsequently arise. A summary of disorders affecting joints is shown in Table 3.3.3.

Dysfunction in the synovial joint

THE SYNOVIAL MEMBRANE: INFLAMMATORY CHANGES

Synovitis is a term used to describe inflammatory changes which are localized within the synovial membrane, and the term **arthritis** describes the extended inflammation which involves other joint structures. Synovitis may be precipitated by any of the following.

Table 3.3.3 Common disorders affecting joints

Classification	Disorder	Aetiology	Joints affected	Joint component affected
1 Degenerative	Osteoarthrosis	Unknown; heredity, trauma, congenital factors, pre-existing diseases are all important	Symmetrical involvement of weight-bearing joints; interphalangeal and metacarpal joints of hand, knee, elbow, shoulder, jaw, and cervical, thoracic and lumbar intervertebral joints	Degeneration of articular cartilage Degeneration of intervertebral discs
2 Inflammatory	Rheumatoid arthritis	Unknown; possibly autoimmune; pre-existing infection or injury, and genetic factors are important HLA-D4 antigen carriers are at risk	Peripheral synovial joints; metacarpophalangeal, metatarsophalangeal and interphalangeal joints of hands and feet; later spread involves the elbow, hip and cervical spine	Synovial membrane Synovial fluid Articular cartilage Capsule Ligaments Tendons
	Juvenile arthritis (Still's disease)	Unknown, may be associated with growth disorders	Distal interphalangeal joints, cervical spine, hip and sacroiliac joints	As above
	Rheumatic fever	Hypersensitivity reaction to Group A streptococci	Wrists, ankles, knees, elbows	All joint components
	(a) Reiter's syndrome	Venereal origin or may follow some forms of dysentery	Peripheral synovial joints, particularly of lower limbs	All joint components
	(b) Psoriatic arthritis	Carriers of certain human leucocyte antigens (HLA) are at risk, e.g. HLA-38	Peripheral, sacroiliac, thoracic and cervical spine	All joint components
	(c) Ankylosing spondylitis	HLA-B27		
3 Metabolic	(a) Gout	*Primary:* enzyme defect in uric acid metabolism (hypoxanthine-guanine phosphoribosyl transferase) *Secondary:* some form of (thiazide) diuretic therapy causes urate retention (impaired renal clearance of urate)	First metatarsophalangeal joints; subsequent involvement of joints in hands, knees, feet	Synovial membrane Deposition of monosodium urate crystals triggers synovitis
	(b) Pseudogout	Associated with diabetes mellitus, hyperparathyroidism	Knee joint, subsequent spread to hip, pelvis, shoulder	Synovial membrane Deposition of calcium pyrophosphate crystals triggers synovitis Synovial fluid alkaline phosphate decreased

Table 3.3.3 Continued

Classification	Disorder	Aetiology	Joints affected	Joint component affected
4 Infective	Bacterial arthritis	Systemic spread of gonococci, meningococci staphylococci and streptococci from other sites May complicate traumatic injuries or rheumatoid arthritis	Symmetrical widespread involvement	All joint components involved
	Viral arthritis	Associated with viral infections of rubella, chickenpox, hepatitis	As above	All joint components involved
	Tuberculous arthritis	Systemic spread from primary focus of mycobacterium TB	Vertebrae, hip joint	All joint components involved
5 Traumatic injuries	Minor trauma	Common in athletes due to overtraining before adequate muscle control gained	Individual joints vary in susceptibility to injury	Synovitis affecting synovial membrane
	Sprain	External injury		Ligaments torn, effusion of fluid into capsule Ligaments ruptured Haemarthrosis may be present
	Dislocation			Fracture of cartilage surfaces

Injury. This may result from minor trauma, e.g. caused by athletic overtraining before adequate muscle control is attained.

Irritation. This may be caused by the synovial deposition of monosodium urate crystals in gout or by bleeding (**haemarthrosis**) into the joint, associated with trauma, haemoglobinopathies, and blood dyscrasias. It may also follow primary degenerative osteoarthrotic changes in cartilage. **Osteoarthrosis**, also referred to as **osteoarthritis**, is not a true arthritis, since inflammation is not the primary event but follows degenerative changes in cartilage.

Infection. This may be acquired following trauma, or precipitated by blood-borne viral or bacterial infections from other foci.

Immunological reactions in rheumatoid disease. The inflamed synovial membrane becomes swollen, hyperaemic and infiltrated by lymphocytes, plasma cells, and polymorphonuclear leucocytes. In rheumatoid arthritism, the synovial membrane proliferates rapidly to form a pannus of tissue which infiltrates and reduces the joint space, and may bind together opposing articular surfaces.

As a result of inflammatory changes, the joint becomes swollen, stiff, and its range of movement may be limited. Although the synovial membrane is not extensively supplied with free nerve endings, pain may be felt, particularly on movement.

SYNOVIAL FLUID CHANGES

The appearance and properties of synovial fluid are markedly altered in a number of disorders; thus, aspiration of fluid from the joint is an essential investigation in the early stages of any joint disorder.

Volume increases in fluid within the joint are characteristic of inflammatory disorders. Increased numbers of leucocytes are present (> 2000 ml) and the fluid may have a turbulent, purulent or bloodstained appearance. Loss of synovial fluid from the joint space may occur during the late stages of degenerative diseases such as osteoarthrosis, or inflammatory disorders such as rheumatoid arthritis. In both, surface fissuring of cartilage, together with the increased pressures generated in the joint during movement, may force fluid down into subchondral bone to form cysts. Changes

in volume and composition do not necessarily accompany traumatic injuries, although bleeding into the joint and effusions of synovial fluid are common features.

Altered viscosity may occur where synovial fluid is diluted and increased in volume. Loss of viscosity may also be brought about by the release of enzymes such as hyaluronidases from synovial cells and leucocytes during the inflammatory changes of rheumatoid arthritis. These enzymes can break down the vital hyaluronic acid–protein complex in synovial fluid.

Clearly, if synovial fluid is lost, or its physical properties are impaired, then great pressures are set up in the joint during movement, and the loss of lubrication results in stiffness and an impaired range of joint movement. This is compounded by inflammatory changes in the synovial membrane. Pain may arise due to pressures exerted on the capsule by an increased volume of synovial fluid and the inflamed, proliferating synovial membrane.

ARTICULAR CARTILAGE AND BONE: DEGENERATIVE CHANGES

Later stages of inflammatory disorders such as rheumatoid arthritis may lead to destruction of the hyaline cartilage, which in turn may lead to collapse of the underlying bone. In rheumatoid arthritis, enzymes are released from synovial cells which cause the breakdown of chondroitin sulphate and collagen in the cartilage matrix. This is probably triggered when synovial cells and leucocytes phagocytose an immune complex of IgG, rheumatoid factor and complement. (Rheumatoid factor is an antibody produced in the rheumatoid synovium.)

A similar mechanism is responsible for the primary degenerative changes which appear in osteoarthrosis. Here, the release of enzymes from chondrocytes produces changes in the structure of chondroitin sulphate. As a result, the healthy, smooth appearance of the cartilage is lost and surface fissuring appears. Regenerating cartilage replaces the damaged tissue, but this is unevenly deposited at the joint margins, and ossifies to form the marginal spurs of bone (**osteophytes**) characteristic of the disease.

Inevitably, all these degenerative changes result in the loss of the compressible properties of hyaline cartilage, and extensive shearing forces are generated on bone during movement, a surface which is ill-suited to cope with this. Wearing of the thickened bone surfaces may be followed by fracture and collapse, so that the bone can no longer function as a component of an efficient lever system. Progressive loss in the range of movement at affected joints occurs and this may be accompanied by deformity and instability of the entire joint structure. Bone pain may be present which is far more severe than that related to elevated intracapsular pressures. Degeneration of vertebral joints may cause nerve root compression, resulting in numbness, pain and sensory impairment, and problems with micturition may arise if the caudae equinae are involved. These problems related to vertebral joints are most likely to present with disorders such as osteoarthrosis.

THE JOINT CAPSULE AND ITS SURROUNDING STRUCTURES

All the pathophysiological changes which have been described in relation to other joint components may eventually extend to involve the fibrous capsule and its surrounding structures – muscles, tendons, ligaments and nerves – which play a vital role in mobility and stability.

In rheumatoid arthritis, distension of the capsule, which is plentifully supplied by pain fibres, by the increased bulk of synovial fluid and pannus causes pain. The pain may be increased due to mechanical changes which produce local ischaemia and muscle spasm, and contractures of the joint may occur if the painful limb is held for a prolonged period in a position which minimizes discomfort.

Deformity may also arise from distension and weakening of the capsule and its surrounding ligaments. The resulting limited movement is further aggravated by muscle spasm and fibrosis, and, together with erosion of the capsular structure, the joint may become unstable and dislocate.

Trauma, dislocation

Dislocations caused by external injury (a direct blow or shearing force) are associated with the rupture of extracapsular ligaments, effusion of fluid into the fibrous capsule, and displacement of articulating bones. A **subluxation** is an incomplete dislocation. Treatment consists of reduction of the dislocation, with surgical repair of the capsule and ligaments if necessary. This is followed by immobilization to promote healing. During this period, exercises must be introduced to maintain the range of motion in a synovial joint and prevent contractures. A traumatic injury which injures the highly vascular synovial membrane may cause bleeding into the joint which in turn causes synovitis. In this situation, aspiration of blood from the joint cavity may be necessary.

A summary of problems: a rationale for planning nursing care

The previous section examined pathophysiological changes in the synovial joint which are brought about by some common disorders. The three major interrelated problems which have emerged – pain, limited movement and deformity – will now be examined in relation to the rationale for planning nursing care. Reference is also made to some aspects of medical treatment and physiotherapy.

PAIN

The causes of pain associated with joint disorders may be summarized as follows.

1 Pressure exerted on the capsule by inflammatory processes within the synovia, and an increased volume of synovial fluid (e.g. inflammatory and infective disorders).
2 Compression of nerve roots which follows trauma or degeneration of intervertebral synovial and cartilaginous joints.
3 Collapse of bone following cartilage erosion (e.g. degenerative disorders such as osteoarthrosis, later stages of rheumatoid arthritis).
4 Spasm in muscle and tendons (any disorder).

Pain caused by inflammation may be relieved in the first instance by the administration of an anti-inflammatory analgesic drug such as indomethacin, aspirin or phenylbutazone as prescribed by the doctor. In the inflamed tissue, there is an increased synthesis of prostaglandins which either initiate or contribute to inflammatory changes such as increases in vascular permeability and leucocyte migration. All three drugs act to prevent prostaglandin synthesis by inhibiting the enzyme prostaglandin synthetase, essential for the formation of prostaglandin precursors, the endoperoxides. Alternatively, immunosuppressive anti-inflammatory drugs, including corticosteroids or azathioprine, may be used in the treatment of rheumatoid disease where the inflammation and pain are severe. Corticosteroids may be given orally or injected into the joint as a medical procedure.

The inflammatory process may be halted by rest, either by resting and immobilizing the affected joint in a splint or by instituting bedrest if several joints are involved in a severe inflammatory disorder such as rheumatoid arthritis. Aspiration of an increased volume of fluid within the joint is a medical procedure which may alleviate pressure on the capsule and thus provide pain relief.

Where cartilage erosion and collapse of the underlying bone have occurred, the use of anti-inflammatory analgesics already described may be inadequate to control bone pain, which is usually severe, therefore the use of more potent opiate-related analgesics, e.g. Diconal or Omnopon, may be necessary. Bone pain is exacerbated by loading, therefore the use of walking aids or other devices to prevent loading stresses on joints, together with appropriate lifting techniques, will do much to alleviate pain.

Pain caused by muscle spasm may be alleviated in a number of ways. Splinting the joint with plaster of Paris or moulded fibreglass splints will enable the muscle to relax and thus reduce pain. Muscle relaxation may also be achieved by physiotherapy techniques such as hydrotherapy and the application of local heat treatments via heating pads or paraffin wax baths. Passive exercises in an acutely inflamed joint may exacerbate or precipitate painful muscle spasm, therefore isometric exercise, i.e. those which develop tension in muscles without moving the joint, are advisable. Although these exercises are initiated on the advice of the physiotherapist, it is important to reinforce them. If the body is constrained in one position for a lengthy period of time, this may increase the severity of muscle and joint pain, and this is particularly true if movement during sleep is restricted. For this reason, it is unusual for potent hypnotic drugs which reduce sleep movements to be prescribed for an individual with joint disease, and a combination of sedative and anti-inflammatory drugs is normally given instead.

Pain produced by compression of a nerve root may be relieved by a combination of traction and local heat treatment, initiated as a medical decision. This is particularly useful in treating joint pain caused by degenerative changes in the cervical spine. Surgical decompression may be considered if these conservative measures fail. (Traction is considered at greater length in Chap. 3.2.)

IMMOBILITY AND STIFFNESS

The causes of limitation of movement associated with joint disorders may be summarized as follows.

1 Loss of joint space due to proliferation of synovial tissue (rheumatoid arthritis).
2 Loss of physical properties of synovial fluid, or loss of fluid per se (inflammatory or degenerative disorders).
3 Degenerative changes in bone and cartilage; loss of the compressive properties of cartilage with subsequent loading of bone (osteoarthrosis, late stages of rheumatoid arthritis).
4 Fibrosis of the capsule, and ligaments and tendons.
5 Spasm of muscle.

A number of nursing actions may alleviate some of these problems. Immobility caused by increases in intracapsular bulk due to inflammation of the synovial membrane may be alleviated by the correct administration of anti-inflammatory analgesics, i.e. regularly as prescribed by the medical staff and after meals to reduce the potential problem of gastrointestinal side-effects, and measures previously described.

Rest of the affected joint, or bedrest if several joints are involved, will help to reduce the great pressures which are generated during movement due to loss of synovial fluid, joint space and cartilage, and which may cause further cartilage erosion. Excessive or abnormal use of the joint will accelerate cartilage fissuring and destruction, therefore loading stresses on weight-bearing joints should be prevented by using walking aids (e.g. tripods, walking frames) and lifting techniques, by supervising exercise periods, and asking the dietician to advise a weight-reducing diet if the individual is obese.

Immobility may be reduced by using physiotherapy exercises which maintain the range of movements in unaffected joints, followed by the introduction of active and passive exercises in affected joints as soon as inflammation and pain have subsided. Exercise also increases muscle mass and tension, thereby promoting stability in the affected joint. In a number of instances it may be advisable to institute, as soon as possible, exercises which attempt to maintain normal joint function, e.g. in ankylosing spondylitis prolonged rest and immobilization may be harmful, resulting in further ankylosis and impairment of mobility. Pain relief is an essential prerequisite here.

DEFORMITY

Joint deformity may be acquired as a result of the following.

1 Distension of the articular capsule by increased intracapsular pressures. This may result in the capsule and ligaments becoming weakened and loose.
2 Collapse of bone and degeneration of cartilage.
3 Prolonged immobilization of a joint in an abnormal position may cause contracture deformity.

Measures which may be implemented to prevent capsular distension and loading collapse of bone and cartilage have been discussed earlier. Contracture deformities may occur in any immobile patient where a limb is held close to the body with the joint flexed for a prolonged period of time. This is followed by muscle atrophy and shortening. The individual with a painful joint is particularly vulnerable, since there may be a tendency to hold the joint in a position which minimizes pain, and a flexion deformity may result. Vital preventive measures here include the effective use of analgesia, frequent changes of position in immobile patients with avoidance of flexion position, and putting joints through a normal range of movement as soon as possible.

In practice, deformity results from the exposure of the joint to a deforming force. Therefore preventing deformity by maintaining good posture is an important aspect of nursing care. Patients with joint disorders should be nursed on firm mattresses, walking aids should be used to maintain a correct upright posture, table heights should be adjusted to prevent continual stooping, and the patient should be positioned with the back straight when sitting up, and with one pillow when lying down.

Splinting is extensively used to correct existing joint deformity; serial splinting may be used to overcome a fixed flexion deformity. Splinting is carried out by a skilled technician or physiotherapist on medical advice.

Surgical intervention

Surgical intervention is indicated when the conservative methods of treatment already described have failed. The aim of surgery is directed at relief of pain, immobility, and correction of deformity. **Synovectomy**, a procedure which excises proliferating synovial tissue, is often successfully employed in rheumatoid disease to reduce pain, and by preventing capsular distension deformity. **Arthrodesis**, or excision of the joint, may be undertaken where cartilage is degenerating and collapse of bone has occurred. This measure alleviates deformity and instability at the expense of loss of movement. An important achievement in the last two decades has been the advent of the **replacement arthroplasty**, in which the joint is reconstructed and a prosthetic implant replaces one or more of the articulating bone surfaces, for example, hip joint replacement. Loss of the compressible properties of cartilage, with subsequent erosion and collapse of the underlying bone, is a fundamental cause of immobility in the diseased joint, and it is an unfortunate fact that once lost, cartilage does not regenerate. At present, the possibility of grafting new cartilage into the joint is under investigation. This may be a viable surgical intervention in the near future.

Review questions

The answers to all these questions can be found in the text. In each case there is at least one correct and one incorrect answer.

1 In a third class lever system

   (a) the point of effort application is situated between the fulcrum and the load
   (b) the lever requires the application of a large effort to produce movement of comparatively small loads
   (c) the fulcrum occupies an intermediary position between the points of application of effort and load
   (d) the point of load application is located between the fulcrum and point of effort application

2 Characteristic features of fibrous joints include

   (a) fibrocartilage sited between opposing bone ends

(b) fibrous connective tissue sited between opposing bone ends

(c) severely limited or absent movement

(d) unlimited movement

3 Synchondroses are

(a) found at the epiphyseal plate in long bones

(b) cartilaginous joints which ossify in adult life

(c) diarthrotic joints

(d) fibrous joints which ossify in adult life

4 The synovial joint found at the elbow is

(a) a gliding joint

(b) a sellar joint

(c) a hinge joint

(d) an ellipsoid joint

5 The range of movements found at a pivot joint includes

(a) rotation

(b) pronation

(c) flexion

(d) gliding

6 Which of the following is/are typical pathological feature(s) of early rheumatoid arthritis?

(a) osteophytes

(b) destruction of articular cartilage

(c) synovitis

(d) proliferation of synovial membrane

Answers to review questions

1 a and d
2 a and c
3 b, c and d
4 c
5 a and b
6 c

Short answer and essay topics

1 List the major function of joints. On what basis are joints classified?

2 Describe the structure, function and location of synovial, cartilaginous and fibrous joints.

3 What is a 'lever'? Explain how lever systems are classified, citing in your answer examples found in the musculoskeletal system in humans.

4 Briefly describe how a lever system may be manipulated to gain a mechanical advantage.

5 Identify the six types of synovial joints and describe the range of movements which they permit.

6 Discuss the structure, function and special properties of (a) synovial fluid, and (b) articular cartilage.

7 List the problems which may arise following prolapse of an intervertebral disc. What treatment is indicated?

8 List the common disorders which affect synovial joints. Which joint components are affected?

9 Describe the common problems which result for individuals afflicted with inflammatory and degenerative joint disease. How have these problems arisen?

10 Discuss the nursing management of pain, immobility and deformity arising from degenerative or inflammatory joint disease.

Suggestions for further reading

Boyle, A. L. (1980) *A Colour Atlas of Rheumatology*. London: Wolfe Medical.

Carlsöö, S. V. (1972) *How Man Moves; Kinesiological Studies and Methods*. (Translated from Swedish by Michael, W. P.). London: Heinemann Medical.

Dagg, A. Innes (1977) *Running, Walking and Jumping: the Science of Locomotion*. Wykeham Science Series.

Dyson, G. (1977) *The Mechanics of Athletics*, 7th edn. London: Hodder and Stoughton.

Freeman, M. A. R. (ed.) (1973) *Adult Articular Cartilage*. London: Pitman.

Gowitzke, B. & Milner, M. (1980) *Understanding the Scientific Bases of Human Movement*, 2nd edn. Baltimore: Williams & Wilkins.

Jaffe, H. L. (1972) *Metabolic, Degenerative and Inflammatory Diseases of Bones and Joints*. Philadelphia: Lea & Febiger.

Kapandji, I. A. (1970–82) *The Physiology of the Joints*, Vol. 1, 5th edn., Vol. 2, 2nd edn., Vol. 3, 2nd edn. Edinburgh: Churchill Livingstone.

Kennedy, P. (1979) *The Moving Body. Faber Anatomical Atlas*. London: Faber & Faber.

Kilgour, O. F. G. (1978) *An Introduction to the Physical Aspects of Nursing Science*, 3rd edn. London: Heinemann Medical.

Kluge, A. (1977) *Chordate Structure and Function*, 2nd edn. London: Collier Macmillan.

Mason, M. & Curry, H. L. F. (eds.) (1975) *An Introduction to Clinical Rheumatology*. London: Pitman Medical.

Panayi, G. S. (ed.) (1980) *Essential Rheumatology for Nurses and Therapists*. Eastbourne: Baillière Tindall.

Pansky, B. (1979) *A Review of Gross Anatomy*, 4th edn. New York: Macmillan.

Sinclair, D. (1975) *An Introduction to Functional Anatomy*, 5th edn. Oxford: Blackwell Scientific.

Sokoloff, L. (1969) *The Biology of Degenerative Joint Disease*. Chicago: University of Chicago Press.

Thompson, C. W. (1984) *Manual of Structural Kinesiology*, 10th edn. St Louis, Miss.: C. V. Mosby.

Section 4
Internal Transport

Chapter 4.1
The Blood

Susan E. Montague

Learning objectives

After studying this chapter, the reader should be able to
1 Describe the appearance and major constituents of blood.
2 Enumerate the major functions of blood as a carrier medium.
3 State the approximate blood volume of a healthy adult.
4 Describe the origin of the various organic and inorganic constituents of plasma and relate these to biochemical estimations performed clinically on serum.
5 List the major groups of plasma proteins and state their functions.
6 Describe the major stages of haemopoiesis and its regulation.
7 Describe the characteristics and functions of mature red blood cells.
8 Write an account of the synthesis, structure, functions and breakdown of haemoglobin A and give examples of other normal and abnormal human haemoglobins and their effects.
9 Define the term anaemia and explain the occurrence of the health problems associated with this state.
10 Outline iron, vitamin B_{12} and folic acid metabolism and the features of the anaemias produced by deficiency of these substances.
11 Describe the functions of the various types of white blood cell and relate these to changes in white cell counts and morphology in illness.
12 Describe the functions of platelets and outline the major health problems associated with platelet deficiency.
13 Relate the process of haemostasis to the major defects of bleeding and clotting, the occurrence of thrombosis and types of antithrombotic therapy.
14 Describe the genetic and physiological bases of the major blood group systems and state their clinical significance.
15 Enumerate and discuss clinical problems associated with the transfusion of blood and apply this knowledge to nursing responsibilities in the care of people undergoing blood transfusion.
16 Describe and explain the occurrence of the major types of fetomaternal blood group incompatibility and discuss the prevention of these.

Introduction

In all complex animals, including humans, the function of internal transport is primarily fulfilled by the specialized tissues which comprise the heart and blood vessels – the cardiovascular system (see Chap. 4.2) – and by the fluid transport medium which is contained within them – the blood. Blood is circulated throughout the body, propelled by the pumping action of the heart, and, in the capillary beds of the tissues, it comes into close proximity with the thin layer of interstitial tissue fluid which directly surrounds the cells.

As a carrier medium, blood transports a wide variety of substances which are involved in all aspects of cellular metabolism. As it passes through the capillaries its composition changes as components are exchanged, across the capillary walls, with those of tissue fluid. This continuous exchange functions to maintain an optimum internal environment for cell and tissue function and is the basis of homeostasis.

APPEARANCE AND COMPOSITION

Newly spilled human blood typically appears as a homogeneous, opaque, somewhat syrupy fluid, which is dark red in colour. The fluid solidifies into a sticky clot which exudes a clear fluid within a few minutes.

If, however, a sample of blood is placed in a glass tube, prevented from clotting by the addition of anticoagulant and allowed to settle, or centrifuged, it separates into its major components (Fig. 4.1.1). These are a pale yellow, slightly opalescent liquid, the **plasma**, which in health constitutes approximately 55% of the total volume, and a lower deposit of formed elements. The latter are the **blood cells** and **platelets**. As can be seen in Fig. 4.1.1, the red cells (erythrocytes) normally constitute nearly all of the cellular deposit. They are topped by a very thin 'buffy

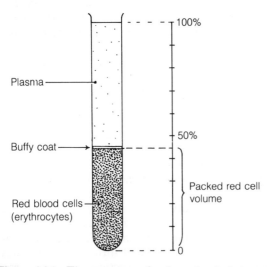

Figure 4.1.1 The appearance of anticoagulated whole blood when allowed to settle

coat', which consists of the white cells (leucocytes) and platelets (thrombocytes). Blood is not a homogeneous substance but a complex fluid containing a variety of cells and platelets in suspension and chemicals in solution.

AN OUTLINE OF THE FUNCTIONS OF BLOOD

Blood provides a medium for the following physiological processes.

Internal transport
Blood carries a wide variety of substances and so subserves the following metabolic functions.

Respiration. The respiratory gases oxygen and carbon dioxide are carried on haemoglobin molecules in red cells and in plasma. Exchange of these with the external environment takes place in the lungs.

Nutrition. Nutrients, absorbed from the intestine, are carried, in the plasma, to the liver and on to other tissues for use in metabolism.

Excretion. The waste products of metabolism are carried in the plasma and passed to the external environment, primarily via the kidneys.

Metabolic regulation. Hormones and enzymes, both of which affect the metabolic activity of cells, are carried in plasma.

Defence against infection by foreign organisms
This is a primary function of white blood cells. Chapters 6.1 and 6.2 describe the protective inflammatory and immune responses, respectively.

Protection from injury and haemorrhage
The protective inflammatory response is the localized response of tissues to injury.

Prevention of haemorrhage is a function of the platelets and a series of clotting and fibrinolytic factors present in plasma. Haemostasis, the process of stopping bleeding from a damaged blood vessel, is described later in this chapter.

Maintenance of water, electrolyte and acid–base balance
Through continuous exchange of its constituents with those of tissue fluid, blood has an essential role in this aspect of homeostasis.

Maintenance of body temperature

Blood carries heat and, because it circulates between body core and periphery, it provides the material basis for heat exchange between the tissues and the external environment.

If any part of the body is functioning abnormally, this is almost always reflected in the composition of the blood. As obtaining a specimen of venous blood is a relatively straightforward procedure, the results of biological and biochemical analysis of such a specimen frequently provide a major aid to medical diagnosis, a guide to treatment, and a means of monitoring a patient's progress.

By the same principle, primary disorders of the blood or loss of blood have widespread effects on body function producing diverse health problems. However, changes in the composition of blood, secondary to disease elsewhere in the body, are far more common than those due to primary disorders of blood.

All personnel involved in handling blood or blood products should take precautions such as wearing protective gloves, in order to avoid inadvertent infection with bloodborne organisms, for example HIV, the virus which causes acquired immune deficiency syndrome (AIDS).

BLOOD WEIGHT AND VOLUME

Blood constitutes approximately 6–8% of an adult's body weight. The proportion is lower for the obese adult as fatty tissue contains relatively little blood. The proportion is slightly higher in infancy and childhood.

A newborn baby weighing 3.4 kg would normally have a total blood volume of approximately 230 ml (8 fl oz), a 70 kg adult male would have approximately 4.8 litres (8.4 pints) of blood, and a 51 kg adult female approximately 3.3 litres (5.8 pints).

CONSTITUENTS OF BLOOD

Plasma

Plasma is a pale yellow, slightly opalescent fluid, comprising approximately 55% of total blood volume – 2.5–3.0 litres in an adult. Its major constituents and properties are summarized in Table 4.1.1.

Table 4.1.1 Major constituents and properties of plasma

Constituent (g/l)	
Water (930)	
Protein (75)	
Inorganic ions (9)	
Organic substances (12)	

Property	Value
Specific gravity	1.026
Viscosity (relative to water)	1.5–1.75 centipoise (cP)
pH	7.35–7.45
H^+ concentration	35–45 nmol/l

Plasma contains the clotting factors and, given an appropriate stimulus such as exposure to air, will form a clot. The clear fluid exuded from clotted whole blood and clotted plasma is called **serum**. The constituents of serum are the same as those of plasma minus the small quantities of clotting factors, such as fibrinogen, which are destroyed during the process of clotting. Clinical investigations of substances being transported in the plasma, for example electrolytes, are usually performed on the serum obtained from a clotted blood specimen, as the unwanted cells can be removed with the clot. However, when cells or clotting factors are to be examined, the blood specimen must not be allowed to clot, and so it is placed in a tube containing an anticoagulant.

Plasma forms about 20% of the body's extracellular fluid and is very similar in composition to the other major extracellular fluid, the interstitial (tissue) fluid. The main difference in composition is in protein content. Interstitial fluid contains only about 2% protein, whereas plasma contains 7%. This is because most of the plasma protein molecules are too large to pass through the capillary walls into the interstitial fluid. The small amount of protein that does 'leak' through, passes eventually into the lymph and is returned to the circulation.

The constituents of plasma reflect overall metabolic function and are frequently the subject of biochemical estimation for clinical purposes.

THE PLASMA PROTEINS

Plasma contains between 60 g and 80 g of protein per litre. Plasma proteins are classified into groups according to their molecular size, water solubility and electrophoretic mobility (migration when placed in electrically charged field). There are two major groups: **albumin**, which moves fastest, and

globulin. Further fractions – alpha (α), beta (β) and gamma (γ) – can be distinguished within the globulin grouping.

Albumin is normally present in larger quantities than any of the globulin fractions. The relative proportions and amounts of plasma proteins can alter in certain diseases, and electrophoretic tracings indicating such changes can aid diagnosis.

Origin of the plasma proteins

Plasma albumin, β globulins and most of the α globulins are synthesized exclusively by the liver. The remaining plasma proteins are produced by other tissues. The γ globulins (immunoglobulins, Ig) are produced by cells of the immune system. Other protein hormones and enzymes, produced by various tissues are present in very small amounts in plasma.

The albumin molecule (mol. wt 69 000) is just too large to pass through the capillary walls. In severe nephrosis, however, the damaged glomerular capillaries 'leak' albumin into the nephron, and the urine can contain large quantities of this protein. The liver has the reserve capacity to synthesize and replace a daily albumin loss of approximately 25 g, provided adequate nitrogen (protein) is obtained from the diet to sustain this extra synthesis. As a result, large quantities of albumin may be lost from the body before a deficiency occurs. Following haemorrhage, plasma proteins are restored to normal levels in a few days.

With the exception of thromboplastin and calcium ions (see Table 4.1.7), all of the clotting factors are proteins, and are synthesized by the liver. Vitamin K is essential in the production of some of these, for example prothrombin. In liver disease, the physiological effects of reduced hepatic synthesis of plasma proteins can produce health problems. These are discussed in Chap. 5.2 and include disordered clotting and the formation of ascites.

Functions of the plasma proteins

Intravascular osmotic effect. Plasma protein molecules do not diffuse readily across the capillary wall because their diameter is greater than that of the pores in the capillary wall. The small quantities of protein, mainly albumin, that do leak through the wall, pass from the interstitial space into the blind-ended lymph capillaries and are returned to the blood via the lymphatic vessels.

Only substances which are unable to pass through the pores of a semipermeable membrane can exert an osmotic effect, and hence it is the dissolved protein in the plasma and interstitial fluid that is responsible for the osmotic effect occuring at the capillary wall. This is often referred to as **colloid osmotic pressure**, or **oncotic pressure**.

The osmotic effect of plasma proteins plays a major role in the formation and resorption of tissue fluid and in determining the normal distribution of water between blood and tissue spaces. This important process is described in Chap. 4.2.

The albumin in plasma is the source of 70–80% of the colloid osmotic pressure. This is because the osmotic pressure of a solution is determined by the *number* of dissolved particles unable to pass through the membrane, rather than their weight and size. Not only is there twice as much albumin as globulin in plasma, but also albumin molecules are considerably smaller than globulin molecules.

When plasma colloid osmotic pressure becomes abnormally low, this affects the distribution of water between blood and tissue spaces, and oedema may result (see Chap. 4.2). The major causes of reduced plasma colloid osmotic pressure are

(a) decreased production of plasma protein, as, for example, in liver disease or protein malnutrition
(b) excessive loss of plasma protein from the body, as, for example, in nephrosis or severe burns
(c) increased porosity of capillary walls, as, for example, in inflammation and allergic reactions.

Contribution to the viscosity of plasma. Plasma and cells contribute almost equally to the viscosity of the blood, which is two to five times that of water. The viscosity of plasma is mainly due to its protein content.

Transport. Many substances that are insoluble in water are carried partially or wholly bound to plasma protein molecules. Substances bound to plasma proteins are unable to enter the cells and take part in metabolism. This is of clinical importance, particularly in the case of drugs, where the bound substance forms a reservoir which must be saturated before the drug can have effect. One substance may also displace another from a binding site, rendering the freed one biologically active. (Further discussion and examples of such interactions can be found in pharmacology texts.) Albumin molecules bind calcium, bilirubin, bile acids and several drugs, for example aspirin, phenylbutazone, thiazide diuretics, digoxin, penicllin, sulphonamides, tolbutamide and tryptophan.

There are specific α globulin molecules which bind the hormones cortisol and thyroxine. Beta globulins carry cholesterol and other lipids, vitamins A, D and K, insulin and iron (transferrin). Gamma globulins bind circulating antigens and histamine.

Protein reserve. Plasma proteins function as a labile protein store, forming part of the body's amino acid pool. They can be broken down into their constituent amino acids by the cells of the mononuclear phagocytic system and, as a last resort, can form a source of replacement amino acids, for tissue protein, in conditions of dietary protein deficiency or intestinal malabsorption. Plasma may be given, intravenously, as a source of protein, but the use of proprietary amino acid preparations for infusion has largely superseded this.

Clotting and fibrinolysis. Clotting and fibrinolysis are functions of plasma, and most of the recognized clotting and fibrinolytic factors are plasma proteins.

Inflammatory response. One of the early features of some types of inflammation is the walling-off of the area by a mesh of fibrin (fibrin clot). This fibrin mesh functions to delay the spread of bacteria and toxins from the initial area of penetration and forms a framework for repair. Fibrin is formed from the soluble plasma protein, clotting factor fibrinogen.

Plasma **kinins**, for example bradykinin, are polypeptides which contribute to the events of the inflammatory response. They cause contraction of smooth muscle, vasodilation and decreased blood pressure, increased vascular permeability, and they stimulate nerve endings producing pain. In high concentrations they promote the migration of white blood cells into the tissues.

Complement (see also Chap. 6.2) consists of a system of plasma proteins which act in a complex sequence (the details of which are beyond the scope of this text) to augment inflammatory and immune defence mechanisms.

Protection from infection. The γ globulins (immunoglobulins, Ig) are produced by the plasma cells of the B-lymphocyte series of cells and function as antibodies, binding and inactivating antigen, in the acquired immune response (see Chap. 6.2).

Maintenance of acid–base balance. Some of the amino acid constituents of proteins have free carboxylic acid groups (COOH). Other amino acids have free basic amino groups (NH_2/NH_3OH). Proteins can, therefore, ionize as acids or bases (that is, they are amphoteric molecules), and so can operate in both acidic and basic buffering systems.

At the pH of blood (7.35–7.45, H^+ concentration 35–45 nmol/l), plasma proteins ionize as anions and act as hydrogen ion acceptors, 'mopping up' excess H^+. However, plasma proteins account for less than one-sixth of the buffering power of the blood, the protein (haemoglobin) in the red cell, being a far more potent hydrogen ion acceptor.

INORGANIC IONS

The inorganic ions (**electrolytes**) of plasma are almost identical to those found elsewhere in the extracellular fluid. Sodium is the major cation and chloride and hydrogen carbonate (bicarbonate) the major anions. Table 4.1.2 presents their normal ranges of concentration. Reference ranges vary between laboratories, depending on the characteristics of the population from which they were derived and the precise measurement techniques employed.

Table 4.1.2 Normal ranges of concentration of inorganic ions in plasma

Ion		Concentration (mmol/l)
Sodium	Na^+	135–146
Potassium	K^+	3.5–5.2
Calcium	Ca^{2+}	2.10–2.70
Chloride	Cl^-	98–108
Hydrogen carbonate	HCO_3^-	23–31
Phosphate	PO_4^{2-}	0.7–1.4

Changes in plasma electrolyte concentration are followed by similar changes in the interstitial fluid. These, in turn, affect cell function. For example, potassium depletion following the increased losses which can occur in severe diarrhoea and vomiting affects tissue excitability causing muscle weakness and abnormalities in the conduction of the cardiac impulse. A fuller discussion of electrolyte disturbances can be found in Chapter 5.4.

The **osmolality** (see Chap. 4.2. for explanation) of plasma is approximately 285 mosmol/kg of water. Solutions used in **intravenous replacement therapy** must be isotonic with normal body fluids or damage to cells may occur; 0.9% (154 mmolal/308 mosmolal) sodium chloride solution and 5% (278 mmolal/278 mosmolal) dextrose solution are examples of commonly used isotonic solutions.

SUBSTANCES IN TRANSIT IN PLASMA

Plasma contains a multitude of substances in transit to various tissues. These may be in simple solution or carried bound to plasma protein molecules. Many are present in very small quantities. They include the following.

Gases
Inspired gases dissolve in plasma to a degree normally dependent only on their solubility in water and their partial pressure at the alveolar membrane.

Oxygen is not very soluble in water, and consequently arterial blood with a PO_2 of 13 kPa (100 mmHg) only carries 0.3 ml/dl (0.13 mmol/l) of oxygen dissolved in the plasma. This amount is physiologically insignificant.

Carbon dioxide is more water soluble than oxygen and approximately 5% of that transported in the blood is carried in simple solution and about 90% as hydrogen carbonate ions in the plasma.

Other gases, such as inhalation anaesthetics, also vary in their solubility and this affects the rate at which they induce anaesthesia.

Nutrients
The most abundant of these is glucose (Table 4.1.3), the primary source of energy for cellular metabolism. Other nutrients in transit include amino acids, fatty acids, triglycerides, cholesterol and other lipids, and vitamins.

Waste products of metabolism
These include urea, uric acid and creatinine, which are excreted by the kidney, and the bile pigment, bilirubin, which is excreted in the bile (Table 4.1.3).

Hormones
Hormones are carred in plasma to their target cells from the endocrine gland that secreted them into the blood.

Enzymes
Except for those enzymes involved in clotting, the removal of intravascular clots and the complement system, most of the enzymes found in plasma derive from the normal breakdown of the cells of blood and other tissues and have no metabolic function in the plasma itself.

Monitoring the activity of certain enzymes in plasma can be a useful diagnostic index for specific abnormalities. Serum is used for the measurement of enzyme activity, in order to avoid contamination by the enzymes of the blood cells. For example, serum amylase (AMS) is raised in people with acute pancreatitis, and acid phosphatase (ACP) is raised in men with carcinoma of the prostate.

Very few enzymes are produced by only one specific tissue and it is often the pattern obtained from the serum level of two or three enzymes, over time, that aids diagnosis, rather than single values. For example, increasing levels of three serum enzymes may be monitored following the cell death which occurs in myocardial infarction. Creatine kinase (CK) tends to peak about 24 hours after an infarction, whereas aspartate transaminase (AST) peaks at 48 hours and hydroxybutyrate dehydrogenase (HBD) after 3–4 days.

Other substances
Other exogenous chemicals may be absorbed into the blood and transported in the plasma. Drugs are probably the major example and these may be the subject of biochemical estimation in cases of poisoning.

Table 4.1.3 summarizes reference ranges of various organic substances in serum. These may vary from those quoted by other sources since the ranges given depend on the characteristics of the population from which they were derived and the precise laboratory techniques employed.

Table 4.1.3 Normal serum levels of various organic substances

Organic substance	Concentration
Glucose	
fasting	3.3–5.5 mmol/l
after a meal	< 10.0 mmol/l
2 hours after glucose	< 5.5 mmol/l
Urea	2.7–8.5 mmol/l
Uric acid (urate)	150–580 μmol/l
Creatinine	40–110 μmol/l
Bilirubin	3–21 μmol/l
Aspartate aminotransferase (AST)	5–30 i.u./l
Alanine aminotransferase (ALT)	5–30 i.u./l
Hydroxybutyrate dehydrogenase (HBD)	150–325 i.u./l
Creatine kinase (CK)	< 130 i.u./l
Amylase (AMS)	150–340 i.u./l
Alkaline phosphatase (ALP)	21–100 i.u./l
Acid phosphatase (ACP)	< 8.2 i.u./l

i.u. = international unit

THE CELLULAR COMPONENTS OF BLOOD

The cellular elements of blood comprise approximately 45% of its volume (see Fig. 4.1.1). There are three major types.

(a) The red blood cells, or **erythrocytes**. These are the most numerous and make up nearly all of the cellular deposit.
(b) The white blood cells, or **leucocytes**.
(c) The platelets, or **thrombocytes**, which together with the leucocytes form the thin buffy coat (see Fig. 4.1.1).

A major part of basic haematological investigation (**haematology** is the study of blood and blood-forming tissues) involves counting the number of each type of cell present per unit volume of blood. In health, the numbers of cells in peripheral blood remain constant within quite narrow limits, that is, they are homeostatically regulated. The morphology (size and shape) of blood cells is investigated using a stained blood film.

Haemopoiesis (the formation of blood cells)

HAEMOPOIETIC SITES

Haemopoietic cells first appear in the yolk sac of the 2-week embryo. By 8 weeks, haemopoiesis has become established in the embryonic liver, and by 12–16 weeks the liver has become the major site of blood cell formation. It remains an active haemopoietic site until a few weeks before birth. The spleen is also active during this period, particularly in the production of lymphoid cells, and the fetal thymus is a transient site for some lymphocytes.

The highly cellular bone marrow becomes an active haemopoietic site from about 20 weeks' gestation and gradually increases its activity until it becomes the major site of production about 10 weeks later.

At birth, actively haemopoietic (red) marrow occupies the entire capacity of the bones and continues to do so for the first 2–3 years of postnatal life. The red marrow is then very gradually replaced by inactive, fatty, yellow, lymphoid marrow. The latter begins to develop in the shafts of the long bones and continues until, by 20–22 years, red marrow is present only in the upper epiphyses (ends) of the femur and humerus and in the flat bones of the sternum, ribs, cranium, pelvis and vertebrae. However, because of the growth in body and bone size that has occurred during this period, the total amount of active red marrow (approximately 1000–1500 g) is nearly identical in the child and the adult. Adult red marrow has a large reserve capacity for cell production. In childhood and adulthood, it is possible for haemopoietic sites outside marrow, such as the liver, to become active if there is excessive demand as, for example, in severe haemolytic anaemia or following haemorrhage. In old age, red marrow sites are slowly replaced with yellow, inactive marrow.

Red marrow forms all types of blood cell and is also active in the destruction of red blood cells, brought about by tissue macrophages which are present in the lining of the marrow blood sinuses. Red marrow is, therefore, one of the largest and most active organs, approaching the size of the liver. About two-thirds of its mass functions in white (myeloid/non-lymphoid) cell production (**leucopoiesis**), and one-third in red (erythroid) cell production (**erythropoiesis**), although there are approximately 700 times as many red cells as white cells in peripheral blood. This apparent anomaly reflects the shorter life span and hence greater turnover of the white blood cells in comparison with the red blood cells.

BONE MARROW BIOPSY (ASPIRATION)

Bone marrow biopsy is usually performed in order to confirm a diagnosis suggested by clinical examination of the patient and investigation of the peripheral blood cells. It provides information about the normality of haemopoiesis and the relative numbers of different types of cell present. It can also demonstrate the presence of foreign cells, as in metastatic cancer and Hodgkin's disease.

Bone must be punctured with a special needle (for example a Klima or Salah needle) or a small piece of bone is removed with a bone-biopsy trephine needle in order to reach the marrow. A suitable site therefore requires red marrow to be present under a relatively thin layer of bone. The tibia is frequently used in young children and the sternum and anterior or posterior iliac crest in adults.

The area is first infiltrated with local anaesthetic and the needle is then pushed through the bony cortex into the medullary cavity and marrow fragments are aspirated using a syringe. The procedure is carried out, aseptically, by a doctor with an assisting nurse. The nurse's responsibilities also include explanation to the patient, skin preparation and assessing the patient for pain and bleeding at the site following the marrow aspiration. Sternal bone marrow aspiration is usually done at the level of the second costal cartilage. It can be an alarming procedure for the patient, who may become distressed at the pressure on his chest and feel that the needle will pierce his heart or lungs. This cannot happen as the needle is fitted with a guard, but prior explanation and reassurance are essential. Haemorrhage may be a particular risk if the patient's clotting mechanism is abnormal due to his illness, as in leukaemia.

BONE MARROW TRANSPLANT

Marrow transplantation, in specialized centres, is now considered as a possible treatment for patients with severe aplastic anaemia, acute leukaemia and the rare, congenital immune deficiency and haemopoietic disorders. This treatment offers the best chance of cure for such patients but also carries with it significant practical difficulties and risks.

Marrow for transplantation is obtained from multiple aspirations performed under general anaesthetic. It is then administered to the recipient, who has previously been given immunosuppressive therapy by intravenous infusion. The stem cells of the donor marrow settle in the recipient's marrow cavity and, if the graft is successful, begin to release cells into the blood after about a fortnight.

Haemopoietic cell types

Whether the different types of mature blood cell arise from one type of stem cell or from separate stem cells has been the subject of much recent scientific debate. However, there is now sufficient evidence for it to be generally accepted that both bone marrow and peripheral blood contain a very

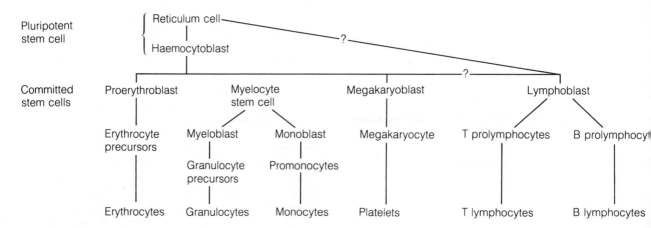

Figure 4.1.2   Summary of the major stages of haemopoiesis

small proportion of 'uncommitted' cells which are capable of mitosis and of differentiation into 'committed' precursors of each of the main types of blood cell (Chanarin *et al.*, 1985).

The development of erythrocytes, leucocytes and thrombocytes from the haemocytoblast is described within the section on that cell type. Figure 4.1.2 diagrammatically summarizes the major stages of haemopoiesis.

Red blood cells (erythrocytes)

ERYTHROPOIESIS (THE PRODUCTION OF RED BLOOD CELLS)

Mature red blood cells develop from haemocytoblasts. This development takes about 7 days and involves three to four mitotic cell divisions, so that each stem cell gives rise to 8 or 16 cells.

The various cell types in erythrocyte development are characterized by

(a) the gradual appearance of haemoglobin and disappearance of ribonucleic acid (RNA) in the cell
(b) the progressive degeneration of the cell's nucleus which is eventually extruded from the cell
(c) the gradual loss of cytoplasmic organelles, for example mitochondria
(d) a gradual reduction in cell size.

The young red cell is called a **reticulocyte** because of the reticular network (reticulum) present in its cytoplasm. This reticulum is composed of ribonucleic acid. As the red cell matures the reticulum disappears. Between 2% and 6% of a newborn baby's circulating red cells are reticulocytes, but this reduces to less than 2% in the healthy adult. However, the reticulocyte count increases considerably in conditions in which rapid erythropoiesis occurs, for example following haemorrhage or acute haemolysis of red cells. A reticulocyte normally takes about 4 days to mature into an erythrocyte.

Regulation of erythropoiesis
In health, erythropoiesis is regulated so that the number of circulating erythrocytes is maintained within a narrow range. Normally, a little less than 1% (2×10^{11}) of these cells are produced per day and these replace an equivalent number that have reached the end of their life span.

Erythropoiesis is stimulated by hypoxia. However, oxygen lack does not act directly on the haemopoietic tissues but instead stimulates the production of a hormone, **erythropoietin (Ep)**. This hormone then stimulates haemopoietic tissues to produce red cells.

Erythropoietin is a glycoprotein with a molecular weight of 46 000. It is inactivated by the liver and excreted in the urine. It is now established that erythropoietin is formed, probably within the kidney, by the action of a renal erythropoietic factor – **erythrogenin (Eg)** – on plasma protein, erythropoietinogen.

$$\text{Erythropoietinogen} \xrightarrow{\text{Erythrogenin}} \text{Erythropoietin}$$

Erythrogenin is present in the juxtaglomerular cells of the kidneys and is released into the blood in response to hypoxia in the renal arterial blood supply.

Various other factors can affect the rate of erythropoiesis by influencing erythropoietin production.

Endocrine factors
Thyroid hormones, thyroid-stimulating hormone, adrenal cortical steroids, adrenocorticotrophic hormone, and human growth hormone (HGH) all promote erythropoietin formation and so enhance erythropoiesis. In thyroid deficiency and anterior pituitary deficiency, anaemia may occur due to reduced erythropoiesis. Polycythaemia (see below) is often a feature of Cushing's syndrome. However, very high doses of steroid hormones seem to inhibit erythropoiesis. Androgens stimulate and oestrogens depress the erythropoietic response. In addition to the effects of menstrual blood loss, this effect may explain why women tend to have a lower haemoglobin concentration and red cell count than men.

Plasma levels of erythropoietin are raised in hypoxic conditions. This produces an **erythrocytosis** (increase in the number of circulating erythrocytes) and the condition is known as **secondary polycythaemia**. A physiological secondary polycythaemia is present in the fetus (and residually in the newborn) and in people living at high altitude because of the relatively low partial pressure of oxygen in their environment. Secondary polycythaemia occurs as a result of tissue hypoxia in diseases such as chronic bronchitis, emphysema and congestive cardiovascular abnormalities associated with right-to-left shunting of blood through the heart, for example

Table 4.1.4 Dietary substances that are essential for normal erythropoiesis

Substance	Utilization
Protein	Synthesis of globin part of haemoglobin and cellular proteins Very low protein intake retards haemoglobin synthesis
Iron	Contained in haem portion of haemoglobin
Vitamin B_{12} (hydroxocobalamin) and folic acid	Involved in DNA synthesis and hence essential to the maturation of red cells
Vitamin C (ascorbic acid)	Necessary for normal folate metabolism Facilitates absorption of iron by reducing ferric iron to ferrous iron Very low levels required before effect seen and the anaemia of vitamin C deficiency (scurvy) is usually the result of blood loss
Vitamin B_6 (pyridoxine), riboflavin and vitamin E	Deficiency of these substances has been associated, occasionally, with anaemia
Trace metals copper and cobalt	There is no evidence that these substances are essential for erythropoiesis in humans Copper is essential for haemoglobin synthesis in other mammals and cobalt is essential for vitamin B_{12} synthesis in ruminants since these animals manufacture this vitamin

Fallot's tetralogy. Erythropoietin is also produced by a variety of tumours of both renal and other tissues.

The oxygen carrying capacity of the blood is increased in polycythaemia but so is the blood viscosity, and the latter produces circulatory problems such as raised blood pressure.

In **primary polycythaemia** (polycythaemia rubra vera), there are increases in the numbers of all the blood cells, and plasma erythropoietin levels are normal. The cause of this condition is unknown.

The underlying cause of secondary polycythaemia is treated with the aim of eliminating hypoxia. Venesection (blood letting) is sometimes employed to reduce red cell volume to normal levels. Frequently blood is removed, centrifuged to remove cells and the plasma returned to the patient (**plasmapheresis**).

In **anaemia** there is a reduction in blood haemoglobin concentration due to a decrease in the number of circulating erythrocytes and/or in the amount of haemoglobin they contain. Anaemia occurs when the erythropoietic tissues cannot supply enough normal erythrocytes to the circulation. In anaemias due to abnormal red cell production, increased destruction and when demand exceeds capacity, plasma erythropoietin levels are increased. However, anaemia can also be caused by defective production of erythropoietin as, for example, in renal disease.

Dietary substances that are essential for normal erythropoiesis are summarized in Table 4.1.4. Requirements for iron, vitamin B_{12} and folic acid are further discussed in the section on anaemia.

CHARACTERISTICS OF MATURE ERYTHROCYTES

Numbers
Erythrocytes are by far the most numerous of the blood cells. The normal **red cell count** (RBC or RCC) is 5.5×10^{12} per litre of blood in men and 4.8×10^{12} per litre in women (see Table 4.1.5 for reference ranges for red cell indices).

Table 4.1.5 Reference ranges for red cell indices

Index	Normal range
Red cell count (RBC or RCC)	
Males	$4.5–6.5 \times 10^{12}/l$
Females	$3.8–5.8 \times 10^{12}/l$
Reticulocyte count	0.2–2.0% of RBC
Haematocrit (Hct)/Packed red cell volume (PCV)	
Males	40–52%
Females	37–47%
Mean cell volume (MCV)	78–93 fl
Haemoglobin concentration (Hb)	
Males	13–18 g/dl
Females	12–15 g/dl
Mean cell haemoglobin (MCH)	27–32 pg
Mean cell haemoglobin concentration (MCHC)	31–35 g/dl
Erythrocyte sedimentation rate (ESR)	Up to 12 mm/h

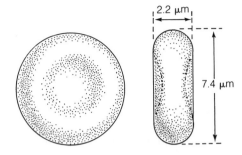

Figure 4.1.3 Diagrams of a mature erythrocyte

When blood is allowed to settle or is centrifuged, under standard conditions, the red cells normally make up almost all the volume of the cellular sediment (see Fig. 4.1.1). The proportion of the blood sample composed of packed red cells (the **packed cell volume, PCV**) is 0.47 in men and 0.42 in women. This measure is also known as the **haematocrit (Hct)**. The haematocrit indicates red cell mass relative to plasma and so values are reduced in anaemia and raised in polycythaemia.

Morphology
Mature circulating erythrocytes are biconcave disc-shaped cells with a mean diameter of 7.4 μm and a thickness of 2.2 μm (Fig. 4.1.3). They have a **mean cell volume (MCV)** of 85 fl (femtolitres – see Appendix 1). Their relative excess of surface area over volume maximizes exchange with their environment and also makes them very deformable, which allows them to pass through the capillaries even though these have a diameter of approximately 6 μm.

Erythrocytes have a complex protein/lipid membrane, the surface of which is the site of various antigens which determine the different blood groups. Many of the cell's enzyme systems are located in the membrane, including those which function in the transport of glucose (which is the red cell's sole source of energy) across the membrane.

Mature erythrocytes lack nuclei and so are unable to divide or to synthesize new proteins. This means that they have a limited life span of 100–120 days. They also lack certain cytoplasmic organelles, for example mitochondria.

The major constituents of a red cell are the protein pigment haemoglobin and a variety of enzymes and other substances which are the products of red cell metabolism. These enzymes and intermediary metabolites are essential to the optimum functioning of haemoglobin and to intracellular stability. The constituents of red cells are surrounded by intracellular fluid and by a cytoplasmic stroma or framework which provides a structural basis for the metabolic function.

HAEMOGLOBIN (Hb)

Haemoglobin is a red-coloured protein pigment found within red blood cells. The normal concentration of haemoglobin in blood is 15.5 g/dl (2.28 mmol/l) in males and 14.0 g/dl (2.06 mmol/l) in females.

Each red cell contains 30 pg (picograms – see Appendix 1) of haemoglobin – **mean cell haemoglobin (MCH)**. The **mean cell concentration of haemoglobin (MCHC)** in an erythrocyte is 32 g/dl.

Functions of haemoglobin
The primary function of haemoglobin is oxygen uptake in the lungs, carriage in the blood and release in the tissues. This function is dependent on the partial pressure of oxygen at these sites. Haemoglobin also functions in carbon dioxide transport, as carbaminohaemoglobin, and as a powerful buffer in the maintenance of blood pH. These functions are further described in Chapter 5.3.

Synthesis
Haemoglobin synthesis takes place during erythropoiesis. Mature red cells are unable to perform this function.

Molecular structure
Haemoglobin has a molecular weight of 68 000. Each molecule consists of two major portions

(a) the iron-containing pigment, **haem**
(b) the protein **globin**.

Haem is composed of ring-shaped organic molecules called pyrrole rings (Fig. 4.1.4a). Four of these are joined by bridges to form a larger ring structure (Fig. 4.1.4b). An iron atom in the ferrous (Fe^{2+}) form is then taken up and held centrally by the nitrogen atom of each pyrrole ring.

Like all proteins, **globin** consists of a long chain of amino acids. There are four types

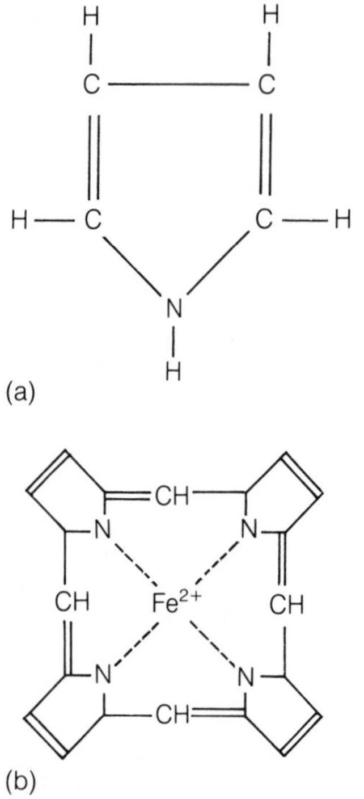

(a)

(b)

Figure 4.1.4   Molecular structure of (a) a pyrrole ring, (b) haem

of globin molecule which occur in normal human haemoglobin. These are distinguished by slight differences in the amino acid composition of the globin chain and are called alpha (α), beta (β), delta (δ) and gamma (γ). Each haemoglobin molecule contains four coiled globin molecules.

There are three important, normal, human haemoglobins.

	Globin chain content
HbA the major adult haemoglobin	2α 2β
HbA$_2$ the minor adult haemoglobin	2α 2δ
HbF fetal haemoglobin	2α 2γ

At birth, HbF makes up two-thirds of the haemoglobin content and HbA the remaining third. By 5 years of age, adult haemoglobin proportions have been obtained. These are HbA > 95%, HbA$_2$ < 3.5%, HbF < 1.5%.

The haemoglobin molecule: its structure and function

Each haemoglobin molecule has four coiled globin chains attached to four haem groups and each haem group contains one ferrous iron atom. This atom has one bond to enter into a loose and reversible combination with a molecule of oxygen, to form **oxyhaemoglobin** (HbO$_2$).

The poisonous gas carbon monoxide (CO) combines with haemoglobin in the same way but more easily and less reversibly because haemoglobin has about 250 times greater an affinity for carbon monoxide than it has for oxygen. The compound formed is called **carbon monoxyhaemoglobin** (HbCO) and the presence of this substance in the blood blocks the availability of haemoglobin for oxygen transport. Carbon monoxide poisoning therefore produces an anaemic hypoxia.

Each haemoglobin molecule can potentially carry four molecules of oxygen. At full saturation this means that one mole of haemoglobin is capable of carrying 4 moles of oxygen. The haemoglobin concentration of the blood is, therefore, a sensitive index of the oxygen-carrying capacity of the blood.

Haemoglobin picks up oxygen molecules one at a time. The reversible binding of each oxygen molecule changes the configuration of the globin and so increases the affinity of the haemoglobin molecule for oxygen. As a result, the affinity of the Hb molecule for the fourth oxygen molecule is 20 times greater than for the first. This phenomenon accounts for the sigmoid shape of haemoglobin's oxygen dissociation curve, which is so important physiologically and is discussed in Chapter 5.3.

Various factors affect the affinity of haemoglobin for oxygen and so shift the oxygen dissociation curve to the right (decreased oxygen affinity), or left (increased oxygen affinity). These factors include pH, temperature and the presence of the substance **2,3-diphosphoglycerate** (2,3-DPG) in the red cell. (2,3-DPG is an intermediary metabolite of glycolysis (glucose breakdown) occurring within the red cell.) Increasing temperature, decreasing pH (i.e. increasing acidity), and increasing levels of 2,3-DPG (precisely the conditions found in the tissues), all reduce haemoglobin's affinity for oxygen, shifting the dissociation curve to the right and so facilitating oxygen delivery in the tissues.

The structure of the haemoglobin molecule also

affects its oxygen affinity. For example, HbF has a higher oxygen affinity than HbA. This allows the fetus to obtain adequate amounts of oxygen from the partial pressure of oxygen in the placental blood supply (which is lower than that occurring in the mother's lungs).

Genetic mutations producing changes in the globin chains can produce abnormal haemoglobin molecules with either high or low oxygen affinities.

ABNORMAL HAEMOGLOBINS (HAEMOGLOBINOPATHIES)

Several abnormal haemoglobins have now been identified and can be distinguished by electrophoresis. The abnormalities always affect the globin part of the molecule and can be divided into two groups.

1 There is substitution of an amino acid in a globin chain by another amino acid, for example HbS which produces sickle cell anaemia (see below).
2 The rate of globin synthesis is subnormal; this produces the disease thalassaemia.

HbS – sickle cell haemoglobin

This is the most commonly occurring abnormal haemoglobin and is found most frequently in the African negro. The molecule differs only slightly from HbA in that the amino acid valine replaces glutamic acid as the sixth amino acid in the beta globin chains. However, this small change in structure has a major effect on molecular function. HbS is less soluble in the reduced form and the molecules link to form tubular structures which distort the red cell into a characteristic sickle shape. Such cells have a reduced life span and so a chronic haemolytic anaemia occurs in people with HbS (**sickle cell anaemia**). Sickle cells also increase the viscosity of the blood, resulting in reduced blood flow and sometimes complete vascular obstruction.

An individual will suffer from sickle cell anaemia if he or she is homozygous for the sickle cell gene, that is, the sickle cell gene has been inherited from both parents. About 1 in 400 African negroes suffers from this disease. The anaemia is severe and typically punctuated by crises of a variety of types, for example a vaso-occlusive, 'painful' crisis. In tropical countries, mortality rate in childhood is high, but in Western countries where sophisticated treatment is possible, patients may live into their sixties (Kenny, 1980).

A heterozygous individual who has inherited one normal and one sickle cell gene has the **sickle cell trait**. Heterozygotes are usually completely well as their red cells do not contain sufficient HbS to produce sickling at normal partial pressures of oxygen. However, they may be at some risk when subjected to hypoxic conditions, for example at altitude and during and following administration of certain anaesthetic gas mixtures. For this reason all negro patients should be routinely screened for HbS before given a general anaesthetic.

Figure 4.1.5 illustrates that heterozygous parents who carry the sickle cell gene have a 1 in 2 chance of producing a child with sickle cell trait, and a 1 in 4 chance of producing an offspring who is homozygous for either the normal HbA gene or the HbS gene for sickle cell anaemia.

Thalassaemia

In thalassaemia there is a genetically determined deficiency in the production of either the alpha or the beta chain of normal adult haemoglobin (HbA). A wide variety of conditions occurs, ranging from some that are so severe that they result in death in utero, to others that are detectable only on laboratory investigation (Gorman, 1980).

All cases of **thalassaemia major** are due to abnormalities of beta globin synthesis. This form of thalassaemia occurs mainly in people from Mediterranean countries and the Middle and Far East. In Britain most patients are Greek, Greek Cypriot and Indian. These individuals suffer from severe anaemia, which is due to their abnormal haemoglobin having a reduced ability to release oxygen and to a decreased life span of their red cells. Current treatment consists of blood transfusions to raise haemoglobin concentration; splenectomy in cases where red cell destruction by the spleen exceeds red cell production; drugs – iron chelators such as desferrioxamine, to reduce iron load; and folic acid administration as the huge haemopoiesis which occurs greatly increases folate requirements.

The carrier state can now be detected for both sickle cell anaemia and thalassaemia major, so affected individuals can be offered genetic counselling. It is also possible to discover, antenatally, if a fetus has the disease and some parents may

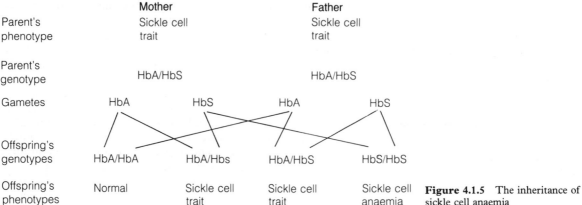

	Mother		Father	
Parent's phenotype	Sickle cell trait		Sickle cell trait	
Parent's genotype	HbA/HbS		HbA/HbS	
Gametes	HbA	HbS	HbA	HbS
Offspring's genotypes	HbA/HbA	HbA/Hbs	HbA/HbS	HbS/HbS
Offspring's phenotypes	Normal	Sickle cell trait	Sickle cell trait	Sickle cell anaemia

Figure 4.1.5 The inheritance of sickle cell anaemia

choose to terminate a pregnancy where it is known that the fetus is affected.

FUNCTION OF RED CELLS

The haemoglobin content and morphology of mature erythrocytes, described in preceding paragraphs, enable them to function efficiently as oxygen and carbon dioxide transporters. Haemoglobin also functions as a powerful buffer.

In addition, red cells contain the enzyme carbonic anhydrase, which catalyses the reaction

$$CO_2 + H_2O \underset{\text{anhydrase}}{\overset{\text{carbonic}}{\rightleftharpoons}} H_2CO_3$$

carbon dioxide · · · · water · · · · · · carbonic acid

$$\rightleftharpoons H^+ + HCO_3^-$$

hydrogen ion · · · · hydrogen carbonate ion

This enzyme has an important function in facilitating carbon dioxide carriage, as hydrogen carbonate ions (HCO_3^-), in plasma. This process is described in Chapter 5.3.

THE ERYTHROCYTE SEDIMENTATION RATE (ESR)

If diluted blood containing an anticoagulant is allowed to stand in a narrow vertical tube, the red cells slowly settle because their density is greater than that of plasma. The erythrocyte sedimentation rate (ESR) is determined by measuring the length of the column of clear plasma, above the cells, after 1 hour.

The red cells form a sediment more rapidly when they have a tendency to pile on top of one another, to form columns (like piles of plates) which are called **rouleaux**. The ESR is primarily a measure of the tendency of the red cells to form rouleaux. This tendency is dependent on the concentration of plasma proteins, particularly fibrinogen and globulins, and hence is directly related to plasma viscosity. Plasma viscosity measurements are replacing ESRs in some centres where appropriate equipment is available. Rouleaux formation is also enhanced by the presence in plasma of proteins which are the byproducts of inflammatory and neoplastic disease.

A raised ESR provides non-specific evidence of disease. It is often used as a screening test when diagnosis is uncertain and to monitor the course of a chronic illness, for example tuberculosis. An ESR of greater than 12 mm/hour is usually considered indicative of disease. The upper limit of the normal range for women is slightly higher than that for men. Exact reference ranges vary as a result of the specific measuring technique employed.

ANAEMIA

Anaemia occurs when the haemoglobin concentration of the blood falls below the lower limit of the reference range for the population (see Table 4.1.5). This happens when, for whatever reason, the erythropoietic tissues cannot supply enough normal erythrocytes to the circulation. Anaemia is therefore not a disease in itself, but a sign of disease. Its general effects on a person are the direct result of a reduction in the oxygen-carrying capacity of the blood. The severity of symptoms is related both to the haemoglobin concentration of the blood and to the length of time over which the anaemia developed.

Young adults with chronic anaemia do not usually experience health problems until their haemoglobin has fallen below 8 g/dl, but older people may have unpleasant symptoms at 11 g/dl. The major problems that occur are

(a) shortness of breath on exertion
(b) palpitations (awareness of the heart beat)
(c) feelings of tiredness and weakness
(d) loss of appetite and dysphagia (difficulty in swallowing)
(e) sore mouth and tongue (glossitis)
(f) pins and needles (paraesthesiae)
(g) intermittent claudication (shooting pains in the legs), particularly on exertion
(h) chest pain – angina pectoris
(i) oedema (swelling) of the ankles at the end of the day (may be due to mild heart failure induced by the anaemia)
(j) in severe anaemia, central nervous system symptoms appear and these include
faintness and giddiness
tinnitus (ringing in the ears)
headache
spots before the eyes.

The anaemic person appears pale because of the lack of the pigment haemoglobin in the blood. This pallor is evident in the skin and mucous membrane of the conjunctiva, mouth and tongue and in the nail beds.

Anaemia may be usefully classified according to its pathophysiology. There are three major types of disturbance which result in anaemia.

1 Decreased production of red cells.
2 Increased loss of red cells from the body.
3 Increased destruction of red cells in the body.

Anaemia is also described in terms of the size and colour (which reflects haemoglobin concentration) of the circulating red blood cells. Red cells may be **microcytic** (of small size), **normocytic** (of normal size), or **macrocytic** (of large size). They may also be **hypochromic** (pale in colour) or **normochromic** (normal in colour). For example, a microcytic, hypochromic anaemia is characteristic of iron deficiency, and a macrocytic,

normochromic or hypochromic anaemia typical of vitamin B_{12} deficiency.

It is important that the underlying cause of the anaemia is identified and treated, if this is possible. The only alternative treatment in severe anaemia is blood transfusion, but this procedure is especially hazardous for people with anaemia (unless the condition is due to acute haemorrhage) because of the possibility of fluid overload. If transfusion is essential, packed cells are preferable to whole blood and units should be given slowly, to minimize the above risk.

Iron metabolism and iron deficiency anaemia
The body of an adult normally contains 4–5 g (75–90 mmol) of iron. The majority of this is in haemoglobin (2.5 g, 45 mmol) and a further 1.5 g (27 mmol) is in storage in the tissues. Small quantities are found in the muscle pigment, myoglobin, and in some intracellular enzymes, such as cytochrome oxidase.

Iron requirements are obtained from dietary sources. Foods vary both in their iron content and in the availability of iron for absorption. Iron is mainly obtained from meat, liver, eggs, green vegetables and fruit. A mixed daily diet typically contains 5–20 mg, which is adequate to meet normal requirements.

In the stomach, gastric acid causes the bound iron to be released and converted to the ferrous form. Ferrous iron is then absorbed in the duodenum and jejunum. Normally only about 10% of the daily intake is absorbed, but a greater amount may be absorbed if body stores of iron are deficient. Absorption of 0.5–1.0 mg (9–18 μmol) of iron is sufficient for the requirements of a healthy man.

The need for iron is greater for women of reproductive age than for men because of the blood loss which occurs during menstruation. Pregnancy increases iron requirements considerably, particularly in the last 2 months when the fetus and placenta are growing rapidly. Growing children also require more iron than a healthy adult man.

Iron is carried in the plasma bound to the protein, **transferrin**. It is stored in the tissues (liver, spleen, bone marrow and cells of the mononuclear phagocytic system) in two forms. About two-thirds of storage iron is attached to the tissue protein **apoferritin** and is stored as **ferritin**. The remaining third is stored as **haemosiderin** which is a conglomerate of ferritin molecules.

Trace amounts of ferritin occur in plasma and the quantity indicates the magnitude of body iron stores. Serum iron concentration is normally 13–32 μmol/l (70–175 μg/dl). If iron deficiency develops, iron stores are depleted first, followed by haemoglobin and then myoglobin and enzymes. During pregnancy, fetal iron requirements take precedence over those of the mother.

Although red cells and their constituent haemoglobin are continually being broken down, the vast majority of the iron released is retained for reutilization in the body. Only small amounts of iron are excreted, in urine and bile, desquamated cells of the skin and gut mucosa, and in menstrual fluid. In healthy adults this amount balances that absorbed from the diet.

Iron poisoning can occur when storage of iron is greatly increased (**siderosis**). The main causes of iron overload are increased intestinal absorption and excessive parenteral administration, as, for example, in multiple blood transfusion. The former may be due to failure to control iron absorption, as in **haemochromatosis**, a condition which may be inherited and is associated with cirrhosis of the liver. Increased absorption may also be due to regular high iron intake. This is usually the result of regularly ingesting large quantities of cheap wine and cider, which contain high levels of iron. Acute iron poisoning is a significant cause of death in young children and the result of accidentally ingesting attractively coloured iron tablets.

Excessive parenteral administration of iron usually occurs in the form of multiple blood transfusions, as, for example, in the treatment of the haemolytic anaemia of thalassaemia.

Iron poisoning, whether acute or chronic, is treated by administration of the chelating agent, desferrioxamine, which both reduces iron absorption from the gut and increases iron excretion in the urine.

The causes of iron deficiency (see below) follow from the physiology described in the preceding paragraphs.

(a) Poor diet.
(b) Excessive requirements.
(c) Lack of gastric acid.
(d) Malabsorption.
(e) Blood loss.

The history of the person with iron deficiency anaemia may contain features which reflect the cause of their iron deficiency. It is important that this cause is found and treated wherever possible. Descriptions of the specific clinical features of this anaemia can be found in medical textbooks.

In most cases of iron deficiency anaemia, the amount of iron required to restore normal amounts of haemoglobin cannot be supplied in the diet and so additional iron preparations are required. Whenever possible these are given orally, for example as ferrous sulphate. Gastrointestinal upsets of various kinds may occur as side-effects. Intramuscular and intravenous iron preparations are available but their use is associated with unpleasant side-effects, such as skin staining and anaphylactic reactions.

Vitamin B_{12} and its deficiency
Vitamin B_{12} (hydroxocobalamin) is synthesized by bacteria and occurs, in significant amounts, only in foods of animal origin, that is, meat, liver, eggs, milk and cheese. Humans obtain all their vitamin B_{12} from their diet: a mixed diet contains 5–30 μg per day and the daily adult requirement is 2–3 μg. This requirement balances the daily loss from the body, largely in urine and faeces.

Vitamin B_{12} is absorbed in the ileum, mainly in combination with **intrinsic factor (IF)**, a glycoprotein secreted by the main gastric glands into the gastric juice (see Chap. 5.1). During absorption, IF is split from the vitamin B_{12}.

Vitamin B_{12} is transported in plasma bound to plasma protein. It is rapidly taken up by the tissues, especially the liver, and liver stores may last for several years, even if no further vitamin B_{12} is absorbed.

Deficiency of this vitamin causes anaemia because the vitamin is necessary for the synthesis of a precursor of thymine. Thymine is a constituent of deoxyribonucleic acid (DNA) and so DNA formation is limited in vitamin B_{12} deficiency; this causes increased size of red cell precursors – a **megaloblastosis**.

The major causes of vitamin B_{12} deficiency are as follows.

1 Lack of gastric intrinsic factor. The commonest reason for this is autoimmune gastric atrophy and it is this condition which produces the classical **pernicious anaemia** of vitamin B_{12} deficiency. Total and partial gastrectomy also result in loss of IF.

2 Malabsorption of the IF/B_{12} complex.
3 Strict vegetarian (vegan) diet, containing no animal products.

Vitamin B_{12} deficiency is treated by intramuscular injections of hydroxocobalamin, administered monthly or even less frequently. Typically, the patient feels significantly better within 48 hours of commencing treatment.

Folic acid (folate) and its deficiency
Folic acid is present in foods of animal and vegetable origin, but is destroyed by heat. The major dietary sources are liver, green vegetables and oranges. A typical Western diet contains 500–800 μg per day and the daily adult requirement is 100–200 μg. This excess of intake over requirement compensates for the poor absorption of some forms of the vitamin and for losses through cooking.

Folates are absorbed in the duodenum and jejunum, and liver stores are sufficient to last a few months if intake ceases. Small amounts are excreted in urine and faeces.

Like vitamin B_{12}, folates promote thymine synthesis and hence DNA synthesis. As a result, folate deficiency causes a megaloblastic anaemia which has essentially the same features as vitamin B_{12} deficiency anaemia, except that neurological deficiency does not occur, and intrinsic factor activity and gastric acid are normal.

The major causes of folic acid deficiency are

(a) poor diet
(b) increased requirements
(c) malabsorption
(d) the presence of folic acid antagnoists, for example the antimitotic drug, methotrexate.

Treatment is with oral folic acid preparations and the response is usually rapid.

THE LIFE SPAN AND BREAKDOWN OF RED CELLS

The life span of normal red cells in a normal circulation is approximately 120 days.

Old red cells are ingested and destroyed by the macrophage cells of the mononuclear phagocytic system (phagocytic cells found in bone marrow, liver, lymph nodes, spleen and subcutaneous tissues). The mechanism(s) by which senescent red cells are recognized are not known. The spleen, in particular, functions in the removal of abnormal red cells, and splenomegaly (enlargement of the spleen) is often a clinical feature of conditions in which abnormalities of red cells occur.

Haemoglobin breakdown
Within the macrophage cells, the ring of the haem in each haemoglobin molecule is opened, by oxidation, so that the haem ring becomes a straight chain. This causes the haem to split from the globin portion of the molecule. The globin is catabolized to its constituent amino acids and these enter the general metabolic pool to be reutilized.

Iron is removed from the opened haem ring and the majority is reutilized in the synthesis of new iron-containing compounds, mainly haemoglobin. The remainder is retained in storage forms in the tissues. Only a very small amount is lost from the body.

The remainder of the haem molecule is converted to the pigment bilirubin and transported to the liver, bound to plasma albumin. In the liver it is conjugated with glucuronic acid to form a water-soluble compound which is excreted in the bile. The excretion of bilirubin is fully described in Chapter 5.2. Figure 4.1.6 summarizes the breakdown of the haemoglobin molecule.

Specific mechanisms exist to prevent the loss from the body of haem and the iron it contains, during red cell destruction. Normally this process occurs within macrophage cells so that haemoglobin is contained. However, if haemolysis occurs in the circulation, haemoglobin is released into the blood. The molecular weight of haemoglobin (68 000) is such that the molecule will pass into the glomerular filtrate and be excreted in the urine.

Excretion of haemoglobin is prevented by the presence of the plasma proteins **haptoglobins** and **haemopexin** in the plasma. These bind to free haemoglobin and haem, respectively, forming molecular complexes which are too large to be excreted and are instead taken up by the mononuclear phagocytic system. Only when both these mechanisms for iron conservation are saturated will free haemoglobin be excreted in the urine (haemoglobinuria).

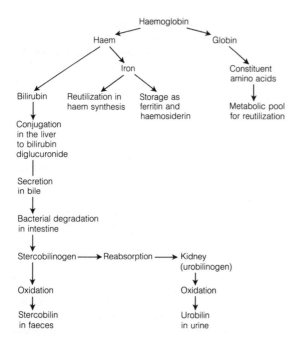

Figure 4.1.6 A summary of haemoglobin breakdown

White blood cells (leucocytes)

The blood of a healthy adult contains $4–11 \times 10^9$ white blood cells per litre. This is the **white cell count (WCC** or **WBC)** which, unlike the red cell count, represents only a small proportion of the total number of white cells in the body. Within these limits, quite large variations can occur in an

Table 4.1.6 Reference ranges for the white blood cell count

White cell type	Count per litre of blood
All white cells (WBC or WCC)	$4–11 \times 10^9$
Differential white cell count, in absolute numbers and as percentage of total count (parentheses)	
Neutrophils	$2.0–7.5 \times 10^9$ (40–75)
Eosinophils	$0.04–0.4 \times 10^9$ (1–6)
Basophils	$0.0–0.1 \times 10^9$ (0–1)
Monocytes	$0.2–0.8 \times 10^9$ (2–10)
Lymphocytes	$1.5–4.0 \times 10^9$ (20–45)

individual, from hour to hour, as various physiological factors, such as exercise and emotion, influence white cells to enter or leave the circulation.

Several types of white cell can be distinguished on examination of a stained blood film. Cells are classified into groups on the basis of various morphological characteristics, such as the presence or absence of a granular cytoplasm, the staining reaction of the cytoplasm and the shape of the nucleus. A **differential white cell count** is obtained by classifying and counting the first 100 white blood cells observed on examining a blood film and expressing the result as a percentage of the total. Reference ranges for the different types of white blood cell are given in Table 4.1.6.

An increase in the white cell count above the normal range (i.e. greater than $11 \times 10^9/l$) is called a **leucocytosis,** and a decrease below the normal range (i.e. less than $4 \times 10^9/l$) a **leucopenia.** If a specific type of white cell is causing the abnormal count, it is described accordingly, for example a neutrophil leucocytosis or a lymphocytosis.

TYPES OF WHITE BLOOD CELL

Granulocytes (polymorphonuclear leucocytes)
These cells are characterized by the presence of granules in the cytoplasm and a lobed nucleus, hence the above names. They are $10–14\,\mu m$ in diameter. Within the group, three cell types can be distinguished from the staining reaction and size of their cytoplasmic granules. They, too, have been named accordingly.

Neutrophils. These contain cytoplasmic granules of varying size that stain violet with a neutral dye. As they are by far the most numerous of the granulocytes (see Table 4.1.6) and have the most lobed nucleus, neutrophils often monopolize the term polymorphonuclear leucocyte (PML) (Fig. 4.1.7a).

Eosinophils. The cytoplasm of these cells contains large, distinctive granules which take up an acidic dye and stain red. The nucleus typically has two lobes (Fig. 4.1.7b).

Basophils. These cells have large, basophilic cytoplasmic granules which take up a basic dye and

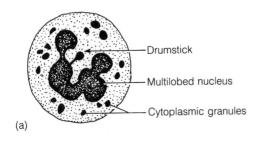

(a)

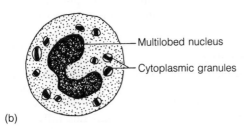

(b)

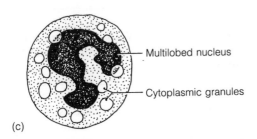

(c)

Figure 4.1.7 Granulocytes: (a) a neutrophil, (b) an eosinophil, (c) a basophil

stain blue/black. The nucleus typically has two to three lobes (Fig. 4.1.7c).

The production of granulocytes (granulopoiesis)
Postnatally, production of granulocytes normally occurs only in red marrow. Granulopoiesis is characterized by progressive condensation and lobulation of the nucleus, loss of RNA and other cytoplasmic organelles, for example mitochondria, and the development of cytoplasmic granules in the cells involved.

The development of a polymorphonuclear leucocyte make take a fortnight, but this time can be considerably reduced when there is increased demand, as, for example, in bacterial infection. The red marrow also contains a large reserve pool of mature granulocytes so that for every circulating cell there may be 50–100 cells in the marrow.

Mature cells pass actively through the endo-thelial lining of the marrow sinusoid into the circulation. In the circulation, about half the granulocytes adhere closely to the internal surface of the blood vessels. These are called **marginating cells** and are not normally included in the white cell count. The other half circulate in the blood and exchange with the marginating population.

Within 7 hours, half the granulocytes will have left the circulation in response to specific requirements for these cells in the tissues. Once a granulocyte has left the blood it does not return. It may survive in the tissues for 4 or 5 days, or less, depending on the conditions it meets.

The turnover of granulocytes is, therefore, very high. Dead cells are eliminated from the body in faeces and respiratory secretions and are also destroyed by tissue macrophages (monocytes).

Control of granulopoiesis
No precise mechanisms for the control of granulocyte production have, so far, been described. However, in health, the count remains relatively constant so it is likely that homeostatic control mechanisms operate.

Functions of granulocytes

Neutrophils. Neutrophils are actively mobile, phagocytic cells and their shape is constantly changing due to amoeboid movement. The granules in their cytoplasm are enzymes which have a digestive, or lytic, action.

The major function of neutrophils is in inflammation of bacterial origin. The vascular endothelium becomes sticky at the site of the inflammation. Neutrophils adhere to this sticky endothelium and to each other, forming clumps. They then actively move into the tissues, probably passing between the junctions of the endothelial cells. This process is known as **diapedesis**.

In the tissues, neutrophils are attracted to the site of inflammation by various substances produced there, for example bacterial toxins and components of the inflammatory and immune responses. This process is known as **chemotaxis**. When a neutrophil comes into contact with a bacterium, **phagocytosis** (cell eating) occurs (Fig. 4.1.8). The cell membrane of the neutrophil becomes invaginated until the bacterium is completely engulfed in a digestive vacuole. Phagocytosis occurs much more easily if the bacterium is coated with IgG antibody.

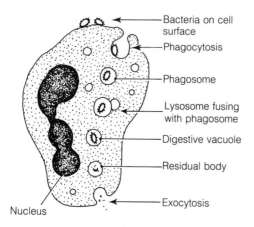

Bacteria on cell surface
Phagocytosis
Phagosome
Lysosome fusing with phagosome
Digestive vacuole
Residual body
Exocytosis
Nucleus

Figure 4.1.8 Diagram of a neutrophil undergoing phagocytosis

Once the bacterium is completely engulfed, the lytic enzymes of the granules are released into the digestive pouch where they digest and kill the bacterium.

Neutrophils also act as scavengers in the tissues, removing small inanimate particles such as antigen–antibody complexes. For this reason they are sometimes called **tissue microphages**, (Greek: *micro* = small, *phage* = eat), whereas monocytes are tissue macrophages. The ability of neutrophils to metabolize without oxygen is crucial to their function of removing bacteria and cell debris in necrotic tissue.

Large numbers of neutrophils may be destroyed during an acute inflammatory response to a bacterial infection. Dead neutrophils, their contents and cell debris form the **pus** which occurs at such a site, for example a boil, abscess or infected cut.

Some bacteria produce toxins which act as pyrogens (fever-producing substances). It is likely that these induce fever by interacting with neutrophils to produce a polypeptide called **leucocyte (or endogenous) pyrogen**. This polypeptide is able to cross the blood–brain barrier and affect the temperature-regulating centre in the hypothalamus.

Bacterial infection can produce a **neutrophil leucocytosis** of up to 40×10^9 per litre. The inflammatory reaction produced by tissue damage and cell death, for example in myocardial infarction, also produces a raised neutrophil count.

The secretion of adrenaline in acute stress produces a transient increase in the number of circulating neutrophils. Increased levels of adrenocorticotrophic hormone and glucocorticoids, for example hydrocortisone, which occur in stress and steroid therapy also increase the number of circulating neutrophils. However, the action of these hormones impairs the ability of neutrophils to migrate into the tissues and so resistance to infection is impaired despite their increased numbers.

Any factor which interferes with the production and/or function of mature neutrophils will reduce resistance to infection. Cytotoxic drug therapy, or deficiency of vitamin B_{12} or folate, for example, may depress granulopoiesis and lead to an **agranulocytosis** (reduction or absence of granulocytes). In contrast, there are large numbers of circulating neutrophils in granulocytic leukaemia. However, a large number of these neutrophils are immature and not fully capable of phagocytosis. Hence resistance to infection is decreased.

Eosinophils (see Table 4.1.6 and Fig. 4.1.7b). Eosinophils are mobile, phagocytic cells, destroying animate and inanimate particles, but less actively so than neutrophils and monocytes. Their cytoplasmic granules contain most of the lytic enzymes found in neutrophil granules.

Eosinophils are involved in immunoglobulin E (IgE)-mediated immune responses and function to neutralize and limit the effect of inflammatory substances, such as histamine and bradykinin. They collect at sites of allergic reaction, for example in the respiratory passages in hay fever and asthma, and the eosinophil count is raised in allergic conditions.

Circulating eosinophil numbers are decreased by adrenocorticotrophic hormone and glucocorticoids.

Basophils (see Table 4.1.6 and Fig. 4.1.7c). The cytoplasmic granules of basophils contain histamine and heparin. Basophils are the circulating counterparts of **tissue mast cells** (i.e. they become mast cells on entering the tissues).

Mast cells are widely distributed in the body and are commonly found in close proximity to the walls of small blood vessels. They are able to bind specific IgE antibody to their surface and subsequent exposure of a basophil that has done this to the antigen for which the bound IgE is specific, leads to rapid breakdown and loss of the basophil's granules.

Basophils are involved in **anaphylactic reac-**

tions, which are mainly due to the effects of the release of **histamine**. In humans, histamine produces vasodilation and hence a fall in blood pressure and rise in heart rate. It increases the tone of most types of smooth muscle and stimulates exocrine secretion. It also produces itching or, in high concentrations, pain. Antihistamine drugs, for example promethazine (Phenergan), are very effective against anaphylactic conditions.

Heparin is a powerful anticoagulant. In anaphylaxis, blood clotting is inhibited due to the release of heparin but the precise physiological function of this is not established.

The degranulation of basophils attracts eosinophils into the region of antigen–IgE interaction by the production of an eosinophil chemotactic factor.

Agranulocytes

These white blood cells are so called because of the non-granular appearance of their cytoplasm

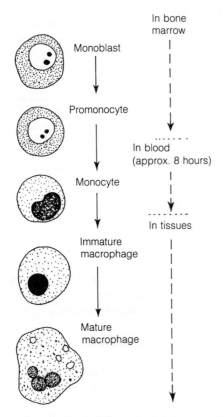

Figure 4.1.9 A summary of monocyte maturation and life cycle

when viewed with a light microscope, although very small granules are present. Two types of cell are distinguished. These are monocytes and lymphocytes.

Monocytes (see Table 4.1.6). Monocytes are large cells with a diameter of 10–18 μm. They have a large oval or indented nucleus which is pale staining and positioned non-centrally. The cytoplasm appears clear and stains pale blue. It contains many mitochondria and other cytoplasmic organelles, and may contain very fine granules, known as azur granules. Monocytes are produced in the bone marrow, developing from nucleated precursors, the monoblast and promonocyte (Fig. 4.1.9).

Mature cells have a half life in blood of approximately 8 hours and, like granulocytes, there is a circulating and marginating pool.

Monocytes are actively phagocytic and, on migration into the tissues, they mature into larger cells called **macrophages** (Greek: *macro* = big, *phage* = eat), which can survive in the tissues for long periods. These cells form the mononuclear phagocytic cells of the mononuclear phagocytic system (reticuloendothelial system) in bone marrow, liver, spleen and lymph nodes. Tissue macrophages (sometimes called **histiocytes**) respond more slowly than neutrophils to chemotactic stimuli. They engulf and destroy bacteria, protozoa, dead cells and foreign matter. They also function as modulators of the immune response by processing antigen structure and facilitating the concentration of antigen at the lymphocyte's surface. This function is essential in order that full antigenic stimulation of both T and B lymphocytes can take place.

Lymphocytes (see Table 4.1.6). Lymphocytes are round cells containing large round nuclei. The cytoplasm stains pale blue and appears nongranular under light microscopy. However, some cytoplasmic granules and organelles are present.

Morphologically, lymphocytes can be divided into two groups: the more numerous **small lymphocytes**, with a diameter of 7–10 μm; and **large lymphocytes**, which have a diameter of 10–14 μm. Lymphocytes are produced in bone marrow from primitive precursors, the lymphoblasts and prolymphocytes. Immature cells migrate to the thymus and other lymphoid tissues, including that found in bone marrow, and undergo further division, processing and maturation.

Functionally, there are two main types of lymphocyte: **T (thymus processed or dependent) lymphocytes**, which are involved in cell-mediated immune responses, and **B (bursa or bone marrow dependent), lymphocytes** (see Chap. 6.2). B lymphocytes produce plasma cells which secrete antibodies (immunoglobulins), and so produce humoral immunity. T and B lymphocytes are morphologically indistinguishable, and their identification is dependent on the detection of surface markers. B lymphocytes have membrane-bound surface immunoglobulins and T cells do not; 60–80% of lymphocytes circulating in the blood are T cells.

Lymphocytes spend a very small proportion of their life cycle in the blood and most of it in the lymphoid tissues. At any one time there are many more lymphocytes in the tissues than in blood. However, unlike granulocytes, lymphocytes recirculate between the blood and lymphatic tissues, returning to the blood via the thoracic duct. Both T and B lymphocytes may have a short or long life span. T cells, though, are generally long lived (2–4 years) and B cells typically have a shorter life span, measured in days or weeks.

As circulating lymphocytes represent only a tiny proportion of the cells in the body's lymphoid tissue, even major abnormalities of lymphoid function may not be apparent from the white cell count. The major causes of an abnormally raised lymphocyte count (**lymphocytosis**) include lymphoid malignancy and viral infection, particularly that caused by the Epstein–Barr (EB) virus, the causative agent of infectious mononucleosis (glandular fever). A decreased lymphocyte count mainly reflects reduced numbers of T lymphocytes. It may be due to immune deficiency disease, or be the result of high levels of circulating glucocorticoids (steroids), cytoxic therapy, or malignant disease, for example Hodgkin's disease.

Leukaemia
Leukaemia is produced by a malignant proliferation of leucopoietic tissue. The disease is classified according to its course and duration and the predominant affected white cell type, for example acute lymphoblastic leukaemia and chronic myeloid leukaemia. The former is mainly a disease of childhood and the latter usually affects middle-aged adults.

Characteristically there is a leucocytosis consisting mainly of immature and abnormal white cells of the type affected. The malignant leucopoietic tissue also infiltrates and displaces other haemopoietic tissues and may infiltrate the liver and spleen.

People who have leukaemia are prone to infection because a large proportion of their white cells is not fully functional due to their immaturity. They may also be anaemic because of depressed erthropoiesis and prone to bleed because of reduced platelet production. The major clinical features of leukaemia are due to these three factors.

Leukaemia is usually diagnosed from an examination of the person's blood count and the results of a bone marrow biopsy. The most effective therapy includes intensive cytotoxic chemotherapy and supportive measures such as the use of broad spectrum antibiotics to prevent infection, and blood transfusion to increase levels of red cells, mature white cells and platelets. Bone marrow transplant may also be considered.

White cell antigens
Antigens on the surface of white blood cells may be either specific to the cell type or also present on other cells of the body. Of the latter type, the most important clinically is the **human leucocyte antigen** (HLA) system. This set of antigens occurs on the surface of most body cells and forms the basis of tissue typing for the purposes of matching prior to organ transplantation. For this reason the HLA system is also called the major histocompatibility complex (MHC).

Platelets (thrombocytes)

Platelets are the smallest cellular elements in blood. They are colourless, non-nucleated, discoid bodies, with a diameter of 2–4 μm and a volume of 7 fl. The number of platelets in a normal adult's blood ranges from 150–400 $\times$ 10^9 per litre. (This is the **platelet count.**)

PRODUCTION (THROMBOPOIESIS)

Platelets are produced in bone marrow. They are formed in the cytoplasm of a very large cell, the **megakaryocyte** (Fig. 4.1.10). The cytoplasm of the megakaryocyte fragments at the edge of the cell. This is called platelet budding. Mega-

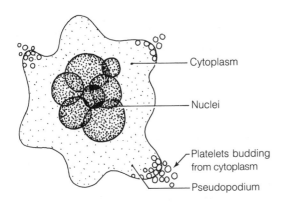

Figure 4.1.10 Diagram of a megakaryocyte showing platelet budding

karyocytes mature in about 10 days, from a large stem cell, the **megakaryoblast**.

It is likely that there are thrombopoietic feedback mechanisms as the platelet count remains fairly constant in health, and platelet production is reduced following an infusion of platelets and increased following removal of platelets. However, these feedback mechanisms have not been elucidated.

At any one time, about two-thirds of the body's platelets are circulating in the blood and one-third are pooled in the spleen. There is constant exchange between the two populations. The life span of platelets is between 8 and 12 days. They are destroyed by macrophages, mainly in the spleen and also in the liver.

STRUCTURE OF PLATELETS

Under the light microscope, platelets are normally disc shaped and their structure appears relatively simple, with a transparent cytoplasm and a central, more darkly staining, granular area. Electron microscopy and biochemical analysis reveal a more complex structure.

The platelet membrane has the dimensions of, and contains the usual components of, a cell membrane. It is the site of many enzymes and also contains receptors for numerous substances, such as collagen, adrenaline and serotonin (5-hydroxytryptamine). The platelet membrane also contains many antigens, some of which are platelet specific and others are also found on different types of cell.

The platelet cytoplasm contains microtubules and microfilaments composed of a contractile protein called **thombosthenin**, which functions in clot retraction. The cytoplasm also contains numerous granules which are composed of enzymes and substances that are released during platelet aggregation. Phospholipids (called platelet factors) are present both in these granules and in the platelet membrane and function in the clotting mechanism.

FUNCTIONS OF PLATELETS

The major functions of platelets are in haemostasis (the arrest of bleeding). Through **adhesion** to the damaged endothelium of a blood vessel and **aggregation** (interaction with each other so that they stick together), they form a **platelet plug**, which can stop blood flow. Haemostasis is initially dependent on the formation of this plug, which occurs very rapidly. The plug is normally stabilized later by the formation of a fibrin clot.

Platelets also function in blood clotting, providing essential phospholipid coagulation factors for various stages in the formation of a clot and in clot retraction.

These important functions of platelets are described in more detail in the next section on haemostasis. Abnormalities of platelet production and/or function are one of the commonest causes of defects in haemostasis. The latter is usually normal until the platelet count falls to $40 \times 10^9/l$ (normal range is $150–400 \times 10^9/l$). When the count falls below $20 \times 10^9/l$, spontaneous bleeding is likely to occur. An abnormally low platelet count is called **thrombocytopenia**. An abnormally high platelet count (**thrombocytosis**) occurs in several diseases, for example myeloid leukaemia, and immediately following splenectomy.

In addition to the above functions, platelets are able to phagocytose relatively small particles, such as viruses and immune complexes. They also store and transport histamine and serotonin, both of which are released when platelets are destroyed and function in the early response to injury, affecting the tone of smooth muscle in blood vessel walls.

Normal platelets probably have a nutritional function, supplying the endothelial cells of blood vessels with nutrients, since these cells atrophy in platelet deficiency. Platelets also secrete a growth factor which stimulates proliferation of smooth muscle cells in artery walls. In excess, the secretion of this factor may play a part in the development of atherosclerosis (Lowe, 1983).

HAEMOSTASIS (THE PHYSIOLOGICAL ARREST OF BLEEDING)

Haemostasis is a fundamental homeostatic process which functions to prevent the loss of blood from the vascular system and to ensure patency of the blood vessels. Normally, it is very effective in controlling bleeding from breached capillaries, arterioles and venules, and even small arteries and veins, but it is insufficient to halt haemorrhage from large vessels. In the last-mentioned situation, non-physiological, first aid, medical and surgical measures are required in order to control blood loss.

Natural haemostasis stops blood loss because constituents of the blood and damaged blood vessel wall react together to form a solid mass which blocks the hole in the vessel. Bleeding will also slow, or cease, if the blood pressure within the vessel approaches or becomes equal to the external hydrostatic pressure. This may occur either if there is a local or general fall in blood pressure, or if inflammation and the accumulation of blood in the surrounding tissues raise the external pressure.

For clarity of description, haemostasis may be divided into four phases. However, it should be realized that the events of these phases overlap and are synergistic; they do not occur in a straight-forward sequence. The four phases of haemostasis are as follows.

1 Contraction of the smooth muscle cells in the wall of the breached vessel, with resultant vasoconstriction – the myogenic reflex.
2 The formation of a platelet plug.
3 The formation of a fibrin clot (blood clotting or coagulation) and retraction of the fibrin clot.
4 Fibrinolysis.

Haemostasis is outlined in Fig. 4.1.11, and the events of each of the above phases are described below.

Phase 1: vasoconstriction (the myogenic reflex)

Immediately following injury, the damaged vessel may dilate. This is probably due to the vasodilator effect of histamine, released from tissue mast cells, in response to trauma, and from platelets. However, after a few seconds, vasoconstriction and retraction of the severed ends of blood vessels occur. This is due to the release from aggregating platelets of serotonin and thromboxane A_2. Both these substances are powerful vasoconstrictors. The vasoconstriction lasts approximately 20 minutes. It reduces blood flow and hence the likelihood of the platelet plug (Phase 2) being washed away.

Capillary walls contain no smooth muscle and

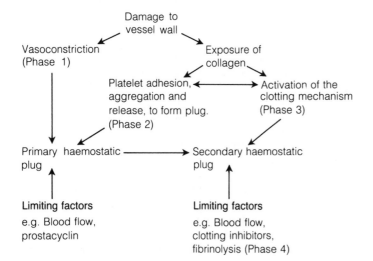

Figure 4.1.11 An outline of the events of haemostasis

so are not stimulated to constrict. In capillary bleeding, blood flow through the capillary bed may be reduced by contraction of precapillary sphincters.

Additionally, adhesion of the endothelial (intimal) surfaces of capillaries and small blood vessels may occur following vasoconstriction, and this may seal the hole.

Phase 2: the formation of a platelet plug (Fig. 4.1.12)

Damage to a blood vessel wall which exposes collagen and elastin microfilaments in the subendothelium causes passing platelets to adhere to the site, since platelets have membrane-based collagen receptors.

Platelet adherence causes the release of adenosine disphosphate (ADP) from the platelet, red cells and vessel wall. The ADP triggers platelet membrane changes which lead to the platelets assuming a spherical shape with pseudopodia, instead of their more characteristic disc shape. These membrane changes are known as contraction and they promote platelet aggregation (interaction with one another so that they stick together). Platelet aggregation is also promoted by many other substances, such as serotonin and certain clotting factors.

The process of platelet adhesion and aggregation is reversible until the platelets discharge or 'release' their granules. After platelet release has occurred, white blood cells begin to adhere to the degranulated platelets and clotting (Phase 3) occurs.

During aggregation the unstable substance **thromboxane A$_2$ (TXA$_2$)** is produced by the platelets from prostaglandin precursors, and this is a powerful inducer of both aggregation and vasoconstriction. The production of TXA$_2$ is inhibited by non-steroidal anti-inflammatory drugs, for example aspirin and phenylbutazone. This action partially explains the antithrombotic effect of these drugs and their ability to prolong bleeding time. The antithrombotic drug sulphinpyrazone (Anturan), also inhibits platelet prostaglandin synthesis and hence adhesion and aggregation.

Within the endothelial cells of the vessel wall, prostaglandin endoperoxides are converted to **prostacyclin**, a substance which powerfully inhibits aggregation and produces vasodilation. Prostacyclin has now been synthesized and is a potential antithrombotic agent for clinical use. The drug dipyridamole (Persantin) exerts its antithrombotic effect by enhancing the action of prostacyclin.

The physiological production, by the platelet and vessel endothelium, of two substances, TXA$_2$ and prostacyclin respectively, which have directly opposing effects, may be a protective mechanism against intravascular thrombosis.

Platelet aggregation also activates platelet phospholipid (platelet factor 3). This substance is not released into the plasma but functions as a surface on to which various blood clotting factors bind and react as a result of their physical proximity. The formation of the prothrombin activator complex occurs in this way. Once thrombin has been formed from activated prothrombin it also acts as a powerful stimulus to platelet aggregation.

In summary, three distinct mechanisms act to produce platelet aggregation and the formation of a platelet plug.

1 Amine release following platelet adhesion.
2 Thromboxane A$_2$ generation.
3 The formation of thrombin in the clotting sequence.

The early stages of haemostasis are dependent on constriction of the damaged vessel wall and the formation of a platelet plug. Blood clotting (Phase 3) plays little or no part in either of these processes. The platelet plug is formed within a few seconds of injury, and can withstand a blood pressure of up to 100 mmHg (13.3 kPa) and so may be sufficient to stop bleeding from small vessels.

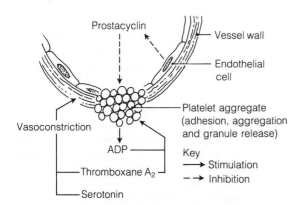

Figure 4.1.12   Summary of events in the formation of a platelet plug

However, if the plug is not stabilized by fibrin fibres around the aggregated platelets, it is likely that it will break down and bleeding will restart after about 20 minutes. This phenomenon may occur following relatively minor injury, such as a dental extraction, in untreated people with a clotting disorder but normal platelet function, for example those with haemophilia.

The normal formation of the platelet plug is, then, as important for haemostasis as blood clotting.

Phase 3: the formation of a fibrin clot (blood clotting or coagulation)

The essential end-point of this process is the formation of the insoluble protein fibrin from the soluble plasma protein fibrinogen. This reaction is catalyzed by the proteolytic enzyme thrombin. Thrombin does not exist in the blood but is formed from its inactive precursor, the soluble plasma protein prothrombin. It is now known that the activation of prothrombin to form thrombin requires the complex interaction of chemical factors derived from damaged tissue, platelets and plasma to produce prothrombin activator (Fig. 4.1.13).

Thrombin splits off two peptides from fibrinogen to form fibrin monomers. These monomers then polymerize to form a soluble fibrin clot which, in the presence of calcium ions and clotting factor XIII (Table 4.1.7), stabilizes into an insoluble clot (Fig. 4.1.14). Insoluble fibrin is laid down as a network of fine white threads which stick to each other and to tissue cells, and which entangle blood cells and platelets. (White fibrin threads can be seen if a blood clot is washed in

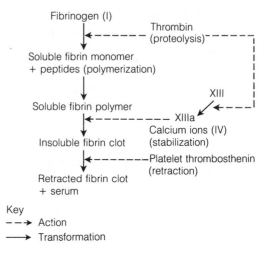

Key
--→ Action
──→ Transformation

Figure 4.1.14 Events in the formation of an insoluble fibrin clot from fibrinogen

water.) A new clot is a soft, jellylike mass. This gradually retracts by more than a half, squeezing out serum. Clot retraction is dependent on normal platelet function, requiring the contractile platelet protein, thrombosthenin. A retracted clot is more solid and elastic and forms a tougher, more efficient haemostatic plug.

The 13 factors involved in blood clotting were discovered mainly during the 1940s and 1950s. In 1961, an International Convention agreed that they should be designated Roman numerals for identification purposes. These numbers indicate the historical sequence of their discovery. The clotting factors are shown in Table 4.1.7.

FORMATION OF PROTHROMBIN ACTIVATOR

The activation of prothrombin (to form thrombin) by the formation of prothrombin activator can occur via two distinct mechanisms. The first of these is triggered by tissue damage. This damage generates thromboplastin (factor III) which is extrinsic to blood and hence this mechanism is called the **extrinsic system**.

The second mechanism is called the **intrinsic system** because only constituents of blood take part in the reactions. It is triggered by contact with a negatively charged surface, such as exposed collagen filaments.

The individual physiological significance of these two separate pathways has not been elucidated.

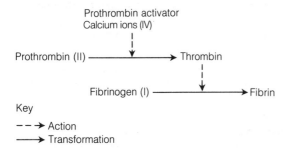

Key

--→ Action
──→ Transformation

Figure 4.1.13 The essential end reactions of the clotting sequence

Table 4.1.7 The clotting factors

Numeral	Name	Site of production	Major function	Deficiency
I	Fibrinogen	Liver	Converted to fibrin by proteolytic enzyme thrombin	Very rare; inherited as autosomal recessive; causes bleeding disorder
II	Prothrombin	Liver: vitamin K dependent	Converted to thrombin by prothrombin activator	Very rare; inherited as autosomal recessive; causes bleeding disorder
III (rarely used)	Thromboplastin (tissue factor or tissue extract)	Damaged cells	Forms complex with VII, in presence of calcium ions, to activate X via extrinsic system	
IV (rarely used)	Ionic calcium	Obtained from dietary sources and from reservoir in bone	Essential for the formation of prothrombin activator, the conversion of prothrombin to thrombin, and the formation of an insoluble fibrin clot	Plasma levels must fall below 1.25 mmol/l to produce delayed clotting; such low levels would cause severe tetany
V	Labile factor (proaccelerin/accelerator globulin)	Liver	Required in the formation of prothrombin activator via intrinsic and extrinsic systems; forms complex with activated X and platelet phospholipid to activate prothrombin	Very rare; inherited as autosomal recessive; causes bleeding disorder
VI	No longer used			
VII	Stable factor (autoprothrombin I, proconvertin)	Liver: vitamin K dependent	Forms complex with thromboplastin (III) to activate X via extrinsic system	As V above
VIII	Antihaemophilic globulin (AHG) (antihaemophilic factor A (AHF$_A$))	Endothelial cells ? Liver	Distinct properties associated with different parts of the molecule. Low molecular weight portion VIIIc is required in the formation of prothrombin activator from blood constituents (intrinsic system). Forms complex with activated IX and platelet phospholipid to activate X	Rare; inherited as sex-linked recessive; absence of activity produces bleeding disorder – classical haemophilia
			High molecular weight protein VIII R-Ag or VIII R-WF, functions as carrier protein and a particular configuration induces platelet aggregation	Very rare; inherited as autosomal dominant; causes complex bleeding disorder – von Willebrand's disease
IX	Christmas factor (antihaemophilic factor B (AHF$_B$)) Plasma thromboplastin component (autoprothrombin II)	Liver: vitamin K dependent	Required in the formation of prothrombin activator from blood constituents (intrinsic system)	Rare; inherited as sex-linked recessive; produces the bleeding disorder haemophilia B – Christmas disease
X	Stuart–Prower factor	Liver: vitamin K dependent	Activation of this factor is key reaction for the production of prothrombin activator via both intrinsic and extrinsic systems	Very rare; inherited as autosomal recessive; causes bleeding disorder
XI	Plasma thromboplastin antecedent (antihaemophilic factor C (AHF$_C$))	Liver	As IX above	Very rare; inherited as autosomal (?) dominant or recessive; causes bleeding disorder

Table 4.1.7   Continued.

Numeral	Name	Site of production	Major function	Deficiency
XII	Hageman factor	Liver	Required for formation of prothrombin activator from blood constituents (intrinsic system); activated by contact with negatively charged surfaces, e.g. collagen, glass; activation is trigger for other plasma enzyme systems such as kinin generation, complement activation and fibrinolysis	Very rare; inherited as autosomal recessive; does not produce a bleeding disorder
XIII	Fibrin Stabilizing factor	Liver	In presence of calcium ions, causes polymerization of soluble fibrin to form insoluble fibrin clot	Very rare; inherited as autosomal recessive; produces bleeding disorder

The key reaction in both systems is the activation of factor X. Later reactions are common to both systems. Once activated, factor X reacts with its cofactor V, in the presence of platelet phospholipid (platelet factor 3) which provides a surface on which the reagents are concentrated. This complex is termed **prothrombin activator** (Fig. 4.1.15).

The extrinsic system for the activation of factor X
This mechanism is activated by tissue damage and the consequent production of tissue extract thromboplastin (factor III). In the presence of calcium ions, thromboplastin forms a complex with factor VII. This complex activates factor X (Fig. 4.1.16). The extrinsic system is rapid, the end-product of a fibrin clot being produced in

10–15 seconds. The normality of the extrinsic system is measured by the prothrombin time test.

The intrinsic system for the activation of factor X
The intrinsic system involves a series of reactions normally initiated by exposure of factor XII to collagen filaments in the subendothelium of a damaged blood vessel wall. This exposure activates factor XII and initiates the intrinsic system of blood clotting.

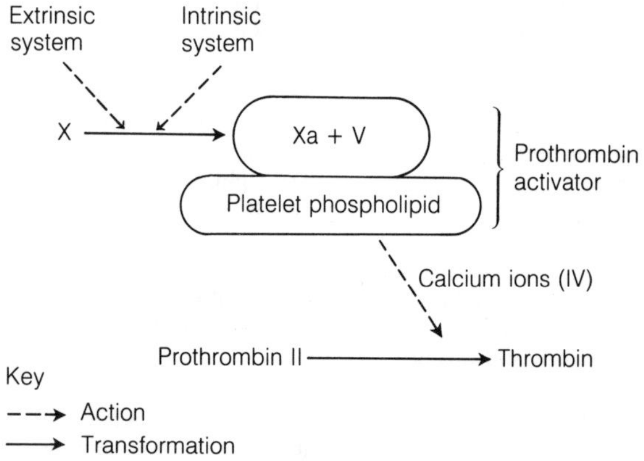

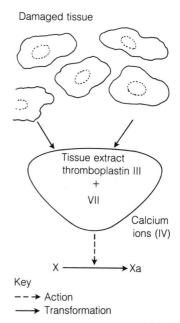

Figure 4.1.15   Production of prothrombin activator

Figure 4.1.16   The extrinsic system for the activation of factor X

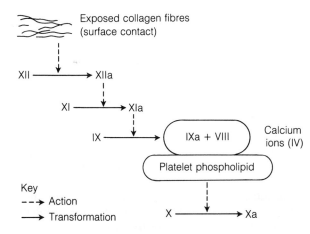

Figure 4.1.17 The intrinsic system for the activation of factor X

verted to an active enzyme (usually identified by the suffix 'a'). The active enzyme then catalyzes the activation of the second proenzyme in the sequence, and so on. A currently accepted sequence for the enzyme cascade, leading to the activation of factor X, is shown in Fig. 4.1.17. Surface contact activates factor XII which then activates factor XI. Activated factor XI activates factor IX. Not all the reactions are of the enzyme–substrate type, as activated factor IX then forms a molecular complex with factor VIII and platelet phospholipid (platelet factor 3). This complex activates factor X.

The cascade sequence is an amplifying system in that minute amounts of the earlier factors can lead to the conversion of large amounts of fibrinogen to fibrin. The entire sequence takes 4–8 minutes (the clotting time). However, once factor X is activated, clotting occurs in seconds, due to the rapid formation and action of thrombin. In other words, in normal clotting, the first appearance of fibrin may take several minutes, but once fibrin has been formed its production is complete within seconds.

In 1964, McFarlane proposed a mechanism for the intrinsic system – the **enzyme cascade** or **waterfall hypothesis**. According to this hypothesis, surface contact initiates a sequence of reactions in which an inactive proenzyme is con-

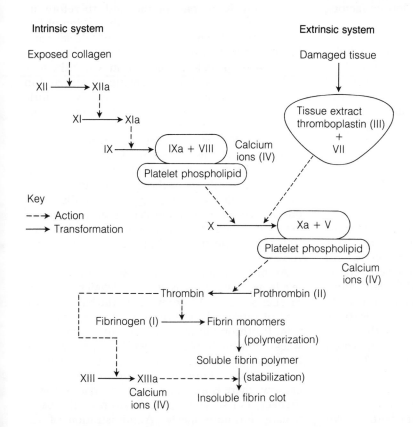

Figure 4.1.18 Summary of the entire clotting sequence

The extrinsic and intrinsic systems of blood clotting are summarized in Fig. 4.1.18. It should be emphasized that although they may appear complex, the mechanisms described here are, in fact, an oversimplification of actual events.

Blood almost certainly contains one or more natural inhibitors of all the clotting factors. These inhibitors regulate the generation and destruction of the active enzymes so that clotting does not spread from the site of blood loss and widespread intravascular coagulation is prevented. Several natural clotting inhibitors (anticoagulants) have been identified. Probably the most important of these is antithrombin III (ATIII). Deficiency of this plasma protein is inherited as a dominant condition. Low levels are associated with an increased tendency to intravascular clotting (thrombosis), particularly during pregnancy. Oral contraceptives containing oestrogen also depress the level of ATIII.

ANTICOAGULANTS

Blood may be prevented from clotting either by inhibiting the normal function of clotting factors, or by inhibiting their production.

Sodium citrate, sodium oxalate or sodium edetate (EDTA: ethylenediamine-tetra-acetic acid)
Outside the body (in vitro) blood can be prevented from clotting by the addition of one of the above salts. These precipitate calcium ions (factor IV) and the clotting sequence is therefore inhibited. For example

$$\text{Calcium ions} + \text{sodium citrate} \rightarrow \text{calcium citrate (insoluble precipitate)} + \text{sodium ions}$$

These substances cannot be used as anticoagulants within the body (in vivo) as the resultant severe reduction in plasma calcium ions would produce clinical problems, such as severe tetany. Blood collected for transfusion and for laboratory analysis of blood cells or clotting factors is usually anticoagulated with citrate phosphate dextrose with adenine (Gunson, 1986). Use of this anticoagulant and preservative solution has increased the storage period of red cells from 21 to 35 days.

Heparin
Heparin is a powerful anticoagulant both in vivo and in vitro. It is ineffective if given orally, being broken down by digestive enzymes, and so is given intravenously.

Heparin is secreted by mast cells in the tissues. It is extracted for clinical use from the lungs and gut of slaughtered cattle. Natural heparin may help maintain the normal fluidity of the blood, but its precise physiological role has not been established.

Heparin acts by inhibiting the action of thrombin. It does so by combining with antithrombin III to produce an even more powerful thrombin inhibitor. As a result, the time taken for blood to clot is greatly increased. Protamine sulphate neutralizes the activity of heparin.

Vitamin K
In vitamin K deficiency, blood clotting time is prolonged and severe bleeding can occur. The vitamin is present in both plant and animal food sources and is also synthesized by bacteria in the large intestine. Deficiency is usually due to reduced absorption rather than to low levels in the gastrointestinal tract.

Vitamin K is fat soluble and therefore its absorption is dependent on the absorption of fats and hence on the normal function of bile salts and lipase enzymes in the gut. Deficiency of the vitamin may occur in any condition in which the flow of bile is decreased, in pancreatic disease and in chronic diarrhoea. The body stores very little vitamin K, so the effects of deficiency become apparent in 3–4 weeks (Fawns, 1978).

Vitamin K is required for the synthesis of clotting factors II (prothrombin), VII, IX and X in the liver. Deficiency of the vitamin therefore leads to reduced synthesis of these factors and hence a prolonged blood clotting time. Liver disease is one of the most common causes of an acquired defect in blood clotting and so the prothrombin time test for clotting is often used as a test of liver function.

At birth, vitamin K and hence prothrombin levels are low. This is due to deficiency in the mother's vitamin K stores and to the baby's initial lack of intestinal bacteria to synthesize the vitamin. Prothrombin levels are usually normal by 2 weeks after birth. If the baby is born prematurely, levels are even lower because of deficient liver cell function due to immaturity. If prothrombin levels are low, there is a greatly increased risk of haemorrhage and so vitamin K is given to the baby, usually intramuscularly. Administration of the vitamin usually produces a rise in prothrombin to normal levels in about 48 hours.

Vitamin K antagonists: the coumarins (e.g. warfarin) and the indanediones (e.g. phenindione). These substances prolong clotting time by blocking the action of vitamin K and so reducing the synthesis of prothrombin and clotting factors VII, IX and X. They are effective when administered orally.

The first vitamin K antagonist to be discovered was dicoumarol. **Dicoumarol** is a product of sweet clover fermentation and was discovered after it caused an epidemic of haemorrhaging in cattle in Alberta, Canada, in 1921. It has a similar molecular structure to vitamin K and acts as an antagonist by replacing the vitamin at its normal site of action. As a result, inert clotting factors are produced.

Other vitamin K antagonists are derived from dicoumarol. They differ mainly in their speed of onset of effect and duration of action. Warfarin and other vitamin K antagonists were first used as rat and mouse poisons. Vitamin K antagonists are widely used for people at risk from thrombosis, and the prothrombin time test is used to assess their therapeutic effect. This should be two to three times normal, reflecting a 5–15% clotting factor activity. Individual response varies widely and so regular monitoring is essential.

The drug is withdrawn slowly at the end of a course of treatment because there is a risk of rebound clotting.

Ancrod (Arvin)

Ancrod is a synthetic form of a glycoprotein extracted from the venom of the Malaysian pit viper. It exerts its anticoagulant effect by converting fibrinogen to an unstable fibrin polymer which is then rapidly broken down by the fibrinolytic system (see next section).

Phase 4: fibrinolysis (the dissolution of a fibrin clot) (Fig. 4.1.19)

Fibrinolysis is the process by which fibrin is degraded into soluble products which are then removed by cells of the mononuclear phagocytic system. Fibrinolysis takes place much more slowly than clotting. In health, the body maintains a homeostatic balance between the continual deposition of small amounts of fibrin and fibrin breakdown. The gradual removal of a blood clot by fibrinolysis is part of the normal process of healing.

Fibrin is broken down by the proteolytic enzyme plasmin. Like thrombin, this enzyme is not present in plasma, but is produced by the activation of the soluble plasma protein, plasminogen. Plasminogen is synthesized in the liver and is present in body fluids as well as in plasma. It has an affinity for fibrin and becomes absorbed on to a fibrin clot as the latter forms.

PLASMINOGEN ACTIVATORS AND FIBRINOLYTIC AGENTS

Plasminogen activators can be produced within the blood, or in the tissues and body fluids. Activation of clotting factor XII leads to the generation of plasminogen activator, and red and white blood cells and other tissues produce plasminogen activators which normally act at sites local to their production. Renal cells, for example, produce the plasminogen activator **urokinase**. A purified form of this substance is used therapeutically as a fibrinolytic agent in antithrombotic therapy.

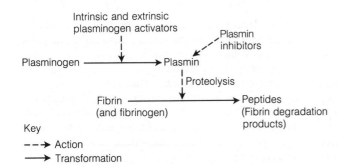

Key

---→ Action

⟶ Transformation

Figure 4.1.19 The fibrinolytic system

In addition, certain bacteria synthesize plasminogen-activating substances. For example, haemolytic streptococci produce **streptokinase**, which is used therapeutically as a fibrinolytic agent.

Several protein plasmin inhibitors are present in plasma, the two most important being antiplasmin and macroglobulin. They inactivate plasmin by forming a complex with the molecule. The substances **epsilon-aminocaproic acid (EACA)** and **tranexamic acid (AMCA)** inhibit fibrinolysis and are used clinically to reduce bleeding at sites where active fibrinolysis occurs, for example the uterus or following dental extraction. They are also used as antidotes to the therapeutic fibrinolytic agents, streptokinase and urokinase.

Tests of haemostasis

Tests of haemostasis can be divided into two groups: screening tests and specific tests which identify the disorder. The latter, which include, for example, specific clotting factor assays, are performed in specialized laboratories. Screening tests assess the function of each of the components of haemostasis, that is

(a) vascular function (Hess test and bleeding time)
(b) platelet function (platelet count and bleeding time)
(c) coagulation (clotting time, activated partial thromboplastin time, prothrombin time and thrombin time)
(d) fibrinolysis (fibrin degradation products).

Details of the conduct of the above tests can be found in haematology texts.

Defects of haemostasis

INCREASED FRAGILITY OF CAPILLARY WALLS AND PLATELET DISORDERS

Increased fragility of capillary walls and disordered platelet function prolong the bleeding time. Bleeding from capillary beds occurs after only minor injury or even apparently spontaneously. Normally such bleeding would be arrested quickly by vessel contraction and platelet plugging. Capillary bleeding shows itself as bruising and as haemorrhagic areas under the skin. Blood loss from nose bleeds and at menstruation is also greater than normal.

The major causes of increased capillary fragility are a reduced platelet count (thrombocytopenia), autoimmune disorder and the effect of drugs, for example penicillin, oxytetracycline, aspirin and thiazide diuretics. Old age and Cushing's syndrome produce atrophy of the supporting connective tissue of capillaries, and bleeding, combined with lack of white cell phagocytic activity, leaves brown areas of haemosiderin under the skin (age spots). Typically these appear on the backs of the hands because of the frequency of minor injury at these sites.

Wherever possible, the cause of the bleeding disorder is treated. Other therapeutic measures include the administration of anti-inflammatory and immunosuppressive agents, splenectomy and platelet transfusion.

CLOTTING DEFECTS

Significant deficiencies of functional forms of clotting factors arise from rare inherited defects in their synthesis (see Table 4.1.7). Almost always, deficiency of a clotting factor leads to a bleeding disorder in which clotting time is prolonged but bleeding time remains normal.

The haemophilias
The haemophilias are the most common of the clotting disorders. **Haemophilia A (classical haemophilia)** is due to deficiency of clotting factor VIII. It is transmitted as a sex-linked recessive disorder (that is, the gene is on the X chromosome), so the disease usually occurs only in men, who inherit an abnormal X chromosome from their unaffected, carrier mothers. Figure 4.1.20 illustrates the inheritance of haemophilia in a family with a carrier mother and normal father. Theoretically, the chances are that half of the children will be normal and, of the other half, the males will have haemophilia and the females will be carriers of the disease.

Where there is evidence that a woman may be a carrier of haemophilia, either through her family history or because she has borne an affected son, assessment of her clotting factor activity may confirm this and genetic counselling should be made available to her.

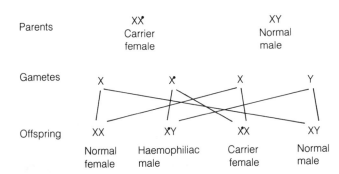

Parents: XX̸ Carrier female / XY Normal male

Gametes: X X̸ X Y

Offspring: XX Normal female / XY Haemophiliac male / XX Carrier female / XY Normal male

Figure 4.1.20 The inheritance of haemophilia

Haemophilia B occurs about six times less frequently than classical haemophilia. It is due to deficiency of clotting factor IX and is also inherited as a sex-linked recessive gene. It is clinically identical to haemophilia A and it was not until factor IX was discovered that the two conditions were distinguished. Factor IX deficiency is commonly called **Christmas disease**, after the first sufferer identified.

In both haemophilia A and Christmas disease the severity of the bleeding disorder depends on the level of active clotting factor present in the plasma. Spontaneous bleeding occurs at very low levels ($< 5\%$).

People with clotting factor deficiencies are usually cared for as out-patients at specialist centres. Therapy now consists of the intravenous administration of a concentrated preparation of the relevant active clotting factor, obtained from normal human or animal plasma. The aim is to raise the patient's plasma level to about 30% of the normal activity of that factor. Through this treatment, it is possible to arrest serious bleeding and hence prevent the development of chronic deformities. Clotting factor therapy also allows people with haemophilia to undergo surgery with far less risk of life-threatening haemorrhage. In most cases, following education, the patient, or a member of his family, is able to administer the clotting factor preparation at home. Since 1984, all UK supplies of factors VIII and IX have been heat treated in order to inactivate HIV, which causes AIDS (Wells, 1986).

It is important to remember that people with disorders of haemostasis and those on antithrombotic therapy have an increased tendency to bleed. Special care should therefore be taken when administering injections to such people. Those with severe haemophilia are not given intramuscular injections because of the risk of bleeding into the muscle. Instead, when necessary, drugs are given via careful subcutaneous or intravenous injections. People at risk of excessive bleeding may be advised not to participate in activities which carry a substantial risk of accidental injury. They should also avoid ingesting irritant substances, such as aspirin and alcohol in excess, which may cause internal haemorrhage.

Thrombosis
Thrombosis is the process of clot formation within the cardiovascular system of a living person. A **thrombus** (plural, thrombi) is an intravascular blood clot formed during life.

Blood does not normally clot unless it is shed from the vascular system. Its intravascular fluidity is dependent on the properties of the normal endothelial lining of the blood vessels, the rate of blood flow and the presence of natural clotting inhibitors. In 1846, Rudolf Virchow described a triad of situations which predispose to intravascular clotting. These were

(a) damaged vessel endothelium
(b) slow blood flow, and
(c) hypercoagulable blood.

Virchow's triad remains valid, and thrombi are most likely to form when these conditions exist. For example, vascular endothelium may be damaged by the presence of atheromatous patches and in diseased heart valves. The presence of atheromata (singular, atheroma) also reduces blood flow, and physical immobility produces venous stasis, particularly in the legs. Thrombosis is initiated by platelet aggregation at the damaged endothelium.

Some thrombi produce no health problems and are gradually broken down by the fibrinolytic system. Others grow large enough to block an artery, producing cell death distal to the blockage (an infarct). Nearly 50% of British adults now die as a result of one of the three major types of thrombosis, and many more suffer the chronic disabling results of non-fatal occurrences (Lowe, 1983). The three major types of thrombosis are:

(a) coronary thrombosis, leading to myocardial infarction (a heart attack)
(b) cerebral thrombosis, leading to cerebral infarction (this type of cerebrovascular accident is the commonest form of stroke)
(c) pulmonary thromboembolism, leading to pulmonary infarction (see below).

People with a previous history of thrombosis or who are known to be at particular risk of its development, may be maintained prophylactically, on long-term anticoagulant therapy, using a drug such as warfarin. Anticoagulant therapy also prevents the spread of an existing thrombus but does not remove it. Problematic thrombi, for example in leg arteries, may be treated with fibrinolytic agents such as streptokinase.

Deep vein thrombosis (DVT). This accounts for the formation of 95% of venous thrombi (Green and Wickenden, 1982). These thrombi usually form in the pocket-like valves of the deep veins of the calf during prolonged periods of lying supine, for example either during or following an abdominal operation. Lying supine increases the pressure in the leg veins and produces venous stasis. The sustained pressure may also damage the venous endothelium.

Deep vein thrombosis may not become clinically apparent. The inflammation associated with its occurrence sometimes produces pain in the calf, especially on dorsiflexion of the foot (**Homans' sign**), and flushing, heat and swelling in the affected area. Superficial leg veins may act as a collateral circulation, enabling blood to bypass the thrombus.

Deep vein thrombi are not a major health hazard in themselves. However, in a small proportion a fragment of the thrombus breaks away. This fragment (an embolus) is transported through the venous system to the right side of the heart and passes into the pulmonary arterial tree where it blocks one of the vessels. The size and severity of effect of the pulmonary infarction thus produced depend on the size of the embolus and the vessel it has blocked. Pulmonary embolism has a mortality rate of 50% and is one of the main causes of sudden death.

It is therefore important to prevent the occurrence of deep vein thrombosis. This may be done by taking action to prevent venous stasis in the leg veins. Teaching a preoperative or bedridden patient the importance of, and how to do, deep breathing exercises (which aid the respiratory pump) and leg exercises (which increase the muscular pump effect) is probably helpful. Prophylactic leg exercises should be performed passively on unconscious and uncooperative patients. Pressure on the leg veins is also reduced by regular changes in position and by the use of bed cradles. Other preventive measures include the application of elastic stockings or crepe bandages to the legs. These support the calf muscles and aid venous tone. Intermittent pneumatic compression of the calves, using electrical apparatus, has also been shown to be beneficial. Low-dose anticoagulant therapy may be prescribed for some high-risk patients.

Disseminated intravascular coagulation (DIC) and defibrination syndrome
In disseminated intravascular coagulation, widespread clotting occurs throughout the circulatory system. This condition may arise in a variety of usually serious illnesses, for example major trauma and surgery, septicaemia, widespread malignancy and mismatched transfusion. Activation of intravascular clotting is precipitated by entry of tissue fluid into the blood or by widespread damage to the vessel endothelium.

Paradoxically, bleeding is the major feature of DIC because clotting factors have been used up intravascularly (defibrination syndrome). Other problems are the result of microthrombi lodging in the arterial blood supply and producing infarctions, in tissues such as brain, lung and kidney. Haemolytic anaemia also occurs as a result of red cells being damaged by intravascular fibrin strands.

Treatment is primarily directed at the usually severe cause of the DIC. Antithrombotic drugs and transfusion of platelets and clotting factor preparations may also be given.

BLOOD GROUPS

Over 400 different, genetically determined, antigens can be identified on the surface of human red blood cells. An **antigen** is a substance that stimulates antibody formation and combines, specifically, with the antibody produced. Some of these antigens are specific to red cells and others are also present on white cells, platelets and on the cells of other tissues. People can be divided into groups (blood groups) according to the presence or absence of particular antigens on the surface of their red cells.

The major blood group systems

More than 12 different blood group systems and subdivisions of these have now been identified. Some of these have been designated letters and others are named after the individual in whom they were first identified. They include the ABO, Rh (CDE/cde), MNS, P, Lutheran (Lu), Kell (K), Lewis (Le), Duffy (Fy) and Kidd (Jk) blood group systems.

Clinical significance

Blood groups become clinically significant when blood transfusion is performed and when a pregnant woman is carrying a fetus which has inherited (from its father) a blood group different from that of its mother. In both these situations the clinical importance of a particular red cell antigen depends on two factors

(a) the frequency of occurrence in the plasma of the corresponding antibody
(b) the ability of this antibody to produce haemolysis of red cells following combination with the antigen.

On this basis, the ABO and Rhesus blood group systems are by far the most important. The other systems usually only cause clinical problems in cases of repeated blood transfusion and, occasionally, in the second and subsequent incompatible pregnancies.

The ABO blood group system: the classical blood groups

In 1900, Landsteiner discovered the ABO blood groups and demonstrated that blood from two individuals could be mixed successfully only if their blood groups matched. This discovery marked the beginning of safe blood transfusion.

The **ABO antigens** occur on the red cell membrane and on most other cells, including white blood cells and platelets. They are glycoproteins, which differ only in one carbohydrate molecular residue and appear on the red cells by about the sixth week of fetal life. At birth they have reached about one-fifth of the adult level.

INHERITANCE OF ABO BLOOD GROUPS

The locus (site) which determines the ABO blood groups is on chromosome nine and there are three alleles, A, B and O, which can occur at this site. Each parent transfers one of his/her two genes to their baby. Table 4.1.8 illustrates how a person's ABO blood group is genetically determined. There are four possible groups – A, B, AB and O – because the A and B genes are codominant and dominant over O (which is amorphic), so that AO and BO react like AA and BB respectively.

THE FREQUENCY OF OCCURRENCE OF THE ABO BLOOD GROUPS

The frequency of occurrence of each of the four ABO blood groups varies widely within different populations. Table 4.1.9 shows the frequency of occurrence of each group in the United Kingdom and Western Europe; groups O and A are by far the most common. However, the percentages of the four blood groups vary considerably in different parts of Britain, and even wider variations in blood group percentages exist between the populations of Western Europe and those of other countries and continents.

Table 4.1.8 The inheritance of the ABO blood groups

If a baby receives	O + O	A + O	A + A	B + O	B + B	A + B
His/her genotype is	OO	AO	AA	BO	BB	AB
His/her phenotype (blood group) is	O	A	A	B	B	AB

Table 4.1.9   The frequency of occurrence of the ABO blood groups in the populations of the United Kingdom and Western Europe

Group	Percentage occurrence
O	46
A	42
B	9
AB	3
All groups	100

ANTIGENS AND AGGLUTINATION

The A and B blood group antigens are sometimes called **agglutinogens** because, in the presence of their specific antibody (an agglutinin), agglutination of the red cells occurs. In **agglutination**, the red cells lose their outline and become massed in clumps. These clumps are sometimes described as having the appearance of paprika pepper. The size of the red cell clumps depends on the strength of the antibody. Agglutination is then usually followed by haemolysis. Agglutination is a quite different process to blood coagulation (clotting), and should not be confused with the latter. O antigen does not normally act as an agglutinogen as it does not usually evoke production of a corresponding antibody.

ANTIBODIES (AGGLUTININS)

The antibodies (or agglutinins) that correspond to the A and B antigens are called anti-A or α, and anti-B or β. Unusually these antibodies are present in the absence of their corresponding antigen and without any immunizing stimulus such as incompatable blood transfusion or pregnancy.

Antibodies are produced by the plasma cells of the immune system (see Chap. 6.2). They are immunoglobins and are found in plasma and in other body fluids. It is thought (Waters, 1983) that certain bacteria and viruses contain antigens so similar to A and B they they trigger production of anti-A and anti-B. Natural exposure probably occurs via the gastrointestinal tract.

Occurrence of antibodies in plasma

If an antigen is present on the red cells of an individual, the corresponding antibody must be absent from the plasma, because if it is present agglutination would occur. If the serum of the four ABO blood groups is examined, it is found that

Group A blood serum contains anti-B
Group B blood serum contains anti-A
Group O blood serum contains both anti-A and anti-B
Group AB blood serum contains neither anti-A nor anti-B.

About 50% of newborn babies have some anti-A and/or anti-B antibodies in their blood, from placental transfer. Actively produced anti-A and anti-B are measurable at about 3 months of age and reach adult levels by about 10 years. There are marked variations in antibody titre between individuals at all ages.

DETERMINATION OF ABO GROUP

In order to determine the ABO group of a blood sample, two test sera, one containing anti-A and one containing anti-B, are used. A variety of techniques is available for ABO grouping (Lockyer, 1982). In the slide technique, a small volume of each serum is placed on a slide and labelled, and then an equal volume of the diluted blood sample is added to each serum. After 10 minutes the slides are observed for agglutination.

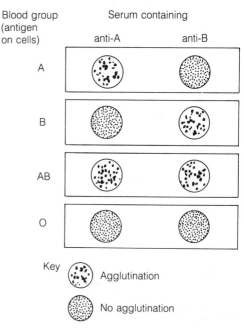

Figure 4.1.21   ABO blood grouping

Figure 4.1.21 illustrates the results for a blood sample of each ABO group.

The Rh (Rhesus) blood group system

After Landsteiner's discovery of the ABO blood group system, the majority of blood transfusions were successful. However, sometimes patients receiving a transfusion of ABO-compatible blood would suffer a transfusion reaction. In addition, some mothers were delivered of an ABO-compatible baby which had severe haemolytic anaemia (haemolytic disease of the newborn). It was believed that this anaemia was caused by incompatible blood group antibodies (agglutinins) present in the mother's plasma, crossing the placenta and destroying her fetus' red cells.

These clinical problems were not clearly understood until Landsteiner joined forces with the American immunologist, Alexander Wiener. In 1939 they discovered the Rh blood group system and went on to discover other blood group systems.

The Rh system is so called because the antigens were discovered on the red cells of Rhesus monkeys. Red cells from a Rhesus monkey were injected into a rabbit and as a result the rabbit's immune system formed an antibody which agglutinated the Rhesus red cells. It was then found that if Caucasian human red cells were mixed with serum from the immunized rabbit, agglutination occurred in 85% of cases. The occurrence of agglutination indicated the presence of Rh antigen on the human red cells. Individuals with Rh antigen on their red cells are termed **Rh-positive**; those with no Rh antigen on their red cells are termed **Rh-negative**.

Rh ANTIGENS AND THEIR INHERITANCE

It is now known that Rh antigens are specific to red cells and that three closely linked genes on chromosome 1 are responsible for their production, or non-production. Each of the three genes has two alleles, called CDE/cde. Alleles C, D and E are dominant over c, d and e. Of these three genes, D is by far the most important clinically. The presence of the D allele gives rise to the Rh antigen, D. The d allele is amorphic and so produces no antigen. The term **Rh-positive**, then, refers to the presence of antigen D.

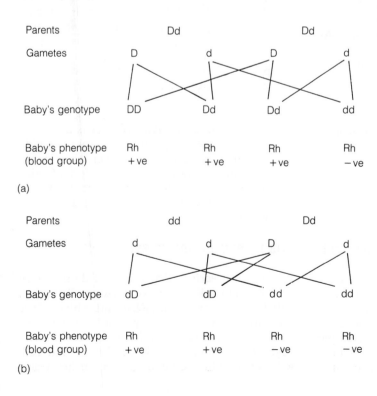

Figure 4.1.22 The inheritance of the D and d Rh blood group antigens: (a) when both parents are Rh-positive heterozygotes, (b) when the mother is Rh-negative and the father a Rh-positive heterozygote

Two examples of the inheritance of the D and d Rh blood group antigens are shown in Fig. 4.1.22. One gene is received from each parent. As D is dominant over d, the genotypes DD and Dd both give rise to Rh-positive red cells and the dd genotype produces Rh-negative red cells.

FREQUENCY OF Rh BLOOD GROUPS

The frequency of the Rh antigens varies in different parts of the world. For example, in the United Kingdom and Western Europe, 83% of people are Rh positive (DD and Dd) and 17% Rh negative (dd); in Japan, 99.7% of the population are Rh positive and only 0.3% Rh negative.

Rh (ANTI-D) ANTIBODY

Unlike the ABO system's anti-A and anti-B, anti-D antibody does not occur naturally in the plasma of Rh-negative individuals. However, immune production of anti-D can be stimulated by contact of an Rh-negative individual's immune cells with the Rh-positive (D) antigen. This **allo-immunization** (immunization against antigens which the individual lacks) can occur in two ways

(a) following transfusion with Rh-positive blood
(b) if the red cells of an Rh-positive fetus enter the circulation of its Rh-negative mother. As this escape of fetal red cells usually only occurs at parturition (delivery), the mother's titre of anti-D is unlikely to be high enough to cross the placenta and destroy her Rh-positive fetus' red cells during a first pregnancy.

Since antigen d is amorphic, there is no corresponding anti-d antibody and so transfusion of an Rh-positive person with Rh-negative blood does not give rise to antibody production.

BLOOD TRANSFUSION

Blood for transfusion has always been and still is a limited and expensive resource. Over the years, techniques have been developed which enable blood to be separated into its various components. These techniques allow modern transfusion practice to aim to give a patient only the blood component he lacks and therefore needs, for example factor VIII in classical haemophilia. Not only does **blood component therapy** represent a more economical use of blood, it also removes many of the health risks associated with whole blood transfusion when this is not required.

Blood groups and blood transfusion

When blood from one individual is transfused into another, it is essential to avoid red cells being agglutinated. Agglutination will occur if the antigens present on the red cells come into contact with their corresponding antibodies.

Usually it is unnecessary to consider the antigens on the cells of the transfusion recipient because the antibodies in the donor's plasma are diluted to such a degree by the much larger volume of the recipient's plasma. This dilution ensures that the donor's antibodies do not cause agglutination of the recipient's red cells. Normally, therefore, it is only *necessary to exclude incompatibility between the antigens on the donor's cells and the antibodies in the recipient's plasma.*

THE ABO BLOOD GROUP SYSTEM

Figure 4.1.23 shows the results of transfusing blood of each of the four ABO groups into recipients of each group. The occurrence of agglutination indicates an incompatible transfusion. It can be seen that intragroup transfusion is always compatible. A recipient of group AB can receive blood from any of the four ABO groups without agglutination occurring. Similarly, group O blood can be given to recipients of all four groups. For this reason, group AB individuals have been termed universal recipients and group O individuals universal donors. However, these terms are not valid as they ignore the effect of other blood group systems, in particular the Rh system, in blood transfusion. Consequently they should no longer be used.

THE Rh BLOOD GROUP SYSTEM

As described earlier, Rh antibodies do not occur naturally but a Rh-negative person will produce anti-D if immunized with Rh-positive blood. If an immunized Rh-negative person subsequently comes into contact with Rh-positive red cells, agglutination and haemolysis of these red cells will occur. This could occur on further trans-

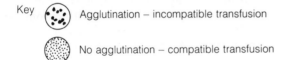

	Gp O	Gp A	Gp B	Gp AB
Donor's red cells	**Recipient's serum contains**			
	anti-A and anti-B	anti-B	anti-A	neither antibody
Gp O	No agg	No agg	No agg	No agg
Gp A	Agg	No agg	Agg	No agg
Gp B	Agg	Agg	No agg	No agg
Gp AB	Agg	Agg	Agg	No agg

Key: Agglutination – incompatible transfusion

No agglutination – compatible transfusion

Figure 4.1.23 The results of transfusing blood of each of the four ABO blood groups into recipients of each group

fusion of incompatible Rh-positive blood, or when an immunized Rh-negative woman is pregnant with an Rh-positive fetus.

It is therefore essential that no Rh-negative person is transfused with Rh-positive blood, except in extreme emergency when no Rh-negative blood is available. This is particularly important for female patients prior to the menopause, as their sensitization to antigen D could adversely affect their subsequent ability to bear a child.

OTHER BLOOD GROUP SYSTEMS

Other blood group systems are of relatively minor clinical significance as antibodies do not occur naturally in the plasma and it is unusual for the antigens of these systems to evoke antibody formation, even after incompatible blood transfusion. However, with repeated transfusions of incompatible blood, antibody formation sometimes occurs and can produce agglutination on subsequent transfusion. Occasionally, too, antibodies from these systems are the cause of haemolytic disease of the newborn.

CROSS-MATCHING

Direct cross-matching of blood for transfusion

with the recipient's blood is routinely performed prior to transfusion for the ABO and Rh systems. It has now also become routine laboratory procedure to test for the presence of atypical antibodies from other blood group systems in the serum of the recipient. This is done prior to crossmatching so that donor blood which does not have the corresponding antigens can be selected.

Cross-matching for the ABO system is performed by mixing a sample of the donor blood, diluted in isotonic saline, with the recipient's serum and observing for agglutination. The latter does not occur if the blood samples are compatible. Separate tests are carried out for crossmatching of the Rh blood group system. Each unit of blood to be transfused is individually cross-matched. Accurate direct cross-matching represents the only safeguard against transfusion complications due to blood group incompatibility.

The effect of storage on blood for transfusion

Donated whole blood can be stored at 4°C for up to 35 days without excessive loss of viability. Red cell metabolism is significantly decreased at 4°C and, as a result, the red cells swell and their fragility is increased. The amount of spontaneous haemolysis that occurs increases with the length of storage. Following transfusion, stored red cells become normal within 48 hours.

Leucocytes and platelets are destroyed within a few days of donation and so these cells cannot be obtained from stored blood. Plasma can be stored, in liquid form, for many months. It can also be frozen successfully. Dried plasma can be preserved for years at a wide range of temperatures. It is reconstituted with sterile water prior to transfusion.

Blood component therapy

The objective of blood component therapy is either to give the patient the particular component of blood that he lacks, for example platelets, or to substitute a normal for an abnormal component, for example exchange of normal red cells for sickle cells in sickle cell anaemia.

The following blood components are available for transfusion.

Whole blood. Whole blood is used to restore blood volume following acute, severe haemorrhage. In an emergency, concentrated red cells and a plasma volume expander, for example dextran, can be used as a substitute if whole blood is not available.

Concentrated (packed) red cells. These constitute the ideal treatment for severe anaemia in which additional blood oxygen carrying capacity but not volume is required.

Washed red cells. Washing red cells with a suitable fluid removes adherent leucocytes, platelets and plasma protein. Washed red cells are given to people who, usually as a result of having received multiple transfusions, have developed antibodies to constituents of blood other than red cells. They are also used to minimize the occurrence of this alloimmunization in people known to require repeated transfusions, for example those with thalassaemia.

Frozen red cells. Red cells can be frozen in liquid nitrogen ($-196°C$) for several years. Glycerol is used as a protective agent. This relatively new procedure is expensive, but frozen red cells are a very useful preparation because they have the lowest leucocyte and platelet content available, and freezing eliminates the hepatitis virus. Freezing is now used as a method of storing selected red cells, for example those which are of a rare blood group.

Platelets. Platelets must be transfused within 24 hours of preparation because of their short survival time. They are mainly required to arrest bleeding due to thrombocytopenia. Platelets from ABO and Rh compatible donors are usually given because of the presence of the ABO antigens on platelets and red cell contamination of the platelet preparation.

Granulocytes. Methods of producing granulocyte preparations for transfusion are as yet experimental. Granulocytes can be obtained in sufficient quantities only by leucapheresis from a single donor. Occasionally patients with chronic granulocytic leukaemia volunteer as donors as they have high levels of circulating granulocytes.

Granulocyte preparations are used fresh as the cells have a short survival time. They are contaminated with red cells and so must be compatible with the recipient's ABO and Rh groups. They are mainly used when antibiotic therapy is unable to control severe infection in patients with neutropenia due to marrow suppression.

Plasma. Plasma is transfused to increase blood volume and to replace plasma which has been lost from the circulation, for example following burns. As in whole blood transfusion, relatively small volumes of plasma can be transfused without considering the effects of the antibodies present. This is because the antibodies are diluted to harmless levels by the recipient's plasma. However, if large volumes of plasma are to be transfused, agglutination may occur. To avoid this, 'conditioned plasma' is used. This is a preparation from which the antibodies have been previously removed by contact with red cells.

Plasma substitutes, such as a solution of the polysaccharide dextran, are sometimes used in emergency and to avoid the risk of transmitting serum-carried diseases. The large dextran molecules are retained in the circulation and so exert an osmotic effect, similar to that of the plasma proteins. This osmotic effect draws fluid into the blood, increasing its volume and hence blood

Table 4.1.10 Plasma components for clinical use

Plasma component	Major uses
Plasma protein fraction (PPF)	Plasma replacement Blood volume expansion
Human albumin	Albumin replacement in hypoalbuminaemia, e.g. in nephrosis
Fresh frozen plasma (FFP)	Clotting factor deficiencies, e.g. DIC, anticoagulant overdose, liver disease
Cryoprecipitate (factor VIII, fibrinogen)	Classical haemophilia von Willebrand's disease Fibrinogen deficiency
Factor VIII concentrate and freeze-dried factor VIII	Classical haemophilia
Factor IX concentrate	Christmas disease
Human immunoglobulin (Ig) (gamma globulin)	Hypogammaglobulinaemia To produce passive immunity to common viral diseases such as rubella (German measles), measles and chickenpox
Human specific globulin (specific antibodies)	Rh anti-D for prevention of Rh immunization To produce passive immunity to rare, life-threatening diseases, e.g. tetanus

pressure. (Plasma components available for clinical use are summarized in Table 4.1.10.)

The risk of transmitting serum-carried infectious diseases in plasma transfusion varies according to the precise component used, since some methods of preparation destroy micro-organisms.

Clinical problems associated with the transfusion of blood

Blood transfusion is a common occurrence in hospital wards. It is also a potentially hazardous procedure for the recipient as a variety of complications can occur. To avoid these, stringent precautions must be taken by the Blood Transfusion Service to ensure that each unit of blood released is suitable for transfusion, and by the laboratory and ward staff involved in administering the transfusion. As a direct result of these safety measures, blood transfusion is a relatively safe procedure, but severe and even fatal reactions do occasionally occur. The large majority of these are due either to administrative errors, for example misidentification of the patient or a blood sample, or to maltreatment of the blood between the time it leaves the blood bank and is transfused (Chanarin *et al.* 1985).

TYPES OF TRANSFUSION REACTION

Immune reactions
Immunization can occur against the antigens which the patient lacks (alloimmunization). Recipients of multiple transfusions may develop antibodies to antigens in the donor blood which have not been grouped and cross-matched. Antigens which trigger alloimmunization are present on red cells, white cells, platelets and plasma proteins. Alloimmunization of the recipient is not in itself a problem, but on subsequent transfusion the antibodies formed may produce an incompatible transfusion reaction (see below).

ABO-incompatible transfusion reactions
A transfusion reaction due to ABO incompatibility is one of the most serious complications of blood transfusion. The clinical problems associated with this reaction are due to the effects of the acute intravascular agglutination and haemolysis that occur. The antigen–antibody reaction probably also triggers the complement and clotting cascades and so may initiate disseminated intravascular coagulation (DIC).

When an ABO-incompatible transfusion is commenced, the recipient often experiences a burning sensation radiating up the arm along the track of the vein from the transfusion site. Within a few minutes he may complain of severe pain in the lumbar region, tightness in the chest and difficulty in breathing. These symptoms are probably due to agglutinated red blood cells blocking small blood vessels in the kidneys and lungs. The patient's temperature and pulse rate rise and his blood pressure falls. He may experience febrile rigors.

The occurrence of such a reaction is an acute emergency and prompt diagnosis and treatment are often lifesaving. For this reason it is important that each nurse caring for a patient undergoing transfusion observes him or her particularly carefully during the first 15 minutes of transfusion. Frequent monitoring of how the patient is feeling, as well as of his temperature, blood pressure and pulse rate, is essential throughout the transfusion. If an ABO-incompatible reaction is suspected, the transfusion is stopped immediately, but the intravenous infusion is maintained for use in treatment. The remaining blood is retained for analysis. The reaction is unlikely to be fatal if less than 350 ml of blood has been transfused.

ABO-incompatible reactions also produce urticarial rashes and diarrhoea and vomiting. The glomerular filtration rate (GFR) decreases due to the fall in blood pressure and to the deposition of thrombi in the renal blood vessels. Free haemoglobin colours the plasma red and is filtered into the renal nephrons where it may be precipitated as acid haematin, obstructing the tubule lumen. Haemoglobin is also excreted in the urine (haemoglobinuria). The fall in glomerular filtration rate and obstruction of the renal tubules lead to a decrease in the volume of urine produced and, in severe cases, to anuria. If renal failure occurs, blood levels of potassium, urea and other nitrogenous compounds increase, with serious metabolic consequences. In order to monitor renal function, it is important that all urine produced by the patient is measured and retained for analysis.

If the patient is anaesthetized or unconscious, he is particularly at risk of a severe reaction as he is unable to complain of the early warning symptoms. The oozing of blood from a surgical wound (due to defibrination syndrome) is sometimes the first sign of a major transfusion reaction in an anaesthetized patient.

Treatment aims to maintain blood pressure and renal perfusion. Anaemia may be corrected by transfusion of compatible packed red cells. If DIC

and defibrination syndrome occur, transfusion of platelets and clotting factors may also be required.

Leucocyte and platelet incompatibility

Transfusion reactions due to leucocyte and platelet antibodies occur most commonly in patients who have received multiple transfusions. The reactions are usually relatively mild, involving fever and anaphylactic responses, such as urticarial rash. Purpura may follow transfusion when platelet antibodies exist. The presence of leucocyte and platelet antibodies reduces the survival of the transfused target cells.

Allergic reactions

Allergic reactions to blood transfusion are quite common. They vary in severity from a minor rise in temperature, itching urticarial rash, through to acute laryngeal and periorbital (around the eyes) oedema. The latter conditions occur only rarely and require urgent treatment.

Pyrogenic reactions

Pyrogenic reactions produce a rise in temperature, as their name suggests. Pyrogenic reactions, usually of unknown origin, used to be one of the commonest types of transfusion reactions, but are now much rarer. This is partly due to the introduction of plastic transfusion packs and partly because of greater understanding of the causes of transfusion reactions, which allow the reactions to be more accurately identified. The severity of a pyrogenic reaction is usually related to the speed of the transfusion. The patient rarely becomes shocked, so although a rise in temperature is unpleasant for him, these reactions are not usually dangerous. Pyrogens may be present in the donor blood as a result of bacterial growth in the transfusion medium prior to sterilization.

Metabolic reactions

Metabolic reactions usually occur only when large volumes of blood (5 units or more) are transfused rapidly.

Hyperkalaemia and hypocalcaemia. Stored blood contains citrate anticoagulant and increased levels of potassium. The latter can produce hyperkalaemia until the transfused red cells regain their normal state (usually within 48 hours). Hyperkalaemia can cause cardiac irregularities and, in severe cases, ventricular fibrillation. The citrate anticoagulant precipitates calcium ions. Normally citrate is converted to hydrogen carbonate in the liver, but this process may be delayed in massive transfusion, and hypocalcaemia resulting in severe muscle spasm (tetany) may occur. Hypocalcaemia may be prevented by giving calcium gluconate to a patient receiving a massive transfusion.

The temperature of and the maltreatment of blood
Blood is stored at 4–6°C. If large volumes are transfused rapidly, this can significantly lower body temperature and may produce cardiac irregularities. Blood should be transfused within half an hour of removal from the blood bank and be stored at room temperature during this time. Blood for transfusion should never be frozen, stored in the ward refrigerator (as the temperature of these is often unreliable), heated, or have anything, for example drugs, added to it. All these procedures can severely damage and haemolyse red cells. If large volumes of blood are to be given rapidly, the blood can be passed through commercially produced blood-warming apparatus, or the blood bag warmed in water at 37°C. The temperature of the water must be measured using a reliable thermometer.

Thrombophlebitis
Thrombophlebitis is inflammation of a vein, often accompanied by thrombus formation. It occurs most commonly as a result of trauma to the vessel wall. The incidence of thrombophlebitis as a complication of blood transfusion has greatly decreased since the introduction of plastic giving sets. If thrombophlebitis occurs, the transfusion will probably need to be resited.

Microvascular obstruction
Microaggregates of platelets, leucocytes and fibrin occur in stored blood. These may obstruct small pulmonary blood vessels and so special filters are often used when large volumes of blood are to be transfused rapidly.

Circulatory overload
Circulatory overload occurs most commonly when anaemic patients are transfused with whole blood. This is because they only need the additional oxygen-carrying capacity of the red cells and not the additional volume. Anaemic patients with existing cardiac disease are particularly at risk. The increased blood volume produced by the transfusion can precipitate cardiac failure and pulmonary oedema.

Abnormal haemostasis
Platelets and some clotting factors do not survive storage. Therefore, massive transfusion with stored blood may significantly reduce the recipient's platelet count and clotting factor titres and thus produce abnormal haemostasis. To prevent this, a unit of fresh frozen plasma may be given for every 4 or 5 units of blood transfused.

Infected blood
It is extremely rare for blood for transfusion to become contaminated with micro-organisms. Although blood is an ideal culture medium, the low temperature at which it is stored limits bacterial growth. Gram-negative *Pseudomonas* and coliform bacilli, capable of reproducing at 4°C, are the most common contaminants.

Occasionally, bacteria are introduced into the blood during the transfusion. These can multiply and reach dangerous levels if the transfusion is slow.

Transfusion of infected blood produces fever, rigors, pains in the chest and abdomen, and hypotension (septicaemic shock). The condition is frequently fatal. If infected blood is suspected, the transfusion is stopped immediately, but the intravenous infusion is maintained. The blood is retained for culture. Treatment includes antibiotic therapy and supportive measures for shock.

Transmission of infectious disease
Despite the screening of donors for hepatitis B surface antigen, viral hepatitis remains a serious complication of blood transfusion. Transmission of other bloodborne diseases such as malaria, syphilis and brucellosis via transfused blood is extremely rare because of donor screening. Universal donor screening (since 1985) for the human immunodeficiency virus (HIV) which causes acquired immune deficiency syndrome (AIDS) and requests to those as high risk of carrying the virus not to donate their blood should mean that people requiring blood transfusion now run practically no risk of HIV infection as a result of undergoing treatment (Wells, 1986).

Iron overload (haemosiderosis)

Haemosiderosis occurs when repeated transfusions of red cells are given, in the absence of haemorrhage, for example in the treatment of thalassaemia major. Excess iron, as haemosiderin, is deposited in the cells of the mononuclear phagocytic system and this eventually damages the liver, heart muscle and endocrine glands. Iron chelation therapy with desferrioxamine is given to reduce the problem of iron overload.

Fetomaternal incompatibility

A fetus' blood cells may carry antigens, inherited from its father, which are not present on its mother's blood cells. If fetal blood cells enter the mother's circulation across the placenta, her immune system will be stimulated to produce IgG antibodies to the foreign antigens. Fetal blood cells most commonly enter the mother's blood during parturition. This alloimmunization of the mother has no harmful effects, in itself, but predisposes her to immune problems if she should subsequently require blood transfusion or tissue transplantation.

If an alloimmunized woman becomes pregnant with another fetus whose blood cells carry the same incompatible antigens, her antibodies will cross the placenta and enter the fetal circulation. The clinical effects on the fetus vary according to the specificity and potency of the maternal antibodies. Table 4.1.11 summarizes the effects of various types of antibody.

Rh HAEMOLYTIC DISEASE OF THE NEWBORN

The most severe form of haemolytic disease of the newborn is caused by maternal Rh anti-D antibodies.

Effects on the fetus

The effects on the fetus are due to agglutination and haemolysis of its Rh-positive red cells by maternal anti-D. The most severe damage occurs when maternal anti-D levels are high throughout the pregnancy. In this situation the fetus develops severe haemolytic anaemia, an enlarged liver, spleen and heart and gross oedema (**hydrops fetalis**). It dies either in utero or soon after birth.

In less severe cases, the baby develops jaundice (**icterus gravis neonatorum**) within 24 hours of birth, because its liver is unable to conjugate the large amounts of bilirubin produced from haemoglobin breakdown following the haemolysis of its red cells. In utero, excess bilirubin is transferred across the placenta and excreted by the mother's liver. The baby's liver may be so severely damaged that death occurs from liver failure. If the baby's serum bilirubin reaches very high levels (over 270 μmol/l), this stains the ganglia in the brain-stem (basal ganglia) yellow and damages them. This condition is called **kernicterus** and produces changes in muscle tone and atheroid cerebral palsy.

The newborn baby may not be anaemic because of its very high rate of erythropoiesis. Its blood contains raised levels of immature red cells

Table 4.1.11   The clinical effects on the fetus of various types of maternal antibodies

Antibody specific for	Clinical effect on the fetus	Comment
Human leucocyte antigens (HLA)	None	The mechanisms which prevent a mother rejecting the foreign HLAs of her fetus have yet to be evaluated
Red cells	Haemolytic disease of the newborn	See text
Granulocytes	Alloimmune neonatal neutropenia	Rare, occur during the first, as well as subsequent, pregnancies This suggests that maternal alloimmunization takes place at an earlier stage of pregnancy than for red cells
Platelets	Alloimmune neonatal thrombocytopenia	

(**erythroblastosis fetalis**). As the rate of red cell haemolysis is highest at birth, anaemia may develop within days.

Prevention
It is now possible to prevent anti-D haemolytic disease of the newborn by preventing Rh-negative women developing anti-D. This can be done in the following ways.

1 By ensuring that an Rh-negative woman does not receive a transfusion of Rh-positive blood.
2 By administering anti-D to every Rh-negative mother giving birth to an Rh-positive baby, provided she has not been immunized previously to the D antigen. The dose of anti-D destroys the leaked fetal red cells before the mother's immune system is stimulated to produce the antibody. For this reason it must be given as soon as possible, and certainly within 72 hours of delivery.

Treatment
The titre of anti-D in the mother's blood is monitored in known affected pregnancies. Intensive plasmapheresis to reduce the anti-D titre can reduce the severity of fetal disease.

The fetus can also be monitored: amniocentesis allows assessment of the bilirubin content of the amniotic fluid and it is also possible, via fetoscopy, to obtain fetal blood samples for analysis.

Treatment for severe fetal disease is exchange transfusion with Rh-negative blood. Intrauterine exchange transfusion can be performed but the procedure is more frequently carried out after birth. The exchange transfusion is given via the umbilical vein. Approximately 10 ml of the baby's blood is removed first and then an equal volume of Rh-negative blood is transfused. The procedure is then repeated. In this way, the baby's Rh-positive blood is progressively removed so that it does not have to metabolize the degradation products of its own blood cells. The transfused Rh-negative cells correct the baby's anaemia. These cells have a normal life span, of approximately 120 days, by which time the maternal anti-D has disappeared.

ABO HAEMOLYTIC DISEASE OF THE NEWBORN

Several factors protect the fetus from the effects of ABO fetomaternal incompatibility so that ABO haemolytic disease of the newborn is uncommon. These factors include

(a) the relative weakness of fetal A and B antigens
(b) the occurrence of A and B glycoproteins on other tissues and in body fluids; this diverts maternal anti-A and anti-B, which crosses the placenta, away from the fetus' red blood cells.

ABO haemolytic disease can occur in a first pregnancy because anti-A and anti-B are naturally present in the plasma. Group A and group B babies of group O mothers are most frequently at risk. Although this situation occurs in about 25% of pregnancies, only about 1% of babies show a degree of haemolysis and jaundice. The condition is usually mild, exchange transfusion being required only in about 1 in 3000 cases (Waters, 1983).

SUMMARY

Blood is the fluid which is normally contained within the heart and blood vessels. It is composed of red cells, white cells, platelets and plasma. Its major function is the carriage of organic and inorganic substances which are continuously exchanged with those contained in the interstitial fluid surrounding the cells. This exchange maintains the composition of the internal environment.

Certain components of blood have a vital function in producing the inflammatory and immune responses and in maintaining the patency of the blood vessels.

The widespread involvement of blood in homeostasis underlies the finding that pathological processes almost always affect its composition. For the same reason, a wide variety of health problems are produced if blood is lost from the circulation and if the blood cells, or the constituents of plasma, are abnormal.

Review questions

The answers to all these questions can be found in the text. In each case there is at least one correct and at least one incorrect answer.

1 The blood volume of a healthy adult male is approximately

(a) 3.3 litres

(b) 4.8 litres
(c) 5.8 litres
(d) 8.4 litres

2 Plasma albumin

(a) is entirely synthesized by the liver
(b) molecules are larger than globulin molecules
(c) is normally present in smaller quantities than plasma globulin
(d) is the major source of the intravascular osmotic effect exerted by plasma proteins

3 The cellular elements of blood

(a) comprise red blood cells, white blood cells and platelets
(b) form the upper portion of a centrifuged anticoagulated blood sample
(c) normally comprise approximately 55% of its volume
(d) are formed by the red bone marrow

4 The production of erythropoietin

(a) probably occurs in the kidney
(b) is stimulated by hypoxaemia
(c) inhibits the production of red blood cells
(d) is inhibited by thryoid hormones

5 Mature red blood cells

(a) can only utilize glucose as an energy source
(b) are unable to synthesize protein
(c) contain the protein pigment myoglobin
(d) have a life span of approximately 12 days

6 Haemoglobin A molecules

(a) are synthesized by mature erythrocytes
(b) contain 2α and 2β globin chains
(c) can each carry one molecule of oxygen
(d) have a higher affinity for oxygen than haemoglobin F

7 Platelets

(a) are the most numerous cellular elements of blood
(b) have a life span of 8–12 days

(c) produce thromboxane A_2
(d) produce prostacyclin

8 Prothrombin

(a) is a soluble plasma protein
(b) requires vitamin K for its synthesis in the liver
(c) is converted to fibrin by the enzyme thrombin
(d) acts as a powerful stimulus to platelet aggregation

9 Heparin

(a) is secreted by mast cells in the tissues
(b) is a powerful anticoagulant when given orally
(c) exerts its anticoagulant effect by inhibiting the action of thrombin
(d) is neutralized by protamine sulphate

10 The A blood group antigen

(a) is produced by the plasma cells of the immune system
(b) occurs in approximately 9% of the United Kingdom population
(c) is present in the plasma of people of blood groups B and O
(d) produces agglutination of red blood cells following combination with antibody anti-A

11 Anti-D

(a) occurs naturally in the plasma of Rh-negative individuals
(b) is produced by Rh-positive individuals following contact with antigen D
(c) is able to cross the placenta and produce agglutination and haemolysis of Rh-positive fetal red cells
(d) is given postnatally to Rh-negative mothers of Rh-positive babies

Answers to review questions

1 b
2 a and d
3 a and d
4 a and b
5 a and b

6 b
7 b and c
8 a and b
9 a, c and d
10 d
11 c and d

Short answer and essay topics

1 List the functions of blood. Choose one of these and discuss it in detail.
2 Describe the origin and functions of the plasma proteins.
3 Describe the structure and function of the haemoglobin A molecule. Name two other forms of human haemoglobin and briefly state the physiological effects of their occurrence.
4 Define the term anaemia. List the health problems associated with severe anaemia and explain their occurrence. Choose one common deficiency anaemia and discuss its causes, prevention and treatment.
5 Write an essay entitled 'The polymorphonuclear leucocytes'.
6 Describe the events occurring in the physiological arrest of bleeding from a small cut.
7 Give an example of three different types of antithrombotic drug. Explain the mode of action of each.
8 Discuss the clinical significance of the ABO and Rh blood group systems.

Suggestions for practical work

WARNING: RISK OF TRANSMISSION OF SERUM-CARRIED DISEASES

Blood samples should always be treated as if they are potentially infectious. In practical work, each individual must only work with samples of his or her own blood. All personnel handling blood or blood products should wear protective gloves in order to minimize infection risk.

1 Using a sterile, disposable lancet, stab a clean fingertip, or earlobe, and obtain a sample of capillary blood. Place a drop of blood on a clean microscope slide. Quickly (i.e. before the blood clots) make a thin smear of blood across the slide by drawing the short edge of a second slide across its surface. Observe red blood cells under a light microscope.
   If facilities are available, stain the blood smear with Leishman's stain and identify the various types of white blood cell.
2 Observe and become familiar with the various types of containers in use in your hospital for the collection of blood samples. Explain why each is required. Match these containers with the forms used for particular blood investigations. Consider the nurse's responsibilities in the collection of blood samples for investigation.
3 Arrange supervised visits to a haemotology laboratory and a chemical pathology laboratory. Observe the range of procedures carried out on plasma and cells, including tests of haemostasis.
   Note how the laboratories produce their reports.
4 With permission, talk to a person being treated for severe anaemia. What health problems has he or she experienced? Over what time period did they develop? Relate the person's answers to your knowledge of the pathophysiology of anaemia.
   What is the cause of this person's anaemia? How is it being treated? Why has this form of treatment been chosen? What are its side-effects, if any?
   Detail the physical nursing care this person requires, relating each aspect to its physiological foundation.
5 With permission, talk to a person who has a clotting disorder (e.g. haemophilia) about his or her condition. Find out when the patient first experienced health problems and the nature and severity of these. What is the cause of this person's clotting disorder? How is it being treated? Does the individual manage his or her therapy at home? What does this involve? Are there any problems associated with the therapy? If the person is in hospital, why has he or she been admitted?
6 Discover the precautions taken to avoid the occurrence of deep vein thrombosis in patients on medical and surgical wards. Explain the pathophysiological basis for these precautions.
7 Arrange a supervised visit to a Regional Blood Transfusion Centre. Observe the precautions associated with the donation of blood for transfusion, the treatment of the blood and the preparation of various blood components. What

are the responsibilities of nurses employed by the Blood Transfusion Service? If possible, talk to donors about their motivation and experience of blood donation.

8 Observe a patient undergoing blood transfusion. What precautions have been and are being taken to ensure the safety of the transfusion? Explain the reasons for these precautions. What is the nurse's responsibility in these? If a component of blood is being transfused, explain the rationale for this.

References

Chanarin, I., Brozovic, M., Tidmarsh, E. & Waters, D. A. W. (1985) *Blood and its Diseases*, 3rd edn. Edinburgh: Churchill Livingstone.

Fawns, H. T. (1978) Vitamin K and blood clotting. *Nursing Times 74*: 1764–1766.

Gorman, A. (1980) Thalassaemia. *Nursing Times 76*: 1348–1350.

Green, S. & Wickenden, A. (1982) Deep vein thrombosis. *Nursing* 1st series *33* 1468–1469.

Gunson, H. H. (1986). Trends in blood transfusion practice in England and Wales. *Health Trends 18*(4): 76–79.

Kenny, M. W. (1980) Sickle cell disease. *Nursing Times 76*: 1582–1584

Lockyer, W. J. (1982) *Essentials of ABO–Rh Grouping and Compatibility Testing*. Bristol: Wright.

Lowe, G. D. O. (1983) Thrombosis. In *Haematology*, Cawley, J. C. (ed.) pp. 68–77. London: Heinemann Medical.

Macfarlane, R. G. (1964) An enzyme cascade in the blood clotting mechanism and its function as a biochemical amplifier. *Nature 202*: 498–499.

Waters, A. H. (1983) Blood transfusion. In *Haemotology*, Cawley, J. C. (ed.) pp. 40–52. London: Heinemann Medical.

Wells, N. (1986) *The AIDS Virus – Forecasting its Impact*. London: Office of Health Economics.

Suggestions for further reading

Cawley, J. C. (ed.) (1983) *Haematology*. London: 19Heinemann Medical.

Chanarin, I., Brozovic, M., Tidmarsh, E. & Waters, D. A. W. (1985) *Blood and its Diseases*, 3rd edn. Edinburgh: Churchill Livingstone.

Clarke, C. A. (1975) *Rhesus Haemolytic Disease* (selected papers and abstracts). Lancaster: MTP.

Cline, M. J. (1975) *The White Cell*. Cambridge, Mass: Harvard University Press.

Guyton, A. C. (1986) *Textbook of Medical Physiology*, 7th edn. London: W. B. Saunders.

Jones, P. (1984) *Living with Haemophilia*. Lancaster: MTP.

Keele, C. A., Neil, E. & Joels, N. (1982) *Samson Wright's Applied Physiology*, 13th edn. Oxford: Oxford University Press.

Miller, J. (1978) *The Body in Question*, London: Jonathan Cape.

Mollison, P. L. (1982) *Blood Transfusion in Clinical Medicine*, 7th edn. Oxford: Blackwell Scientific.

Race, R. R. & Sanger, R. (1975) *Blood Groups in Man*, 6th edn. Oxford: Blackwell Scientific.

Titmuss, R. M. (1970) *The Gift Relationship*. Harmondsworth: Pelican/Penguin.

Chapter 4.2
Cardiovascular Function

Rosamund A. Herbert
Jennifer A. Alison

Rosamund A. Herbert
Jennifer A. Alison

Learning objectives

After studying this chapter the reader should be able to
1 Discuss the vital significance of maintaining an adequate circulation of blood to the tissues at all times.
2 Describe the normal structure and function of the heart and vascular system.
3 Demonstrate how disruption of normal function in any part of the cardiovascular system can lead to abnormal states or disease and interferes with tissue nutrition.
4 Discuss the common forms of treatment used in cardiac and vascular disease, e.g. rest, surgery, drug therapy, and relate them to the abnormal physiology.
5 Base his or her assessment, planning, delivery and evaluation of nursing care of patients with cardiovascular problems on sound physiological principles.
6 Discuss the various ways of assessing the cardiovascular system and be aware of the significance of parameters being monitored.
7 Give adequate and comprehensive explanations of normal and abnormal conditions to patients and other colleagues.

Introduction

The functioning of the human body as a whole depends on the individual and collective function-
ing of all the cells. For optimal function, each cell depends on being in a stable internal environment with a constant supply of nutrients and removal of unwanted substances; it is the role of the circulatory and lymphatic systems to perform these functions. Blood, the fluid within the circulatory system, has been considered in Chapter 4.1, and the lymphatic system in Chapter 6.2.

As maintaining an adequate circulation is vital to health, many of the nurse's observations are concerned with assessing, either directly or indirectly, the state of an individual's circulatory system, e.g. taking the pulse rate, recording blood pressure and pallor of the skin. Pressure area care can even be included: pressure sores develop because pressure (usually the individual's own weight) occludes the circulation to tissues, which become ischaemic, and without adequate perfusion the integrity of the tissue is destroyed and the tissue breaks down (Chap. 6.1).

GENERAL DESCRIPTION OF THE CARDIOVASCULAR SYSTEM

The function of the cardiovascular system is to ensure an adequate circulation of blood to the tissues of the body at all times and thus to transport substances to and from the individual cells as required. The cardiovascular system is dynamic and is capable of maintaining an adequate flow to the tissues under varying circumstances, for example at rest, when exercising, and when standing on one's head. As there is a finite volume of blood, this flexibility is achieved by directing blood to where it is most needed and away from less active areas. Conflicts can, and do, arise; for instance, if strenuous exercise is undertaken shortly after eating a large meal, both the gastrointestinal system and the muscles require an increased blood flow, which may not be feasible. It is not possible to supply both systems with large blood flows at the same time. After a meal and during digestion, the gut has priority and strenuous exercise cannot usually be maintained.

However, at all times, maintaining an adequate blood flow to vital organs such as the brain and heart has priority. The cardiovascular system as a whole is controlled and coordinated by centres in the brain, although local influences and reflexes are also very important. Tissues that do not have an adequate blood flow, that is, are **ischaemic**, show signs of adverse effects quickly.

The main role of the cardiovascular system is that of general transport and can be subdivided into the following.

(a) Delivery of substances to all body tissues to maintain their nutrition and metabolic function, i.e. oxygen, nutrients and other substances manufactured in the body such as hormones, amino acids, defence cells.
(b) Removal of carbon dioxide and metabolic end-products from tissues and delivery to the appropriate organs for breakdown and elimination, e.g. lungs, liver, kidneys.
(c) The dissipation of heat away from active tissues and its redistribution around the body to maintain normal body temperature.

In humans, as in all mammals, there are two distinct circuits within the cardiovascular system, known as the systemic and pulmonary circulations (Fig. 4.2.1). Both circulations originate and terminate in the heart, which is itself functionally divided into two pumps.

The **systemic circulation** supplies all the body tissues, and is where exchange of nutrients and products of metabolism occurs; all the blood for the systemic circulation leaves the left side of the heart via the **aorta**. This large artery then divides into smaller arteries and blood is delivered to all tissues and organs. These arteries divide into smaller and smaller vessels (Fig. 4.2.2), each with its own characteristic structure and function. The smallest branches are called **arterioles**. The

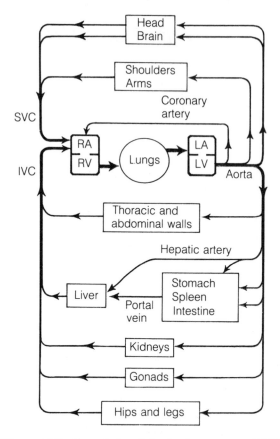

Figure 4.2.1 General plan of the circulatory system (SVC superior vena cava; IVC inferior vena cava; RA right atrium; RV right ventricle; LA left atrium; LV left ventricle)

arterioles themselves branch into a number of very small thin vessels, the **capillaries**, and it is here that the exchange of gases, nutrients and waste products occurs. Exchange occurs by diffusion of substances down concentration and pressure gradients. The capillaries then unite to form larger vessels, **venules**, which in turn unite to form fewer and larger vessels, known as **veins**. The veins from different organs and tissues unite to form two large veins. The **inferior vena cava** (from the lower portion of the body) and the **superior vena cava** (from the head and arms), which return blood to the right side of the heart. Thus there are a number of parallel circuits within the systemic circulation (see Fig. 4.2.1).

The **pulmonary circulation** is where oxygen and carbon dioxide exchange between the blood and alveolar air occurs. The blood leaves the right

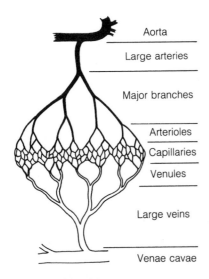

Figure 4.2.2 Relationship between the various blood vessels in the circulatory system

side of the heart through a single artery, the **pulmonary artery**, which divides into two – one branch supplying each lung. Within the lung, the arteries divide, ultimately forming arterioles and capillaries; venules and veins return blood to the left side of the heart.

All arteries carry blood away from the heart and all arteries, except the pulmonary artery, carry blood with a high oxygen content and deliver this to the tissues. Veins carry blood back to the heart. Figures 4.2.3 and 4.2.4 show the major arteries and veins in the body.

Normally there is only one capillary bed for each branch of a circuit; however, there are a few instances where there are two capillary beds, one after each other, in series. These are known as **portal systems** or portal circulations. One example of this is in the liver. Part of the blood supply to the liver is venous blood coming directly from the gastrointestinal tract and spleen via the hepatic portal vein. This arrangement enables the digested and absorbed substances from the gut to be transported directly to the liver, where many of the body's metabolic requirements are synthesized. Thus there are two microcirculations in series, one in the gut and the other in the liver. (For fuller details see Chap. 5.2.)

The force required to move the blood through the blood vessels in the two circulations is provided by the heart, which functions as two pumps, the left side of the heart supplying the

systemic circulation and the right side the pulmonary circulation. The systemic circulation is much larger than the pulmonary circulation and thus the force generated by the left side of the heart is much greater than that of the right side of the heart. However, as the circulatory system is a closed system, the volume of blood pumped through the pulmonary circulation in a given period of time must equal the volume pumped through the systemic circulation – that is, the right and left sides of the heart must pump the same amount of blood. In a normal resting adult, the average volume of blood pumped simultaneously is approximately 5 l per min. As there are approximately 5 l of blood in an adult, this means that the blood circulates around the body approximately once every minute. During heavy work or exercise, the volume of blood pumped by the heart can increase up to 25 l per min (or even 35 l per min in elite athletes).

THE HEART

Location and structure (Fig. 4.2.5)

The heart lies in the **mediastinum** which is the central part of the thorax, lying between the two pleural sacs which contain the lungs. The mediastinum extends from the sternum to the vertebral column and contains all the thoracic organs except the lungs.

The heart is shaped like a blunt cone that lies with its base towards the head. Approximately two-thirds of its mass lies to the left of the body's midline. The heart is enclosed in a fibrous sac, the **pericardium**, which provides a tough protective membrane around it and anchors it in the mediastinum by its attachments to the large blood vessels entering and leaving the heart, to the diaphragm and to the sternal wall of the chest. Between the pericardium and the heart is a potential space called the **pericardial cavity**. This cavity contains watery fluid, known as **pericardial fluid**, which prevents friction as the heart moves. Inflammation of the pericardium is called **pericarditis** and is often associated with a pericardial friction rub which can be heard on auscultation of the pericardial area.

THE WALLS

The walls of the heart consist mainly of muscle

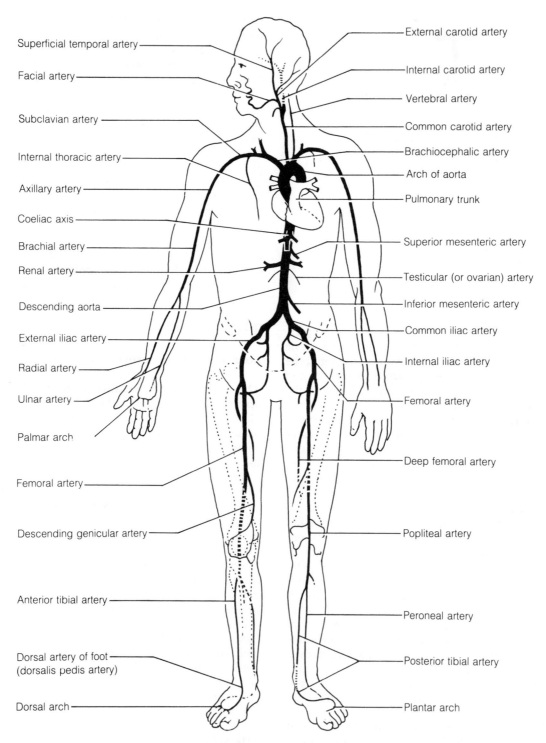

Superficial temporal artery

Facial artery

Subclavian artery

Internal thoracic artery

Axillary artery

Coeliac axis

Brachial artery

Renal artery

Descending aorta

External iliac artery

Radial artery

Ulnar artery

Palmar arch

Femoral artery

Descending genicular artery

Anterior tibial artery

Dorsal artery of foot
(dorsalis pedis artery)

Dorsal arch

External carotid artery

Internal carotid artery

Vertebral artery

Common carotid artery

Brachiocephalic artery

Arch of aorta

Pulmonary trunk

Superior mesenteric artery

Testicular (or ovarian) artery

Inferior mesenteric artery

Common iliac artery

Internal iliac artery

Femoral artery

Deep femoral artery

Popliteal artery

Peroneal artery

Posterior tibial artery

Plantar arch

Figure 4.2.3 Major arteries of the human body (anterior view)

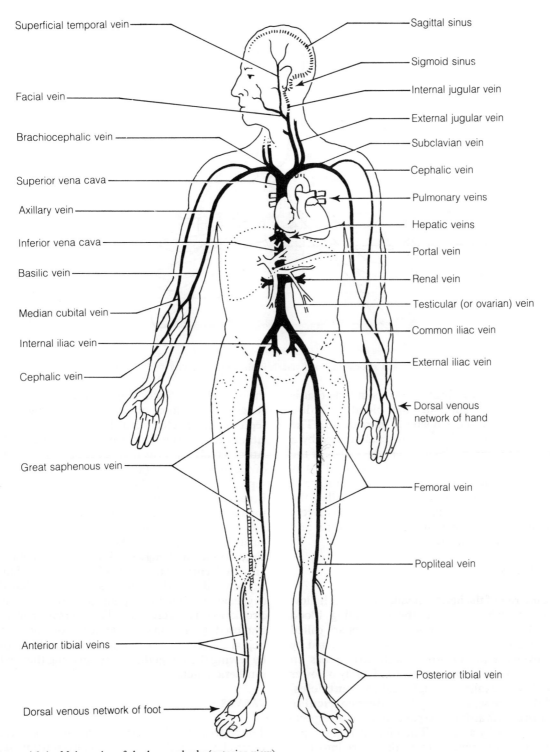

Superficial temporal vein

Facial vein

Brachiocephalic vein

Superior vena cava

Axillary vein

Inferior vena cava

Basilic vein

Median cubital vein

Internal iliac vein

Cephalic vein

Great saphenous vein

Anterior tibial veins

Dorsal venous network of foot

Sagittal sinus

Sigmoid sinus

Internal jugular vein

External jugular vein

Subclavian vein

Cephalic vein

Pulmonary veins

Hepatic veins

Portal vein

Renal vein

Testicular (or ovarian) vein

Common iliac vein

External iliac vein

Dorsal venous network of hand

Femoral vein

Popliteal vein

Posterior tibial vein

Figure 4.2.4 Major veins of the human body (anterior view)

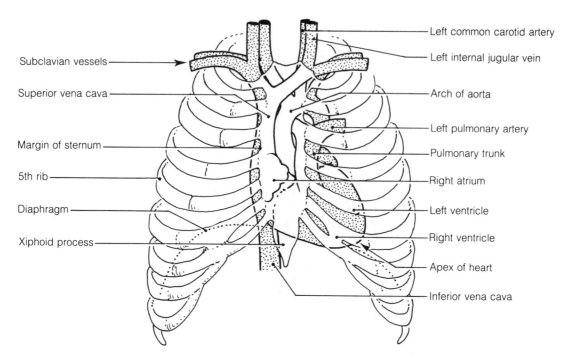

Subclavian vessels

Superior vena cava

Margin of sternum

5th rib

Diaphragm

Xiphoid process

Left common carotid artery

Left internal jugular vein

Arch of aorta

Left pulmonary artery

Pulmonary trunk

Right atrium

Left ventricle

Right ventricle

Apex of heart

Inferior vena cava

Figure 4.2.5 Location of the heart and associated blood vessels in the thoracic cavity

called the **myocardium**. It is this muscle that is responsible for pumping the blood. (Cardiac muscle will be discussed later in more detail.) The myocardium is lined on its internal surface by a layer of epithelium which covers the valves of the heart and the tendons that hold them. This is called the **endocardium**. On the outer surface of the myocardium is an external layer called the **epicardium**. Inflammation of the various layers is referred to as myocarditis, endocarditis and epicarditis, respectively.

CHAMBERS

The interior of the heart is divided into four hollow chambers which receive the circulating blood (Fig. 4.2.6). The two upper chambers are called the right and left **atria,** and they are separated by the **interatrial septum**. (Each atrium has an appendage called an **auricle**.) The two lower chambers are called the right and left **ventricles,** and these are separated by the **interventricular septum**. Each atrium is separated from its respective ventricle by a valve. Therefore it is the septa and the valves which divide the heart into four chambers.

Abnormalities can occur in the septa that divide the chambers. Such abnormalities are usually congenital and most often consist of a hole in the septum (septal defect) that allows a communication either between the atria or between the ventricles. Atrial septal defects usually occur because of failure of the foramen ovale (an opening in the interatrial septum of the fetal heart) to close after birth. Because the pressure in the left atrium is higher than in the right, blood flows from the left atrium into the right. This overloads the pulmonary circulation and decreases the flow in the systemic circulation, which may result in inhibition of body growth. Ventricular septal defects allow mixing of oxygenated and deoxygenated blood so that the oxygenation of the blood pumped into the systemic circulation is decreased. The person becomes cyanosed (a bluish discoloration of the skin). Septal defects can be corrected surgically, either by sewing them together or by covering them with synthetic patches.

VALVES

The valves that lie between the atria and their ventricles are called the **atrioventricular (AV) valves**. The AV value between the right atrium and the right ventricle is called the **tricuspid**

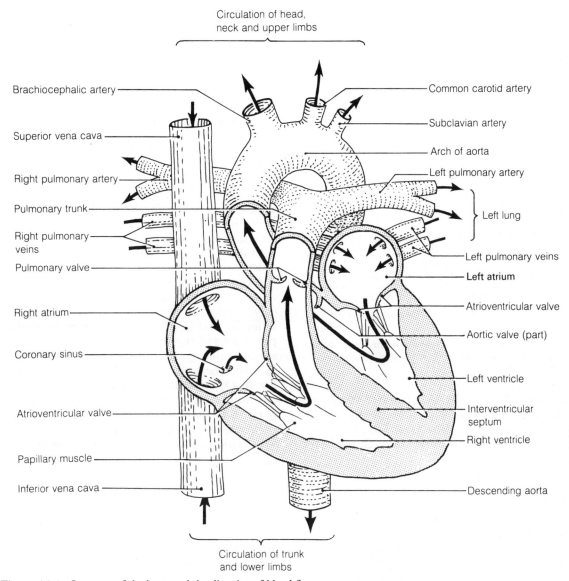

Circulation of head,
neck and upper limbs

Brachiocephalic artery

Superior vena cava

Right pulmonary artery

Pulmonary trunk

Right pulmonary
veins

Pulmonary valve

Right atrium

Coronary sinus

Atrioventricular valve

Papillary muscle

Inferior vena cava

Common carotid artery

Subclavian artery

Arch of aorta

Left pulmonary artery

Left lung

Left pulmonary veins

Left atrium

Atrioventricular valve

Aortic valve (part)

Left ventricle

Interventricular
septum

Right ventricle

Descending aorta

Circulation of trunk
and lower limbs

Figure 4.2.6   Structure of the heart and the direction of blood flow

valve because it consists of three cusps, or flaps. The AV valve between the left atrium and the left ventricle is called the **mitral** or **bicuspid valve** because it has two cusps. When the ventricles contract, the pressure exerted on these valves by the blood forces them to close, thus preventing backflow of blood into the atria. To stop the valves themselves being forced backwards into the atria, they are attached to muscular projections of the ventricular wall (**papillary muscles**) by fibrous strands called **chordae tendineae** (Fig. 4.2.7).

The valves that lie between the ventricles and the major arteries that leave the heart are called the **semilunar valves,** so named because of their half-moon or crescent shape (Fig. 4.2.7). The **pulmonary semilunar valve** lies where the pulmonary trunk leaves the right ventricle; the **aortic semilunar valve** lies where the aorta leaves the left ventricle. The semilunar valves are forced open when the ventricles contract, allowing blood to be ejected into the aorta and pulmonary artery. As soon as the ventricles relax, there is a back

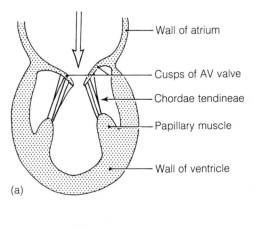

Wall of atrium

Cusps of AV valve

Chordae tendineae

Papillary muscle

Wall of ventricle

(a)

Wall of artery

Cusp of semilunar valve

(b)

Figure 4.2.7   Structure of (a) the atrioventricular valves (AV), tricuspid and mitral, and (b) the semilunar valves, pulmonary and aortic

pressure of blood from the aorta and pulmonary artery that closes the valves. This prevents backflow of blood into the ventricles. Thus the valves ensure one-way flow of blood through the heart.

The closure of the valves defines the period of ventricular contraction since the AV valves close when the ventricles start contracting and the semilunar valves close when the ventricles stop contracting (i.e. relax). Thus, during each cardiac cycle two **heart sounds** can be heard through a stethoscope applied to the chest wall. The **first heart sound**, which can be described as a 'lubb' sound (long and booming), is associated with the closure of the AV valves at the beginning of ventricular contraction. The **second heart sound**, which can be described as a 'dupp' sound (short and sharp), is associated with the closure of the semilinar valves at the beginning of ventricular relaxation. The sounds result primarily from turbulence in the blood flow created by the closure of the valves. They may be represented phonetically as in Fig. 4.2.8.

Although heart sounds are associated with the closure of the valves, the sounds produced by each valve tend to be heard more clearly in a slightly different position from the anatomical surface projections of the valves. Figure 4.2.8 shows the surface projections of the valves and the circles indicate where the sounds they produce are best heard. Occasionally, sounds additional to these two heart sounds are heard over a normal healthy heart (see below).

As well as auscultation of the chest for heart sounds, the heart rate is often counted by listening

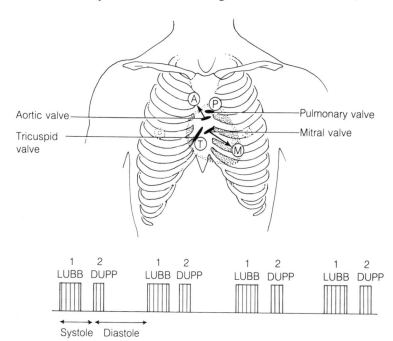

Aortic valve

Tricuspid valve

Pulmonary valve

Mitral valve

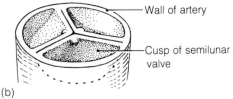

| 1 | 2 | | 1 | 2 | | 1 | 2 | | 1 | 2 |
| LUBB | DUPP | | LUBB | DUPP | | LUBB | DUPP | | LUBB | DUPP |

Systole   Diastole

Figure 4.2.8   (a) Anatomical surface projections of the heart valves (the circles indicate the areas at which sounds produced by each valve are best heard) (b) Phonetical representation of the heart sounds

to the **apex beat**. The apex of the heart lies close to the anterior chest wall, usually in the fifth intercostal space in the mid-clavicular line. It produces a pulsation against the chest wall with each ventricular contraction. This pulsation can often be seen and felt in thin individuals and can be heard by auscultation.

Damage to the heart valves can disrupt the flow of blood through the heart: rheumatic fever is the most common cause of valvular damage. It is an autoimmune or allergic disease which can affect many different tissues of the body but especially the heart valves. The damage to the valves can cause the cusps of the valve to stick together, thus narrowing the opening. This is called valvular **stenosis**. At the same time, damage to the edges of the cusps means that they cannot close completely and there is backflow, or regurgitation, of blood through the valve. The valve is said to be **incompetent**. Stenosis and incompetence often co-exist but usually one predominates, so the valve is said to be either stenosed or incompetent.

Stenosed or incompetent valves cause sound additional to the normal heart sounds. Normally, when blood flows smoothly in a streamlined manner it makes no sound, but turbulent flow produced either by blood flowing rapidly in the usual direction through an abnormally narrowed (stenosed) valve or backwards through a leaky (incompetent) valve produces a **murmur** or sloshing sound. (Murmur may also be heard if blood is able to move between the two atria or the two ventricles via a septal defect.)

Damaged valves decrease the efficiency of the heart as a pump as more work has to be done either to pump blood through a stenosed valve or to pump the extra blood that flows back through an incompetent valve. In the long term, the effects of valvular damage can result in failure of the pump.

Heart valves can be replaced surgically by artificial valves or tissue valves. The most commonly used artificial valve is the tilting disc valve. If an artificial valve is used, the patient will be placed on long-term anticoagulant therapy to prevent clot formation on the valve.

DIRECTION OF BLOOD FLOW

As already stated, valves ensure one-way flow of blood through the heart (see Fig. 4.2.6). Blood enters the right atrium from the **superior vena cava**, which brings venous blood from the upper portion of the body (i.e. head and arms), and from the **inferior vena cava**, which brings blood from the lower portions of the body (i.e. trunk and legs). This blood passes from the right atrium through the tricuspid valve into the right ventricle whence it is pumped through the pulmonary valve into the **pulmonary trunk**. The pulmonary trunk divides into the **right** and **left pulmonary arteries** which carry the blood to the right and left lung respectively, where gas exchange takes place. These are the only arteries to carry deoxygenated blood. This blood returns to the left atrium via four **pulmonary veins**. This is the only place in the body where veins carry highly oxygenated blood. The blood passes from the left atrium into the left ventricle. The left ventricle pumps blood through the aortic valve into the aorta. Arteries branch from the aorta and carry blood to all parts of the body.

Cardiac muscle

Cardiac muscle combines certain properties of skeletal and smooth muscle. Like skeletal muscle, it is striated and its myofibrils contain actin and myosin filaments (see Chap. 3.1). However, cardiac muscle differs from skeletal muscle in that the cell membranes that separate individual muscle cells have a very low electrical resistance. These membranes are called **intercalated discs** (Fig. 4.2.9). The action potentials travel from one cardiac muscle cell to another through the intercalated discs without significant hindrance. In this respect, cardiac muscle is similar to smooth muscle. The ability of the action potential to spread to all cells if one cell becomes excited means that cardiac muscle acts as a functional whole. This ability is further enhanced by the shape of the action potential. After an initial spike, the membrane has a plateau of depolarization which lasts for 0.15–0.3 seconds (Fig. 4.2.10). This long duration of the action potential (relative to the velocity of conduction) ensures that the impulse has time to travel over the whole atrial or ventricular muscle mass, causing complete contraction of each muscle mass as a unit, before any portion of the muscle can repolarize and relax. This is essential for efficient pumping.

The action potential is followed by a period in which the muscle is refractory to restimulation. This prevents tetanic contractions and fatigue and

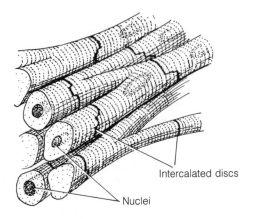

Figure 4.2.9 Structure of cardiac muscle showing lattice arrangement and intercalated discs

ensures that the muscle relaxes between action potentials, permitting it to fill with blood. If there were no relaxation phase, i.e. if tetanic stimulation were possible, no blood would enter the heart and thus the heart's action as a pump would be defunct.

Therefore, three factors affect the way cardiac muscle contracts: the intercalated discs, the long duration of the action potential, and the long refractory period. These factors, combined with sequential depolarization of the muscle due to the organization of the conducting system (see later), ensure that the heart muscle contracts in a coordinated and effective manner.

The way the actual excitation of the cardiac muscle cell results in contraction (i.e. excitation–contraction coupling) is similar to that of skeletal muscle (see Chap. 3.1). However, in cardiac muscle more of the calcium required for muscle contraction comes from the transverse tubules (T-tubules) rather than from the sarcoplasmic reticulum as in

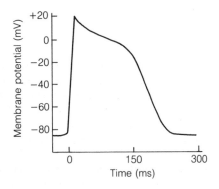

Figure 4.2.10 The action potential of cardiac muscle

skeletal muscle. The importance of this is that the T-tubules are filled with extracellular fluid, and so the concentration of calcium in the extracellular fluid can alter the strength of cardiac muscle contraction.

As in skeletal muscle, the strength of contraction or tension that a cardiac muscle fibre can generate is related to its length at rest. This is known as the **length–tension relationship** and is discussed in Chapter 3.1 in regard to skeletal muscle. The more the cardiac muscle fibre is stretched before contraction, the greater will be the force of contraction or energy output. This is true up to a critical point, beyond which further increases in muscle length result in a decrease in the strength of contraction because the muscle is stretched beyond its physiological limits.

The length of cardiac muscle fibres is affected by the amount of blood in the chambers of the heart before contraction. Normally, during sedentary activities, the resting length of cardiac muscle (i.e. the length of the muscle fibres during the relaxation phase of the cardiac cycle) does not yield maximum force of contraction, thus leaving a margin for coping with increasing amounts of blood returning to the heart. When venous return to the heart increases, as during exercise, there is a consequent increase in the amount of blood in the ventricles, termed the ventricular end-diastolic volume (VEDV). This stretches the muscle fibres and results in a greater force of contraction and more blood is expelled.

This is the basis of **Starling's law of the heart** which basically states that the force of contraction of cardiac muscle is proportional to the resting length of the muscle fibres. Therefore, within physiological limits, the more the heart is filled with blood during its relaxation phase, the greater will be the force of contraction, and hence the quantity of blood pumped out during contraction will increase. Other factors affecting the quantity of blood ejected by each ventricle with each contraction (stroke volume) will be discussed later.

Blood supply – the coronary circulation

When the body is at rest, the oxygen consumption of the heart muscle is approximately 8 ml/100 g/min, which is greater than that of any other tissue.

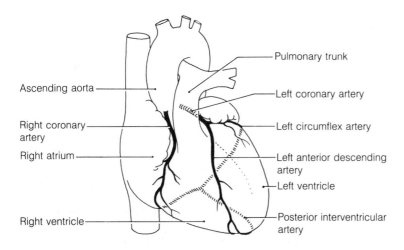

Ascending aorta

Right coronary artery

Right atrium

Right ventricle

Pulmonary trunk

Left coronary artery

Left circumflex artery

Left anterior descending artery

Left ventricle

Posterior interventricular artery

Figure 4.2.11 The coronary circulation

Oxygenated blood is carried to the cardiac muscle by the right and left **coronary arteries** and their branches (Fig. 4.2.11). The coronary arteries originate from the aorta just beyond the aortic valve. The right coronary artery supplies the right atrium and ventricle and portions of the left ventricle. The left coronary artery divides near its origin into the left anterior descending branch and the circumflex branch. The former supplies blood to the anterior part of the left ventricle and a small part of the anterior and posterior portions of the right ventricle. The circumflex branch supplies the left atrium and upper front and rear of the left ventricle. The major coronary arteries lie on the surface of the heart and the branches penetrate deep into the cardiac muscle.

Most of the venous blood from the left ventricular muscle returns to the right atrium through the **coronary sinus**, which is a thin-walled vein without any smooth muscle to alter its diameter. Most of the venous blood from the right ventricular muscle flows through the small anterior cardiac veins into the right atrium.

FACTORS THAT INFLUENCE CORONARY BLOOD FLOW

Aortic pressure

The primary factor responsible for maintaining blood flow to the cardiac muscle is the aortic pressure, which is generated by the heart itself. Generally, an increase in aortic pressure will result in an increase in coronary blood flow, and vice versa. However, under normal conditions blood pressure is kept within relatively narrow limits by the baroreceptor mechanism (see later) so that changes in coronary blood flow are primarily caused by changes in the resistance of the coronary vessels.

Demand for oxygen

The major factor affecting the resistance of the coronary vessels is the demand of the myocardium for oxygen. At rest, oxygen extraction from the coronary circulation is almost three times greater than in the normal circulation. Therefore, when the demand for oxygen is increased, little additional oxygen can be removed from the blood unless the flow is increased. Blood flow increases almost directly in proportion to oxygen need due to dilation of the blood vessels. It is believed that oxygen lack causes this dilation of the coronary arterioles, however the exact autoregulating mechanism by which the heart is able to alter its blood-flow has not been determined. Some possible mechanisms are discussed later under control and coordination of the cardiovascular system.

Nervous control

The autonomic nervous system has an indirect effect on coronary blood flow. Sympathetic stimulation increases heart rate and contractility and this increases the demand for oxygen, resulting in dilation of the coronary arterioles which enhances coronary blood flow. Conversely, parasympathetic stimulation slows the heart rate and has a slight depressive effect on contractility, which results in a decreased cardiac oxygen consumption and therefore coronary blood flow decreases.

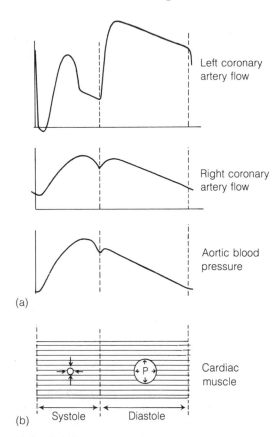

(a)

(b) Systole | Diastole

Left coronary artery flow

Right coronary artery flow

Aortic blood pressure

Cardiac muscle

Figure 4.2.12 (a) Schematic representation of blood flow in the left and right coronary arteries during systole and diastole (based on Gregg, 1965) (b) Diagram of a blood vessel within the myocardium showing compression during systole due to the contraction of the cardiac muscle, and opening during diastole due to the internal blood pressure (P)

Compression of vessels during systole
Unlike other arteries in the body, the majority of blood flow in the coronary arteries occurs during diastole rather than in systole because when ventricular cardiac muscle contracts (systole) it compresses the vessels within it, thus reducing blood flow. As higher pressures are generated in the left ventricle than the right, this compression effect is more marked in this area. During diastole, the cardiac muscle relaxes and no longer obstructs blood flow. Figure 4.2.12 shows the effects of systole and diastole on coronary blood flow.

Coronary artery disease

The integrity of the cardiac muscle depends on the maintenance of an adequate oxygen supply for all levels of activity. Failure to maintain an adequate blood flow to cardiac muscle results in the muscle becoming ischaemic and may lead to death of the muscle tissue.

One of the major causes of cardiac muscle becoming ischaemic is the development of atheromatous plaques in the coronary arteries. This is called **coronary artery disease**, or ischaemic heart disease, and is one of the major causes of disability and death in developed countries. The atheromatous plaques cause the coronary arteries to be narrowed, thus decreasing or, sometimes, completely blocking blood flow to the cardiac muscle. (How these plaques form and factors that might predispose an individual to develop such plaques are dealt with under Atherosclerosis later in this chapter.)

If constriction of the coronary arteries occurs over many years, **collateral vessels** (small communications between arteries) can develop which help to supply blood to the heart muscle beyond the narrowed artery. The symptoms of coronary artery disease depend on the degree of obstruction of the vessels, and range from ischaemic pain due to moderate obstruction to death in some instances of complete obstruction.

ANGINA PECTORIS

Often the first sign that the obstruction of the coronary arteries is great enough to cause ischaemia of the heart muscle is when activities that increase the oxygen demand of the heart cause pain. This pain is called **angina pectoris**. It is usually felt beneath the upper sternum and is often referred to the left arm and shoulder and neck, or even to the opposite arm and shoulder. This distribution of pain is due to the fact that the heart and arms originate during embryonic life in the neck, therefore both these structures receive pain nerve fibres from the same spinal cord segments.

Generally, the pain occurs on exercise when the increased demand of the cardiac muscle for oxygen cannot be met due to the narrowing of the coronary arteries. The pain usually causes the person to stop the activity. It disappears a few minutes after the cessation of activity when blood flow can again match the oxygen requirements of the muscle. People who suffer from angina pectoris may also feel the pain when they experience emotions that increase the metabolism of the heart.

Sometimes an ECG taken during exercise can help in the diagnosis of coronary artery disease.

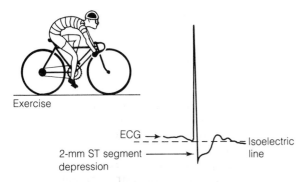

Exercise

ECG →

2-mm ST segment depression →

Isoelectric line

Figure 4.2.13 ST segment depression on the ECG due to ischaemia of the cardiac muscle during exercise

Typically, when the muscle becomes ischaemic, the ST segment on the ECG is depressed (Fig. 4.2.13).

Treatment for angina pectoris is aimed at decreasing the workload of the heart so that the oxygen requirement of the cardiac muscle is less, thus preventing ischaemic episodes.

Nitrates, e.g. **glyceryl trinitrate**, taken before activity that is known to cause angina can often prevent it, or if taken during an angina attack will often give immediate relief. Some forms of glyceryl trinitrate are given sublingually as this allows rapid absorption through the buccal mucosa. It can also be given via pads on the skin to allow for continuous slow absorption.

Nitrates dilate blood vessels everywhere in the body. Dilation of the arterioles lowers blood pressure, which reduces the resistance that the heart has to pump against and therefore decreases the cardiac work. Consequently, the requirement for oxygen by the cardiac muscle is decreased and hence symptoms are reduced. Venous dilation increases the capacity of the venous circulation and therefore reduces venous return to the heart. The lowered blood volume in the heart decreases the stretch on the ventricular muscle fibres in diastole, resulting in a reduced force of contraction. This again lowers the cardiac work and decreases oxygen consumption. The decreased diastolic stretch on the ventricles allows a greater diastolic coronary blood flow and this also improves oxygenation of the myocardium. Because of their action as vasodilators, one side-effect of nitrates is headaches.

Another group of drugs, the **β-adrenergic sympathetic blocking drugs**, is also used in the management of angina pectoris to decrease the workload of the heart. Sympathetic nervous activity increases heart rate and contractility, thus increasing cardiac oxygen consumption. Therefore, blocking the sympathetic stimulation of the heart by β-adrenergic blocking drugs, e.g. propranolol, reduces oxygen consumption and thus relieves angina.

Advice should be given to angina sufferers on ways of adapting their lifestyle to reduce attacks without becoming an invalid. They should be encouraged to stop smoking as nicotine is a vasoconstrictor and so opposes the action of nitrates. Part of discharge planning for such patients should involve health education on diet and stress management.

MYOCARDIAL INFARCTION

If there is complete occlusion of a coronary artery, the blood flow to the muscle beyond the blockage ceases and the muscle dies. The wedge-shaped area of dead cardiac muscle is termed an **infarct**, and the overall process is called a **myocardial infarction**. Often the person is said to have had a 'coronary occlusion' or a 'heart attack'.

Complete blockage of an artery can be caused by a local blood clot, called a **thrombus**, which develops on an atherosclerotic plaque and grows so large that it completely blocks the lumen of the artery. Alternatively, a clot forming on an atherosclerotic plaque can break away (i.e. embolize) and flow downstream until it wedges in a more peripheral branch of the artery causing a blockage at that point.

Sometimes a sudden increase in myocardial oxygen consumption, e.g. during strenuous exercise, in the presence of severely atherosclerotic arteries can have the same effect as complete occlusion of the artery. When a reasonably large coronary artery is occluded, various changes take place within the area of ischaemic muscle. The muscle fibres in the very centre of the area die, and this usually causes severe, continuous, crushing retrosternal chest pain which radiates to the arms, neck and jaw. Unlike angina pectoris, this pain is not relieved by rest and is often associated with nausea, dyspnoea and sweating accompanied by fear, shock and weakness. Around the area of dead muscle is a non-functional area which does not contract, called a zone of injury. Next there is an area that is mildly ischaemic and therefore contracts only weakly (Fig. 4.2.14). This last area is

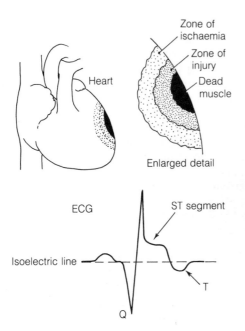

Zone of
ischaemia

Zone of
injury

Dead
muscle

Heart

Enlarged detail

ECG

ST segment

Isoelectric line

T

Q

Figure 4.2.14 Grades of muscle damage following a myocardial infarction and related ECG changes, namely, enlarged Q wave, ST segment elevation and T wave inversion

supplied by small collateral arteries (i.e. small branches from other arteries) which are sufficient to provide enough blood to the muscle at rest but not during exercise. Therefore, to prevent an increase in the size of the infarct, i.e. to stop this part of the muscle dying, the patient should be kept at rest, and all nursing care should be planned with this in mind.

In the days following myocardial infarction, more of the muscle fibres immediately around the dead area die because of prolonged ischaemia. At the same time the non-functional area becomes smaller because the size of the collateral vessels increases. Eventually the dead muscle is replaced by fibrous tissue which forms a scar that becomes smaller over a period of years.

The diagnosis of a myocardial infarction can often be made on the basis of clinical features. However, a number of tests help to confirm the diagnosis. Muscle cells that die release enzymes into the serum, raised levels of which indicate muscle death. The enzymes most commonly measured are creatinine kinase (CK), aspartate transaminase (AST) and lactate dehydrogenase (LD). Following myocardial infarction, there is usually a seven- to tenfold increase in CK, a five-

fold increase in AST, and a threefold increase in LD (Moss, 1981). As changes in these enzyme levels occur over 7–9 days, serial measurements are taken. These enzymes are not exclusive to cardiac muscle, and can be raised as a result of damage to other muscle, e.g. by recent exercise, muscle injury or even intramuscular injection. More recently, some laboratories have been able to measure an isoenzyme of CK (CK-MB isoenzyme) which is only found in heart muscle. Measurement of this isoenzyme ensures that levels are not confused with injury of other muscle but, more importantly, helps to give an idea of the extent of muscle damage.

The ECG can aid in the diagnosis of a myocardial infarction. Alterations in the ECG result from the changes that are occuring within the muscle of the infarcted area. Most commonly, in the lead that overlies the infarct, there are large Q waves, ST segment elevation and T wave inversion (Fig. 4.2.14).

Approximately 25% of patients with myocardial infarction die within the first hour, usually due to **ventricular fibrillation**. When the ventricles fibrillate they do not pump blood, there is no cardiac output and the person dies within 2–3 minutes. Ventricular fibrillation can develop for a number of different reasons, all related to the changes that occur as a result of the muscle damage. It commonly occurs during the first 10 minutes after an infarct, although it can develop 3–5 hours after one. This increased risk of the ventricles fibrillating can last for several days. Therefore, following myocardial infarction, the cardiac rhythm is continuously monitored by an ECG projected onto an oscilloscope so that if the patient goes into ventricular fibrillation, appropriate action, i.e. electrical defibrillation (see later), can be taken immediately.

The coronary arteries, as well as supplying blood to the cardiac muscle, also supply the tissue of the specialized conducting system (see later). For example, the atrioventricular node is supplied by the right coronary artery in 90% of individuals. Therefore, sometimes following coronary occlusion the conducting system is damaged by ischaemia and oedema which can lead to arrhythmias such as heart block (see later).

Immediately following an acute myocardial infarct, treatment aims to: (i) relieve pain by the use of analgesic drugs, e.g. diamorphine, (ii) promote oxygenation of the myocardial tissue

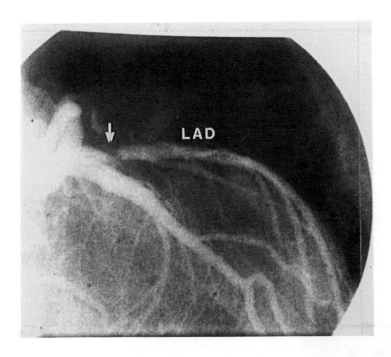

Figure 4.2.15 Coronary angiography showing 90% diameter obstruction (arrowed) in the proximal part of the left anterior descending (LAD) artery (courtesy Hallstrom Institute of Cardiology, Sydney, Australia)

by rest and administration of supplemental oxygen, (iii) detect any complications, such as ventricular fibrillation, by continuous monitoring.

SURGERY IN CORONARY ARTERY DISEASE

Since the late 1960s it has been possible surgically to improve the blood supply to the myocardium in some cases of coronary artery disease (either before or after myocardial infarction) by **aortic–coronary bypass grafts**. In this operation vein grafts, usually from the long saphenous vein in the leg of the recipient, are used to bypass the obstruction in the coronary artery. The graft is attached to the aorta and then to a more peripheral vessel beyond the obstruction. Often as many as three vessels can be grafted.

Before such an operation the extent of coronary artery disease is assessed by **coronary angiography**. The heart is catheterized (therefore the overall procedure is sometimes called **cardiac catheterization**) via the femoral artery. A contrast dye is injected into each of the coronary arteries and continuous cine-radiography allows visualization of the blood vessels. Figure 4.2.15 shows a typical picture of obstruction of the left anterior descending artery. Dye is also injected into the left ventricle (called a **ventriculogram**)

to assess left ventricular function by looking for areas of uncoordinated contraction due to scarring from previous infarction (Fig. 4.2.16). The best candidates for bypass grafts are those with proximal severe obstruction with the development of a rich collateral circulation, while maintaining a relatively normal left ventricular function.

It should be remembered that cardiac catheterization is not without risks, namely haemorrhage from the insertion site, infection and arrhythmias. Nursing care should be directed towards the recognition of these complications. The patient should rest in bed, the insertion site should be inspected for bleeding every 15 minutes for 4 hours, and the pulse and blood pressure should be monitored frequently, as should temperature.

CARDIOGENIC SHOCK

About 15% of patients who suffer an acute myocardial infarction have such massive ischaemic damage to the heart that the muscle cannot pump effectively. The cardiac output is reduced, resulting in a decreased blood flow to all the organs of the body, and subsequently to a reduction in their functions. This is known as **cardiogenic shock**. It may also occur after cardiac surgery, massive pulmonary embolism and pericardial tamponade

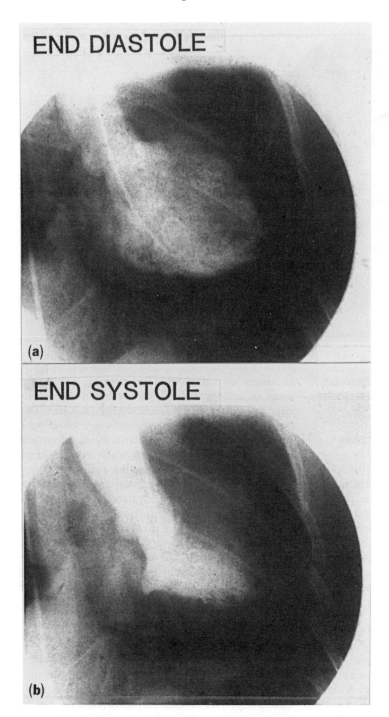

Figure 4.2.16 Ventriculogram showing a normal left ventricle (a) at end-diastole and (b) vigorous contraction at end-systole (courtesy Hallstrom Institute of Cardiology, Sydney, Australia)

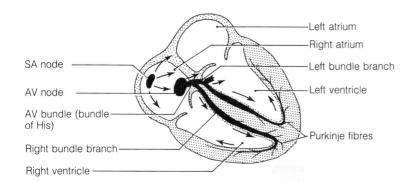

SA node

AV node

AV bundle (bundle of His)

Right bundle branch

Right ventricle

Left atrium

Right atrium

Left bundle branch

Left ventricle

Purkinje fibres

Figure 4.2.17 The conducting system of the heart. The direction in which the impulse travels is indicated by arrows. The impulse arises in the sinoatrial (SA) node and travels to the atrioventricular (AV) node, the atrioventricular (AV) bundle, the left and right bundle branches, and the Purkinje fibres

(where effusion of fluid into the pericardium causes severe compression of the heart).

The conducting system

The heart muscle must contract in an orderly and coordinated fashion to ensure efficient pump action. This is achieved by the conducting system of the heart which is a network of specialized muscle fibres concerned with the initiation and propagation of the wave of excitation that results in cardiac muscle contraction.

The conducting system consists of the sinoatrial (SA) node, the atrioventricular (AV) node, the atrioventricular (AV) bundle (also called the bundle of His), the left and right bundle branches, and the Purkinje fibres (Fig. 4.2.17).

The impulse is initiated in the SA node which lies in the right atrium near the superior vena cava. It spreads through the muscle mass of both atria causing them to depolarize and contract. It then enters the AV node which lies at the base of the right atrium. The AV node and AV bundle provide the only conduction link between the atria and the ventricles as all other areas are separated by non-conducting connective tissue. This means that malfunction of the AV node or bundle may cause complete dissociation of atrial and ventricular contractions. Conduction of the impulse through the AV node is delayed for approximately 0.1 seconds since these nodal fibres have a low conduction velocity. This allows time for the atria to finish contracting and empty their contents into the ventricles before ventricular contraction begins. Once through the AV node the impulse travels rapidly through the remainder of the conducting system. It enters the AV bundle and then travels along the left and right bundle

branches that lie on either side of the interventricular septum. These branches spread downward towards the apex of the respective ventricles. They then divide into small branches, the Purkinje fibres, which carry the impulse to the many unspecialized ventricular muscle fibres. The impulse then spreads to the remaining muscle fibres via the intercalated discs, resulting in contraction of the ventricular muscle.

The rapid conduction of the impulse once it leaves the AV node means that depolarization of the whole right and left ventricular muscle mass occurs more or less simultaneously, thus ensuring a coordinated contraction. (For a full discussion of the general concept of depolarization, see Chap. 2.1.)

The conduction of these electrical impulses in the heart that cause the cardiac muscle to contract produces weak electrical currents that spread throughout the body. By placing electrodes in various positions on the body surface, this electrical activity can be recorded; the recording is called an **electrocardiogram (ECG)**. (Figure 4.2.22 shows a typical ECG.) The P wave represents the depolarization of the atria that results in atrial contraction. The QRS waves together represent the depolarization of the ventricles which results in ventricular contraction. The ECG will be examined in more detail later in this section.

AUTORHYTHMICITY

The SA node has the ability to generate an impulse that results in the contraction of the cardiac muscle, without any external stimulus, an ability called **autorhythmicity**, which is due to spontaneous changes in the permeability of the cell membrane to potassium and sodium. There is a gradual decrease in the permeability of the membrane to

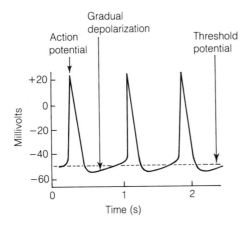

Figure 4.2.18 Spontaneous rhythmical discharge of the sinoatrial (SA) node

potassium ions and a natural leakiness of the membrane to sodium ions. This causes a gradual depolarization of the membrane towards threshold, at which point an action potential occurs (Fig. 4.2.18). This action potential is transmitted through the conducting system and results in contraction of the cardiac muscle, i.e. the electrical changes precede the muscular contraction. Following the action potential, the membrane potential in cells of the SA node returns to the initial resting value and gradual depolarization begins again. Thus rhythmical, repetitive, self-excitation of the cells occurs which causes rhythmical and repetitive cardiac muscle contraction. The SA node discharges at a rhythmical rate of about 100 per min. Other areas of the heart also exhibit autorhythmicity, particularly the AV node and the remainder of the conducting system. The AV node discharges at a rhythmical rate of 40–60 per min and the remainder of the conducting system discharges at a rate somewhere between 15 and 40 per min. Normally these tissues do not have the opportunity to initiate an action potential because they are depolarized by the impulse from the SA node before they reach their own threshold of self-excitation. Therefore the SA node controls the pace of the heart because its rate of rhythmical discharge is greater than that of any other part of the heart. Thus the SA node is known as the normal **pacemaker** of the heart. If for some reason the SA node were inactive, then the tissue with the next fastest autorhythmical rate would take over the pacing of the heart.

ABNORMALITIES IN CONDUCTION

Abnormalities in conduction which result in variations from the normal rhythm, i.e. **arrhythmias** (sometimes called dysrhythmias), can affect the pumping ability of the heart. (These are summarized in Table 4.2.1.)

Heart block
Occasionally the transmission of the impulse through the heart is blocked at a critical point in the conduction system. Such a phenomonen is called heart block. The most common heart block is atrioventricular block in which a disturbance in the conducting tissue of the AV node, the bundle of His (AV bundle) or the bundle branches, delays or completely prevents transmission of the impulse from the atria into the ventricles. This can occur as a result of damage or destruction of the conducting tissue by ischaemia or as a result of depression of conduction caused by various drugs, e.g. digitalis toxicity.

If transmission of the impulse is completely blocked, as in third degree heart block (Table 4.2.1), atrial and ventricular contractions become totally dissociated. The atria continue to contract at their normal rate, stimulated by the SA node, but the impulse now cannot be transmitted to the ventricles. Therefore a site in the conducting tissue in the ventricles with the fastest rate of autorhythmicity begins to act as the ventricular pacemaker. Usually the ventricular rate is less than 40 beats/min as this is the highest autorhythmical rate of the conducting tissue within the ventricles. This is usually insufficient to maintain an adequate blood flow from the heart and results in poor cerebral, coronary and renal perfusion.

Occasionally complete heart block can occur intermittently. There is normal AV conduction followed by no conduction at all through the AV tissue. It can take 5–10 seconds for the ventricles to start contracting at their own inherent rate of 15–40 beats per min. During this time the brain is without blood supply and the person faints. This is known as **Stokes–Adams syndrome**.

Complete heart block can be rectified by the insertion of an **artificial pacemaker** which is an electronic device that delivers an electrical stimulus to the heart through a pacing wire electrode normally inserted into the right ventricle (Fig. 4.2.19). All pacemakers used today are 'demand' pace-

Table 4.2.1 Cardiac arrhythmias

Normal sinus rhythm
(included for reference)

Description
Each P wave is followed by a QRS complex
Rate 60–100 beats/min – in this example the rate is 63 beats/min

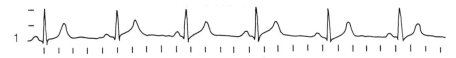

Sinus arrhythmia

Description
Normal variation of sinus rhythm
Heart rate increases with inspiration and decreases with expiration

Treatment
None

Sinus bradycardia

Description
Sinus rhythm (i.e. each P wave is followed by a QRS complex); however, the heart rate at rest is less than 60 beats/min – in this example the rate is 50 beats/min

Causes
May be normal in athletes
Hypothermia
Increased vagal tone due to bowel straining, vomiting, intubation, mechanical ventilation, pain
β-blocking drugs

Treatment
Atropine, isoprenaline
If unresponsive to these, may require temporary ventricular pacemaker

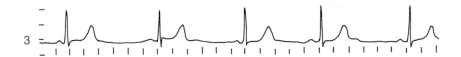

Sinus tachycardia

Description
Sinus rhythm (i.e. each P wave is followed by a QRS complex); however, the heart rate at rest is greater than 100 beats/min – in this example the rate is 125 beats/min

Causes
Physiological response to fever, exercise, anxiety, pain
May accompany shock, left ventricular failure, cardiac tamponade, hyperthyroidism, pulmonary embolis
Sympathetic stimulating drugs

Treatment
Correction of the underlying cause

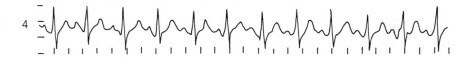

Table 4.2.1 Continued

Atrial flutter

Description
Characteristic atrial 'saw tooth' waves that occur at a regular interval and at a rate of 250–400 min
Only every 2nd, 3rd or 4th atrial impulse reaches the ventricles; in this example every 2nd atrial impulse is followed by a QRS complex (2:1 block)

Causes
Heart failure, pulmonary embolis, valvular heart disease, digitalis toxicity

Treatment
Propranolol; quinidine or digitalis (unless flutter is due to digitalis toxicity)
Direct current shock (i.e. cardioversion)

Atrial fibrillation

Description
Atrial activity is seen as rapid, small, irregular waves
Ventricular rate is completely irregular

Causes
Congestive cardiac failure, chronic obstructive lung disease, pulmonary embolis, hyperthyroidism, mitral stenosis, post-coronary artery bypass or valve replacement surgery

Treatment
Digitalis to slow ventricular rate
May require elective direct current cardioversion

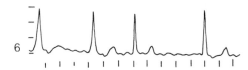

Premature ventricular contraction (ventricular ectopic beat)

Description
Beat occurs prematurely
The QRS complex is wide and distorted

Causes
Myocardial infarction, heart failure, stimulants (e.g. nicotine, caffeine), drug toxicity (e.g. digitalis, aminophylline), electrolyte imbalance (especially hypokalaemia), anxiety, catheterization of ventricle (e.g. as in a pacemaker)

Treatment
None if benign
Lignocaine, procainamide
Stop digitalis if induced by this drug
Potassium chloride if cause is hypokalaemia

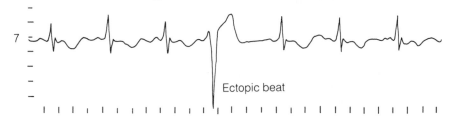

Ectopic beat

Table 4.2.1 Continued

Ventricular tachycardia

Description
QRS complexes are wide and bizarre
No visible P waves
Ventricular rate 140–220 beats/min – in this example ventricular rate is 166 beats/min

Causes
Myocardial infarction, drug toxicity (digitalis or quinidine), hypokalaemia, hypercalcaemia

Treatment
Lignocaine
If no effect, direct current cardioversion

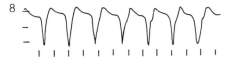

Ventricular fibrillation

Description
Ventricular rhythm is rapid and chaotic
QRS complexes are wide and irregular
No visible P waves

Causes
Myocardial ischaemia or infarction, hypo- or hyperkalaemia, congestive cardiac failure, untreated ventricular tachycardia, drug toxicity (e.g. digitalis), electrocution, hypothermia

Treatment
Cardiopulmonary resuscitation
Direct current shock 200–400 joules
Sodium bicarbonate
Antiarrhythmic drugs after resuscitation

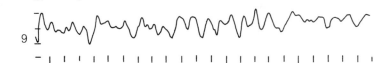

Asystole

Description
Ventricular standstill
No QRS complexes

Causes
Myocardial ischaemia or infarction, acute respiratory failure, aortic valve disease, hyperkalaemia

Treatment
Cardiopulmonary resuscitation
Calcium chloride
Isoprenaline
Adrenaline

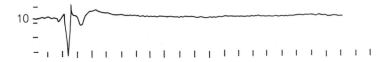

Table 4.2.1   Continued

Heart block
(i)   First degree AV block

Description
P-R interval is prolonged
QRS complex normal

Causes
Inferior myocardial ischaemia or infarction (i.e. right coronary artery disease), potassium imbalance, digitalis toxicity, hypothyroidism

Treatment
Correct underlying causes
Discontinue digitalis

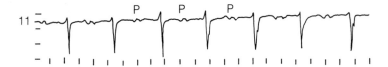

(ii)   Second degree AV block

Description
Atrial rate regular but only every 2nd, 3rd or 4th atrial impulse results in a ventricular contraction – in this example every 2nd atrial impulse is followed by a QRS complex (2:1 block)

Causes
Inferior myocardial infarction, digitalis toxicity, vagal stimulation

Treatment
Atropine
Discontinue digitalis
May require temporary pacemaker

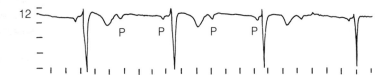

(iii)   Third degree AV block (complete heart block)

Description
Atrial rate is regular
The ventricular rate is slow and regular; however, there is no relationship between the atrial (P) waves and the ventricular (QRS) complexes

Causes
Myocardial ischaemia or infarction, digitalis toxicity, complication of mitral valve replacement

Treatment
Pacemaker

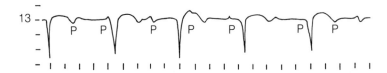

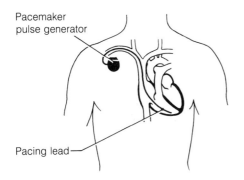

Figure 4.2.19 Artificial pacemaker implanted under the skin in the upper chest with leads to the heart

makers, which means that the electronic circuitry within them is able to detect the heart's normal electrical activity and will only stimulate the heart when the natural rate falls below the preset rate of the pacemaker. Pacemakers inserted for heart block normally only activate the ventricles because the SA node fires to activate the atria. Sometimes a pacemaker lead is placed in the right atria as well as in the right ventricle to provide more physiological pacing. These pacemakers are called dual chamber pacemakers.

Insertion of a pacemaker is a simple operation, often performed under local anaesthetic. The pacemaker is usually implanted under the skin in the upper chest and the leads to the heart are inserted via the subclavian vein and passed into the muscle of the right ventricle. There is little specific postoperative care necessary and patients usually return home after 2–3 days. Patients are advised not to move the arm nearest the pacemaker too vigorously for about 1 month after insertion in case the wire into the ventricle is dislodged. After this there are no restrictions on activity.

Theoretically, electromagnetic radiation may affect the performance of a pacemaker, but the metal containers in which the pacemakers are housed provide a high degree of protection against this hazard. On the other hand, a pacemaker can trigger off the metal detectors at airports, shops and libraries.

The batteries in the pacemakers can last up to ten years, after which replacement is necessary.

Premature beats
Sometimes a small area of cardiac muscle becomes much more excitable than normal, possibly due to a local area of ischaemia, overuse of stimulants

such as caffeine or nicotine, electrolyte imbalance or drug toxicity (e.g. digitalis). This area may initiate an impulse prior to that from the SA node. A wave of depolarization spreads outward from the irritable area and initiates a contraction of the heart before one would be expected from the normal rhythm of the SA node. This is called **premature contraction** or **beat**. The focus at which the abnormal impulse is generated is called an **ectopic focus** and may be in the atria or ventricles, and the premature contraction that it initiates is often called an **ectopic beat** (Greek: *ektopos* = displaced). If an ectopic focus becomes so irritable that it establishes rhythmical impulses of its own at a more rapid rate than that of the SA node, this ectopic focus becomes the pacemaker of the heart.

The output of blood from a premature contraction is decreased because the ventricles have not had normal filling time. Therefore the pulse wave passing to the periphery may be so weak that it cannot be felt at the radial artery. Thus, when the apex beat is counted simultaneously with the radial pulse rate, there is a deficit in the number of pulses felt in relation to the number of beats counted. This is known as **pulse deficit**.

Ventricular premature beats may be treated by giving lignocaine or procainamide, both of which reduce the autorhythmicity of the cells and their responsiveness to excitation. If the ectopic beats are caused by digitalis toxicity, then this drug is stopped. Premature beats induced by hypokalaemia are treated by giving intravenous potassium chloride.

Fibrillation
In normal conduction, once the impulses leave the specialized conducting tissue they travel around the heart stimulating the cardiac muscle to contract. The impulses eventually die away because when they return to the originally stimulated muscle, this muscle is still in a refractory state.

However, if the length of the pathway along which the impulse travels is lengthened (as might occur in a dilated heart due to congestive cardiac failure), by the time the impulse returns to the starting position, the originally stimulated muscle might be out of the refractory state and able to be restimulated. A similar situation occurs if conduction is prolonged, as might happen if there is a blockage in the conduction system, ischaemia of the muscle, or hypo- or hyperkalaemia. Restimulation can also occur if the refractory period is shortened, as in response to various drugs, e.g. adrenaline, or in electrocution. Therefore if an impulse constantly meets an area which is no longer refractory, it keeps travelling around and around the heart in a so-called circus movement. This leads to continuous and completely disorganized contractions called **fibrillation**.

If the ventricles fibrillate, different parts of the ventricle no longer contract simultaneously and therefore no blood is pumped. The person dies within 2–3 min if no action is taken. An ECG tracing of ventricular fibrillation is shown in Table 4.2.1. The electrical activity in the ventricles is completely disorganized so there are no distinctive wave patterns.

In an attempt to stop the ventricles fibrillating, a very strong direct current can be passed through them which throws all the ventricular muscle into a refractory period simultaneously. This is called **electrical defibrillation** or **direct current cardioversion**. It is done by placing two large electrodes ('paddle') on the chest, one just to the right of the sternum below the clavicle and the other just to the left of the apex of the heart or the left nipple (Fig. 4.2.20). The direct current (commonly 200–400 joules) passes through the heart causing all impulses to stop for 3–5 seconds, after which it is hoped that the SA node fires and resumes the pacing of the heart.

When a patient is being defibrillated it is important that nobody is touching either the patient or his bed otherwise they too will receive the shock.

CARDIAC ARREST

A fibrillating ventricle is totally inefficient as a pump and therefore fails to maintain an adequate cardiac output. This failure of the heart to maintain an adequate circulation, particularly to the brain, is called **cardiac arrest**, a state which may also result from ventricular tachycardia and asystole (see Table 4.2.1). At normal body temperatures, irreversible changes occur in the brain if circulation fails for more than 2–3 min.

The two primary signs of cardiac arrest are

unconsciousness
absent pulses.

Other signs are

convulsion, often at the onset of cardiac arrest
absence of respiration or gasping respiration –

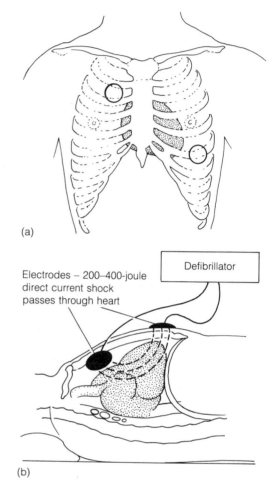

(a)

Electrodes – 200–400-joule direct current shock passes through heart

Defibrillator

(b)

Figure 4.2.20 Electrical defibrillation: (a) the circles show the position of the large electrodes ('paddles') across the long axis of the heart, (b) when the defibrillator discharges, a direct current shock passes through the heart

this occurs within 20–30 seconds from the onset of cardiac arrest
dilation of the pupils – this occurs within 30–60 seconds
no audible heart sounds
cyanosed pallor.

There is no time to wait for an absolute diagnosis. If someone is unconscious and no carotid or femoral pulse is palpable, he or she should be considered to have had a cardiac arrest. In order to save the person's life, cardiopulmonary resuscitation should begin immediately. If circulation is restored within 2–3 min, complete recovery is possible.

Cardiopulmonary resuscitation consists of ventilating the patient to ensure that the blood is oxygenated, and external cardiac massage to circulate this oxygenated blood. The ABC of cardiopulmonary resuscitation is as follows.

A – The Airway must be opened by tilting the head backwards and pulling the lower jaw forwards in order to stop the tongue falling back. Any debris in the mouth should be cleared out quickly and any false teeth removed.

B – Breathing. If the person is breathing spontaneously, artificial ventilation may not be required. However, in most instances breathing ceases within 20–30 seconds of the onset of cardiac arrest. The person should be artifically ventilated by

(a) mouth-to-mouth or mouth-to-nose ventilation
(b) mask and bag with oxygen-enriched air, or
(c) endotracheal intubation and ventilation with oxygen-enriched air.

The person should be ventilated approximately 15 times a minute. The efficacy of ventilation is checked by observing if there is adequate chest expansion.

C – Circulate. Since the heart is no longer efficiently pumping blood, its action is taken over by external cardiac massage in which the heart is rhythmically compressed between the sternum and the spine to cause blood to be ejected. To perform external cardiac massage, the heel of one hand is placed on the lower half of the sternum and the other hand over the first (avoiding the xiphisternum). The sternum is depressed approximately 4–5 cm by using the operator's body weight through straight arms (Fig. 4.2.21). This compresses the heart between the sternum and the spine and causes blood to be expelled. The sternum should be depressed every second. For effective compression the patient should be lying on a firm surface. (A board is often put under the patient if he or she is in bed.) The efficacy of the compression should be checked by an assistant feeling the femoral pulses.

If the person has been seen to collapse and is reached within 1 minute, a **precordial thump** may be given, i.e. the lower part of the sternum is given a firm thump, once, with the ulnar border of a clenched fist. This manoeuvre is only of value when blood is still circulating in the heart and may be enough to 'shock' the heart into beating again.

For cardiopulmonary resuscitation, if there is

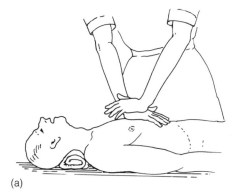

(a)

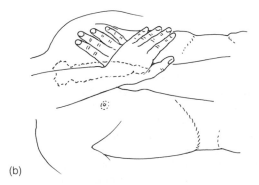

(b)

Figure 4.2.21 External cardiac massage: (a) position of the operator, (b) position of the hands on the lower half of the sternum

only one operator, then 2 breaths should be given to every 15 compressions of the heart; if there are two operators, 1 breath should be given to every 4–5 compressions.

Once cardiopulmonary resuscitation is established, further action can be taken to get the heart pumping effectively by itself again. If facilities are available

(a) An ECG should be recorded, remembering that artificial ventilation and cardiac massage must be continued throughout. If ventricular fibrillation is present, electrical defibrillation should be performed. If there is no defibrillator available, intravenous lignocaine as a bolus might be given. If asystole is present, it is vital to get some electrical activity in the heart. Calcium chloride may be given followed by electrical defibrillation. If asystole persists, adrenaline might be injected directly into the heart.

(b) A sodium bicarbonate infusion should be put up as soon as possible. This is to correct acidosis, since during cardiac arrest poor perfusion of tissue induces metabolic acidosis. The acidosis depresses myocardial contractility; induces intracellular loss of potassium which depresses cardiac response to catecholamines; reduces the threshold for ventricular fibrillation; and induces asystole. If the acidosis is not corrected, resuscitation will not be successful.

The electrocardiogram

The electrical impulses that precede cardiac muscle contraction arise in the conducting system of the heart. These impulses excite the muscle fibres of the heart and cause them to contract, and it is this coordinated contraction of cardiac muscle that pumps the blood into the pulmonary and systemic circulations.

The impulse formation and conduction produce weak electrical currents that spread through the entire body. By applying electrodes to various positions on the body and connecting these electrodes to an electrocardiograph, an electrocardiogram is recorded. Therefore the electrocardiogram (ECG) is a graphic recording of the electrical processes that initiate the contraction of the cardiac muscle. A typical ECG wave pattern is seen in Fig. 4.2.22.

The value of such recordings is that abnormalities in the conduction system or abnormalities in the cardiac muscle itself will result in changes on the ECG and so can be diagnosed. For diagnostic purposes a 12-lead ECG is used as this looks at the heart from 12 different angles.

Sometimes it is necessary to monitor the cardiac rhythm of patients with cardiac abnormalities continuously so that any untoward changes can be immediately observed. For this the ECG signal from just one lead is used.

LEADS OF THE ECG

The 12-lead ECG includes 6 recordings from the extremities and 6 recordings from the chest. Each of the 12 leads looks at the same phenomenon (i.e. the conduction of the electrical impulse through the heart) from a different point. Since the electrical currents picked up at each point of recording are slightly different, the waveform patterns will vary between leads. However, for each lead there is a normal waveform pattern.

The six extremity leads are as follows.

Three **bipolar leads**, I, II and III (known as **standard leads**), that record the difference in electrical potential between two points electrically equidistant from the heart.

Lead I records the difference in potential between the left arm and right arm.
Lead II records the difference in potential between the left leg and right arm.
Lead III records the difference in potential between the left leg and the left arm.

These three leads form an Einthoven triangle whose apices are the two arms and the left leg. The heart is considered to lie in the centre of this triangle (Fig. 4.2.23).

Three **unipolar leads** aVR, aVL, aVF (a = augmented; since deflections are small the recordings are augmented 1.5 times). These three leads record the difference in electrical potential between a zero reference electrode and the limb potential.

aVR records the difference between the zero potential and the right arm.
aVL records the difference between the zero potential and the left arm.
aVF records the difference between the zero potential and the left leg (Fig. 4.2.23).

The above six leads can be thought of as looking at the heart in a vertical plane. Leads I, II and aVL look at the left lateral surface of the heart;

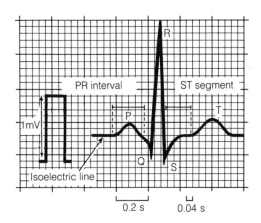

Figure 4.2.22 A normal ECG complex

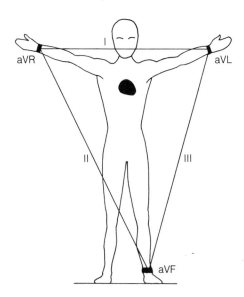

Figure 4.2.23 Position of the electrodes to record the six extremity (limb) leads of the ECG

leads III and aVF at the inferior surfaces; and lead aVR looks at the atria.

The six **chest leads** are unipolar leads and are recorded from the following positions (Fig. 4.2.24).

V_1 4th right intercostal space at the sternal border.
V_2 4th left intercostal space at the sternal border.
V_3 Midway between V_2 and V_4.
V_4 5th left intercostal space in the midclavicular line.

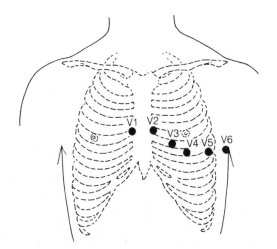

Figure 4.2.24 Position of the chest electrode to record the six chest leads of the ECG

V_5 Left anterior axillary line at the same horizontal level as V_4.
V_6 Left midaxillary line at the same horizontal level as V_4.

These leads look directly at the heart in a horizontal plane. Leads V_1 and V_2 look mostly at the right ventricle; leads V_3 and V_4 look at the septum between the ventricles and the anterior wall of the left ventricle; leads V_5 and V_6 look at the anterior and lateral walls of the left ventricle. However, a unipolar chest lead does not only record the electrical potential from a small area of the underlying myocardium, it records all the electrical events that occur with each heart beat as viewed from the selected lead site.

For diagnostic purposes it is necessary to look at the heart from these 12 different positions since abnormalities in certain parts of the heart will be more clearly visible (i.e. will cause greater changes in the waveform pattern) in the leads that look most directly at this area of the heart.

Figure 4.2.25 shows the normal waveform pattern for each lead of the 12-lead ECG.

COMPONENTS OF THE ECG

Figure 4.2.22 shows a typical ECG complex.

The first wave seen is the **P wave**, which represents atrial depolarization that precedes atrial contraction. It is normally in a positive (upwards) direction. After a slight pause, the next three waves, Q, R, S, appear very close to each other. These waves together are called the **QRS complex** and represent depolarization of the ventricles that precedes ventricular contraction. Within the QRS complex, the **Q wave** is the first downward deflection, the **R wave** is the first upward deflection, and the **S wave** is the first downward deflection following the R wave. There is a pause after the QRS complex and the next positive wave is the **T wave**, which represents ventricular repolarization. Occasionally there is a small positive wave after the T wave known as the U wave.

The distance from the beginning of the P wave to the beginning of the QRS complex is known as the **P–R interval**. This represents the time taken for the impulse to travel from the SA node through the atria to the ventricular fibres. It includes the normal delay of excitation in the AV node. The **QRS interval** is measured from the onset of the Q wave to the termination of the S

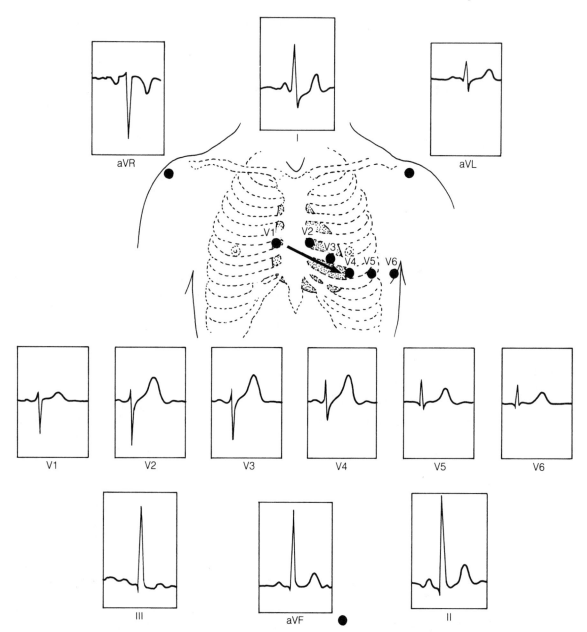

Figure 4.2.25 Normal complexes for each lead of the 12-lead ECG (the arrow shows the major direction of the wave of depolarization)

wave and represents the total time for ventricular depolarization.

The **ST segment** is the period between the S and T waves, and is part of ventricular repolarization.

The connections of the electrodes to the electro-graphic apparatus are such that if the wave of depolarization is moving towards the active electrode, the resultant deflection on the ECG will be in an upward direction; if the wave of depolarization is moving away from the electrode, the resultant deflection will be in a downward

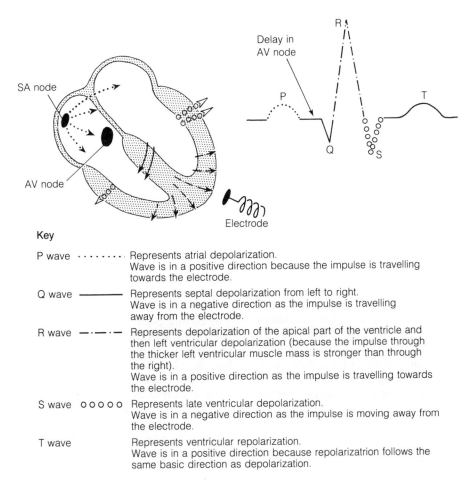

Key

P wave ········ Represents atrial depolarization.
Wave is in a positive direction because the impulse is travelling towards the electrode.

Q wave ———— Represents septal depolarization from left to right.
Wave is in a negative direction as the impulse is travelling away from the electrode.

R wave —·—·— Represents depolarization of the apical part of the ventricle and then left ventricular depolarization (because the impulse through the thicker left ventricular muscle mass is stronger than through the right).
Wave is in a positive direction as the impulse is travelling towards the electrode.

S wave ooooo Represents late ventricular depolarization.
Wave is in a negative direction as the impulse is moving away from the electrode.

T wave Represents ventricular repolarization.
Wave is in a positive direction because repolarizatrion follows the same basic direction as depolarization.

Figure 4.2.26 Genesis of a lead II ECG; each symbol of the wave of depolarization represents a different waveform on the ECG

direction. Figure 4.2.25 shows the major direction of a normal wave of depolarization. If leads aVR and V_5 are taken as an example, in lead aVR the wave of depolarization is mostly travelling away from the electrodes so that most of the deflections are downward; in lead V_5 the wave of depolarization is mostly travelling towards the electrode so that most of the deflections are upward. To clarify this further, the genesis of a lead II ECG is followed in more detail in Fig. 4.2.26.

The ECG is recorded on graph paper in which horizontal and vertical lines are present at 1 mm intervals. A heavier line occurs every 5 mm. The vertical lines show the time interval (1 mm = 0.04 s, 5 mm = 0.2 s). The horizontal lines show the amplitude (10 mm = 1 mV) (see Fig. 4.2.22). In routine electrocardiograph practice, the recording speed is 25 mm/s.

CALCULATION OF HEART RATE

The **heart rate** is the number of times the heart beats each minute and is represented on the ECG by the number of atrial depolarizations (P waves) or ventricular depolarizations (QRS complexes) that occur each minute. Heart rate can also be measured indirectly by counting the pulse rate. The normal heart rate is between 60 and 100 beats/min. It is fastest at birth and decreases with age. It is also slightly faster in females than in males.

RHYTHM (see Table 4.2.1)

The normal rhythm of the heart is called **sinus rhythm**. This denotes that the impulse arises in the SA node and follows the normal conduction pathway. Thus it implies that a P wave precedes every QRS complex. Variations of sinus rhythm can occur in a normal heart. For example, there may be changes in rate which are related to respiration, the rate increasing with inspiration and decreasing with expiration (see Chap. 5.3). This is known as **sinus arrhythmia** and is more common in children than in adults. These changes in rate can be felt while taking the pulse. Other variations of sinus rhythm are **sinus bradycardia** (Greek: *brady* = slow) in which the heart rate at rest is below 60 beats/min, as may occur in well trained athletes or when the body temperature is lowered; and **sinus tachycardia** (Greek: *tachos* = speed) in which the heart rate at rest is above 100 beats/min, as may occur when there is an increased body temperature, e.g. in fever.

POTASSIUM IMBALANCE

The ECG changes that occur with potassium imbalance are important since the ECG can give a guide to serum potassium levels when direct determination of serum potassium is not possible. The importance of detecting changes in potassium levels is that, in the heart, both hypokalaemia

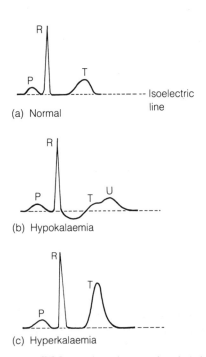

(a) Normal
(b) Hypokalaemia
(c) Hyperkalaemia

Figure 4.2.27 ECG complexes in potassium imbalance

(potassium deficiency) and hyperkalaemia (potassium excess) can diminish excitability and reduce the conduction rate of cardiac muscle. This may lead to cardiac arrest. Therefore the ECG changes indicative of a raised or lowered serum potassium should be reported immediately. The effects of hypokalaemia and hyperkalaemia on the ECG are seen in Fig. 4.2.27. In hypokalaemia there is ST segment depression (i.e. the ST segment is below the isoelectric line) and there is a prominent U wave immediately following the T wave. In hyperkalaemia there is peaking of the T wave, the more severe effects including widening of the QRS complex and lengthening of the P–R interval.

The cardiac cycle

The cardiac cycle is the period from the end of one heart contraction to the end of the next. An understanding of how all the events in the cycle fit together is essential to understanding how the heart acts as an efficient pump.

In a normal cycle, the heart muscle contracts and then relaxes. **Systole** refers to the phase of contraction and **diastole** to the phase of relaxation. Since pumping of the blood is mainly achieved by the contraction and relaxation of the ventricles, systole usually refers to the phase of ventricular contraction and diastole to the phase of ventricular relaxation.

First the events that occur in late diastole will be described. Throughout the description reference should be made to Fig. 4.2.28. It is important to note from this figure that the sequences of events in the right and left sides of the heart are the same. The difference between the figures of the right and left side is simply due to the fact that the pressures generated by the right ventricle during systole are considerably lower than those generated by the left ventricle. This is because the total resistance to flow in the pulmonary circulation is less than in the systemic circulation. The difference in the amount of pressure generated by the right and left ventricle is clearly reflected in the size of the muscular ventricular wall, the right ventricular wall being much thinner than the left. Despite the lower pressure, the right ventricle ejects the same amount of blood as the left with each contraction.

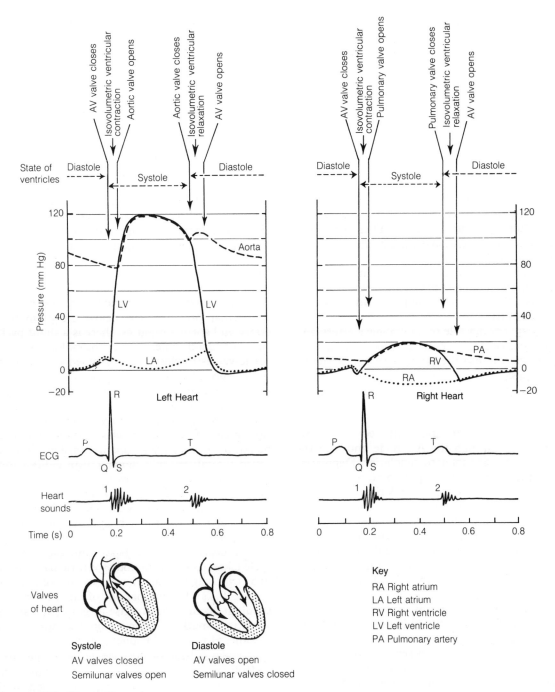

Figure 4.2.28 The cardiac cycle

DIASTOLE (LATE)

In diastole, the atria and ventricles are relaxed. Blood enters the right atrium from the superior and inferior venae cavae, and enters the left atrium from the pulmonary veins. This blood entering the atria causes the pressure within them to be slightly higher than that in the ventricles. This forces the AV valves to be open and blood passes from the atria into the ventricles throughout diastole. During all this time the pulmonary and aortic valves are closed because the pressures in the pulmonary artery and the aorta are higher than those in the ventricles. This ensures that the blood collects in the ventricles.

Towards the end of diastole the SA node discharges and the atria depolarize, causing a P wave on the ECG. This depolarization of the atria results in atrial contraction which pumps blood from the atria into the ventricles. It is important to note that approximately 80% of ventricular filling occurs before atrial contraction. Atrial contraction merely 'tops up' the ventricles. Therefore atrial dysfunction does not have a great effect on ventricular filling.

The amount of blood in each ventricle at the end of diastole, i.e. just before the ventricles contract, is called the **ventricular end–diastolic volume (VEDV)**.

Throughout diastole the pulmonary artery and aortic pressures fall because blood is moving out of arteries into the pulmonary and systemic vasculature.

SYSTOLE

The wave of depolarization initiated by the SA node passes through the AV node and into the ventricles, causing them to depolarize. Depolarization of the ventricles is represented on the ECG by the QRS complex. This depolarization triggers contraction of the ventricles, and the pressures within them rises. Almost immediately the pressures in the ventricles exceed those in the atria, causing the AV valves to close and thus preventing backflow of blood into the atria. The turbulence created by the closure of the AV valves generates the first heart sound.

At the very beginning of ventricular contraction, the pressures in the pulmonary artery and aorta exceed those in the ventricles, therefore the pulmonary and aortic valves remain closed and no blood leaves the ventricles. Each ventricle is contracting but the blood volume within them remains constant. This is known as **isovolumetric ventricular contraction**. This phase of contraction ends when the pressures in the ventricles exceed those in the pulmonary artery and aorta, the semilunar valves open and **ventricular ejection** occurs. The ejection of blood is at first rapid and then tapers off. The amount of blood ejected by each ventricle with each contraction is called the **stroke volume**.

The pulmonary artery and aortic pressures rise as the blood flows in from the ventricles. Peak pressures in the pulmonary artery and aorta occur before the end of ventricular ejection. This is because the rate of ejection of blood from the ventricles during the last part of systole is quite low and is less than the rate at which the blood is leaving the arteries via the arterioles. Therefore the volume of blood, and hence the pressure within the pulmonary artery and aorta, begin to decrease.

Throughout the entire period of ventricular ejection, the atrial pressures rise slowly because of the continued flow of blood from the veins.

DIASTOLE (EARLY)

Almost as soon as the ventricular muscle relaxes, ventricular pressure falls below the pulmonary artery and aortic pressures, there is a slight back-surge of blood which causes the pulmonary and aortic valves to close. The turbulence created by the closure of these valves generates the second heart sound. For a short time the AV valves are also closed because the pressure in the ventricles still exceeds that in the atria. Therefore in the early phase of ventricular relaxation the volume of blood in the ventricles remains constant. This is called **isovolumetric ventricular relaxation**. It ends as ventricular pressure falls below atrial pressure, causing the AV valves to open and ventricular filling to begin. Ventricular filling occurs rapidly at first so that it is almost completed in early diastole.

At a heart rate of 72 beats/min, each cardiac cycle takes 0.8 s. The period of ventricular diastole (0.5 s) is longer than the period of ventricular systole (0.3 s). During periods of rapid heart rate, as might occur with exercise or emotional stress, there is a marked reduction in the duration of diastole. However, this does not seriously impair ventricular filling as most filling occurs in early diastole.

Cardiac output

Cardiac output is the rate at which the heart pumps blood. Therefore it is the amount of blood available for the transport of nutrients to the tissues. For this reason, cardiac output is central to the function of the circulatory system.

Cardiac output is often defined as the volume of blood pumped out of the left ventricle into the aorta each minute. It is usually expressed in litres per minute (l/min) and can be calculated by multiplying **stroke volume** (which is the volume of blood pumped out of each ventricle per contraction) by the number of contractions per minute (i.e. the heart rate). Therefore

cardiac output = stroke volume × heart rate

For example, if an adult at rest has a heart rate of 72 beats/min and each ventricle ejects about 70 ml of blood with each beat, then

cardiac output = 72 beats/min × 0.07 litres/beat

= 5.0 litres/min

At rest, a normal cardiac output is about 5.0 l/min. This can increase to as much as 25 l/min in a normal person performing strenuous exercise and to as much as 35 l/min in a well-trained athlete. The maximum percentage that the cardiac output can increase above the normal resting cardiac output is called the **cardiac reserve**.

An increased cardiac output, as might be required to supply various tissues and organs with a greater blood flow, e.g. muscles during exercise or the gut during digestion, is achieved by either an increase in heart rate or an increase in stroke volume, or both.

The mechanisms by which heart rate and stroke volume can be altered are as follows.

CONTROL OF HEART RATE

As was described earlier when discussing the conducting system of the heart, the SA node, in the absence of any nervous or hormonal influences, discharges spontaneously and rhythmically at about 100 beats/min. If uninfluenced by other factors, this rate set by the SA node would never vary. However, the rate of the heart does vary to meet the changing requirements of the tissues for oxygen. There are various factors that alter heart rate.

Nervous control
By far the most important factor that alters heart rate is the influence of the autonomic nervous system. The heart is supplied by both sympathetic and parasympathetic nerves (Fig. 4.2.29). The sympathetic nerves to the heart originate in the cardiovascular centre which is a group of neurones within the medulla, innervating the SA node, the AV node and portions of the myocardium. Stimulation of these sympathetic nerves causes an increase in heart rate. The parasympathetic nerve to the heart (vagus nerve) originates in the nucleus ambiguus within the medulla. These fibres innervate the SA node and the AV node. Parasympathetic stimulation causes a decrease in heart rate.

At rest a normal heart rate is about 60–80 beats/ min. This is less than the inherent rate of discharge of the cardiac pacemaker, the SA node. Therefore, in the resting state, parasympathetic influence is dominant (sometimes known as the **vagal brake**).

Primarily, heart rate is regulated by a balance between the slowing effects of parasympathetic discharge and the accelerating effects of sympathetic discharge.

Hormonal control
Adrenaline, released from the adrenal medulla, is a sympathetic mediator and therefore causes an increase in heart rate by stimulating the β_1 receptors in the cardiac muscle conduction system (see Chap. 2.4).

Stretch
Stretch of the right atrial wall by an increased venous return can increase the heart rate by as much as 10–15%. This is because there are **stretch receptors** in the wall of the right atrium (and also in the superior and inferior venae cavae) which send impulses that stimulate the sympathetic output. This is known as the **Bainbridge reflex**.

Temperature
A raised body temperature causes an increased heart rate because it increases the rate of discharge of the SA and AV nodes. Decreased body temperature, as might occur in exposure to cold or in deliberate cooling of the body during cardiac surgery, decreases heart rate.

Drugs
Administration of certain drugs can alter the heart rate. Such drugs that alter the heart rate are called **chronotropic drugs** (Greek: *chronos* = time). Those that increase the heart rate are called positive chronotropic drugs, e.g. isoprenaline, adrenaline,

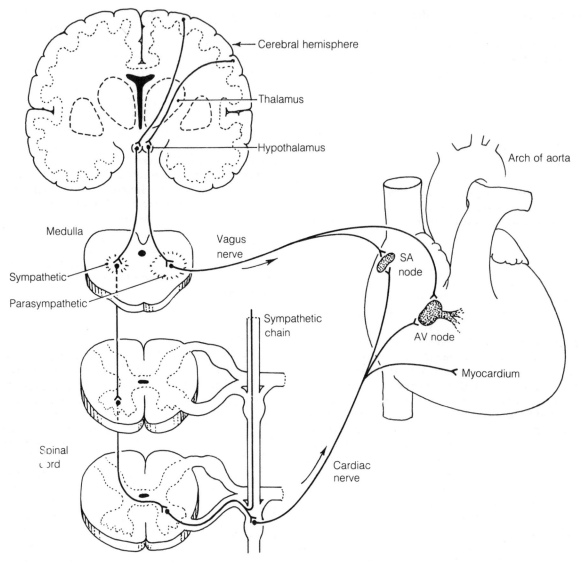

- Cerebral hemisphere
- Thalamus
- Hypothalamus
- Arch of aorta
- Medulla
- Vagus nerve
- SA node
- Sympathetic
- Parasympathetic
- Sympathetic chain
- AV node
- Myocardium
- Spinal cord
- Cardiac nerve

Figure 4.2.29 Sympathetic and parasympathetic innervation of the heart (based on Tortora and Anagnostakos, 1981)

and those that decrease it are called negative chronotropic drugs, e.g. β-adrenergic blocking drugs such as propranolol.

Heart rate is sensitive to other factors, including plasma electrolyte concentrations and hormones other than adrenaline, but these are less important. Generally, heart rate is somewhat faster in females than males and it declines slightly with age.

For a given stroke volume, all the factors that increase heart rate will increase cardiac output and all the factors that decrease heart rate will decrease cardiac output.

CONTROL OF STROKE VOLUME

Venous return and ventricular end-diastolic volume

In a normal heart, the most important factor that controls the amount of blood pumped from a ventricle with each beat (i.e. the stroke volume) is the amount of blood that is in the ventricle immediately before contraction (i.e. the ventricular end-diastolic volume). The more blood in the ventricle before contraction the greater the amount of blood pumped (see p. 324).

Since stroke volume is so dependent on ventricular end-diastolic volume, it is important to know what determines ventricular end-diastolic volume. The major determinant is **venous return**. The heart will pump as much blood as is returned to it from the veins, and this is principally due to Starling's law of the heart, which states that the force of contraction is a function of the length of the muscle fibre. The force of contraction adjusts according to the volume of blood in the chambers of the heart, and so the heart is said to **auto-regulate**.

Since venous return alters ventricular end-diastolic volume, and hence stroke volume, it will also alter cardiac output, as it will be remembered that CO = SV × HR. Hence venous return and the related ventricular end-diastolic volume are crucial determinants of cardiac output. Factors that affect venous return are discussed later in this chapter.

Nervous regulation
Sympathetic nerves not only supply the SA node and conducting system, but also innervate the atrial and ventricular myocardium (see Fig. 4.2.29). Sympathetic stimulation causes the release of the sympathetic mediator, noradrenaline, which increases ventricular (and atrial) contractility. **Myocardial contractility** is the strength of contraction at a given degree of stretch (stretch is synonymous with muscle fibre length). It has been shown that muscle fibre length is dependent on ventricular end-diastolic volume, therefore myocardial contractility is the strength of contraction at a given ventricular end-diastolic volume.

Increased contractility results in an increased stroke volume even though ventricular end-diastolic volume (VEDV) does not change. This is possible because, normally, when a ventricle contracts it does not empty completely; the amount of blood that remains after ejection is the **ventricular end-systolic volume (VESV)**. When contractility increases, the ventricle empties more completely, thereby increasing stroke volume (SV). Thus

$$SV = VEDV - VESV$$

where the end-systolic volume is dependent on myocardial contractility. The effect of sympathetic stimulation of the heart on stroke volume is seen in Fig. 4.2.30.

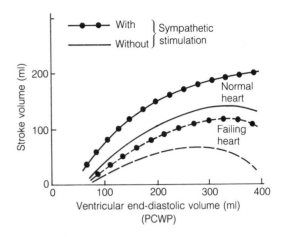

Figure 4.2.30 The effect of sympathetic stimulation on the normal and failing heart. Ventricular end-diastolic volume (VEDV) is measured by pulmonary capillary wedge pressure (PCWP). For the same VEDV, sympathetic stimulation in both the normal and the failing heart increases contractility, thus increasing stroke volume (based on Vander *et al.*, 1975)

Sympathetic stimulation not only causes a more forceful contraction but also a more rapid contraction. This is important when an increase in heart rate (also the result of increased sympathetic activity) reduces the time available for diastolic filling of the ventricles. If contraction is more rapid, a larger fraction of the cardiac cycle will be available for filling.

Certain drugs can alter myocardial contractility. Drugs that affect contractility are called **inotropic drugs** (Greek: *inos* = fibre). Those that increase contractility are called positive inotropic drugs, e.g. digitalis, adrenaline, and those that decrease contractility are known as negative inotropic drugs, e.g. β-adrenergic blocking drugs such as propranolol.

Hormonal regulation
Circulating catecholamines, adrenaline and noradrenaline, released from the adrenal medulla produce similar changes in myocardial contractility, and hence stroke volume, to those induced by the sympathetic nerves to the heart via the β_1 receptors.

Arterial blood pressure
An increase in the resistance to the ejection of blood from the ventricles can decrease stroke volume if the force of contraction stays constant.

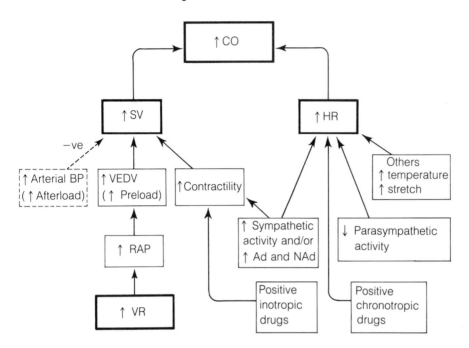

Figure 4.2.31 Main factors that can alter cardiac output. (CO = cardiac output, SV = stroke volume, HR = heart rate, VEDV = ventricular end-diastolic volume, RAP = right atrial pressure, VR = venous return, Ad = adrenaline, NAd = noradrenaline.) An increased arterial blood pressure (i.e. an increased afterload) causes a decrease in stroke volume and consequently a decrease in cardiac output

Resistance to ejection is determined by the distensibility of the large arteries and the total peripheral resistance. This resistance to ejection is called the **afterload**.

The arterial blood pressure gives an indication of the degree of afterload. If the arterial blood pressure is raised (i.e. increased afterload) and the force of contraction stays the same, the stroke volume will be reduced. This is because, for the same force of contraction, the muscle fibres are less able to shorten because they are pushing against a greater load. The ventricle does not empty as much and so stroke volume is decreased.

In a normal heart, changes in blood pressure, up to a certain point, do not affect overall cardiac output. This is due to the self-adjustment of the heart. When the stroke volume falls due to an increased afterload, the ventricular end-systolic volume increases. This end-systolic volume is added to the normal diastolic filling in the next cycle, therefore increasing the volume of blood in the ventricle. This causes a greater stretch on the ventricular muscle which enhances the force of contraction, with a consequent increase in stroke volume. This process continues with each contraction until the original stroke volume is achieved. However, the end-systolic volume and the end-diastolic volume will remain elevated

Vasodilator drugs, e.g. sodium nitroprusside, phentolamine or nitroglycerine, reduce afterload.

Figure 4.2.31 summarizes the main factors that can alter cardiac output. A normal functioning heart will pump out all the blood that is returned to it from the veins. Therefore venous return is the primary factor determining cardiac output. Any factors that increase stroke volume or increase heart rate can conceivably increase the cardiac output, and vice versa.

MEASUREMENT OF CARDIAC OUTPUT

Cardiac output can be assessed indirectly by observing related variables, e.g. urine output, peripheral toe and limb temperatures, and can be classed as high, normal or low. Often this is adequate for many clinical situations.

However, in critically ill patients it is often necessary to have an accurate, direct and repeatable measurement of cardiac output to assess the effects of treatment. The most commonly used direct measure of cardiac output is **thermodilution**. In this method a triple lumen Swan–Ganz catheter is inserted into a peripheral vein and advanced into the right atrium. A bolus of cold saline of known temperature is injected through the catheter into the right atrium; it mixes with the blood flowing through the right side of the heart and the temperature change is sensed by a thermistor in the tip of a flotation catheter lying in the pulmonary artery. The temperature changes by an amount proportional to the blood flow and thus cardiac output can be calculated.

HEART FAILURE

It can be seen from the previous discussion that venous return largely determines cardiac output. However, this is only true while the heart is an effective, adjustable pump. If the ability of the heart to pump is impaired, it may not be able to provide an adequate cardiac output to meet the needs of the body, despite a satisfactory venous return. In such a situation the heart is said to have failed. This is known as **heart failure, cardiac failure** or **pump failure**.

In the discussion of heart failure we will first consider the pressure and volume changes that occur and then discuss mechanisms that compensate for a decreased cardiac output.

The left and right side of the heart can be considered as two separate pumping systems connected in series. It is possible that one side of the heart may fail independently of the other. For example, myocardial infarction may only affect the left side, or pulmonary disease may predominantly affect the right side. A consideration of the pressure and volume changes that occur when one side of the heart fails independently of the other gives a good understanding of the importance of the integrity of the pumping system.

If the left side of the heart fails, the output of this side is less than the total volume of blood received from the right side. Effects occur in both the vessels behind the pump and beyond the pump. Blood is backlogged behind the left ventricle. This increases the volume and pressure of blood in the left atrium and consequently in the pulmonary veins and the pulmonary capillary bed. The accumulation of blood in the pulmonary capillary bed causes the pulmonary capillary pressure to rise. When it rises about 28 mm Hg (3.7 kPa) – the colloid osmotic pressure of plasma – fluid filters out of the capillaries into the interstitial spaces and alveoli of the lungs. Fluid in the interstitial spaces and alveoli is called **pulmonary oedema** and this compromises gas exchange in the lungs (see Chap. 5.3). If severe, pulmonary oedema will be life threatening since the fluid in the alveoli prevents adequate oxygenation of the blood.

The effects beyond the left ventricle are due to the decreased cardiac output. There is a decreased perfusion of the body tissues, and effects will depend on how severely impaired is the cardiac output. Renal function especially is depressed, resulting in fluid retention.

Symptoms of left-sided heart failure are related to the high pressures in the pulmonary circulation and the low cardiac output. For example, dyspnoea on exertion (i.e. shortness of breath or difficult breathing during exercise) is due to the low cardiac output failing to provide adequate oxygenation to the tissues plus the increased venous return pooling in the pulmonary circulation, causing pulmonary oedema and consequent reduction in gas exchange. **Orthopnoea** (difficulty in breathing when lying down) results from the effects of a sudden increase in venous return (which occurs on lying down) not being pumped out by the left side of the heart. Blood accumulates in the pulmonary circulation and has the same effects as that following exercise. Orthopnoea is relieved by sitting up because this decreases venous return due to the changes in hydrostatic pressure. The nurse should assist the patient into a comfortable, well-supported sitting position which will improve chest expansion and decrease venous return. **Paroxysmal nocturnal dyspnoea** is another symptom of left-sided heart failure, i.e. periods of difficult breathing that occur during the night, probably due to a sudden increase in right ventricular output or in the body's need for oxygen. It is thought that factors such as dreams, nightmares or a full bladder could trigger such occurrences, which may cause the patient typically to rush to breathe out of an open window.

Left-sided heart failure most often results from dysfunction of left ventricular muscle due to myocardial infarction. Other causes may be haemodynamic, such as volume overload that occurs with incompetence of the mitral or aortic valves, or pressure overload as occurs with stenosis of the aortic valve or with systemic hypertension where there is an increased resistance to ventricular ejection (i.e. an increased afterload). In such instances the left ventricle may hypertrophy because of the increased work of pumping extra blood (as occurs with incompetent valves) or of pumping against a resistance (as in hypertension).

If the right side of the heart fails, the output of the right ventricle is less than the total volume of blood being returned to it from the systemic circulation. Blood backlogs behind the right ventricle. This increases the volume and pressure in the systemic venous circulation. (This increased pressure can often be observed because it causes the jugular vein in the neck to be distended or bulging rather than flat.) The systemic capillary pressure is increased and this causes movement of fluid into the interstitial spaces. This fluid is visible as swelling or oedema of the dependent parts of the body where the hydrostatic pressure is greatest, e.g. the ankles, and the sacrum when lying, and this may predispose to the formation of pressure sores. The liver and spleen are also distended as they act as reservoirs for the backlogged blood.

The increased venous pressure of right-sided heart failure can be measured by inserting a catheter into the right atrium and connecting it to a pressure transducer. This measurement is called central venous pressure and is described later.

The effects beyond the right ventricle are related to the decreased output. The reduction of blood pumped from the right ventricle means that less blood returns to the left ventricle and subsequently there is a decrease in the cardiac output, which has the same results as a decreased cardiac output from left ventricular failure.

Right-sided heart failure can occur independently of left-sided failure, usually as a result of lung disease when increased resistance to blood flow in the pulmonary circulation eventually causes the right ventricle to fail. When right-sided heart failure is precipitated by lung disease it is called **cor pulmonale**. Most commonly, right-sided heart failure occurs as a result of left-sided failure because the increased back pressure in the pulmonary circulation means that the right ventricle is working against an increased load and eventually fails.

When both sides of the heart fail together, the condition is called **congestive cardiac failure**.

We will now consider the mechanisms that compensate for a decreased cardiac output due to heart failure. Some of these mechanisms are seen to act when the heart failure occurs acutely, others occur when heart failure develops slowly (i.e. chronic heart failure).

When failure is acute, as in myocardial infarction in which the ventricular muscle is damaged, the ventricle cannot pump out all the blood that is returning to it. The cardiac output falls immediately and the blood dams up in the heart, causing the right atrial pressure to rise. The fall in cardiac output leads to a reduction in arterial pressure

which activates the baroreceptor reflex via the carotid and aortic baroreceptors. This results in strong sympathetic stimulation that has two effects: (i) it increases myocardial contractility and therefore the efficiency of the heart as a pump; and (ii) at the same time, it causes constriction of the veins which increases the amount of blood returning to the heart and therefore further increases the right atrial pressure. This increase in right atrial pressure increases the ventricular end-diastolic volume which enhances the ability of the heart to pump (i.e. increases stroke volume) by stretching the muscle fibres. Both these sympathetic effects help to restore the cardiac output towards normal.

Increased sympathetic activity also causes constriction of the arterioles of less vital organs such as the gut, skin and kidney, and redirects blood to the more immediately essential organs such as the brain and heart.

Figure 4.2.30 shows the effect of sympathetic stimulation on the failing heart. Pulmonary capillary wedge pressure (see later) is used to measure the left ventricular end-diastolic volume. A raised pulmonary capillary wedge pressure at rest signifies left ventricular dysfunction.

In chronic heart failure a third compensatory mechanism, i.e. expansion of the extracellular fluid volume, occurs. The low cardiac output reduces the blood flow to the kidneys. This is compounded by the effect of the increased sympathetic activity diverting blood flow away from the kidneys. Consequently there is a reduction in the glomerular filtration rate, the renin–angiotensin system is activated and antidiuretic hormone is secreted, which increases tubular reabsorption of salt and water. This results in expansion of the extracellular fluid volume and therefore increases total blood volume. More blood flows back to the heart causing additional stretch on the muscle fibres which increases the force of contraction, thus helping to return cardiac output towards normal. This is sometimes called **compensated heart failure**. This mechanism can eventually have a detrimental effect if the heart failure becomes more severe. The volume of blood in the ventricles can become so great that the muscle fibres are stretched beyond their physiological limit and cannot produce as much force. More blood remains in the ventricles after ejection, causing the heart to dilate. This is seen on chest x-ray as an enlarged heart. The effectiveness of the heart as a pump decreases and the cardiac output is again reduced. A vicious circle ensues of further increasing fluid retention, greater stretch of the muscle beyond the physiological limit, still less effective pumping and a further reduction in cardiac output which eventually leads to death. This is sometimes called **decompensated heart failure**.

Certain measures can be taken to prevent this occurring. Cardiac glycosides, e.g. digitalis, are used to increase contractility of the ventricles. This gives an increased emptying of the ventricles in a shorter time which allows a longer diastolic rest for a given cardiac output. Therefore cardiac glycosides increase the work done by the dilated and failing heart without increasing its oxygen consumption. The increased cardiac output helps to improve renal function and the longer diastole allows a greater filling time for the coronary arteries, therefore improving myocardial oxygen supply. Cardiac glycosides also slow the heart rate by their action on the AV node and by increasing vagal activity which slows SA node firing. Slowing of the heart rate allows a longer diastolic rest and increased filling time of the coronary arteries. However, the tachycardia in heart failure is mostly a sympathetic reflex response to a reduced cardiac output. The slowing of heart rate that occurs with cardiac glycosides is chiefly due to the restoration of normal output which decreases the sympathetic activity (Laurence, 1980).

The nurse should monitor carefully patients receiving digitalis for undue slowing of their heart rate, coupling of beats and nausea, all of which may complicate its administration. The patient is likely additionally to be prescribed a diuretic, e.g. frusemide, and the nurse should bear in mind when planning care the rapidity of action of this drug.

BASIC PRINCIPLES OF FLUIDS, PRESSURE, FLOW AND RESISTANCE

Before discussing the structure and function of blood vessels in detail, it is necessary to consider briefly some of the properties of fluids and the principles that govern the flow of fluids through vessels.

All fluids (when in a confined space) exert a pressure. The term **hydrostatic pressure** refers to the force that a liquid exerts against the walls of

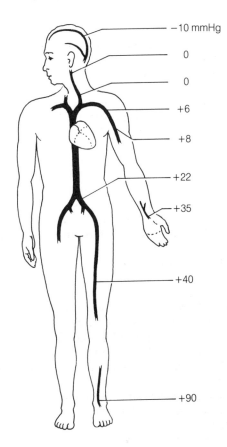

Figure 4.2.32 Effect of hydrostatic pressure on the venous pressures throughout the body in an individual standing absolutely still (all figures for pressure are in mmHg and are approximate only)

its container. The pressure that blood exerts in the vascular system is known as **blood pressure**. Pressure varies with the height of the liquid column and this can be observed in the veins of a person standing up (Fig. 4.2.32): the venous pressures in the feet are considerably greater than in the head (this is, of course, related to the effect of gravity). The effect of density on hydrostatic pressure is shown by the fact that 1 mm of mercury (mm Hg) exerts the same pressure as 13 mm of water (mm H_2O), because mercury is more than 13 times as heavy as water for an equal volume.

If pressure is exerted on a confined fluid, the pressure will be transmitted equally in all directions – this is known as Pascal's principle. If there is a weak point in the container's wall and the pressure exerted is great enough, the container wall may burst. This is what happens when an aneurysm bursts. When an individual is hypertensive, the blood vessels harden or undergo sclerotic changes (**arteriosclerosis**) to prevent the vessels bursting with the elevated blood pressure.

The distensibility of the container also influences the hydrostatic pressure that develops: if the container is distensible, the pressure in the fluid is less than in a rigid container.

Flow of fluids

The flow of a fluid through a vessel is determined by the pressure difference between the two ends of the vessel and also the resistance to flow.

PRESSURE DIFFERENCE

For any fluid to flow along a vessel there must be a pressure difference otherwise the fluid will not move. In the cardiovascular system the 'pressure head' or force is generated by the pumping of the heart and there is a continuous drop in pressure from the left ventricle to the tissues and also from the tissues back to the right atrium. Without this drop in blood pressure, no blood would flow around the circulatory system.

RESISTANCE TO FLOW

Resistance is a measure of the ease with which a fluid will flow through a tube: the easier it is, the less the resistance to flow, and vice versa. In the circulatory system the resistance is usually described as the **vascular resistance**; as it mainly originates in the peripheral blood vessels, it is also known simply as the **peripheral resistance**.

Resistance is essentially a measure of the friction between the molecules of the fluid, and between the tube wall and the fluid. The resistance depends on the viscosity of the fluid and the radius and length of the tube.

Radius of the tube
The smaller the radius of a vessel, the greater is the resistance to the movement of particles; this increased resistance results from a greater probability of the particles of the fluid colliding with the vessel wall. When a particle collides with the wall, some of the particle's kinetic energy (energy of movement) is lost on impact, resulting in the slowing of the particle. Thus, in a smaller diameter

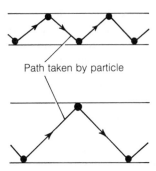

Figure 4.2.33   Diagram showing the effect of the diameter of a vessel on the flow of fluid. In a narrow vessel, particles collide more frequently with the wall and with each collision lose energy; thus the resistance to particle movement is greater in narrow vessels than in wide ones

vessel, there will be a greater number of collisions and a reduction in the energy content and speed of the particles moving through the vessel (Fig. 4.2.33). This results in a decrease in the hydrostatic pressure.

Small alterations in the size of the radius of the blood vessels, particularly of the more peripheral vessels, can greatly influence the flow of blood. Atheromatous changes in the walls of large and medium-sized arteries cause narrowing of the lumen of the vessels and result in an increased vascular resistance.

Length of the tube
The longer the tube, the greater the resistance to the flow of liquid through it. A longer vessel will require a greater pressure to force a given volume of liquid through it than will a shorter vessel. However, the length of the blood vessels in the body is not altered significantly and the overall length is kept to a minimum because of the parallel circuits in the systemic circulation (see Fig. 4.2.1).

Viscosity of the fluid
Viscosity is a measure of the intermolecular or internal friction within a fluid or, in other words, of the tendency of a liquid to resist flow. The rate of flow varies inversely with the viscosity: the greater the viscosity of a fluid, the greater is the force required to move that liquid.

Thus, changes in blood viscosity affect flow. Normally the viscosity of blood remains fairly constant, but in polycythaemia, in which there is an increased red cell content, the viscosity of the blood can be considerably increased and the blood flow reduced. Severe dehydration, where there is a loss of plasma, can also lead to increased viscosity. Cooling of the blood similarly increases its viscosity.

The nature of the lining of the tube or vessel also influences the way fluids flow. If the lining of the blood vessel, the endothelium, is smooth, the fluid will flow evenly; this is known as streamline or **laminar flow**. However, if the lining is rough or uneven or the fluid flows irregularly, **turbulent flow** is set up. Laminar flow is characteristic of most parts of the vascular system and is silent, whereas turbulent flow can be heard, e.g. during blood pressure measurements with a sphygmomanometer.

It is sometimes necessary to measure blood flow in patients and it is usual simply to measure the quantity of blood that passes a given point in the circulation over a given period of time. One method used in the clinical situation is by means of an **ultrasonic flowmeter** applied to the surface of the skin over a blood vessel. This makes use of the Doppler effect (a shift in the frequency of the ultrasonic waves when they are reflected off the moving blood cells). It is a useful and non-invasive method of assessing the condition of the peripheral arteries, in peripheral vascular disease or after vascular surgery for example.

THE VASCULAR SYSTEM

In order to fulfil its role the vascular system must

ensure delivery of blood to all tissues
be flexible and adaptable so that blood flow can be varied according to the metabolic requirements of individual tissues or the body as a whole
convert a pulsatile blood flow in the arteries into a steady flow in the capillaries to facilitate optimum transfer of substances to and from the cells
return blood to the heart

The structure of the vessels in the different parts of the vascular system varies and the differences relate directly to the function of each type of vessel. The walls of all the blood vessels, except the capillaries, have the same basic components (Fig. 4.2.34), but the proportion of the components varies with function (Fig. 4.2.35).

The innermost layer, or **tunica intima**, is a single layer of extremely flattened epithelial cells, called the endothelium, which is supported by a basement membrane and some connective and elastic tissue. Capillaries are formed from endothelial tissue and do not have the middle and outer layers. The middle layer, or **tunica media**, is predominantly smooth muscle and elastic tissue. The outer layer, or **tunica adventitia**, is composed of fibrous connective tissue, collagen and fibroblasts. The tunica media exhibits the greatest variation throughout the vascular tree, for example it is absent in the capillaries but comprises almost the whole mass of the heart.

The arterial system

The **aorta** and **large arteries** are highly elastic and distensible vessels with a relatively large diameter (the lumen of the aorta is approximately 2.5 cm in diameter). As they have relatively wide diameters, they are low resistance vessels and

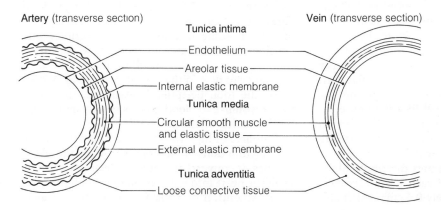

Artery (transverse section)

Tunica intima
─Endothelium─
─Areolar tissue─
─Internal elastic membrane

Tunica media
─Circular smooth muscle and elastic tissue─
─External elastic membrane

Tunica adventitia
─Loose connective tissue─

Vein (transverse section)

Figure 4.2.34 Structure of blood vessel walls, comparing that of a large artery with that of a large vein

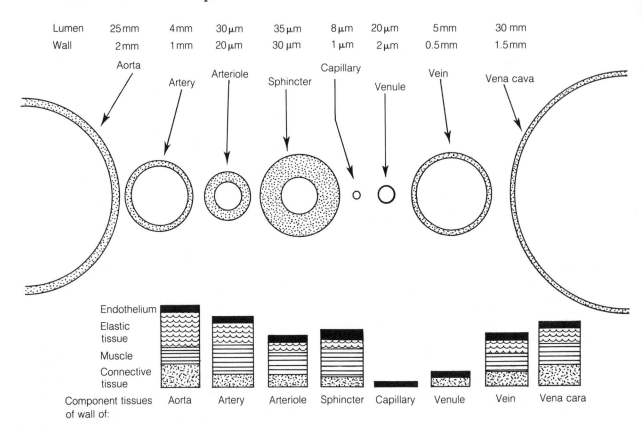

| Lumen | 25 mm | 4 mm | 30 μm | 35 μm | 8 μm | 20 μm | 5 mm | 30 mm |
| Wall | 2 mm | 1 mm | 20 μm | 30 μm | 1 μm | 2 μm | 0.5 mm | 1.5 mm |

Figure 4.2.35   The variations in size and components of the walls of the various blood vessels in the circulatory system

conduct blood flow through them easily (see previous section). When the heart contracts and forces blood into the aorta, the elastic fibres are stretched and so the vessel is distended. At the end of ventricular contraction, the force generated by the heart is reduced, and so the force stretching the elastic fibres is removed and they tend to return to their initial smaller size. The effect of this elastic recoil in between contractions is to sustain the pressure head: as the arteries return to their original size, the blood in the lumen is forced onwards around the vascular tree. Thus potential energy stored in the elastic fibres during ventricular contraction is converted into kinetic energy moving the blood onwards during the diastolic phase.

This mechanism contributes to the conversion of a pulsatile ejection from the heart to a steady flow through the arterial system.

The **medium-sized arteries** distribute blood to all parts of the body and each has a relatively large diameter (approximately 0.4 cm) which aids

blood flow. The walls of these arteries are distensible, but as the vessels become smaller with each further branching or subdivision, the amount of elastic tissue decreases whilst the smooth muscle component increases (Fig. 4.2.35).

The small arteries, more often referred to as **arterioles**, have a much smaller diameter (20 μm) and a thicker wall with muscle tissue predominating; there can be up to six concentric layers of muscle in some arterioles. The arterioles offer considerable resistance to blood flow because of their very small radius, and are the major site of resistance to flow in the vascular tree (Fig. 4.2.36). Thus the total peripheral resistance, that is the total resistance to blood flow, is mainly determined by the radius of the arterioles.

This area of high resistance to blood flow serves several functions: first, together with the elastic arteries, it converts the pulsatile ejection of blood from the heart into a steady flow through the capillaries; second, if no resistance were present and a

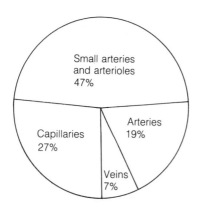

Figure 4.2.36 Distribution of vascular resistance (based on Despopoulos and Silbernagl, 1981)

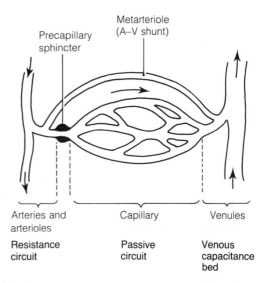

Figure 4.2.38 Diagrammatic representation of the microcirculation

high pressure persisted into the capillaries, there would be a considerable loss of blood volume into the tissue by transudation of fluid across the capillary wall. The pulmonary circulation is a *low* pressure circulation partly to prevent this happening; if pulmonary pressures increase for some pathological reason (as discussed earlier), pulmonary oedema may develop.

The arterioles are also important in determining the blood supply to different tissues and regions.

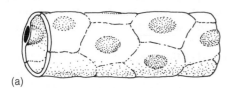

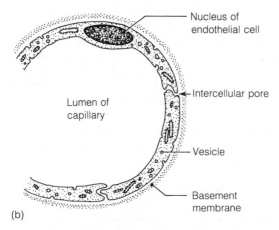

Figure 4.2.37 Structure of the capillary wall, showing (a) the surface and (b) the cross-sectional view

There are specialized regions near the junction between the terminal (smallest) arterioles and the capillaries known as **precapillary sphincters**, which consist of a few smooth muscle cells arranged circularly (see Fig. 4.2.38). If the sphincters are relaxed and the lumen patent, the capillary beds distal to the sphincter are open and perfused. If the sphincters are partially constricted, blood flow to the capillaries will be reduced, and if fully contracted, no blood will flow through. In active muscle, for instance, many more capillaries are patent due to relaxation of the sphincters and thus blood flow is increased; this has the effect of greatly increasing the surface area available for exchange of substances and at the same time reduces the distance across which substances have to diffuse to reach the cells. The mechanisms that control the arteriolar radius and precapillary sphincters will be considered later. Suffice it to say here that altering the radius is the normal mechanism for controlling the resistance and altering blood flow, and that both the sympathetic nervous system and local factors are involved.

The capillaries

The capillaries form the part of the circulatory system where the exchange of gases, fluids, nutrients and metabolic waste products occurs between the blood and the individual cells; capillaries are thus sometimes known as the **exchange vessels**. The capillaries form a dense network of narrow, short tubes; they can be as little as 3–4 μm in diameter (i.e. half the diameter of red blood cells), and up to 30–40 μm (these large blood spaces are usually known as *sinusoids*). On average, capillaries have a diameter of 6–8 μm and are approximately 750 μm long (Fig. 4.2.37). The total number of capillaries in the body has been estimated to be of the order of 40 000–50 000 million. In the resting state, probably only about 25% of the capillary beds are patent. For exchange of substances to be efficient, it is necessary to have short distances for substances to diffuse, a large surface area (the total cross-sectional area of all the capillaries is about 700 times larger than that of the aorta), and a slow steady flow of blood, about 0.3–0.5 mm/s (the flow velocity is about 700 times lower in the capillaries than in the aorta because of the narrower vessels). This part of the circulatory system is often referred to as the **microcirculation** (Fig. 4.2.38).

The structure of the microcirculation is modified in different tissues to meet specific functional requirements. Different tissues have varying abundance of capillaries, e.g. dense connective tissue has a poor capillary network as compared to cardiac muscle. Electron microscopy has shown that the nature of the endothelium is not the same in all parts of the circulation. Three different kinds of capillary walls have been identified, and the terms continuous, fenestrated and discontinuous are used to describe them, according to the size of the intercellular gaps or pores present in each.

Another modification in the structure of the microcirculation in tissues is the presence of **arteriovenous shunts** or **arteriovenous anastomoses**. These are direct connections between the arterial and venous systems that bypass the capillary beds (Fig. 4.2.38). If these shunts are patent, blood can flow rapidly through the vessels, but does not serve any nutritive purpose. These short connecting vessels have strongly developed muscular control and are under sympathetic nervous control. They are found in many tissues and organs. In the skin, for example, they enable cutaneous blood flow to be increased to allow dissipation of heat from the body surfaces when exercising or in high environmental temperatures (see Chap. 6.1).

The capillary network, whatever its form, drains into a series of vessels of increasing diameter to form venules and veins.

The venous system

The venous system acts as a collecting system, returning blood from the capillary networks to the heart passively down a pressure gradient. The capillaries merge to form venules, which in turn unite to form larger, but fewer, veins which amalgamate finally into the venae cavae. The walls of veins consist of the same three layers as arteries, but the elastic muscle components are much less prominent; the walls in general are thinner and more distensible than those of arteries (see Fig. 4.2.35). The vessels have a relatively large diameter (the vena cava is 2–3 cm in diameter) and thus offer low resistance to blood flow. Some veins, especially in the arms and legs, have internal folds of the endothelial lining (Fig. 4.2.39) that function as valves and allow blood to flow in one direction only, towards the heart. These valves can be damaged if overstretched by high venous pressures for long periods, for example during pregnancy or in people who stand for extended periods; the valves become incompetent, lose their function, and varicose veins develop. As a result of this, oedema and varicose ulcers can develop (Fig. 4.2.39).

A major part of the blood volume, approximately 60% (Fig. 4.2.40), is contained within the venous system and for this reason veins are sometimes referred to as **capacity vessels**. The capacity of the venous system can be modified by altering the lumen size of the muscular venules and veins; the changes are mediated by altering the **venomotor tone**, that is, the degree of contraction of the smooth muscle in the tunica media. Venomotor tone is mainly under the control of the sympathetic nervous system. Changes in the venomotor tone can increase or decrease the capacity of the venous circulation and therefore can partially compensate for variations in the effective circulating blood volume.

VENOUS RETURN

Venous blood flow occurs along relatively small pressure gradients and even small variations in resistance and vessel radius affect the return flow

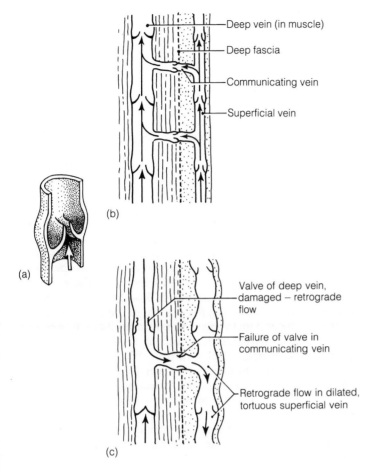

Figure 4.2.39 Diagrams showing: (a) a single normal venous valve, (b) the normally functioning valves in the superficial and deep veins of the leg, and (c) the formation of varicose veins when valves in the deep vein become incompetent

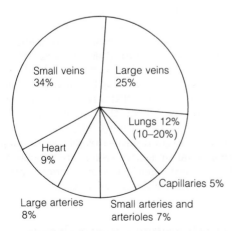

Figure 4.2.40 Percentage of the total blood volume in each portion of the circulatory system (based on Despopoulos and Silbernagl, 1981)

to the heart. As flow can only occur when there is a pressure gradient, the pressure in the venules must at all times be greater than the pressure in the right atrium, and this is normally the situation (see Fig. 4.2.44).

However, the effect of gravity retards venous return: when upright, as the veins are distensible and due to the hydrostatic pressure of a column of blood in the veins below the level of the heart, blood tends to collect or pool in the feet and legs. When vertical, the leg veins take on a circular form which has a greater capacity; when horizontal the veins take on an elliptical shape with a lower capacity. Increased venomotor tone, reducing the diameter and hence capacity of the veins, helps to reduce venous pooling. Venous pooling is a useful term but it suggests stagnation which does not occur; venous pooling simply indicates that the veins accommodate a greater volume of blood.

One can see the effect of gravity on the veins in the neck: when sitting or standing the neck veins above a level 5–10 cm higher than the heart are not prominent, but when lying down the veins distend. This is due to the fact that, in contrast to venous return from the feet, blood from the head returns to the heart aided by gravity when upright. However, the blood supply to the head has to overcome the effect of gravity; failure of this phenomenon can be observed when someone stands up too quickly after bending down and feels dizzy due to a temporary reduction in the effective pressure head delivering blood to the brain.

It is vital that an adequate venous return to the heart is maintained at all times because the cardiac output depends on the venous return – in most instances the cardiac output equals the venous return. Thus, if the venous return falls, cardiac output and blood pressure may also drop. Several mechanisms exist to help maintain the venous return at all times. Increasing the venomotor tone is an important mechanism as it decreases the capacity of the venous system and so aids venous return. After a long period of bedrest when the body is not constantly being exposed to the force of gravity and the veins do not have to compensate, venomotor tone is reduced, and this method of reducing the effect of gravity is temporarily less efficient. This should be remembered when helping someone to get up after a period of bedrest. It is essential to move slowly and steadily and to support the person in case he or she becomes dizzy and faint.

Venous return is also assisted by two systems sometimes referred to as the **skeletal muscle pump** and **respiratory pump**. Contraction of the skeletal muscles, especially in the limbs, squeezes the veins and this pushes blood in the extremities towards the heart; backflow is prevented by the presence of numerous valves. There are also many communicating channels which allow emptying of blood from the superficial limb veins into the deep veins when rhythmic muscular contractions occur (Fig. 4.2.39). Consequently, every time a person moves his or her legs or tenses the muscles, a certain amount of blood is pushed towards the heart. The more frequent and powerful such rhythmic contractions are, the more efficient their action. (Sustained continuous muscle contractions, unlike rhythmic contractions, impede blood flow due to the veins being continuously 'blocked'.) The muscle pump mechanism is an efficient system: the venous pressure in the feet of someone walking is of the order of 25 mmHg (3.3 kPa), whereas in the feet of an individual standing absolutely still it is of the order of 90 mmHg (12 kPa); see Fig. 4.2.32. So when an individual stands still for long periods of time, the muscle pump cannot operate and venous return is decreased. This can result in people fainting due to an inadequate cerebral blood flow. e.g. soldiers fainting on parade, people fainting in operating theatres after standing still for long periods. Thus it is advisable to contract the muscles of the legs and buttocks voluntarily to aid venous return if standing still for long periods.

Respiration produces cyclical variations in intrapleural and intrathoracic pressure (Chap. 5.3). With each inspiration, the pressure is lowered with the thorax and hence also within the right atrium of the heart; this increases the pressure gradient and aids blood flow back to the heart. Simultaneously, the descent of the diaphragm into the abdomen raises the intra-abdominal pressure and increases the gradient to the thorax, again favouring venous return. With expiration, the pressure gradients are reversed and blood tends to flow in the opposite direction; fortunately this tendency is prevented by the valves in the medium-sized veins.

Thus venous return is maintained by changes in venomotor tone, altering the capacity of the venous system, and by the skeletal muscle and respiratory pumps. Obviously it is also necessary to maintain an adequate circulating blood volume. If the blood volume is depleted for some reason, e.g. dehydration or haemorrhage, in the short term venoconstriction and vasoconstriction in the body's blood reservoirs, such as the skin, liver, lungs and spleen, can increase the effective circulating blood volume. However, the blood volume must be restored eventually by fluid replacement. The pressures in the central regions of the venous system directly reflect the blood volume; thus central venous pressure (CVP), or right atrial pressure, is a good indicator of blood volume, unlike arterial pressures which are reflexly regulated and controlled. (Central venous pressure measurements are described later.)

Arteriosclerosis and atherosclerosis

Structural changes in the walls of blood vessels are

very common and lead to changes in the properties and normal functioning of the vessels; but perhaps more seriously, in the long term these changes are also responsible for, or linked with, several of the major diseases prevalent today. **Arteriosclerosis** refers to the hardening of the arteries: muscle and elastic tissue are replaced with fibrous tissue and calcification may occur. Due to the lack of distensibility, arteriosclerotic vessels are less able to change their radius and lumen size. Arteriosclerosis includes atherosclerosis.

Atherosclerosis refers to hardening and obstruction of the arteries, but is due to the deposition of lipids and other substances, in the form of an atheromatous plaque, in the intima of the medium and large-sized arteries. In affected vessels the intima becomes thickened, smooth muscle cells, collagen and elastic fibres accumu-

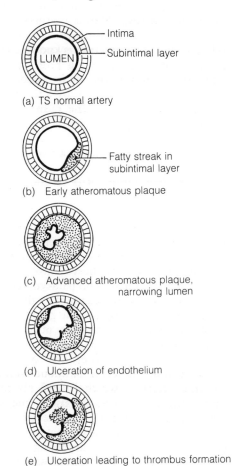

(a) TS normal artery

(b) Early atheromatous plaque

(c) Advanced atheromatous plaque, narrowing lumen

(d) Ulceration of endothelium

(e) Ulceration leading to thrombus formation

Figure 4.2.41 Atherosclerotic changes that may occur in arteries (the stages illustrated are not necessarily sequential)

late, and lipids (especially cholesterol) are deposited in the arterial wall (Fig. 4.2.41). Though much research has been carried out to determine how the plaque is formed, the mechanism is still not fully understood. The outer part of the plaque is fibrous but the centre is soft and has the consistency of gruel or porridge, hence the use of *athera*, a Greek word meaning gruel. Atherosclerosis is not a simple lesion; a variety of pathological changes occur.

In many cases there is damage to the endothelium of the arteries. Platelets circulating in the blood adhere to the damaged portions of the endothelium; it is thought that a substance released from platelets, known as platelet factor, promotes the proliferation of the smooth muscle cells. Muscle cells, probably from the media, migrate into the intima of the artery wall. The plaque appears above the normal endothelial surface and serum lipids accumulate within it, especially cholesterol and low density lipoprotein (LDL) which carries cholesterol in the plasma.

Clinical problems can ensue from several aspects of plaque formation and atherosclerosis. The vessels can actually become blocked or stenosed; also, as a result of the endothelial damage and platelet adhesion, an ulcer-like site can develop that can lead to thrombus formation. Sometimes pieces of dead tissue from the damaged arterial wall can break off and lead to an embolus. Stenosis and thrombus and/or embolus formation can occur together. If the process progresses, the intima is 'eaten away' and, especially if the blood pressure is elevated, the vessel can develop an aneurysm.

Atherosclerosis is responsible for most coronary artery and ischaemic heart disease, for much cerebrovascular disease (e.g. strokes), peripheral vascular disease and for most abdominal aortic aneurysms. The occurrence of atheroma varies in different sites in the body and, as would be expected from the diseases it is associated with, is common in the proximal coronary arteries, the aorta, the iliac, popliteal, femoral and internal carotid arteries and in the circle of Willis in the brain. There is some evidence to suggest that atheroma formation is enhanced in areas where the arteries branch or divide and turbulence in blood flow is increased.

RISK FACTORS

Atherosclerosis almost certainly has a multifactorial aetiology. Some of the accepted risk factors – for instance, smoking and hypertension – may have their own adverse effect by damaging the endothelium. As far as smoking is concerned, nicotine is thought to increase platelet adhesion, and carbon monoxide may increase the permeability of the arterial endothelium, thus enhancing plaque formation.

Many studies show a positive relationship between elevated serum cholesterol levels and the incidence of atherosclerosis, especially associated with coronary artery disease. Elevated cholesterol levels may be the result of endogenous (metabolic disease, e.g. diabetes, or genetic) factors or exogenous (high fat, especially saturated fats and high cholesterol diet) factors. There is no definitive causal relationship between saturated fat intake in the diet and the serum cholesterol levels, but there is evidence that unsaturated fatty acids lower blood cholesterol whereas saturated fats tend to raise it.

There is also clear evidence that lowering very high plasma concentrations of cholesterol and low density lipoprotein cholesterol lowers the incidence of coronary heart disease. In a study conducted by the Lipid Research Clinics (1984) in the USA, a group of men who were given a drug that lowers plasma cholesterol levels, cholestyramine, suffered 20% fewer episodes of angina and 19% fewer coronary attacks than a control group matched for dietary fat intake.

Most researchers believe that the process of atherogenesis begins early in life and that fatty streaks observed in the arterial walls of children may, in some instances, be precursors of atherosclerosis. Thus, although there is no proof as to the causal relationship between fats and atherosclerosis, it does seem sensible to reduce the amount of saturated fat in the diet. It is also perhaps important that the body manufactures cholesterol in amounts up to four to six times those found in the diet. There is some evidence to suggest that hypercholesterolaemia itself can alter the normal endothelium.

Many other factors have been associated with the development of arteriosclerosis. It is more common in the industrialized areas of the world. It has been suggested that atherosclerosis is a result of a more sedentary lifestyle; indeed, regular physical exercise has been found to promote a favourable ratio of lipoproteins in the blood (a higher high density lipoprotein (HDL) and lower

LDL content). There are sex differences too: males are affected more than women until the time of the menopause, when the sex distribution is equal. The incidence of atherosclerosis also increases with increasing age.

Blood vessel diameter

Blood vessels play an important role in determining local blood flow to tissue and in central reflexes to maintain blood pressure. The diameter of all vessels (except capillaries) is altered by changing the degree of contraction of the smooth muscle in the tunica media of the vessel wall. The muscle is arranged in a circular pattern and when it contracts the diameter of the vessel lumen becomes smaller (a process known as **vasoconstriction**), and when it relaxes the vessel diameter increases (known as **vasodilation**). Normally, the smooth muscle is in a state of partial contraction all the time, and this level of contraction is known as the **vasomotor tone**; arteriolar tone refers to the tone of the arterioles and venous or venomotor tone to the tone of the veins. So, by altering the degree of muscle contraction, the radius of the vessel can be increased or decreased.

The smooth muscle of the vessel walls is influenced by both nervous and chemical factors.

NERVOUS CONTROL OF BLOOD VESSEL DIAMETER

Sympathetic nerves innervate the smooth muscle of all arteries and veins, but the arterioles, precapillary sphincters and postcapillary venules are more densely supplied with nerve endings than are the other vessels. (There is no parasympathetic nerve supply to blood vessels, except in the nervi erigentes, which supplies part of the genital tract, and in the salivary glands.) The effect of sympathetic discharge is to cause muscle contraction and hence vasoconstriction. However, due to the fact that the muscle of the vessels is always in a state of partial contraction (the vasomotor tone), the same sympathetic constrictor nerves can also accomplish dilation simply by decreasing the frequency of discharge of impulses along the sympathetic nerve fibres (Fig. 4.2.42). Thus, by increasing the discharge rate, the vessel is constricted, whereas by decreasing the discharge rate the same vessel can dilate. Part of the cardiovascular centre in the

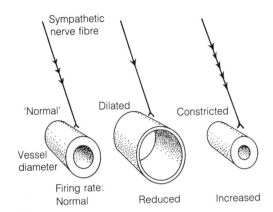

Figure 4.2.42 Effect of sympathetic nerve firing rate on blood vessel diameter

medulla oblongata of the brain, described as the vasomotor centre (discussed later), controls the rate of discharge down the sympathetic nerves.

The vasoconstriction that results from increased sympathetic activity causes an increase in the total vascular resistance. An increased sympathetic discharge has other consequences, too. The capillary hydrostatic pressure drops due to the increased arteriolar resistance; this favours the uptake of interstitial fluid from the tissues into the blood and tends to increase the circulating blood volume. Another consequence of the increased sympathetic discharge is to increase the venomotor tone, so the capacity of the venules and veins decreases and venous return increases. These mechanisms are particularly useful in conditions such as haemorrhage when there is a decrease in the circulating blood volume. (The opposite effects will occur if the sympathetic discharge is reduced.)

The blood vessels in different parts of the body are not all equally affected by the sympathetic output; this is due partly to the variation in distribution of sympathetic nerves to the vessels and also to the varying distribution of α- and β-adrenoreceptors in the smooth muscle (see Chap. 2.4). For example, blood vessels to the skin will vasoconstrict strongly with sympathetic stimulation whereas cerebral vessels show only slight vasoconstriction.

The postganglionic sympathetic fibres normally release noradrenaline (norepinephrine) and this acts upon the smooth muscle to produce the contraction and vasoconstriction. Adrenaline (epinephrine) released from the adrenal medulla causes vasoconstriction in a similar way to the sympathetic nervous discharge (see Chap. 2.4).

The fact that altering blood vessel calibre has several physiological effects is used therapeutically and there are many drugs available that alter blood vessel calibre. For instance, drugs that mimic the action of the sympathetic nervous system (known as sympathomimetics), e.g. dopamine, adrenaline and noradrenaline, cause vasoconstriction (they also have varying effects on the heart). Adrenaline influences both α- and β-receptors and affects both the heart and peripheral vessels, whereas noradrenaline mainly affects α-receptors and primarily the vascular system, causing an increase in peripheral resistance. Nicotine increases adrenaline output from the adrenal medulla and so causes vasoconstriction.

Adrenergic blocking drugs 'block' the activity of adrenaline and noradrenaline at the smooth muscle receptor sites and so cause vasodilation. For example, α-receptor blocking drugs are used in the treatment of peripheral vascular disease in order to increase blood flow to ischaemic tissues; the drugs tend to increase blood flow to the skin rather than to the muscles and so are particularly useful in some instances to assist in the treatment of varicose ulcers. Drugs can also be given that directly affect the smooth muscle in the blood vessel wall, e.g. glyceryl trinitrate (discussed earlier) is a potent vasodilator as it relaxes the vascular muscle, and is particularly effective in the treatment of angina.

CHEMICAL CONTROL OF BLOOD VESSEL DIAMETER

As well as being affected by sympathetic nervous control, the vascular smooth muscle is also directly influenced by chemical factors such as hormones and locally produced metabolites. The effects of adrenaline and noradrenaline have already been mentioned. A number of other agents have a role, too: for instance, histamine and plasma kinins, released as part of the inflammatory response when tissues are injured, cause vasodilation of the small vessels, whereas angiotension II, formed by the action of renin on angiotensinogen, is a potent vasoconstrictor. The response of vascular smooth muscle to local metabolites is important, and is discussed below.

Response to local metabolites
The coordination of metabolic needs when tissues are active and require an increased blood supply does not depend on the presence of nerves or

hormones, but is controlled locally. When a tissue is active, there is a local increase of cell metabolic products in the interstitium and these chemicals directly influence the smooth muscle of the pre-capillary vessels causing vasodilation, and thus increasing the blood flow to the active tissue. This mechanism of increasing blood flow in response to local demand, known as **active hyperaemia**, is highly developed in tissues such as the heart and skeletal muscle, and to a lesser extent in the gas-trointestinal tract.

The precise nature of the metabolites that produce this relaxation of the smooth muscle still has to be resolved, but the likely factors include

(a) a rise in the potassium ion concentration in the vicinity of the precapillary sphincters
(b) hyperosmolarity of the interstitial fluid
(c) adenosine
(d) increased pH
(e) lack of oxygen
(f) high levels of carbon dioxide and lactic acid.

The smooth muscle of the smaller arterioles and precapillary sphincters shows inherent myogenic activity, and the metabolites, whichever they prove to be, suppress this myogenic activity – the

sphincters relax and the capillaries governed by them open (see Fig. 4.2.38). With the increased blood flow, the accumulated metabolites are dissi-pated and the inherent tone can be re-established and the sphincters close again. Thus there is auto-regulation of the blood flow in the sense that the increase in blood flow is directly in proportion to the increased activity of the tissue.

Locally mediated vasodilation can be easily demonstrated. If the blood supply to the arm is occluded for a couple of minutes (for example using a sphygmomanometer cuff inflated to 200 mmHg), the arm goes pale as it is ischaemic. While the cuff is inflated the cells are still metab-olizing; the metabolites accumulate as there is no blood flow to remove them. This causes the pre-capillary vessels to dilate. As the cuff is deflated, once the pressure in the cuff is below systolic pressure, blood can again flow into the arm. As a result of the vasodilation there is an enlarged capacity in the capillary system and so the arm flushes considerably 'redder' than the other arm due to the increased blood content. In this instance, the ischaemia was responsible for producing the arteriolar dilation and so the process is known as **reactive hyperaemia**.

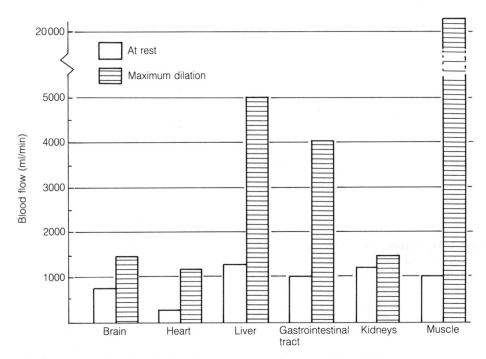

Figure 4.2.43 Approximate blood flow in selected tissues at rest and at maximal vasodilation

So, by a combination of nervous reflexes and chemical changes, the blood flow to most tissues is regulated according to the metabolic needs (Fig. 4.2.43 shows the possible extent of the blood flow changes). The chemical control is of overwhelming importance in conditions of increased metabolic activity of tissues. For instance, resting skeletal muscle flow is maintained by the influence of sympathetic vasoconstrictor fibres, but when exercising, massive vasodilation is achieved by the influence of local metabolites produced by the active tissue.

The control of two of the vital circulations, those of the heart and brain, is predominately influenced by chemical rather than nervous factors. The coronary vessels are very sensitive to oxygen lack and dilate when oxygen levels in the blood drop. Blood flow to the brain under normal circumstances varies very little. However, if the chemical composition of the blood or tissue fluid changes markedly, blood flow does alter; the main determinants of the cerebral vascular diameter are the local carbon dioxide and hydrogen ion levels (see Chap. 5.3).

If an individual hyperventilates and expires extra carbon dioxide, the hydrogen ion concentration is reduced and the cerebral vessels constrict; the individual may become dizzy and light-headed due to cerebral hypoxia. Conversely, if the plasma carbon dioxide or hydrogen ion levels increase, the cerebral vessels dilate; this can be observed if an individual breathes air with a high carbon dioxide level, say 7% when he or she may experience an intense pounding headache due to cerebral vasodilation. A decreased oxygen tension or a rise in body temperature also produces dilation of the cerebral vessels.

The blood flow to the skin, in contrast to the heart and brain, is mainly determined by sympathetic nervous activity and local factors are much less important. Cutaneous blood flow depends not so much on the metabolic activity of the skin itself as on the requirements for maintenance of body temperature, and thus skin blood flow is centrally coordinated. Local reflexes do operate if part of the body is exposed to extremities of temperature (e.g. a hand in hot water vasodilates). Under maximum heat load, total skin blood flow can be as much as 3–4 l/min whereas under cold stress the total skin blood flow can be as little as 50 ml/min.

BLOOD PRESSURE WITHIN THE CARDIOVASCULAR SYSTEM

Blood pressure refers to the pressure exerted by the blood on the blood vessel walls. All fluids, and blood is no exception, exert pressure on the walls of the vessel that they are held in. The pressure exerted by a liquid against the walls of its container is known as the **hydrostatic pressure** and thus blood pressure is a hydrostatic pressure.

Each blood vessel has its own blood pressure value, for instance arterial blood pressure, capillary blood pressure, venous blood pressure, right atrial pressure, and so on. As can be seen from Fig. 4.2.44, the pressure in the blood vessels falls continuously from the aorta to the end of the systemic circulation in the right atrium. The pressures in the pulmonary circulation are considerably lower than in the systemic circulation but there is still a pressure gradient from the right ventricle to the left atrium. The pressure must drop in order for blood to flow.

When we talk clinically about 'blood pressure' per se, we are usually referring to systemic arterial blood pressure. The emphasis on arterial blood pressure is logical and important because it is this pressure that ensures an adequate blood flow to the tissues and to vital organs such as the brain and heart. If the blood pressure falls too far, tissue flow is reduced and so nutrient and gas supplies may become inadequate. A person fainting is an example of this, with a reduced blood supply to the brain.

As we shall see later, there are several reflexes operating to regulate and maintain arterial blood pressure within normal limits. Blood pressures in other regions of the circulatory system should similarly be kept within normal ranges: for instance, abnormal capillary blood pressures will alter the exchange of fluids to and from the tissues (see later).

Blood pressure is a function of blood flow and vascular resistance.

Blood flow
The circulatory system is a closed system and thus the total flow leaving and returning to the heart will be the same. Thus blood flow is equivalent to cardiac output.

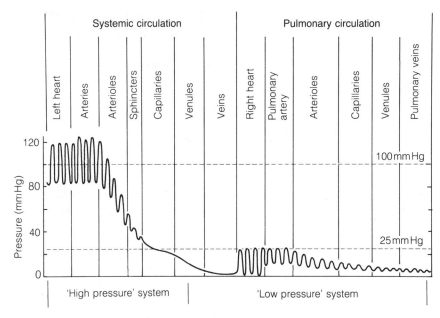

Figure 4.2.44   Blood pressures in each section of the systemic and pulmonary circulations

Pressure difference

This is the difference between the mean pressure in the aorta and the pressure in the vena cava just before the blood enters the heart (this latter value is almost zero). Since the blood pressure is essentially the same in the aorta and all *large* arteries, the pressure difference can be said to be equivalent to mean arterial pressure.

Resistance to flow

This is the total resistance to blood flow. As the majority of the resistance is found in the peripheral vessels, especially the arterioles, it is often described as the total peripheral resistance.

Thus we have the equation

mean arterial pressure = cardiac output

× total peripheral resistance

or

$$BP = CO \times TPR$$

This is one of the fundamental equations of cardiovascular physiology. You can see from the equation that blood pressure can be maintained by altering cardiac output and/or total peripheral resistance. The cardiac output itself is changed by altering the heart rate and stroke volume.

Arterial blood pressure

Arterial blood pressure fluctuates throughout the cardiac cycle. The contraction of the ventricles ejects blood into the pulmonary and systemic arteries during systole and this additional volume of blood distends the arteries and raises the arterial pressure. When the contraction ends, the stretched elastic arterial walls recoil passively and this continues to drive blood through the arterioles. As the blood leaves the arteries the pressure falls; the arterial pressure never falls to zero because the next ventricular contraction occurs whilst there is still an appreciable amount of blood within the arteries (Fig. 4.2.45). Thus the pressure in the major arteries rises and falls as the heart contracts and relaxes. The maximum pressure occurs after ventricular systole and is known as the **systolic pressure**. When the blood pressure in the aorta exceeds that in the ventricle, the aortic valve closes; this accounts for the dicrotic notch (Fig. 4.2.45). Once the aortic valve has closed, the blood pressure in the aorta and large arteries falls as blood flows through the arterioles and capillaries to the veins. The level to which the arterial pressure has fallen before the next ventricular systole, that is the minimum pressure, is known as the **diastolic pressure**.

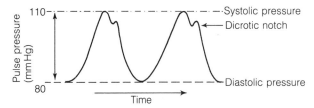

Figure 4.2.45 Waveform showing that pressure in the main arteries rises and falls as the heart contracts and relaxes

The systolic pressure is determined by the amount of blood being forced into the aorta and arteries with each ventricular contraction, i.e. the stroke volume, and also by the force of contraction. An increase in either will increase the systolic pressure. Similarly, if the arterial wall becomes stiffer, as happens in arteriosclerosis, the vessels are not able to distend with the increased blood volume and so the systolic pressure is increased.

Diastolic pressure is also influenced by several factors. The diastolic pressure provides information on the degree of peripheral resistance: if there is increased arteriolar vasoconstriction, this will impede blood flowing out of the arterial system to the capillaries, and the diastolic pressure will rise. Conversely, if the peripheral resistance is reduced by vasodilation, more blood will flow out of the arterial system and thus diastolic pressure will fall. Drugs that modify the degree of arterial vasoconstriction and alter the peripheral resistance will obviously affect the diastolic pressure, and vasodilator drugs, for example hydrallazine, are sometimes used in the treatment of hypertension.

The diastolic pressure also depends on the level of the systolic pressure, the elasticity of the arteries and the viscosity of the blood. Alterations in the heart rate will also affect diastolic pressure: with a slower heart rate, the diastolic pressure will be lower as there is a greater time for blood to flow out of the arteries, and vice versa.

PULSE PRESSURE AND MEAN ARTERIAL PRESSURE

Each ventricular contraction initiates a pulse, or wave, of pressure through the arteries and these pulses can be palpated (felt) wherever an artery passes near the skin and over a bony or firm surface. (Figure 4.2.46 shows the common sites where the pulse is felt.) Thus there is one pulse per heart beat, and so the pulse rate is used as an easy

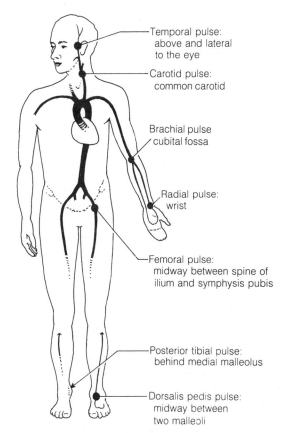

Figure 4.2.46 Common sites in the body where the pulse can be felt

method for counting the heart rate. These palpable pulses represent the difference between the systolic and diastolic pressures and this difference is known as the **pulse pressure**, e.g. the pulse pressure in an individual whose blood pressure is 120/70 mmHg is (120−70) = 50 mmHg. Two of the main factors that alter the pulse pressure are the stroke volume and decreased arterial compliance.

It is sometimes useful to have an average, or mean, value for the arterial pressure, rather than maximum and minimum (systolic and diastolic) pressures, as it is the mean pressure that represents the pressure driving blood through the systemic circulation. The **mean arterial pressure** is not a simple arithmetical mean; it is estimated by adding one-third of the pulse pressure to the diastolic pressure. So, for a blood pressure of

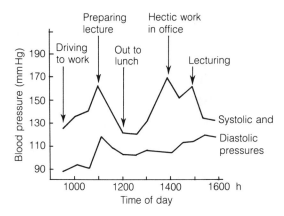

Figure 4.2.47 Variation of arterial blood pressure in one 'normal' individual over a 7-hour period (redrawn from O'Brien and O'Malley, 1981)

120/70 mmHg

$$\text{mean pressure} = 70 + (1/3 \times 50)$$

$$= 87\,\text{mmHg}$$

From the point of view of actual tissue perfusion, it is generally the mean arterial pressure that matters, rather than the precise values of systolic and diastolic pressures.

Blood pressure values

There is no such value as a 'normal' blood pressure for the population as a whole; there is a usual or 'normal' value for any particular individual, but even that value varies from moment to moment under different circumstances (Fig. 4.2.47) and over longer periods of time. Many factors, both physiological and genetic, have an influence on blood pressure and thus it is not surprising that individuals have significantly different, but 'normal', blood pressure values. Therefore it is more

Table 4.2.2 Some average blood pressure values, taken from 250 000 healthy individuals (from Durkin, 1979)

Age (years)	Blood pressure (mmHg)	
	Systolic	Diastolic
Newborn	80	46
10	103	70
20	120	80
40	126	84
60	135	89

appropriate to refer to a normal range of blood pressure than to a single value. Normal blood pressures are said to range from 100/60 mmHg to 150/90 mmHg (O'Brien and O'Malley, 1981).

Parameters such as age, sex, and race influence blood pressure values. In Western societies, blood pressure values tend to increase with advancing age (Table 4.2.2), therefore a blood pressure which would be 'normal' for a 70 year old might be considered 'abnormal' for a 40 year old. This is not universal, for example South Sea Islanders show little, if any, increase in mean blood pressure with increasing age (Keele *et al.*, 1982). The elevation in blood pressure with age may be due either to genetic or environmental factors and is likely to be a result of arteriosclerosis.

Men generally have higher blood pressures than women. Race also seems to influence blood pressure levels, e.g. in the USA negroid races tend to have higher blood pressures than whites.

Most authorities agree that a resting diastolic pressure persistently exceeding 90 mmHg or 95 mmHg indicates **hypertension**, that is, a raised blood pressure; this is an arbitrary definition but proves to be useful for clinical practice. A persistently low blood pressure, **hypotension**, is relatively rare, although temporary or transient hypotension is more common, e.g. in haemorrhage or fainting.

HYPERTENSION

One of the reasons that clinicians are so concerned about the level of an individual's blood pressure is that there is a significantly increased mortality in those with untreated hypertension when compared with individuals with a 'normal' blood pressure (**normotensive**): a 35-year-old man with a diastolic pressure of 100 mmHg can expect a 16-year reduction in life expectancy (Bannan *et al.*, 1980). O'Brien and O'Malley (1981) estimate that nearly one-quarter of the adult population have an elevated blood pressure.

Individuals who are hypertensive usually have few, if any, symptoms and often the hypertension is only diagnosed as part of a routine medical screening, for example for insurance purposes. The effects of a raised blood pressure are insidious and develop over many years: the heart has to increase in size (detectable on x-ray) and strength to overcome the increased resistance caused by the increased blood pressure. The arteries respond to the increased pressure by hypertrophy of the smooth muscle in their walls, so that they are able to withstand exposure to the higher pressures. Atherosclerosis formation is also potentiated. The blood vessels most commonly affected are the cerebral, coronary and renal vessels; cerebrovascular accidents (strokes) and myocardial infarctions are the commonest clinical manifestations, followed by renal disease.

There has been much research and discussion into the causes of hypertension. In a few instances, hypertension is secondary to renal or endocrine disease, but in the majority of cases the cause of primary or **essential hypertension** is not fully understood. The aetiology of essential hypertension is almost certainly multifactional and it is likely to prove to be a combination of genetic and environmental factors. Mechanisms that seem to be involved include some that affect the extracellular fluid volume and expand the circulating blood volume, e.g. excessive renin secretion and angiotensin production, increased sympathetic activity and excessive dietary salt intake, possibly associated with a low potassium intake (MacGregor, 1983). Some of the treatments prescribed for hypertension relate to these mechanisms, i.e. diuretics (e.g. a thiazide) to increase sodium and water loss; methyldopa, β-adrenoreceptor blocking drugs (e.g. propranolol), and relaxation techniques to reduce sympathetic activity; restriction of salt intake. One drug, captopril, inhibits angiotensin-converting enzyme in the lungs and reduces the production of angiotensin II. Raised peripheral resistance is linked with hypertension and so drugs that produce vasodilation are useful.

Thre are many risk factors associated with the development of hypertension, including obesity, high alcohol and salt intakes and some drugs (e.g. oral contraceptives, corticosteroids, monoamine oxidase inhibitors). There is also often a positive family history of hypertension: if both parents are hypertensive, there is a significantly greater risk that their children will also develop high blood pressure.

If hypertension is diagnosed and effectively treated, usually by drug therapy, much of the cardiovascular-related disease can be prevented.

Measurement of arterial blood pressure

The first documented measurement of blood pressure dates back to the eighteenth century. In 1773, Stephen Hales, an English theologian and scientist, directly measured mean blood pressure in an unanaesthetized horse by inserting an open-ended tube directly into the animal's neck; the blood entered the tube and rose upwards (to a height of 2.5 m) towards the tube opening until the weight of the column of blood was equal to the pressure in the circulatory system of the horse. This is the basis of a simple pressure manometer which is still used for measuring blood pressure (see CVP monitoring). It is the basis too for measuring cerebrospinal fluid pressures during a lumbar puncture.

Catheters can be inserted directly into an artery (the radial artery is often used) to give direct arterial pressure measurements. The indwelling catheter is now usually attached to small electronic transducers, and pressures can be monitored continuously. Figure 4.2.48 shows an arterial pressure waveform. (For a more comprehensive discussion on transducers and direct pressure measurements, see section on venous pressure.)

However, in most instances it is not desirable or practicable to use invasive techniques to measure arterial pressures. In the eighteenth century, an Italian physiologist, Scipione Riva Rocci, invented the **sphygmomanometer** (*sphygmo* = pulse) which enabled a non-invasive measurement of systolic pressure. A rubber inflatable cuff is placed over the brachial artery and the pressure in the cuff is raised until the cuff pressure exceeds that of the blood in the artery. At this point the artery collapses and no radial pulse can be felt as blood is not able to flow through the brachial artery. The pressure in the cuff is then slowly released and the radial pulse reappears. The pressure at which the pulse reappears corresponds to the systolic pressure as it is the point at which the peak pressure

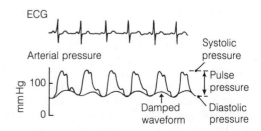

Figure 4.2.48 Normal and damped arterial pressure waveforms when recorded continuously with a pressure transducer

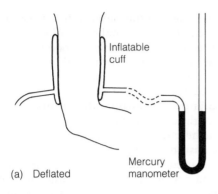

Inflatable cuff

Mercury manometer

(a) Deflated

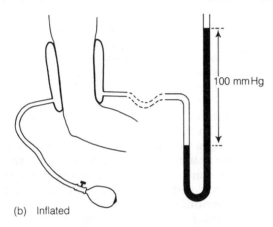

100 mmHg

(b) Inflated

Figure 4.2.49 Principle of sphygmomanometry; (a) the cuff is deflated; there is no pressure in the cuff, and the mercury levels in the two arms of the U-tube manometer are the same (in actual clinical sphygmomanometers mercury is stored in a reservoir and not in a U-tube as drawn here); (b) the cuff is inflated; the pressure in the cuff is equal to the difference in height between the two mercury levels

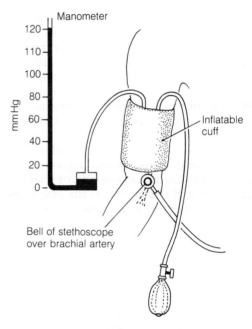

Manometer

mmHg

Inflatable cuff

Bell of stethoscope over brachial artery

Figure 4.2.50 The auscultatory method for measuring systolic and diastolic pressures

The method was developed further a few years after Riva Rocci by a Russian surgeon, Dr Nicolai Korotkov. Korotkov reported a method for measuring both systolic and diastolic pressures by auscultation, that is, by listening, using a stethoscope placed over the brachial artery and the sphygmomanometer (Fig. 4.2.50). Various sounds were audible and Korotkov classified the sounds into five phases, which are now known simply as the **Korotkov sounds** (Fig. 4.2.51).

(i.e. the systolic) in the brachial artery exceeds the occluding pressure in the cuff.

The mercury sphygmomanometer is used as the standard reference for measuring blood pressure and it still forms the basis for our present-day indirect method of assessing arterial pressure (Fig. 4.2.49). Traditionally, blood pressure is measured in millimetres of mercury (written as mmHg); this means that if the blood pressure is 100 mmHg, the pressure exerted by the blood is sufficient to push a column of mercury up to a height of 100 mm. (The SI unit for pressure is the pascal (Pa) or kilopascal (kPa) and so sometimes blood pressure may be written as, say, 13.3 kPa instead of 100 mmHg.)

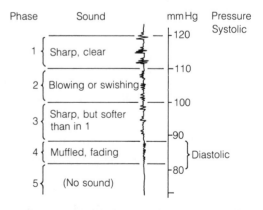

Phase	Sound		mmHg	Pressure
				Systolic
1	Sharp, clear		120	
			110	
2	Blowing or swishing		100	
3	Sharp, but softer than in 1			
			90	
4	Muffled, fading			Diastolic
			80	
5	(No sound)			

Figure 4.2.51 The Korotkov sounds (based on O'Brien and O'Malley, 1981)

The precise origin of the various Korotkov sounds is not fully understood, but they are due primarily to turbulent flow and vibratory phenomena in the brachial artery as it opens and closes with each beat and as the blood flows through the semi-occluded vessel. When the pressure in the blood pressure cuff is greater than that in the artery, the vessel is completely occluded and there is no blood flow and no turbulence, and hence no sound.

There is considerable controversy as to whether phase 4, the muffling of the sounds, or phase 5, the disappearance of the sounds, is the best measure of the diastolic pressure. In the UK, phase 4 is favoured in clinical practice, whereas in the USA, phase 5 is used. O'Brien and O'Malley (1981) suggest that both phases 4 and 5 should be recorded, for example 120/72/64 mmHg. Phase 5 correlates better with direct arterial measurement of blood pressure and there is also often better agreement among observers when using the disappearance of sounds rather than muffling. The main problem with using phase 5 is that in some individuals, especially when the cardiac output is high, the sounds do not disappear (although they do muffle), and sometimes persist right down to zero. However, in most people muffling and disappearance of the sounds usually occur within 10 mmHg of each other and may even occur together. When transferring between hospitals, nurses should always check on local policy regarding this.

Nurses should ensure that blood pressures measurements are taken under standardized conditions and using the correct technique. Blood pressure values vary according to the situation that the individual is in (see Fig. 4.2.47) and many physiological variables influence them. The individual should rest for at least 5 minutes before measurement, and should avoid exertion and not eat or smoke for 30 minutes beforehand. Both systolic and diastolic pressures are reported to rise by 10–33 mmHg within 15–45 minutes after the subject has eaten a meal (Jensen, 1982). Blood pressure also rises as the bladder fills.

The emotional state of the patient, e.g. whether anxious or in pain, will affect blood pressure values, but this is often difficult to avoid in clinical situations. A study, performed in Italy, illustrates this point clearly: the simple arrival of doctors at the bedside of patients induced an immediate rise in blood pressure (and heart rate), with mean values increasing by approximately 27 mmHg for systolic and 15 mmHg for diastolic above the previsit

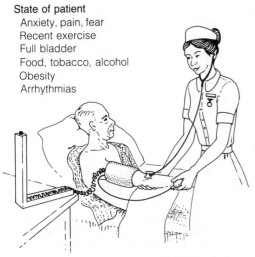

State of patient
  Anxiety, pain, fear
  Recent exercise
  Full bladder
  Food, tobacco, alcohol
  Obesity
  Arrhythmias

Nurse
  Training
  Observer bias
  Preferred digit
  Lack of concentration
  Sight and hearing
  Distance from sphygmo-
    manometer
  Diastolic dilemma

Cuff
  Correct application
  Dimensions of bladder
  Positioning of bladder

Patient
  Position
  Right or left arm
  Arm supported or not

Surroundings
  Room temperature
  Noise
  Other distractions

Sphygmomanometer
  Height
  Upright/vertical scale
  Maintenance
  Clogging of vent
  Level of mercury

Figure 4.2.52 Potential sources of error in blood pressure measurement (based on O'Brien and O'Malley, 1981)

values (Mancia *et al.*, 1983). The fact that anxiety influences blood pressure readings should be remembered, especially when intending to use observations taken at the time of admission to hospital as a baseline for subsequent observations.

There are also many potential sources for error in the actual measurement technique (Fig. 4.2.52); for instance, the arm should be at heart level or, more specifically, the arm should be horizontal with the fourth intercostal space at the sternum. The observer should also support the patient's arm, otherwise the patient will have to perform isometric muscle contractions which can increase the diastolic pressure. Raising or lowering the arm away from heart level causes significant changes in blood pressures, the error can be as large as 10 mmHg. The same arm should be used each time as in some individuals there are differences in the right and left brachial artery pressures. An appropriate size cuff should also be used; ideally the bladder of the cuff should encircle the arm, but if this is not feasible the centre of the bladder must be placed directly over the brachial artery. Tight or constricting sleeves of clothes pushed up to allow application of the cuff will also give false readings.

Observer bias, especially by looking at previously charted values and expectations of individuals' values, e.g. older people having higher blood pressures, is also a potential source of inaccuracy. The observer should also be at eye level with the mercury manometer scale when reading off the values. For some reason observers show a strong preference for the terminal digits 0 and 5, e.g. 125/75 mmHg, even though a 5-mmHg mark does not appear on many scales!

It is advisable to record an approximate value for systolic pressure by palpation, before auscultation, because in some people the Korotkov sounds appear normally giving the systolic pressure, but then disappear for a short time before returning above the diastolic pressure. This period of silence is known as the **auscultatory gap** and, although nothing can be heard, the pulse can be felt. (For a detailed discussion of the correct procedure see O'Brien and O'Malley, 1981.)

On many occasions it may not be possible to obtain all the optimum conditions, and if this is the case the qualifying factor(s) should be recorded on the chart, e.g. '150/94 mmHg – patient in severe pain'.

If it is not possible to use the arms for blood pressure readings, it is possible, using special large leg cuffs, to record the blood pressure using the Korotkov sounds from the popliteal artery in the popliteal fossa (at the back of the knee). The technique is more cumbersome but is useful in some instances, e.g. for patients with suspected coarctation of the aorta or with arm injuries. Pressure in the arteries of the legs is normally the same as that in the arms.

Anaeroid sphygmomanometers that work on a bellows system rather than on a mercury column, and also semiautomated systems that detect Korotkov sounds using a microphone or detect arterial blood flow using ultrasound, may be used clinically to measure blood pressure. However, the values obtained do not exactly correlate with direct arterial measurements.

A crude value for mean arterial pressure can be obtained using the method described earlier to demonstrate reactive hyperaemia. The mean arterial pressure is the pressure level when the arm flushes bright red as blood returns to the arm. This is described as the 'flush method' and is sometimes used in children or in shocked patients when other methods are not possible.

Measurement of venous pressures

Venous pressures reflect venous return to the heart and cardiac function. As venous pressures are not reflexly maintained at any specific or predetermined level, fluctuations in venous return (especially the circulating fluid volume) and cardiac function are reflected in the venous pressures. Observations of the jugular vein in the neck gives a crude indication of venous pressure: a raised jugular venous pressure (JVP) may indicate cardiac failure, for instance.

The venous pressure most frequently monitored is the **central venous pressure (CVP)**, which is the pressure in the central veins (the superior and inferior venae cavae) as they enter the heart. This is an invasive procedure and the catheter is usually inserted via the subclavian, internal jugular or, less often, the antecubital veins. Figure 4.2.53 shows a direct pressure monitoring system. As the tip of the catheter used to measure the CVP lies in the right atrium, CVP is synonomous with right atrial pressure (RAP). (The catheter is radio-opaque and its position is confirmed after insertion with a chest x-ray.)

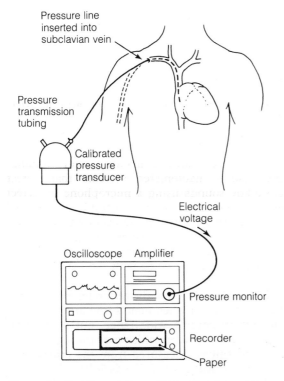

Figure 4.2.53 Direct pressure monitoring system (diagram shows monitoring of central venous pressure)

Figure 4.2.54 shows the factors that determine the CVP value. If the tricuspid valve is normal, the CVP equals the end-diastolic pressure in the right ventricle and as such it is an index of right ventricular function; impaired right ventricular function would lead to a back pressure that would raise the pressure in the atrium and hence give a higher CVP reading.

The volume of blood returning to the heart, i.e. the venous return, is the other major determinant of the CVP. Changes in the circulating fluid

volume and venomotor tone will alter venous return: an increase in the circulating fluid volume or venomotor tone will increase the venous return and give a higher CVP reading, and vice versa. One of the major clinical uses of CVP measurements is to monitor the circulating blood volume, so it is used in the management of fluid replacement in hypovolaemia, e.g. after burns, haemorrhage or surgery. Sequential measurements give a good indication of adequate fluid replacement and also help to prevent fluid overload. The technique and conditions must be consistent each time a measurement is taken. The CVP can be measured intermittently using a simple fluid manometer (Fig. 4.2.55) or continuously using a pressure transducer.

The fluid manometer is usually a branch of a normal giving-set linked to a venous catheter via a three-way tap and mounted against a centimetre reference scale. The central line is often used for giving clear parenteral fluids (e.g. saline or dextrose) when not monitoring pressure values. When measurements are being made, the tap is turned so that the pathway between the branch of the giving-set and the patient's vein are in direct communication (and the flow of fluid from the intravenous infusion is temporarily stopped). Therefore the fluid level in the manometer directly reflects the CVP. With a simple manometer the CVP is recorded in centimetres of water (strictly

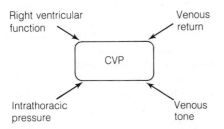

Figure 4.2.54 Factors that determine the central venous pressure (CVP)

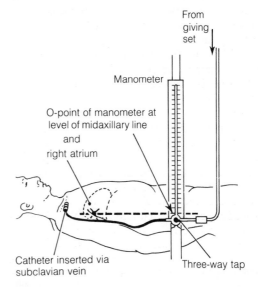

Figure 4.2.55 Measurement of central venous pressure using a fluid manometer system

this should be centimetres of saline or dextrose). A pressure of 10 cmH$_2$O means a pressure sufficient to raise a column of water to a height of 10 cm. CVP pressure is normally in the range of 0–10 cmH$_2$O (1.36 cmH$_2$O is equivalent to 1 mmHg), depending on the zero reference point used (see below).

The pressure transducer
A transducer is basically a small electronic device which converts one form of energy to another, and in this instance the transducer is converting the physiological variable of venous blood pressure into an electrical signal (it can equally well be used to measure arterial pressures). The electrical signal can then be processed, displayed and recorded at convenience; the values are often displayed on the oscilloscope screen in a waveform (Fig. 4.2.56) or a single value can be given. Before use, a transducer must be calibrated against a known pressure source, usually a column of mercury.

The intravascular catheter must remain fully patent for accurate monitoring; thrombus formation or kinking will cause a reduction or damping of the signal (Fig. 4.2.48 shows a damped arterial waveform). The tendency for thrombus formation is greatest with arterial lines because the relatively high pressure tends to force blood back into the catheter. The likelihood of this happening is reduced by using some form of continuous infusion device between the catheter and the transducer.

All direct blood pressure measurements, whether recorded via a transducer or manometer, must be recorded from a definite zero reference point and the same point must be used in each series of measurements. **Zero referencing** is the technique whereby the effects of both atmospheric and hydrostatic pressures are considered on the ultimate reading.

For arterial, central venous and pulmonary wedge pressures (see below), the common reference points are the sternal angle and the midaxillary line in the fourth intercostal space. The midaxillary line is preferred because it is anatomically in line with the right atrium. Placing the transducer or manometer at different levels relative to the patient alters the reading and so it is useless to compare measurements that are made with either the patient or manometer/transducer in different positions. Accurate alignment is achieved by using a spirit level, often attached to a telescopic arm. The zero reference point is often marked on the patient's skin. The patient should ideally be lying flat or, if not flat, at an angle of 45°, and all readings should be taken with the patient in the original position.

The exact position of the zero reference point is not so important when monitoring arterial pressures which cover a wide range, but is essential with venous pressures which vary through a small range of values. Measured from the midaxillary line, the normal range of CVP is 5–10 cmH$_2$O (0.5–1.0 kPa); from the sternal angle it is 0–5 cmH$_2$O (0–0.5 kPa). For a transducer, the normal range is 3–8 mmHg (0.4–1.1 kPa). CVP readings alter during inspiration and expiration as the intrathoracic pressure changes. Intermittent positive pressure ventilation increases the mean thoracic pressure by up to 5 mmHg and this is reflected by a rise in CVP; this must be remembered when interpreting the readings of ventilated patients.

WEDGE PRESSURES

Sometimes it is useful to have more specific information on the function of the left ventricle since this directly influences blood flow to the systemic circulation. In many instances, right arterial pressure (RAP) reflects left atrial pressure and so information on the left side of the heart can be gained by inference from CVP readings. However, there are many situations, for instance patients with pulmonary or cardiac problems (e.g. myocardial infarction, pulmonary oedema, after cardiopulmonary bypass surgery), when CVP is an unreliable indication of left ventricualr function.

This problem was overcome in the early 1970s with the development of a flotation, or Swan–Ganz, catheter which enables measurement of the

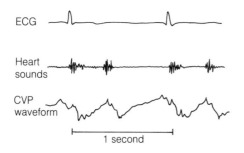

Figure 4.2.56 Waveform obtained when central venous pressure (CVP) is recorded continuously with a pressure transducer (electrocardiogram (ECG) and heart sound recordings are also shown)

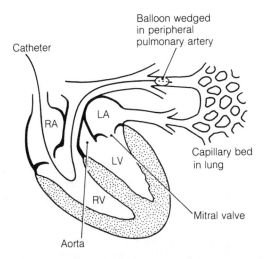

Figure 4.2.57 Position of Swan–Ganz catheter when 'wedged' in small pulmonary artery

pulmonary capillary wedge pressure (PCWP) or pulmonary wedge pressure (PWP). The technique involves inserting the catheter, which has a small inflatable balloon at its tip; it is usually inserted, like a CVP line, via the subclavian or antecubital vein. When the catheter tip reaches the superior vena cava the balloon is inflated, either with air or carbon dioxide, and the catheter 'sails' or 'floats' through the right side of the heart into the pulmonary artery where it eventually wedges in a small branch of the pulmonary circulation (Fig. 4.2.57). The catheter is attached to a pressure transducer and thus records a 'wedge pressure'.

With the balloon inflated and wedged, the pulmonary artery behind the wedged catheter is occluded and so the pressure recorded is a reflection of the pressure in front of the catheter tip, that is, from the left ventricle. At the end of ventricular diastole, all the valves between the left ventricle and pulmonary arteries are open so that

there is a continuous column of blood between these two points. Thus, in these circumstances

left ventricular end-diastolic pressure (LVEDP)

= pulmonary artery diastolic pressure (PADP)

= pulmonary capillary wedge pressure (PCWP)

Thus, use of the Swan–Ganz catheter is a relatively simple means of estimating the pressure in the left ventricle at the end of diastole, just before contraction.

It is possible to assess the position of the catheter, because wedging produces a characteristic change in the pulmonary artery pressure trace (Fig. 4.2.58). The normal PCWP range is from 4 mmHg to 12 mmHg (0.5–1.6 kPa). If the balloon were to remain inflated, damage would occur to the pulmonary capillary. Thus the PCWP is recorded intermittently. With the balloon deflated the catheter floats back into the main pulmonary artery and allows the pulmonary artery pressure (PAP) to be measured continuously.

Flotation catheters are now being used much more frequently and in some units PCWP is replacing CVP as a measure of cardiac function (but not for fluid replacement therapy) in many critically ill patients.

OSMOSIS AND THE FORMATION OF TISSUE FLUID

Osmosis

Osmosis is defined as the flow of solvent (water) across a semipermeable membrane, from a dilute to a more concentrated solution (or from an area of high water concentration to an area of low water

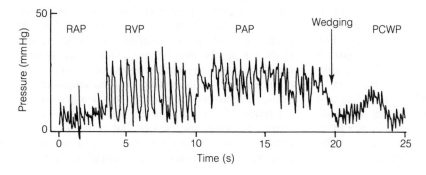

Figure 4.2.58 Pressure and waveform changes as a flotation catheter passes through the right side of the heart and into the pulmonary artery (RAP = right atrial pressure; RVP = right ventricular pressure; PAP = pulmonary artery pressure; PCWP = pulmonary capillary wedge pressure)

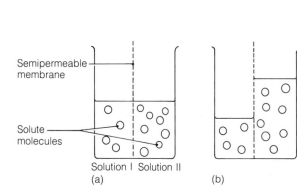
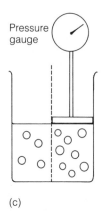

Semipermeable membrane

Solute molecules

Solution I | Solution II
(a)   (b)   (c)

Pressure gauge

Figure 4.2.59   The process of osmosis

concentration). This process occurs until the concentrations on either side of the membrane are equal (Fig. 4.2.59).

In Fig. 4.2.59a, two solutions of equal volume but differing concentrations are separated by a semipermeable membrane. Solution I is less concentrated than solution II. Solute molecules are too large to pass through the pores in the semipermeable membrane, but solvent molecules (not illustrated) can pass through freely.

In Fig. 4.2.59b, solvent has moved across the semipermeable membrane, from solution I to solution II, until the concentration of the two solutions is equal. This movement of solvent is called osmosis.

Osmotic pressure is the pressure required to stop the movement of solvent by osmosis (Fig. 4.2.59c). The greater the difference in concentration between the solutions on either side of the semipermeable membrane, the greater is the pressure required to halt the osmotic movement of solvent across the membrane.

A solution's ability to induce osmosis and hence osmotic pressure is expressed by the unit the **osmole**. The number of osmoles in a solution is calculated by multiplying the molar concentration of the solution by the number of ions present in each molecule of solute. The unit thus reflects the number of particles present in a solution. For example, a one molar solution of sodium chloride contains two osmoles, as sodium chloride dissociates into one sodium and one chloride ion.

The terms osmolality and osmolarity describe the osmole concentration of a solution. **Osmolality** is the number of osmoles present per kilogram of solvent, and **osmolarity** expresses the number of osmoles per litre of solution. When, as

in living systems, water is the solvent, the two values are practically identical since 1 litre of water weighs approximately 1 kilogram (1 ml water weighs 1 g). As the concentration of physiological solutions is low, this is expressed in milliosmoles (mosmol).

The term **tonicity** refers to a solution's osmotic concentration compared with that of intracellular and extracellular body fluids, which are normally in osmotic equilibrium. Hence a **hypertonic solution** is more concentrated than body fluids, and cells placed in such a solution shrink and shrivel (crenate) due to movement of water, by osmosis, to the more concentrated hypertonic solution. Conversely, a **hypotonic solution** is less concentrated than body fluids, and cells placed in such a solution swell, as water enters by osmosis, and may rupture, depending on the magnitude of the osmotic gradient created.

An **isotonic solution** is one which has the same osmotic activity as body fluids. Cells placed in such a solution neither gain nor lose water.

THE CAPILLARY WALL

In the tissues, blood is contained within the capillaries and is separated from the cells by the capillary wall, which is approximately $0.5\,\mu m$ in diameter, and the fluid-filled interstitial space. Cells are rarely more than $20\,\mu m$ away from a capillary. Capillary walls consist of a single layer of endothelial cells, resting on a basement membrane. Electron microscopy has suggested the existence of slit-like spaces, with a diameter of 10 nm, between the endothelial cells. These are known as pores. They represent only a very small proportion of the total surface area of the capillary wall. It is through these pores that water and solutes diffuse to and from the blood and interstitial fluid.

Unlike the cell membrane, the capillary wall is not a single semipermeable membrane. However, if its pores are thought of as the 'pores' in a single membrane, it is possible to examine osmotic effects across the capillary wall.

Formation and resorption of tissue fluid

Under normal conditions, the volume of the plasma and interstitial fluids changes very little over time, despite very high rates of diffusion of various substances both into and out of capillaries.

At the end of the last century, the physiologist Starling pointed out that the mean forces tending to move fluid out through capillary walls are in near equilibrium with those forces which tend to move fluid from the interstitial spaces into the capillaries. As a result, the volume of fluid passing out of the capillaries, by filtration, almost exactly equals the volume returned. This phenomenon is known as the Starling equilibrium of capillary exchange.

Starling proposed that the direction and rate of transfer between plasma in the capillaries and fluid in the tissue spaces depended on three factors (Fig. 4.2.60).

(a) The hydrostatic pressure on each side of the capillary wall.
(b) The osmotic pressure of protein in the plasma and in the tissue fluid.
(c) The properties of the capillary wall. (These have been discussed already.)

The capillary blood pressure (P_{cap})
This pressure forces fluid and solutes through the capillary walls into the interstitial spaces. The average pressure in the systemic capillaries is about 25 mmHg (3.3 kPa) but this is a broad generalization, for it represents the midpoint in a capillary where the pressure may be, say, 35 mmHg (4.7 kPa) at the arteriolar end and 15 mmHg (2 kPa) at the venous end. Blood pressure in any capillary is not constant over time; it depends on the arteriolar tone and the resistance offered and, as blood pressure is a hydrostatic pressure, it also depends on the position relative to the heart. The capillary pressure characteristically differs in certain tissues too. For instance, the blood pressure in the renal glomeruli is approximately 70 mmHg (9.3 kPa) and in the lungs 8 mmHg (1.1 kPa).

The interstitial fluid pressure (P_{if})
Tissue fluid, like any fluid, exerts a hydrostatic pressure. The value for this pressure must be subtracted from the capillary hydrostatic pressure to arrive at the effective transmural pressure, that is, the 'driving' pressure across the wall. The experimental work measuring interstitial fluid pressure gives conflicting values according to the precise methodology used: some measurements suggest a negative or subatmospheric pressure whereas others indicate pressures between 0 mmHg (atmospheric pressure) and 2 mmHg (0.3 kPa) – the latter figures have been adopted for this text.

The plasma colloid osmotic pressure (OP_{cap})
This produces fluid movement, by osmosis, into the capillary with a force of approximately 28 mmHg (3.7 kPa).

The interstitial fluid colloid osmotic pressure (OP_{if})
This produces fluid movement, by osmosis, out of the capillary and has a magnitude of approximately 5 mmHg (0.7 kPa).

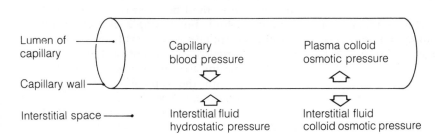

Figure 4.2.60 Forces affecting fluid movement across the capillary wall

Lumen of capillary

Capillary wall

Interstitial space

Capillary blood pressure

Interstitial fluid hydrostatic pressure

Plasma colloid osmotic pressure

Interstitial fluid colloid osmotic pressure

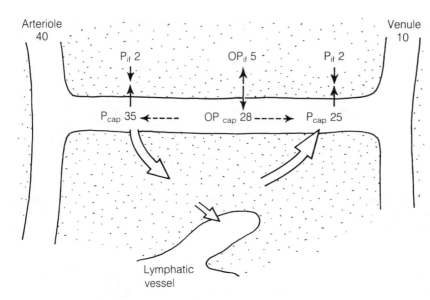

Arteriole
40

Venule
10

P_{if} 2 OP_{if} 5 P_{if} 2

P_{cap} 35 ◄---- OP_{cap} 28 ----► P_{cap} 25

Lymphatic
vessel

Figure 4.2.61 Diagram summarizing the forces contributing to the formation and reabsorption of tissue fluid in the systemic circulation (for full explanation see text; all figures refer to pressures in mmHg). (P_{cap} = capillary blood pressure; P_{if} = interstitial fluid pressure; OP_{cap} = plasma colloid osmostic pressure; OP_{if} = interstitial fluid colloid osmotic pressure.) *NB.* P_{cap} is the only pressure for which magnitude alters

Filtration and absorption of tissue fluid

As already discussed, the pressures involved vary in different tissues and under different circumstances, therefore hypothetical figures are used to illustrate this example (see also Fig. 4.2.61).

P_{cap} (arterial end)	=	35 mmHg
P_{cap} (venous end)	=	15 mmHg
P_{if}	=	2 mmHg
OP_{cap} (tending to retain fluid in capillary)	=	28 mmHg
OP_{if} (tending to retain fluid in interstitial spaces)	=	5 mmHg

At the arterial end of a capillary

Transmural pressure:

$$P_{cap} - P_{if} = 35 - 2 = 33\,\text{mmHg}$$

Net osmotic pressure:

$$OP_{cap} - OP_{if} = 28 - 5 = 23\,\text{mmHg}$$

Therefore the forces tending to move fluid out of the capillaries exceed those tending to retain fluid by 10 mmHg (1.3 kPa). Thus at the arterial end, fluid leaves the capillaries.

At the venous end of a capillary

Transmural pressure:

$$P_{cap} - P_{if} = 15 - 2 = 13\,\text{mmHg}$$

Net osmotic pressure:

$$OP_{cap} - OP_{if} = 28 - 5 = 23\,\text{mmHg}$$

Therefore the forces tending to move fluid into the capillary exceed those tending to move fluid out of the capillary by 10 mmHg (1.3 kPa). Thus at the venous ends of capillaries, due to the fall in capillary blood pressure which occurs along the length of the capillary, fluid enters the capillaries. Thus a circulation of tissue fluid is established.

As a result of these forces, a small percentage of the plasma volume entering the capillaries is filtered into the interstitial space and flows through the tissues. The vast majority of the volume filtered is reabsorbed at the venous ends of the capillaries. The remaining fluid passes into the lymph capillaries. Lymph drainage plays an important part in maintaining local fluid equilibrium in some tissues.

If these forces become abnormal, this affects the distribution of fluid between the plasma and the interstitial fluid. For example, if mean capillary blood pressure falls considerably, as might occur in severe haemorrhage, the outward forces become less than the inward forces and so more fluid is reabsorbed into the capillary than is filtered. Initially, therefore, following haemorrhage plasma volume increases at the expense of interstitial fluid volume.

OEDEMA

More commonly, though, the imbalance which occurs is such that the outward forces exceed the inward forces and this produces an excessive accumulation of interstitial fluid which is called **oedema**. If oedema is present for some time, the tissue spaces become stretched and this increases the ease with which further oedema can develop.

Most oedema demonstrates the phenomenon of **pitting**. If the skin over an oedematous area is pressed with the index finger, a pit appears in the tissues and remains for up to 30 seconds after the finger is removed. This is because the oedema has been pushed away through the tissue spaces, by the pressure from the finger, and it takes some time to flow back once the pressure is released.

Causes of oedema

There are many causes of oedema, but there are only four major physiological mechanisms underlying its formation; all these result in the outward forces exceeding the inward forces in the affected capillaries. Tissues do not become oedematous until there is quite a large imbalance in these forces acting at the capillary walls. Table 4.2.3 classifies the causes of oedema according to the four physiological mechanisms involved.

The accumulation of oedema is subject to the force of gravity and so generalized oedema tends to become apparent first in the soft tissues of dependent parts of the body, such as the ankles, wrists and fingers and around the sacrum. **Ascites**, an abnormal accumulation of fluid in the peritoneal cavity, occurs due to physiological mechanisms similar to those of oedema.

Oedema is treated according to its cause. Diuretics, such as frusemide, are frequently prescribed to promote the excretion of sodium and water. The effectiveness of treatment may be monitored with the help of accurate fluid balance recordings and by recording the weight of the patient, daily, at the same time of day. The skin of an oedematous person requires special care, as waterlogged tissues tend to be deprived of oxygen and nutrients, because of the increased barrier to diffusion, and hence vulnerable to trauma and less able to heal.

In an individual who has varicose veins the valves in the veins are not functional and the

Table 4.2.3 Causes of oedema and their physiological mechanisms

Physiological mechanism	Cause	Examples
Increased mean capillary blood pressure	Increased venous pressure	Back pressure due to reduced cardiac output in cardiac failure Venous obstruction due, for example, to pressure from plaster casts/bandages/garters which are too tight Fluid overloading with intravenous fluids
	Sodium and water retention	Reduced renal blood flow Renal failure
	Raised aldosterone secretion	Mineralocorticoid effect of corticosteroid therapy
	Inability to destroy aldosterone	Liver damage, e.g. cirrhosis
Decreased plasma colloid osmotic pressure	Decreased production of plasma protein Loss of plasma protein from the body	Liver damage Malnutrition (kwashiorkor) Nephrosis, nephrotic syndrome Burns, draining wounds and fistulae
Increased permeability of capillary walls to protein	Increased porosity of capillary walls	Inflammation, e.g. due to burns and infections Allergic reactions, e.g. hives
Decreased lymphatic removal of protein and tissue fluid (leads to increased interstitial fluid pressure and interstitial fluid colloid osmotic pressure)	Congenital absence of lymphatic vessels Blockage of lymphatic vessels	Malignant disease Surgical removal of lymph nodes Infestation by parasitic worm, filaria

venous and capillary pressures are high. Thus fluid tends to leak out of the capillary and oedema develops. As described above, the normal exchange of gases and nutrients is disrupted; the muscles in the leg can become painful and weak and the skin can become gangrenous and varicose ulcers develop. When a limb is oedematous, as a result of venous stasis, elevation of that limb and the application of an elastic stocking will facilitate venous return.

Interstitial fluid formation and reabsorption in the pulmonary circulation

The mechanics of interstitial fluid formation and reabsorption in the low pressure pulmonary circulation are essentially the same as for the systemic circulation. The major difference is that the mean blood pressure in pulmonary capillaries is only approximately 8 mmHg (1.1 kPa), much less than the plasma colloid osmotic pressure of 28 mmHg (3.7 kPa).

Pulmonary oedema occurs via the same mechanisms as in the peripheral tissues. The major cause is failure of the left side of the heart to maintain its output of blood. This can lead to great increases in pulmonary capillary blood pressure. Pulmonary oedema increases the diffusion barrier between alveoli and blood and, almost invariably, fluid enters the alveoli. It thus interferes with the diffusion of the respiratory gases and this produces hypoxia and breathlessness. In very severe cases, pulmonary oedema can produce death, from asphyxiation, in less than an hour.

As with peripheral oedema, the cause of the condition must be treated. Symptomatic relief can be obtained by nursing the patient sitting up. This maximizes the efficiency of his respiratory effort. Oxygen is usually required to relieve breathlessness and a bronchodilator, such as aminophylline, may also be prescribed. In very severe cases, intermittent positive pressure ventilation may be necessary. Diuretic therapy promotes fluid loss. In acute cases, morphine may be given to allay pain and emotional distress. This drug also produces vasodilation and so helps to reduce blood pressure. Vasodilators, such as sodium nitroprusside, also play a major role in the treatment and prevention of acute pulmonary oedema.

The breathlessness and discomfort produced by this condition are extremely frightening and the patient's distress tends to aggravate his condition. The ability of nurses and others in the care team to reassure and calm him is therefore also an important aspect of care.

CONTROL AND COORDINATION OF THE CARDIOVASCULAR SYSTEM

The function of the cardiovascular system is to ensure an adequate blood flow to all tissues at all times. All homeostatic mechanisms, and the cardiovascular system is no exception, need an integrating and control centre which receives 'information' from receptors via afferent pathways and coordinates the effector responses.

The coordinating centres that link cardiac and vascular responses are in the medulla and pons of the brain. The major parameter that these centres regulate is the systemic arterial blood pressure, as it is this that ensures blood flow to the tissues, especially to the vital organs of the brain and heart.

As seen earlier, mean arterial pressure can be regulated by altering cardiac output and/or the total peripheral resistance (i.e. $BP = CO \times TPR$). Thus, any factor which influences either cardiac output or peripheral resistance will affect blood pressure, for instance heart rate, stroke volume, radius of arterioles, venomotor tone, circulating blood volume. Many systems are involved in controlling these factors, including nervous, hormonal and local factors; some of the systems act rapidly to maintain blood pressure on a short-term basis (e.g. nervous reflexes) and others operate in the longer term (e.g. hormonal fluid balance mechanisms). Thus the role of the cardiovascular control centres is to integrate the function of the heart and vascular system.

Medullary cardiovascular centres

In the medulla and pons there is a large diffuse area of interconnected neurones concerned with the regulation of the heart and vascular system. This region, known as the **cardiovascular centre (CVC)**, receives afferent inputs from other parts of the central and peripheral nervous systems and sends out efferents through sympathetic

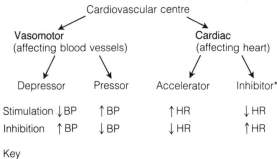

Stimulation ↓BP | ↑BP | ↑HR | ↓HR
Inhibition ↑BP | ↓BP | ↓HR | ↑HR

Key
BP = blood pressure
HR = heart rate

Figure 4.2.62 Subdivisions of cardiovascular control centres (*cardiac inhibitory centre corresponds to nucleus ambiguus)

and parasympathetic fibres to the heart and blood vessels. In this way the CVC is able to regulate heart function and blood pressure in accordance with the body's physiological needs.

The CVC is sometimes subdivided into discrete units (Fig. 4.2.62), e.g. vasomotor centre, cardiac accelerator centre, but there is no clear anatomical separation between these various centres and their function is not always discretely divided up. For

instance, the vasomotor centre, the region supposedly exerting control over sympathetic vasoconstrictor activity, also contains neurones that affect the heart and produce positive chronotropic (increase heart rate) and positive inotropic (increase stroke volume) responses. For these reasons, as there are no definite anatomical or functional divisions between the control centres, it is appropriate to use the term **medullary cardiovascular centre** (Keele *et al.*, 1982). Thus, for the purposes of this discussion, the controlling centres will be described simply as the cardiovascular centre (CVC). The afferent inputs to the CVC are from the baroreceptors and chemoreceptors as well as from the cortex, hypothalamus and other areas of the central nervous system. The efferent pathways are mainly via the sympathetic nerves to the heart (altering the force and rate of contraction) and the blood vessels (altering blood vessel diameter). Sympathetic vasoconstrictor activity in the arterial system increases the peripheral resistance whilst in the venous system it increases the venous return to the heart. The parasympathetic outflow via the vagus nerve arises from the nucleus ambiguus (which corresponds to the centre previously known as the cardiac inhibitory centre; Keele *et al.* 1982). Figure 4.2.63 summarizes the afferent and efferent pathways involved.

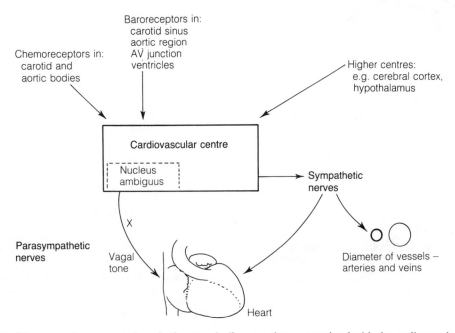

Figure 4.2.63 Diagrammatic representation of afferent and efferent pathways associated with the cardiovascular control centre

The baroreceptor mechanism

The baroreceptor mechanism is the most important of the feedback systems involved in the maintenance of arterial blood pressure on a short-term
basis, e.g. during changes in posture, different
activity levels, etc. As the parameter being regulated is the arterial pressure, it is appropriate that
the major receptors for the feedback loop, i.e. the
baroreceptors, are located within the arterial system. The baroreceptors are situated in the tunica
adventitia of both the internal carotid artery
(especially in the carotid sinus) and the transverse
part of the aortic arch (Fig. 4.2.64). These areas
are linked to the CVC by the sinus nerve (part of
the glossopharyngeal (IX) cranial nerve) and the
aortic nerve (part of vagus (X) cranial nerve)
respectively.

The **baroreceptors** are specialized nerve endings that respond to the degree of stretch of the
arterial wall, that is, they are stretch receptors.
The degree of stretching in the arterial wall is
directly related to the blood pressure in the carotid
sinus and aortic arch; if there is an increase in the
blood pressure, there will be more stretch, and if
blood pressure falls, less stretch. The stretching
alters the rate of firing of impulses along the nerves
to the CVC. The baroreceptors are sometimes
known as pressoreceptors.

There is always a basal rate of discharge to the
CVC from the baroreceptors along these afferent
fibres which relates directly to the degree of
stretching of the arterial wall; when these endings
are stretched there is an increased firing rate in the
afferent fibres which serves to communicate the
degree of stretch to the CVC in the brain (to the
region often referred to as the vasomotor centre).
So, when the baroreceptors are further stretched,
the firing rate increases, and vice versa (Fig.
4.2.65). The discharge from the baroreceptors is
pulsatile due to the change in the arterial pressure
during the cardiac cycle.

The normal action of the CVC on this afferent
input is *inhibitory* and so an increased blood press-

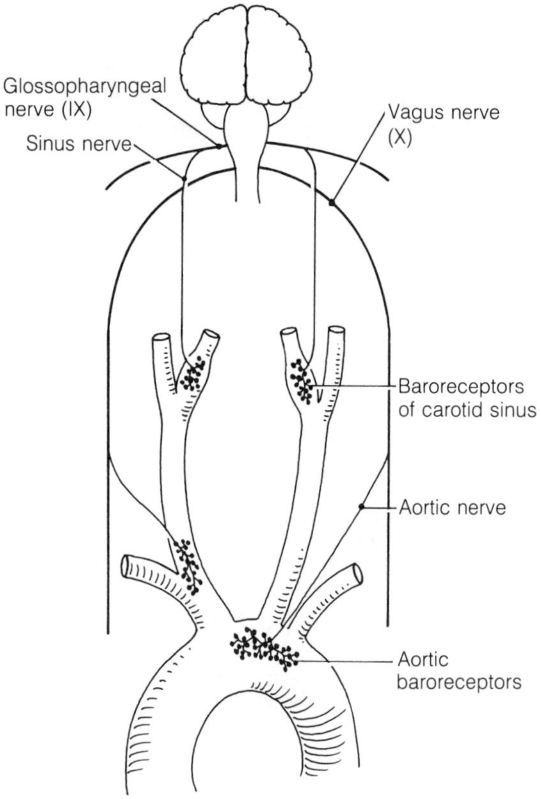

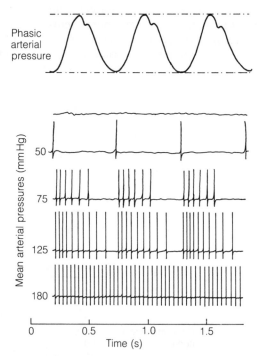

Figure 4.2.64 Arterial baroreceptor system

Figure 4.2.65 Relationship between mean arterial blood
pressure and the impulse firing rate in a single afferent
nerve fibre from the baroreceptors in the carotid sinus with
different levels of mean arterial pressure. The baroreceptors
discharge at a greater frequency as the pressure is increased
(see text)

ure produces an increased firing rate, and this leads to a reduced sympathetic discharge rate to the blood vessels (producing vasodilation) and reduced sympathetic discharge to the heart (producing a decrease in the force and rate of contraction). Simultaneously, there is an increased parasympathetic vagal discharge to the heart which also reduces the heart rate. The effect of the vasodilation, reducing the total peripheral resistance, and the drop in heart rate and stroke volume, reducing cardiac output, is to cause a drop in the arterial blood pressure. Thus, the initiating stimulus – the increased blood pressure – has been negated, and the blood pressure returned to 'normal'.

$$CO \times TPR = BP$$

$$(HR \times SV) \times TPR = BP$$

$$(If \downarrow HR \ and \downarrow SV \ and \downarrow TPR, \ then \downarrow BP)$$

If there is a fall in blood pressure, the firing rate will decrease and, by a reverse of the mechanisms described above, heart rate and force of contraction will increase and vasoconstriction occurs (Fig. 4.2.66).

Other mechanoreceptor reflexes

There are also stretch receptors, similar to baroreceptors, in other parts of the thorax, for instance at the junction between the vena cava and right atrium, at the junction of the pulmonary vein and left atrium, sparse fibres scattered over the atria and ventricles and probably at other sites within the thorax. All the afferent nerve endings are responsive to stretch and mechanical distortion and so are often referred to as **mechanoreceptors**. The impulses from these mechanoreceptors are all carried in the vagus to the CVC and so often referred to simply as vagal endings.

The mechanism of action of the cardiac vagal mechanoreceptors is not exactly like the classic baroreceptor system in the systemic arteries; their response is complex and their role is not fully understood. Different receptors seem to have different responses. Some of the receptors seem to play an important role in the control of body sodium and water by acting as blood volume receptors and influence the output of antidiuretic hormone from the posterior pituitary. Other

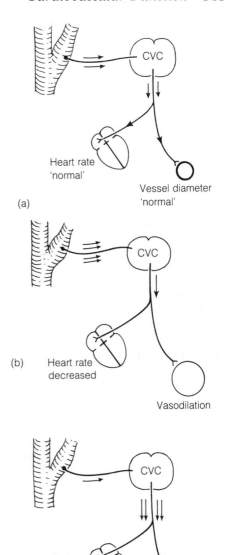

Figure 4.2.66 Diagrammatic representation of the inverse relationship between baroreceptor firing rate and sympathetic outflow to the heart and blood vessels: (a) pressure 'normal', (b) pressure high, (c) pressure low (CVC, cardiovascular centre)

receptors seem to cause reflex vasodilation, e.g. in the renal circulation.

Activation of some of the fibres from the ventricles seem reflexly to inhibit sympathetic discharge and produce a profound bradycardia. Patients having coronary angiography often show bradycardia when the contrast medium is injected, and it has been suggested that this is due to chemical excitation of these ventricular fibres. The vasovagal syncope and bradycardia of myocardial infarction may also be ascribed to reflexes aroused by an increase in activity of these ventricular fibres (Keele *et al.*, 1982).

Stretch of some of the atrial vagal endings causes an increase in heart rate and an increase in the strength of contraction (the Bainbridge reflex). The likely function of this reflex is to help prevent the accumulation of blood in the veins, atria and pulmonary circulation.

Thus the effects of these other mechanoreceptor reflexes are varied, and although they may not be primarily concerned with blood pressure regulation, they play an important role in other cardiovascular reflexes.

Chemoreceptor input

The chemoreceptors in the carotid and aortic bodies (see Chap. 5.3) can also influence the CVC, although they are mainly involved with the regulation of respiration. The chemoreceptors are responsive to oxygen lack, hypercapnia, acidaemia and asphyxia. Under normal circumstances the chemoreceptors only have a small effect on the CVC, but under 'emergency' conditions of anoxia or asphyxia they can and do provoke powerful reflex sympathetic effects; for example, a decrease in oxygen or increase in carbon dioxide (or hydrogen ion concentration) causes an increase in heart rate and blood pressure (Fig. 4.2.67).

Influence of other areas of the central nervous system

Emotions such as excitement, fear and anxiety can cause an increased heart rate and blood pressure, whereas a sudden shock may induce a bradycardia. Stimuli from the hypothalamus influence vasomotor tone and this is one of the mechanisms used to control heat loss and maintain normal body temperature. Pain produces changes in heart rate and blood pressure. The respiratory centres also influence the CVC.

Illustrations of cardiovascular control

The activity of the cardiovascular system is thus coordinated by centres in the brain. As far as the normal control of arterial pressure is concerned, the baroreceptor mechanism is the most important. The baroreceptors respond immediately to a change in arterial pressure and their response is fully active within a minute or so. Besides these direct nervous reflexes, via the CVC, there are other moderately rapidly acting (in terms of minutes to hours) systems that operate for the short-term control of blood pressure, namely hormonal effects, particularly adrenaline and noradrenaline, and intrinsic physical mechanisms altering the shift of fluids from interstitium to plasma.

The baroreceptors are not effective for the long-term control of blood pressure; it seems that the baroreceptor/CVC tonic discharge rate becomes reset at a different level when blood pressure is chronically elevated. In the long term, renal and body fluid mechanisms become the dominant pressure-determining factors.

A few illustrations of the cardiovascular system 'in action' will now be given. (Most of the separate components of these changes have been discussed earlier in the chapter, so for the details of mechanisms see the relevant part of the chapter.)

CHANGES IN POSTURE

When an individual is supine, the arterial and

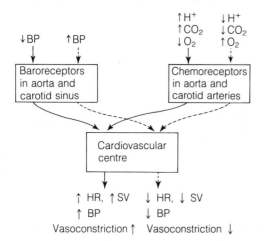

Figure 4.2.67 Role of baroreceptors and chemoreceptors in regulating heart function and blood pressure

venous pressures are not significantly influenced by the hydrostatic pressure of the blood. However, on standing, due to the effects of gravity, the arterial and venous pressures in the head drop whilst the pressure in the feet and lower parts of the body rises (see Fig. 4.2.32). Due to the high compliance of the veins, the limb veins distend and blood tends to pool in the arms and legs. As the venous capacity has increased, the venous return to the heart falls. This is the sequence of events that occurs every time an individual changes from a horizontal to a vertical position. The decreased venous return reduces the stretching of the cardiac muscle fibres and the stroke volume decreases (Starling's law of the heart). As a consequence of the decrease in stroke volume, the cardiac output falls and the arterial blood pressure falls.

This drop in arterial blood pressure is detected by the baroreceptors; the baroreceptors in the carotid sinus are particularly affected. Thus there is decreased baroreceptor activity and the CVC initiates sympathetic activity to increase heart rate, stroke volume and vasoconstriction. The arterial vasoconstriction increases peripheral resistance and the increased venomotor tone enhances venous return to the heart. These changes together correct the blood pressure and compensate for the effect of gravity. The response is complete, in normal healthy individuals, literally within seconds. Figure 4.2.68 summarizes these changes.

If the body does not correct the blood pressure or if the compensatory mechanisms are sluggish, postural or orthostatic hypotension develops with dizziness and even fainting if severe. You can observe this mechanism sometimes if you get up too quickly after bending down.

Patients who are in bed for long periods of time, and thus remain more or less horizontal, do not have to overcome the effects of gravity. (As an aside, this is rather like astronauts in space, in fact much of the work carried out to study the effects of prolonged weightlessness have been done by confining subjects to bed!) Once bed-bound patients start to get up, they often suffer from weakness, dizziness, giddiness or can faint, and this is again often a result of orthostatic hypotension. Whilst confined to bed, patients lose general muscle tone, including that of the blood vessels, especially veins, and also suffer a decreased efficiency of the cardiovascular reflexes, e.g. the

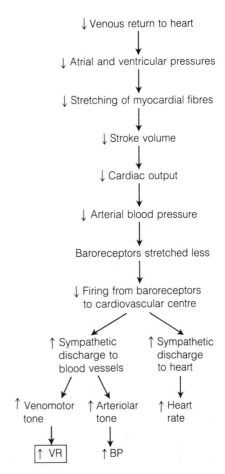

Figure 4.2.68 Sequence of events that occurs immediately after changing posture from lying down to standing up. *NB.* Once venomotor tone has increased and the venous return (VR) to the heart has increased, the heart rate (HR) and blood pressure (BP) can return to 'normal'

baroreceptor mechanism. The decrease in muscle tone also means that the muscle pump mechanism that normally aids venous return in the vertical position is less efficient.

It has been suggested that bedrest may increase the work load of the heart by increasing the venous return, and hence cardiac output, due to the horizontal position. (Olsen (1967) reviews the hazards of immobility and bedrest.) Certainly there is a decline in general cardiovascular fitness, or a 'detraining effect', due to the reduction in activity levels, which will certainly be deleterious to any patient.

HAEMORRHAGE

When an individual suffers a haemorrhage, there is a decrease in the circulating blood volume; this leads to a reduction in the venous return to the heart, and to a reduced stroke volume and cardiac output, and to a drop in blood pressure. (This sequence of events is the same as discussed above when changing from a horizontal to a vertical posture.) The drop in blood pressure sets off the baroreceptor mechanism, and intense sympathetic activity is established which causes an increased force and rate of contraction of the heart (so attempting to maintain cardiac output) and the vasoconstriction increases peripheral resistance and venous return. The venoconstriction reduces the size of the blood reservoir in the veins and so increases the effective circulating blood volume.

The intense arteriolar constriction results in a drop in the capillary blood pressure, which has the effect of increasing fluid shifts *from* the tissue spaces *into* the capillaries, and this also increases the effective circulating blood volume.

Under conditions of shock or haemorrhage the circulation of blood to the vital organs (brain and heart) has priority and so blood is 'shunted' away from less vital areas, i.e. from the skin, skeletal muscle, gut, renal system and spleen, by intense vasoconstriction.

The sympathetic arousal causes the release of adrenaline and noradrenaline from the adrenal medulla, and these hormones augment the effects of the sympathetic nerves. The decreased renal blood flow activates the renin–angiotensin system which causes more vasoconstriction, and also leads to the release of aldosterone to conserve sodium. This in turn leads to increased release of ADH from the posterior pituitary and so tends to conserve water (see Chap. 2.5). The volume receptors in the atria also increase ADH production.

Thus there is a wide range of responses to haemorrhage. The above physiological changes explain the common signs and symptoms observed in a shocked patient:

pale due to peripheral vasoconstriction
tachycardia ⎫
sweaty ⎬ due to sympathetic arousal
weak pulse due to low blood volume
oliguria due to reduced renal blood flow and fluid-conserving mechanisms.

The blood pressure may well be low; however, it is often the last 'sign' to develop as the cardiovascular reflexes may be 'strong' enough to maintain the arterial blood pressure for some time (this obviously depends on the rate and nature of the haemorrhage). An indication of how effective these compensatory mechanisms are is seen by considering a case of 'controlled haemorrhage' when donating blood: the body is normally easily able to compensate for the loss of 500 ml or so of blood (almost 10% of the total blood volume).

However, in 'clinical' haemorrhage, if the circulating blood volume is not restored fairly quickly, the tissues deprived of blood, i.e. the ischaemic tissues, will start to necrose. Thus the priority in the treatment of these patients is to replace fluids – often plasma expanders or whole blood are given (see Chap. 4.1).

For a detailed discussion on shock see Kelman (1980).

The effect of exercise on the cardiovascular system

The 'mental' anticipation of exercise can result in sympathetic arousal and lead to an increase in heart rate. Once the exercise is underway, the muscles that are active require an increased blood flow to provide the necessary oxygen and nutrients and to remove waste products. This increased blood flow is achieved by local factors: vasodilation is produced by the accumulation of local metabolites that directly affect the smooth muscle of the arteriolar walls and precapillary sphincters. Thus a greater number of capillary beds are opened up and the distance substances have to diffuse to and from the cells is reduced.

As a consequence of the vasodilation, the peripheral resistance in the muscle drops, and in order to maintain the total peripheral resistance at adequate levels (so that blood pressure does not fall too far), compensatory vasoconstriction occurs, especially in the kidneys and gut. Blood flow to the skin may increase in order that the body can lose heat, and arteriovenous anastomoses open up.

In order for the blood flow to the muscles to increase, the cardiac output must also increase, and this is partly brought about by sympathetic activity. The venous return to the heart is increased by an increased venomotor tone and especially by the action of the skeletal muscle and respiratory pump mechanisms. The increased venous return (by Starling's law) results in an increased stroke volume. The sympathetic influence also increases the force and rate of cardiac contractions.

The maximum attainable heart rate varies with sex, age and state of physical fitness. As a useful guide, the maximum attainable heart rate is given by the formula 200 − (0.65 × age in years).

The cardiac output increases from the resting value of approximately 5 l min and can reach 25 l min (or even 35 l min in elite athletes).

The systolic blood pressure increases due to the increased stroke volume. The diastolic pressure may increase, but if the total peripheral resistance falls, the diastolic pressure may fall too.

The ability of the cardiovascular system to cope with exercise increases considerably with training. Athletes have a slower heart rate at rest, because they have a greater stroke volume – resting heart rates of 40 are not uncommon in very fit people. This increases their reserve when they undertake exercise.

Assessment of the adequacy of the circulatory system

There are many ways of assessing the adequacy of the circulation in the body and of the circulatory system as a whole; some are very simple and others more sophisticated, requiring technical equipment and invasive procedures (Fig. 4.2.69).

For instance, simple observations include assessment of the peripheral circulation by observing the colour and temperature of the skin: with a greater blood flow, the skin appears flushed and warmer (e.g. with exercise, cellulitis), and with a decreased blood flow, the skin is pale and cold (e.g. peripheral vascular disease, ischaemia, low blood volume). The temperature of the skin in the extremities is often used as an indication of peripheral perfusion in intensive care units.

Information can also be gained by assessing the symmetry of colour, temperature and pulses on both sides of the body or in both limbs. The development and early signs of pressure sores are indications that the circulation to a particular area is inadequate. The character of the peripheral

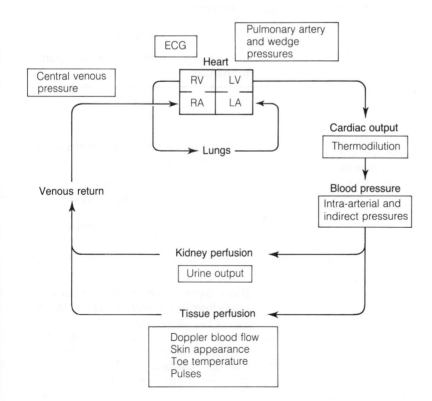

Figure 4.2.69 Summary of some of the techniques which can be used to monitor and assess the cardiovascular system

pulses, e.g. whether weak and thready or bounding, gives qualitative information on the circulation; the pulse obviously gives an easy method of counting heart rate.

The efficiency of the circulatory system as a whole can be assessed by monitoring the function of vital organs, and kidney function is often used. A decreased blood flow to the kidney can lead to oliguria: regular half-hourly or hourly urimeter readings made on catheterized patients who are critically ill are sometimes used to assess renal perfusion which can give information on the fluid volume or progress of, say, shock.

Other non-invasive methods but which require some technical apparatus are commonly used, for instance sphygmomanometers to measure blood pressure, ECG machines to monitor rate, rhythm and conduction patterns in the heart, or Doppler equipment to monitor blood flow in the peripheral circulation.

Invasive techniques are useful in giving information on venous return to the heart and cardiac function, e.g. central venous pressure measurements, wedge pressures and cardiac output estimations. The circulatory system can also be visualized by injecting radio-opaque substances into the blood and then taking x-rays of the arteries. This technique is called angiography or arteriography and the film an arteriogram. Similar procedures can outline the ventricles, e.g. a ventriculogram. Angiography is used in the diagnosis of atherosclerosis. However, invasive procedures do carry some risk to the patients.

These are just some of the numerous possible investigations and ways of monitoring the cardiovascular system. Although the more complex technical investigations are useful, the simple and safe ones should be used first.

Review questions

The answers to all these questions can be found in the text. In each case there is at least one correct and at least one incorrect answer.

1 The following statements concern the circulatory system.

   (a) The blood vessels with the greatest elasticity are the arteries.
   (b) About two-thirds of the circulatory blood volume is located in the systemic veins.
   (c) The portion of the circulatory system with the largest total cross-sectional area is made up of the capillaries.
   (d) A portal system carries blood from capillaries to venules.

2 The following statements are concerned with blood pressure.

   (a) The highest pressure in an artery develops during the systolic phase of the cardiac cycle.
   (b) The diastolic pressure can be estimated by palpation of the radial artery (when using a sphygmomanometer).
   (c) Arteriosclerosis, which increases the stiffness of the arterial walls, tends to raise the systolic arterial pressure.
   (d) The capillary blood pressure is the same in all tissues.

3 Vasodilation of the arterioles is caused by

   (a) nicotine
   (b) noradrenaline
   (c) histamine
   (d) local metabolites.

4 Movement of blood in the veins is

   (a) assisted by smooth muscle pump mechanism
   (b) assisted by respiratory movements
   (c) kept moving forward by valves
   (d) increased in exercise.

5 The following factors will increase the venous return to the heart.

   (a) Increased venomotor tone.
   (b) An increase in right atrial pressure.
   (c) Lying down after standing up.
   (d) Increased skeletal muscle activity.

6 The following statements concern tissue fluid formation.

   (a) Blood hydrostatic pressure is the most important factor in moving substances out of the capillaries.
   (b) Blood oncotic pressure is the most important factor in moving fluid back into the capillaries.
   (c) Loss of fluid from the capillaries is decreased by constriction of the veins.
   (d) Loss of fluid from the capillaries is increased when arterioles constrict.

7 Oedema formation may occur when

    (a) plasma osmotic pressure is increased
    (b) the capillary permeability to plasma proteins is increased
    (c) large volumes of isotonic saline are given intravenously
    (d) lymph flow is blocked.

8 The following statements concern cardiac muscle and its blood supply.

    (a) The Purkinje fibres are specialized nerve cells.
    (b) The length of the cardiac muscle fibres affects the force of contraction.
    (c) The blood supply to the myocardium is via the coronary arteries which lead directly off the pulmonary artery.
    (d) The coronary arteries dilate in response to lack of oxygen.

9 According to Starling's law of the heart, increased venous return

    (a) increases the heart rate
    (b) decreases the end-diastolic volume
    (c) increases the stroke volume
    (d) decreases the cardiac output.

10 In an electrocardiogram

    (a) the P wave is associated with atrial repolarization
    (b) the QRS complex represents depolarization of the ventricles
    (c) the QRS complex immediately follows contraction of the ventricles
    (d) the T wave is due to repolarization of the ventricles.

11 The following statements concern the autonomic nerve supply to the heart.

    (a) Stimulation of the vagus nerve slows the heart rate.
    (b) Sympathetic stimulation increases the rate of depolarization of the sinoatrial node.
    (c) Parasympathetic stimulation increases the force of myocardial contraction.
    (d) Cutting all the autonomic nerves to the heart will produce an increase in the resting heart rate.

12 Cardiac output (in litres per minute) divided by the heart rate (in beats per minute) equals

    (a) mean arterial pressure
    (b) end-diastolic volume
    (c) mean stroke volume
    (d) peripheral resistance.

13 Which of the following is/are correct? (BP = blood pressure; CO = cardiac output; SV = stroke volume; HR = heart rate; TPR = total peripheral resistance.)

    (a) BP/HR = SV
    (b) CO = BP/TPR
    (c) BP = HR × SV × TPR
    (d) SV = CO/HR

14 The following statements concern the cardiovascular control centre.

    (a) It is located in the cerebral cortex.
    (b) It controls the heart rate only.
    (c) All 'information' from the baroreceptors passes to the vasomotor centre, which is inhibitory in nature.
    (d) It receives 'information' from higher centres and the hypothalamus.

15 The following statements concern the baroreceptor mechanism.

    (a) Baroreceptors monitor both the magnitude and rate of change of arterial blood pressure.
    (b) Baroreceptors are located in the carotid sinus and aortic arch.
    (c) The frequency of signals from the baroreceptors is inversely proportional to the blood pressure.
    (d) A high frequency of signals from the baroreceptors reduces vasoconstriction and slows the heart rate.

16 The following statements concern the clinical assessment of the cardiovascular system.

    (a) The electrocardiogram gives information on the conducting patterns of the heart.
    (b) Atrial fibrillation results in an irregular radial pulse.
    (c) Measurements of BP using the sphygmomanometer give exactly the same values as direct intraarterial pressure measurements.

(d) Pulmonary wedge pressure measurements are useful in monitoring fluid replacement therapy.

17 During a period of moderate exercise, the following cardiovascular adjustments take place.

(a) Heart rate increases due to increased activity of the parasympathetic nerves.
(b) Skin blood flow increases.
(c) Sympathetic vasoconstriction of visceral arterioles increases.
(d) Muscle blood flow increases due to an accumulation of local vasodilator metabolites.

18 During a severe haemorrhage, the following cardiovascular adjustments take place.

(a) Heart rate increases in an attempt to maintain an adequate cardiac output.
(b) Loss of fluid from capillaries increases.
(c) Systolic and diastolic pressures increase.
(d) The patient will be pale and sweaty due to arousal of the sympathetic nervous system.

19 If the heart suddenly stops beating

(a) the individual will slowly lose consciousness in a few minutes
(b) the pupils of the eye will constrict
(c) the physical signs are similar to those of ventricular fibrillation
(d) there will be no peripheral pulses.

Answers to review questions

1 a, b and c
2 a and c
3 c and d
4 b, c and d
5 a, c and d
6 a and b
7 b, c and d
8 b and d
9 c
10 b and d
11 a, b and d
12 c
13 b, c and d
14 c and d
15 a, b and d
16 a and b
17 b, c and d
18 a and d
19 c and d

Short answer and essay topics

1 Explain how the structure of the heart is related to its function. In your answer suggest how damage to heart muscle might interfere with its function.
2 Describe the normal cardiac cycle, linking the electrical and mechanical changes that occur.
3 Discuss the factors that regulate cardiac output. In your answer define the terms: cardiac output; stroke volume; heart rate; Starling's law.
4 State Starling's law of the heart and explain why this mechanism is important.
5 Why is the maintenance of arterial blood pressure so important? Discuss the factors that determine blood pressure and explain the changes that occur when a patient gets up after a long period of bedrest.
6 Discuss the use of central venous pressure monitoring in the critically ill patient.
7 Discuss the pathophysiology of congestive cardiac failure and explain why oedema formation occurs.
8 Explain how the circulatory function is altered by (a) atherosclerosis; (b) hypertension; and (c) haemorrhage.

Suggestions for practical work

Ask a colleague if you may carry out the following on him or her.

1 Using a stethoscope, listen to the heart sounds at various positions on the chest (both front and back). Which heart sound is loudest when listening to the apex beat? Why is this?
2 Count the heart rate (using either apex beat or radial pulse) when lying down and immediately after standing up. Account for the change in rate that you have observed. (For maximum effect, count the heart rate for the first 15 seconds after standing, and multiply by 4 to get the rate per minute.)
3 Compare the neck veins when lying down, sit-

ting up at 45° and then sitting upright. Explain any changes you observe.
4 To assess the importance of the correct procedure for recording accurate blood pressure measurements
   (a) record the BP in both arms
   (b) ask the subject to hold an arm out straight and then record the BP (the subject will be performing isometric muscle contractions which may alter your reading)
   (c) record the BP with the arm above and below heart level (easier to do if the subject is horizontal)
   (d) compare Korotkov phase 4 and phase 5 values for diastolic pressure. Is there any difference, and if so, is it clinically significant?
5 Take BP readings under varying circumstances and compare the values recorded: for instance after exercise, hot bath, large meal, 'frightening/ surprising' your subject, 15 minutes resting (add any other circumstances to this list that might apply to hospital patients having regular blood pressure measurements taken).
6 Demonstrate reactive hyperaemia. Hold an arm up to drain it of some blood. Apply the sphygmomanometer cuff and inflate it to 200 mmHg. The subject can then lower his or her arm. Keep the pressure in the cuff elevated for approximately 2 minutes (the subject will experience a 'pins and needles' sensation, which is normal – why?). Then slowly reduce the pressure in the cuff. At what pressure does colour return to the arm? Why does the arm go redder than the other control arm?
7 Record the heart rate and blood pressure of a group of subjects (the larger the group the better). Try to include subjects of different ages. Compare these values and discuss any differences. What are 'normal' values?
8 With a group of young subjects, compare resting blood pressure and heart rate values with values recorded after 5 minutes' exercise (e.g. running on the spot). Why do these parameters alter? Explain the different responses of subjects in your group.

References

Bannan, L. T., Beevers, D. G. & Wright, N. (1980) ABC of blood pressure reduction: the size of the problem. *British Medical Journal* 281: 821–923.
Despopoulos, A. & Silbernagl (1981) *Colour Atlas of Physiology*. Stuttgart: Georg Thieme Verlag.
Durkin, N. (1979) *An Introduction to Medical Science*. Lancaster: MTP Press.
Gregg, D. E. (1965) In *Physiology and Biophysics*, 19th edn, Buck, T. C. and Patton, H. D. (eds), Philadelphia: Saunders.
Jensen, J. T. (1982) *Physics for the Health Professions*. New York: Wiley.
Keele, C. A., Neil, E. & Joels, N. (1982) *Samson Wright's Applied Physiology*. Oxford: Oxford Medical Publications.
Kelman, G. R. (1980) *Physiology: A Clinical Approach*. New York: Churchill Livingstone.
Laurence, D. R. (1980) *Clinical Pharmacology*. Edinburgh: Churchill Livingstone.
Lipid Research Clinics Program (1984) The Lipid Research Clinics Coronary Primary Prevention Trial. *Journal of American Medical Association* 251, 351–64, 365–74.
MacGregor, G. A. (1983) Dietary sodium and potassium intake and blood pressure. *Lancet* 750–753.
Mancia, G. et al (1983) Effects of blood pressure measurement by the doctor on patient's blood pressure and heart rate. *Lancet* 695–698.
Moss, D. W. (1981) Diagnostic enzymology: some principles and applications. *Hospital Update October*: 999–1010.
O'Brien, E. & O'Malley, K. (1981) *Essentials of Blood Pressure Measurement*. Edinburgh: Churchill Livingstone.
Olsen, E. V. (1967) The hazards of immobility. *American Journal of Nursing* 67, (4): 781–796.
Tortora, G. J. & Anagnostakos, N. P. (1981) *Principles of Anatomy and Physiology*. New York: Harper & Row.
Vander, A. J., Sherman, J. H. & Luciano, D. S. (1975) *Human Physiology – The Mechanisms of Body Function*. New Delhi: Tata/McGraw Hill.

Suggestions for further reading

Astrand, P. O. & Rodahl, K. (1977) *Textbook of Work Physiology: Physiological Bases of Exercise*. New York: MacGraw-Hill.
Bassey, E. J. & Fentem, P. H. (1981) *Exercise: the Facts*. Oxford: Oxford University Press.
DHSS (1981) *Prevention and health – Avoiding Heart Attacks*. London: HMSO.
George, R. J. D. & Banks, R. A. (1983) Bedside measurement of pulmonary capillary wedge pressure. *British Journal of Hospital Medicine March*: 286–291.
Joseph, S. (1982) *Living with a Pacemaker*. London: Chest, Heart and Stroke Association.
McGurn, W. C. (1981) *People with Cardiac Problems*. Philadelphia: J. B. Lippincott.
Office of Health Economics (1982) *Coronary Heart Disease – The Scope for Prevention*. London: OHE.

Section 5
The Acquisition of Nutrients and Removal of Waste

Chapter 5.1
The Acquisition of Nutrients

Susan M. Hinchliff

Learning objectives

After studying this chapter, the reader should be able to

1 Relate the following oral structures to their functions: lips, tongue, taste buds, teeth, salivary glands.
2 List the constituents of saliva and describe the regulation of its secretion.
3 Describe how and where peristalsis, segmentation and mass movements occur in the gastrointestinal tract.
4 Discuss how the generalized structure of the gut wall is adapted throughout its length, according to function.
5 Write an essay entitled 'The neural and hormonal control of gastrointestinal motility and secretion'.
6 List the functions of the stomach and explain fully the mechanism for gastric juice production.
7 State how the exocrine pancreas contributes to digestion.
8 Draw a diagram to illustrate the formation, secretion and function of bile.
9 Describe how the intestinal enzymes complete digestion, and summarize the digestion of carbohydrates, proteins and fat.
10 Teach a junior nurse how nutrients are absorbed from the small intestine.
11 Discuss the importance of the colon and relate this to stoma formation.
12 Describe how defaecation occurs, and relate this to diarrhoea and constipation.
13 Assess a patient's bowel habits.
14 Relate types of gastrointestinal pain to their source.

Introduction

Every day, the average adult consumes approximately 1 kg of solids and 1.2 kg of fluid. In general, the form in which we eat food is unsuitable for immediate use by the body for growth, repair and the production of energy for the performance of physical work. During its passage though the gastrointestinal, or alimentary (from the Latin *alimentum*, meaning nourishment), tract the food undergoes six processes: ingestion, mastication, digestion, absorption, assimilation of useful components, and elimination of non-usable residues. The effect of the first three of these – ingestion, mastication and digestion – is to render the food into a state in which it can be absorbed. It is only at stage 5 – assimilation – that food becomes of use to the body's cells. Substances which are not of use are eliminated at the final stage.

The adult gastrointestinal tract consists of a fibromuscular tube, approximately 4.5 m long and of variable diameter, that extends from the mouth to the anus (Fig. 5.1.1). Throughout, the structure of the tract shows variation according to function.

The tract is in contact with the external environ-

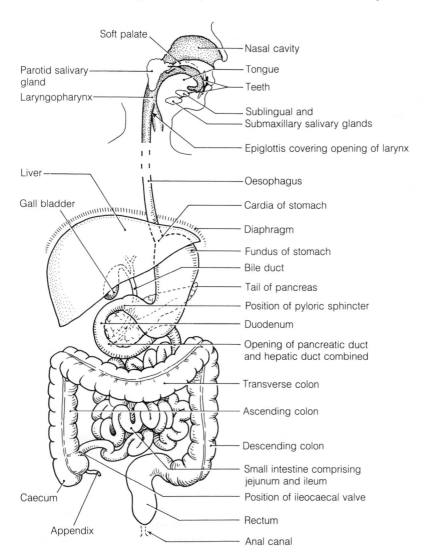

Figure 5.1.1 Diagrammatic representation of the gastrointestinal tract

ment at both ends, and as a result the potential problem exists of infective agents entering the body via this route. This problem is normally dealt with by (a) patches of lymphatic tissue throughout the tract, and (b) marked changes in the pH of the medium which most micro-organisms find it difficult to withstand.

THE MOUTH

Food is ingested by the mouth (sometimes called the oral or buccal cavity), which is the only part of the gastrointestinal tract to have a bony skeleton.

In the mouth, food is mixed with saliva, broken up into small chunks by the teeth, formed into a bolus and propelled backwards into the oesophagus by the tongue.

The mouth is divided into two parts.

(a) The **vestibule**: that part between the teeth and jaws, and the lips and cheeks. The parotid salivary glands open here.

(b) The **oral cavity**; this is the area that lies inside and is bounded by the teeth. The epithelium here is typically 15–20 layers of cells thick, and is thus structurally adapted to the amount of friction to which it is subjected during mastication.

The Lips

The lips form a muscular entrance to the mouth. They are necessary for ingestion as they form a muscular passageway for food and fluids and help to grasp particles of food. The act of drinking is accomplished by forming the lips into a tube and simultaneously inhaling – it is, of course, vital to stop the inspiratory process before swallowing fluid in order to prevent it entering the respiratory tract. Once food and fluids are in the mouth, the lips guard the entrance during chewing and swallowing.

Just how necessary the lips are for distinct enunciation becomes clear when watching a ventriloquist. Clear speech in those with injured lips is extremely difficult. A further function of the lips is to convey information about the mood of the person – smiling, grimacing, pursing, kissing, etc.

The lips are covered by squamous, keratinized epithelial tissue which is very vascular and very sensitive. In fact, the lips form one of the body's erogenous zones, with a large area of the sensory cortex of the brain in relation to actual size of the lips devoted to receiving sensations from them.

As there are no sebaceous glands present on the lips, they can quite quickly become chapped and cracked. In patients who are dehydrated, the lips become very dry and the keratinized tissue may peel off; this may lead to bleeding and can be painful. It is for this reason that careful attention must be paid to lubricating the lips in such circumstances.

Some individuals carry the herpes simplex virus in the cells of the lip area. The virus lies dormant until, for some reason, the individual's resistance is lowered, when the characteristic 'cold sore' lesion develops. This can cause discomfort and is cosmetically disfiguring.

The inside of the lip is formed of non-keratinized mucous epithelial tissue, and here the profuse blood capillaries lie very close to the surface. It is for this reason that anaemia can be assessed, albeit only roughly, from the coloration of this tissue.

The teeth

A child has 20 deciduous (milk) teeth, and an adult 32 permanent teeth. The teeth perform several functions during mastication – tearing (the canines), cutting (the incisors) and grinding (the

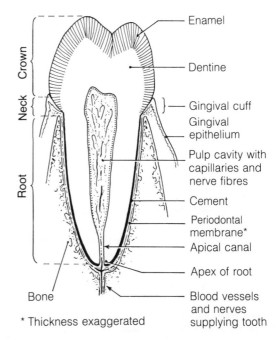

Figure 5.1.2 Diagram to show generalized structure of a tooth

molars and premolars). However, each tooth roughly conforms to the basic structure shown in Fig. 5.1.2, with a visible **crown** projecting above the gum and a **root** embedded in the jaw.

Enamel covers the crown and the root. This is an extremely hard substance which, when intact, protects the underlying **dentine**, which has a structure rather similar to that of bone. Once the enamel is breached, the dentine can be attacked by acids. Bacteria can feed and multiply on sugary residues left in such areas when the teeth are not properly cleaned. These attacks will eventually result in dental caries and if untreated, loss of the affected teeth. Proprietary soft drinks, such as colas, can be detrimental to teeth since they are both sugary and acidic; some drinks having a pH as low as 3.5. Those who drink such products should be encouraged to clean their teeth afterwards.

The **pulp cavity** lies within the dentine and contains nerve fibres. Once the damage caused by dental caries affects this area, intense pain will result when the nerve fibres are stimulated.

Nurses should pay careful attention to the care of the teeth of those unable to do this for themselves. Edentulous patients (that is, those who have lost their teeth) may find eating extremely difficult, and this may be a contributory factor to the nutritional problems of the elderly emaciated or anaemic patient. As well as their practical functions, teeth have a cosmetic value and this should be borne in mind when a patient has a dental clearance – a procedure which may have psychological implications. Clear speech may pose problems initially in such a patient.

In the elderly, the gum tissue covering the jaw tends to shrink and thus dentures that were once well fitting may become loose, making the gums sore and eating difficult. This is a factor which should be considered when assessing the mouth of an elderly patient.

Candida albicans (thrush) is a fungus which can affect the gums and oral mucosa of the very old, the very young, the debilitated and those on long-term antibiotic therapy. White, ulcerated plaques form over the affected area. An antifungal substance, such as nystatin in the form of lozenges, mouthwash or tablets, is usually prescribed.

The tongue

The tongue is formed of striated (voluntary) muscle which has only one insertion, to the hyoid bone. It is, however, anchored to the anterior floor of the mouth, behind the lower incisor teeth, by a fold of skin called the **frenulum**.

The tongue is very vascular, and bleeds profusely when cut. It is also extremely sensitive, as it is supplied by a large number of nerve endings. We become aware of the tongue's sensory function when it develops a lesion or ulcer, and what may in fact be quite small feels enormous. As with the lips, the tongue is represented by a much larger area of the sensory and motor cortex of the brain in relation to its size than one might expect. This relatively large cerebral representation in the motor cortex allows innervation of the numerous fine movements of the tongue necessary for speech. If the tongue is injured in any way, then lisping and speech defects can result. Even a tiny lesion on the tongue can cause speech problems.

Apart from its function in speech, the tongue is necessary for swallowing ingested and masticated food. The tongue is very mobile and readily extensible and distensible. It also helps to mix food with saliva.

Examination of the tongue can give an indication of the health of the individual. Normally, the tongue should be pink, moist and neither smooth nor cracked. It should not appear dry, which indicates a general state of dehydration,

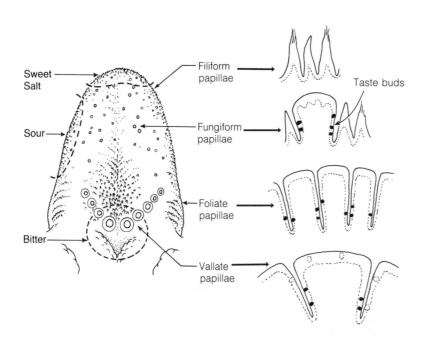

Figure 5.1.3 Diagram of tongue to show papillae

nor should it appear furred or brown. A smooth tongue may be seen in pernicious anaemia.

The tongue and the inside of the mouth should be virtually odourless. A faeculent smell may indicate intestinal obstruction; an ammoniacal, uriniferous smell (fetor hepaticus) may indicate liver failure; and a smell of new-mown hay may indicate uncontrolled diabetes mellitus. The mouth may retain for some while odours of food and drink that have recently been ingested, e.g. garlic and alcohol. It is always a good idea to smell the breath of a patient who has been admitted unconscious, as this may supply clues to the nature of his condition.

An important function of the tongue is that it allows us to taste, and therefore derive enjoyment from food. On the superior surface of the tongue, there are numerous **papillae** – variously called filiform, fungiform, circumvallate and foliate papillae according to their shape (Fig. 5.1.3). These areas contain some 10 000 **taste buds** which allow us to differentiate between the four taste modalities – sweet, sour, salt and bitter. Taste receptor cells (chemoreceptors) lie in these papillae. When chemicals in solution (e.g. dissolved in saliva) stimulate these cells, an impulse is generated which travels along the nerve fibre supplying each receptor cell, eventually arriving at the sensory cortex or the brain (see Chap. 2.2). Here, the particular taste sensation is distinguished and recognized. It used to be thought that each of the different types of papillae on the tongue (illustrated in Fig. 5.1.3) responded to a single taste modality. However, this is no longer thought to be so; it is now postulated that each receptor is potentially capable of responding to all types of taste, although each is associated particularly with one modality. Each of the four possible types of taste results in different and distinguishable firing patterns in the nerve fibres, and these are interpreted in the cerebral cortex. Interpretation is aided by impulses received from the olfactory nerve, which supplies the olfactory epithelium in the nose. The fact that taste is potentiated by smell becomes evident when one has a cold: inflammation and hypersecretion of the nasal mucosa impair the ability to smell, and this subsequently impairs the ability to taste food effectively.

The salivary glands

In terms of their contribution to digestion, these exocrine glands are non-essential; in practical terms, however, they are most important as their secretion aids speech, chewing, swallowing and general oral comfort.

There are three main pairs of glands: the parotid glands, submaxillary and sublingual glands. Other smaller glands exist over the surface of the palate, inside the lips and the tongue. The latter secrete saliva in response to local mechanical stimuli, and are not under the control of the parasympathetic nervous system.

PAROTID GLANDS

These glands are situated by the angle of the jaw, lying posterior to the mandible and inferior to the ear. They are the largest of the three pairs of salivary glands, and are supplied by parasympathetic fibres of the glossopharyngeal nerve (the IXth cranial nerve). The parotids produce a watery secretion which forms 25% of the total daily salivary secretion. This saliva is rich in both enzymes and antibodies (IgA). In infectious parotitis (that is, inflammation of the parotids), more commonly called mumps, these glands swell and produce the characteristic appearance of mumps.

SUBMAXILLARY GLANDS

This pair of glands lies below the maxilla (i.e. the upper jaw). The submaxillary glands produce a more viscid saliva which forms 70% of the total daily production. Their nerve supply comes from the facial nerve (the VIIth cranial nerve).

SUBLINGUAL GLANDS

These lie in the floor of the mouth below the tongue, and produce a scanty secretion (only 5% of the total produced). They also are supplied by the facial nerve.

Saliva

Each day, adults produce approximately 1–1.5 l of saliva. Saliva is formed mainly of water (99% of total), and normally has a pH of 6.8–7.0. It does, however, become more alkaline as the secretory rate increases during chewing.

Saliva has several functions.

(a) It cleanses the mouth. There is a constant

production of saliva with a backwards flow directed towards the oesophagus. Saliva contains lysozyme, which has an antiseptic action, and also immunoglobulin (IgA), which has a defensive function. In general, a normal flow of saliva helps to prevent dental caries and halitosis. A sudden flush of saliva into the mouth ('waterbrash') often heralds vomiting.

(b) Saliva is necessary for oral comfort, and the lubrication it provides reduces the friction that would otherwise be produced by speech and chewing. A dry mouth is extremely uncomfortable, and does not permit easy speech. This condition may occur normally when the subject is nervous, frightened or anxious and when activity of the sympathetic nervous system or the action of drugs inhibits salivary secretion.

In a patient who is dehydrated, the mouth may be coated with thick secretions. In order to remove these, frequent, repeated and thorough cleansing of the mouth may be necessary.

Certain drugs, e.g. atropine and hyoscine (scopolamine), block parasympathetic action and therefore cause a dry mouth – that is, they inhibit salivary secretion. It is for this reason that these drugs are often given as part of a premedication prior to surgery. During surgery, it is necessary to reduce the volume of saliva produced, since in the absence of the cough reflex while the patient is anaesthetized the saliva may be inhaled.

(c) It is necessary for chemicals in food to be in solution (i.e. mixed with saliva) for them to stimulate the taste receptors in the papillae on the tongue. A person who has a dry mouth cannot taste food effectively (and therefore fails to enjoy meals).

(d) Saliva is necessary for the formation of a **bolus**, that is, a ball of partly broken-up food which is ready to be swallowed. The mucins present in saliva (viscid glycoproteins) help in the moulding and lubrication of the bolus. The role of saliva in swallowing becomes evident if one tries to eat a dry biscuit when anxious or nervous, and when consequently one's salivary secretion is decreased. This task is exceedingly difficult.

(e) Saliva contains a digestive enzyme, **salivary** or **alpha (α) amylase**, which was formerly referred to as ptyalin. Salivary amylase acts upon cooked starch (e.g. bread, pastry) and converts the polysaccharide starch to disaccharides (maltose and dextrins). The longer the starch remains in the mouth, and the more it is chewed, the greater will be the effect of this enzyme. Usually in digestion, salivary amylase plays only a minor and non-essential role.

In addition to the substances already mentioned, saliva contains calcium, sodium, chloride, hydrogen carbonate and potassium ions. If for some pathological reason, for example inflammation, infection or neoplasm, the ducts from the salivary glands become blocked, then these electrolytes (that is, the solution of the ions referred to) may become even more concentrated within the gland which leads to the formation of salivary stones.

CONTROL OF SALIVARY SECRETIONS

Saliva is produced in response to the following.

(a) The thought, sight or smell of food. This is a conditioned reflex. When a substance which is consciously recognized as food is anticipated by thought, sight or smell, impulses travel from the receptor to the cerebral cortex and thence to the medulla in the brain-stem. The salivary nuclei are situated in the reticular formation in the floor of the fourth ventricle. Impulses travel from there to the salivary glands and secretion is effected.

(b) The presence of food in the mouth. This produces mechanical stimulation of the salivary glands, and this response represents an unconditioned reflex (i.e. it is not learned). The impulse generated goes straight to the salivary nuclei in the medulla, and does not travel via the cerebral cortex.

Secretion of copious watery saliva occurs as a result of excitation of the parasympathetic nervous system, which also increases blood flow to the salivary glands. The glossopharyngeal and facial nerves (referred to earlier), which supply the salivary glands, form part of the parasympathetic cranial outflow.

Acetylcholine (ACh) is the parasympathetic neurotransmitter which brings about salivary secretion; atropine and hyoscine block the receptor sites for acetylcholine and thus inhibit secretion. Neostigmine, conversely, inhibits **acetylcholine esterase (AChE)** which destroys acetylcholine once it has brought about secretion. The administration of neostigmine therefore increases salivary secretion.

Stimulation of the sympathetic nervous system results in vasoconstriction of the blood vessels to the glands, hence only small amounts of concentrated saliva are produced.

When carrying out a nursing assessment of a patient's mouth, one may obtain relevant information from observation of the state of the lips, gums and tongue. The nurse should notice the condition of the patient's teeth and whether or not dental caries are present; the smell of the mouth; any bleeding, thrush (candidiasis) or overt problems such as cracked lips or herpes simplex.

Before digestion can occur, the ingested and masticated food (now in the form of a bolus) must be swallowed. This process is sometimes referred to as **deglutition**. The food is gathered into a bolus by the tongue, and before swallowing takes place the mouth is normally closed. To gain an idea of how difficult it is to swallow with the mouth open, think of the problems one experiences in the dentist's chair; under these circumstances, it is possible to swallow only small quantities of saliva. This needs to be remembered when performing mouth care for a patient. The tongue

contracts and presses the bolus of food against the hard palate in the roof of the mouth (Fig. 5.1.4). It then arches backwards and the bolus is propelled towards the pharynx.

In order that the bolus is not propelled upwards into the nasopharynx, this area is shut off by the soft palate rising. In order, too, that the bolus is not propelled into the larynx and respiratory tract, the larynx rises to a position under the base of the tongue and the epiglottis closes over the trachea. Respirations are inhibited so that food and fluids are not inhaled into the trachea by the negative pressure generated in the respiratory tract during inspiration.

Once the bolus reaches the posterior pharyngeal wall, the musculature there contracts around it. The food, often including up to 100 ml of air, enters the oesophagus and after this point the process becomes involuntary. The earlier stages described are all under voluntary control, but once the bolus touches the posterior pharynx and tonsillar area, impulses are generated which travel to the medulla and trigger off the **swallowing reflex**. This reflex is absent in the deeply unconscious or anaesthetized patient.

THE OESOPHAGUS

Once in the oesophagus, ring-like contractions

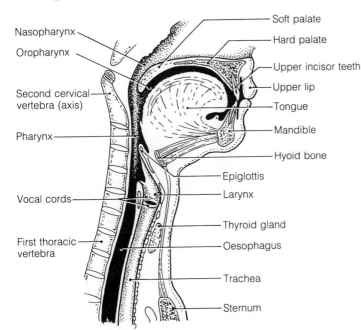

Figure 5.1.4 Diagram to illustrate mouth, pharynx and oesophagus

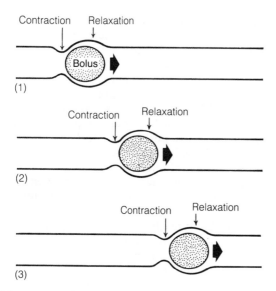

Figure 5.1.5 Diagram to illustrate peristalsis

propel the food towards the cardiac sphincter of the stomach. This movement is called **peristalsis** (Fig. 5.1.5) and is found throughout the gastro-intestinal tract from the oesophagus to the colon. It occurs as a response to stretching of the smooth muscle layer of the tract, and thus is referred to as a myenteric reflex. Contraction of circular muscle fibres occurs, followed by a contraction of the adjacent distal longitudinal muscle fibres while the circular fibres relax. These movements occur progressively throughout the tube in a coordinated manner, and give the appearance of smooth waves of contraction. The process can be initiated by stimulation of the vagus nerve (Xth cranial nerve) and also by stimulation of parasympathetic nerve fibres in the myenteric

(Auerbach's) plexus situated in the muscular layer of the tract.

The time taken for food to travel the 20–25 cm of the oesophagus depends on the consistency of the food and the position of the body. The patient who has to eat while flat on his back may experience some discomfort and difficulty in swallowing – a process which is normally aided by gravity, although, incidentally, it is possible to swallow whilst standing on one's head! Liquids take only 1–2 seconds to reach the cardiac sphincter, but a more solid bolus takes longer, perhaps up to 2–3 minutes. The small amount of air swallowed with the food is usually expelled some time later by belching.

The lower 4 cm of the oesophagus is referred to as the **gastro-oesophageal**, or **cardiac sphincter** and is normally in a state of tonic contraction until swallowing occurs, when the circular muscle fibres relax. This is a physiological rather than an anatomical sphincter; the circular muscles in the area act to prevent reflux of the gastric contents into the oesophagus.

The pressure within the stomach is normally 5–10 mmHg above atmospheric pressure, due to the pressure exerted on the stomach from below by the contents of the abdominal cavity. Since the pressure within the upper oesophagus is less than this – usually some 5–10 mmHg below atmospheric pressure as the upper oesophagus lies within the thoracic cavity – it can be seen that without the presence of a functional sphincter reflux could easily occur. The gastro-oesophageal region lies partly within the abdominal cavity, and this helps to account for the above-atmospheric pressure in this area.

Any condition which causes the pressure within

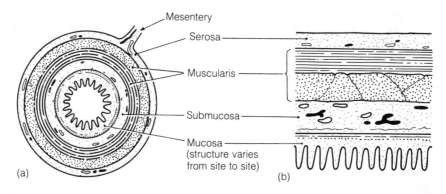

Figure 5.1.6 Diagrams to show generalized structure of the gut wall: (a) cross-section (b) longitudinal section

the stomach to rise (e.g. pregnancy, an abdominal tumour or ascites, all of which would increase the pressure on the stomach) could potentially give rise to problems of reflux and consequent oesophagitis. When the gastro-oesophageal region rises into the thoracic cavity, a condition known as hiatus hernia, the pressure within this region may fall and reflux occur.

GENERALIZED STRUCTURE OF THE GASTROINTESTINAL TRACT

The basic structure of the tract follows the same pattern from the mouth to the anus, with some functional adaptations throughout (Fig. 5.1.6).

The wall of the tube generally consists of four layers.

The mucosa. This is the innermost layer, that is, the layer nearest to the lumen of the tube, and it exhibits a great deal of variation throughout. Mucous stratified epithelial cells line the lumen, and it is from this layer that all the glands develop. Mucus-secreting cells are situated throughout the epithelium. These cells are subjected to an enormous amount of frictional wear and tear, and hence their rate of mitosis is high. The epithelial cells lie on a sheet of connective tissue called the **lamina propria**. Distal to this, there is a thin layer of smooth muscle tissue called the **muscularis mucosae** which allows the foldings and distension that characterize the mucosa. Throughout the tract, the mucosal layer contains patches of lymphoid tissue which have a defensive function.

The submucosa. This lies distal to the mucosa, and consists of loose connective tissue which supports the blood vessels, lymphatics and nerves carried in this layer. The nerve fibres in this layer form the **submucosal** or **Meissner's plexus**.

The muscularis. As its name suggests, this layer is formed of muscle fibres. The muscle fibres in the gastrointestinal tract are referred to as **smooth, involuntary, unstriated** or **visceral muscle fibres**. They exhibit the following characteristics.

(a) They respond to stimulation by the autonomic nervous system or certain hormones.
(b) They exhibit continuous rhythmic and

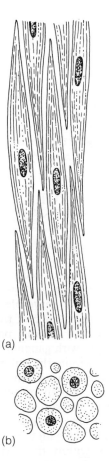

Figure 5.1.7 Diagrams to show gross structure of smooth muscle: (a) surface view, (b) cut section

inherent contractions that may be modulated by the above stimuli.
(c) They do not contract as forcefully as striated muscle fibres.
(d) They do not exhibit the same finely controlled contractions as striated muscles; their contractions bring about rather more diffuse movements of the whole muscle mass.

The fibres consist of elongated spindle-shaped cells whose contractile proteins are not arranged in myofibrils, with tapered ends (Fig. 5.1.7), and with a single centrally located, elongated nucleus. The fibres are bound together in sheets called **fasciculi**, and it is the sheet of fibres that forms the contractile unit (see (d) above).

The smooth muscle fibres in the gastrointestinal tract are arranged in two layers – circular (inner) and longitudinal (outer) – in all areas except the stomach, where there is an additional layer of oblique muscle fibres. Between the two layers of muscle fibre there is a network of nerve fibres called the **myenteric** or **Auerbach's plexus**.

Adventitia or serosa. This is the outermost, protective layer, and is formed of connective tissue and squamous serous epithelium. It is continuous with the mesentery in the abdominal cavity, and carries blood vessels and nerves.

NEURAL CONTROL OF THE GASTROINTESTINAL TRACT

Internal nerve plexuses

The gastrointestinal tract receives nerve fibres from the autonomic nervous system. The submucosal and myenteric nerve plexuses form the local internal nervous system of the tract.

In the **submucosal plexus**, parasympathetic nerve fibres synapse with ganglion cells which lie in small clusters within the submucosal tissue. Postganglionic fibres arise from here and travel to the glands and smooth muscle of the tract. Sympathetic fibres also run within this plexus.

In the **myenteric plexus**, parasympathetic nerve fibres travel to ganglion cells which lie in large clusters within the muscularis layer, between the circular and longitudinal muscle fibres. Postganglionic fibres arise here and innervate the smooth muscle within this layer. This muscle is also supplied with sympathetic fibres from the myenteric plexus.

Wingerson (1980) suggested that **Substance P** (P for 'powder'), a peptide made up of 11 amino acids and found in high concentrations in the gut, may be a chemical mediator, that is, it acts like a neurotransmitter in reflexes at myenteric level. It can therefore be referred to as a **regulatory peptide** or **neuropeptide**. Previous work on Substance P has shown it to be involved in sending pain impulses along sensory nerve fibres. In addition, it brings about vasodilation and contraction of non-vascular smooth muscle.

Serotonin (5-hydroxytryptamine, or 5H-T) is also thought to be synthesized within the myenteric plexus and may have a role as an interneuronal transmitter.

Both submucosal and myenteric plexuses run from the oesophagus to the anus, and receive both sympathetic and parasympathetic nerve fibres. There are neuronal connections between the two internal plexuses, and activity in one can therefore affect the other. The arrangement of the neurones within the plexuses allows stimulation at the upper end of the tract to be transmitted to distal portions of the tract. Thus, entry of food into the upper oesophagus can lead to gastric and intestinal secretion.

Areas of the gut, for example the stomach, receive direct 'external' autonomic fibres from the vagus (a branch of the parasympathetic nervous system) and the sympathetic nervous system; they also receive autonomic stimulation via the local internal plexuses.

Autonomic nervous system (see Chap. 2.4)

Secretion and motility throughout the gastrointestinal tract are under the control of both autonomic nervous and hormonal factors. The effect of the hormonal factors will be discussed later in relation to each relevant area of the tract.

With regard to neural control: in general, parasympathetic activity leads to an increase in both motility and secretion and to relaxation of the gut sphincters. The parasympathetic supply to the oesophagus, stomach, pancreas, bile duct, small intestine and proximal colon is via the **vagus nerve** (Xth cranial nerve). The term vagus comes from the Greek, meaning to wander (cf. vagrant), and this nerve is indeed very widespread in its connections, supplying areas of the cardiovascular, respiratory and gastrointestinal tracts.

The parasympathetic supply to the salivary glands is via the facial and glossopharyngeal nerves (VIIth and IXth cranial nerves), with the parasympathetic ganglia lying near the glands. The parasympathetic nerve supply to the distal colon is via the nervi erigentes in the sacral outflow from the spinal cord.

Activity of the sympathetic nervous system in general leads to a decrease in the blood flow to the gut, and a consequent decrease in secretions and lessening of motility and excitation (i.e. contraction) of the gut sphincters. Within the gastrointestinal tract, as elsewhere in the body, there are two types of receptors present on the postsynaptic muscle fibres which initiate responses to the catecholamines released as a result of sympathetic stimulation. These are termed α and β_2

Table 5.1.1 The effects of autonomic stimulation of the gastrointestinal tract

Area of gastrointestinal tract	Parasympathetic stimulation	Sympathetic stimulation (receptors indicated where appropriate)
Salivary glands	Profuse watery secretion	α-receptors Sparse viscid secretion
Stomach		
Motility	↑	β_2-receptors ↓
Secretion	↑	Usually inhibits. α-receptors → contraction
Pancreas (secretion of acinar cells)	↑	↓
Gall bladder and bile ducts	Contraction	Relaxation
Small and large intestines		
Motility	↑	α and β_2-receptors ↓
Secretion	↑	Usually inhibits
Sphincters	Relaxation	α-receptors → contraction

↑ increases; ↓ decreases.

(β_1 receptors being present only in cardiac muscle).

When stimulated, **α receptors** mediate contraction of smooth muscle in the walls of the gastrointestinal tract and the walls of the blood vessels. When stimulated, **β_2 receptors** bring about relaxation of smooth muscle fibres.

Table 5.1.1 shows the details of autonomic stimulation of the gastrointestinal tract.

The amount of food we take in daily is regulated only to a certain extent by hunger.

Hunger refers to a physiological sensation of emptiness, usually accompanied by contraction of the stomach. There is no direct relationship between the blood sugar level and hunger – diabetics who are hyperglycaemic often feel hungry. To a large extent, what and how much we eat are dictated by social customs and the pattern of our day; it depends not only on hunger but also on appetite. **Appetite** refers to a pleasant feeling of anticipation of forthcoming food. The two sensations of hunger and appetite are therefore related, but quite different. Appetite is affected by one's emotional state, for example a patient who is nervous or apprehensive may well lose his appetite, that is, become anorexic.

The hypothalamus is important in regulating food intake, although the exact mechanism for this is still uncertain. There appear to be two hypothalamic areas that interact in regulating food intake – the **feeding centre** and the **satiety centre**. Stimulating the feeding centre in rats leads to eating, whereas destruction leads to anorexia which is eventually fatal. Stimulation of the satiety centre leads to cessation of eating. Rats in whom this area has been ablated (destroyed) exhibit a voracious appetite, which results in what is called hypothalamic obesity. It is thought that the satiety centre is influenced by the level of glucose utilization in receptor cells in this area of the hypothalamus. These cells are therefore termed **glucostats**.

For details of how to assess a patient's nutritional status, the reader is urged to consult Goodinson (1986).

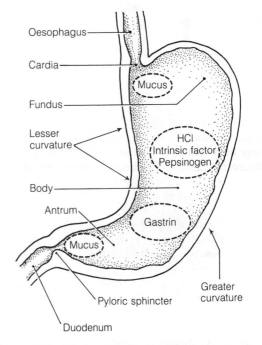

Figure 5.1.8 Diagram of stomach to show gross structure

THE STOMACH

The stomach, continuous with the oesophagus above and the duodenum below, is the most dilated area of the gastrointestinal tract (Fig. 5.1.8). It is roughly J-shaped, although the size and shape vary between individuals and with its state of fullness. Its internal surface area has visible folds called **rugae**, which together with the muscle layer allow distension. The epithelium of the body of the stomach (about 75% of the total) is further folded, being composed of **gastric pits** containing microscopic gastric glands (Fig. 5.1.9). This arrangement results in a total surface area of some 800 cm^2.

Functions of the stomach

A RESERVOIR FOR FOOD

This aspect of the stomach's function allows us to eat large meals at quite widely spaced intervals. The importance of this function is demonstrated in patients who have undergone a gastrectomy,

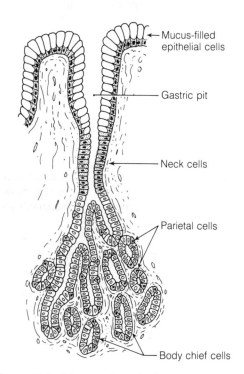

Figure 5.1.9   Diagram of gastric pit

Labels on figure:
- Mucus-filled epithelial cells
- Gastric pit
- Neck cells
- Parietal cells
- Body chief cells

and who therefore can cope only with small, frequent meals. At rest, the volume of the 'empty' stomach in the adult is approximately 50 ml. However, by **receptive relaxation** of the muscle in the stomach wall as food enters, it can accommodate up to 1.5 litres of food and fluids. The stomach of the neonate is about the size of a hen's egg – a fact to be remembered when feeding a baby.

This ability of the stomach to store food, together with the presence of the **pyloric sphincter**, allows for the controlled release of gastric contents into the duodenum, and so prevents rapid overloading of the small intestine. In patients who have had all or a large amount of their stomach resected, this controlled release into the duodenum is not possible. As these patients eat, relatively large, undiluted and therefore highly concentrated amounts of foodstuffs are poured rapidly into their duodenum. This causes distension and stimulates sympathetic nerve fibres in the duodenum and jejunum. Furthermore, since the duodenal contents are hypertonic in relation to plasma and extracellular fluid, fluid moves from the interstitial spaces and from the blood vessels in the gut wall into the lumen of the small intestine to dilute the contents. This results in a quite rapid fall in blood volume and a consequent fall in the cardiac output. As a result, the patient complains of feeling weak, dizzy and faint during or after eating food. This is often referred to as the **dumping syndrome**.

A further problem for these patients may occur when glucose is absorbed very rapidly after a meal. The consequent hyperglycaemia leads to a rapid increase in insulin secretion, and as a result the blood sugar level drops dramatically, causing the patient to become hypoglycaemic.

Not all patients after a gastrectomy experience these problems; but for those in whom they do occur, the symptoms are most unpleasant. Such patients may decrease their food intake in an effort to avoid these episodes, with consequent nutritional problems. It is therefore important that patients who have had radical gastric surgery receive dietary advice regarding the necessity to take small, frequent meals as part of their care and discharge planning.

PRODUCTION OF INTRINSIC FACTOR

This glycoprotein is necessary for the absorption

of **vitamin B$_{12}$ (cyanocobalamin)** in the terminal ileum. Intrinsic factor binds to vitamin B$_{12}$ in the small intestine to form a complex, which appears to bind to receptors in the ileal wall, allowing vitamin B$_{12}$ to transfer across the gut wall into the blood. Trypsin is also thought to be necessary for efficient vitamin B$_{12}$ absorption to occur.

Vitamin B$_{12}$ is required for healthy functioning of nerve fibres and their myelin sheaths in the spinal cord. It is also necessary for the formation of the red blood cell stroma during erythropoiesis in the red bone marrow. Lack of vitamin B$_{12}$ leads to a megaloblastic (pernicious) anaemia and, if untreated, to subacute combined degeneration of the spinal cord. Since our diet normally contains sufficient vitamin B$_{12}$ to supply our needs, megaloblastic anaemias are usually the result of failure to produce intrinsic factor, or of failure to absorb the vitamin in the small intestine. This latter situation may occur following an ileostomy or when chronic diarrhoea exists.

Intrinsic factor is produced by the gastric parietal cells, which also produce gastric acid. Patients who lack intrinsic factor often also lack gastric acid – that is, they have **hypochlorhydria** or **achlorhydria**. Tests used in the diagnosis of megaloblastic anaemia may therefore attempt to elicit gastric acid secretion by, for example, the administration of histamine or, more commonly, pentagastrin. Some patients with megaloblastic anaemia are thought to produce antibodies to their own intrinsic factor, and in these individuals the cause of the disease is autoimmune.

GASTRIC ABSORPTION

Ingested food is only partly broken down in the stomach, and the resulting molecules are still, in general, too large to cross the gastric wall. In addition, carrier systems for the transport of molecules across the gastric wall have not been found. Hence only a small amount of absorption can occur in the stomach. It is possible to absorb the following.

(a) A small amount of water.
(b) Alcohol – the bulk of alcohol absorption occurs in the small intestine, but about 20% may be absorbed in the stomach. If foods containing fat (e.g. milk, cheese etc.) are taken before drinking alcohol, they may help to delay alcohol absorption in the small intestine by slowing down gastric emptying. This will not, however, prevent gastric alcohol absorption.
(c) Some drugs – in particular aspirin. Aspirin (acetylsalicylic acid) is a weak acid, and as such can cross the stomach wall. In so doing, it increases the hydrogen ion concentration of the cells in the wall, thus lowering their cellular pH. This may cause cell damage. Prolonged aspirin ingestion may therefore lead to gastric irritation (gastritis) and bleeding. Aspirin and other salicylates should therefore never be administered to patients with lesions of the gastric mucosa or dyspepsia, to those on anticoagulant therapy or those with haemophilia. Patients who are prescribed aspirin (e.g. for the treatment of rheumatoid arthritis) are usually given enteric-coated aspirin; the coating on the drug ensures that it is not absorbed in the stomach, and this therefore avoids problems of gastric irritation, bleeding and haematemesis (vomiting of blood).

THE STOMACH ACTS AS A CHURN

It converts ingested food to a thick minestrone soup-like consistency, by mixing it with gastric secretions. This serves to dilute the foodstuffs, and thus prevents the entry of a solution into the duodenum that is hypertonic in relation to extracellular fluid. The resulting semiliquid substance is called **chyme**.

Mixing is achieved by gastric peristalsis. Rhythmic waves of contraction of the three layers of smooth muscle in the stomach wall pass from cardia to pylorus about three times a minute, each wave lasting approximately half a minute. These waves of contraction allow the more liquid contents to leave the pylorus of the stomach quite rapidly and enter the duodenum. As the pyloric muscle contracts, the lumen of the stomach partially closes, and this causes the more solid food in the antrum to pass backwards towards the body of the stomach. This forwards and backwards movement increases the efficiency of gastric mixing. **Motilin** is a regulatory peptide secreted by cells in the duodenum and jejunum in response to acid chyme, which increases gastric motility

SECRETION OF MUCUS

Mucus consists of a gel, formed of the protein mucin and glycoproteins, which adheres to the gastric mucosa. It is secreted by cells in the necks of the deep gastric glands in both the cardia and the pylorus. Mucus protects the stomach wall from being digested by the proteolytic enzyme pepsin, which is produced in the stomach (i.e. it prevents autodigestion). It also contains some hydrogen carbonate (bicarbonate) and this partially neutralizes gastric acid and so prevents this acid from damaging the gastric cells. It further helps to lubricate the food in the stomach.

In order to carry out these protective functions, the layer of mucus covering the rugae needs to be at least 1 mm thick. Mucus hypersecretion occurs in gastric inflammation (gastritis) and as a response to irritant or toxic agents (e.g. alcohol and micro-organisms). Additionally, to help patients who have gastric ulcers, mucus secretion can be increased by the administration of carbenoxolone (a liquorice derivative). Protection of the gastric cells does not depend solely on mucus secretion, but also on there being intact gastric cell membranes with a low permeability to hydrogen ions and tight junctions between these cells.

Certain substances (e.g. aspirin, alcohol and bile salts) disturb the cellular arrangement of the gastric mucosa, and increase the permeability of the cell membranes to hydrogen ions. If the cell membrane is breached, hydrogen ions enter the cell and lower the cellular pH. The subsequent cell damage may lead to hypoxic and eventually necrotic areas of gastric mucosa, and hence to the formation of an ulcer. Healing of such areas is made difficult by the gastric acid in which they are bathed, and by the autodigestive effect of pepsin.

Histamine is a substance, produced by circulating mast cells and basophils, which increases gastric acid secretion. It achieves this effect by binding to histamine receptors (H_2-receptors) on the gastric parietal cells. H_2-receptor blocking agents can therefore be used to decrease gastric acid secretion; such drugs, e.g. cimetidine (Tagamet), have been used successfully in the treatment of gastric ulcers.

SECRETION OF GASTRIC JUICE

Gastric juice consists of a mixture of secretions from two types of cells, both of which are absent from the pylorus.

(a) **Parietal** or **oxyntic cells** – these secrete hydrochloric acid (HCl), and also the intrinsic factor referred to earlier.

(b) **Chief** or **zymogen cells** – these secrete enzymes.

Some 2–3 litres of juice with a pH of 1.5–3.0 are secreted each day in the adult.

Hydrochloric acid secretion
The parietal cells, lying in the gastric pits (see Fig. 5.1.9), produce gastric acid, and it is likely that these cells may have fine channels (canaliculi) leading from the cytoplasm of the cell to the lumen of the gastric pits and through which the acid is discharged into the stomach. There are about 1000 million parietal cells in the healthy adult stomach.

Hydrogen ions are secreted by the parietal cells *against* a concentration gradient; that is, they are produced by the parietal cells and move into the lumen of the stomach where the hydrogen ion concentration is high, rather than into the blood which has a low concentration of hydrogen ions. There is evidence that the hydrogen ions are transported actively by a pump mechanism in the cell membrane. Chloride ions, too, are actively secreted against a concentration gradient, and the presence of a similar pump for this action is consequently postulated. The energy for this active transport of both hydrogen and chloride ions is obtained from the aerobic breakdown of glucose.

Blood leaving the stomach via the gastric vein has a lower partial pressure of carbon dioxide than blood arriving at the stomach via the gastric artery. It is therefore postulated that CO_2 diffuses

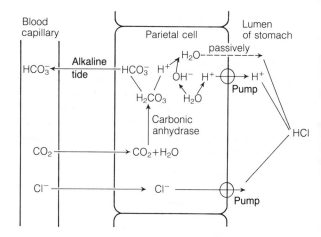

Figure 5.1.10 Diagrammatic representation of hydrochloric acid production within the parietal cell

into the parietal cells from arterial blood. Within the parietal cell, CO_2 combines with water under the influence of the enzyme carbonic anhydrase (CA) to form H_2CO_3 (carbonic acid). This dissociates to form a hydrogen ion (H^+) and a hydrogen carbonate ion (bicarbonate, HCO_3^-) (Fig. 5.1.10). The hydrogen carbonate so produced enters the capillaries and hence the venous blood draining the stomach. This movement of base into the venous blood is referred to as the **alkaline tide**. Thus, in a sense, the parietal cells act as both endocrine and exocrine glands.

The alkaline tide is of significance in disorders in which there is prolonged vomiting (e.g. hyperemesis gravidarum in pregnancy). In such conditions, there is an absolute loss from the body of gastric acid, while the venous blood continues to become more alkaline. This may result in a metabolic alkalosis. It is thought that within the parietal cell, H_2O dissociates into a hydrogen ion (H^+) and an hydroxyl ion (OH^-). The hydrogen ion produced by the dissociation of carbonic acid neutralizes the hydroxyl ion produced by the dissociation of water. This leaves the hydrogen ion produced by the dissociation of water to be secreted actively into the lumen of the gastric pits. Equal numbers of chloride ions are similarly secreted actively into the lumen of the gastric pits; water follows passively to dilute the HCl so formed.

Acid secretion is increased when stimulated by histamine or the hormone gastrin. Pentagastrin, a synthetic gastrin and hence a potent stimulant for acid secretion, is used clinically to investigate gastric acid secretion.

Functions of gastric acid
1 It inactivates salivary amylase.
2 It is bacteriostatic and therefore protective.
3 It tenderizes proteins by denaturing them, that is, it alters their molecular structure.
4 It curdles milk. The hydrochloric acid acts on casein (a soluble protein in milk) and converts it to paracasein. This combines with calcium ions to form curds preparatory to milk digestion. In animals, this function is performed by the enzyme rennin which is absent in the human stomach.
5 It converts pepsinogen to pepsin.

Gastric enzyme secretion
The chief (or zymogen) cells produce a scanty secretion rich in **pepsinogen**, an enzyme precursor. While gastric pH is below 5.5, this pepsinogen is converted into **pepsin** by hydrochloric acid. Once this conversion has occurred, the pepsin so formed can itself convert pepsinogen into more pepsin. Pepsin is a proteolytic enzyme, that is, it acts on proteins and starts their digestion. It converts proteins into polypeptides (long chains of amino acids) by breaking bonds between specific amino acids. Pepsin is most active when the gastric pH is below 3.5, and it is responsible for between 10% and 15% of all protein digestion. Once chyme leaves the stomach, the activity of pepsin stops due to the change to an alkaline medium.

The secretion of pepsinogen is closely linked to the secretion of gastric acid. Stimulation of the vagus nerve leads to the release of acetylcholine at the vagal axon terminations. This stimulates both the chief cells and the parietal cells, and provides the rationale for using agents which block acetylcholine receptors in the treatment of gastric ulcers. Such anticholinergic agents (e.g. atropine, belladonna, propantheline and scopolamine) prevent the secretion of both gastric acid and pepsinogen.

Some authorities (Sanford, 1982) cite the presence of a lipase in the stomach which acts on triglycerides containing short chain fatty acids. It is not particularly active at the low pH of the stomach, however.

Control of gastric juice secretion
This is both neural and hormonal (humoral) (Fig. 5.1.11).

Neural control. It is usual to describe two phases in the neural control of gastric juice secretion, although these two phases are not distinct.

In the **cephalic** or **psychic phase**, secretion is brought about in response to the sight, thought or smell of food. This is a conditioned (i.e. learned) reflex, and is mediated by the vagus nerve. Additionally, the presence of food in the mouth leads to gastric secretion. Stimulation of the vagus nerve by such cephalic impulses leads to the release of acetylcholine at the axon terminations. This chemical (as described earlier) stimulates both parietal and chief cells. Surgical section of branches of the vagal nerve (selective vagotomy) will prevent vagal stimulation, resulting in reduced secretion of gastric juice.

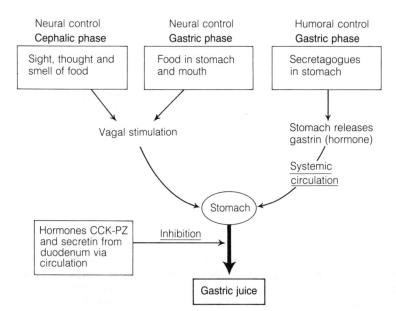

Figure 5.1.11   Flow chart to show gastric juice secretion

The **gastric phase** is not mediated by the vagus nerve. The presence of food in the stomach produces mechanical stimuli which result in gastric juice secretion. Stretch receptors in the stomach wall respond to distension of the stomach wall, and chemoreceptors to protein molecules within the stomach. Impulses travel from these stretch and chemoreceptors via afferent nerve fibres to the submucosal plexus. Here, the nerve fibres synapse with parasympathetic neurones, and excitatory impulses are conveyed to the parietal cells.

Hormonal control. Hormonal (humoral) influences also contribute to the gastric (i.e. non-vagal) phase of gastric juice secretion. The word 'hormone' is derived from the Greek and means 'arouse to activity', and throughout the gut there are several such regulatory hormones (or peptides), many of which are also active as transmitters in the central nervous system. For this reason they are sometimes referred to as **neurohormones**, **neuropeptides** or **neurotransmitters**.

Gastrin is the singular term, which in fact refers to a group of chemically similar hormones discovered in 1905 and produced by the G-cells in the lateral walls of the gastric glands in the antrum of the stomach. There are two main molecular variations of gastrin – G17, with 17 amino acids, present in quite large amounts, and a smaller quantity of G34, with 34 amino acids. It appears

that the terminal five of the amino acids in the molecule (that is, the terminal pentapeptide) form the critical segment for the stimulation of gastric acid secretion. Hence the synthetic gastric acid stimulant is called pentagastrin.

A small amount of gastrin is produced by the duodenal mucosa. This production is sometimes referred to as the **intestinal phase** of the control of gastric juice secretion. Pathologically, gastrin-secreting tumours may occur in the pancreas, but it is unlikely that gastrin is normally produced in the pancreas. Such tumours result in a condition termed Zollinger–Ellison syndrome in which excessive amounts of gastric acid are produced.

Gastrin is secreted in response to the presence in the stomach of certain foodstuffs referred to as **secretagogues**, examples of which are meat, alcohol, tea, coffee and colas, and also in response to the acetylcholine release as a result of vagal stimulation. **Bombesin** is a regulatory peptide which is also called **gastrin-releasing peptide**. It is produced by P cells throughout the gastrointestinal tract and is a potent releaser of most of the regulatory peptides. It is also found in the brain and pancreas.

Once produced, gastrin enters the gastric capillaries and then the systemic circulation. When it reaches the stomach via the bloodstream it

(a) stimulates the parietal cells to produce gastric acid by liberating histamine

(b) has a minor role in stimulating the chief cells to produce pepsinogen

(c) stimulates the growth of the gastric and intestinal mucosa

(d) brings about enhanced contraction of the gastrointestinal sphincter, and hence prevents reflux during gastric activity

(e) stimulates the secretion of insulin and glucagon when it reaches the pancreas via the systemic circulation.

Gastrin secretion is inhibited by the presence of gastric acid in the antrum of the stomach via a negative feedback mechanism. When gastric acid secretion falls, more gastrin is secreted. Thus, when hypochlorhydria or achlorhydria exists, blood gastrin levels are permanently raised.

Circulating secretin from the duodenum similarly inhibits gastrin release (Table 5.1.2), as do CCK, GIP, VIP (see later in chapter) and **somatostatin**. Somatostatin is a regulatory peptide which inhibits the release and action of many gut peptides, since it is thought to be antagonistic to bombesin. It is found in the gut, thyroid, pancreas and brain. It inhibits the secretion of growth hormone and so is also called growth hormone release inhibitory hormone (GHRIH).

Control of gastric motility and emptying

The time foods remain in the stomach depends upon their consistency and composition. Carbohydrate foodstuffs, together with liquids, leave fastest, then protein-based foods, and finally fatty foods are the slowest to leave the stomach. Hence foods rich in fats, such as fried fish and chips and cheese, have the highest satiety value and give a feeling of fullness.

When products of protein digestion and acids enter the duodenum, they initiate the **enterogastric reflex**, and this results in a slowing of gastric motility. Gastric contents may start to enter the duodenum some 30 minutes after entering the stomach. Gastric emptying is usually complete after 4–5 hours. This, though, will depend on the individual's emotional state; fear and anxiety, and states of generalized sympathetic nervous system stimulation will result in delayed gastric emptying. This may explain why, in the operating theatre, it is sometimes found that the stomach still contains food even though the patient has fasted preoperatively.

As already indicated, the structure of the stomach differs from that of the rest of the tract in that the gastric muscle coat consists of three layers and not two, i.e. it contains circular, longitudinal and oblique muscle fibres. It is this arrangement of muscle fibres that facilitates gastric churning. When food enters the stomach, the muscle layer relaxes reflexly – termed **receptive relaxation**. Peristalsis then occurs, which allows mixing of the gastric contents. This peristaltic activity is most marked at the pylorus (from the Greek for gatekeeper), and indeed may be visible after feeding in a baby with congenital pyloric stenosis or in the adult with secondary pyloric obstruction.

During gastric emptying, the antrum, pylorus and duodenal cap function as one unit. First the antrum contracts, then the pylorus and finally the duodenal cap. This sequence results in squirts of chyme entering the duodenum, and this is sometimes referred to as the **gastric pump mechanism**. Reflux of chyme into the antrum is prevented by the contraction of the pylorus persisting after the relaxation of the duodenal cap.

Once the stomach contents have entered the duodenum, mild contractions occur that persist and increase in intensity over a period of hours. If no food is received, these waves form hunger contractions, each of which can last up to half an hour and may be painful. Patients who are required to fast may experience these contractions, and become anxious about their origin. It is helpful to warn fasting patients that they may experience such pains, and that this is quite normal.

Table 5.1.2 Factors influencing gastric juice secretion

Factors stimulating secretion	Factors inhibiting secretion
Meat products	Catecholamines (adrenaline
Alcohol	and noradrenaline)
Caffeine-containing drinks	Vagotomy
(e.g. coffee and colas)	Fear
Vagal stimulation	Anxiety
Anger	Depression
Hostility	Atropine
Histamine	Prostaglandins
Acetylcholine	CCK ⎫ produced when
Hypoglycaemia	Secretin ⎬ chyme enters the
Gastrin	⎭ duodenum
Bombesin	Somatostatin
	GIP (gastric inhibitory
	peptide)
	VIP (vasoactive intestinal
	polypeptide)

When glucose and fats enter the duodenum, a regulatory peptide called **gastric inhibitory peptide (GIP)** is secreted by the K cells of the duodenal and jejunal mucosa. The GIP decreases gastric secretion and motility and so is considered to be a physiological enterogastrone; it also stimulates the secretion of insulin and is thus sometimes referred to as **glucose-dependent insulin releasing peptide** (Daggett, 1981).

Vasoactive intestinal polypeptide (VIP) similarly inhibits gastric motility. It is produced by the D_1 cells in the duodenum and colon and is widely distributed throughout the body. It is vasodilatory and acts as a smooth muscle relaxant.

NAUSEA

This term refers to the unpleasant sensation which precedes vomiting. It may be accompanied by one or more of the following features

pallor
sweating
waterbrash (sudden and profuse secretion of saliva into the mouth)
antiperistalsis (reverse waves of peristalsis from pylorus to cardia).

VOMITING

This occurs as a result of a reflex, and can be defined as the forceful expulsion of gastric and intestinal contents through the mouth. During the process the larynx is closed and the soft palate rises to close off the nasopharynx and so prevent the inhalation of vomitus. The diaphragm and abdominal wall contract strongly, the pylorus closes and this results in a sharp rise in the intragastric pressure, which causes the sudden expulsion of the gastric contents. The muscles of the stomach itself, and of the oesophagus, play a relatively passive role.

The **vomiting centre** is situated in the reticular formation of the medulla oblongata of the hindbrain. Impulses travel from the vomiting centre to the muscles of the abdominal wall and diaphragm via the extrapyramidal tract, and to the stomach via the vagus nerve.

Vomiting may be stimulated by any of the following.

(a) Irritation of any part of the gastrointestinal tract. Such irritation may be the result of chemical (e.g. alcohol), microbiological (e.g. staphylococci) or mechanical stimuli (e.g. handling of the viscera during surgery); in this respect vomiting can be regarded as an important protective reflex.

(b) Impulses from the semicircular canals in the ear, for instance in sea sickness.

(c) Cerebral tumours or a rise in intracranial pressure.

(d) Higher cerebral centres, as a response to intense anxiety, fear, unpleasant sights or smells etc.

(e) Some drugs, for example digitalis, anaesthetic agents, opiates and emetics such as ipecacuanha. Apomorphine is a potent emetic, and as such is used in alcohol aversion therapy.

Useful information can be obtained from an examination of the patient's vomitus, e.g. if it is feculent, this may indicate an intestinal obstruction; if altered blood is present, then a gastric ulcer may be suspected; if undigested food is present, then pyloric obstruction may be a problem.

Vomiting can present the following problems.

(a) Loss of fluid, and also a loss of those electrolytes present in gastric juice, mainly sodium, chloride and potassium.

(b) Loss of gastric acid, which can result in a metabolic alkalosis. Similar problems can result from long-term nasogastric drainage, when the gastric aspirate is not replaced.

(c) Exhaustion and soreness of those muscle groups (described earlier) used in vomiting.

(d) Weight loss and nutritional disturbances if the vomiting is prolonged.

(e) If the vomiting is chronic, metabolic acidosis may be more of a problem than the alkalosis just described. Metabolic acidosis will result when the patient starts to utilize body fat as a source of energy once his glycogen stores are depleted. Under these circumstances, ketones (e.g. acetoacetic acid) are produced.

(f) Inhalation of vomitus. This can lead to aspiration pneumonia. Unconscious patients should always be placed carefully in the semiprone position, so that if they do vomit, the vomitus will drain out of the mouth by gravity and so is less likely to be inhaled into the respiratory tract. Inhalation of vomitus can be a cause of death in the unaccustomed drinker who drinks alcohol heavily (for example, at a party), collapses into an unrousable stupor, vomits without waking and inhales the vomitus.

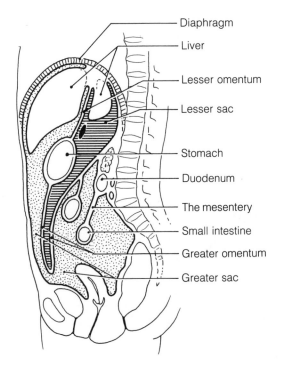

Figure 5.1.12 Diagram to show peritoneum of female in lateral view

THE SMALL INTESTINE

This consists of a long coiled tube, the structure of which closely follows the generalized pattern

described earlier. In life, the tube is approximately 3.0–3.5 m long in the adult; in the cadaver, it appears longer due to relaxation of the smooth muscle layer.

Most of the small intestine is attached to the posterior abdominal wall by a double fold of serous membrane called peritoneum (Fig. 5.1.12). The peritoneum reflected off (folded away from) the posterior abdominal wall looks rather like a fan, and supports the blood vessels, lymphatics and nerves which supply the small intestine; this portion of peritoneum is called the **mesentery**. A further fold of peritoneum extends from the liver to the stomach, and is called the **lesser omentum**. The **greater omentum** hangs rather like an apron in front of the intestines, and is reflected off the stomach. The peritoneum, as well as carrying blood vessels and nerves, serves a protective function and can 'wall off' areas of infection or inflammation, and so prevent the spread of peritonitis.

The small intestine consists of three sections (Fig. 5.1.13).

(a) The C-shaped and wider **duodenum**, lying mostly behind the peritoneum. This part is 20–25 cm long. (The name duodenum derives from the Latin, meaning 12 fingers; in the adult, the length of the duodenum is about equal to the width of 12 fingers.) The loop of this C-shaped area surrounds the pancreas.

(b) The **jejunum** (meaning empty) forms

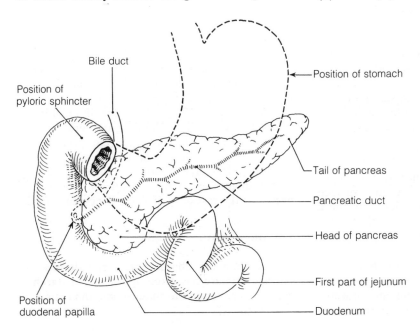

Figure 5.1.13 Diagram to show duodenum, jejunum and ileum

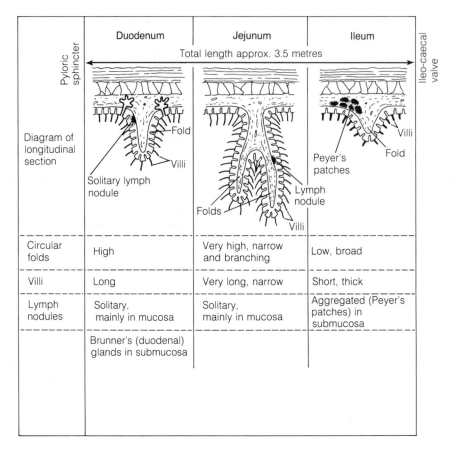

	Duodenum	Jejunum	Ileum
Diagram of longitudinal section		Total length approx. 3.5 metres	
Circular folds	High	Very high, narrow and branching	Low, broad
Villi	Long	Very long, narrow	Short, thick
Lymph nodules	Solitary, mainly in mucosa	Solitary, mainly in mucosa	Aggregated (Peyer's patches) in submucosa
	Brunner's (duodenal) glands in submucosa		

Figure 5.1.14 Diagram to show differences between areas of the small intestine

approximately 40% of the remainder of the small intestine.

(c) The **ileum** (meaning twisted) is a slightly longer tube making up the final 60%.

The jejunum and ileum are both suspended by mesentery. The two areas are, in general, not morphologically distinct. However, some differences are evident between the proximal jejunum and the terminal ileum, for example the jejunum has somewhat thicker walls which are more vascular and folded, whereas the ileum is more sparsely folded on its luminal surface (Fig. 5.1.14). Furthermore, in the jejunum there are solitary protective lymph nodes; in the ileum, however, these are aggregated into groups called **Peyer's patches**.

The function of the small intestine is to complete the digestion of ingested food and to absorb practically all the nutrients and most of the water from the chyme which enters this area from the gastrointestinal tract.

About 50% of the small intestine can be removed surgically before there is an appreciable effect upon digestion and absorption. Patients who undergo surgery, for whatever reason, which leaves them with 25% or less of their small intestine can survive only with total parenteral feeding, i.e. the infusion into a large vein (usually the superior vena cava) of amino acids, lipids, glucose, vitamins, electrolytes and trace elements in sufficient quantities to meet their full nutritional needs for the rest of their lives (Breckman, 1981).

The duodenum

By the time the chyme enters the duodenum, salivary amylase has acted upon cooked starch to begin its conversion to maltose and dextrins, and pepsin has acted upon proteins starting their breakdown into polypeptides. The duodenum does not itself secrete digestive enzymes. It does, though, secrete hormones, some of which have already been referred to (e.g. GIP, VIP, CCK and motilin).

The duodenum receives the secretions of the pancreatic and common bile duct at a point called the sphincter of Oddi, which lies about 10 cm below the pylorus at the level of the first to third lumbar vertebrae (L1–3). The secretions produced by both the pancreas and the liver and delivered to the duodenum are alkaline, having a pH in the range 7.8–8.4. There is therefore a sharp change in the pH of the intestinal contents (from the gastric pH of 1.5–3.0) after the addition of bile and pancreatic juice to the duodenal pH of approximately 7.0. The digestive enzymes in the small intestine are, like all enzymes, pH specific, i.e. there is a critical pH for their optimal activity, and act best at a pH of 6.5–7.0. This change in the intestinal pH in the area between the stomach and the duodenum does not favour the reproduction of pathogenic micro-organisms.

The first few centimetres of the duodenum (the duodenal cap) receive acid chyme from the stomach, and this tissue therefore (like that of the stomach) must be protected to prevent ulcer formation. In the first part of the duodenum, there are a large number of mucus-secreting glands in the submucosa, called **Brünner's glands**, which are characteristic of the duodenum. The mucus is secreted into the bases of the intestinal crypts, that is, into the areas between the mucosal folds.

Exocrine pancreas

The pancreas is a gland which has both endocrine and exocrine functions, and is therefore described as a mixed gland. Only the exocrine function will be described here (see Chap. 2.5 for details of pancreatic endocrine function).

Structurally, the pancreas resembles the salivary glands. It is soft, friable and pink, consisting of a head enclosed within the loop of the duodenum, a body and a tail. The whole lies horizontally below the stomach, the tail extending towards the right of the abdomen (see Fig. 5.1.13).

The pancreas has a major role in digestion. Its exocrine digestive function is served by **acinar cells**, which structurally resemble cells of the salivary glands. These cells form and store **zymogen**

granules which consist of protein-based digestive enzymes. These enzymes are discharged in response to stimulation, mainly hormonal, and are secreted into the pancreatic duct. This duct joins with the common bile duct to form a slightly dilated area called the ampulla of Vater, which then empties into the duodenum via the sphincter of Oddi.

PANCREATIC JUICE

Between 1.5 and 2 litres of juice are secreted daily (most of which is later reabsorbed), with a pH of 8.0–8.4. (It should be remembered that this does not represent the final pH of the duodenal contents, since mixing with acid chyme will lower the ultimate duodenal pH.) In theory, the pancreas produces two secretions depending on hormonal stimuli – a copious watery secretion and a scantier one rich in enzymes; in practice, a mixture of the two secretions is always released.

Profuse watery secretion
This contains large amounts of hydrogen carbonate ions, which are manufactured by the cells of the pancreatic duct in the presence of the enzyme carbonic anhydrase by a process similar to that described earlier for the production of gastric acid. Hydrogen ions are formed in the process, and blood leaving the pancreas is therefore more acid than the blood entering it. This is to some extent balanced by the hydrogen carbonate added to the blood during the formation of gastric acid by the stomach. In addition, this watery secretion contains ions of sodium, potassium, calcium, magnesium, chloride, sulphate and phosphate, plus some albumin and globulin.

Scanty enzyme-rich secretion
Three proteolytic enzymes are produced by the acinar cells of the pancreas, and for the reason described earlier in relation to pepsin – namely, to prevent autodigestion – they are secreted in the inactive form. In addition, it is thought that there is a trypsin inhibitor present in the pancreas. Figure 5.1.15 summarizes the activity of the pancreatic proteolytic enzymes.

Trypsinogen is converted into the active form **trypsin** by the enzyme **enterokinase** (also called enteropeptidase) secreted by the duodenal mucosa. The amount of enterokinase produced is increased by the presence of circulating CCK-PZ (cholecystokinin-pancreozymin).

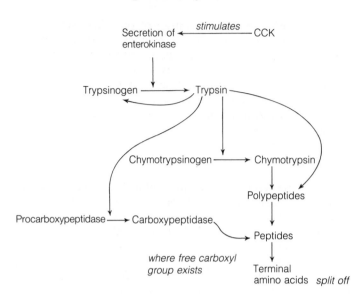

Figure 5.1.15 Summary of activity of the pancreatic proteolytic enzymes

Trypsin acts on proteins and polypeptides, breaking bonds between the amino acids and thus forming peptides (short sections of amino acid units). Trypsin, once formed, has three further actions.

(a) It activates trypsinogen to form more trypsin.
(b) It activates chymotrypsinogen to form chymotrypsin, which has the same proteolytic functions as trypsin.
(c) In addition, trypsin activates procarboxypeptidase to form active carboxypeptidase.

Carboxypeptidase acts on peptides. Whatever their individual molecular structure, all amino acids have an amine group and a carboxyl group (see Fig. 5.1.20). Single amino acids bond together by means of the amine group of one amino acid joining to the carboxyl group of another forming a peptide linkage. Carboxypeptidase splits off the terminal amino acid from the end of a peptide, acting on the free carboxyl group. It thus produces free amino acids.

Pancreatic juice also contains **pancreatic**

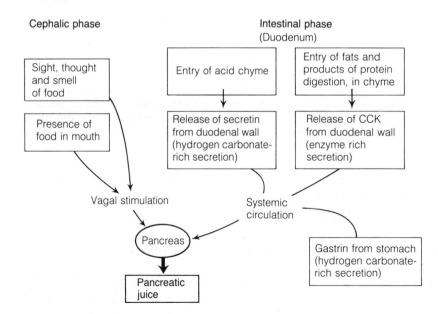

Figure 5.1.16 Flow chart to illustrate pancreatic juice secretion

amylase (sometimes called diastase). This acts on starch and converts it to maltose.

The enzyme **pancreatic lipase** acts on neutral dietary fats (triglycerides: molecules made up of three fatty acids and glycerol) and converts them to single fatty acids and glycerol once they have been emulsified by the action of bile salts.

Ribonuclease and **deoxyribonuclease (RNAase and DNAase)** act on RNA and DNA, respectively, and convert them into free nucleotides.

Control of pancreatic juice secretion
The secretion of pancreatic juice is primarily under hormonal control, although the vagus nerve has a minor influence (Fig. 5.1.16).

Secretin. This hormone is produced by S cells deep in the mucosal glands in the wall of the duodenum and upper jejunum in response to acid chyme entering that area. Secretin enters the systemic venous circulation and eventually arrives back at the pancreas via the pancreatic artery, where it brings about the production of a copious watery secretion which is rich in hydrogen carbonate (bicarbonate) ions but low in enzymes. Hydrogen carbonate levels in pancreatic juice may be up to five times greater than in plasma, and the concentration rises as secretion increases.

The actions of secretin can be summarized as follows.

(a) It increases the secretion of water and hydrogen carbonate by the pancreas.
(b) It augments the actions of CCK.
(c) It decreases gastric secretion and motility.

Secretin was the first hormone ever identified, being described by Bayliss and Starling in 1902. It was after this discovery that Starling introduced the term hormone. In molecular terms, secretin is similar to VIP, GIP and glucagon.

CCK. This hormone is usually referred to simply as CCK, an abbreviation which stands for cholecystokinin. It used to be thought that there was a hormone, pancreozymin (PZ), which stimulated the secretion of pancreatic enzymes, and another hormone named cholecystokinin which stimulated the gall bladder to contract and thus release bile. It is now clear that a single hormone has both actions; however, this misconception provides the historical explanation for the fact that some authorities refer to CCK as CCK-PZ.

The molecular structure of CCK is very similar to that of gastrin. It acts not only in the gut but also as a neurotransmitter in the central nervous system.

It is secreted by the columnar cells of the duodenal and jejunal mucosal crypts in response to products of protein digestion and fats entering the duodenum. The CCK circulates via the systemic circulation and arrives back at the pancreas, where it stimulates the acinar cells to discharge their zymogen granules; hence it leads to the release of an enzyme-rich secretion. In summary, CCK

(a) stimulates the secretion of an enzyme-rich secretion from the pancreas
(b) augments the activity of secretin
(c) slows gastric emptying and inhibits gastric secretion
(d) increases the secretion of enterokinase
(e) stimulates glucagon secretion
(f) stimulates the motility of the small intestine and colon
(g) causes the gall bladder to contract and therefore to release bile.

The secretion of secretin and CCK is referred to as the **intestinal phase of pancreatic secretion**.

Gastrin. The release of gastrin into the circulation by the stomach stimulates the production of a pancreatic secretion rich in bicarbonates. This is sometimes called the **gastric phase of pancreatic secretion**. This is of only minor importance in terms of pancreatic secretion.

Vagal stimulation. The **cephalic phase of pancreatic juice secretion** refers to the production of pancreatic juice as a result of vagal stimulation brought about by the sight, thought or smell of food, or to the presence of food in the mouth.

Acetylcholine, released as a result of vagal stimulation, acts directly on the acinar cells, bringing about the release of an enzyme-rich secretion similar to that produced in response to stimulation by CCK. Vagal stimulation is not of major importance in pancreatic secretion.

The role of bile in digestion

The formation of bile will be further referred to in Chapter 5.2 (on liver physiology); however, the role of bile in digestion will be discussed here.

Bile contains no digestive enzymes; neither does the liver secrete any digestive hormones. The main importance of bile in digestion lies in its role in the emulsification of fats and the consequent absorption of lipids and of fat-soluble vitamins and iron.

Bile is a slightly syrupy fluid, with a colour range of greeny-yellow to brown. It consists of water (97%) and bile salts (0.7%). The water contains mucin and hydrogen carbonate. Electrolytically, bile is very similar to pancreatic juice, with a pH of 7.8–8.0. The total volume secreted daily is between 0.5 and 1.0 litre.

BILE SALTS

The bile salts (see Chap. 5.2) make up 0.7% of the total volume of bile, there being some 2–4 g in total in the body. Bile salts are formed from cholic and deoxycholic acid. Both these acids are steroids manufactured by the liver from cholesterol. In the liver, cholic acid is conjugated with taurine and glycine, which are acid derivatives of the amino acid cystine, to form **taurocholic acid** and **glycocholic acid**. The term conjugation refers to the process during which the substances are joined together chemically with the elimination of water. The bile acids form bile salts with sodium and potassium in solution in the bile, e.g. sodium taurocholate and sodium glycocholate.

Bile salts

(a) deodorize faeces
(b) activate lipase (in the small intestine) and other proteolytic enzymes. (Lipase is water soluble, and for its optimal activity fats must be broken down, that is, emulsified, into small droplets.)
(c) have a detergent-like action on fats, that is, they reduce the surface tension of fat droplets and therefore contribute to the emulsification of fats.

The steroid part of the bile salt molecule is lipid soluble, and will therefore dissolve in fat droplets. The bile salt molecule also has a water-soluble carboxyl portion, which remains on the surface of the fat droplet and which carries a negative charge. This dissolves in the watery portion of bile. As negative charges repel one another, the charges on the surfaces of the fat droplets cause them to remain separated into small droplets, i.e. they emulsify. Large fat droplets are broken

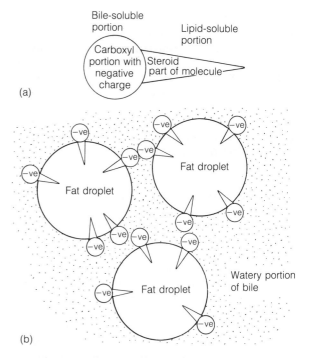

Figure 5.1.17 Diagrammatic representation of the emulsification of fats

into smaller droplets mechanically as well as chemically due to intestinal activity (Fig. 5.1.17).

Bile salts combine with lipids plus lecithin and cholesterol to form **micelles**, which are water-soluble complexes about 1 μm in diameter. Micelles allow fats to be absorbed more easily. If bile salts do not reach the intestine, e.g. in posthepatic jaundice, are lost, e.g. following an ileostomy, or not produced for some reason, then about 25% of ingested fat may be lost in the stools, which will consequently be bulky and have an offensive odour.

BILE PIGMENTS

These make up 0.2% of the total composition of bile. Bile pigments – mainly **bilirubin** and a small amount of **biliverdin** – are formed from the breakdown of old red blood cells in the reticulo-endothelial system – the mononuclear phagocytic system (see Chap. 4.1 for fuller details).

The pigments travel to the liver bound to plasma albumin, where they are conjugated, usually with glucuronic acid, in the presence of the enzyme glucuronyl transferase to form water-soluble bilirubin diglucuronide. This substance enters the bile, giving it its golden colour, and during its passage through the intestine is subjected to oxidation and acted upon by bacteria. Eventually, **stercobilinogen** is formed, some of which is absorbed into the bloodstream and ultimately excreted by the kidneys, where it is called **urobilinogen**. The remainder within the bowel is converted into **stercobilin** by bacterial action. This is a brown pigment which colours the faeces.

Bile contains cholesterol, which forms 0.06% of its total composition. Excess cholesterol from the body is excreted in the bile. When the normal concentration of cholesterol rises, crystals of cholesterol form within the biliary system and act as nuclei for the deposition of calcium and phosphate salts. This eventually leads to the formation of **gall-stones**. About 10% of the middle-aged population of the UK have some gall-stones, even though these may not be clinically apparent. Approximately 85% of all gall-stones formed are cholesterol based.

In addition, bile contains the following:

Fatty acids.
Lecithin.
Inorganic salts.
Alkaline phosphatase.
Excretory products of steroid-based hormones.

Patients having T-tube drainage of their bile duct after cholecystectomy and exploration of the common bile duct may lose a significant amount of electrolytes, especially sodium, and these will require replacement.

Control of bile secretion and release

The secretion of bile is under the influence of both hormonal and nervous stimuli (Fig. 5.1.18). The major stimulus for bile release is CCK, which is liberated into the circulation in response to fats and products of protein digestion entering the duodenum. It circulates to the gall bladder and there causes contraction of the wall and simultaneously relaxation of the sphincter of Oddi. Substances which bring about contraction of the gall bladder are referred to as **cholagogues**. Once the gall bladder has discharged its contents into the duodenum, further bile will flow directly into the gut from the hepatic cells via the hepatic

ducts. This happens continuously in patients who have undergone a cholecystectomy. Stimulation of the vagus nerve during digestion has an action similar to that of CCK.

The amount of bile secreted by the hepatic cells largely depends upon the blood levels of circulating bile salts. After passage through the intestine in the bile, about 97% of bile salts are reabsorbed in the ileum into the portal circulation, and recycled back to the liver. This is referred to as the **enterohepatic circulation of bile salts** (see Fig. 5.2.11). The remainder is lost from the body in the faeces. After a meal, therefore, once the products of digestion have reached the ileum, bile salts are reabsorbed into the blood, resulting in a high blood level of bile salts. By some mechanism not yet understood, high blood levels of bile salts stimulate the hepatic cells to increase their own secretion. This is an example of autostimulation. Substances which increase the secretion of bile are referred to as **choleretics**.

A further stimulus to hepatic production of bile salts is thought to be an increase in liver blood flow following a meal.

Intestinal juice

This is secreted by the jejunum and ileum. The enzymes present in intestinal juice are responsible for the completion of digestion. Up to 3 litres of this secretion, with a pH of 7.8–8.0, are normally produced each day by the small intestine in response to local mechanical stimuli and to the chemical stimuli of partially digested food products on the intestinal mucosa. In addition, secretion of intestinal juice is stimulated by circulating VIP (vasoactive intestinal polypeptide), and also to some extent by emotional disturbances.

Intestinal juice is rich in mucus, some of which is added by the Brünner's glands in the first few centimetres of the duodenum. The remainder of the watery juice, rich in mucus, is secreted by Lieberkühn's glands in the jejunum and ileum.

The surface area of the small intestine is increased by visible folding of the mucosa into valvulae conniventes, and by the presence of finger-like processes, each about 0.5 mm long, called **villi**. These features serve to increase the surface area to 600 times that of a non-folded tube of the same length. The total surface area is further increased by the presence of microvilli on the villous surfaces. This large surface area, of

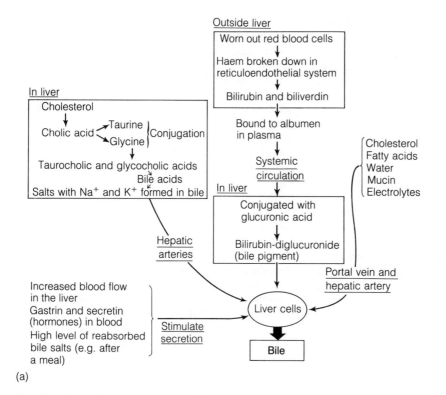

(a)

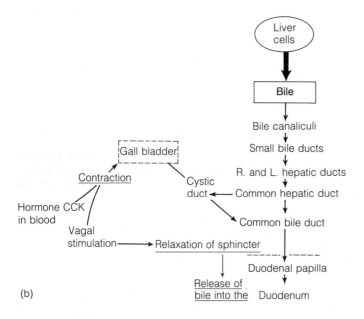

(b)

Figure 5.1.18 Flowcharts to illustrate (a) formation, and (b) secretion of bile

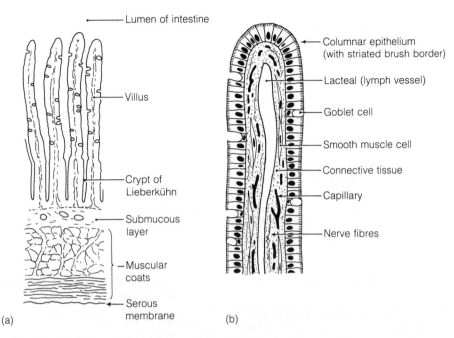

Figure 5.1.19 Diagram to show (a) villi in small intestine, and (b) a single villus

about 200 m², increases the efficiency of digestion and absorption carried out in the jejunum and ileum. The mucosa of the small intestine is invaginated between adjacent villi to form pits, called the **crypts of Lieberkühn**, where the mucus-secreting glands are located.

VILLI (FIG. 5.1.19)

There are 20–40 villi/mm² of the intestinal mucosa. Each villus consists of a finger-like process, the surface of which is covered by simple columnar epithelium continuous with that of the crypts. Within each villus there is a central **lacteal** which contains lymph and which empties into the local lymphatic circulation; each also has a capillary blood supply which links with both the hepatic and portal veins. The surface of the villus is usually covered with a layer of mucus which prevents autodigestion by proteolytic enzymes.

Villi contain two main cell types: the **goblet cells**, secreting mucus and situated mainly in the crypts, and the **enterocytes**, which are involved in both digestion and absorption. Enterocytes are tall columnar cells, their nuclei lying towards their bases. Their surfaces are each covered with up to 3000 **microvilli** (microscopic finger-like processes) which increase the villous surface area.

These microvilli give what is known as a brush border to the villus. Enterocytes have many mitochondria to provide for the high-energy demands of enzyme secretion and absorption.

Cell division in the villi is rapid (that is, these cells have a high rate of mitosis) and occurs in the crypts. Cells gradually migrate up from the crypts over a period of about 30 hours to replace those enterocytes being shed from the tips of the villi. There is a very high rate of enterocyte turnover, mainly as a result of the area being subjected to a great deal of friction by the gut contents.

Between enterocytes, at intervals, are situated lymphocytes and plasma cells. These latter secrete IgA, an immunoglobulin which protects the gastrointestinal tract from pathogens. There are also cells in the intestinal wall which secrete 5-HT (serotonin). The action of this substance in the gut is not fully understood, although it is known to stimulate the contraction of smooth muscle and may thus have a role in intestinal motility (Wingerson, 1980).

The villi contain a small number of smooth muscle fibres originating from the muscularis coat of the small intestine. Contraction of these muscle fibres assists lymphatic drainage in the central lacteals. The villi also move in response to the mechanical stimulation of food passing along the tract.

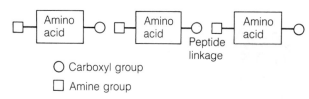

O Carboxyl group

□ Amine group

Figure 5.1.20 Diagrammatic representation of amino acid linkage

In starvation, the enterocytes shrink, the net result being a decrease in the height of the villi by up to half. The effect of this is to decrease the absorptive powers of the small intestine. Conversely, villi increase in height when actively absorbing nutrients.

INTESTINAL ENZYME SECRETION

It is the outer membrane (i.e. that which projects into the lumen of the gut) of the enterocytes which contains the remaining enzymes necessary to complete digestion. The juice produced by the glands in the crypts is practically enzyme free, contrary to past theories. Any enzymes which are present in the actual intestinal juice are probably released from disintegrating shed enterocytes.

Enzymes present in the enterocyte membrane

Aminopeptidases. These act on peptides, splitting off amino acids by acting on the amine group at the ends of peptide chains, breaking the peptide bond, and thus releasing free terminal amino acids (Fig. 5.1.20).

Dipeptidases. These act on dipeptides (Fig. 5.1.20), (i.e. units of two joined amino acids) and break them into single amino acids.

Maltase. This acts on maltose to convert it to glucose.

Lactase. This acts on lactose to convert it to glucose and galactose.

Sucrase. This acts on sucrose to convert it to glucose and fructose.

Digestion is now complete.

1 Proteins have been broken down into amino acids.

2 Fats have been broken down into fatty acids and glycerol.
3 Carbohydrates have been broken down into monosaccharides – glucose, fructose and galactose.

Ingested foodstuffs have been digested and rendered into a form in which they can be absorbed.

ABSORPTION

Each day, approximately 8–9 litres of water and 1 kg of nutrients pass across the wall of the gut from its lumen into its blood supply. This process requires energy, which is derived from the oxidation of glucose and fatty acids. The energy demands of the gastrointestinal tract are extremely high for both secretion and absorption, and also to provide for the rapid rate of mitosis in the epithelial cells lining the tract.

Transport of nutrients
Transport of nutrients across the cell membranes of the gastrointestinal epithelial cells can be either active or passive.

Active transport. This requires the expenditure of energy. It occurs when the concentration of the substance in the gut is less than the concentration of the substance in the plasma; transport must thus occur against a concentration gradient.

Vitamin B_{12} and iron are actively absorbed into the bloodstream from the ileum, as are sodium ions, glucose, galactose and amino acids. These substances all require **carrier molecules** to facilitate their absorption. Water follows the passage of these substances passively along an osmotic gradient.

Passive transport. This requires no energy consumption. Water, lipids, drugs and some electrolytes and vitamins are examples of substances transported passively from the gut into the blood. Passive transport is influenced by concentration and electrical gradients. Some substances require a carrier molecule to assist their passage across the cell membranes of the gut wall. When this carrier-mediated transport occurs passively, it is referred to as **facilitated diffusion**, an example of which is the transport of glucose molecules into cells under the influence of insulin.

Osmotic transport of water
The 8–9 litres of water which are transported

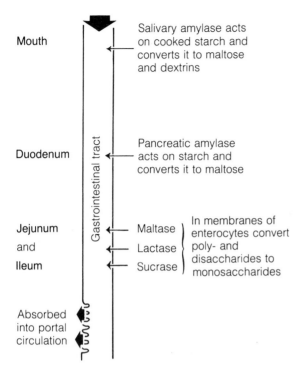

Figure 5.1.21 Summary of carbohydrate digestion

transport). Glucose and sodium ions are thought to share the same carrier molecule, which facilitates their transport across the cell membrane of the enterocyte. The concentration of sodium ions *inside* the enterocyte is always low. Sodium moves into the cell along a concentration gradient and glucose moves with it; sodium is then pumped actively out of the enterocyte into the extracellular spaces of the villi, and glucose moves with it; then both substances pass into the capillary network. The active absorption of sodium ions into the blood is therefore crucial to glucose absorption. Galactose and glucose share a common carrier; fructose, however, requires a different carrier, and its transport is not influenced by sodium.

Metabolism of monosaccharides

Monosaccharides are transported via the hepatic portal vein to the liver. Here, galactose and fructose are converted to glucose (Fig. 5.1.22). Some glucose is converted by the liver under the influence of insulin into **glycogen**; this is referred to as **glycogenesis**. About 100 g of glycogen (sufficient to maintain blood glucose levels for up to 24 hours) are stored in the liver. Glycogen is also stored within skeletal muscle, and this glycogen, through the action of muscle phosphorylase, can be reconverted to glucose to provide the energy for muscle contraction (see Chap. 3.2).

Glucose surplus to the body's needs for the maintenance of blood glucose levels and glycogen stores is converted by the liver into fat, and stored in fat (adipose) depots throughout the body. Blood glucose is maintained at a resting level of 3.5–5.5 mmol/l by a series of mechanisms described in Chapter 2.5. When the blood glucose level falls, liver glycogen can be broken down (**glycogenolysis**) to reform glucose under the influence of glucagon and adrenaline in the presence of the liver enzyme phosphorylase. If glycogen stores are depleted, the liver can manufacture glucose from fats and proteins. This process is referred to as **gluconeogenesis**.

When circulating glucose arrives at the tissues, insulin facilitates its uptake by the cells. This process was referred to earlier as an example of facilitated diffusion. The role of insulin is described fully in Chapter 2.5.

Oxidation of glucose to form energy occurs in the mitochondria of the cells. The glucose is first broken down into pyruvic acid, a process which does not require oxygen (i.e. it is anaerobic).

daily across the gut wall and reabsorbed into the blood passively follow the passage of actively or passively transported water-soluble substances to restore osmotic balance. If the transport of water required energy, our present average dietary intake would be insufficient to meet this need, and we would have to increase our intake considerably in order to reabsorb water.

Absorption of nutrients and minerals

MONOSACCHARIDES (FIG. 5.1.21)

Each day, approximately 500 g of monosaccharides are absorbed; this will of course vary according to the individual's dietary habits. All such absorption has occurred by the time the terminal ileum is reached.

Monosaccharides pass from the gut lumen across the enterocytes on the villi into the villous capillaries, and thence into the hepatic portal vein. Their active transport is facilitated by a high concentration of sodium ions on the surface of the enterocytes (low sodium levels at this point inhibit

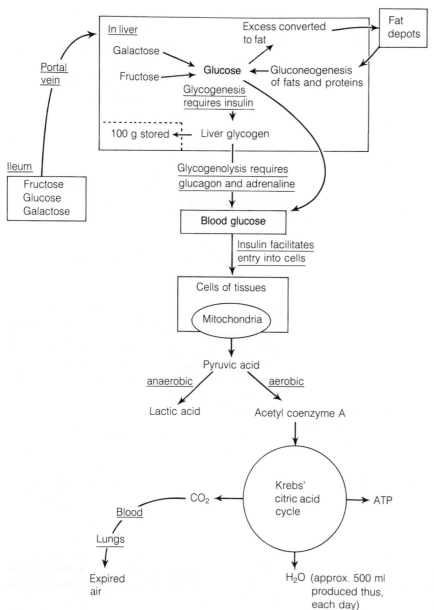

Figure 5.1.22 Diagram to illustrate metabolic pathways for glucose

Pyruvic acid is converted aerobically (i.e. in the presence of oxygen) into acetyl co-enzyme A (acetyl CoA). Acetyl CoA then undergoes a series of biochemical changes catalyzed by enzymes; these changes are referred to as the **Krebs' citric acid cycle**. They result in the formation of adenosine triphosphate (ATP), water and carbon dioxide. Energy is 'stored' in the ATP molecule.

A relatively large amount of energy is released when one of the phosphate bonds in ATP is broken to form ADP (adenosine diphosphate) and free phosphate.

If insufficient oxygen is available for the conversion of pyruvic acid to acetyl CoA, then lactic acid is formed. Thus, in hypoxic conditions such as shock, blood lactic acid levels may rise and

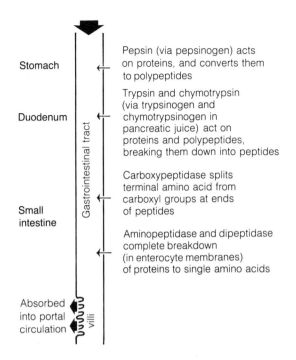

Figure 5.1.23 Summary of protein digestion

result in a metabolic acidosis. The oxidation of 1 g of glucose results in the release of 16 kJ (4 kcal) of energy in the form of heat.

AMINO ACIDS

Fig. 5.1.23 summarizes the digestion of proteins. Each day, approximately 200 g of amino acids are absorbed from the ileum. Absorption of at least 50 g per day is required in order to remain in positive nitrogen balance, and to meet the needs of the adult body for protein for growth and repair of tissues. The amount required to maintain nitrogen balance varies during the life cycle. It is greater, for example, during childhood and in pregnancy.

The mechanisms for the absorption of amino acids have not yet been fully elucidated by research. However, there appear to be three separate mechanisms for acidic, basic and neutral amino acids involving different carrier molecules. As described earlier for glucose, absorption of amino acids seems to be linked to, and facilitated by, sodium absorption. Sodium on the surface of the enterocyte membrane appears to increase the affinity of the carrier molecules for amino acids.

Once the amino acids have entered the enterocyte they appear to move passively into the blood capillaries of the villus and hence pass to the hepatic portal vein.

Most of the absorption of amino acids occurs in the first part of the small intestine. Some amino acids may enter the colon, where they are metabolized by the colonic bacteria.

The 10–20 g of protein present in the faeces is derived from dead bacteria and shed gut epithelial cells.

Metabolism of amino acids (Fig. 5.1.24)
Proteins, unlike fats and carbohydrates, cannot be stored by the body. Once absorbed into the blood, amino acids enter a common circulating pool, from where appropriate acids are taken to build up proteins for cell reproduction and growth, the formation of enzymes and hormones, and plasma proteins.

The liver can interconvert amino acids, i.e. it can use the eight amino acids essential in adults to synthesize other non-essential amino acids. The essential amino acids must be present in the diet for this to occur.

Amino acids can be used to meet energy demands once stores of glycogen are depleted. The oxidation of 1 g of amino acids results in the production of 16 kJ (4 kcal) of heat energy. Amino acids can be converted by the liver into glucose (gluconeogenesis), and any excess amino acids are broken down by the liver by the process of **deamination**. The nitrogen portion of the amino acid is converted into ammonia, which is then converted into urea, via a series of biochemical reactions termed the **Krebs' urea cycle**. Thus the more protein we take in our diet, the more urea will be produced. Urea enters the blood and is excreted in the urine. In renal failure, the kidneys are unable to excrete the urea and so blood levels rise, resulting in uraemia. In liver failure, the liver is unable to form urea although it can form ammonia. High levels of this toxic substance build up in the blood, resulting in hepatic coma. In both renal and liver failure, therefore, treatment involves limitation of the dietary intake of protein.

FATS (FIG. 5.1.25)

Each day, about 80 g of fat are absorbed from the small intestine, mostly in the duodenum. Micelles are formed in the duodenum by the action of bile

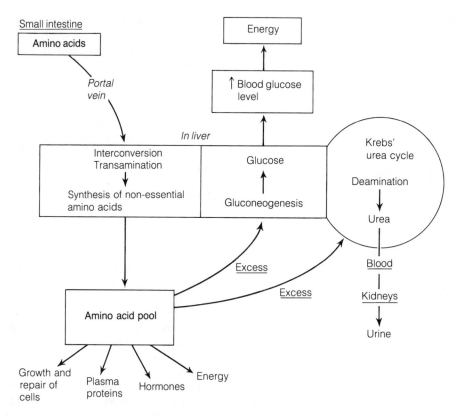

Figure 5.1.24   Diagram to illustrate metabolic pathways for amino acids

salts on lipids. Micelles are spheres of 3–10 nm in diameter, comprising fatty acids, monoglycerides and cholesterol, which form the transport mechanism for fats. They move to the microvilli on the enterocytes and there they discharge their contents, which enter the enterocytes by passive diffusion. The bile salt portion of the micelle remains within the gut where it is available for further micelle formation.

Short-chain fatty acids (i.e. those with fewer than 10–12 carbon atoms) pass from the enterocyte into the capillary network and thence to the hepatic portal vein, travelling as free fatty acids (Fig. 5.1.26). This route accounts for about 20% of fat transport. Longer-chain fatty acids (i.e. those with more than 12 carbon atoms) are resynthesized within the enterocyte to triglyceride. They become coated with a layer of lipoprotein, cholesterol and phospholipid to form **chylomicrons**. These complexes enter the central lacteals of the villi. The creamy substance so formed in the lacteal is termed **chyle**. This enters the lymphatic circulation and hence, eventually, the bloodstream.

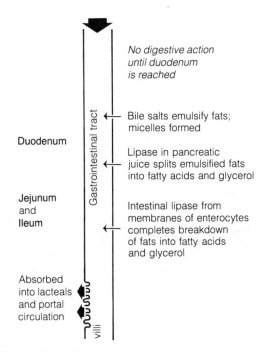

Figure 5.1.25   Summary of fat digestion

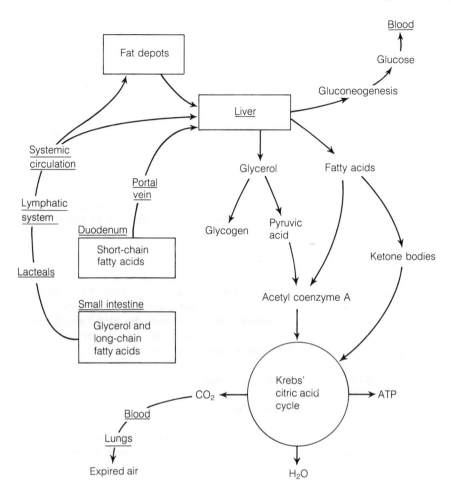

Figure 5.1.26 Diagram to illustrate metabolic pathways for fats

Faeces contain about 5% fat – most of which is derived from bacteria. Infants lose 10–15% fat in their stools as their ability to absorb fats is not well developed at birth.

Cholesterol is found in the blood mainly as

(a) **high density lipoproteins (HDLs)**, formed of a large amount of protein combined with a little cholesterol; cholesterol is carried to the liver for excretion in this form, and

(b) **low density lipoproteins (LDLs)** and **very low density lipoproteins (VLDLs)**, formed of a little protein combined with a large amount of cholesterol, which represent cholesterol on its way to the tissues.

It is thought that it is in the latter form – low and very low density lipoproteins – that cholesterol is laid down in atheromatous plaques in the arterial wall. The ratio of HDLs to LDLs and VLDLs has been shown to be increased in the blood of vegetarians, in those whose fat intake is largely polyunsaturated, and in those who take regular exercise to keep fit; this ratio is reduced in those who smoke cigarettes. It is possible that a high ratio of HDLs to LDLs and VLDLs may actually afford protection in some way from ischaemic heart (Tudge, 1985).

Fat metabolism (Fig. 5.1.26)

The oxidation of 1 g of fat results in the release of 38 kJ (9 kcal) of heat energy. Absorbed fat not required for the production of energy is stored in the adipose tissue of fat depots as neutral fat, i.e. triglycerides. These depots are situated in subcutaneous and retroperitoneal tissue.

When fats are required for energy, they are mobilized from the depots under the influence of hormones, e.g. growth hormone or cortisol, and are carried as triglycerides in the blood to the liver. Here they are broken down to fatty acids and glycerol, and released into the bloodstream for use by the cells. The liver can convert fats to glucose by gluconeogenesis.

The fatty acids are converted to acetyl CoA in the presence of oxygen and glucose. This acetyl CoA then enters the Krebs' citric acid cycle (mentioned earlier under glucose metabolism). If glucose is not present (as might be the case in starvation or uncontrolled diabetes mellitus), acetyl CoA metabolism is deranged. In the liver, molecules of acetyl CoA pair, resulting in the formation of ketone bodies, namely acetoacetic acid and β-hydroxybutyric acid. These can undergo oxidation to release energy; accumulation of these acids in the blood, however, can lead to metabolic acidosis.

Glycerol is converted either to glycogen for storage or pyruvic acid, which enters the Krebs' citric acid cycle.

SODIUM AND WATER

Each day, depending on thirst and social habits, approximately 2 litres of fluid are ingested. The secretion of digestive juices adds a further 8–9 litres of fluid to the gut contents (Table 5.1.3). Of this 10–11 litres of fluid which daily passes through the gastrointestinal tract, only some 50–200 ml are lost from the body in the faeces; the rest is absorbed at a rate of 200–400 ml/hour from the small and large intestines, as follows.

Jejunum   5–6 litres reabsorbed in 24 hours
Ileum   2 litres reabsorbed in 24 hours
Colon   1.5–2 litres reabsorbed in 24 hours

From this, it is possible to estimate the amount of absolute fluid loss from the body if an ileostomy is

Table 5.1.3 Summary of daily secretion of digestive juices

Secretion	Volume (ml)	pH
Saliva	1000–1500	6.8–7.0
Gastric juice	2000–3000	1.5–3.0
Pancreatic juice	1500–2000	8.0–8.4
Bile	500–1000	7.8–8.0
Intestinal juice	3000	7.8–8.0

Most of this daily secretion is reabsorbed. If this were not so, 8–9 litres would represent an exceedingly high rate of fluid and electrolyte loss. Quite large losses may occur in conditions such as gastroenteritis, cholera and typhoid and following an ileostomy.

performed (Breckman, 1981). Sodium ions are actively absorbed in the jejunum, ileum and colon; chloride ions follow passively, as does water.

Water can either move out of the gut lumen into the blood, or it can move from the blood to dilute the gut contents when these are hypertonic. It is for this reason that hypertonic enema solutions (e.g. magnesium sulphate) result in the production of large watery stools. Movement of water from the blood into the gastrointestinal tract occurs (as described earlier) in dumping syndrome following major gastric resection or total gastrectomy. The movement of a large volume of water into the gastrointestinal tract can, in some cases, lead to severe shock.

POTASSIUM

Some potassium is actively secreted into the gut, particularly in mucus. Usually, though, potassium is passively absorbed into the blood along a concentration gradient from the ileum and colon. In patients suffering from diarrhoea, or in those who have an ileostomy, hypokalaemia can be a potential problem.

VITAMINS

Most water-soluble vitamins (those of the B group, except vitamin B_{12}, and vitamin C) are absorbed passively with water. Vitamins A, D, E and K are fat soluble, and their absorption depends on efficient micelle formation and subsequent entry into the enterocytes. The production of bile salts and the secretion of lipase are thus necessary for the efficient absorption of these vitamins.

Vitamin B_{12} is absorbed in the terminal ileum. As described earlier, it forms a complex with intrinsic factor from the gastric parietal cells. The complex is thought to bind to receptors on the ileal wall, and vitamin B_{12} is then able to transfer across the gut wall into the blood.

CALCIUM

About 30–80% of ingested calcium is actively absorbed in the upper part of the small intestine under the influence of parathyroid hormone and calcitonin (see Chap. 2.5).

Calcium absorption is facilitated by the active metabolite of vitamin D, formed in the kidney under the influence of parathyroid hormone, called 1,25-dihydroxycholecalciferol. This substance brings about the synthesis of a protein in the gastrointestinal mucosa which binds to calcium ions and is necessary for their transport across the gut wall.

When serum calcium levels fall, more 1,25-dihydroxycholecalciferol is formed, and so more calcium can be absorbed from the small intestine. Calcium absorption is facilitated by lactose and proteins, and inhibited by oxalates and phytic acid (found, for example, in cereals and rhubarb), and phosphate.

IRON

In the UK, the average daily intake of iron is 15–20 mg. Only about 5–10% of the total dietary iron intake is absorbed into the blood from the gastrointestinal tract. Each day, about 1 mg of iron is lost from the body through desquamation of skin and via the faeces and urine. In females of child-bearing age there is an additional absolute loss of some 25 mg of iron in an average menstrual flow.

Most of the dietary intake of iron is in the ferric form. However, iron is more readily absorbed in the ferrous form, and reduction from the ferric form to the ferrous form is facilitated by gastric juice and also by vitamin C. Patients who have undergone radical gastric surgery may therefore have problems with iron absorption and may become anaemic.

Iron is actively absorbed in the upper part of the small intestine; thus patients with colostomies and ileostomies should not experience problems with iron absorption. The enterocytes store iron, and more iron is absorbed from the lumen only when these cellular stores are depleted. The enterocytes discharge their iron stores into the blood when serum iron levels fall.

Iron travels in the blood mostly bound to apoferritin, a globular protein, and while in the blood it is referred to as **transferrin**. Once iron binds to apoferritin, **ferritin** is formed. This is the principal storage form of iron, although a small amount is stored as **haemosiderin**. About 70% of the body's iron is in haemoglobin; 3% is in myoglobin, a muscle protein; the rest is stored in the liver as ferritin or haemosiderin.

If the passage of chyme through the small intestine is hastened in any way, there will not be efficient absorption of nutrients. It normally takes about 9 hours from the time of ingestion for nutrients to reach the terminal portion of the small intestine; patients who suffer from chronic conditions resulting in 'intestinal hurry' may well show signs of malabsorption of some nutrients.

THE LARGE INTESTINE

In an adult, the large intestine is approximately 1.5 m long, and consists of the caecum, appendix, colon and rectum (Fig. 5.1.27).

The large intestine has five functions.

(a) Storage of food residues prior to their elimination.
(b) Absorption of most of the remaining water, electrolytes and some vitamins.
(c) Synthesis of vitamin K and some B vitamins by colonic bacteria.
(d) Secretion of mucus, which acts as a lubricant for the elimination of faeces.
(e) Elimination of food residues as faeces.

Each day, about 1 litre of chyme (with a consistency like that of thin porridge) enters the large intestine laterally via the ileocaecal valve. The ileocaecal valve is normally closed, as a result of back pressure from the colonic contents. It has two horizontal folds which project into the caecum, and are formed of circular muscle fibres. The valve acts like a sphincter to prevent chyme leaving the small intestine rapidly, i.e. before adequate time has elapsed for full absorption to occur. The ileocaecal valve opens in response to peristaltic waves which bring chyme into contact with it. In addition, when food enters the stomach, a reflex is set up via the vagus nerve which stimulates peristalsis in the colon. This causes the caecum to relax and the ileocaecal valve to open. This reflex is called the **gastrocolic reflex**; it is particularly evident after breakfast, when food enters an empty stomach. The consequent colonic peristalsis causes the rectum to fill with faeces, and this results in the urge to defaecate.

The large intestine has a diameter of 5–6 cm. Apart from the stomach, it is the widest part of the gastrointestinal tract and allows the storage of large amounts of food residues, which move slowly through this section of the tract. The large intestine has no villi, and hence has a much smaller

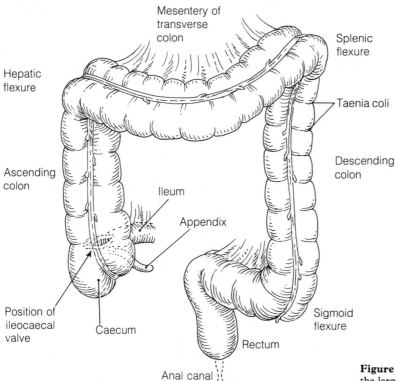

Figure 5.1.27   Diagram to show the large intestine

internal surface area – approximately one-third that of the small intestine.

Throughout the length of the large intestine there are many patches of lymphoid tissue in the muscularis layer of the tube. These have a protective function.

The caecum

This is a blind-ending pouch about 7 cm long, entered via the ileocaecal valve and leading to the colon. In humans it has no significant function, but in herbivorous animals it is concerned with cellulose digestion.

The appendix

This is a vermiform (worm-like) blind-ending sac projecting from the end of the caecum, and about the size of an adult's little finger. It has no function in humans. It is made up mainly of lymphoid tissue, and can enlarge in infection or inflam-

mation. Inflammation (i.e. appendicitis) can occur if the opening of the appendix to the caecum is blocked by a hard mass of faeces (a faecolith). If this inflammation causes the enlarged appendix to rupture, then faecal material containing bacteria will enter the abdominal cavity and peritonitis is likely to follow.

The colon

This dilated area of the large intestine is anatomically divided into three main regions (Fig. 5.1.27).

1 *The ascending colon.* This is on the right of the abdomen, extending upwards as far as the lower border of the liver.
2 *The transverse colon.* The ascending colon turns at the hepatic flexure, or right colic flexure, into the transverse colon which lies slightly 'higher' on the left than the right, transversely below the liver and the stomach.
3 *The descending colon.* At the left colic or splenic flexure, the transverse colon turns to form the

descending portion, on the left of the abdomen. This gives rise to the sigmoid (s-shaped) colon, which empties into the rectum.

The colon differs from the generalized structure of the gastrointestinal tract described earlier in that the longitudinal muscle bands in the muscularis layer are incomplete. As a result, the wall is gathered into three longitudinal flat bands called the **taenia coli**. Because these bands are shorter than the rest of the colon, the wall pouches outwards to form **haustrations** (derived from the Latin word for bucket) between the taeniae when the circular muscle fibres contract. As the haustrations fill and empty, they aid in kneading the colonic contents.

FUNCTIONS OF THE COLON

Storage
The main function of the colon is to store unabsorbed and unassimilable food residues. The colon thus acts as a reservoir; 70% of the residue of food is excreted within 72 hours of ingestion, and the remaining 30% can stay within the colon for a week or longer. As it does so, progressively more water is reabsorbed.

Absorption of sodium and water
Sodium is actively transported from the colon to the hepatic portal vein, and water and chloride ions follow passively. The amount of water reabsorbed from the colonic contents depends on the length of time the residue remains in the colon. In the constipated person, the food residue remains within the colon for several days and hence most of the water is reabsorbed, resulting in hard pellets of faeces which are difficult to eliminate.

The bulk of the food residue, and hence potential faeces, is made up of cellulose – a substance which humans are unable to digest since they lack the vital enzyme cellulase.

Some drugs, for example aspirin, prednisone and some anaesthetics, and also amino acids can be absorbed by the colonic mucous membrane. Hence steroid retention enemas can be used successfully to reduce inflammation of the colon in patients suffering from ulcerative colitis.

Secretion of mucus and electrolytes
Mucus contains hydrogen carbonate and hence colonic mucus gives the contents of the colon a pH of 7.5–8.0. In addition, some potassium ions and some hydrogen carbonate ions may be secreted actively into the colon.

Incubation of bacteria
Many of the bacteria which colonize the large intestine are anaerobic species (i.e. they do not need oxygen for their survival). *Bacteroides fragilis* and *Clostridium perfringens (welchii)* are both anaerobic; *Enterobacter aerogenes* is aerobic, as its name implies, and there are also some streptococci and lactobacilli and *Escherichia coli*. The relatively sluggish movements of the colon are conducive to colonization by bacteria. Many of these bacteria, which compose the gut flora, exhibit a modified symbiotic relationship with humans; that is, each derives mutual benefit from the other and they live together harmoniously. It should be noted, though, that these commensals (literally, 'those who eat at the same table', from the Latin) can become pathogenic, especially if introduced into another part of the body, for example gut bacteria may cause cystitis if introduced into the bladder during catheterization.

The bacteria synthesize vitamin K, thiamine, folic acid and riboflavin in small amounts. The amount produced is not normally nutritionally significant; however, in vitamin deficiency or starvation this contribution may be of some benefit. These bacteria also synthesize a small amount of vitamin B_{12}, but since this vitamin can be absorbed only from the ileum, the amount thus synthesized is normally excreted.

Patients who are on long-term antibiotic therapy may lose these commensal bacteria, and this loss provides the opportunity for colonization by pathogenic, antibiotic-resistant bacteria, a potential problem which one should be aware of in such patients. The first sign of such a problem is usually diarrhoea.

Bacterial fermentation of food residues produces quite large amounts of gas, called flatus, which consists of nitrogen, carbon dioxide, hydrogen, methane and hydrogen sulphide. Between 500 ml and 700 ml of flatus may be produced each day, although the amount will show considerable variation depending on the food eaten; foods such as baked beans, onions, cauliflower and pulses lead to an increase in flatus production due to fermentation of their residues by the colonic bacteria. The production of flatus may also result from air swallowing in anxiety states (aerophagia).

Normal bowel movements allow the expulsion of gases so produced.

In patients who have had abdominal surgery, the smooth muscle activity of the intestine may be inhibited because of trauma resulting from operative handling of the gut. This results in cessation of movements of the small intestine, a condition termed **paralytic ileus**. This may occur not only after abdominal surgery, but also as a response to intestinal obstruction. In paralytic ileus, peristalsis stops and thus movement of the gut contents stops; this results in the formation of pockets of gas and fluid. This gas cannot be passed as flatus, and the consequent accumulation of gas and fluid leads to abdominal distension with increase in girth and the production of considerable abdominal discomfort.

If paralytic ileus occurs as a result of abdominal surgery, after a few days the abdominal smooth muscle starts to contract again, peristalsis is once more evident, and flatus can be expelled.

In a patient suffering from paralytic ileus, no food or fluids should be given until flatus is passed – a sign of returning peristalsis. Aspiration of a nasogastric tube helps to remove gases produced and digestive fluids secreted until the paralysis has passed; this may prevent some of the discomfort experienced by the patient.

THE ROLE OF DIETARY FIBRE IN THE LARGE INTESTINE

The time taken for food residues to be expelled is directly related to the amount of dietary fibre ingested. Dietary fibre, which used to be termed 'roughage', decreases the mouth-to-anus transit time.

Dietary fibre is made up largely of **cellulose**, the substance found in cell walls. Humans do not product cellulase, the enzyme found in herbivorous animals such as cows and rabbits, which is necessary to digest and utilize cellulose, and since cellulose cannot therefore be absorbed, it stays in the bowel where it exerts a hygroscopic effect, i.e. it attracts water to it. Thus, stools high in fibre tend to be bulkier and softer in consistency, and this makes them easier to expel.

Diverticular disease is a condition more prevalent in omnivores (meat and vegetable eaters) than vegetarians, whose diet is always high in fibre; and there is a higher incidence of the condition (about 10% of men and women over the age of 40) in the Western world than in the Third World, where less meat and refined food and more vegetable fibre are eaten.

Diverticula are pouches or sacs that occur in the walls of the intestine as a result of weakness of the muscle layer at that point. They may be congenital. Eating a diet high in fibre does not prevent such diverticula forming (**diverticulosis**); it will, however, help to prevent **diverticulitis** (inflammation of the diverticula). The latter condition results when hard masses of faeces collect in the diverticula and cause inflammation. This results in increased peristalsis accompanied by discomfort and diarrhoea. Stools high in fibre are softer and pass through the bowel at a speed which is not conducive to pockets of faeces being trapped within the diverticula.

Mann (1981) has estimated that vegetarians eat approximately 41.5 g of fibre daily, whereas omnivores eat only 21.4 g of fibre daily; 33% of his sample of omnivores were found to be suffering from diverticular disease, compared to only 12% of his vegetarian sample.

Burkitt and Trowell (1975) have suggested that the longer mouth–anus transit times found in people consuming a low-fibre (refined) diet allow bacterial toxins and metabolites to remain in contact with the gut wall for a longer period of time; this may be linked to the higher incidence of carcinoma of the large intestine and rectum in the Western world compared to the Third World. In addition, the faeces of omnivores contain a higher proportion of *Bacteroides* than do those of herbivores. It is now thought possible that these bacteria may act on bile acids to form carcinogenic products. The slower the bowel transit time, the longer such carcinogens may have to exert their effects on the intestinal wall.

A further benefit of a high-fibre diet is that the softer stool produced is easier to expel; thus the necessity to strain at defaecation is eliminated, and this may reduce the incidence of haemorrhoids.

The report of the National Advisory Committee on Nutrition Education (1983) recommends that fibre intake should be increased by 33%, to 30 g per head per day – mainly by increasing consumption of wholegrain cereals.

MOVEMENTS OF THE COLON

Although the colon has only incomplete bands of longitudinal muscle fibres, it does have complete

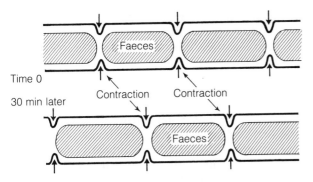

Figure 5.1.28 Diagram to show segmentation in the colon

bands of circular muscle fibres. When these latter bands contract, segmentation results. Segmentation allows mixing of the colonic contents and facilitates colonic absorption. When the circular muscle bands contract, they divide the colon into segments. Contractions occur about once every half-hour, after which the circular muscle fibres relax and adjacent bands of muscle contract, thus breaking up the first segment of faeces (Fig. 5.1.28).

Segmentation is a non-propulsive movement, and in the constipated person it is the main movement that occurs in the large bowel. Peristalsis, described earlier, also occurs in the colon, propelling faeces towards the rectum.

Two or three times a day, usually after meals, an increase in activity occurs in the colon. After meals, especially after breakfast (which is eaten when the stomach is empty), the gastrocolic reflex occurs; this brings about increasing contractions of the terminal ileum, relaxation of the ileocaecal valve and colonic peristalsis. The gastrocolic reflex therefore allows filling of the colon.

Associated with the gastrocolic reflex, the phenomenon of **mass movement** occurs in the colon. Mass movement propels the faeces in the mid-colon towards the rectum; the haustrations in the mid-colon disappear, the tube becomes shortened and flattened by waves of rapidly advancing powerful contractions, and the colonic contents are moved at speed, within a few seconds, into the sigmoid colon. The subject only becomes aware of mass movements when faeces enter the rectum.

Analgesics such as morphine, codeine and pethidine decrease these movements in the colon; ganglion-blocking agents and aluminium-based antacids have a similar effect. Persons taking these drugs therefore may become constipated.

The rectum

The rectum is a muscular tube about 12–15 cm long, capable of great distension. It is usually empty until just before defaecation. Mass movement of the colon leads to sudden distension of the rectal walls, which brings about what is often termed the "call to stool", when the subject becomes aware of the need to defaecate.

The rectum opens to the exterior via the anal canal, which has both internal and external anal sphincters. The **internal anal sphincter** is composed of smooth muscle fibres, and is not under voluntary control. Fibres from the sympathetic nervous system to the internal sphincter are excitatory, i.e. when active they bring about contraction; fibres from the parasympathetic system are inhibitory, and stimulation of these causes relaxation of the sphincter.

The **external anal sphincter** is composed of striated muscle fibres and is under the control of the will from about the age of 18 months. This control, or continence, is learned, and is not present in the baby and young infant. The external sphincter is supplied by the pudendal nerve and is maintained in a state of tonic contraction. This voluntary sphincter control may be lost in people with damage to their pudendal nerve or spinal cord or after a cerebrovascular accident.

DEFAECATION

When faeces enter the rectum, afferent impulses travel to the sacral spinal cord. If it is convenient to defaecate, the sacral spinal cord will reflexly initiate defaecation. Impulses travel from the sacral spinal cord not only to the terminal colon and anal sphincter, but also to the cerebral cortex, which can inhibit spinal cord activity if defaecation is not convenient. In patients who have had a cerebrovascular accident or who have senile deterioration of their cerebral functions, such inhibition may not be possible and faecal incontinence results. Similarly, in babies or young infants, or in patients with sacral spinal cord lesions, reflex efferent impulses from the sacral spinal cord immediately cause contraction of the terminal colon and relaxation of the sphincters. Defaecation is thus a reflex response to faeces entering the rectum.

Defaecation is accompanied by some measure of straining. The degree of effort necessary will depend upon the consistency of the faeces. A deep breath is taken and expired against a closed glottis (sometimes called **Valsalva's manoeuvre**). This forced expiration is aided by contraction of the muscles of the abdominal wall, which further serves to raise the intra-abdominal pressure. This, together with contraction of the levator ani muscles, causes the pressure in the rectum to rise to about 200 mmHg (26 kPa) in the adult. This increase in pressure, together with relaxation of the anal sphincters, leads to the expulsion of the rectal contents. During straining, there is a sharp rise in blood pressure followed by a sudden fall. For this reason, defaecation can sometimes precipitate cerebrovascular accidents.

FAECES

Approximately 100–150 g of faeces are usually eliminated every day, consisting of 30–50 g of solids and 70–100 g of water. The solid portion is made up largely of cellulose, epithelial cells shed from the lining of the gastrointestinal tract, bacteria, some salts and the brown pigment stercobilin. It seems that a small amount of potassium is actually secreted into the faeces in the colon – a fact which becomes significant in diarrhoea when quite large amounts of potassium may be lost. Indole and skatole are two products of amine breakdown, arising from bacterial decomposition, which give faeces their characteristic odour.

DIARRHOEA

This results when movements of the intestine occur too rapidly for water to be absorbed in the colon. Stools are therefore produced in large amounts, and may range from being loose to being entirely liquid, such as the 'rice water' stools associated with cholera. Diarrhoea, depending on its severity and duration, can be a mere nuisance or can be fatal. If diarrhoea is severe, large amounts of water (consisting of ingested fluids and digestive juices, together with sodium and potassium) are lost in the stools. This rapidly results in dehydration and electrolyte imbalance. In chronic diarrhoea, hypokalaemia is often found, and loss of alkaline digestive juices may cause a metabolic acidosis. Infants with diarrhoea and vomiting can quickly become severely dehydrated.

The problems that the sufferer is aware of are exhaustion and abdominal pain, possibly embarrassment, anxiety or fear and excoriation of the perineum. While urgent medical management of the condition is essential, nursing activites should in addition be directed towards the relief of these problems.

Causes of diarrhoea

Diet. Certain foods, notably fruits such as prunes, rhubarb and gooseberries, and highly seasoned dishes can result in diarrhoea. Similarly, diarrhoea may be the result of allergies to some foods, and to a high intake of alcohol.

Diverticulosis. This has been discussed in some detail in the section on the role of dietary fibre.

Drugs. Certain drugs, for example antibiotics and iron preparations, cause diarrhoea in some people. Laxatives, by definition, result in the passage of a large, loose stool.

Infections. Organisms such as *Salmonella*, *Shigella* and staphylococci all produce an inflammatory enteritis when present in sufficient numbers. Such intestinal infections are commonly accompanied by nausea and abdominal pain, and usually follow the ingestion of contaminated food.

Inflammatory conditions of the gastrointestinal tract. These include, for example, ulcerative colitis and regional ileitis (Crohn's disease). The inflammation leads to an increase in peristalsis and also excess mucus production; intestinal contents are therefore moved rapidly through the large bowel.

Malabsorption syndrome. In conditions in which, for example, gluten, lactose or fats are not absorbed from the ileum, bulky, offensive stools are produced.

Neoplasms. Malignant growths in the bowel may result in a change of bowel habits, such as alternating periods of diarrhoea and constipation.

Stress. Diarrhoea may be a physiological response to stress.

Thyrotoxicosis. In this condition, there is a general speeding up of body activities, and intestinal hurry commonly occurs.

In treating diarrhoea, the predisposing cause

must be sought and treated and dehydration and electrolyte imbalances must be corrected. Concurrently, provision must be made for symptomatic relief of the problems presented to the patient by the condition, for example the position of the patient's bed in the ward must be carefully considered. If the patient is ambulant, it would be helpful to him if his bed were situated not too far from the lavatory. If the patient is on bedrest, then provision must be made for adequate ventilation of the bed area and for deodorant sprays. Soft lavatory paper or tissues are more comfortable for the patient, and sometimes a barrier cream for the perineal area is helpful to prevent excoriation. Diarrhoea is an unpleasant condition which causes many patients embarrassment and worry, and reassurance that they are not a nuisance is frequently necessary.

CONSTIPATION

This term refers to the difficult passage of hard stools. Many people may wrongly regard themselves as being constipated if they do not defaecate every day. However, it may be normal for one person to have two bowel actions every day whereas for another it is normal to defaecate only two or three times each week. In this latter case, so long as the stools are of normal consistency and are not difficult to pass, such a person could not be regarded as being constipated.

Constipation is the opposite of diarrhoea in that the food residues become hard, due to the reabsorption of most of the water when they remain in the colon for a long time. They thus become difficult and often painful to eliminate. Constipation frequently occurs when the diet contains insufficient fibre. Food residues tend to remain in the colon until eventually both the colon and the rectum are full of faecal material. This results in the sufferer complaining of a feeling of fullness or of feeling 'bloated'. Abdominal distension may well be evident, and this may lead to a feeling of nausea. In addition, halitosis, a furred tongue, headache, irritability and flatulence may also occur.

When constipated stools are passed, they may be so hard that an anal fissure (a tear in the anal mucosa) occurs with consequent bleeding and pain. The degree of straining necessary to pass such hard stools may result over a period of time in the development of varicosed rectal veins, that

is, haemorrhoids or piles. These varicosities occur as a result of the rise in pressure in the rectal veins which accompanies prolonged straining, and this leads to incompetence of the rectal venous valves. Venous return along the rectal veins becomes sluggish, and the veins become distended. The degree of distension may be such that the veins assume a grape-like appearance, and in severe cases the varicosed rectal veins may prolapse through the anal sphincter. Haemorrhoids further serve to aggravate the problem of constipation as the sufferer tends to delay defaecation in an attempt to avoid the consequent pain.

Causes of constipation

Avoidance of defaecation. Some examples of conditions in which this may occur have already been described. In addition, embarrassment at having to use a bedpan or commode in the close vicinity of other patients may lead to constipation in any patient in hospital. If a patient is required to delay defaecation as a result of a commode, lavatory or bedpan not being available at the time when it is required, then as a result he may suffer extra discomfort by adding constipation to his existing problems (Wright, 1974).

At home, the patient who is too weak or who is in too much pain (for example, from arthritis) to move to the lavatory may avoid defaecation.

Dehydration. A decrease in fluid intake, or an increase in fluid loss, can cause constipation.

Depression and dementia. Both of these conditions result in a general slowing down of both physical and mental activities. This would include the slowing down of colonic movements and constipation may result. Antidepressant drug therapy may further serve to worsen the condition.

Drugs. Certain analgesics, notably codeine, all narcotics (e.g. morphine and heroin), some antihypertensive agents (e.g. methyldopa), anticholinergics (e.g. the antispasmodic propantheline), aluminium antacids (e.g. Aludrox), and iron preparations may all cause constipation.

Haemorrhoids. Varicosities and prolapse of rectal veins can be exceedingly painful, and sufferers may attempt to avoid the passage of stools.

Hypothroidism (myxoedema). Patients suffering from this condition tend to have general depression of all their body activities. Faeces therefore pass through the colon slowly and constipation results.

Inactivity. Exercise tends to stimulate peristalsis and thus defaecation. Patients who are on prolonged bedrest may therefore suffer from constipation. In addition, such patients may suffer from a decreased appetite and may therefore decrease their dietary intake. For many reasons, too, they may be reluctant to ask for a bedpan or commode, and may therefore delay defaecation.

Insufficient dietary fibre. Dietary fibre is hygroscopic, i.e. attracts water. It therefore provides bulk to the stool and aids elimination. Elderly patients without their own teeth or with badly fitting dentures may tend to eat a soft, low fibre diet, and thus aggravate the problems arising due to weak musculature of the pelvic floor.

Neoplasms. Change in bowel habits brought about by intestinal growths can lead to alternating bouts of diarrhoea and constipation.

Weak musculature of the pelvic floor. (i.e. the levator ani muscles). In the elderly, or in multiparous women (i.e. those who have had several babies), the muscles of the pelvic floor tend to become weak and therefore less able to contract efficiently during defaecation, which therefore becomes inefficient.

Complications of constipation

In order to initiate the call to stool, a faecal mass of 100–150 g is necessary. If the faecal mass is less than this, then straining is necessary in order to eliminate it. During straining, momentary circulatory stasis occurs, with a sharp increase in the thoracic, intra-abdominal and blood pressures. This can lead to the propagation of thrombi as emboli which may occlude either the cerebral or pulmonary circulation, depending on their site of origin. The increase in blood pressure can result in the rupture of an existing aneurysm in the cerebral circulation or aorta.

It should be possible to prevent constipation by adding bran to the diet and by encouraging an adequate amount of exercise. The daily ingestion of about 30 g of bran together with a diet high in fibre (e.g. fruit, cereals and vegetables) are sufficient to prevent constipation in most subjects (Harris, 1980). Bran is cheap, non-habit forming and more effective than aperients. People eating bran, though, should increase their fluid intake in order to prevent its hygroscopic action leading to dehydration. Bran is not palatable on its own, but can be added to cereals, soups and gravy to make it more acceptable. However, once constipation has occurred, the following measures, in addition to the above, may be helpful.

Management of constipation

The position adopted for defaecation. This affects the efficiency of the mechanism. A comfortable squatting position is more efficient than an upright one. Sitting on a bedpan is uncomfortable and therefore does not aid defaecation; it may in fact increase the amount of straining required. For patients who have suffered a myocardial infarction, the use of a commode probably causes less overall stress than that associated with the use of a bedpan. Additionally, the assurance of privacy is a psychological help.

Lubricants. The administration of oral or rectal lubricants, such as liquid paraffin or dioctyl orally, or glycerine suppositories rectally, serves to soften the faeces. Such lubricants are not absorbed. However, if taken frequently, they may interfere with the absorption of fat-soluble vitamins A, D, E and K.

Bowel stimulants. The commonest aperients in this group are the senna derivatives, bisacodyl (Dulcolax), Senokot and cascara. It is thought that these substances irritate the colonic mucosa and thus aid defaecation. The now seldom-used soap and water enema is an example of an irritant administered rectally.

Osmotic aperients. Magnesium sulphate (Epsom salts), which can be given either orally or as an enema, oral milk of magnesia, and phosphate enemas are examples. These substances are hygroscopic, that is, they draw water into the lumen of the gut from the surrounding blood capillaries. A large watery stool will therefore follow their administration.

Bulking agents. Methylcellulose derivatives, examples of which are dietary fibre, Isogel, Normacol and Celevac, reduce mouth-to-anus transit time by attracting water to the gut contents and thus providing a bulky but relatively soft stool.

If the above attempts to manage the problem of constipation fail, then it may be necessary to carry out a manual removal of faeces. This is a painful and embarrassing procedure, and should only be attempted by a doctor or trained nurse experienced in the technique. The patient will usually need analgesia or sedation before this procedure.

Nursing assessment of a patient's bowel habits

Whether an individual is admitted to hospital or cared for in the community, it is important to assess his or her bowel function. Bowel problems may not be central to the patient's need for care, but they may aggravate the primary problems if allowed to develop. In addition, the topic in general is one in which the nurse may, with effect, attempt some health education. Assessment of a patient's bowel habits links with the assessment of his state of hydration, the condition of his tongue, the smell of his breath, assessment of his food and fluid intake and activity levels.

The following points relating to bowel habits should be assessed by the nurse, either from direct observation or by questioning.

(a) Frequency of the bowel actions.
(b) Quantity of stool produced and variation from normal: stools will be increased in volume in patients taking bran, large quantities of fruit and vegetables or in those on a vegetarian diet.
(c) Consistency of the stool passed: the stool will be softer in those on a high-fibre diet and should float in the lavatory.
(d) Inexplicable changes in bowel habits in terms of amount, frequency or consistency.
(e) Presence of mucus in the stool: this may indicate an inflammatory condition in the bowel.
(f) Presence of blood: whether it is fresh or altered, blood may give clues about where in the gastrointestinal tract bleeding is occurring.

(g) Presence of undigested food: this will occur in conditions of intestinal hurry.
(h) Colour of the stool: dark stools may result from oral iron preparations; tarry stools may indicate melaena (i.e. bleeding from the gastric or upper intestinal region); pale stools may be the result of obstructive jaundice.
(i) Offensive odour of stool: this may occur with malabsorption states.
(j) If the stool is a response to an enema or to aperients, is it an adequate response?
(k) Pain on defaecation: this may indicate the presence of haemorrhoids, anal fissure, a perineal lesion of constipation.

Gastrointestinal pain

Pain experienced as a response to gastrointestinal (i.e. visceral) stimuli is different in kind from that experienced as a response to cutaneous stimuli. The nerves in the viscera are not able to convey discrete sensations of touch, temperature, etc, to the cerebral cortex; instead, they convey impulses which result from distension of the gastrointestinal wall, and the pattern of these impulses is interpreted in the cerebral cortex. The effect on the individual of pain resulting from gastrointestinal stimuli will vary with, among other factors, that individual's pain threshold, degree of preparation for, or anticipation of, pain and the duration of the pain.

Areas of the gastrointestinal tract will be considered in anatomical order.

MOUTH

Oral pain can range from that of an inflamed tongue or gums to toothache. With toothache, the nerve endings themselves in the pulp cavity are stimulated, and pain impulses travel in the trigeminal nerve to the sensory cortex. Toothache is an example of a pain which can be acute or chronic in nature.

OESOPHAGUS

Pain here is often referred to by the sufferer as 'heartburn'. It usually results from reflux of acid gastric contents into the oesophagus, causing irritation of the epithelial tissue. Impulses so produced travel in the lateral spinothalamic tract to the thalamus and thence to the sensory cortex.

STOMACH

Pain here may result from hunger contractions of the stomach; from distension when the stomach is overfull after a heavy meal; and from the formation of a gastric ulcer. This last example results in what sufferers often term a 'gnawing' pain. The inflammation produced around the ulcerated region leads to oedema and an increase in tension in the area. Pain impulses travel to the central nervous system from the stomach along the gastric sympathetic fibres.

INTESTINE

Intestinal pain is usually referred to as **colic**, i.e. it results from prolonged contraction (spasm) of smooth muscle. Mechanical obstruction is one condition which may result in colic pain; this occurs when the area of intestine above the obstruction dilates due to the accumulation of gas and fluid. Biliary colic occurs when gall-stones obstruct the common bile duct. This leads to an increase in tension in that area, and also to a degree of local ischaemia. Most nerve fibres carrying painful stimuli pass from the intestine to the spinal cord in the region of T11–L2, via sympathetic nerves. Impulses then pass to the thalamus via the lateral spinothalamic tract of the spinal cord, and hence to the sensory cortex.

APPENDIX

When this organ becomes inflamed and oedematous, pain is initially felt in the periumbilical region. The explanation for this apparent anomaly is that embryologically the appendix develops from the midgut, lying in the periumbilical area, and its nerve supply reflects this development. A few hours after the pain is first felt, it commonly localizes over the appendix itself. The pain of appendicitis is usually severe.

PERITONEUM

The pain of peritonitis is usually severe, and the patient typically lies very still with a rigid abdominal wall 'guarding' the underlying inflammation. When the visceral (inner) layer of peritoneum is stimulated, a diffuse sensation of pain results that is poorly localized. This diffuse sensation results because the peritoneum is insensitive to local mechanical stimuli and contains relatively few pain receptors. In contrast, stimulation of the parietal (outer) layer of the peritoneum by inflammation results in a well-localized sensation of pain.

RECTUM AND LARGE INTESTINE

Pain fibres from the rectum accompany the pelvic parasympathetic nerve fibres. Pain from the proximal part of the large intestine is typically experienced as periumbilical; pain from the distal end of the large intestine is experienced as hypogastric.

Review questions

The answers to all these questions can be found in the text. In each case there is at least one correct and at least one incorrect answer.

1 Which of the following statements concerning saliva is/are correct?

(a) Saliva plays an essential role in carbohydrate digestion.
(b) Over 2 litres of saliva are secreted each day.
(c) Saliva contains lysozyme.
(d) Secretion of saliva is stimulated by the administration of hyoscine.

2 During swallowing

(a) the epiglottis shuts off the nasopharynx
(b) respirations are inhibited
(c) segmentation occurs in the oesophagus
(d) the tongue pushes against the soft palate.

3 Stimulation of the parasympathetic nervous system leads to

(a) an increase in gastric motility
(b) stimulation of the alpha receptors in the pyloric sphincter
(c) relaxation of the pyloric sphincter
(d) sparse salivary secretion.

4 Secretion of gastric juice is stimulated by

(a) secretagogues
(b) prostalglandins
(c) secretin
(d) acetylcholine.

5 Which of the following is/are function(s) of the stomach?

(a) Absorption of all alcohol.
(b) Conversion of ingested foodstuffs to chyle.
(c) Secretion into the gastric juice of carbonic anhydrase.
(d) Formation of polypeptides.

6 Which of the following is/are secreted in the duodenum?

(a) Motilin.
(b) Gastric inhibitory peptide (GIP).
(c) Carbonic anhydrase.
(d) Trypsinogen.

7 Pancreatic juice secretion is stimulated by

(a) enterokinase
(b) chyle entering the duodenum
(c) gastrin from the stomach
(d) procarboxypeptidase.

8 The secretion of bile results in

(a) the formation of cholic acid
(b) activation of intestinal lipase
(c) micelle formation
(d) synthesis of cholesterol.

9 The enterocytes

(a) synthesize IgA
(b) decrease in size in starvation
(c) synthesize aminopeptidase
(d) absorb maltose.

10 Which of the following is/are actively absorbed from the ileum?

(a) Fructose.
(b) Chloride.
(c) Alcohol.
(d) Acetyl co-enzyme A.

11 Which of the following is/are not a function of the colon?

(a) Absorption of sodium ions.
(b) Secretion of hydrogen carbonate ions.
(c) Synthesis of vitamin D.
(d) Digestion of cellulose.

12 Which of the following is/are an example of an osmotic aperient?

(a) Milk of magnesia.
(b) Isogel.
(c) Senokot.
(d) Dulcolax.

Answers to review questions

1 c
2 b
3 a and c
4 a and d
5 d
6 a and b
7 c
8 b and c
9 b and c
10 a
11 c and d
12 a

References

Breckman, B. (1981) *Stoma Care: a Guide for Nurses, Doctors and Other Health Care Workers*. Beaconsfield: Beaconsfield Publishers.

Burkitt, D. & Trowell, J. (1975) *Refined Carbohydrate Foods and Disease*. London: Academic Press.

Daggett, P. (1981) *Clinical Endocrinology*. Physiological Principles of Medicine Series. London: Edward Arnold.

Goodinson, S. M. (1986) Assessment of nutritional status *Nursing*, 3 No. 7, 252–258.

Harris, W. (1980) Bran or aperients. *Nursing Times*, 8 May: 811–813.

Mann, J. (1981) The well nourished vegetarian. *New Scientist 89*, No. 1239.

National Advisory Committee on Nutrition Education (1983) *Proposals for Nutritional Guidance for Health Education in Britain*. London: HMSO.

Sanford, P. (1982) *Digestive System Physiology*. London: Edward Arnold.

Tudge, C. (1985) *The Food Connection – The BBC Guide to Healthy Eating*. London: BBC.

Wingerson, L. (1980) Gut feelings about neuropeptides. *New Scientist* 3 April: 16–18.

Wright, L. (1974) *Bowel Function in Hospital Patients*. Royal College of Nursing Research Project, Series 1, No. 4. London: Royal College of Nursing.

Suggestions for further reading

Coates, V. (1975) *Are They Being Served?* London: Royal College of Nursing.

Jones, D. (1977) *Food for Thought – a descriptive study of the nutritional and nursing care of unconscious patients in general hospitals.* Royal College of Nursing Research Project, Series 2, No. 4. London: Royal College of Nursing.

McMinn, R. M. H. & Holdell, M. H. (1974) *Functional Anatomy of the Digestive System.* London: Pitman Medical.

Nurses' Clinical Library (1985) *Gastrointestinal Disorders.* Nursing 85 Books. Springhouse, Pennsylvania: Springhouse.

Nursing (1982) Nutrition. *Nursing 2*, No. 4 (August).

Nursing (1984) Faecal elimination. *Nursing 2*, No. 30 (October)

Nursing (1986) Nutrition. *Nursing 3*, No. 7 (July).

Nursing (1986) Nutrition. *Nursing 3*, No. 8 (September).

Nursing (1980) Nutrition and health. *Nursing 1*, No. 11 (March).

Nursing (1980) Nutrition in illness. *Nursing 1*, No. 12 (April).

Nursing (1980) Faecal elimination. *Nursing 1*, No. 17 (September).

Polak, J. & Bloom, S. (1983) Gut regulatory peptides. *British Medical Journal 286*: 1461–1466.

Watson, J. F. & Royle, J. R. (1987) *Watson's Medical-Surgical Nursing and Related Physiology*, 3rd edn. London: Baillière Tindall.

Whitehead, S. A. (1981) The puzzle of peptides – neurotransmitters or gut hormones? *Nursing Times* 15 January: 122–123.

Chapter 5.2
The Liver

Jennifer Storer

Learning objectives

After studying this chapter, the reader should be able to

1 Describe the role of bile salts in the digestion and absorption of fats.
2 Discuss the significance of the enterohepatic circulation.
3 Explain the metabolism of bilirubin and the causes of jaundice.
4 Name the constituents of bile and list their functions.
5 Describe how bile is secreted from the liver cell to the intestine.
6 Explain the causes and effects of cholestasis.
7 List the factors associated with the formation of gall-stones.
8 Explain the formation of ascites and list the principles of treatment.
9 Teach a junior nurse the importance of preparation and care for the patient undergoing liver biopsy.
10 State how the liver is involved in the maintenance of blood glucose levels.
11 Discuss the detoxication of substances within the liver.
12 List the effects of cellular damage of the liver.

Introduction

The liver is vital to life. It is an organ that is metabolically active in the synthesis and catabolism (breakdown) of fats, proteins, carbohydrates and vitamins. It also metabolizes and detoxifies hormones, steroids and exogenous substances, such as drugs and alcohol, and secretes bile. This continuous biochemical activity gives rise to considerable heat production, second only to that of muscular activity. Under basal conditions the liver is responsible for most of the body heat. The wide variety of reactions taking place in the liver allows for integration and regulation of its various functions. Other body tissues do not demonstrate such a wide functional ability. Although liver tissue has a considerable regenerative capacity, it can malfunction under certain conditions. The diversity of reactions taking place within the organ is then made apparent in the widespread bodily effects of abnormal liver function.

STRUCTURE

The liver, situated in the upper right quadrant of the abdominal cavity, is the largest glandular organ of the body, weighing, on average, 1.5 kg in men and 1.3 kg in women. It consists of four anatomical lobes, the largest being the right which lies under the right dome of the diaphragm; the smaller left lobe lies under the left dome. Two lesser segments of the right lobe, the caudate and the quadrate lobes, are located on the undersurface (Fig. 5.2.1). The liver is encased by the rib cage so that the organ is not normally palpable

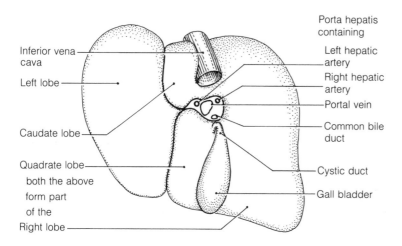

Porta hepatis
containing

Left hepatic
artery

Right hepatic
artery

Portal vein

Common bile
duct

Cystic duct

Gall bladder

Inferior vena
cava

Left lobe

Caudate lobe

Quadrate lobe
  both the above
  form part
  of the
Right lobe

Figure 5.2.1   The inferior surface of the liver showing the position of the four lobes

(Fig. 5.2.2). During a deep inspiration, the lower edge moves 1–3 cm downwards and may be palpated at 'two fingers depth' below the ribs on the right side of the abdomen.

The peritoneum covering the liver forms several ligaments attaching the organ to the diaphragm and the abdominal wall. The ligament attaching the liver to the abdominal wall is known as the **falciform ligament**.

Blood supply

The blood supply to the liver derives from two sources. The **hepatic artery**, a branch of the coeliac artery, carries arterial blood from the systemic circulation. The **portal vein**, supplied by the splenic and superior mesenteric veins, carries blood drained from the stomach and upper intestine (Fig. 5.2.3). These vessels enter the liver on the undersurface at the **porta hepatis** (see Fig. 5.2.1), and divide immediately into right and left branches which subdivide further through the hepatic tissue. Blood leaves the liver in the left, right and central **hepatic veins** which open directly into the inferior vena cava as they leave the liver. There are sphincters in all the vascular compartments of the liver regulating the local supply from the hepatic artery and portal vein as well as the total liver blood flow and capacity of the portal venous bed. In certain conditions, for example in cardiac failure, the liver can accommodate up to one-third of the total body blood volume; thus the splanchnic vessels (i.e. those to the viscera) play a major part in the regulation of the general circulation. The vascularity of the liver is responsible for the problems of trauma to the organ caused by stabbing, gunshot wounds or, more frequently, car accidents. A considerable volume of blood may be lost, necessitating rapid repair and supportive measures.

Histology

Microscopically, liver tissue is divided into **lobules**, 1–2 mm in diameter (Fig. 5.2.4). These constitute the functional units of the liver. The approximately hexagonal lobules consist of a central vein from which single columns of **hepatocytes** (liver cells) radiate towards the surrounding thin layer of connective tissue. Within the connec-

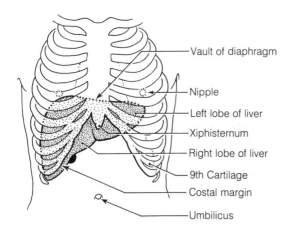

Vault of diaphragm

Nipple

Left lobe of liver

Xiphisternum

Right lobe of liver

9th Cartilage

Costal margin

Umbilicus

Figure 5.2.2   The position of the liver in relation to the rib cage

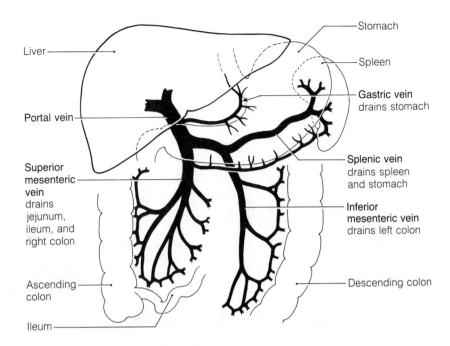

Figure 5.2.3 The main features of the portal vein

tive tissue are situated portal canals, each containing a branch of the hepatic artery and portal vein and an interlobular bile duct. Between the layers of cells lie sinusoids receiving blood from both artery and vein and draining into the central vein. The system of veins runs approximately perpendicular to the portal canals. Surrounding the liver cells and in direct contact with them is a network of minute tubules called the **bile canaliculi** which carry bile produced in the liver cells. The canaliculi drain into larger ductules and terminate in the interlobular bile ducts of the portal canals. The flow of blood and bile is shown schematically in Fig. 5.2.5.

The sinusoids are lined with flat phagocytic cells of the mononuclear phagocytic system

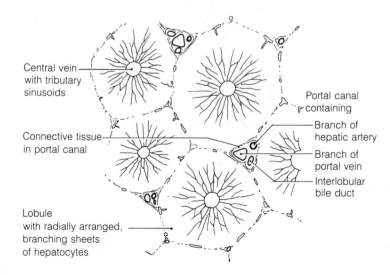

Figure 5.2.4 The general features of the liver lobules at low magnification

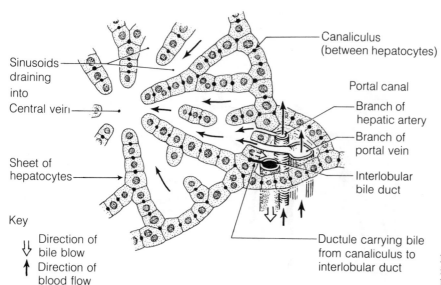

Sinusoids draining into Central vein

Sheet of hepatocytes

Canaliculus (between hepatocytes)

Portal canal

Branch of hepatic artery

Branch of portal vein

Interlobular bile duct

Ductule carrying bile from canaliculus to interlobular duct

Key

⇊ Direction of bile blow

↑ Direction of blood flow

Figure 5.2.5 Diagram indicating the flow of blood and bile within the liver lobule

(reticuloendothelial system), called **Kupffer cells**. These are important in phagocytosis and also in the production of antibodies. Behind these cells is the space of Disse (Fig. 5.2.6) which contains tissue fluid bathing the microvillous border of the hepatocytes. The sinusoidal lining is apparently freely permeable to nutrients and other molecules contained in the plasma. The microvilli maximize the area of the liver cell available for absorption of these substances and fluids from the plasma. The area of hepatocyte adjacent to the bile canaliculus also has microvilli projecting into the lumen.

Biliary drainage

The biliary drainage of the liver is completed as the interlobular ducts join with one another until the left and right hepatic ducts emerge from their respective lobes at the porta hepatis and unite to form the common hepatic duct (Fig. 5.2.7). This is joined by the cystic duct from the gall bladder,

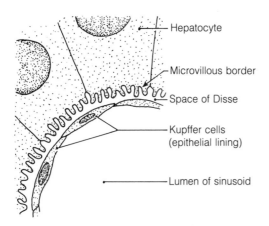

Hepatocyte

Microvillous border

Space of Disse

Kupffer cells (epithelial lining)

Lumen of sinusoid

Figure 5.2.6 Diagram indicating the position of the space of Disse

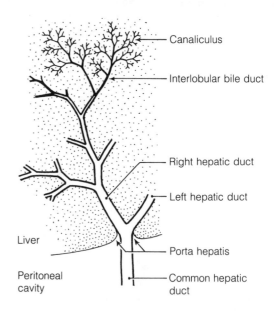

Canaliculus

Interlobular bile duct

Right hepatic duct

Left hepatic duct

Liver

Peritoneal cavity

Porta hepatis

Common hepatic duct

Figure 5.2.7 Diagram showing the intrahepatic biliary drainage (the biliary tree)

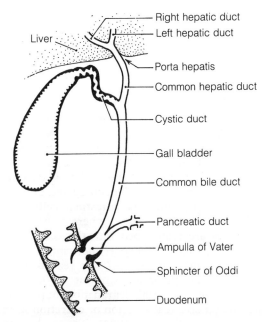

Figure 5.2.8 The drainage of bile from the liver to the intestine (the biliary tract)

forming the common bile duct. The total length of the common bile duct is 7–8 cm and it opens, in common with the pancreatic duct, via the ampulla of Vater, a spindle-shaped dilation, into the second part of the duodenum 10 cm from the pylorus (Fig. 5.2.8). This common opening of the ducts is surrounded by a circular muscle called the **sphincter of Oddi** which controls the flow of bile into the duodenum. Each duct also possesses its own sphincter so that bile and pancreatic juice may be discharged independently.

THE GALL BLADDER

The gall bladder is a pear-shaped sac with a capacity of about 50 ml, lying under the right liver lobe, to which it is bound by connective tissue and small blood vessels. The walls consist of a network of elastic and non-striated muscular tissue. The mucous membrane is thrown into folds, producing a honeycomb appearance in the body of the gall bladder, and is lined with columnar epithelium. There are no glands within the gall bladder. The cystic duct joins the gall bladder to the common bile duct.

BILE

The production of bile is the main exocrine function of the liver, providing an excretory route for several substances. Bile is a yellowish-green viscous fluid, slightly alkaline in reaction (pH 7–8) and having a bitter taste. About 500–1000 ml are produced daily by the liver and pass into the common bile duct. There is a diurnal rhythm of secretion, more being produced during the day. In the common hepatic duct, above the cystic duct, the bile is 97–98% water and contains 2–3% solids. The main constituents of bile contribute to its functions.

Bile acids (salts)

Bile acids constitute approximately half the solid matter of bile. In solution, they form salts, usually with sodium. They are important in promoting the absorption of fats and fat-soluble substances from the intestine. There are 2–4 g of bile salts in the body pool and approximately 0.3–0.6 g are lost in the faeces per day. Since the pool remains constant, the liver synthesizes 0.3–0.6 g bile salts per day (Jones and Meyers, 1979).

The two primary acid forms, cholic acid and chenodeoxycholic acid (CDCA), are synthesized in the liver from cholesterol (Fig. 5.2.9). The secondary forms, deoxycholic and lithocholic acids, are produced in the intestine as a result of bacterial action. Only small quantities of the latter are produced, which are excreted in the faeces. The deoxycholic acid is absorbed with the remaining cholic and CDCA by the terminal ileum and

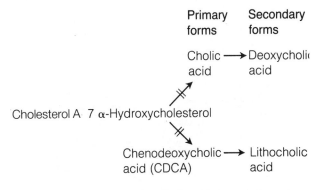

Figure 5.2.9 The formation of bile acids from cholesterol

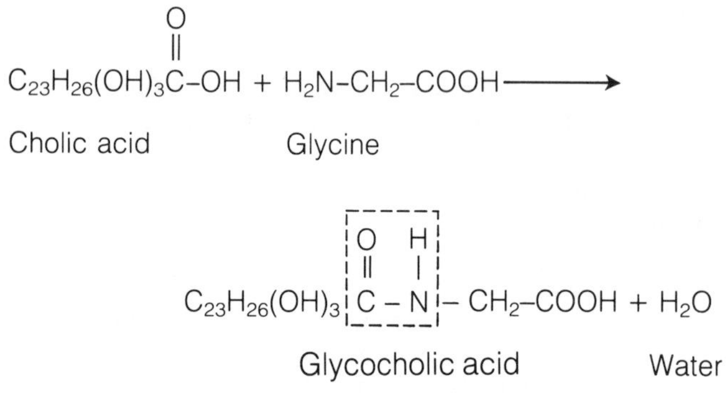

$$C_{23}H_{26}(OH)_3\overset{\overset{\displaystyle O}{\|}}{C}-OH + H_2N-CH_2-COOH \longrightarrow$$

Cholic acid   Glycine

$$C_{23}H_{26}(OH)_3\boxed{\overset{\overset{\displaystyle O}{\|}}{C}-\overset{\overset{\displaystyle H}{|}}{N}}-CH_2-COOH + H_2O$$

Glycocholic acid   Water

Figure 5.2.10   The conjugation of cholic acid with glycine occurring in the liver cell

proximal colon. The synthesis of bile acids is controlled by a feedback mechanism operating at the first reaction (see A in Fig. 5.2.9).

Within the liver cells, the bile acids are conjugated with the amino acid glycine or taurine to form glycocholic or taurocholic acid. A **conjugation reaction** involves the formation of an amide linkage (similar to that of amino acid links to form proteins) with the elimination of water (Fig. 5.2.10). The salts (usually sodium) of these conjugated acids are secreted from the liver cells into the canaliculi. A carrier-mediated mechanism

involving active transport against a concentration gradient occurs (Jones and Meyers, 1979).

In the bile, the salts form **micelles**, aggregates of molecules having both polar and non-polar characteristics which keep non-polar, insoluble substances such as cholesterol in solution (see Fig. 5.1.17).

In the intestine the micelles are disrupted, allowing the cholesterol to be released and excreted in the faeces while the bile salts are operative in the absorption of fat molecules from the lumen (see also Chap. 5.1). Some of the cholate is deconjugated by bacterial action. Unconjugated bile salts passively diffuse into the intestinal cells at all levels of the small intestine, whereas the conjugated ones are reabsorbed into the cells of the lumen by an active transport mechanism in the terminal ileum. A total of 97% is reabsorbed in this way. The remainder continue to the colon where further bacterial degradation occurs followed by passive reabsorption or excretion in the faeces (Campbell *et al.*, 1984). The reabsorbed bile salts enter the portal venous system draining the intestine. Most are bound to serum proteins and returned to the liver where, at the sinusoids, the salts are taken up by the hepatocytes via a carrier-mediated system. Thus only small amounts enter the systemic circulation. The bile salts taken up by the liver are available for further secretion

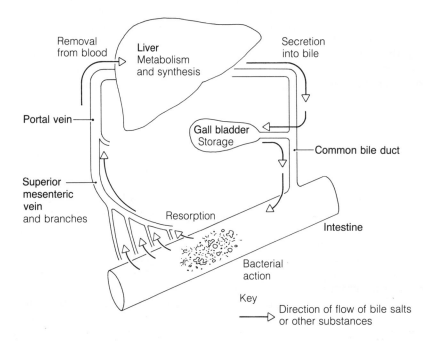

Key

⟶ Direction of flow of bile salts or other substances

Figure 5.2.11   A diagrammatic representation of the enterohepatic circulation

from the hepatocytes into the bile canaliculi. This cycle is called the **enterohepatic circulation**. It provides a mechanism for the conservation and reutilization of the bile acids and also of other compounds secreted in the bile, such as oestrogenic hormones, progesterone, the thyroid hormones, vitamin A and its metabolites as well as various drugs (Fig. 5.2.11).

The total bile salt pool may circulate between the liver and intestine two to three times during each meal, so that 15–30 g bile salts per day are secreted by the hepatocytes into the canaliculi. Various factors affect the enterohepatic circulation of compounds. These include the extent and rate of secretion of the compound into the bile; the activity of the gall bladder; the fate of the substance in the intestine, particularly the consequences of bacterial deconjugation reactions; and the fate of the substance after reabsorption (Smith, 1973).

The importance of bile acids is threefold (see Chap. 5.1). First they activate enzymes involved in the absorption of nutrients such as pancreatic lipase. Second, as emulsifiers they are involved in the digestion and absorption of fats. And finally, high blood levels of bile acids themselves stimulate further bile and bile acid production. They also stimulate a variety of hepatic secretory functions such as phospholipid and cholesterol secretion.

ABNORMALITIES OF BILE SALT METABOLISM

Abnormalities of bile salt metabolism and turnover may occur in two ways. Obstruction of the biliary tree results in a decreased quantity of bile salts reaching the intestine while there is an increased proportion within the liver and serum. Since the bile salts are required for the digestion and absorption of fat from the intestine, this process is impaired, resulting in an increased excretion of fat within the faeces, known as **steatorrhoea**. In conditions of hepatocellular damage such as cirrhosis there is a significant decrease in the total bile acid pool due to a decreased synthesis of bile acids, especially cholic acid in the liver, and an absence of deoxycholic acid due to changes in the colonic flora or intestinal contents such that the normal production of deoxycholic acid from cholic acid does not occur. This also results in steatorrhoea due to poor digestion and absorption of fats (Schwarz *et al.*, 1979).

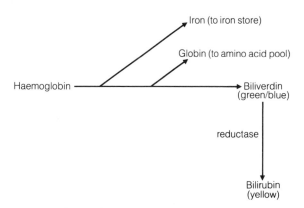

Figure 5.2.12 Breakdown of haemoglobin in mononuclear phagocytic cells of the liver

Bile pigments

These are the excretory products of haem, which are responsible for the colours of bile and faeces (see also Chap. 4.1). About 260–300 mg of bilirubin are produced in adults per day, although the liver has the reserve capacity to excrete five to ten times this amount. About 75% is derived from the haemoglobin of mature red cells, the breakdown of which takes place in the mononuclear phagocytic (reticuloendothelial) cells of the liver and spleen. The remainder is released in the liver from tissue cytochromes and other haem products. The breakdown of haemoglobin can be summarized as shown in Fig. 5.2.12.

BILIRUBIN

Bilirubin, the predominant pigment in bile, is yellow in colour, weakly acid, soluble in lipid and sparingly soluble in water. Within cells it interferes with vital metabolic functions. Unconjugated bilirubin is transported in the plasma tightly bound to albumin. Drugs such as sulphonamides or salicylates compete for binding and so facilitate the diffusion of bilirubin into the liver and other tissues. In neonates such drug administration would facilitate the entry of unconjugated bilirubin to the brain, increasing the risk of kernicterus. Unbound (lipid-soluble) bilirubin is taken up by the liver cells with ease, whereas the bound complex utilizes a carrier mechanism in the cell membrane. Within the cell, bilirubin is bound

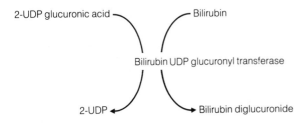

Figure 5.2.13 Conjugation of lipid-soluble bilirubin to form water-soluble bilirubin diglucuronide

to one of two soluble proteins of low molecular weight called the Y and Z carrier proteins. Competition with other substances may occur. At the smooth endoplasmic reticulum the lipid-soluble unconjugated bilirubin is rendered water soluble, and thus easily excreted, by conjugation to form bilirubin diglucuronide. The reaction is catalyzed by the microsomal enzyme bilirubin uridine diphosphate (UDP) glucuronyl transferase (Fig. 5.2.13). This reaction can be induced (increased)

by drugs such as phenobarbitone or inhibited (decreased) by others – novobiocin, for example.

The secretion of the conjugated bilirubin to the bile canaliculi is poorly understood. It occurs against a large concentration gradient and is probably carrier mediated. The conjugated bilirubin is part of the micellar complex in the bile and thus passes into the intestine (Fig. 5.2.14).

In the terminal ileum and colon, bacterial activity releases unconjugated bilirubin which is then reduced to urobilinogen. Small quantities of urobilinogen absorbed by the intestine enter the enterohepatic circulation and are re-excreted by the liver and the kidney; 0.5–5.0 μmol per day enter the systemic circulation to be excreted as urinary urobilinogen, which may be detected using Ehrlich's aldehyde reagent. Increased amounts occur in various conditions, for example haemolytic jaundice. Urobilinogen is colourless, but oxidizes on exposure to air to an orange-red urobilin. It may be detected visually if urine is allowed to stand.

Some 150–500 μmol of urobilinogen is excreted

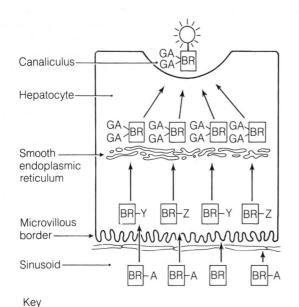

Key
BR   Bilirubin
A   Albumin
Y and Z   Carrier proteins
GA   Glucuronic acid
☀   Micelle

Stage
4: Incorporation of diglucuronide into micelle in the bile

3: Conjunction with glucoronic acid to form diglucuronide

2: Binding to carrier protein

1: Bound or free bilirubin

Figure 5.2.14 A diagrammatic representation of the passage of bilirubin through the hepatocyte

in the faeces. Here it is known as stercobilinogen. It is also oxidized on exposure to air, giving rise to stercobilin and the characteristic faecal coloration. In the neonate bilirubin only is found in the faeces, as the bacterial reduction mechanism for its conversion to urobilinogen does not develop fully for some months. This explains the yellow colour of babies' stools. In the adult, oral antibiotics may also inhibit the bacterial reactions.

Serum bilirubin levels are measured using the Van den Bergh test. The normal ranges are as follows.

Total bilirubin	5.0–17.0 μmol/l
Conjugated ('direct')	< 3.0 μmol/l
Unconjugated ('indirect')	2.0–15.0 μmol/l

JAUNDICE

Jaundice occurs when the normal metabolism of bilirubin is altered resulting in an increased serum bilirubin level. Clinically it may be identified by a yellow coloration of the skin, sclera and other elastic tissues which is visible in daylight at serum levels greater than 34 μmol/l. The urine may be dark due to the presence of excess bilirubin and the faeces may be pale due to its absence. Traditionally jaundice has been classified as haemolytic, hepatocellular or obstructive (cholestatic). This, however, is an oversimplification and the following classification may be more helpful.

Jaundice due to increased bilirubin load
An increased production and/or breakdown of red cells may occur. As these are the precursors of haem and thus bilirubin, an increased amount of the latter is formed. The liver is unable to conjugate all the pigment so that a rise in the level of unconjugated pigment in the serum takes place. This situation occurs in various haemolytic conditions both congenital, such as hereditary spherocytosis, and acquired, such as reaction to the drug phenacetin (formerly present in several analgesic preparations but now withdrawn from use because of renal problems).

Serum levels of bilirubin are usually less than 85 μmol/l and the jaundice is mild and lemon yellow. The faeces are dark due to increased quantities of stercobilinogen, but the urine rarely darkens although the urinary urobilinogen is raised. Anaemia also accompanies the condition, because of the increased breakdown of red cells.

Jaundice due to deficiency of the transferase enzyme
Deficiency of the conjugation enzyme, bilirubin UDP glucuronyl transferase, also leads to raised levels of unconjugated bilirubin. This occurs in various hereditary conditions, the commonest being Gilbert's disease. A mild, benign, fluctuating jaundice occurs with no change in colour of the urine or faeces.

The liver of a neonate may be immature, giving rise to a temporary deficiency of transferase. The decreased conjugation rate leads to a retention of unconjugated bilirubin and the development of an orange-yellow jaundice after the first 24 hours in a full-term infant and after approximately 48 hours in a preterm infant. This 'physiological' jaundice, common in normal babies, usually subsides within 2 weeks.

Unconjugated bilirubin, being lipid soluble, has an affinity for nervous tissue, particularly in the newborn, and at high levels passes into the basal ganglia and other areas of the brain and spinal cord causing disturbance of cellular metabolism and resulting in **kernicterus**. The mortality from this is high, and surviving infants are usually mentally handicapped. Treatment of rising bilirubin levels (estimated from heel-prick blood samples by a bilirubinometer) by phototherapy, which converts bilirubin to a colourless compound with no known long-term deleterious effects on the infant, is used for more severe 'physiological' jaundice. Serum bilirubin levels exceeding 340 μmol/l in full-term infants or 270 μmol/l in preterm infants are considered dangerous, and exchange transfusion may be considered (Valman, 1980). Kernicterus may also develop when haemolytic disease of the newborn due to rhesus incompatibility between the mother and infant occurs (see Chap. 4.1).

An abnormal progesterone in the breast-milk of some mothers inhibits the transferase reaction, giving rise to 'breast-milk jaundice' in the infant. This may continue for up to 2 months after delivery.

Jaundice due to liver cell damage
General damage to the liver by toxic or infective agents causes disruption of the cells with a regurgitation of bilirubin, both conjugated and unconjugated, into the blood. Infective agents include, most commonly, the viruses responsible for hepatitis A and B and also yellow and glandular fevers.

Table 5.2.1   The characteristics of different types of jaundice

	Haemolytic	Hepatocellular		Obstructive	
Abnormality	Increased bilirubin load	Deficiency of transferase	Cellular damage	Intrahepatic fibrous obstruction	Extrahepatic obstruction
Example	Hereditary spherocytosis	Gilbert's disease	Infective hepatitis	Biliary cirrhosis	Carcinoma of head of pancreas
Urine					
Colour	Normal	Normal	Dark	Dark	Dark
Urobilinogen	Increased	Variable	Variable	Absent	Absent
Bilirubin	Absent	Absent	Increased	Increased	Increased
Faeces					
Colour	Dark	Normal	Paler than normal	Pale	Pale
Stercobilinogen	Increased	Normal	Low	Absent	Absent
Plasma					
Unconjugated bilirubin	Increased	Increased	Decreased	Increased	Increased
Conjugated bilirubin	Normal	Decreased	Increased	Increased	Increased

Toxic agents include drugs such as halothane (anaesthetic) and the monoamine oxidase inhibitors (antidepressants) to which some individuals develop sensitivity reactions.

Serum bilirubin levels are variable, often greater than 340 µmol/l, and show a rise during the icteric (jaundiced) period followed by a fall during convalescence. The urine is dark due to the presence of conjugated water-soluble bilirubin, while the faeces are paler than normal as little bilirubin reaches the intestine due to the microscopic obstruction of the bile canaliculi.

Jaundice due to intrahepatic obstruction
Microscopic intrahepatic obstruction can occur which leads to a regurgitation of conjugated bilirubin to the blood. In the condition of biliary cirrhosis, fibrous obstructions to the secretion of bile at the canaliculi occur. The onset of jaundice is gradual but high levels of serum bilirubin (greater than 510 µmol/l) may be reached after prolonged illness. The urine is dark and the faeces pale. In prolonged jaundice the skin may appear greenish in colour due to the presence of biliverdin.

Some drugs may cause intrahepatic obstruction by forming plugs within the canaliculi, thus leading to jaundice. Chlorpromazine (Largactil) is an example of this, 1–2% of patients receiving it developing jaundice.

Jaundice due to extrahepatic obstruction
Extrahepatic obstruction to bile flow may be caused within the biliary tree by gall-stones or by external obstructions causing occlusion of the ducts such as carcinoma of the head of the pancreas. Gall-stones often give rise to an intermittent jaundice. The urine is dark and the faeces pale.

The characteristics of the different types of jaundice are shown in Table 5.2.1. Treatment includes bedrest, especially in acute conditions, in order to minimize liver metabolism. Generally, restriction of dietary fats is not necessary unless the patient cannot tolerate them. Jaundiced patients may be sensitive about their appearance and may wish to wear dark glasses to prevent others seeing their yellow eyes, and may be nursed in a room with soft lighting. Other aspects of treatment are specific to the underlying cause.

Lipids

The major lipids in bile are cholesterol and lecithin (a phospholipid).

CHOLESTEROL

Cholesterol is a chemically unreactive hydrocarbon, insoluble in water. It occurs in cell membranes and circulates in the plasma. Cholesterol is synthesized in the liver and is also absorbed from the intestine. The latter, i.e. that absorbed, derives from endogenous cholesterol secreted in the bile, cholesterol from shed epithelial cells and from exogenous sterols in food, which are hydrolyzed in the intestine. The size of the body pool is variable, normal total cholesterol for the whole body being $3.5-7.0 \mu mol/l$. Under normal conditions the rate of synthesis shows a diurnal rhythm. A high cholesterol intake decreases the rate of synthesis in the liver whereas loss of cholesterol through a biliary fistula increases it. The biliary content is $1.6-4.4 mmol/l$. Cholesterol absorbed from the intestine is esterified – i.e. its alcohol (–OH) group is linked to an organic acid (usually a fatty acid), eliminating water – within the intestinal cells and carried as a lipoprotein or chylomicron in the lymph to the thoracic duct. From there it travels in the systemic circulation to the liver where it is rapidly assimilated. The free unesterified form is excreted from the liver cells into the bile micelles. Cholesterol is also metabolized to bile acids within the liver.

When an imbalance of the main constituents of bile occurs, namely a reduction of bile salts and phospholipids and an increase of cholesterol, the latter may be precipitated, leading to the formation of gall-stones. When prolonged raised serum cholesterol levels occur, as in biliary cirrhosis, flat or slightly raised soft yellow areas called **xanthomas** appear on the face, neck, chest or back. These disappear as the cholesterol level falls.

PHOSPHOLIPIDS

Lecithin accounts for 90% of the phospholipids in bile; others are lysolecithin (3%) and phosphatidyl ethanolamine (1%). Lecithin (or phosphatidyl choline, which is an alternative name) is important in fatty acid metabolism.

Fatty acids are delivered to the liver in a free form or esterified with glycerol or cholesterol. Others are synthesized within the liver from carbohydrate or amino acid precursors. In turn, these fatty acids are metabolized to the forms in which they leave the liver, either as lipoproteins or as phospholipids containing choline, for example lecithin. Deficiency of choline or constituent components leads to a decrease in phospholipid synthesis and a consequent increase in the fatty acid content of the liver. The resultant accumulation of lipid within the tissue is known as **fatty liver**. It may be prevented or corrected by the administration of the so-called lipotrophic substances, including the amino acids methionine and cysteine.

Within the bile, lecithin and its partially hydrolyzed derivative lysolecithin, which has a detergent action, are incorporated into the micelles. They aid the emulsification of dietary lipids and undergo further hydrolysis within the intestine, to be excreted in the faeces. They do not enter the enterohepatic circulation.

Other constituents of bile

ELECTROLYTES

The following ions are secreted into the bile: sodium, potassium, chloride, hydrogen carbonate, calcium and magnesium. These pass into the bile at the hepatocytes as a result of the osmotic effect of the actively transported bile salts. The hormones secretin and cholecystokinin (CCK), both released by the duodenal mucosa, and insulin and glucagon, both from the pancreas, also stimulate the secretion of chloride and bicarbonate (Jones and Meyers, 1979). Within the gall bladder, reabsorption of water and the electrolytes occurs so that their respective concentrations are diminished disproportionately. Whereas sodium is the dominant electrolyte in hepatic bile, chloride predominates in gall bladder bile (Smith, 1973). The alkalinity of the bile aids the neutralization of the acidic food (chyme) within the intestine.

ALKALINE PHOSPHATASE

This enzyme is synthesized in the liver while another related form (i.e. an isoenzyme) is synthesized in bone. The two forms can be separated by electrophoresis. The normal serum level is 21–100 i.u./l. The biological role of hepatic alkaline phosphatase is not known, although involvement in the transport of compounds into the bile has been postulated (Haroff and Hardison, 1979). Increased serum levels occur in various abnormal hepatic conditions and therefore its measurement may be useful in diagnosis, particularly to distinguish between different forms of jaundice. An increase to greater than 200 i.u./l indicates an intra- or extrahepatic biliary obstruction. Raised levels of one of the retained biliary products, possibly bile acids, stimulate the production of the enzyme. There is only a moderate increase, up to 150 i.u./l, in viral hepatitis where liver cells are damaged. Levels are also raised in conditions which increase the production of the bone isoenzyme, e.g. during periods of active bone growth.

MINOR CONSTITUENTS

The following substances are found in small quantities in the bile: vitamin B_{12}, nucleoproteins, mucin, triglycerides, free fatty acids, plasma proteins such as albumin, and free amino acids.

Vitamin B_{12} is stored in the liver. It has a lipotrophic effect and is necessary for protein metabolism and for formation of the erythrocytic stroma (see Chap. 4.1). It is reabsorbed into the enterohepatic circulation.

Secretion of bile from the liver cell to the duodenum (see also Chap. 5.1)

The secretion of bile from the hepatocytes into the canaliculi takes place in different ways. First, there is a **bile acid-dependent secretion** which, as its name suggests, is determined by the secretion of bile acids. This involves their active transport into the canaliculi where they exert an osmotic effect which draws water, electrolytes and bile pigments across the cell membrane. The volume of bile produced is determined by the rate at which bile salts are returned to the hepatocytes by the enterohepatic circulation.

There is also a **bile acid-independent secretion** which involves an active sodium pump mechanism carrying water, small solutes and electrolytes into the bile.

Third, there is a **ductular secretion** which modifies the canalicular flow of bile. This is stimulated by the hormone secretin, released from the duodenal mucosa when the acid food chyme enters the duodenum. The enzyme acts on the biliary ductules giving rise to a 'watery' bile due to increased secretion of water, sodium bicarbonate and sodium chloride. Since this secretion takes place mainly during a meal, it does not result in storage of the bile in the gall bladder where water may be reabsorbed. Of the approximately 600 ml of bile secreted in 24 hours, 225 ml are bile acid dependent, 225 ml are bile acid independent and 150 ml are ductular.

The hepatic bile, produced continuously, is stored in the gall bladder between meals. Here reabsorption of some constituents occurs, particularly of water and electrolytes, which leads to an increased concentration of five to ten times of other constituents, particularly bilirubin, bile salts, cholesterol, fatty acids and lecithin (Smith, 1973). Muscular contraction of the gall bladder, stimulated by the vagal nerve and the hormone cholecystokinin, produced by the intestinal wall, releases bile into the duodenum via the cystic duct and the relaxed sphincter of Oddi. Two or three cycles of bile secretion occur at each meal, resulting in approximately 500–1000 ml total secretion in 24 hours. Evacuation of the gall bladder takes place only once during a meal. The bile flow subsequent to that is directly from the liver into the intestine.

CHOLESTASIS

Interference of the bile flow from the liver cell to the duodenum results in a syndrome called cholestasis – an accumulation of bile in the liver cells with a resultant retention in the blood of substances normally excreted in the bile. The effects vary with the type and duration of the obstruction. **Extrahepatic obstruction** is mechanical; examples are gall-stones in the biliary ducts or carcinoma of the pancreas occluding the duct. **Intrahepatic obstructions** are due to damage of the liver itself by agents including viruses (hepatitis), alcohol and drugs. Jaundice (described earlier) will be evident, as will steatorrhoea. **Malabsorption** of other nutrients apart from fats occurs, particularly of the fat-soluble vitamins A, D, E and K and calcium which is absorbed with fats in the form of a soap. On a long-term basis this can lead to a state of malnutrition, with vitamin and calcium deficiencies. The diet supplied to the patient should therefore be strictly controlled. Restricting the intake of fat to about 30–40 g per day makes life more tolerable as the patient passes fewer bulky stools. Medium-chain triglycerides (smaller parts of fat molecules) can be given orally to increase the energy intake as they are water soluble and absorbed directly into the blood without the need for the presence of bile salts. Intramuscular preparations of fat-soluble vitamins and calcium supplements prevent deficiency conditions (Bateman, 1979).

Pruritus (itching of the skin) is associated with

cholestasis and for many years this was thought to be due to the presence of bile salts irritating cutaneous sensory nerves. However, pruritus may well be due to an as yet unidentified serum factor, since there is a low correlation between serum bile acid levels and the presence of pruritus (Freedman *et al.*, 1981). Treatment involves the use of antihistamines or cholestyramine, an exchange resin said to bind bile salts in the intestine causing their faecal excretion and consequent lowering of the serum bile acid level. There is often a reluctance by the patient to take this preparation because it can produce nausea. Scrupulous skin hygiene is necessary for affected patients by the use of soft sheets, lotions such as calamine, bathing in tepid water and keeping the nails short so that scratching does not damage the skin.

Clotting abnormalities also occur in cholestasis. In particular, the prothrombin time is prolonged because vitamin K is not absorbed, and spontaneous bruising may occur. Collection of blood by venepuncture should be as infrequent as possible. Dental care should be gentle since a toothbrush may cause bleeding of the gums; mouthwashes and the use of swabs may be more appropriate.

GALL-STONES

Gall-stones occur in both men and women, with a higher incidence among the latter; 12–15% of adult females in the United Kingdom are affected, the incidence increasing with childbearing (Kupfer and Northfield, 1981).

The predominant constituent of gall-stones found in westerners is cholesterol. At present, understanding of gall-stone formation remains incomplete, but a major factor involves an alteration in the 'cholesterol-carrying capacity' of the bile. Normally the bile is saturated with cholesterol and has sufficient capacity within the micelles to carry the amount present. In certain circumstances it may become lithogenic (potentially gall-stone forming), when there is insufficient cholesterol-carrying capacity and the bile is supersaturated with cholesterol. This crystallizes out and aggregates to form a stone if an appropriate (but unknown) nucleating agent is present.

Gall-stones are formed in the gall bladder since the bile accumulates there between meals. Factors associated with gall-stone formation are obesity, the use of the drug clofibrate (to treat certain hyperlipidaemias), and use of the oestrogen-containing contraceptive pill which increases the amount of cholesterol secreted in the bile (Kupfer and Northfield, 1981). A reduced bile salt pool (and thus cholesterol-carrying capacity) occurs with prolonged cholestyramine therapy for pruritus and following ileal resection when the enterohepatic circulation of bile acids is broken. There is an increased incidence of gall-stones in these conditions.

Other constituents of gall-stones include calcium salts of bilirubin and trace quantities of fatty acids, phospholipids, bile acids and glycoproteins. When large amounts of calcium are present the stones are radio-opaque. Only about 10% of gall-stones can be seen on x-ray. Pigment stones, without cholesterol, occur associated with haemolytic conditions when excessive quantities of bilirubin, particularly the less soluble unconjugated form, are present in the bile.

Cholecystitis is an inflammation of the gall bladder, usually in association with obstruction of the cystic duct or the neck of the gall bladder by gall-stones. It may be acute or chronic, the former being the third commonest cause of acute pain requiring hospital admission in the United Kingdom (Gunn, 1981). Biliary colic, the pain of acute or chronic cholecystitis, is severe. It arises in the epigastrium and moves to the right side and back or shoulder. The pain lasts several hours and may be accompanied by nausea and vomiting. Jaundice may be slight or latent, that is, revealed only by bilirubin estimations. The attack may be precipitated by the ingestion of fatty foods. The pain is due to the gall bladder contracting in an attempt to overcome the blocked cystic duct. The walls are also stretched to accommodate the accumulating inflammatory exudate. Thus the splanchnic and phrenic nerve endings in the gall bladder wall are stimulated. Irritation of the overlying peritoneum, which is innervated by spinal nerves, gives rise to superficial pain. The diaphragm lies close to the gall bladder and stimulation of the sensory nerves supplying it, that is, the phrenic and some intercostal nerves, gives rise to the referred pain. Treatment includes bedrest, suitable analgesia such as pethidine, antispasmodics, antibiotics if infection is present, and maintenance of fluid and electrolyte balance by intravenous therapy. Medical treatment of gall-stones involves the administration of chemicals which dissolve the stone.

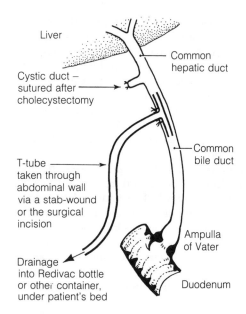

Liver

Common
hepatic duct

Cystic duct –
sutured after
cholecystectomy

T-tube
taken through
abdominal wall
via a stab-wound
or the surgical
incision

Common
bile duct

Drainage
into Redivac bottle
or other container,
under patient's bed

Ampulla
of Vater

Duodenum

Figure 5.2.15   T-tube in position

If **cholecystectomy** (surgical removal of the gall bladder) is to take place immediately, the stomach is kept empty by nasogastric suction. During cholecystectomy, exploration of the common bile duct is undertaken to detect the presence of further stones. A T-tube is inserted to maintain the patency of the bile duct during healing and to enable observation of the biliary flow (Fig. 5.2.15). Observable drainage occurs via the stem portion of the T-tube which is brought out through the abdominal wall via either the incision or a stab wound and attached to a drainage bag. During the first 24 hours, frequent observation for haemorrhage is necessary. Small amounts of blood mixed with bile are lost during the first few hours and thereafter normally bile alone in variable quantities. The tube is clamped, so that all the bile flows through the bile duct, and gradually shortened prior to removal. Observation of the remaining dressing is necessary to detect bile seepage. Large quantities may damage the skin and leakage into the abdominal cavity may cause peritonitis.

Chronic cholecystitis may not be due to the presence of gall-stones. Bouts of biliary colic occur at intervals accompanied usually by general digestive complaints such as nausea, flatulence and dyspepsia.

SYNTHETIC PROCESSES OF THE LIVER

In addition to the production of bile, which is the excretory route for some substances, the liver is also the site of many synthetic processes, some of which are unique to it.

Plasma proteins

All plasma proteins except the immune (γ) gamma-globulins are manufactured within the liver.

ALBUMIN

This is the smallest plasma protein, with a molecular weight of about 69 000. About 175 mg albumin per kg body weight per day are made, or 12 g per day in an adult. About one-sixth of dietary nitrogen is utilized in this way. The synthesis is dependent on prevailing amino acid levels so that the nutritional status of the individual is important. Protein-deficient diets such as those which lead to malnutrition states also lead to a lowered serum albumin. The total body pool of albumin is about 300 g, of which approximately one-third is in the blood at a level of 35–50 g/l.

Albumin is the protein mainly responsible for the maintenance of the colloidal osmotic effect of the plasma which influences the exchange of water between the tissues and blood. The osmotic effect is due to its lower molecular weight and the quantity of molecules present compared with other plasma proteins.

GLOBULINS

The total serum globulin level is 18–32 g/l, of which 7–15 g/l are γ-globulins. The α_2, β and some α_1-globulins are synthesized in the liver. The γ (immune) globulins are made in the mononuclear phagocytic (reticuloendothelial) system outside the liver. The globulin molecules are much bigger than albumin, their molecular weights ranging from 90 000 to over a million. They are involved in the transport of various substances such as cortisol, oestrogen, copper (ceruloplasmin) and iron (transferrin). As with albumin, their synthesis is dependent on the amino acid level.

Changes in plasma protein levels occur in liver disease, with usually a decrease in serum albumin and an increase in serum globulins. These changes are detected using flocculation tests or electrophoresis. Flocculation (or turbidity) tests determine quantitative changes in the amounts of plasma albumin and globulins and measure the immunoglobulin response to liver cell damage rather than the damage itself. More accurate estimations of the proportions of plasma protein can be made by electrophoresis, a technique based on the fact that different proteins move at different rates in an electric field. This separates the albumin, α_1, α_2, β and γ-globulins. Different patterns are obtained in different conditions.

The altered albumin:globulin ratio of the plasma caused by defective hepatocellular function leads to **ascites**. This is a condition of abdominal distension due to the accumulation of fluid in the peritoneal cavity, and may be acute or chronic in onset. It generally occurs when both hepatocellular failure and portal hypertension are present. The exact mechanism is not understood and the following is a possible explanation. When the liver cells are damaged, the synthesis of albumin is decreased, leading to lowered plasma levels. Since this protein is so important in the maintenance of the osmotic pressure of the plasma, a decrease leads to a lowered pressure with a consequent movement of fluid and electrolytes, particularly sodium, into the peritoneal cavity from the circulation. Portal hypertension also contributes to this movement, with a resultant depletion of effective intravascular fluid and electrolytes. This situation stimulates sodium and water retention in the kidney via the renin–aldosterone system (see Chap. 5.4.) as a homeostatic mechanism to restore the blood volume.

The treatment of ascites requires bedrest to decrease the metabolic activity of the liver. Restriction of fluids (to approximately 1.5 litres per day) and a low-sodium diet are given and strict attention is paid to fluid balance. Daily weighing to detect the amount of water gained or lost is particularly important and it is the nurse's responsibility to ensure that this is carried out at the same time each day, preferably after the patient has emptied his bladder. Diuretic therapy may be required if weight loss is insufficient. The prevention of pressure sores is a priority goal in the patient's care plan as the skin is stretched and more liable to breakdown.

Abdominal paracentesis (removal of fluid from the peritoneal cavity via a trochar and cannular) may be used as a diagnostic procedure to detect protein and electrolyte levels and the type of any cells or micro-organisms present in the ascitic fluid. Generally, large quantities of fluid are not removed because of the danger of complications such as hypokalaemia, hyponatraemia, encephalopathy, or renal failure due to a reduction in body fluids.

Acute ascites may occur as a result of a precipitating factor such as shock, infection or a large intake of alcohol when hepatocellular function is depressed.

Serum transaminases

Two enzymes of medical significance – aspartate aminotransferase (AST, previously glutamic oxaloacetate transaminase, or GOT) and alanine aminotransferase (ALT, previously glutamic pyruvate transaminase, or GPT) – are both synthesized in the liver as well as in other tissues. The reaction catalyzed by these enzymes is a transamination in which the deamination of an amino acid is coupled with the simultaneous amination of a keto acid (Fig. 5.2.16).

Glutamic acid and α-ketoglutaric acid are almost always involved in any transamination reaction. The enzymes named catalyze the following reactions with the formation of two non-essential amino acids from products of carbohydrate and fat metabolism.

Aspartate aminotransferase

$$\text{Glutamic acid} + \text{oxaloacetic acid} \rightleftharpoons \alpha\text{-ketoglutaric acid} + \text{aspartic acid}$$

Alanine aminotransferase

$$\text{Glutamic acid} + \text{pyruvic acid} \rightleftharpoons \alpha\text{-ketoglutaric acid} + \text{alanine}$$

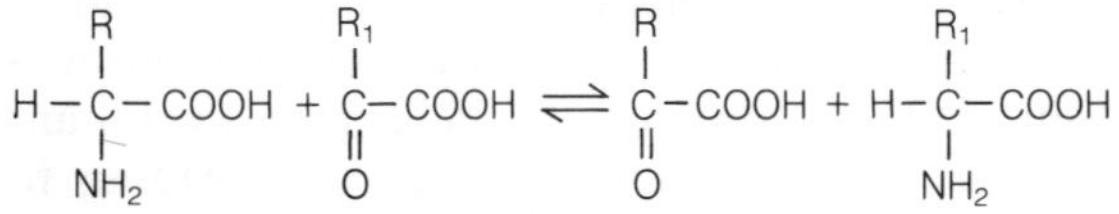

Glutamic acid α-Ketoglutaric acid

Figure 5.2.16 A transamination reaction

There are other transaminase enzymes which catalyze reactions involving other amino acids. Reactions may be coupled so that non-essential amino acids (i.e. those that can be made in the body) can be formed from essential ones (i.e. those not made in the body and therefore necessary in the diet).

Aspartate aminotransferase is a mitochondrial enzyme. Raised serum levels may be indicative of myocardial infarction (although they may also result from damage to other organs) since the cells of the heart synthesize the enzyme, and so myocardial cell damage leads to release of the enzyme into the circulation. Normal serum levels are < 18 i.u./l. Alanine aminotransferase is a cytoplasmic enzyme present in the heart, skeletal system and liver. In the liver the amount is less than aspartate aminotransferase but proportionately greater than that in the heart or skeletal system, thus an increase in serum level is specific for liver damage. The normal serum level is < 20 i.u./l. The degree of increase is indicative of the cause and thus aids diagnosis.

Blood clotting factors

Many of the protein clotting factors involved in the blood coagulation mechanism (see Chap. 4.1) are made in the liver.

FIBRINOGEN (FACTOR I)

This is the soluble precursor of insoluble fibrin which forms fine strands as the basis of a blood clot.

PROTHROMBIN (FACTOR II)

This is another plasma protein and the precursor of thrombin which is required for the conversion of fibrinogen to fibrin. The synthesis of prothrombin is dependent on the amount of vitamin K reaching the liver cells as well as on their functional state. In cholestasis, there is a decreased absorption of the fat-soluble vitamin K, with a resultant decrease in prothrombin production leading to delayed clotting, estimated by the measurement of the prothrombin time. Administration of intramuscular vitamin K improves the clotting mechanism.

The neonate has a low prothrombin level. In the adult vitamin K is synthesized in the intestine by bacteria but this does not occur until the intestinal flora are established during the first months of life (see Chap. 5.1). Thus, in certain circumstances in which babies are prone to bleed, such as prematurity or following an operative or traumatic delivery, a synthetic analogue of vitamin K (Konakion) is given to reduce the risk of prolonged bleeding.

OTHER CLOTTING FACTORS

The following are synthesized in the liver: V, XI, XII, and XIII and also VII, IX and X which require the presence of vitamin K (see Chap. 4.1 for full names).

HEPARIN

This is mucopolysaccharide made in the liver which is involved in the balance of fibrinolysis, preventing coagulation of plasma.

The liver also clears the active clotting factors from the blood, thus preventing excess clotting.

In conditions of hepatocellular damage, the synthesis of clotting factors may be impaired, resulting in a prolonged prothrombin time and spontaneous bleeding, bruising and purpura; this should be borne in mind particularly if **liver biopsy** is considered. The liver, being a vascular organ, is liable to bleed profusely, especially if the clotting mechanism is not functioning properly. It is thus vitally important to know the prothrombin time. If it is prolonged, treatment with vitamin K for several days prior to the biopsy is necessary. As a precaution in case of haemorrhage, the patient's blood group should be known and cross-matched blood made available. Following a biopsy it is necessary for the patient to remain in bed for 24 hours, for the first 4 hours lying on his right side, thus putting pressure on the biopsy site to reduce the likelihood of bleeding. Observations of pulse and blood pressure are made, usually every 15 minutes initially, to detect possible haemorrhage. Observation of vital signs is maintained regularly for 24 hours.

Glycogen formation and the maintenance of blood glucose levels

The liver has a short-term store of glycogen, a complex carbohydrate compound.

Glucose, absorbed from the intestine, readily diffuses from the portal circulation into the liver cells. Within the hepatocytes it is phosphorylated (that is, a phosphate group is added) under the influence of the enzyme glucokinase to form glucose-6-phosphate. An enzyme converts this to glucose-1-phosphate which can then be added on to the glycogen molecule by glycogen synthetase.

Liver glycogen is broken down by phosphorylase to release glucose into the blood to maintain the blood glucose level, which is vital to life (see Chap. 2.5). Alternatively, glycogen may be broken down to glucose-6-phosphate. A phosphatase enzyme, found only in the liver, catalyzes the hydrolysis of glucose-6-phosphate to release glucose to the blood. The reactions involved are summarized in Fig. 5.2.17.

Other sources from which liver glycogen may be derived via the formation of glucose-6-phosphate include fructose and galactose, pyruvic and lactic acids (products of carbohydrate metabolism), deaminated amino acids and the hydrolysis of neutral fats. Gluconeogenesis (new glucose formation from protein or fat precursors) occurs only in the liver. It is important since fasting exhausts the liver glycogen store of approximately 100 g in 24 hours.

Blood glucose is required by the cells of the body as an energy source. When glucose is withdrawn from the blood by tissues, the decreased level stimulates the release of glucose from liver glycogen in order to maintain the blood level. This is stimulated by the diabetogenic group of hormones including adrenaline, thyroxine, the diabetogenic factor of the anterior pituitary and glucagon. These hormones stimulate the activity of liver phosphorylase, thus promoting hepatic glycogenolysis (breakdown of glycogen). This action is opposed by insulin. (Chapter 2.5 contains a full discussion of the endocrine control of blood glucose levels.) Changes in blood glucose level as a result of hepatic malfunction occur in acute liver failure rather than in chronic conditions. Resultant hypoglycaemia (low blood glucose level) may lead to coma and death.

The **galactose tolerance test** has been used as a test of liver function because this sugar is metabolized only in the liver. Abnormally high blood concentrations an hour or two following its ingestion indicate impaired liver function.

Red cell production and destruction (see Chap. 4.1)

The liver is an important site of red cell formation after the third month of fetal life. However, as the bone marrow begins to act as a blood-forming organ from about $4\frac{1}{2}$ months of fetal life, the erythropoietic activity of the liver declines. In later life the liver resumes erythropoietic activity if severe marrow damage occurs.

In the adult, the Kupffer cells of the liver break down the mature erythrocytes. The haemoglobin is released and broken down as described in the section on bile pigments.

DEGRADATION REACTIONS PRODUCING EXCRETORY PRODUCTS

As do the synthetic reactions, metabolism occurring in the liver also involves the breakdown of various substances.

Deamination of proteins

When amino acids are broken down, usually in the liver, they are deaminated so that ammonia is released, and the remaining non-nitrogenous part is further metabolized to other amino acids, glucose, fat or ketone bodies. Some ammonia is incorporated into newly formed amino acids.

The liver is the only tissue which converts the surplus toxic ammonia into **urea** which is water soluble, non-toxic and can be excreted via the

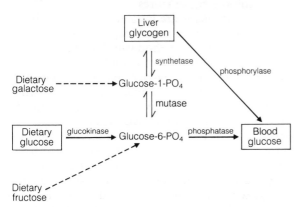

Figure 5.2.17 Processes by which glucose is released into the blood

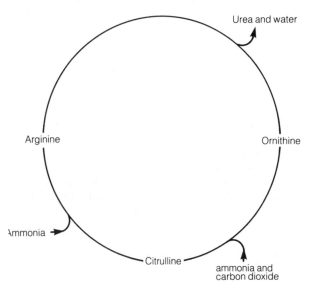

Figure 5.2.18   The Krebs' urea cycle

kidney. The formation of urea is a cyclical reaction in which two molecules of ammonia unite with one of carbon dioxide with the elimination of water. The amino acid ornithine is involved in the Krebs' urea cycle (Fig. 5.2.18).

Another source of ammonia is from the breakdown of protein material by the intestinal flora. The ammonia is absorbed into the portal circulation and taken up by the liver, where it is detoxified by urea formation.

It has been suggested that ammonia toxicity may be responsible for the varied changes of **hepatic encephalopathy**. The exact cause of this condition is not known, but ammonia toxicity is the most widely investigated possibility. When severe liver damage occurs, the portal blood may bypass the remaining healthy liver cells by means of a collateral circulation or by passing directly through damaged hepatocytes, as in acute liver failure. This means that ammonia enters the systemic circulation without undergoing metabolism in the liver. Ammonia may exert toxic effects on tissues, particularly nervous tissue. Blood levels of up to 5000 $\mu g/l$ ammonia (upper limit of normal is 800–1000 $\mu g/l$) have been found in encephalopathy. The symptoms involve mental and neurological dysfunction such as disturbed consciousness, personality changes, intellectual deterioration, slurred speech and a characteristic 'flapping' tremor of the hands. Encephalopathy develops more frequently in chronic liver conditions and is reversible in its early stages.

Treatment of encephalopathy involves removal of the precipitating factor, a low-protein diet and intestinal antibiotics (in order to prevent ammonia being formed from protein in the gut by the intestinal flora). When treatment is commenced, a gradual increase in the amount of dietary protein given is made with the aim of establishing the upper level of protein tolerance. It is important that the nurse ensures that the correct diet is received and consumed by the patient. The nature and quantity of leftovers and the consumption of any other foods need to be recorded, so that an accurate assessment of dietary intake can be made. Neomycin is the antibiotic of choice as it is non-absorbable. The synthetic dissaccharide lactulose may also be given. This is hydrolyzed in the colon to form lactic, acetic and formic acids and carbon dioxide. The acidic medium thus produced traps ammonia and possibly other toxic compounds.

Breakdown of fats

The liver stores some fat in the form of neutral fat or triglycerides and also the fat-soluble vitamins A, D, E and K. The fat is derived from three sources: triglycerides absorbed from the intestine and transported in chylomicrons via the lymphatics and systemic circulation, free fatty acids from the adipose tissue stores, and fat synthesized in the liver.

Within the liver, newly arrived fats are converted to fatty acids which may be metabolized in several different ways (Fig. 5.2.19). Triglycerides are reformed by the esterification of glycerol. These form lipoprotein by the addition of protein and pass into the systemic circulation to be deposited in the adipose tissue stores.

Other fatty acids are involved in phospholipid formation and in the esterification of cholesterol. Fatty acids are broken down by oxidation to provide energy in the form of adenosine triphosphate (ATP), an energy-rich compound, with the release of carbon dioxide. An intermediary product in this breakdown is acetyl coenzyme A (acetyl CoA). An alternative metabolic path for acetyl CoA occurs when two molecules join to form acetoacetyl CoA which is metabolized irreversibly in the liver to the **ketone bodies** acetoacetic acid, β-hydroxybutyric acid and acetone. Very low levels of these are found in the blood of normal individuals. They can be metabolized to produce energy in tissues other than the liver.

In abnormal conditions such as fasting or

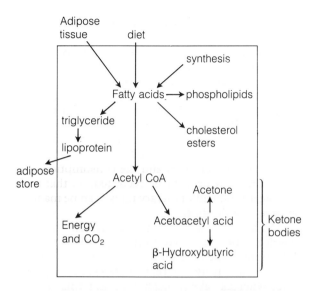

Figure 5.2.19 An outline of fatty acid metabolism in the liver

diabetes, when carbohydrate is not available as an energy source, the oxidation of fatty acids is increased to meet the energy demands. Since more acetyl CoA is formed than can be metabolized to carbon dioxide with the production of energy an accumulation of the ketone bodies occurs, leading to increased blood levels and the condition of **ketosis**. The ketones are excreted in the urine and there is often a sickly sweet odour of acetone in the expired breath, sometimes described as being akin to the smell of new-mown hay.

Fatty acids are synthesized in the liver cells from non-lipid substances, a process known as **lipogenesis**. Carbohydrates and some amino acids as well as fatty acids may be broken down to acetyl coenzyme A, which is then involved in a complex series of reactions resulting in a step-by-step increase of the chain length of the fatty acids.

In the condition of **fatty liver**, abnormal accumulations of fat of up to 30% of the total liver weight are found. Fatty liver occurs when there is an absence of lipotrophic substances, such as methionine, in the diet. This causes a reduction in the amount of phospholipids made in the liver and results in an accumulation of fatty acids. Other conditions in which fatty changes are found in the liver include kwashiorkor, obesity, and in association with excessive alcohol consumption. The exact mechanism of fat accumulation in each of these is not completely understood.

Elimination of copper

The liver is the main site of copper metabolism in the body. Dietary copper is rapidly cleared from the plasma by the liver and incorporated into proteins such as cytochrome oxidase (a respiratory electron-transfer enzyme) and ceruloplasmin (a blue-coloured glycoprotein transporting copper in the blood). Copper is excreted in the bile bound to a carrier, usually a bile acid or pigment, which prevents its reabsorption by the intestine. Excess copper is toxic. In Wilson's disease, a congenital condition involving the accumulation of copper, it produces structural and functional changes to the cells of the liver, nervous system, eyes and kidneys.

Detoxification of endogenous and exogenous compounds

The metabolism of steroid and thyroid hormones and drugs, particularly those administered orally, in the liver and their consequent excretion in the bile clear these substances from the blood and tissues. The biological effects of the compounds are curtailed and the body is protected against the intoxicating results of their accumulation.

A two-phase reaction occurs to render the substances more polar so that they can be excreted in the bile. The first phase, which takes place in the endoplasmic reticulum of liver cells, is an oxidative reaction which provides an available hydroxyl (–OH), carboxyl (–COOH) or amine (–NH₂) group for the second phase of conjugation – that is, combination with another molecule. Substances which may be utilized in conjugation reactions include the sulphate group, the amino acid glycine, glutathione (a tripeptide of three amino acids – glutamic acid, cysteine and glycine), and acetyl and methyl derivatives. The most important conjugate metabolically is the carbohydrate derivative glucuronic acid. This can attach to a variety of chemical groups giving them a versatile ability to conjugate many different molecules. Drugs conjugated by glucuronic acid include imipramine, morphine, chloramphenicol, salicylic acid and indomethacin.

Certain drugs and chemical agents are harmful to the liver and cause disruption of liver function in various ways. *Paracetamol in overdose* is increasingly used as a suicide agent. Taken in large quantities, paracetamol produces hepatic necrosis which can result in acute liver failure and possibly death. It is more dangerous when taken in conjunction with dextropropoxyphene, as in Distalgesic. The ingestion of at least 10–15 g (20–30 500-mg tablets) of paracetamol is necessary to produce damage of the liver. In addition to hepatic damage, there is also myocardial and renal damage. The damage to the liver is caused by a highly reactive intermediate metabolite of paracetamol, which is normally inactivated by conjugation with hepatic glutathione. However, when the stores of this are depleted, the metabolite is free to bind with vital liver cell macromolecules, with consequent disruption of the tissue. Since it takes time for the glutathione store to be depleted, damage to the liver does not become evident for 3 or 4 days, hence the apparent recovery at 48 hours prior to deterioration.

Many other drugs may cause liver damage by different mechanisms. For some the extent of damage is dependent on the amount of the drug taken, whereas for others it is not.

ALCOHOL CONSUMPTION

Consumption of alcohol has long been associated with the incidence of cirrhosis of the liver. Alcohol is not stored in the body but undergoes obligatory oxidation, predominantly in the liver. A healthy individual cannot metabolize more than 160–180 g per day, but since alcohol induces (i.e. increases the activity of) the enzymes involved in its catabolism, alcoholics may be able to metabolize more than this.

The Health Education Authority currently recommends that women drink no more than 13 units of alcohol per week, and men no more than 20 units per week. One unit of alcohol is equivalent to $\frac{1}{2}$ pint (0.3 litres) of beer, one measure of spirits or one glass of wine.

Ethanol is metabolized by alcohol dehydrogenase within the liver cells to acetaldehyde, which is toxic and causes membrane damage and cell necrosis. Acetaldehyde can be further broken down to acetyl coenzyme A, which may be metabolized in the same way as it is when produced from fats or carbohydrates, to give energy, carbon dioxide and water. Hydrogen is produced when ethanol is converted to acetaldehyde. This hydrogen replaces fatty acids as a fuel so that the fatty acids accumulate, giving rise to ketosis, an increase

in triglycerides, hyperlipidaemia and the formation of a fatty liver.

Alcohol can also be metabolized by a system of enzymes in the microsomes of the cell which metabolize other drugs as well. Alcohol induces this system, which may explain the increased tolerance of the alcoholic to alcohol and also to sedative drugs. Alcohol also stimulates the formation of collagen, which disrupts the cellular structure, and the resulting fibrosis impedes the passage of substances between the blood and liver cells. The damage is reversible at first if the causative agent, alcohol, is removed but progresses to an irreversible state.

Alcohol is prohibited for patients who are convalescent or recovered from hepatitis or other infections of the liver for a period of 6–12 months. This allows the liver to regenerate as fully as possible, whereas taking alcohol during this period might lead to permanent damage.

Not only do drugs and chemical agents cause disorders of liver function, as demonstrated above, but altered liver function, as in conditions of hepatocellular failure, may affect the metabolism of drugs. Detoxification reactions may be reduced so that the pharmacological action of the drug is prolonged. Drugs which may be metabolized abnormally in liver disease include tolbutamide (an antidiabetic agent), phenytoin (an anticonvulsant), theophylline (for bronchospasm) and diazepam (a tranquillizer and sedative).

PARENCHYMAL DYSFUNCTION OF THE LIVER

We have seen that a wide variety of reactions takes place within the liver cell. When the cells are damaged there is an alteration in the metabolism of many substances, affecting different functions and with various effects. Almost every system of the body is affected in some way. **Cirrhosis**, a process of fibrosis and nodule formation following hepatocellular necrosis, demonstrates the wide variety of effects of liver cell failure. There are many different causes of cirrhosis, including viral hepatitis, certain metabolic conditions such as Wilson's disease, and prolonged cholestasis; consumption of alcohol is more commonly associated, especially in Western countries, whereas protein malnutrition may be responsible in Third World populations. The liver cell structure is destroyed by the causative agent. Following this, fibrous tissue is laid down, distorting the lobular structure of the liver. The healthy tissue attempts regeneration but the new growth is irregularly shaped, which distorts and thus disrupts the blood supply to the lobules. As a result of this, pressure in the portal vessels increases, leading to impaired perfusion and a decrease in the nutrient supply to the liver tissue. The effects of this type of damage to the liver are many and varied.

Portal hypertension and collateral circulation

Nodules formed as part of the cirrhotic process exert pressure on the hepatic vessels, thus tending to decrease the rate of blood flow within the liver. The pressure of the portal system, normally 5–10 mmHg (0.7–1.3 kPa), increases with the additional resistance to the flow of blood in the vessels of the liver. When the pressure is increased above 14 mmHg (1.9 kPa), a collateral circulation occurs via tributaries of the portal vessels to the systemic circulation, with consequent bypassing of the liver by the portal blood. The main sites of joining with the systemic circulation occur in the submucosa of the stomach and lower oesophagus, the rectum and anterior abdominal wall.

Varices

The presence of a collateral circulation in the oesophagus gives rise to varices, a number of dilated vessels which may rupture causing bleeding which presents as haematemesis or malaena. The precise factors responsible for the rupture are unknown, but an elevated portal pressure appears relevant. Bleeding in this way may precipitate liver cell failure, coma, jaundice and ascites, and may be fatal.

Initial treatment for bleeding involves the replacement of blood volume by blood transfusion. Oral neomycin is given to prevent bacterial breakdown of protein in the intestine which would otherwise increase the load at the liver and so increase the likelihood of the development of coma. The blood from the rupture itself constitutes an increased amount of protein in the intestine. Vasopressin (Pitressin) may be given intravenously. This reduces the portal pressure and contracts the oesophageal muscle. It also stimulates intestinal contractions which help to remove the blood from the bowel. Compression of the varices by means of a Sengstaken–Blakemore oesophageal compression tube is used infrequently as there are various difficulties associated with its use, for example obstruction of the pharynx and resultant asphyxia. Surgical intervention, creating a shunt from the portal to the systemic circulation and thus reducing portal pressure, may be performed. This procedure has varying survival rates depending on the type of shunt formed.

Ascites

This condition has been discussed previously.

Circulatory changes

The general circulation becomes hyperdynamic. An increased blood flow through the skin is observed with consequent flushing of the extremities (palmar erythema) and bounding pulses. The cardiac output is also increased, as evidenced by tachycardia. Blood flow to the spleen is increased while that to the kidney is reduced, which may result in abnormal renal function. The blood pressure is low. The cause of these changes is uncertain but they may be due to the presence of a vasodilator substance in the circulation.

Pulmonary changes

Some patients with cirrhosis may be cyanosed and it is found that their arterial blood has a reduced oxygen saturation compared with normal. Within the lungs there are microscopic arteriovenous fistulae through which the blood may be shunted, thus bypassing the alveoli (i.e. the functional units). There is also an increase in the diffusion barrier which may be related to the dilation of the blood vessels and the thickening of their walls by a layer of collagen.

Jaundice

This is caused by the inability of the damaged liver

cell to metabolize bilirubin. In chronic conditions, such as cirrhosis, regeneration of liver tissue compensates for the functional loss of necrosed tissue. Jaundice may be absent or mild. However, in conditions of acute liver failure, as in viral hepatitis, the extent of the jaundice parallels the cell damage.

Loss of tissue

There is a loss of flesh and muscle-wasting in patients with prolonged liver cell failure, due to the inability of the liver to synthesize tissue protein. Anorexia and poor dietary habits add to the malnutrition.

Skin changes

The appearance of vascular **spider naevi** on the face, neck, forearm and back of the hand is attributed to oestrogen excess. The arterial spider (a synonym) consists of a central arteriole from which numerous small vessels radiate, resembling spiders' legs. When sufficiently large it may be seen to pulsate.

   Palmar erythema, a bright red coloration of the palms of the hands (often called liver palms), which are warm, also occurs in some cirrhotic patients. This may be due to oestrogen excess or to the circulatory changes.

Endocrine changes

A fall in the circulating testosterone levels of cirrhotic males due to decreased production and increased binding by globulins has been associated with loss of body hair and testicular atrophy. **Gynaecomastia** (development of breast tissue) occurs in some males. Raised levels of oestrogens, which are normally inactivated by the liver, may be responsible for this but the mechanism is not known.

Changes in nitrogen metabolism

Amino acids are excreted in the urine, and plasma levels show a characteristic change: there is an increase in methionine, tyrosine and phenylalanine and a decrease in valine, isoleucine and leucine.

Urea production is impaired but there is a considerable reserve capacity for synthesis which maintains the normal blood level. The failing liver cannot convert ammonia to urea.

Neurological changes

These have been discussed previously.

Disordered blood coagulation

Failure to synthesize many of the proteins involved in blood coagulation results in prolonged bleeding. Spontaneous bruising and the appearance of purpura may occur. Care needs to be taken when operative procedures are considered, including venepuncture and liver biopsy, as well as large-scale surgery. Administration of vitamin K to decrease a prolonged prothrombin time is essential treatment.

Fetor hepaticus

The breath of patients with severe hepatocellular disease or with an extensive collateral circulation may have a sweetish, slightly faecal odour. This is presumed to be due to the exhalation of methyl mercaptan. This compound is apparently metabolized from the amino acid methionine by the intestinal flora. Since the normal demethylation reactions are reduced in a damaged liver, the methyl mercaptan passes into the systemic circulation. It is excreted in the urine as well as exhaled in the breath. The presence of fetor hepaticus may precede coma.

INVESTIGATION OF THE LIVER

A series of biochemical tests on the serum is made when investigating hepatic disease. These tests have been referred to in the relevant parts of this chapter, and are listed in Table 5.2.2.

LIVER TRANSPLANTATION

The first liver transplant in a human was carried out in 1963. Since then, several hundred have

Table 5.2.2 The major biochemical liver function tests

Biochemical test	Normal range	Importance
Alkaline phosphatase	21–100 i.u./l	Diagnosis of jaundice
		Presence of tumours
Bilirubin		Diagnosis of jaundice
Total	5–17 μmol/l	Degree of liver cell damage
Conjugated	< 3 μmol/l	
Plasma proteins		Diagnosis of jaundice
Albumin	35–50 g/l	Degree of liver cell damage
Globulin	7–15 g/l	Course of chronic hepatitis and cirrhosis
Serum transaminases		
Aspartate aminotransferase (AST)	5–30 i.u./l	Early diagnosis of liver disease
Alanine aminotransferase (ALT)	5–30 i.u./l	Degree of liver cell damage
Prothrombin time	10–15 s	Degree of liver cell damage

been performed, mainly in America and England. The operation is rarely undertaken as the survival rate is low at present. When liver transplantation is performed, the donor liver must function immediately otherwise the patient will die. There is no artificial process analogous to haemodialysis utilized in renal failure and transplantation. Conditions for which transplants have been given include primary malignant hepatic tumours, cirrhosis and biliary atresia, particularly in children. Immediate postoperative problems to be corrected are hypoglycaemia, failure of blood clotting, and electrolyte abnormalities. Long-term problems include cholestasis, the formation of fistulae, infection and rejection of the transplanted organ.

The physiology of the liver involves all aspects of metabolism and their interrelationships. The functions of the liver are numerous and have diverse effects throughout the body, effects which are particularly noticeable when the liver is not functioning in its normal manner. The importance of the liver to the well-being of the individual cannot be overemphasized.

Review questions

The answers to all these questions can be found in the text. In each case there is at least one correct and at least one incorrect answer.

1 Bile is secreted by the hepatocyte into the

   (a) sinusoid
   (b) portal canal
   (c) canaliculus
   (d) hepatic duct.

2 Micelles in the bile are made up of

   (a) alkaline phosphatase and bilirubin
   (b) cholesterol, bile salts and phospholipid
   (c) cholesterol, bilirubin and vitamin K
   (d) bile pigments and phospholipid.

3 Within the liver cell, bilirubin is conjugated with

   (a) glucuronic acid
   (b) glutathione
   (c) albumin
   (d) urobilinogen.

4 Kernicterus occurs as a result of

   (a) a deficiency of glucuronyl transferase
   (b) damage to the nervous tissue by unconjugated bilirubin
   (c) the presence of an abnormal progesterone in the breast-milk
   (d) microscopic plugs within the bile canaliculi.

5 The hormone stimulating contraction of the gall bladder is

   (a) cholestyramine
   (b) secretin
   (c) cholecystokinin
   (d) lecithin.

6 Which statement(s) about albumin is/are true?

   (a) Its synthesis is dependent upon the amount of fat in the diet.
   (b) It transports iron in the plasma.
   (c) It is normally present in an equal amount to that of the globulins.

(d) It is important in the maintenance of the osmotic pressure of the plasma.

7 The production of urea

(a) prevents a build-up of toxic ammonia
(b) is responsible for the changes of encephalopathy
(c) takes place in the intestine
(d) is decreased following myocardial infarction.

8 Paracetamol damages the liver by

(a) producing an immunological reaction
(b) forming an active metabolite
(c) inducing an enzyme
(d) forming acetaldehyde.

9 Which statement(s) about a collateral circulation is/are true?

(a) It enables substances secreted to the bile to be reabsorbed and transported to the liver.
(b) It consists of tributaries from the portal vessels to the systemic circulation.
(c) It may be responsible for carcinoma at a hepatobiliary site.
(d) It carries nutrients absorbed from the intestine to the liver.

10 Cirrhosis of the liver is

(a) found only in alcoholics
(b) caused by hormonal changes
(c) a complication of oesophageal varices
(d) a process of fibrosis and nodule formation.

Answers to review questions

1 c
2 b
3 a
4 b
5 c
6 d
7 a
8 b

9 b
10 d

References

Bateman, E. C. (1979) Dietary management of liver disease. *Proceedings of the Nutrition Society 38*: 331.

Campbell, E. J. M., Dickinson, C. J., and Slater, J. D. H. (eds) (1984), *Clinical Physiology*. 5th edn. Oxford: Blackwell Scientific Publications.

Freedman, M. R., Holzbach, R. T. & Ferguson, D. R. (1981) Pruritis in cholestasis: no direct causitive role for bile acid retention. *American Journal of Medicine 70*: 1011.

Gunn, A. A. (1981) Acute cholecystitis. *The Practitioner 225*: 491.

Haroff, D. E. & Hardison, W. G. M. (1979) Induced synthesis of alkaline phosphatase by bile acids in rat liver cell culture. *Gastroenterology 77*: 1062.

Jones, R. S. & Meyers, W. C. (1979) Regulation of hepatic biliary secretion. *Annual Review of Physiology 41*: 67.

Kupfer, R. M. & Northfield, T. C. (1981) Gallstones. *The Practitioner 225*: 499.

Schwarz, C. C., Almond, H. R., Vlahcevic, Z. R. & Swell, L. (1979) Bile acid metabolism in cirrhosis and determination of biliary lipid secretion rates in patients with advanced cirrhosis. *Gastroenterology 77*: 1177.

Smith, R. L. (1973) *The Excretory Function of Bile (The Elimination of Drugs and Toxic Substances in Bile)*. London: Chapman and Hall.

Valman, H. B. (1980) Jaundice in the newborn. *British Medical Journal 280*: 543.

Suggestions for further reading

Bouchier, I. A. D. (1982) *Gastroenterology*, 3rd edn. London: Baillière Tindall.

Nurses' Clinical Library (1985) *Gastrointestinal Disorders*. Nursing 85 Books. Springhouse, Pennsylvania: Springhouse.

Sherlock, S. (1985) *Diseases of the Liver and Biliary System* 7th edn. Oxford: Blackwell Scientific.

Watson, J. E. & Royle, J. R. (1987) *Watson's Medical–Surgical Nursing and Related Physiology*, 3rd edn. London: Baillière Tindall.

Wheeler, P. & Williams, R. (1980) A better outlook: the investigation of liver disease. *Nursing Mirror* 24th July: 42.

Chapter 5.3
Respiration

Janet Stocks

Learning objectives

After studying this chapter, the reader should be able to

1 Discuss the ways in which the respiratory, cardiovascular and neuromuscular systems are interrelated to ensure a continuous exchange of oxygen and carbon dioxide between the atmosphere and the tissues.
2 Describe the role of the diaphragm and other respiratory muscles in breathing.
3 Assess an individual's respiratory status and recognize the significance of these observations.
4 Relate the structure of the respiratory system (lungs, thorax and airways) to the function of gas exchange.
5 State why the composition of alveolar air differs from that of atmospheric or expired air.
6 Describe how a pneumothorax may occur and the principles of treatment for this condition.
7 Discuss the importance of even distribution of gas and blood flow through the lungs and describe the causes and effects of ventilation-perfusion imbalance.
8 List factors which may increase the work of breathing.
9 Describe methods commonly used to assess lung function.
10 List causes of inspiratory failure.

11 Discuss the effects of changes in dead space and breathing patterns on the magnitude of alveolar ventilation.
12 Describe the ways in which oxygen and carbon dioxide are transported around the body.
13 Discuss the physiological significance of the sigmoid shape of the oxygen–haemoglobin dissociation curve.
14 Enumerate the causes of (a) hypoxia and (b) hypercapnia.
15 Describe how the basic rhythm of respiration is generated.
16 Describe how alveolar ventilation is regulated according to body needs.

Introduction

In an adult, the body cells consume about 250 ml oxygen (O_2) per minute under resting conditions, this demand increasing up to thirtyfold during strenuous exercise. The end-product of the energy-producing (metabolic) processes is carbon dioxide (CO_2), of which approximately 200 ml per minute is produced at rest. While we can survive without food or water for days, we can live without O_2 for only a matter of minutes – primarily because the brain cells are incapable of functioning without it. Furthermore, any significant build up of CO_2 in the body would be highly toxic.

The primary function of the respiratory system is to provide an adequate supply of O_2 to the tissues and to remove the metabolically produced CO_2. To do this, several complex and inter-related processes must occur. Not only must the lungs be supplied with air (ventilation) and blood (perfusion), but the two must also be evenly matched to ensure that efficient gas exchange by diffusion can occur. Complex mechanisms are involved in the transport of O_2 and CO_2 around the body in blood, due to the relative insolubility of O_2 and the toxic effects that large quantities of dissolved CO_2 would have on the acid–base balance in the body. Finally, the rate and depth of breathing must be carefully regulated accord-ing to body requirements. In this chapter, each of these processes (summarized in Fig. 5.3.1) will be considered in detail. However, it is first nec-essary to appreciate the relationship between the structure and functions of the respiratory system.

THE RELATIONSHIP BETWEEN THE STRUCTURE AND FUNCTIONS OF THE RESPIRATORY SYSTEM

Exchange of air between lungs and atmosphere

The respiratory system consists of the lungs and the chest structures responsible for moving air in and out of the lungs (Fig. 5.3.2). The two cone-shaped lungs lie in the thoracic cage (chest), which consists of the ribs, sternum, thoracic vertebrae, diaphragm and intercostal muscles, which form an airtight protective cage around the lungs.

The basic rhythm of respiration is generated by cyclical nerve impulses passing from the respirat-ory centre in the medulla of the brain, down the phrenic and intercostal nerves to the diaphragm and intercostal muscles respectively. Cyclical excitation of these respiratory muscles results in alternate expansion and relaxation of the thoracic cage which in turn causes air to be drawn into (**inspiration**) and expelled from (**expiration**) the lung. To understand how this process, known as **ventilation** (or breathing!), occurs, it is important to remember how gases behave and to appreciate the structural relationship between the lungs and thoracic cage.

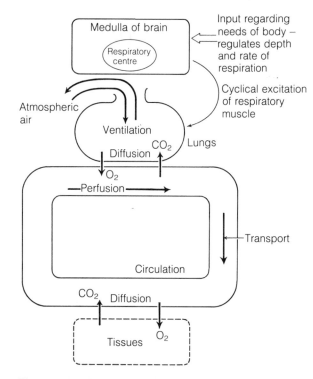

Figure 5.3.1 Schematic representation of the processes involved in respiration

BEHAVIOUR OF GASES

Whenever a substance exists as a gas, its molecules are free to move about independently in space. This perpetual motion of gas molecules results in numerous collisions which exert a certain pressure. Any factor that increases the number of collisions occurring (such as a rise in temperature which increases the speed at which the molecules travel) will cause a rise in gas pressure, and vice versa.

If temperature remains constant, the pressure of a gas varies inversely with the volume in which it is contained (decreased volume leads to increased concentration of gas molecules and hence more collisions). Thus, if the same volume of air (say 50 ml) is drawn into three syringes (see Fig. 5.3.3), the pressure within them can be altered, relative to atmospheric pressure, by clos-ing the ends of the syringes and then moving the plungers to decrease (Fig. 5.3.3b) or increase (Fig. 5.3.3c) the internal volume of the syringe.

Although gas molecules are always in perpetual motion, there will only be a net flow of gas from

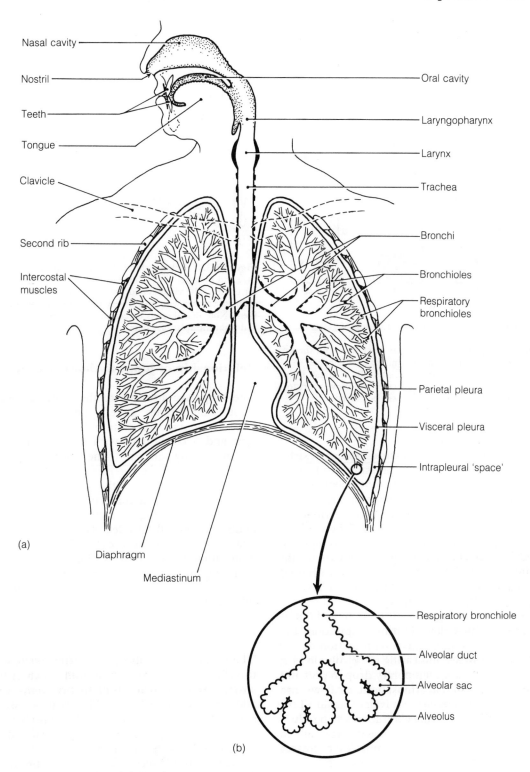

Nasal cavity

Nostril

Teeth

Tongue

Clavicle

Second rib

Intercostal muscles

Oral cavity

Laryngopharynx

Larynx

Trachea

Bronchi

Bronchioles

Respiratory bronchioles

Parietal pleura

Visceral pleura

Intrapleural 'space'

(a)

Diaphragm

Mediastinum

Respiratory bronchiole

Alveolar duct

Alveolar sac

Alveolus

(b)

Figure 5.3.2 Organization of the respiratory system

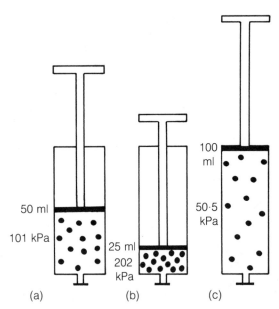

Figure 5.3.3 Pressure–volume relationship of gases. Atmospheric pressure has been taken to be 101 kPa (760 mmHg). By moving the plunger down as in (b), the gas in the syringe is compressed to 25 ml (volume is halved), with a corresponding rise in gas pressure. By contrast, if the plunger is withdrawn, the gas molecules will move apart to fill the 100-ml space, but pressure will be halved (c)

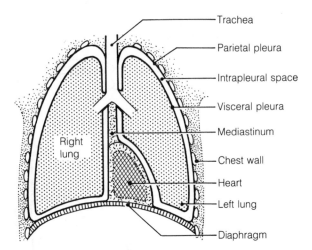

Figure 5.3.4 Relationship between the lungs, thoracic cage and pleurae. The volume of pleural fluid is exaggerated for clarity. The visceral and parietal pleurae normally lie in intimate contact with only a very thin layer of fluid between them. Note that there is no communication between the left and right intrapleural fluids

one area to another if a pressure gradient exists. Movement of gases along a pressure gradient is said to occur by **bulk flow**, movement always being from an area of high to low pressure until an equilibrium is achieved. Thus, if the syringes in Fig. 5.3.3 are opened to the atmosphere, there would be no net exchange of gas from syringe (a), air would flow out of syringe (b), and air would flow into syringe (c), until the pressure within all returned to atmospheric (which at sea level is equal to 101 kPa or 760 mmHg).

During a breathing cycle air flows in and out of the lungs by bulk flow. During inspiration, the thoracic cage expands due to contraction of the respiratory muscles, increasing lung volume as it does so. This causes a temporary drop in pressure inside the lungs so that atmospheric air flows into them until pressures are equivalent again. This process is reversed during expiration when reduction in the volume of the thoracic cage due to relaxation of the respiratory muscles decreases lung volume and temporarily causes pressures within the lung to exceed that of the atmosphere.

Relationship of lungs to thoracic cage

The lungs are attached to the thoracic cage by serous membranes called the **pleurae**. The visceral pleura covers the outer surface of each lung and is continuous with the parietal pleura, which is firmly adherent to the inner surface of the thoracic cage. These two membranes normally lie in intimate contact with one another, being separated only by a thin layer of pleural fluid. In effect, each lung is enclosed within its own double-walled, fluid-filled sac, by which it is attached to the thoracic cage. There is no communication between the intrapleural fluid surrounding the right and left lung since the mediastinum effectively divides the thoracic cavity into two halves (Fig. 5.3.4)

PLEURAL FLUID

The two pleural layers are effectively sealed together by a film of pleural fluid which exerts a strong surface tension force that prevents separation of the membranes. This seal is essential since it enables the lungs (which themselves contain no skeletal muscle) to be expanded and relaxed by movements of the chest wall.

Pleural fluid is secreted by the membranes and is basically a capillary filtrate. Anything that upsets the normal balance of forces occurring at the capillaries such as

(a) increased capillary permeability (as in inflammatory conditions)
(b) increased pulmonary capillary pressure (as in left ventricular failure)
(c) reduced flow through pulmonary lymphatics (as in tumours and infections)

may result in excess pleural fluid formation, known as **pleural effusion**. This condition may develop gradually and not be recognized until the accumulation of fluid is great enough to compress the lung, causing difficulty in breathing (**dyspnoea**). This unpleasant symptom can be relieved by draining the pleural effusion by **chest aspiration (thoracentesis)**. This aseptic procedure is performed by a doctor using local anaesthetic. It involves passing a biopsy needle through the chest wall, between the ribs, at the affected site. A sample of the fluid is sent for analysis and medical management can then be directed towards treating the causative disease.

As well as assisting the doctor, nursing responsibilities include: ensuring that the patient knows what to expect and how he can cooperate; helping him to remain still, in a supported sitting position, during the procedure; and observing for signs of respiratory distress and bleeding both during the aspiration and after it has been completed.

INTRAPLEURAL PRESSURE

The relaxed volume of the lung is considerably smaller than that of the thoracic cage (Fig. 5.3.5). Attachment of the lung to the inside of the thorax by the pleura results in the thoracic cage being pulled inwards while the lung is expanded (Fig. 5.3.5c). Both the thorax and lung are elastic structures, that is, they will spring back (or recoil)

to their original size as soon as any distending pressure (or force) is removed. While the pleural fluid seal prevents the lung actually separating from the thorax, the tendency of the thorax to spring out and that of the lung to collapse produces forces pulling in opposite directions. This produces a negative pressure (relative to atmosphere) inside the intrapleural space. The resting volume adopted by the lung at the end of a normal expiration is known as the **functional residual capacity (FRC)**. This volume is usually about 3 litres in an adult male and represents the lung volume which is achieved when the outward pull of the thorax is exactly balanced by the inward recoil of the lung. At the end of expiration, intrapleural pressure is about $-0.5\,\text{kPa}$ ($-4\,\text{mmHg}$). The more the lungs are expanded, the greater will be their tendency to spring back to their relaxed volume. Consequently, intrapleural pressure becomes increasingly negative during inspiration (Fig. 5.3.6).

Intrapleural pressure is sometimes referred to as **intrathoracic pressure** since it is transmitted to all structures in the thorax including the heart. The negative intrathoracic pressure assists venous return by exerting a slight 'sucking' force, sometimes called the 'respiratory pump'. This force is increased during exercise when deeper breathing is accompanied by greater negative intrathoracic pressure. Conversely, venous return may be impeded in some patients during mechanical ventilation as a result of the applied positive pressure which is transmitted not only down the airways but also around all the thoracic blood vessels.

Pneumothorax

If the pleural fluid seal is broken (e.g. by a spontaneous intrapulmonary air leak, during chest surgery or by a stab wound), air will be drawn into

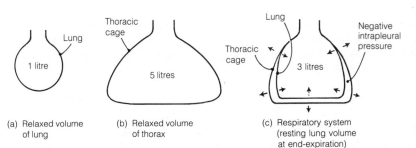

(a) Relaxed volume of lung

(b) Relaxed volume of thorax

(c) Respiratory system (resting lung volume at end-expiration)

Figure 5.3.5 Diagrammatic representation of the relative sizes of the lung, thoracic cage and total respiratory system at rest. When the lungs are attached to the thoracic cage by the pleural fluid seal, the thorax is pulled inward, while the lung is expanded. At end-expiration the tendency of the lung to recoil to its relaxed volume is exactly matched by the tendency of the thorax to recoil outwards

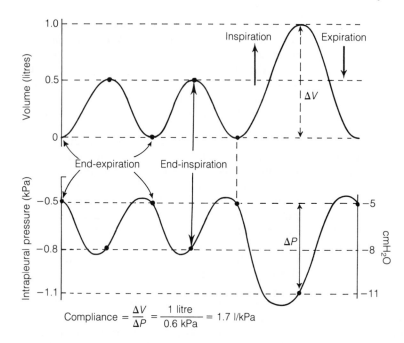

$$\text{Compliance} = \frac{\Delta V}{\Delta P} = \frac{1 \text{ litre}}{0.6 \text{ kPa}} = 1.7 \text{ l/kPa}$$

Figure 5.3.6 Changes of intrapleural pressure during the respiratory cycle

the intrapleural space, creating a space between the lungs and thorax and abolishing the negative intrapleural pressure. This is called a pneumothorax. When a pneumothorax occurs, the lung in the affected area collapses, the chest wall bows out and, since movement of the chest wall is no longer effective in expanding the lung, gas exchange is seriously impaired. Fortunately, since there is no continuity between the pleural fluid seal around the right and left lung, only one side will collapse unless both sides are damaged simultaneously.

Rapid diagnosis and treatment of a pneumothorax are essential. Decreased movement of one side of the chest on inspiration (except following pneumonectomy, when this is to be expected), difficulty in breathing, sudden deterioration in condition, cyanosis and chest pain may all suggest the presence of a pneumothorax and should be reported immediately. Diagnosis is usually confirmed by a portable chest x-ray.

Treatment is aimed at restoring the lung to its original size as rapidly as possible. This is usually achieved by the insertion of a tube to draw air from the pleural space into a water seal drainage apparatus.

Water seal chest drainage

Various methods may be used to achieve drainage of a pneumothorax or pleural effusion, but the basic principle is always the same – to allow air and excess fluid to escape from the pleural cavity while preventing any reflux. The basic apparatus is illustrated in Fig. 5.3.7.

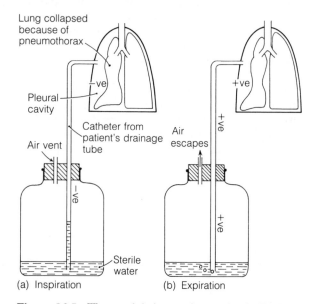

Figure 5.3.7 Water seal drainage using one bottle. During inspiration (a), the negative pressure in the pleural cavity sucks water up the tube. During expiration (b), pressure in the pleural cavity and tubing becomes positive, forcing air out of the cavity

The drainage tube from the patient is connected to a long catheter, the end of which is submerged below a few centimetres of sterile water in a calibrated drainage bottle. The decrease in pressure inside the pleural cavity during inspiration causes air to be sucked up the tube, usually to a height of about 10–20 cm.

Consequently, drainage bottles must always be kept well below the level of the patient's chest, preferably on the floor, to prevent water being sucked into the chest. The danger of this occurring is sometimes prevented by separating the drainage bottle from the underwater seal. During expiration the rise in pressure forces air out of the pleural cavity. The water level in the tube will fall and some air will bubble out through the water and escape from the bottle via the air vent. Re-entry of air during the following inspiration is prevented by the water seal.

The magnitude of pressure swings in the pleural cavity (as reflected by the swings of water level in the tube) will depend on the amount of air present and the depth of breathing. Coughing and deep breathing promote drainage. Drainage may be assisted by mild suction, in which case there will be no swings in water level as the applied negative pressure will be held at a constant level.

A patient undergoing chest drainage requires regular observation. If he experiences chest pain or difficulty in breathing, or has a rising pulse rate, this requires immediate investigation. During drainage, the amount of air and the type and volume of fluid drained are observed and recorded.

It is also essential that the patency of the drainage system is regularly checked. Blockage is indicated by the absence of oscillations in the water level in the closed tube (suction having first been disconnected if it is being used). Blockage may be caused by the drainage tube becoming kinked, by the patient lying on it, or by a clot or other debris inside the tube. Precautions should also be taken to ensure that drainage bottles are not accidentally moved or knocked over, and a spare set of equipment should always be available in case of emergency.

If either the patient needs to be moved or the bottle must be moved or replaced, the drainage tube must first be sealed with two clamps placed close to the chest wall. Sealing in this way ensures that neither fluid nor air can accidentally enter the chest.

Having drainage tubes in situ is uncomfortable and care should be taken to support them so that they do not pull on the chest wall or become dislodged.

When no air remains in the pleural space, the bubbling and swinging of water levels will cease. Re-expansion of the lung is confirmed by chest x-ray before removal of the tubes.

When the tube is removed the wound is immediately sealed to prevent recurrence of pneumothorax, and the patient is closely observed afterwards for signs of leakage of air into the chest.

Inspiration

The most important muscle of inspiration is the **diaphragm**. This is a dome-shaped sheet of muscle attached to the lower ribs. When the diaphragm contracts during inspiration, it flattens, pressing down on the abdominal contents and lifting the rib cage, thus enlarging the thoracic cage and lung, both from top to bottom and from front to back (Fig. 5.3.8). Since the air in the lungs now occupies a greater volume, alveolar pressure temporarily falls below atmospheric pressure. This causes air to be drawn into the lungs by bulk flow until alveolar pressure again equals atmospheric at the end of inspiration (termed end-inspiration). By contrast, intrapleural pressure becomes increasingly negative throughout inspiration, due partly to the increase in the elastic recoil of the lung and partly to the fall in alveolar pressure (Fig. 5.3.9).

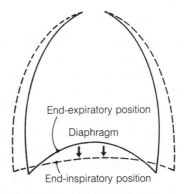

Figure 5.3.8 Diaphragmatic movements during respiration. The dome-shaped diaphragm contracts during inspiration, pressing down on the abdominal contents and lifting the rib cage, resulting in an increase in the volume of the thorax

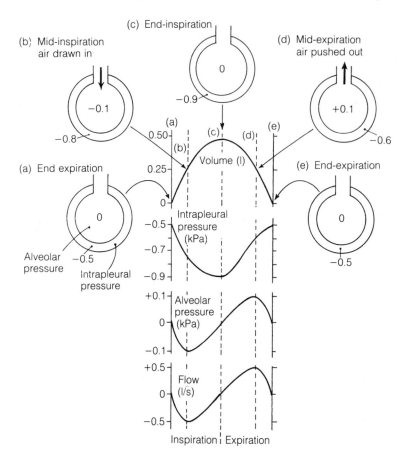

Figure 5.3.9 Comparison of changes in alveolar and intrapleural pressure during the respiratory cycle. All pressures are given in kPa (multiply by 10 to obtain values in cmH$_2$O). Note that intrapleural pressure normally remains negative throughout the respiratory cycle whereas alveolar pressure is atmospheric at end-expiration (points a and e), and end-inspiration (c), (where there is no flow of gas), but becomes negative during inspiration (b), and positive during expiration (d), thereby providing the driving force to move air in and out of the lungs, as shown by the recording of airflow

Recent work has shown that the external intercostal muscles play virtually no part in expanding the rib cage during quiet breathing. They do, however, play a vital role in stabilizing (stiffening) the rib cage. If, for example, the intercostal muscles are paralysed, much of the effort expended during breathing will be spent on distorting the rib cage rather than on effecting gas exchange. During exercise, when requirements for gas exchange increase, or in the presence of an upper airway obstruction, which impedes the flow of gas into the lungs, the external intercostals, together with other accessory muscles of inspiration such as the sternomastoids, scalene and pectoralis muscles, do assist in the expansion of the rib cage. Use of these accessory muscles to aid ventilation may also be observed in patients with obstructive airway disease, such as chronic bronchitis.

Expiration

During quiet breathing, expiration is a passive process requiring no active muscle contraction. At end-inspiration, the respiratory muscles relax allowing the elastic lung and thorax to recoil to their original resting volume (functional residual capacity). This reduction in volume compresses the air in the lungs so that alveolar pressure exceeds atmospheric, providing the necessary driving force to push air out of the lung. Expiratory flow ceases when alveolar pressure is again equal to atmospheric pressure. During expiration, intrapleural pressure becomes progressively less negative, partly due to the progressive reduction in elastic recoil of the lung as it returns to its resting position, but also because of the positive alveolar pressure during expiration (Fig. 5.3.9).

Active expiratory efforts may occur if requirements for gas exchange increase (during exercise) or if the airways are narrowed. The most important accessory muscles of expiration are those of the abdominal wall. When these muscles contract, they increase abdominal pressure which forces the diaphragm higher up into the thorax. This causes a greater compression of alveolar gas and thus facilitates expiratory flow.

Organization of the airways

For air to reach the lungs it must pass through a series of branching airways, which become progressively more numerous and smaller in diameter (Fig. 5.3.10). Air may enter the respiratory passages via the nose or mouth, although very young infants, as a result of anatomical differences, are preferential nose breathers and rarely breathe through their mouths until about 3–6 months of age. It is therefore essential that their nasal passages are always kept as clear as possible.

The **nasal cavity** consists of a large, irregular cavity divided by a septum. It is lined with ciliated epithelium, the surface area of which is greatly increased by the presence of bony projections, called turbinates or conchae, which project into the cavity. The ciliated epithelium ensures that air is warmed, filtered and moistened as it passes through the nose. The mucous membrane covering the turbinates is vascular and swells rapidly when inflamed or irritated, to the extent that the entire nasal cavity may become blocked during head colds or allergic reactions.

Having passed through the nose or mouth, air enters the **pharynx** (throat), which is also a common passageway for food and water. The pharynx divides into the oesophagus (along which food is directed on its way to the stomach) and the larynx, through which air passes on its way to the trachea, bronchi, bronchioles and alveoli.

The **larynx** is composed of cartilages, connected by ligaments and moved by various muscles. It produces the bump in the neck called the Adam's apple and, after puberty, is larger in men than in women. It is lined with mucous membrane that is continuous with that of the pharynx and trachea. As well as being part of the airway, it contains the elastic vocal cords which function in sound production. The space between the vocal cords, through which the air passes, is called the **glottis**.

Inflammation of the mucous membrane lining

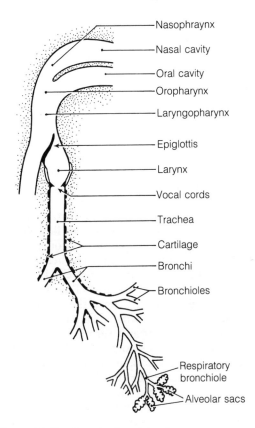

Figure 5.3.10 Organization of the airways

of the larynx produces hoarseness and loss of voice (**laryngitis**). This may be relieved by removing irritants, avoiding smoking, resting the voice, steam inhalations and cough medication. Persistent hoarseness is usually the first symptom of laryngeal cancer. The small larynx of a child under 5 is particularly susceptible to spasm when inflamed and may become partially or totally obstructed. This produces a barking cough and inspiratory stridor (an abnormal, high-pitched sound heard on inspiration) and can cause severe respiratory distress (**croup**). Croup requires immediate treatment by allowing the child to breathe air saturated with water vapour (cooled steam). This moistens secretions and reduces inflammation, hence relieving the obstruction.

Valsalva's manoeuvre involves forced expiratory effort against a closed glottis, such as occurs when straining to move a heavy object, to change position in bed or to defaecate, especially when constipated. Such straining reduces venous return to the heart, due to increased intrathoracic pressure, and, on relaxation of the muscles, venous return is correspondingly increased. While the performance of Valsalva's manoeuvre creates no problems for healthy people, those with cardiovascular disease are usually advised to avoid straining in this way by exhaling instead of holding their breath. This decreases the risk of the heart suddenly becoming overloaded.

The **trachea** is a cylindrical tube, approximately 10 cm long, which is composed of 16–20 incomplete (C-shaped) rings of cartilage joined together by fibrous and muscular tissue. The cartilage, which forms the anterior and lateral walls of the trachea, provides rigidity to the trachea and prevents it collapsing and blocking air flow during inspiration when airway pressure becomes negative. The muscular portion of the trachea lies over the oesophagus, providing a flattened surface which can be stretched slightly. This facilitates movement of food boluses down the oesophagus. The trachea is lined with ciliated epithelium which assists in filtering and warming inspired air.

The trachea extends to the level of the 5th thoracic vertebra, where it divides (bifurcates) into the two **primary bronchi**. The right bronchus is a shorter, wider tube than the left and lies in a more vertical position. Consequently any foreign bodies that enter the trachea are more likely to be inhaled into the right main bronchus where they may become lodged, requiring removal through a bronchoscope.

The primary bronchi enter the right and left lungs at the **hilum**. The bronchi are composed of tissues similar to those of the trachea, but as they become progressively smaller by subdivisions inside the lungs, the cartilage becomes less well defined and irregular in shape.

The volume of the nasal cavity, pharynx, larynx, trachea and bronchi (which are collectively known as the conducting airways) is approximately 150 ml in the adult. Since no gas exchange can occur across these thick-walled tubes, the space within them is called the **anatomic dead space**.

Smaller branches of the bronchi which are less than 1 mm in diameter are called **bronchioles**. Many of these are lined with thin respiratory epithelium across which gas exchange can occur. The walls of the bronchioles, like those of the larger airways, are composed of rings of smooth muscle. Changes in smooth muscle tone are particularly effective in altering the calibre of the bronchioles since, unlike the trachea and bronchi,

they have no cartilage in their walls to maintain rigidity. The tone of the smooth muscle is under autonomic control. It contains β receptors, which cause relaxation of the muscle in response to sympathetic stimulation (e.g. during exercise) and to drugs such as adrenaline (epinephrine), noradrenaline and salbutamol, resulting in generalized **bronchodilation**. Since this facilitates air entry to the lungs, these drugs are frequently administered to patients with bronchoconstriction. Bronchodilation also occurs in response to certain hormones and to local increases in the level of carbon dioxide.

Bronchoconstriction is caused by parasympathetic activity, acetylcholine, histamine and stimulation of receptors in the trachea and large bronchi by irritants such as cigarette smoke. In asthma there is a generalized hypersensitivity of the airways, and widespread bronchoconstriction may occur in response to allergies, infections or irritants, causing severe respiratory distress. Parasympathetic activity is also stimulated by breathing very cold air.

Functions of the airways

Inasmuch as they constitute a dead space in which no gas exchange can occur and offer a resistance to breathing, which increases in the presence of any form of obstruction, it might be thought that the airways are nothing but a hindrance to efficient respiration. However, they perform the vital function of protecting the delicate respiratory tissues in the alveoli by filtering, warming and humidifying the air during its passage to the lungs.

The structure of the epithelial lining is well suited to its function as an air filter (Fig. 5.3.11). The epithelial glands secrete a thick, sticky substance called mucus which lines all parts of the respiratory tract above the bronchioles. This layer of mucus waterproofs the epithelial surface (thereby diminishing water loss from the body), acts as a protective barrier against irritants and traps foreign particles. Hypertrophy of the mucous glands and excessive mucus production, such as occurs in chronic bronchitis, will occur in response to prolonged irritation of the mucosa by cigarette smoke and other pollutants.

The epithelial lining of all large airways also contains hair-like projections called cilia. These beat between 600 and 1000 times each minute,

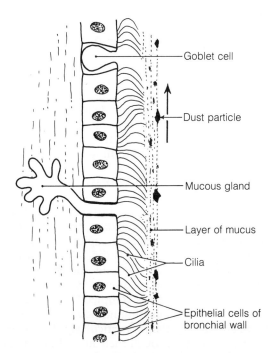

- Goblet cell
- Dust particle
- Mucous gland
- Layer of mucus
- Cilia
- Epithelial cells of bronchial wall

Figure 5.3.11 Epithelial lining of the respiratory tract. Mucus is secreted by mucous glands and goblet cells and traps inhaled particles. The cilia propel the mucus towards the pharynx (the mucociliary escalator) where it can be swallowed

sweeping the mucus and any trapped particles in it towards the pharynx where it can be swallowed and eliminated in the faeces. This mechanism also contributes to defence against infection since many bacteria enter the body on dust particles and can be 'swept out' again in this manner. The large numbers of phagocytic cells in the epithelial lining also contribute to defence by engulfing debris, dust and bacteria. The cilia may be unable to clear the mucous blanket adequately if

(a) the secretions become too thick and tenacious, as in dehydration or cystic fibrosis
(b) the cilia are damaged or paralysed by inadequate humidification of inspired air (this occurs especially during artificial ventilation) or exposure to irritant gases such as those in cigarette smoke
(c) there is excessive production of mucus, as in upper respiratory tract infections.

In any of these situations, there will be a retention of secretions which are likely to contain bacteria and may therefore lead to respiratory infections.

In addition, excess mucus may drain into the lower airways causing partial or complete airways obstruction and seriously interfering with gas exchange.

Pneumonia is an acute inflammatory process caused by bacteria or viruses which usually begins in the large bronchi and then spreads towards the periphery of the lung. The inflammatory process causes vasodilation of the pulmonary arterioles and capillaries with consequent leakage of plasma, fibrin and blood cells. This exudate frequently collects in the alveoli where its presence prevents gas exchange – the affected part of the lung is said to be **consolidated**. The pneumococcal bacteria which are the commonest cause of pneumonia are normally present in the upper respiratory tract in about 50% of the 'healthy population'. The excellent defence mechanisms afforded by the respiratory tract in health prevent these bacteria entering the lower respiratory tract and initiating the inflammatory process. It is only when the body's resistance to disease is lowered in old age, or by malnutrition or other diseases, that the risk of contracting pneumonia is increased. Intoxication with alcohol or drug abuse also increases the risk, both by reducing the phagocytic ability of the macrophages and by depressing the cough reflex which would normally facilitate removal of bacteria and mucus from the lungs.

COUGH AND SNEEZE REFLEXES

Coughing is the body's reflex mechanism for attempting to remove excess mucus or other irritants from the air passages. This reflex originates in receptors in the lining of the respiratory tract which, when activated, cause stimulation of the respiratory centre in the medulla of the brain, which in turn regulates the special breathing pattern used during coughing. After an initial deep inspiration, a forced expiration is made. The glottis is closed during initial expiration so that no air can leave the lungs. This enables the pressure beneath the glottis to build up rapidly and, when it is suddenly re-opened, air rushes out at a speed which may approach 500 mph, carrying liquid matter up and clearing the airways. Sneezing is a similar protective reflex which is stimulated by irritation of the nasal mucosa. It involves a series of short inspirations followed by an explosive expiration, usually with the mouth closed. The rapid spread of respiratory infections is largely due to the speed at which infected droplets of water are expelled from the body during coughing and sneezing, which enable them to travel considerable distances outside the body.

If a cough results in expulsion of mucus (sputum), it should be encouraged since it will help prevent accumulation of secretions within the lungs. Sputum is an abnormal excess of secretions from the airways. Laboratory analysis of sputum can be used to assist in the diagnosis of both respiratory and cardiac disorders. Specimens of sputum should be collected into a sterile container soon after waking from a night's sleep, as the overnight accumulation of secretions makes it easier for the patient to produce sputum rather than saliva. The amount of sputum produced, its colour, consistency and the degree of effort required to produce it should be observed. Many patients may feel embarrassed when asked to spit in order to produce a specimen of sputum and so it is important that the nurse ensures the patient's privacy and shows no personal distaste. Precautions should also be taken to avoid cross-infection from a potentially infected sputum specimen.

Following intrathoracic surgery, the intercostal nerves are sometimes injected with a local anaesthetic to reduce pain. If this is not done, a patient may adopt a pattern of shallow breathing and inhibit his cough reflex in an attempt to minimize pain. This may delay re-expansion of the lungs or lead to obstruction of the smaller airways with subsequent alveolar collapse (atelectasis). Morphine and narcotic drugs should be used sparingly since they tend to depress respiration and the cough reflex. A nurse may provide support by placing her hands on the anterior and posterior chest walls in the painful areas while encouraging deep breathing and coughing.

Other methods of preventing accumulation of retained secretions include the following.

(a) Early mobilization and/or frequent changes of position.
(b) Maintenance of adequate fluid balance (without which secretions become thick and tenacious).
(c) Administration of inhalations, expectorants or mucolytic enzymes.
(d) Physiotherapy and postural drainage.
(e) Administration of antibiotics if infective organisms are present in sputum.

If the cough reflex is lost (during anaesthesia or unconsciousness) or if a tracheostomy or endotracheal tube is present, the patient will be unable to raise his own secretions. Under these circumstances, it may be necessary to use a small sterile catheter attached to a suction pump to remove secretions. During suction, care should be taken to avoid (a) traumatizing the delicate respiratory epithelium, (b) introducing any infection, and (c) creating too great a negative pressure in the airways which could lead to lung collapse.

HUMIDIFICATION AND WARMING

As air passes along the upper respiratory tract it is exposed to a large surface area of highly vascular epithelium and undergoes a rapid exchange of heat and water. Lack of humidification can cause severe damage to the cilia and thereby initiate or aggravate respiratory disease. Consequently, compressed O_2 (which is completely dry) should always be warmed and humidified before being administered to patients, particularly if they are unable to breathe through their nose. Inadequate humidification of inspired gases was a major cause of lung damage in mechanically ventilated patients before the significance of humidification was appreciated, and even now the efficiency of some ward humidifiers is highly questionable.

Under normal circumstances, approximately 150–200 ml of water are lost from the body in expired gas each day. If ventilation increases due to exercise, fever or lung disease or if an individual mouth-breathes, this loss will be greater and must be compensated for by increased fluid intake.

If very cold air is breathed, parasympathetic stimulation of the bronchiolar smooth muscles causes bronchoconstriction, which slows the passage of air into the alveoli and allows more time for it to be warmed up. This reflex may be evident from the 'tight' feeling in the chest often experienced when out walking in cold weather and, in patients with hypersensitive airways, may be sufficient to provoke an attack of asthma.

SWALLOWING REFLEX

The presence of food or water in the pharynx normally stimulates the swallowing reflex, which involves a temporary cessation of breathing and the closure of the larynx (glottis) by the lowering of a leaf-shaped piece of tissue called the **epiglottis**. If this reflex is absent (e.g. in an unconscious patient or during and after anaesthesia), food or water may be inhaled into the lungs, resulting in choking and possible airway obstruction, lung collapse or pneumonia. To minimize this risk, atropine is given preoperatively to reduce secretions, and oral food and fluid are withheld from patients postoperatively until the swallowing reflex is regained.

SPEECH

The airways play a vital role in the production of sound and speech. Sound is produced by the vibration of air as it passes through the vocal cords, which are two strong bands of elastic tissue stretched across the lumen of the larynx, covered by membranous folds. The glottis can be varied in shape and size to produce different levels of pitch in sound production. Normally, sound is produced during expiration, and inspiration occurs silently. Inflammation, infection or the presence of a tumour in the larynx may lead to hoarseness and loss of voice. The nasal cavities and sinuses act as resonating chambers which alter the quality of sound.

SMELL

The olfactory receptors which are concerned with the sense of smell are found in the posterior portion of the nasal cavities. Their efficiency is greatly diminished by any obstruction (e.g. oedema) of the nasal passages and sinuses. It is for this reason that the common cold affects one's ability to smell and, since the sense of smell enhances the sense of taste, food seems tasteless and less appetizing.

Structure and functions of the alveoli

The epithelial lining of the airways becomes progressively less ciliated as it approaches the **alveolar ducts**. These are thin tubes of squamous epithelium branching off the terminal bronchioles and opening into the alveoli, which are tiny cup-shaped hollow sacs. Three types of cells are found in the alveolar epithelium.

(a) Thin squamous epithelial cells which cover most of the alveolar surface.
(b) Larger cuboidal epithelial cells (Type II) which are metabolically active and respon-

sible for both epithelial cell renewal and synthesizing **surfactant**, a phospholipid which reduces surface tension forces in the lung. Infants born prematurely may have a deficiency of surfactant due to immaturity. The lungs of such infants are very stiff, difficult to inflate and tend to collapse on expiration, resulting in a condition known as the **respiratory distress syndrome**. Steroid therapy given to mothers prior to impending premature delivery may accelerate fetal lung development, thus helping to prevent this condition.

(c) Alveolar macrophages. These are active phagocytic cells thought to be derived from bone marrow. They provide the chief defence mechanism against any debris or bacteria that reaches the alveoli. They migrate through the epithelium, engulf any foreign matter and are eliminated mainly in the sputum. There are no ciliated or mucus-producing cells in the alveolar epithelium.

The lungs are divided into approximately 300 million alveoli. Similarly, the pulmonary artery, which brings venous blood to the lungs for gas exchange, branches and subdivides into millions of thin-walled capillaries forming a dense network of blood vessels wrapped around each alveolus (Fig. 5.3.12).

Gas exchange between the pulmonary capillary blood and the air within the alveoli occurs by diffusion. To understand this process it is again necessary to remember how gases behave and to appreciate the fine structure of the alveolar–capillary (or pulmonary) membrane.

BEHAVIOUR OF GASES

Partial pressures
When different kinds of molecules are present in a gas mixture, e.g. as in air, where the main components are O_2 and nitrogen (N_2), each gas exerts its own pressure (depending on the number of molecules present etc.) as if it were completely filling any available volume. The total pressure of a gas mixture is merely the sum of all the individual (or partial) pressures (Dalton's law of partial pressures). Since air consists of approximately 21% O_2, and atmospheric air has a total pressure of 101 kPa (760 mmHg) at sea level, the partial pressure of O_2

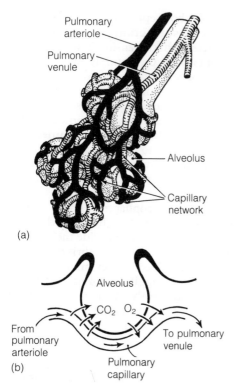

(a)

Pulmonary arteriole

Pulmonary venule

Alveolus

Capillary network

(b)

Alveolus

CO_2 O_2

From pulmonary arteriole

To pulmonary venule

Pulmonary capillary

Figure 5.3.12 Relationship between alveoli and blood vessels. Gas exchange can occur across the vast surface area provided by the dense network of capillaries

(referred to as PO_2) in air is

PO_2 atmospheric air at sea level

$$= \frac{21}{100} \times 101 = 21.2 \, \text{kPa} \, (160 \, \text{mmHg})$$

Similarly, the PN_2 of this air is $79 \, \text{kPa}$ ($590 \, \text{mmHg}$).

Diffusion of gases
As discussed on p. 468, movement of gases by bulk flow (e.g. in and out of the lungs) can only occur if there is a difference in total pressure between two areas. However, even if no gradient of total pressure exists, a net movement of any individual gas in a gas mixture can occur by a process called diffusion, providing a partial pressure gradient for that particular gas exists. Providing no specific barrier to diffusion exists, a gas will move from an area of high to lower partial pressure until an equilibrium is achieved. The constant consumption of O_2 and production of CO_2 by the cells and the continual renewal of gas in the alveoli during ventilation maintain a perpetual pressure gradient for O_2 and CO_2 both within the lungs and at tissue level, thereby enabling gas exchange to occur by passive diffusion. The rate at which a gas can diffuse, and hence the amount of gas exchange that can occur in the lungs or tissues, is proportional to

(a) the magnitude of the partial pressure gradient
(b) the solubility of the gas
(c) the thinness of the membrane across which the gases must move
(d) the surface area available for diffusion.

THE ALVEOLAR–CAPILLARY MEMBRANE

Mixed venous blood (i.e. blood that has supplied the tissues with O_2 and removed their excess CO_2) is continually brought to the lungs via the pulmonary artery (right ventricular output). Since this blood has a lower PO_2 and higher PCO_2 than the air in the lungs, gas exchange can occur by the simple process of diffusion, with O_2 diffusing into the blood for transportation to the tissues, and CO_2 diffusing out of the blood into the air for expulsion during expiration.

For gas exchange to occur in the lungs or between the blood and tissues, gases have to cross cellular membranes. Surfaces of the lung that are thin enough to permit rapid diffusion of gases include the alveoli, the alveolar ducts and the respiratory bronchioles, and are collectively known as the pulmonary (or alveolar–capillary) membrane (Fig. 5.3.13b). This membrane consists of the alveolar epithelium, the interstitium and the capillary endothelium. It is exceedingly thin, being under $0.4 \, \mu\text{m}$ thick in most parts, which is less than the diameter of a red blood cell. The blood travelling through the capillaries is thus brought into intimate proximity with the alveolar air, thereby facilitating diffusion (Fig. 5.3.13a). The function of the lung as an organ of gas exchange is also favoured by its subdivision into millions of alveoli. This arrangement vastly increases the surface area available for diffusion to about $70 \, \text{m}^2$ (which is some 40 times larger than the entire external body surface area).

Under normal conditions, gas exchange in the lungs is so efficient that venous blood entering the capillaries can equilibrate with alveolar air in about 0.2 seconds. Since it usually takes the red blood cell approximately 0.7 seconds to pass

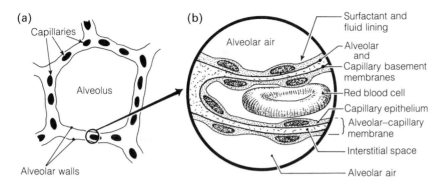

Figure 5.3.13 (a) Cross-section through an alveolus (b) Higher magnification showing histology of part of the alveolar–capillary membrane. The dense network of capillaries forms an almost continuous sheet of blood in the alveolar walls, providing a very efficient arrangement for gas exchange

through the pulmonary capillaries at rest, much of the diffusing capacity of the lung is normally held in reserve. The surface area of the pulmonary membrane can be increased during exercise by the opening up, or 'recruitment', or additional capillary units. The commonest cause of a reduction in surface area is poor matching of the distribution of air and blood in the lungs, but it may also be caused by collapse or surgical removal of part of the lung. Thickening of the pulmonary membrane may result from pulmonary oedema, or deposition of fibrous tissues in the alveolar wall. This, like a reduction in surface area, rarely causes any problems at rest, but may seriously limit gas exchange during exercise when, due to the increased blood flow through the lungs, the time available for gas exchange may be reduced to 0.3 seconds.

Having considered how the structure of the lung is suited to its primary role as an organ of gas exchange, the various physiological processes that are involved in providing the tissues with O_2 and eliminating excess CO_2 will be considered.

VENTILATION OF THE LUNGS

Pulmonary ventilation

The volume of air that passes in and out of the lungs during each breath is called the **tidal volume**. The total volume of air exchanged between the respiratory system and atmosphere each minute is referred to as the **minute volume** or **pulmonary ventilation**. This is determined by the tidal volume and respiratory rate, both of which vary enormously in health and disease, and according to the age of the subject. However, in a resting adult with an 'average' tidal volume of about 500 ml and respiratory rate of 12 breaths/

min, pulmonary ventilation will be approximately 6000 ml/min. However, not all the air that is inspired actually reaches the alveoli for gas exchange to occur. Approximately 150 ml of each tidal breath is 'trapped' within the dead space and is breathed out again during the subsequent expiration with its composition unchanged.

Alveolar ventilation

The volume of fresh air entering the alveoli each minute is called the **alveolar ventilation**, and is calculated as follows.

Respiratory rate × (tidal volume − dead space)

= alveolar ventilation

(e.g. 12 breaths/min × (500 − 150 ml)

= 4200 ml/min)

Thus, although pulmonary ventilation may be 6000 ml/min, only 4200 ml/min would actually be available for gas exchange. The minimal alveolar ventilation compatible with life would be about 1200 ml/min. By contrast, during maximal breathing efforts, it can temporarily be increased to as high as 100 l/min. Shallow rapid breathing is generally very inefficient since the smaller the tidal volume, the greater the proportion that will be wasted in the dead space (Table 5.3.1). Similarly, the efficiency of ventilation will be impaired if the dead space increases, either due to attachment to some external apparatus (such as a snorkel or mechanical ventilator) or to lung disease. In either situation pulmonary ventilation will have to be increased if an adequate level of alveolar

Table 5.3.1   Effects of changing patterns of breathing on alveolar ventilation

Pattern	Tidal volume (ml)	Respiratory rate/min	Total ventilation (ml/min)	Dead space ventilation (ml/min)	Alveolar ventilation (ml/min)
A Rapid, shallow breathing	200	30	6000	4500	1500
B 'Normal'	500	12	6000	1800	4200
C Deep, slow breathing	1000	6	6000	900	5100

Calculations based on a constant dead space of 150 ml. Note that total ventilation remains constant throughout, but that alveolar ventilation shows dramatic changes.

ventilation is to be maintained. For this reason patients should not be left connected to a mechanical ventilator for prolonged periods if the machine is to be switched off. When patients are being weaned from mechanical ventilation, they are usually left connected to the machine while it is switched off for gradually increasing periods of time. During these periods it is particularly important that they are observed carefully for signs of respiratory distress.

TRACHEOSTOMY

A tracheostomy is an artificial opening into the trachea at the level of the second or third cartilaginous ring which is kept patent by the insertion of a metal or Portex tube. A tracheostomy decreases the work of breathing by eliminating the resistance of the upper airway and reduces the dead space by about 50%, thereby greatly increasing the efficiency of breathing. It may be performed

(a) if there is severe upper airway obstruction
(b) if there is paralysis of the vocal cords
(c) to improve the efficiency of ventilation and facilitate the removal of secretions from the airways, particularly in patients who require mechanical ventilation.

Although a tracheostomy is frequently a life-saving measure, it does have certain disadvantages, in that

(a) air passes directly to lower airways without being filtered, warmed and humidified in the nasal passages
(b) it interferes with speech production and coughing
(c) there is an increased danger of water entering the airways so that care is required even when taking a shower.

Composition of atmospheric, expired and alveolar air

Atmospheric air consists primarily of 21% O_2 and 79% N_2, with negligible quantities of CO_2 and water vapour. The composition of alveolar air differs from that of atmospheric air for several reasons.

(a) As air passes along the airways it is humidified with water vapour which causes a slight reduction in the partial pressure of the other gases present.
(b) Alveolar air is continually losing O_2 as it diffuses into the blood, and this O_2 is replaced by CO_2 which diffuses out of the blood.
(c) Only a fraction of alveolar gas is renewed with each breath.

Thus alveolar air contains considerably less O_2 and more CO_2 and water vapour than atmospheric air. Since expired alveolar air is 'diluted' with atmospheric air from the dead space, the composition of mixed expired gas falls between that of alveolar and of atmospheric air (Table 5.3.2).

Exchange of alveolar air with atmospheric air

The functional residual capacity (FRC) may be regarded as a reservoir of air in the lungs which is diluted by successive inspirations of tidal air. Only about 12% of alveolar gas can be exchanged with fresh gas during a single breath and it usually takes about 23 seconds to renew half the gas in the lungs.

The mechanism of partial dilution means that the composition of alveolar gas remains very stable. Thus, the FRC acts as a buffer, preventing the marked fluctuations in alveolar (and hence

Table 5.3.2 Composition of atmospheric, expired and alveolar airs, expressed as partial pressures at sea level (kPa and mmHg) and as percentage concentrations

Gas	Atmospheric air			Expired air			Alveolar air		
	Partial pressure			Partial pressure			Partial pressure		
	kPa	mmHg	%	kPa	mmHg	%	kPa	mmHg	%
Oxygen	21.2	159	20.8	15.6	117	15.4	13.3	100	13.2
Carbon dioxide	0.04	0.3	0.04	3.8	29	3.8	5.3	40	5.3
Nitrogen	79.6	597	78.6	75.6	567	74.6	76.4	573	75.3
Water	0.5	3.9	0.5	6.3	47	6.2	6.3	47	6.2
Total	101.3	760	100	101.3	760	100	101.3	760	100

$1 kPa = 7.5 mmHg$

Note that since atmospheric pressure at sea level is 101.3 kPa, partial pressures of each gas expressed in kilopascals are virtually identical to their percentage concentrations within each gas mixture.

The water content of atmospheric air varies daily according to prevailing conditions.

arterial blood) gases that would otherwise occur during the respiratory cycle.

CHANGES IN BAROMETRIC PRESSURE

Low atmospheric pressure (high altitude)
There is a progressive decrease in atmospheric pressure as the distance above sea level increases, so that although atmospheric air maintains a relatively constant composition, there is a fall in partial pressure of each of its constituent gases. At an altitude of 3000 metres (10 000 feet), atmospheric pressure is reduced to 70 kPa (523 mmHg). Consequently atmospheric PO_2 is reduced to 21% of 70 kPa (i.e. 15 kPa or 110 mmHg) while alveolar PO_2 decreases to around 9 kPa (68 mmHg), i.e. 13% of 70 kPa (Table 5.3.2). This causes a marked reduction in the normal pressure gradient for O_2 across the alveolar–capillary membrane, which reduces the rate at which O_2 can diffuse into the blood, resulting in a general deficiency of O_2 in the body (hypoxia) unless augmented O_2 is breathed. Rapid ascent to altitude generally results in some degree of mountain sickness due to hypoxia. However, the body can gradually acclimatize to chronic exposure to low inspired PO_2 (by changes in ventilation, cardiac output, and composition of the blood), as demonstrated by the remarkably normal lives led by the inhabitants of the Andes and Himalayas who permanently live at an altitude of about 5500 m.

High atmospheric pressure
Atmospheric pressure increases below sea level to the extent that 10 m below the surface it is twice as great as normal (202 kPa or 1520 mmHg). This can cause increased diffusion of O_2 and N_2 into body fluids with subsequent interruption of normal cellular activity. More dangerously, on return to sea level excess N_2, which has been forced into the body fluids under pressure, re-expands forming bubbles in body tissues. These cause severe pain, gastrointestinal distension and possible paralysis or death due to embolism. Deep-sea divers may breathe reduced concentration of O_2 to prevent excess uptake of O_2 on descent, and use decompression chambers to allow the excess dissolved N_2 to escape gradually on return to the surface.

The work of breathing

The amount of energy expended on breathing depends on

(a) the rate and depth of ventilation
(b) the ease with which the lungs and thorax can be expanded
(c) the resistance to airflow offered by the airways.

Normally less than 1% of the resting metabolic rate is spent on breathing, but this may rise to 3% during heavy exercise and up to 50% in severe respiratory disease, resulting in exhaustion of the patient and possible respiratory failure.

COMPLIANCE

The term compliance is used to describe the distensibility of the lungs and thorax. Under normal conditions the lungs are very compliant, requiring an inflating pressure (i.e. intrapleural pressure change) of only about 0.3 kPa (2 mmHg) to produce a tidal volume of 500 ml (see Fig. 5.3.6), whereas 100 times more pressure may be required to inflate a child's balloon to the same volume. The lungs will become stiffer (decreased compliance) if there is

(a) a reduction in lung size (atelectasis, pneumonectomy)
(b) increased pulmonary fluid or blood (e.g. pulmonary congestion in heart failure)
(c) a deficiency of surfactant; this phospholipid, which is secreted by the alveolar epithelial cells, normally decreases surface tension forces, facilitating lung expansion.

In contrast, compliance will increase (the lungs will be more distensible) if there is a loss of elastic tissue such as occurs with ageing and in emphysema due to overstretching and destruction of alveolar walls. This is not as beneficial as it might at first appear since the loss of elastic recoil results both in overinflation of the lungs (which can no longer 'spring back' as efficiently as normal during expiration) and a narrowing of the airways (which no longer have the normal degree of traction exerted on their outer walls). Consequently, active expiratory efforts, accompanied by a marked rise in the work of breathing, may be required to force air out of the lungs during expiration.

RESISTANCE OF THE AIRWAYS

Airways resistance can be defined as the pressure required to produce a flow of gas of 1 litre/s through the airways. In a healthy adult, resistance is approximately 0.2 kPa/l/s (2 cmH$_2$O/l/s) during mouth breathing. (Note that in lung function testing, pressures are usually measured in kPa or cmH$_2$O: 1 kPa = 7.5 mmHg = 10.20 cmH$_2$O.) The magnitude of airways resistance is influenced by several factors, the most important of which is the calibre of the airways. The narrower the airways, the higher their resistance and the greater the pressure change (and hence the work of breathing) required to drive the airflow.

The calibre of the airways may be affected by physical, chemical or neural factors, including the following.

(a) Changes in lung volume: resistance falls during inspiration and rises during expiration due to the variable intrathoracic pressure surrounding the airways (see Fig. 5.3.9).
(b) Smooth muscle tone: as discussed previously, sympathetic stimulation causes bronchodilation and a fall in resistance, while parasympathetic stimulation has the reverse effect.
(c) Airway obstruction: resistance increases if the airways become partially blocked by secretions, scar tissue or external compression (e.g. in chronic bronchitis or by neoplasm or enlarged thyroid).

DYSPNOEA

One of the most important symptoms of lung disease is that of dyspnoea. Dyspnoea is a subjective feeling of 'air hunger' or shortness of breath which occurs when the demand for ventilation is out of proportion to the patient's ability to respond to that demand and, as a result, breathing becomes difficult, laboured and uncomfortable.

Abnormalities of both the lung and chest wall may increase the demand for ventilation (e.g. if there is mismatching of the air and blood supply in the lung), and decrease the ability to respond to this demand (due to abnormalities of resistance or compliance).

Since there are no objective signs of dyspnoea, which is something that only the patient can feel, it is very difficult to assess how severe it is. However, standard questionnaires may be used to grade breathlessness according to how far a person can walk on the level or upstairs without pausing for breath. These can be of particular value when following a patient's progress.

To experience difficulty in breathing is frightening. In severe cases, for example in asthma or acute pulmonary oedema, the patient is likely to be very distressed and may panic. If they are able, such individuals will, almost certainly, want to sit upright, perhaps leaning slightly forward and supporting themselves against another piece of furniture, for example the back of another chair or a bed-table. This upright position maximizes their breathing efforts (see Body position and ventilation, p. 486).

Severely dyspnoeic patients expend a great deal of energy in their fight to breathe effectively and soon become very weak and tired. They need to use their accessory muscles of respiration to aid ventilation and will probably breathe through their mouths, since resistance to airflow is less through the mouth than through the nose. Continual mouth breathing dries the oral mucous membranes and so such patients require frequent mouth care as soon as they are able to tolerate this.

The acute stress of dyspnoea is produced by sympathetic nervous stimulation (see Chap. 2.4), and the signs and symptoms associated with this, plus the physical exertion of breathing produce a clinical picture of a frightened, tired patient, fighting to breathe. He may, for example, be pale or flushed, cold and clammy, or hot and sweating. His respiratory rate and heart rate will be raised and his pupils may be noticeably dilated.

It is essential that nurses caring for dyspnoeic patients take effective steps to relieve their physical problems and to prevent or alleviate unnecessary mental distress. The patient can be helped by being supported in a comfortable, upright sitting position and by being encouraged to relax and to breathe less rapidly and more deeply, since this will tend to maximize alveolar ventilation. He will be reassured, and hence more able to relax, by not being left alone, by appropriate explanations of what is being done medically to help him and, of course, by experiencing evidence of this. In every situation it is immensely important that by their words and actions, those caring for a patient instil in him a confidence in their ability to help him.

Actual medical treatment will depend on the cause of the dyspnoea and may include oxygen therapy, bronchodilator drugs in asthma and chest aspiration when pleural effusion is present.

INSPIRATORY FAILURE

Like all skeletal muscles, the respiratory muscles may fatigue if their workload is too high. During severe respiratory disease, when the work of breathing is greatly increased, the respiratory muscles may not be able to sustain the necessary effort required to maintain adequate gas exchange and respiratory failure may occur, requiring artificial ventilation until recovery occurs. Mechanical ventilation of the lungs will also have to be employed if the respiratory muscles are paralysed by drugs (e.g. tubocurarine), disease (poliomyelitis) or trauma to the respiratory centre.

Mechanical ventilation
Artificial ventilation is most commonly performed using a machine that intermittently applies a positive pressure to the lungs (IPPV) during inspiration. Expiration is achieved by passive recoil of the lungs. Inflation pressures are rarely allowed to exceed 3 kPa (30 cmH$_2$O) since higher pressures predispose to lung rupture (pneumothorax) and may impede venous return. Frequent determinations of blood gases and observations on the patient help determine optimum settings on the ventilator. Suctioning of the tracheal tube, by which the patient is generally attached to the ventilator, may be necessary to ensure that the airway remains free of excess secretions.

Emergency resuscitation
If respiration suddenly ceases (**respiratory arrest**), immediate action is required to re-establish the O$_2$ supply to the brain. It must first be ascertained whether the heart is still beating, without which no O$_2$ can be transported to the brain. The resuscitation techniques of mouth-to-mouth respiration and external cardiac massage are described in Chap. 4.2.

Assessment of lung function

Lung function tests are being increasingly performed in patients of all ages with respiratory disease. Some tests, such as peak flow rate, are simple enough to perform anywhere, whereas others require complex apparatus and highly skilled staff and are therefore restricted to pulmonary function laboratories.

These investigations require the patient's full participation and often need to be repeated in order to obtain accurate results. If the patient is anxious, he is less likely to be able to cooperate fully. It is therefore important that he is prepared in advance, so that he understands what to expect and how he can help.

Lung function tests are rarely diagnostic in their own right since the same functional abnormalities can occur in several different diseases. However, they can provide valuable information on the type and severity of the abnormality and enable a subject's progress, response to therapy or fitness for surgery to be assessed objectively. The measurement of arterial blood gases gives the most pertinent information on lung function and is described in a later section. A brief description of some of the commoner methods of assessing lung and airway function is given below.

LUNG VOLUMES AND CAPACITIES

As a result of the negative intrapleural pressure surrounding them, the lungs still contain a considerable volume of air at the end of a normal expiration, commonly known as the functional residual capacity (FRC). This cannot be assessed by spirometry, which measures only *changes* in volume not absolute values, but can be measured using a gas dilution technique (West, 1977) or by whole body plethysmography (DuBois *et al.*, 1956a).

Once the FRC is known, all standard lung volumes can be computed by adding or subtracting the appropriate inspired or expired volumes measured, using a spirometer. This consists of a drum inverted into a tank of water. With his nose clipped, the subject breathes in and out of the tube which passes up inside the apparatus, the drum moves down and up respectively. The drum is usually counterbalanced so that inspiration causes an upward deflection of the pen, while expiration causes a downward deflection. All respiratory movements are recorded by the pen on a variable speed kymograph, enabling volume changes to be measured directly from the chart. A spirogram showing differing depths of inspiration and expiration is shown in Fig. 5.3.14.

If, at the end of normal expiration, the subject breathes in as deeply as possible, he will inflate his lungs up to the **total lung capacity (TLC)**. The volume of air that is drawn into the lungs beyond that already present at end-expiration is called the **inspiratory capacity (IC)** and usually amounts to about 3000 ml. The TLC (= FRC + IC) of a healthy young adult male is approximately 6000 ml. Measurement of TLC gives the best indication of lung size and is particularly useful in diseases which cause a reduction in lung volume,

such as pulmonary fibrosis. An increase in TLC is most frequently associated with obstructive airway disease, in which gas becomes trapped in the lungs behind narrowed airways.

If at the end of normal expiration the subject breathes out as far as possible, a further 1000–1500 ml of air may be expelled from the lungs by active contraction of the abdominal and other expiratory muscles. This is known as the **expiratory reserve volume (ERV)**. However, no matter how much effort is exerted, some air will always remain in the lungs at the end of a maximal expiratory effort. This is known as the **residual volume (RV)** (RV = FRC − ERV) and normally amounts to approximately 1500 ml. The RV tends to be decreased in pulmonary fibrosis, where the lung is stiff and tends to recoil to a smaller volume, but increased in the presence of airway obstruction (due to gas trapping), or as the result of respiratory muscle weakness.

The total volume of air that an individual can forcibly expel from his lungs following a maximal inspiratory effort is called the **vital capacity (VC)** (VC = IC + ERV). In a normal adult, the VC is about 4500 ml. The VC is a measure of an individual's ability to inspire and expire air (i.e. of the maximum **stroke volume** of the lungs), and is determined primarily by the strength of the respiratory muscles and the amount of effort that is required to expand the lungs and thoracic cage.

ASSESSMENT OF AIRWAY FUNCTION

Although the resistance of the airways can be measured directly by plethysmography, the technique is complicated and the equipment costly (DuBois *et al.*, 1956b). Consequently, airway function is most commonly assessed by measuring forced expiratory flow rates. Generally, the higher

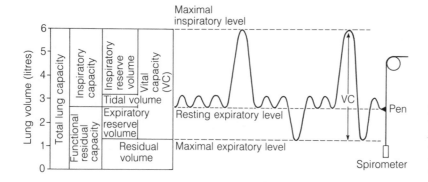

Figure 5.3.14 Lung volumes. Note that only *changes* in lung volume can be measured using the spirometer. Other methods are required for the assessment of functional residual capacity

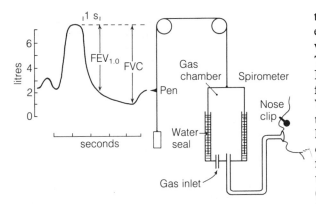

Figure 5.3.15 Measurement of timed forced expiratory volume ($FEV_{1.0}$) and forced vital capacity (FVC)

the resistance the slower that air can be forced out of the lungs.

Peak flow
The simplest of such measurements is peak flow rate (PFR), in which an individual simply takes a full inspiration and blows out as forcibly as possible into an instrument that registers the maximal rate of exhalation (i.e. peak flow rate) in litres per minute on a dial. PFR readings are usually taken at intervals in order to monitor a patient's progress, and nurses are often responsible for supervising the patient whilst taking this measurement. It is essential that the patient understands that it is not the *volume* of air expired that is crucial, but the greatest rate/force with which he can expire it!

Forced vital capacity
Alternatively, by using a spirometer and asking

the subject to inhale as deeply as possible and then exhale as far and as forcibly as possible, a forced vital capacity (FVC) can be recorded (Fig. 5.3.15). The volume exhaled in the first second is called the $FEV_{1.0}$. Normally about 80% of the FVC can be forcibly expired in 1 s, i.e. the $FEV_{1.0}/FVC = 0.8$. Variations in this ratio are often helpful in distinguishing obstructive from restrictive types of lung disease (Fig. 5.3.16). In restrictive lung diseases such as kyphoscoliosis and pulmonary fibrosis, inspiration is limited by weakness of the respiratory muscles or reduced compliance (increased stiffness) of the lung or chest wall, thereby reducing the VC. Such lungs also tend to recoil to lower volumes, resulting in reduction of FRC and RV, and hence an overall fall in TLC. However, the increased traction applied to the outer airway walls holds them more widely open, thereby facilitating expiratory air flow and resulting in an $FEV_{1.0}$ which, while reduced compared to normal values (due to the diminished total volume of gas in the lungs), is normal or even elevated when related to the subject's actual FVC. (Fig. 5.3.16c). A similar pattern is seen in patients who have lost some of their lung tissue following pneumonectomy.

In obstructive lung disease (bronchitis, asthma), the FVC is also reduced, primarily because airway narrowing or closure during expiration limits the amount of gas able to leave the lungs. Hence TLC is typically abnormally large due to the gas trapping. However, the increased airways resistance (or decreased elastic recoil as in emphysema) causes a much greater reduction in $FEV_{1.0}$ than FVC, giving a low $FEV_{1.0}/FVC$ ratio (Fig. 5.3.16b).

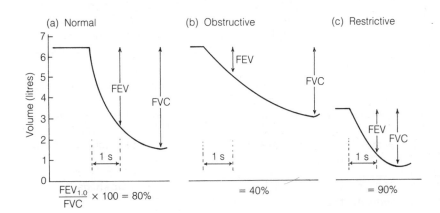

Figure 5.3.16 Normal and abnormal patterns of forced expiration

BODY POSITION AND VENTILATION

Breathing tends to be most efficient when the person is sitting or standing since FRC is approximately 0.5 litres greater when in an upright position than when lying down. This is because the doming movement of the diaphragm into the chest is less when upright than it is when supine, and because the pulmonary blood volume also decreases on moving from lying down to an upright position.

Unless a patient's condition contraindicates an upright position, as for example in unconsciousness or shock, he should be helped and encouraged to sit comfortably and, where feasible, to stand and take gentle exercise, in order to minimize the risk of chest infection resulting from respiratory stasis.

Positioning is of particular importance for people who are dyspnoeic. Here, a supported, upright sitting position maximizes their breathing efforts.

PERFUSION OF THE LUNGS

The pulmonary circulation begins at the main pulmonary artery which receives mixed venous blood pumped by the right ventricle. This artery then branches and divides, like the bronchial tree, into a series of pulmonary arterioles and finally into the dense capillary network that surrounds the alveoli. Following gas exchange across the pulmonary membrane, the blood is collected from the capillary bed into small pulmonary veins which eventually unite to form the four large pulmonary veins which drain into the left atrium.

Pressures within the pulmonary circulation

The primary function of the pulmonary circulation is to bring the entire cardiac output into intimate contact with alveolar air so that gas exchange can occur. This is facilitated not only by the structural arrangements of alveoli and pulmonary capillaries but by the very low resistance of the pulmonary circulation. This means that considerably less pressure has to be exerted to pump blood round the pulmonary circuit than around the systemic circulation. Indeed, the systolic/diastolic blood pressure in the pulmonary

artery is only about 25/8 mmHg (3.3/1 kPa), compared with 120/80 mmHg (16/10.6 kPa) in the aorta. Mean pulmonary capillary pressure is normally about 15 mmHg (2 kPa). Since the osmotic pressure of plasma proteins is 25 mmHg (3.3 kPa), there is a net force of about 10 mmHg (1.3 kPa) keeping fluid within the capillaries and preventing the formation of tissue fluid within the alveoli.

PULMONARY OEDEMA

If pulmonary capillary pressure rises above 25 mmHg (3.3 kPa), the net flow of fluid out of the capillaries into the interstitial space will cause thickening of the pulmonary membranes. Fluid may also enter the alveoli causing dyspnoea and hypoxia. This condition is known as pulmonary oedema or cardiac asthma (see also Chap. 4.2). The patient's distress is increased when lying down (**orthopnoea**) because vital capacity is reduced and the greater pulmonary blood volume increases pulmonary congestion. Such patients need to be nursed in a relatively upright position, even during sleep, in order to avoid paroxysmal nocturnal dyspnoea.

A rise in left atrial pressure (such as may occur in mitral stenosis or during the development of left heart failure) will precede the development of pulmonary oedema, and may be detected by measuring the pulmonary capillary wedge pressure (PCWP). This closely reflects left atrial pressure due to the low resistance of the pulmonary circuit and is measured by passing a fine catheter from the vena cava through the right side of the heart and into the pulmonary circulation until it becomes 'wedged' due to the diminishing calibre of the vessels (see also Chap. 4.2).

A severe acute onset of pulmonary oedema may lead to death from respiratory failure within a couple of hours if left untreated. Management is usually aimed at treating the primary cause and may include

(a) administration of digitalis to increase contractibility of the heart
(b) increased concentrations of inspired O_2
(c) fluid restriction and diuretic therapy
(d) nursing the patient in the upright position.

Ventilation–perfusion relationships

For adequate gas exchange in the lung, not only

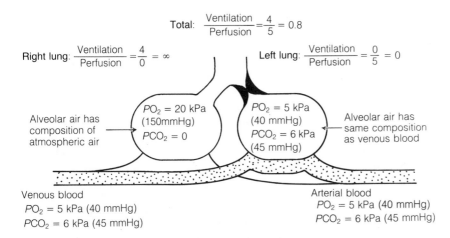

Total: $\dfrac{\text{Ventilation}}{\text{Perfusion}} = \dfrac{4}{5} = 0.8$

Right lung: $\dfrac{\text{Ventilation}}{\text{Perfusion}} = \dfrac{4}{0} = \infty$

Left lung: $\dfrac{\text{Ventilation}}{\text{Perfusion}} = \dfrac{0}{5} = 0$

Alveolar air has composition of atmospheric air

$PO_2 = 20$ kPa (150mmHg)
$PCO_2 = 0$

$PO_2 = 5$ kPa (40 mmHg)
$PCO_2 = 6$ kPa (45 mmHg)

Alveolar air has same composition as venous blood

Venous blood
$PO_2 = 5$ kPa (40 mmHg)
$PCO_2 = 6$ kPa (45 mmHg)

Arterial blood
$PO_2 = 5$ kPa (40 mmHg)
$PCO_2 = 6$ kPa (45 mmHg)

Figure 5.3.17 Extreme imbalance between ventilation and perfusion. Note that despite an adequate overall ventilation–perfusion ratio of 0.8, all the ventilation goes to one lung through which there is no perfusion, whereas all the blood goes to the other lung where there is no ventilation. Consequently no gas exchange occurs. Such a situation would obviously be incompatible with life (values are given in SI units, with mmHg in parentheses)

must sufficient air and blood be delivered to the alveoli each minute, but they must be delivered in the right proportions. In a healthy adult at rest, alveolar ventilation is about 4.0 litres/min, while pulmonary capillary blood flow is about 5 litres/min. Consequently the ventilation–perfusion ratio at rest is

$$\frac{\text{alveolar ventilation}}{\text{pulmonary capillary blood flow}} = \frac{4\,\mathrm{l/min}}{5\,\mathrm{l/min}} = 0.8$$

During severe exercise, when there is a greater increase in ventilation (up to 15-fold) than cardiac output (up to 6-fold), this ratio may rise to about 2.0, but it is generally kept fairly constant under widely varying conditions.

However, it is not enough to have an acceptable overall ratio of ventilation to perfusion. If 5 litres of blood flowed through a completely unventilated lung each minute, while 4 litres of air were delivered to the other unperfused lung, the overall ratio would still be 0.8 and yet no gas exchange could occur (Fig. 5.3.17)

Even in the healthiest lungs, some regional differences in ventilation–perfusion ratios occur. For example, both ventilation and perfusion (blood flow) are preferentially distributed to the base of the lung due to the effect of gravity. Certain compensatory measures are available to facilitate matching of ventilation and perfusion. These depend on the sensitivity of the smooth muscle in the bronchioles and pulmonary arterioles to local changes in gas tensions. For example, gas in relatively unventilated areas of the lung will tend to have an elevated PCO_2 which stimulates bronchodilation, enhancing ventilation to the area, whereas

a decrease in PO_2 stimulates pulmonary vasoconstriction, thereby diverting some of the blood flow away from the area.

In health, adequate gas exchange occurs despite regional differences, and blood leaving the lungs is virtually in equilibrium with alveolar gas. However, in the presence of lung disease, the distribution of ventilation and perfusion often becomes far more uneven, which diminishes the efficiency of gas exchange and increases the work of breathing.

Inequalities of the ventilation–perfusion ratio may occur for several reasons, some of which are summarized below.

AREAS OF HIGH VENTILATION–PERFUSION RATIO

If intrathoracic pressure should exceed pulmonary artery pressure, either due to a rise in intrathoracic pressure (e.g. during IPPV) or a fall in pulmonary artery pressure (shock, haemorrhage), the pulmonary capillaries will be flattened, with subsequent reduction of blood flow to an area. Gas exchange will be severely impaired in such ventilated but relatively unperfused areas, which are referred to as the **alveolar dead space**.

AREAS OF LOW VENTILATION–PERFUSION RATIO

Certain areas of the lung may take longer to fill and empty than others, for example if they are supplied by obstructed or narrowed airways. They therefore receive a disproportionately small share of the ventilation so that only limited gas exchange can occur. Such regions act as a physiological right-to-left shunt, enabling venous blood to pass through the lungs and return to the heart virtually unchanged, resulting in a fall in arterial PO_2.

Mismatching of ventilation and perfusion is the commonest cause of hypoxia in lung disease. It may or may not be accompanied by hypercapnia (increased arterial PCO_2), depending on the individual's ability to increase his ventilation, but it always increases the work of breathing.

TRANSPORT OF GASES AROUND THE BODY

In an earlier section, it was seen how the structure of the lung is well suited to its functions in that it provides a vast surface area for gas exchange together with a very thin pulmonary (alveolar–capillary) membrane across which gases can easily diffuse. Similar structural arrangements are available in the tissues where the thin-walled capillaries branch and divide to bring the arterial blood into intimate proximity with the cells, being separated from the intracellular fluid by only the

capillary membrane, the interstitial fliud and the cell membrane, none of which normally presents any barrier to diffusion. Gas exchange cannot occur across the thicker wall blood vessels. Consequently, the composition of systemic arterial or venous blood remains unchanged as it travels to the tissues or to the lungs, respectively.

It has already been mentioned that the rate of diffusion of a gas is proportional not only to the surface area and thinness of the membrane but also to the partial pressure gradient and solubility of that gas. The constant consumption of O_2 and production of CO_2 in the tissues combined with ventilation of the lungs with fresh air provides the necessary pressure gradients (shown in Fig. 5.3.18) to ensure adequate diffusion in the lungs and tissues. However, gas exchange would be seriously limited if O_2 had to be transported around the body in simple solution since only 0.23 ml O_2 can dissolve in each litre of blood for each kPa change in PO_2 (or 0.03 ml O_2/litre blood per mmHgPO_2). Consequently, arterial blood with a PO_2 of 13 kPa (100 mmHg) can only carry 3 ml O_2/l blood in simple solution. Even if the tissues were capable of extracting all this O_2, with a cardiac output of 5 l/min, only 15 ml O_2 would be supplied each

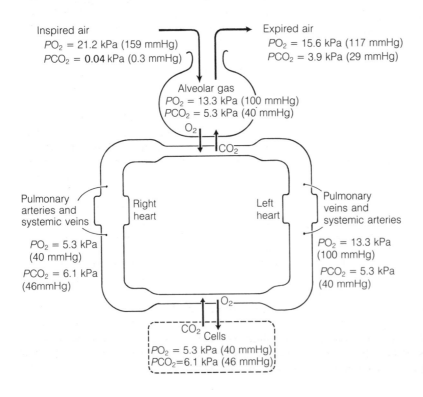

Inspired air
PO_2 = 21.2 kPa (159 mmHg)
PCO_2 = 0.04 kPa (0.3 mmHg)

Expired air
PO_2 = 15.6 kPa (117 mmHg)
PCO_2 = 3.9 kPa (29 mmHg)

Alveolar gas
PO_2 = 13.3 kPa (100 mmHg)
PCO_2 = 5.3 kPa (40 mmHg)

O_2

CO_2

Pulmonary arteries and systemic veins

Right heart

Left heart

Pulmonary veins and systemic arteries

PO_2 = 5.3 kPa (40 mmHg)
PCO_2 = 6.1 kPa (46mmHg)

PO_2 = 13.3 kPa (100 mmHg)
PCO_2 = 5.3 kPa (40 mmHg)

O_2

CO_2 Cells
PO_2 = 5.3 kPa (40 mmHg)
PCO_2=6.1 kPa (46 mmHg)

Figure 5.3.18   Summary of O_2 and CO_2 pressures in inspired and expired air and throughout the body. (Partial pressures are given in SI units with conversion to mmHg in parentheses.) Venous blood arriving at the lungs is 'arteriolized' as it passes through the pulmonary capillary network – picking up O_2 and losing CO_2. It is then pumped back to the tissues for further gas exchange

minute, whereas even at rest the body requires 250 ml O_2/min.

Transport of CO_2 would be less limited by solubility, since CO_2 is approximately 20 times more soluble than O_2. Hence, despite the smaller pressure gradients for PCO_2 than PO_2 at lung and tissue level (Fig. 5.3.18) – approximately 0.8 kPa (6 mmHg) for PCO_2 compared with 8 kPa (60 mmHg) for PO_2 – similar quantities of both CO_2 and O_2 can diffuse. However, since CO_2 dissolves in water to form carbonic acid, the carriage of large quantities of CO_2 in simple solution would cause a dangerous increase in the acidity of the blood and body fluids. Consequently, both O_2 and CO_2 require fairly complex mechanisms to facilitate their transport around the body. These will now be considered.

Transport of oxygen

Under normal conditions, about 99% of O_2 in the blood is bound to haemoglobin (Hb), the remainder being carried in simple solution. Despite the small contribution of dissolved O_2 to the total O_2 content of the blood, it plays a vital role in that it is this dissolved O_2 that determines the PO_2 of the blood and maintains the necessary pressure gradients for diffusion to occur, and not that bound to Hb, which is no longer 'free' to exert a pressure.

HAEMOGLOBIN

Haemoglobin is a unique conjugated protein found in red blood cells (see also Chap. 4.1). Each molecule of Hb consists of 4 haem groups (the iron-containing pigment which gives blood its characteristic colour) attached to 4 polypeptide chains (which make up the protein, globin). Each haem group is capable of combining with 1 molecule of O_2, by a process known as oxygenation, to give a bright red compound, oxyhaemoglobin (HbO_2) which gives arterial blood its characteristic colour. By contrast, reduced haemoglobin (Hb) is a purple compound (hence the colour of venous blood). The normal Hb complement ranges from 12 to 15 g Hb/dl blood in women, and from 13 to 18 g Hb/dl blood in men. (Hb concentration is expressed per decilitre of blood rather than per litre by clinical convention.) Each gram of Hb is capable of combining with up to 1.34 ml O_2. Consequently, with an Hb comple-

ment of 15 g/dl blood, the total amount of O_2 that could be carried (known as the **O_2 capacity** of the blood) would be around 20.1 ml/dl blood (15 × 1.34 ml). (In SI units, normal Hb content is 2.2 mmol/l blood, and since each molecule of Hb can combine with 4 molecules of O_2, the O_2 capacity would be 8.8 mmol/l (1 mmol O_2 = 22.4 ml).) By contrast, the O_2 capacity of an anaemic individual with only 7.5 g Hb/dl will be reduced to 10 ml O_2/dl blood.

However, the actual quantity of O_2 in the blood (**O_2 content**) is determined not only by the Hb complement but also by the PO_2 of the blood. On exposure to increasing levels of PO_2, the O_2 content of the blood will gradually rise as more and more O_2 combines with Hb. At a certain PO_2, when O_2 content equals O_2 capacity, the Hb will be unable to take up any more O_2 and is said to be fully (100%) saturated, i.e.

$$\text{Hb saturation} = \frac{O_2 \text{ content}}{O_2 \text{ capacity}} \times 100$$

THE OXYGEN DISSOCIATION CURVE

The degree to which O_2 combines with Hb under varying conditions can be measured in the laboratory by exposing blood in several containers to gas mixtures with varying concentrations of O_2. After equilibration, the Hb saturation is calculated by dividing the O_2 content of each sample by the O_2 capacity of the blood. The latter is taken as the maximum O_2 content obtained in any of the samples despite any further increases in PO_2. For each blood sample, Hb saturation is plotted against the PO_2, as shown in Fig. 5.3.19, where Hb saturation is plotted on the left-hand axis.

It can be seen that the relationship between Hb saturation and PO_2 is not a linear one, but that the curve is sigmoid (S-shaped). The reason for this is that, although each of the 4 haem groups in a Hb molecule can combine with 1 molecule of O_2, they do so with varying affinities. The first haem group combines with relative difficulty (but conversely 'holds on' to its O_2 more strongly when a fall in PO_2 occurs); the second and third haem groups have a far greater affinity for O_2, as seen from the steep part of the curve where Hb saturation increases rapidly from 25% to 75% for a relatively small change in PO_2; whereas the fourth haem group combines with the greatest difficulty.

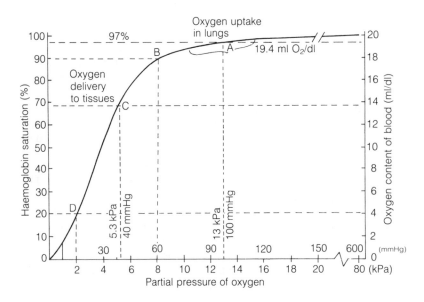

Figure 5.3.19 The oxygen–haemoglobin dissociation curve. This oxygen dissociation curve applies when pH is 7.4, PCO_2 is 5.3 kPa (40 mmHg) and blood is at 37°C. The total blood oxygen content is shown, assuming a haemoglobin concentration of 15 g/dl blood (i.e. O_2 capacity of 20 ml/dl)

The shape of this curve is physiologically advantageous for several reasons. The blood arriving in the lungs has a PO_2 of 5.3 kPa (40 mmHg) or lower, and is exposed to the alveolar PO_2 of 13.3 kPa (100 mmHg). Consequently, O_2 diffuses along the pressure gradient across the alveolar–capillary membrane and into the plasma. The rise in plasma PO_2 creates a pressure gradient between it and the inside of the erythrocytes, enabling O_2 to diffuse into the red blood cell. However, PO_2 within the erythrocyte rises more slowly since the Hb rapidly loads O_2, thereby removing it from free solution. This maintains the pressure gradient and facilitates continuing rapid diffusion of O_2 until the Hb is fully saturated. As PO_2 within the erythrocyte increases from 5.3 kPa (40 mmHg) to 8 kPa (60 mmHg), there is rapid loading of O_2, with Hb saturation increasing from 70% to 90% (this is on the steep portion of the O_2–Hb dissociation curved). Thereafter relatively less O_2 is taken up per unit change in PO_2, with 97% Hb saturation not being achieved until PO_2 reaches about 13.3 kPa (100 mmHg). This flat upper portion of the curve provides an excellent safety factor in the supply of O_2 to the tissues since even if alveolar PO_2 should fall to 8 kPa (60 mmHg), as might occur in lung disease or when breathing at altitude, 90% of the Hb leaving the lungs will still be saturated with O_2.

As the blood enters the tissues, still with a PO_2 of 13.3 kPa (100 mmHg) in a healthy individual, it is exposed to a PO_2 of 5.3 kPa (40 mmHg). Oxy-

gen therefore diffuses from the plasma across the capillary membrane, the interstitial fluid and the cell membrane. The resultant drop in plasma PO_2 creates a pressure gradient between it and the erythrocyte so that the oxyhaemoglobin begins to dissociate. The release of previously bound O_2 into solution enables the pressure gradient to be maintained, facilitating diffusion of O_2 into the tissues until an equilibrium is achieved, and the PO_2 of blood equals that of the tissues. The level of tissue PO_2 (5.3 kPa) falls on the steep portion of the O_2–Hb curve (point C, Fig. 5.3.19), meaning that O_2 is readily released to the tissues. If the level of tissue activity increases, its PO_2 may fall as low as 2 kPa (15 mmHg) (point D, Fig. 5.3.19). This will enable the Hb to release 80% of its O_2 to the tissues, demonstrating a considerable reserve capacity above resting levels. Oxygen is loosely bound to Hb (steep curve) down to about 1.3 kPa (10 mmHg); below this the affinity of Hb for O_2 increases (flatter curve). However, PO_2 seldom falls this low except in working muscles, where a special O_2-carrying protein, called myoglobin, is capable of extracting all the O_2 that the blood delivers.

The venous blood then returns to the lung, where the O_2 released in the tissues can be replaced by exposure of the pulmonary blood to alveolar air. Of course, the lower the PO_2 of the blood returning to the lung, the greater the pressure gradient will be for O_2 diffusion and the faster the loading of the Hb will be.

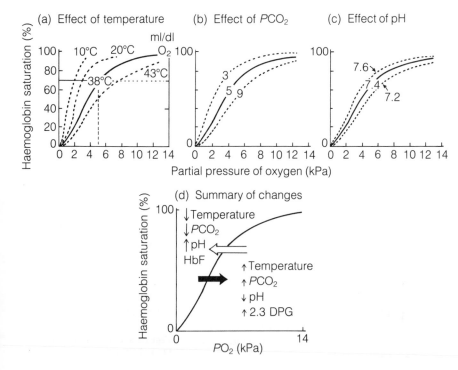

Figure 5.3.20 Factors influencing the position of the O_2 dissociation curve. A shift of the curve to the right ($\rightarrow$) facilitates unloading of O_2. Thus, at any given PO_2, more O_2 is released to the tissues if there is an increase in temperature or PCO_2 or a fall in pH (such as may occur in exercising muscle). Conditions in the lung (fall in temperature, PCO_2 and rise in pH) shift the curve to the left ($\Leftarrow$), thereby facilitating loading of Hb with O_2 at any given PO_2

FACTORS INFLUENCING THE OXYGEN–HAEMOGLOBIN DISSOCIATION CURVE

The affinity of Hb for O_2 at any given PO_2 is influenced by several factors which assist in maintaining a constant intracellular PO_2 under varying conditions. Increases in the amount of CO_2 and hydrogen ions (H^+) reduce the ability of Hb to bind O_2. Thus, for any given PO_2, blood entering the tissues with a PCO_2 of 6.1 kPa (46 mmHg) will release more of its O_2 than blood with a PCO_2 of 5.3 kPa (40 mmHg) (Fig. 5.3.20). This is sometimes expressed by saying the curve has 'shifted to the right' with respect to the standard O_2–Hb dissociation curve (see Fig. 5.3.19), which is itself based on arterial blood with a PCO_2 of 5.3 kPa and pH of 7.4.

In the tissue capillaries, CO_2 and H^+ bind with Hb, facilitating the release of O_2, whereas in the lungs release of CO_2 and H^+ from Hb facilitates O_2 uptake. A rise in temperature also reduces the O_2-binding ability of Hb, which again facilitates O_2 delivery to active tissues.

A substance called **2,3-diphosphoglycerate** (2,3-DPG), which is a product of red blood cell metabolism, combines reversibly with Hb, causing it to release more of its O_2 at any given PO_2. An increase in 2,3-DPG concentration occurs in response to chronic hypoxia (e.g. in anaemia, at altitude) and also during pregnancy, which facilitates release of O_2 across the placenta to the fetus. Maternal–fetal O_2 transfer is also facilitated by the fact that the fetus has a slightly different type of Hb, known as HbF, which has an increased affinity for O_2. This is gradually replaced by adult Hb (HbA) during the first year of life.

The gas carbon monoxide (CO) is a poison because it binds very tightly with Hb to form carboxyhaemoglobin. Since CO binds to exactly the same sites on the Hb molecule as O_2, but with a far greater affinity, it prevents O_2 transport and can lead to death.

CHANGES IN OXYGEN CONTENT

So far we have only considered the changes in Hb saturation as the blood is exposed to changes in PO_2 around the body. However, if the Hb complement of the blood is known, the O_2 capacity can be calculated (Hb g/dl $\times$ 1.34 ml O_2). Since O_2 capacity is equivalent to 100% Hb saturation, an

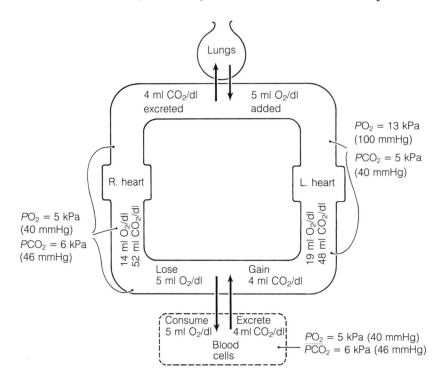

Figure 5.3.21 Transport of O_2 and CO_2

axis representing O_2 content of the blood can be added to the Hb–O_2 dissociation curve (see Fig. 5.3.19, right-hand axis). Thus, with an Hb complement of 15 g/dl, O_2 capacity (i.e. 100% saturation) = 20 ml O_2/dl, whereas 50% saturation represents an O_2 content of 10 ml/dl blood, and so on. Changes in O_2 content (assuming a Hb of 15 g/dl) as blood travels around the body are summarized in Fig. 5.3.21. It can be seen that under resting conditions, blood returning to the lung contains about 14 ml O_2/dl blood at a PO_2 of 5.3 kPa (40 mmHg). During passage through the lung, an additional 5 ml O_2/dl are loaded, due to the rise in PO_2, which are then released at tissue level. During exercise the Hb may release up to 15 ml/dl O_2 (if PO_2 falls to 1.3 kPa), and therefore returns to the lung with only 4 ml/dl. In order to meet this increased O_2 consumption by the tissues, there must be a marked increased in both ventilation and cardiac output.

A muscle requires up to 50 times more O_2 during exercise than when at rest. This is achieved by

(a) an increase in cardiac output (and pulmonary blood flow) from 5 l/min to 30 l/min (a sixfold increase)

(b) redistribution of blood, with three times as much of the cardiac output going to the active muscle than when at rest

(together these mechanisms increase muscle blood flow 18-fold)

(c) three times as much O_2 being extracted by the muscle, with Hb releasing 15 ml O_2/dl blood during exercise instead of the 5 ml/dl blood released at rest.

This gives a total increase in blood flow to the muscle of 54-fold, which is accompanied by up to a 30-fold increase in ventilation to supply the necessary additional O_2.

Measurement of arterial blood gases
Of all the laboratory tests that are relevant to respiratory diseases, measurement of arterial blood gases is the most important. Knowledge of the levels of O_2, CO_2 and H^+ concentration (pH) in the arterial blood can help to determine the nature and severity of the disease and the response to various forms of energy. The alveolar–arterial O_2 difference (A–aDO_2), which in health should be very small, may be measured to assess the efficiency of gas exchange across the pulmonary membrane.

By using suitable electrodes, PO_2, PCO_2 and pH, or $[H^+]$, of blood can be determined electronically using very small samples. Blood samples may be taken from the radial or brachial arteries and should immediately be taken to the laboratory in a heparinized, iced receptacle, ensuring that no contamination with room air occurs.

Oxygen and CO_2 electrodes have been built into the tips of catheters so that a continual reading of arterial blood gases can be obtained. Recently, considerable advances have been made in the development of skin electrodes for measuring transcutaneous blood gases. Despite their limitations, these are proving to be valuable under certain circumstances and avoid the need to take arterial blood samples.

Respiratory failure
Respiratory failure is said to occur when the lung fails to oxygenate the arterial blood adequately and/or prevent undue retention of carbon dioxide. There is no absolute definition of the levels of arterial blood gases that indicate respiratory failure, but in general an arterial PO_2 of less than 8 kPa (60 mmHg) or a PCO_2 of more than 6.7 kPa (50 mmHg) are values that may be used.

HYPOXIA

Whenever tissues do not get sufficient O_2 to meet their needs or are unable to utilize it, the condition is known as hypoxia. This may be due to any of the following, e.g.:

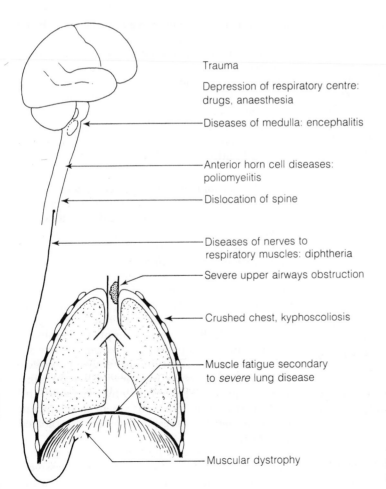

Trauma

Depression of respiratory centre: drugs, anaesthesia

Diseases of medulla: encephalitis

Anterior horn cell diseases: poliomyelitis

Dislocation of spine

Diseases of nerves to respiratory muscles: diphtheria

Severe upper airways obstruction

Crushed chest, kyphoscoliosis

Muscle fatigue secondary to *severe* lung disease

Muscular dystrophy

Figure 5.3.22 Potential causes of alveolar hypoventilation

Any factor which reduces arterial PO_2. Factors such as breathing at altitude or alveolar hypoventilation will reduce the pressure gradient between the lungs and tissues and hence impede diffusion.

Alveolar hypoventilation is most commonly caused by diseases outside the lung (Fig. 5.3.22), indeed the lungs may be completely normal. However, in the presence of very severe respiratory disease, the work of breathing may be increased to such an extent that the respiratory muscles fatigue and are no longer capable of sustaining an adequate level of alveolar ventilation.

Reduced blood supply to the tissues. This may result from a fall in cardiac ouput (e.g. during heart failure or shock), and is known as ischaemic or stagnant hypoxia.

Interstitial oedema. This increases the distance across which gases must diffuse and may result in local areas of hypoxia, necrosis and possibly gangrene.

Anaemic hypoxia. Oxygen delivery to the tissues may be impeded if there is a reduction in available Hb (the various causes of which are summarized in Chap. 4.1). Severe hypoxia at rest may result from a sudden loss of available Hb (haemorrhage, carbon monoxide poisoning). In addition, individuals suffering from anaemia tend to have limited exercise tolerance and suffer from excessive tiredness.

An increased demand for gas exchange by the tissues. This may occur, for example, as a result of neoplasms or during severe exercise. If the increased demand cannot be met by the normal adjustments of the respiratory and cardiovascular systems, hypoxia will occur.

Histotoxic hypoxia. Even if the supply and composition of arterial blood are adequate, hypoxia may occur if there is inactivation of the enzyme, cytochrome oxidase, by cyanide poisoning. This prevents uptake of O_2 by the cells and rapidly leads to death.

Various compensatory changes occur in the presence of chronic hypoxia, including (a) an increase in 2,3-DPG, and (b) an increase in the number of red cells (polycythaemia) and Hb complement (unless the hypoxia is due to anaemia). Signs and symptoms of hypoxia may include: (a) systemic oedema due to pulmonary vasoconstric-

tion, giving the patient a bloated appearance; (b) rapid pulse rate; (c) cyanosis (see below); (d) increased respiratory rate if arterial PO_2 is low; (e) excessive tiredness and limited exercise tolerance; (f) various effects on the brain resulting in headache, nausea, restlessness, disturbances of visual acuity, reduced mental efficiency, and, if severe, possible stupor or coma. If the brain is totally deprived of O_2, death occurs in about 2–3 min.

Cyanosis

If arterial blood contains more than 5 g reduced Hb/dl, it will take on a bluish hue which is most clearly visible in the mucous membranes. Although central cyanosis is a common clinical sign of respiratory insufficiency, it is not a reliable guide to the degree of hypoxia since it is influenced by an individual's Hb complement. Thus an anaemic patient with only 7.5 g Hb/dl blood would have to have a PaO_2 of 3 kPa (23 mmHg) – i.e. an Hb saturation of 33% – before 5 g Hb/dl were reduced (see Fig. 5.3.19). By contrast, a polycythaemic patient will show frank cyanosis with only mild hypoxia.

Peripheral cyanosis is more likely to be seen when blood flow through the skin is slow, as for example in cold weather or when viscosity of the blood is increased. Generally, it is easier to observe cyanosis in fair-skinned individuals and in good lighting (see also Chap. 6.1).

Blue bloaters and pink puffers. Generalized pulmonary vasoconstriction in response to hypoxia is in part responsible for the marked differences seen between patients with obstructive airways disease who tend to fall into two distinct categories: either 'blue bloaters' or 'pink puffers'. The blue bloaters tend to be those whose predominant problem is one of chronic bronchitis. They do not usually complain of dyspnoea but, due to widespread yet variable airways obstruction, have very poor matching of the blood and air supply in their lungs which results in impaired gas exchange, hypoxia and cyanosis. In addition, generalized hypoxic vasoconstriction results in a large rise in pulmonary artery pressure (pulmonary hypertension) and right heart failure. The subsequent development of systemic oedema is responsible for their bloated appearance.

By contrast, the pink puffers, whose problem is primarily one of emphysema, are excessively dyspnoeic (puffing) on exercise, but by increasing their level of pulmonary ventilation, generally manage to maintain their blood gases within reasonable limits. Consequently they stay 'pink' and do not usually suffer from cor pulmonale (right heart failure secondary to pulmonary disease, as described above).

OXYGEN THERAPY

Although the administration of increased concentrations of inspired O_2 may be of great benefit in certain types of hypoxia, it may be of limited value or even potentially dangerous in other types. It is therefore essential that the cause of hypoxia is diagnosed by the doctor before O_2 is prescribed. The O_2 may be administered via: (a) face masks, (b) nasal catheters, (c) oxygen tents, or (d) endotracheal or tracheostomy tubes. The method of choice usually depends on the concentrations required and the age and comfort of the patient, for example some patients feel claustrophobic behind face masks and may prefer nasal catheters.

Patients receiving oxygen therapy require frequent nose and mouth care because although the oxygen passes through a humidifier before reaching the patient, humidification may be inadequate and breathing dry gas dries and irritates mucous membranes. Since oxygen masks fit closely over the nose and around the chin, they should be moved frequently to relieve pressure on the underlying tissues, which should be observed for inflammation. Lack of care here can result in the formation of pressure sores under a tight-fitting mask.

Oxygen therapy may be of benefit to a patient in three different ways.

To correct a decreased alveolar PO_2. Alveolar PO_2 can be increased, enabling adequate O_2 uptake by the blood despite alveolar hypoventilation or decreased barometric pressure.

To elevate alveolar PO_2. Patients suffering from a reduction in diffusing capacity across the pulmonary membrane will benefit if augmented O_2 is breathed since the increased alveolar–arterial pressure gradient will facilitate O_2 diffusion and partially compensate for the increase in membrane thickness or reduction in surface area available for gas exchange.

To increase dissolved O_2. In the presence of an anaemic crisis or CO poisoning, Hb is no longer available to transport O_2. Under these circumstances, the amount of O_2 physically dissolved in the blood will be critical. By breathing 100% O_2, PaO_2 increase to 80 kPa (600 mmHg) and at this pressure 2 ml O_2 will be dissolved per decilitre of blood. This effect can be augmented by administering O_2 at increased pressures (hyperbaric O_2 therapy), but this therapy is potentially dangerous and is limited to treating emergencies over a brief (less than 3 hours) period of time. Hyperbaric O_2 therapy is therefore rarely used to treat respiratory failure or any chronic form of hypoxia, but is valuable in treating CO poisoning and decompression sickness. It may also be used in the treatment of infections with anaerobic bacteria, such as gas gangrene, and as an adjacent to radiotherapy where the higher tissue PO_2 increases the radiosensitivity of tissues with a relatively poor blood supply.

Oxygen therapy is of very limited value if:

(a) blood is bypassing the lungs (R–L intracardiac shunt, such as may occur in 'blue babies' with congenital heart defects)
(b) there is generalized reduction in circulation
(c) there is any interference with O_2 uptake by the tissues (cyanide poisoning)
(d) there is chronic anaemia.

Hazards of oxygen therapy
Despite the numerous benefits of O_2 therapy, it is not without its dangers, which are summarized below.

(a) Oxygen is a colourless, odourless gas which supports combustion and therefore presents a fire risk.
(b) Compressed O_2 is very dry and must be humidified before reaching the patient (see above).
(c) By abolishing the hypoxia, O_2 therapy may completely suppress respiration in patients with chronic CO_2 retention (see Control of ventilation, p. 504).
(d) High concentrations of O_2 over prolonged periods may cause severe lung damage (O_2 toxicity). Inspired O_2 concentrations are therefore always kept as low as possible while maintaining reasonable blood gases.
(e) Newborn infants, especially those born prematurely, may develop fibrosis behind the lens of the eye leading to blindness if treated

with O_2 (**retrolental fibroplasia**). This condition can be prevented by constant monitoring of arterial blood gases and by keeping the arterial PO_2 below 8 kPa (140 mmHg).

Transport of carbon dioxide

Carbon dioxide is transported in the blood in three different ways.

(a) About 5% is carried in simple solution (0.6 ml CO_2 will dissolve in each decilitre of blood for every kPa of PCO_2, or 0.075 ml/dl/mmHg PCO_2).
(b) About 5% is carried in combination with Hb.
(c) Approximately 90% is transported in the form of hydrogen carbonate (bicarbonate) ions.

The constant production of CO_2 in the cells means that the PCO_2 of intracellular fluid always exceeds that of the blood flowing into the tissue capillaries (see Fig. 5.3.21). This creates the necessary pressure gradient to enable CO_2 to diffuse out of the cells into the interstitial fluid, across the capillary membrane and into the plasma, where a small quantity will dissolve in simple solution, according to the equation

$$CO_2 + H_2O \rightleftharpoons H_2CO_3$$

carbon dioxide + water $\rightleftharpoons$ carbonic acid

However, this reaction is very slow unless catalyzed by the enzyme carbonic anhydrase, of which there is very little present in the plasma, but abundant supplies inside the red blood cell. Most of the CO_2 that diffuses out from the tissues diffuses into the red blood cell, where the constant and rapid production of carbonic acid 'soaks up' the CO_2, thereby keeping the PCO_2 of the erythrocyte relatively low and maintaining a pressure gradient along which CO_2 can move.

The total amount of carbonic acid in the erythrocyte itself remains relatively low since it almost completely ionizes (dissociates) into hydrogen (H^+) and hydrogen carbonate (bicarbonate, HCO_3^-) ions. Thus

$$H_2O + CO_2 \underset{\text{anhydrase}}{\overset{\text{carbonic}}{\rightleftharpoons}} H_2CO_3 \rightleftharpoons H^+ + HCO_3^-$$

water + carbon dioxide $\rightleftharpoons$ carbonic acid

$\rightleftharpoons$ hydrogen ion + hydrogen carbonate ion

The reaction shown above is a reversible one (i.e. it can proceed in either direction) and obeys the law of mass action, in that any increases in the concentration of reacting substances on the left-hand side (i.e. CO_2) will drive the reaction towards the right, and vice versa. Consequently at tissue level, where there is a continuous addition of CO_2 into the blood, the reaction will be driven to the right, resulting in the continuous production of H^+ and HCO_3^-. (By contrast, when CO_2 leaves the blood as it passes through the lungs, the reaction will be driven to the left.)

Hydrogen carbonate ions can pass freely across the red cell membrane. Consequently, as their concentration rises there is a net diffusion of HCO_3^- out of the cell and into the plasma along the concentration gradient. Here they react with sodium chloride (NaCl) to form sodium hydrogen carbonate ($NaHCO_3$) and chloride ions (Cl^-). The HCO_3^- that remains in the erythrocyte is transported as potassium hydrogen carbonate (bicarbonate). This net movement of negative HCO_3^- into the plasma leaves the inside of the red blood cell relatively positive. Electrical neutrality is maintained by diffusion of negatively charged chloride ions (Cl^-) from the plasma into the red blood cells. This is sometimes known as the **chloride shift**. In addition, Hb is a very effective buffer which combines with and neutralizes many of the H^+ released on ionization of carbonic acid.

$$HbO_2 + H^+ \rightleftharpoons HHb + O_2$$

oxyhaemoglobin　　　　　　reduced haemoglobin

Not only does this prevent any marked rise in acidity in the red blood cell, but it assists the unloading of O_2 from oxyhaemoglobin for use by the tissues and facilitates the combination of CO_2 with Hb. The Hb can carry both O_2 and CO_2 simultaneously since CO_2 is attached to the amine groups in the globin part of the molecule while O_2 is carried on the haem group. However, reduced Hb can carry more CO_2 than oxyhaemoglobin so that the unloading of O_2 in the peripheral capillaries facilitates loading of CO_2. Approximately 5% of the total CO_2 content in blood is carried in combination with Hb, forming a compound known as **carbaminohaemoglobin**. The chemical reactions involved in CO_2 transport at tissue level are shown in Fig. 5.2.23.

When venous blood reaches the lung capillaries it is exposed to the lower PCO_2 of alveolar air so that dissolved CO_2 diffuses out of the red blood

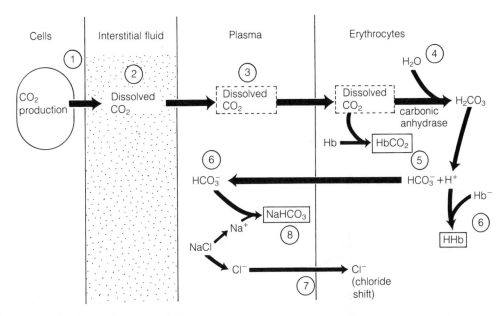

Figure 5.3.23 Summary of the chemical reactions involved in CO_2 transport at tissue level. Although there is a continuous production of CO_2, the numbers refer to the approximate sequence of events; 90% of the CO_2 is carried in the blood in the form of hydrogen carbonate (HCO_3^-) ions, 5% in simple solution, and 5% attached to haemoglobin as $HbCO_2$ (for key to symbols, see text)

cells and plasma and across the pulmonary membrane to be excreted during ventilation. The resultant fall in blood PCO_2 causes a reversal of all the processes that occur at tissue level (Fig. 5.3.24). The constant production of CO_2 in the erythrocyte keeps the PCO_2 of the red cells relatively high, thus maintaining a pressure gradient between blood and alveolar air along which CO_2 can diffuse. In this way all the CO_2 released by the tissues (about 4 ml/dl blood/min at rest, or 200 ml/min when cardiac output is 5 l/min) can be delivered into the alveoli and expired. The changes in PCO_2 and CO_2 content of the blood around the body are summarized in Fig. 5.3.21. During exercise, much larger quantities of CO_2 are produced by the tissues. However, the automatic increase in alveolar ventilation and cardiac output that occurs during exercise, together with the increased rate of diffusion of CO_2 both at tissue level and in the lungs as a result of the higher pressure gradients that are created, normally ensure that arterial PCO_2 remains relatively constant, between 4.9 kPa and 5.7 kPa (37 and 43 mmHg) in an adult.

HYPOCAPNIA

A decrease in arterial PCO_2 to below normal levels is known as hypocapnia. This will occur if alveolar ventilation is increased above metabolic needs (hyperventilation), which results in a lowering of alveolar PCO_2, a rise in the alveolar–capillary PCO_2 gradient and an increase in the rate at which CO_2 can diffuse out of the blood. Hyperventilation can cause a marked decrease in arterial PCO_2 but has a far less marked effect on PO_2 due to the shape of the O_2-dissociation curve (see Fig. 5.3.19).

Hyperventilation may result from: (a) central nervous system stimuli (pain, temperature, anxiety, salicylate poisoning); (b) hypoxia (high altitude); or (c) excess acid in the body (metabolic disorders). Any lowering of arterial PCO_2 generally decreases the stimulus to breathe, and may be followed by deep, slow breathing or even apnoea, resulting in marked hypoxia of the individual and disturbances of cerebral function. For this reason, voluntary attempts to hyperventilate should never be attempted unless properly supervised. In

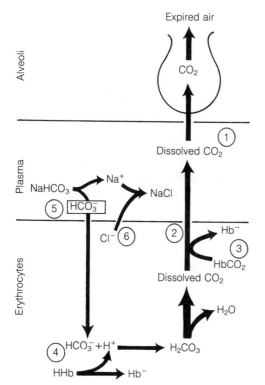

Figure 5.3.24 Summary of chemical reactions involved in CO_2 transport as blood passes through the pulmonary capillaries

addition, the fall in arterial PCO_2 will result in respiratory alkalosis. The effects of hypocapnia can be eliminated by getting the individual to rebreathe from a paper bag.

HYPERCAPNIA

Any increase in arterial PCO_2 above normal levels is called hypercapnia. The two main causes of CO_2 retention are alveolar hypoventilation and severe ventilation–perfusion imbalance. Retention of CO_2 may also occur if there is either a general or local impairment of circulation allowing CO_2 to accumulate in the tissues. It is important to remember that whereas hypercapnia is *always* accompanied by hypoxia (unless O_2 therapy is being given), the reverse is not true.

Any increase in arterial PCO_2 acts as a powerful stimulus to breathing. The rate and depth of breathing are increased and the patient may experience severe dyspnoea in his attempt to eliminate the excess CO_2. If PCO_2 rises above about $10\,kPa$ ($75\,mmHg$), the central nervous system becomes depressed as a result of dilation of cerebral blood

vessels, which causes cerebral oedema and increased intracranial pressure. The individual will suffer from headaches, lethargy and may progress to a stuporous or comatose state. Total anaesthesia and death will occur if arterial PCO_2 rises to 13–$20\,kPa$ (100–$150\,mmHg$).

The only way of treating hypercapnia is to increase ventilation. Airways obstruction should be relieved if this is a primary problem, by using bronchodilators if appropriate, by removal of any foreign bodies or excess secretions, or by performing a tracheostomy. Mechanical ventilation may be required.

CONTROL OF VENTILATION

Respiration is largely an involuntary act resulting from the automatic generation of rhythmic breathing by the respiratory centre in the brainstem. The respiratory muscles are innervated by the phrenic nerve to the diaphragm (which originates from the IIIrd, IVth and Vth cervical nerves) and the intercostal nerves (which originate from the thoracic portion of the spinal cord, T1–12) to the intercostal muscles. Spinal transection above C3 results in total respiratory paralysis, whereas if the damage occurs in the thoracic portion of the spinal cord, although the intercostal muscles will by paralysed resulting in a loss of stability of the chest wall, diaphragmatic breathing can continue.

Respiratory rhythm generated in the brain is conveyed to the spinal cord through three anatomically separate pathways:

(a) The voluntary pathway, which runs in the dorsolateral region of the spinal cord. The voluntary control of breathing is accomplished by descending pathways from the cerebral cortex to the medullary respiratory centre.

(b) The involuntary pathway, which runs in the ventrolateral part of the cord. Interaction between voluntary and involuntary breathing occurs at spinal level. Certain lesions of the brainstem or spinal cord may destroy voluntary control of breathing but leave automatic control unaffected, or vice versa, depending on the exact location.

(c) Tonic influences on the respiratory motor neurones arise from a nucleus in the upper medulla and help to determine the degree to which various respiratory muscles are relaxed or contracted. The axons from this nucleus run in close proximity to the ventrolateral rhythmical pathways.

Even during passive expiration, nerve impulses pass to the respiratory muscles which maintain their tone, thereby helping to maintain the stability of the chest wall and posture. However, as inspiration starts, there is a rapid increase in the number of impulses arriving at the respiratory muscles, so that the force of inspiration gradually increases and thoracic expansion occurs. Then, at the end of inspiration, a sudden reduction in the number of impulses results in relaxation of the inspiratory muscles, and expiration usually follows passively by elastic recoil of the lungs and thoracic cage. After a given period of time (usually about 2–3 seconds in the resting adult), the barrage of impulses returns and the cycle is repeated.

The level of ventilation must be adapted according to changes in body requirements or atmospheric conditions if adequate oxygenation is to be maintained, e.g. during exercise or when breathing at high altitude. It has long been thought that powerful chemoreceptor reflexes are of primary importance in the response to such influences on the respiratory system. (A chemo-receptor is a receptor that responds to a change in the chemical composition of the blood or fluid surrounding it.) However, the complexity of the physiological mechanisms involved is such that it is still not known exactly how chemoreceptors achieve this response. It is likely that several mechanisms are involved and that at any given moment in time, respiration is being controlled not by any single factor but by several interrelated events.

Although chemoreceptors seem to be primarily involved in changing the *level* of ventilation, receptors in the lungs and airways are more important in regulating the *pattern* of breathing. Stretch receptors in the airways may be important in producing a combination of tidal volume and respiratory rate that achieves adequate alveolar ventilation while minimizing the work of breathing. Very rapid, shallow breathing is inefficient both because of the wasted energy spent on ventilating the dead space and because the energy required to overcome airways resistance increases with increasing flow rates. Similarly, very deep breaths result in a disproportionate increase in the energy required to stretch the lungs and chest wall. These reflexes can be overcome by cortical control, for example during speech, and by other reflexes, such as the protective cough and sneeze reflexes. Other respiratory reflexes, such as those from pain, temperature, the viscera (e.g. swallowing, straining), diaphragmatic muscle spindles and baroreceptors, have also been described but are of lesser importance. Most reflexes operate eventually through the central rhythm-generating mechanisms in the brain.

Central rhythm-generating mechanisms

The respiratory nerves receive their impulses via synaptic connections in the spinal cord from neurones whose cell bodies lie in the medulla and pons in a portion of the lower brain-stem known as the **respiratory centre**. These neurones control the automatic system responsible for the periodic nature of inspiration and expiration, and so transection of the brain-stem above this level leaves the mechanism for generating rhythmic breathing essentially intact.

It has been found that certain neurones in the medulla record in perfect synchrony with inspiration, and stimulation of such neurones results in sustained inspiratory effort. These have been called the **inspiratory neurones**. Similarly, other groups of neurones (the **expiratory neurones**) discharge synchronously with expiration. Interaction between these two groups of neurones is apparently responsible for the inherent rhythmicity of impulses in the medulla which continues, albeit somewhat irregularly, even when all known afferent stimuli have been abolished. It therefore appears that these cells do have the capacity for self-excitation, although they are dependent on various inputs if a smooth and regular respiratory rhythm is to be maintained.

It is believed that the inspiratory neurones may be arranged in a loop in such a way that when one of these neurones becomes excited, it sends a signal around the circuit stimulating all other neurones and finally becoming restimulated itself (Fig. 5.3.25). Fibres from this circuit carry impulses down the spinal cord where they synapse with the nerves supplying the inspiratory muscles to cause contraction. The faster the inspiratory neurones transmit their impulses around the circuit, the more frequent will be the impulses arriving at the respiratory motor units, thereby increasing the force of muscle contraction and hence the depth of inspiration. Not only do excitatory nerve impulses pass out of the circuit to the

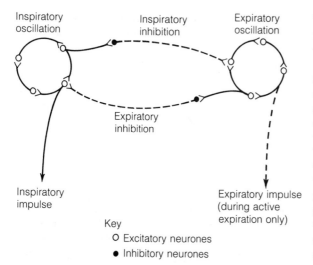

Key
O Excitatory neurones
● Inhibitory neurones

Figure 5.3.25 Diagrammatic representation of possible interactions between inspiratory and expiratory neurones

respiratory muscles, but inhibitory impulses are transmitted to the expiratory neurones. This reduces or stops any expiratory activity during inspiration.

The expiratory neurones are probably organized in a similar manner, although impulses are usually only transmitted down the spinal cord to the expiratory muscles during active expiration, since no muscle contraction is necessary to effect passive expiration. Oscillation of the impulse around the expiratory circuit causes self-excitation of all the expiratory neurones and transmission of inhibitory impulses to the inspiratory neurones.

Thus, these two groups of neurones mutually inhibit one another so that when one is oscillating the other stops, and vice versa.

At the transition between inspiration and expiration, large numbers of respiratory neurones suddenly either stop or start discharging and this is accompanied by equally rapid changes in muscle activity. The current explanation for this rapid transition between inspiration and expiration is based on some inhibitory threshold mechanism which, once activated or inactivated, rapidly 'turns off' either the inspiratory or expiratory neurones and hence the activity of the appropriate muscles (D'Angelo, 1980). It is thought that once this inhibitory threshold is reached, inspiration is suddenly switched off, enabling expiration to occur. During expiration, the inhibition on the inspiratory cells will gradually decrease until a new lower threshold is reached, enabling the 'on switch' for inspiration to be reactivated. The level at which the thresholds are set may be altered according to various inputs from the brain and rest of the body, thereby allowing deeper respiration to occur before inspiration is switched off.

A schematic representation of the way in which the basic rhythm of respiration may be controlled is shown in Fig. 5.3.26.

INFLUENCE OF THE PONTINE RESPIRATORY CENTRES

The medullary neurones receive an input from two separate areas in the pons: (a) the **apneustic**

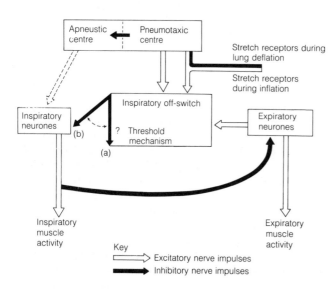

Key
⇨ Excitatory nerve impulses
➡ Inhibitory nerve impulses

Figure 5.3.26 Hypothetical representation of the system involved in the generation of rhythmic breathing. Inspiration will only occur if the 'inspiratory off-switch' is *not* activated, i.e. when it is at position (a). Once the combined inspiratory inhibitory effects reach a certain threshold level, inspiration is switched off (moves to position b). Once inspiration has ceased, expiratory neurones are no longer inhibited and active expiratory effort can occur if necessary

centre, which stimulates the inspiratory centre (but which appears to play a relatively small role in the intact animal), and (b) the **pneumotaxic centre** (or nucleus parabrachialis) which has an inhibitory effect on the inspiratory neurones, either directly or by its inhibitory influence on the apneustic centre. If the pneumotaxic centre is destroyed, the unopposed action of the apneustic centre will cause prolonged inspiration with only occasional brief expiratory efforts. However, if the entire pons is destroyed, an essentially normal balance between inspiration and expiration occurs, providing vagal input is intact.

VAGAL INFLUENCES

There are numerous stretch receptors in the lung that are sensitive to the degree of lung expansion. Impulses from these receptors pass via the vagus to the respiratory centres in the medulla, where they have an inhibitory effect on inspiration during lung inflation, and an inhibitory effect on expiration during lung deflation. These reflexes therefore help to maintain the rhythmicity of breathing and to prevent excessive changes in lung volume.

Vagal input is thought to play a much larger part in controlling breathing and maintaining lung volume in young babies, in whom the brainstem (and hence pontine control) is more poorly organized. The **Hering–Breuer reflex** is vagally mediated and results in temporary apnoea if the lung is overinflated. This reflex is easily elicited in babies and during general anaesthesia but appears to have no physiological significance in older, awake subjects.

If the vagi are cut, a highly characteristic pattern of deep, slow breathing occurs. This is because, if the inhibitory effects of the vagus on inspiration are removed, it will take longer to reach the necessary threshold before inspiration can be 'switched off'. The activity of the pons (which normally acts as a central relay station for vagal input) becomes more important if there is any decrease in vagal input.

VOLUNTARY CONTROL OF BREATHING

Voluntary control over breathing patterns can be demonstrated during breath-holding or hyperventilation, although the duration of such efforts is usually limited by the intensity of the involuntary chemical stimuli that they induce. Nevertheless, in the absence of proper supervision, or as a result of anxiety states, hyperventilation may be carried to extremes, resulting in a marked reduction of arterial PCO_2, a slowing or cessation of respiration, and respiratory alkalosis.

Complex control of the respiratory system is necessary during speech and singing. Whenever nerve impulses are transmitted from the brain to the vocal cords, simultaneous impulses are transmitted to the respiratory centre to control the flow of air between the vocal cords. Marked changes in respiration are also seen in emotional states (e.g. laughing and crying). Finally, cortical factors may play an important role in preparing the body for additional activity and may contribute to the rise in ventilation that occurs during exercise.

Mental alertness and wakefulness have a stimulating effect on breathing, whereas sleep, sedatives, some anaesthetics and alcohol usually reduce both the rate and depth of ventilation.

FAILURE OF THE RESPIRATORY CENTRE

Despite the duplication of neurogenic mechanisms which help to ensure that respiration will continue even if one part of the respiratory centre is damaged, occasionally all these mechanisms fail simultaneously. This may be caused by a number of factors, such as barbiturate overdose, cerebral contusion and severe hypoxia. The respiratory muscles may also be paralysed by:

(a) damage to anterior horn cells which relay nerve impulses to the respiratory muscles (poliomyelitis)
(b) drugs which block the transmission of nerve impulses at the motor end-plates (e.g. tubocurarine which is used during general anaesthesia as a muscle relaxant).

Management of respiratory centre failure is extremely difficult, and in general the only effective treatment is to use artificial ventilation until recovery occurs. However, in some cases the damage may be irreversible and recovery, therefore, impossible.

Chemoreception

The powerful influence of arterial PO_2 and PCO_2 on ventilation is not surprising when one con-

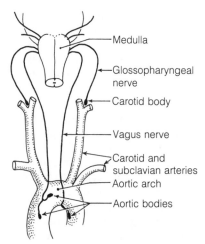

Figure 5.3.27   Peripheral chemoreceptor system involved in the control of breathing

siders that the main function of breathing is to maintain homeostasis by providing the body with O_2 and eliminating CO_2.

PERIPHERAL CHEMORECEPTORS

The carotid bodies and similar vascular structures around the aortic arch contain chemoreceptors that are important in the response to blood gases (Fig. 5.3.27). These receptors sense the levels of PO_2 and PCO_2 and hydrogen concentration $[H^+]$ in the blood and relay the information (via the glossopharyngeal and vagal nerves) to the respiratory areas in the brain. Denervation of these structures abolishes the response to hypoxia and decreases the response to $[H^+]$ and hypercapnia. While homeostasis of arterial O_2 depends on peripheral arterial chemoreceptors, the response to hypercapnia is primarily dependent on other receptors which are found in the brain-stem and are known as central chemoreceptors.

CENTRAL CHEMORECEPTORS

For many years it was believed that the central response to hypercapnia was due to the direct effect of CO_2 on the rhythm-generating mechanisms in the brain-stem. However, there is increasing evidence that specific receptor sites, known as central chemoreceptors, are situated just under the ventral surface of the medulla. The actual mechanism by which they exert their control has been the subject of much controversy in

the past. However, it now appears that when arterial PCO_2 rises, CO_2 diffuses from the cerebral blood vessels across the blood–brain barrier and into the cerebrospinal fluid (CSF) which bathes the entire brain, including the central chemoreceptors. The blood–brain barrier is relatively impermeable to $[H^+]$ or hydrogen carbonate (bicarbonate) ions (HCO_3^-), but dissolved CO_2 crosses easily. However, once in the CSF, the CO_2 liberates H^+ according to the reaction

$$CO_2 + H_2O \rightleftharpoons H_2CO_3 \rightleftharpoons H^+ + HCO_3^-$$

It is the rise in $[H^+]$ that causes the stimulation of the chemoreceptors, which in turn send excitatory impulses to the inspiratory neurones to increase their rate of discharge. The resultant hyperventilation reduces arterial PCO_2 in the blood and hence PCO_2 of the CSF, so that homeostasis can be maintained. Conversely, any fall in arterial PCO_2, and hence fall in $[H^+]$ within the CSF, inhibits the rate of discharge of the respiratory neurones, leading to a reduction in ventilation. Thus, it can be seen that the PCO_2 in the blood regulates ventilation by its effect on the $[H^+]$ of the CSF.

Chemoreceptor reflexes

HYPOXIA

Under normal circumstances, the ventilatory response to hypoxia is weak. Increased levels of inspired O_2 will only reduce ventilation by about 20%, whereas no significant increase in ventilation occurs until arterial O_2 drops below 8 kPa (60 mmHg), when the PCO_2 is within normal limits. The hypoxic drive can only increase ventilation to 50% above resting values. The central chemoreceptors are not sensitive to changes in arterial PO_2. Indeed, in the absence of the peripheral chemoreceptors, hypoxia depresses ventilation – presumably by direct action on the respiratory centre.

The relatively weak hypoxic drive under normal conditions is not surprising since: (a) increasing ventilation does little to increase O_2 delivery to the tissues due to the shape of the O_2-dissociation curve (see Fig. 5.3.19), and (b) increasing ventilation in response to hypoxia, in the presence of a normal PCO_2, would eliminate excess CO_2 from the body which would consequently depress breathing.

The only two circumstances in which the hypoxic reflex assumes particular importance are:

(a) at altitude when, after a few days acclimatization, there is a decrease in the buffering capacity of the blood and CSD (p. 507). This enables the hypoxic drive to act unopposed by the accompanying fall in PCO_2, which would normally have a depressant effect
(b) in chronic lung disease, when the hypercapnic drive has been depressed.

HYPERCAPNIA/CHANGES IN [H$^+$]

The most powerful factor regulating alveolar ventilation is arterial PCO_2. Whenever arterial PCO_2 rises, respiration is stimulated, whereas a fall in arterial PCO_2 below the normal level of 5.3 kPa (40 mmHg) results in a reduction in alveolar ventilation. In healthy subjects, if a mixture of air containing 5% CO_2 and normal levels of O_2 is breathed, a three- to fourfold increase in ventilation occurs. Ventilation may be increased up to 15-fold if the inspired concentration of CO_2 is raised to 15%. However, no further increases occur above this level, since very high levels of arterial PCO_2 depress the entire central nervous system, including the respiratory centre, and may prove lethal. The sensitivity of this feedback mechanism is such that, despite marked variations in tissue activity throughout the day, arterial PCO_2 rarely changes by more than 0.4 kPa (3 mmHg), except during sleep when it may rise a little more. The increase in ventilation that occurs in response to a rise in PCO_2 results in increased excretion of CO_2, thereby enabling arterial PCO_2 to return to normal levels, and vice versa.

Any rise in the [H$^+$] in arterial blood (fall in pH) acts as a powerful stimulus to breathing. However, the maximum increase that will occur purely in response to a fall in pH is about fivefold (i.e. only about a third of that which can be induced in response to CO_2). Although it is difficult to separate the ventilatory response to [H$^+$] from that caused by any accompanying change in PCO_2, the response to [H$^+$] can be seen in patients with metabolic disorders, such as diabetes mellitus. Such patients frequently hyperventilate in response to their low blood pH (caused by excess acids in the blood) despite the fact that this 'blows off' excess CO_2, resulting in a low arterial PCO_2 ($PaCO_2$), which would depress respiration under normal circumstances.

Undoubtedly the predominant effect of changing $PaCO_2$ is on the central chemoreceptors, whereas hypoxia stimulates the peripheral receptors. Quantitative assessment of their relative importance and interactions is difficult because $PaCO_2$ and [H$^+$] stimulate both peripheral and central mechanisms, whereas hypoxia stimulates the peripheral but depresses the central mechanisms. However, it is clear that the peripheral chemoreceptors are responsible for about one-third of the ventilatory response to hypercapnia, and are also the chief site at which interaction between the effects of PCO_2, PO_2 and [H$^+$] occurs.

EXERCISE

During exercise, ventilation increases in almost direct proportion to the amount of work being performed by the body. There are two distinct components in the ventilatory response to exercise: an immediate increase when exercise first starts, and a slower rise to plateau level as exercise is maintained up to 120 l/min in a fit individual (i.e. a 20-fold increase over resting levels), (Fig. 5.3.28). The physiological mechanisms involved in this are still poorly understood. The initial increase in ventilation is thought to be too rapid to involve the chemoreceptor system, especially since no significant changes in blood gases occur during exercise due to the increased ventilation and cardiac output. The speed of this initial reaction strongly suggests a neurological mechanism, which could be due either to a direct drive from the cerebral cortex, or to stimulation of the respiratory centre by

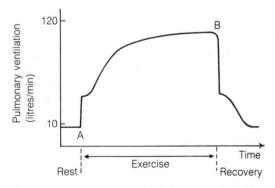

Figure 5.3.28 Changes in ventilation during exercise. There is a sudden increase in ventilation at the very onset of exercise (A) and an equally rapid, though larger, decrease in ventilation immediately the exercise ceases (B)

stimulation of proprioceptors in the exercising muscles and joints.

The cause of the subsequent slow rise in ventilation and the factors sustaining ventilation when the plateau has been reached remain unexplained. There is some evidence that peripheral O_2 chemosensitivity is increased with exercise. Alternatively, the deeper, more rapid breathing that occurs during exercise could cause more marked oscillations in blood gases during the breathing cycle than normal, which may stimulate the peripheral chemoreceptors even though the mean levels of $PaCO_2$, PaO_2 and $[H^+]$ remain constant.

CHRONIC LUNG DISEASE

Some patients with chronic lung disease have a poor response to hypercapnia. This is partly due to mechanical obstruction to breathing in that they are incapable of increasing their work of breathing any further. However, in some patients, the 'blue bloaters', there is a true reduction in response to $PaCO_2$. At one time this was thought to be due to changes in the buffering capacity of the blood and CSF in response to chronic CO_2 retention, but this is no longer considered to be an adequate explanation. It is known that there is a wide variation in normal response to CO_2 and it may be that the peripheral chemoreceptors in 'blue bloaters' develop a reduced responsiveness to CO_2, similar to the fall in peripheral chemosensitivity seen in people living at altitude. These patients no longer respond to rising PCO_2, and become dependent on the hypoxic drive to breathe. Administration of augmented O_2 in this situation can eliminate this drive and cause a total cessation of breathing.

This risk is reduced by administering O_2 in minimal dosage compatible with adequate oxygenation of arterial blood. The Venturi-type mask is usually used for patients with chronic lung disease since this enables the flow of O_2 to be continuously diluted with air. These masks are designed to provide a specified fractional concentration of O_2 which cannot be exceeded by the patient or his attendants increasing the flow rate of oxygen. A mask supplying 27% O_2 is generally recommended in these cases, although masks supplying only 24% O_2 may be used in acutely ill patients who cannot be closely monitored. The response of patients to O_2 therapy should be noted at frequent intervals by observing their state of consciousness, the frequency and depth of breathing and by measuring the PCO_2 and pH of their arterial blood.

Other patients with chronic lung disease, the 'pink puffers', retain their CO_2 sensitivity, remaining relatively well oxygenated at the expense of considerable, often distressing, effort.

Cheyne–Stokes respiration

Severely hypoxic patients, especially those in whom respiration is being driven primarily by a lack of O_2, frequently exhibit a striking pattern of periodic breathing known as Cheyne–Stokes respiration. This is characterized by periods of apnoea lasting approximately 15–20 seconds, alternating with periods of hyperventilation of about the same duration. It seems to occur because of the relative insensitivity of the respiratory control mechanisms to small changes in PO_2. The period of hyperventilation raises arterial PO_2 so that peripheral chemoreceptors are no longer stimulated, and breathing then stops until the PO_2 has fallen sufficiently to reactivate the receptors, when the cycle begins again. Cheyne–Stokes breathing is frequently seen at high altitude (especially during sleep), in anaesthetized individuals who are allowed to breathe spontaneously, and in some patients with severe cardiorespiratory disease or brain damage.

The regulation of respiration by the interaction of multiple factors is summarized in Fig. 5.3.29. Under normal circumstances, this system ensures that homeostasis of the blood gases and $[H^+]$ can be maintained at all times, despite widely varying circumstances.

Assessment of respiratory rate and rhythm

The rate of respiration varies according to the size of the individual, the metabolic demands of the body and the mechanical properties of the lung and chest wall. As described above, the respiratory control centre in the medulla receives a constant barrage of information from receptors all over the body and normally adjusts the rate and depth of breathing to achieve the necessary level of alveolar ventilation with the least investment of energy.

Respiratory rates generally tend to decrease with increasing body size. Thus, newborn infants

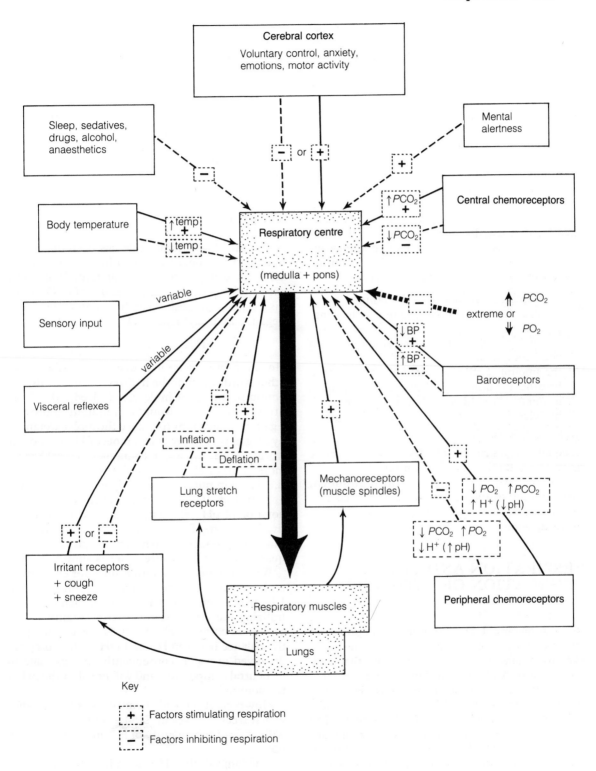

Figure 5.3.29   Summary of factors controlling respiration

have a 'normal' rate of about 40 breaths per minute, whereas in adults it is generally between 10 and 15 breaths per minute. However, it should be stressed that there is considerable individual variation, and changes in respiratory rate can only provide vital warning signs of respiratory (or cardiac) problems if an accurate record has previously been kept for that particular patient. The depth of respiration (tidal volume) normally increases with increasing body size.

Although respiration is generally an involuntary act, it is easily influenced by subjective factors, particularly if a patient is aware that his pattern of breathing is being observed. Consequently, it is usually more satisfactory if respirations are assessed while the patient thinks his pulse rate is still being counted. While counting respiratory rate it is also vital to observe

(a) the depth and regularity of breathing
(b) the presence of any breath sounds (wheezing, stridor)
(c) the presence of any chest recession or uneven chest movements
(d) the amount of effort required by the patient to breathe, including whether or not the patient is mouth breathing and/or using accessory muscles to aid ventilation, (see section on dyspnoea, p. 482).

Adequate assessment of the above usually requires a full minute's observation.

A normal respiratory rate is termed **eupnoea**, whereas **tachypnoea** refers to an increased respiratory rate, and a cessation of respiration is called **apnoea**.

RESPIRATION AND THE REGULATION OF HYDROGEN IONS

Respiration and CO_2 transport have profound effects on the acid–base status (or pH) of the body. Hydrogen ions are far more reactive than any other positively charged ions in the body and it is the concentration of hydrogen ions $[H^+]$ that determines the acidity of blood and body fluids.

Even small changes away from the normal $[H^+]$ will affect the structures of proteins. Enzymes are particularly susceptible to changes in the pH of the surrounding medium, and even small changes may drastically reduce their efficiency, whereas

larger alterations will result in a life-threatening disruption of metabolism. Consequently, the intake and production of H^+ in the body, which may vary considerably according to diet, level of exercise, drugs or disease, must be matched by an efficient system of buffering the H^+ in the body fluids and controlled excretion of H^+ by the lungs and kidneys, to ensure output equals input and that homeostasis is maintained.

Before considering the regulation of $[H^+]$ it is important to understand the notation of pH.

The pH notation

A standard logarithmic scale has been developed to denote the concentration of hydrogen ions in a solution, using water as a standard. Water can exist in two forms: (a) as molecules of H_2O, or (b) dissociated into positive hydrogen ions (H^+) and negative hydroxyl ions (OH^-)

$$H_2O \rightleftharpoons H^+ + OH^-$$

However, pure water ionizes only very slightly, in other words most of the hydrogen remains tightly bound in the water molecule. Indeed, 10 000 000 litres of pure water would be required to provide 1 g of hydrogen ions (or, since the molar weight of hydrogen is 1, to produce 1 mole of H^+). In other words, 1 litre of water contains only 0.0000001 mole (mol) of H^+. Thus the concentration of H^+ (i.e. $[H^+]$ in pure water could be expressed as log -7. In fact, $[H^+]$ is usually expressed using the notation of pH, where pH is defined as the negative logarithm of the hydrogen ion concentration. Since the product of two negative signs is a positive quantity, the pH of pure water is simply 7.

Water is a neutral solution since for every molecule which dissociates, 1 hydrogen ion (H^+) and 1 hydroxyl ion (OH^-) are formed, each one neutralizing the other. Many other substances, such as sodium chloride, glucose and urea, do not upset the balance between H^+ and OH^- when they are dissolved in water, consequently they are said to be neutral compounds, and will not alter the pH of the solution.

However, if an acid (defined as any substance which frees or gives up H^+) is added to water, it will increase the quantity of H^+ and cause a fall in pH.

The range of the pH scale is from 0 to 14. If the pH of a solution is 0, this means that 1 litre

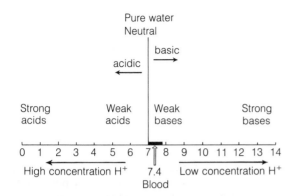

Figure 5.3.30 The pH scale

contains 10^0 or 1 mole of hydrogen ions. (Rather confusingly, the logarithm of 1 is 0, since $10^{-1} = 0.1$, $10^0 = 1$, $10^1 = 10$.) At the other end of the scale, pH is 14 (Fig. 5.3.30). Thus, if pH is below 7, it is said to be acidic, whereas if it is greater than 7 (i.e. with a lower concentration of H^+ than in water), the solution is called basic.

Although all acids are alike in that they donate H^+, they differ in their ability to do so. Some acids such as hydrochloric acid ionize completely, donating all their H^+, and are said to be strong acids (pH $\simeq$ 1.0), i.e.

$$HCl \rightleftharpoons H^+ + Cl^-$$

whereas others, such as carbonic acid, dissociate much less freely, and therefore cause a smaller fall in pH ($\uparrow$ [H^+] of a solution) (pH $\simeq$ 6.0).

$$H_2CO_3 \rightleftharpoons H^+ + HCO_3^-$$

When a base (any compound which accepts H^+) is added to a solution, the concentration of H^+ falls with respect to OH^-, causing a rise in pH. This may be achieved either by the donation of hydroxyl ions (OH^-), which combine with H^+ to form H_2O, or by the direct acceptance of H^+.

Concentration of hydrogen ions in the blood

The normal pH of blood is 7.4 (i.e. it has $0.000\,000\,04$ mol H^+/l) and it is therefore a slightly alkaline fluid. The range of pH compatible with life is only 7.0–7.8, and in a healthy adult arterial blood pH is almost always maintained between 7.35 and 7.45.

Despite the advantages of the pH scale, the

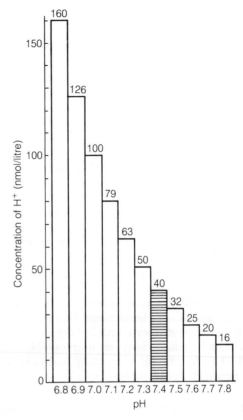

Figure 5.3.31 Diagram to show the effects of changing pH on hydrogen ion concentration

logarithmic instead of linear relationship between pH and the actual [H^+] can cause confusion. Looking at Fig. 5.3.31, it can be seen that the change in [H^+] per unit fall in pH is proportionately much greater than per unit rise in pH. Thus, if pH changes from 7.4 to 7.2, [H^+] increases by 23 nmol/l, whereas an increase in pH from 7.4 to 7.6 represents a reduction of only 15 nmol/l of [H^+]. Consequently, the clinical use of the pH scale may be replaced in future by expressing [H^+] directly in nanomoles per litre (nmol/l), where 1 nanomole $= 10^{-9}$ moles. Arterial blood normally has a [H^+] of 40 nmol/l (range 36–45 nmol/l). The range compatible with life lies between 16 and 100 nmol H^+/l.

Buffer systems

The pH of blood is normally kept within fairly narrow limits by two buffer systems

(a) the haemoglobin in red blood cells
(b) the carbonic acid–hydrogen carbonate system in plasma.

Buffers are systems which minimize any change in overall pH, by accepting H^+ when $[H^+]$ rises (i.e. pH falls) and by donating H^+ when $[H^+]$ falls (i.e. pH rises).

HAEMOGLOBIN

The plasma proteins and haemoglobin play a vital role in maintaining a constant blood pH despite the continuous addition of CO_2 into the blood at tissue level, and its removal as blood passes through the lungs.

At tissue level, the following reaction occurs as CO_2 enters the red blood cell

$$CO_2 + H_2O \rightleftharpoons H_2CO_3$$

If the carbonic acid were allowed to dissociate into free H^+ and HCO_3^-, there would be a marked rise of H^+ in the erythrocyte and consequent disruption of its function. However, haemoglobin reacts with H_2CO_3 forming potassium hydrogen carbonate and reduced haemoglobin

$$H_2CO_3 + KHb \rightleftharpoons HHb + KHCO_3$$

carbonic + potassium $\rightleftharpoons$ reduced + potassium
acid   haemoglobin   haemoglobin   hydrogen
                                     carbonate

Reduced haemoglobin acts as a much weaker acid than H_2CO_3 and therefore binds its hydrogen more tightly, preventing a rise in $[H^+]$.

As the blood passes through the lungs, CO_2 is released into alveolar air. Potentially this could result in a marked decrease of $[H^+]$ in the blood, since the reaction

$$H^+ + HCO_3^- \rightleftharpoons H_2CO_3 \rightleftharpoons H_2O + CO_2$$

is driven to the right by the continuous excretion of CO_2. However, under these circumstances some of the reduced haemoglobin dissociates to release H^+, thereby maintaining a constant level of H_2CO_3 (and hence H^+) in the blood.

$$HHb + KHCO_3 \rightleftharpoons KHb + H_2CO_3$$

In addition, this 'releases' the potassium haemoglobin to buffer more H^+ when the blood passes through the tissues again. The efficiency of this buffer system depends on alveolar ventilation

being regulated according to body needs, so that only the correct amount of CO_2 is excreted at any given time.

CARBONIC ACID–HYDROGEN CARBONATE SYSTEM

Although carbonic acid (H_2CO_3) is a stronger acid than reduced haemoglobin (Hb), it is still considerably weaker than many of the acids ingested or produced by the body during metabolism. Consequently, together with its salts, potassium and sodium hydrogen carbonate (bicarbonate) ($KHCO_3$ and $NaHCO_3$), it can buffer acid produced during metabolism, ingested in the diet, or released in diabetes. In addition, the HCO_3^- has the advantage that, since it can be converted to H_2CO_3 and then excreted as CO_2 and H_2O, any excess acid can be removed from the body, providing ventilation can be adjusted adequately.

If excess H^+ are added to the body, e.g. lactic acid formed during exercise, the HCO_3^- will 'mop up' excess H^+ to form H_2CO_3 and a neutral salt. The H_2CO_3 can then be prevented from releasing any of *its* H^+ by the action of the haemoglobin buffering system, i.e.

$$H_2CO_3 + KHb \rightleftharpoons HHb + KHCO_3$$

Once the blood reaches the lung, these reactions are reversed, so that the HCO_3^- is converted into H_2CO_3, resulting in excretion of CO_2 and production of H_2O. However, to do this, ventilation must be increased above normal levels and the additional excretion of CO_2 (which came not from the tissues, but from the 'pool' of HCO_3^- in the blood) will result in a reduction in blood HCO_3^- (base deficit) and a temporary reduction in its buffering power. Although the level of blood HCO_3^- is initially affected by dietary intake, secondary changes may occur as a result of its role as a buffer. The $[HCO_3^-]$ will decrease if: (a) additional H^+ is added to the blood, or (b) hyperventilation occurs resulting in excessive loss of CO_2 and H_2O from the body. The $[HCO_3^-]$ will rise if there is: (a) any loss of H^+, or (b) CO_2 retention.

The level of HCO_3^- (which is ultimately regulated by the kidney) plays a vital role in maintaining blood pH at its normal level. The ratio of HCO_3^- to H_2CO_3 must be maintained at 20:1 if the pH of the blood is to remain at 7.4. The PCO_2 of blood determines the amount of H_2CO_3 (i.e. dissolved CO_2).

The way in which blood pH is determined by the ratio of bicarbonate to dissolved CO_2 (i.e. base to acid) can be demonstrated by the Henderson–Hasselbach equation, which states

$$\text{blood pH} = 6.1 + \log \frac{[\text{hydrogen carbonate}]}{[CO_2 \text{ in solution}]}$$

The figure 6.1 merely represents a logarithmic constant describing the degree to which carbonic acid dissociates into HCO_3^- and H^+ (i.e. how 'strong' an acid it is). Logarithms are used on the right-hand side to compensate for the fact that $[H^+]$ on the left is also expressed logarithmically (i.e. as pH). At a normal arterial PCO_2 of 5.3 kPa (40 mmHg), there are usually 26 mmol/l of HCO_3^- and 1.3 mmol/l of dissolved CO_2, resulting in a hydrogen carbonate : carbonic acid ratio of 20 : 1

$$\text{pH} = 6.1 + \log \frac{26}{1.3}$$
$$= 6.1 + \log 20$$
$$= 6.1 + 1.3$$
$$= 7.4$$

Any increase in $[HCO_3^-]$ or fall in H_2CO_3 (i.e. PCO_2) will cause an increase in the ratio and a rise in pH (i.e. increased alkalinity of the blood, or alkalosis). Conversely, any reduction in the ratio, whether by a fall in $[HCO_3^-]$ or a rise in PCO_2, will decrease pH (increased acidity of the blood, known as acidosis).

The buffer systems in the blood provide a highly efficient mechanism for responding to immediate changes in $[H^+]$ and help to prevent any marked fluctuations in pH despite the variations in intake or production of acids in the body. However, they are a temporary expedient, and can only continue to function if there is some means by which excess acid or base can actually be excreted from the body. The two organs most active in this capacity are the lungs, which generally respond very rapidly to short-term changes, and the kidneys, which function in more long-term control (see Chap. 5.4).

The role of the respiratory system

The respiratory system plays a vital role in maintaining blood pH in that alveolar ventilation is normally adjusted very precisely to the level of PCO_2 and $[H^+]$ in arterial blood (see Control of ventilation).

Any increase in PCO_2 or $[H^+]$ – and hence fall in pH – is sensed by central and peripheral chemoreceptors and results in a rapid increase in alveolar ventilation, thus speeding the reaction

$$H^+ + HCO_3^- \rightarrow H_2CO_3 \rightarrow CO_2 + H_2O$$

which causes an increased excretion of CO_2 and H^+ from the blood, whereas any fall in the level of PCO_2 or $[H^+]$ depresses the respiratory drive. In this way, the respiratory system, acting with the blood buffers, normally provides a highly efficient mechanism for maintaining blood pH within a narrow range.

However, the overall increase or decrease in $[HCO_3^-]$ resulting from such changes in the level of alveolar ventilation has to be compensated for by the kidney. The role of renal control also assumes great importance in the presence of respiratory disease when alveolar ventilation cannot be increased sufficiently to excrete sufficient CO_2. Renal control of acid–base balance and acid–base disturbances is described in Chapter 5.4.

Review questions

The answers to all these questions can be found in the text. In each case, there is at least once correct and at least one incorrect answer.

1 Which of the lung volumes given below can be measured using a simple spirometer?

(a) Functional residual capacity.
(b) Tidal volume.
(c) Residual volume.
(d) Vital capacity.

2 Which of the following conditions in the red cell shift(s) the O_2-dissociation curve to the right?

(a) Rise in temperature.
(b) Reduction in PCO_2
(c) Reduction in pH.
(d) Rise in 2,3-diphosphoglycerate.

3 When CO_2 is given off by the blood in the lungs

(a) chloride ions move in to the red cells
(b) hydrogen carbonate ions diffuse into the red cells
(c) the loading of O_2 onto haemoglobin is assisted
(d) carbonic acid in the red cells and plasma breaks down to form CO_2 and H_2O.

4 When a pneumothorax occurs, the chest wall

(a) collapses inwards
(b) expands outwards
(c) remains where it was.

5 Which of the following conditions may cause a rise in airways resistance?

(a) Bronchoconstriction.
(b) Increased lung volume.
(c) Emphysema.
(d) Airway obstruction.

6 Which of the following factors in blood exert(s) the most important control on ventilation?

(a) PCO_2
(b) pH
(c) PO_2

7 At any given arterial PO_2, cyanosis will be more apparent if the subject is

(a) anaemic
(b) polycythaemic
(c) hot
(d) cold.

8 In which of the following hypoxic conditions would oxygen therapy be of most benefit?

(a) Right-to-left cardiac shunt.
(b) Ventilation–perfusion imbalance.
(c) Pulmonary oedema.
(d) Chronic anaemia.

9 If arterial pH is to remain at 7.4, the ratio of hydrogen carbonate to carbonic acid in arterial blood must be maintained at

(a) $1:20$
(b) $20:1$
(c) $1:200$
(d) $200:1$

10 In a healthy adult, the 'average' size of the functional capacity is

(a) 500 ml
(b) 1500 ml
(c) 3000 ml
(d) 4500 ml

11 At end-expiration

(a) alveolar pressure is equal to atmospheric pressure
(b) intrapleural pressure is equal to atmospheric pressure
(c) alveolar pressure is slightly greater than atmospheric pressure
(d) intrapleural pressure is slightly less than atmospheric pressure.

12 During inspiration

(a) alveolar pressure is more negative than atmospheric pressure
(b) alveolar pressure is higher (less negative) than intrapleural pressure
(c) alveolar pressure is equal to atmospheric pressure
(d) alveolar pressure is lower (more negative) than intrapleural pressure.

Answers to review questions

1 b and d
2 a, c and d
3 b, c and d
4 b
5 a, c and d
6 a
7 b and d
8 b and c
9 b
10 c
11 a and d
12 a and b

Short answers and essay questions

1 What is meant by the ventilation–perfusion ratio? What would be the causes and effects of areas of lung with
  (a) a high ventilation–perfusion ratio?
  (b) a low ventilation–perfusion ratio?

2 Discuss the factors which influence the transport of oxygen from the atmosphere to the tissue cells.

3 Describe the ways in which carbon dioxide is transported from the cells to the lungs for excretion. What advantages do these methods have over transport in simple solution?

4 What respiratory adaptations occur in response to
(a) chronic hypoxia
(b) exercise.

5 How is alveolar ventilation regulated according to the body's needs?

6 Define the term 'dyspnoea'. Discuss the nursing care of a young woman experiencing an acute attack of asthma.

7 In what circumstances may oxygen therapy
(a) benefit a hypoxic patient
(b) be of limited value to a hypoxic patient.
How would you assist a junior learner to care for a patient receiving 27% oxygen via a Venturi mask?

Suggestions for practical work

1 Using a Douglas bas, collect the expired minute volume, and compare volumes per minute collected (a) at rest, (b) while walking briskly, and (c) while running up and down stairs.

2 Using a spirometer, make a recording of tidal breathing, inspiratory and expiratory reserve volumes and vital capacity.

3 With a nose clip in position, breathe through tubes of various dimensions. Notice how much harder it is to breathe through a narrow straw than through wider bore tubes due to the higher resistance (small radius) of the former.

4 (a) See how long you can hold your breath before the resultant increase in arterial PCO_2 forces you to take a breath.

(b) Under supervision, breathe as deeply and rapidly as possible for about 30 seconds. Notice the sensation this produces, particularly the diminished desire to breathe during the next minute or so due to the reduced arterial PCO_2 produced by hyperventilating.

References

D'Angelo, E. (1980) Control of breathing: central mechanisms and peripheral inputs. *Bulletin Europeen de Physiopathologie Respiratoire 16*: 111–122.

DuBois, A. B., Botelho, S. Y., Bedell, G. N., Marshall, R. & Comroe, J. H. (1956a) A rapid plethysmographic method for measuring thoracic gas volume. *Journal of Clinical Investigation 35*: 322–326.

DuBois, A. B., Botelho, S. Y. & Comroe, J. H. (1956b) A new method for measuring airway resistance in man using a body plethysmograph. *Journal of Clinical Investigation 35*: 327–334.

West, J. B. (1977) Ventilation–perfusion relationships. *American Review of Respiratory Diseases 116*: 919–943.

Suggestions for further reading

Cherniack, R. M. (1977) *Pulmonary Function Testing*. Philadelphia: W. B. Saunders.

Cotes, J. E. (1979) *Lung Function: Assessment and Application in Medicine*, 4th edn. Oxford: Blackwells.

Scalding, J. G., Cumming, G. and Thurlbeck, W. M. (1981) *Scientific Foundations of Respiratory Medicine*. London: Heinemann Medical.

Watson, J. E. (1979) *Medical-Surgical Nursing and Related Physiology*, 2nd Edn. Philadelphia: W. B. Saunders.

West, J. B. (1981) *Pulmonary Pathophysiology: the Essentials*. Baltimore: Williams and Wilkins.

West, J. B. (1985) *Respiratory Physiology – the Essentials*, 3rd edn. Baltimore: Williams and Wilkins.

Chapter 5.4
Renal Function

Susan M. Goodinson

Learning objectives

After studying this chapter, the reader should be able to

1 Describe the anatomical features of the kidney and lower urinary tract.
2 List the functions of the kidneys in humans.
3 Review the regional structures and mechanisms which subserve filtration, reabsorption and secretion in the nephron.
4 Discuss the mechanisms involved in the formation of concentrated and dilute urine.
5 List the causes of acute and chronic renal failure, and relate these to the pathophysiological changes which occur.
6 Identify the major problems which arise during renal failure and review their management.
7 Explain the mode of action of diuretic drugs.
8 Briefly discuss the role of the kidney in the genesis of hypertension.
9 Describe the major influences in the regulation of body water, sodium and potassium balance.
10 Discuss the role of the kidney in the regulation of acid–base balance.
11 Describe the process of micturition.
12 Identify problems which result from micturition disorders and review their nursing management.
13 Describe the volume and composition of the body fluid compartments and explain how major imbalances can arise.

14 Identify the symptoms of fluid and electrolyte imbalance and describe how they may be alleviated.

Introduction

The kidney is an organ which fulfils a vital role in maintaining the volume and composition of the body fluids and is thus a major regulator of the internal environment; indeed, it is frequently referred to as 'the ultimate regulator of homeostasis'.

In this chapter, the physiological mechanisms which subserve normal renal function, and thus homeostasis are examined. Consideration is also given to the problems which arise when normal renal mechanisms are altered by disease.

GROSS ANATOMY OF THE KIDNEY

In humans, the kidneys are paired, compact organs which lie one on either side of the vertebral column at the level of the 12th thoracic to the 3rd lumbar vertebrae. Situated behind the peritoneum, they are attached by adipose tissue to the posterior abdominal wall.

Supporting connective tissue covers the anterior surfaces of each kidney, renal blood vessels, aorta and the adrenal glands. Figure 5.4.1 represents a coronal section of the kidney and upper ureter. Each kidney is covered by an external capsule of

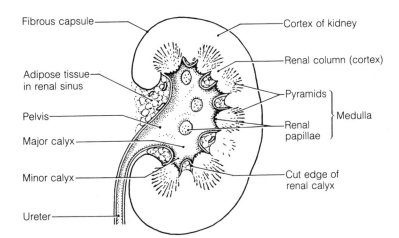

Figure 5.4.1 Coronal section through a kidney

fibrous connective tissue. Renal arteries, veins, lymphatics and nerves enter and leave at a medial indentation known as the **hilum**. Also entering at the hilum is the funnel-shaped upper end of the ureter which expands to form an internal cavity known as the **renal pelvis**. Beneath the capsule lie two distinct areas of tissue, the outer **cortex** and inner **medulla**. The medulla consists of 8–18 wedge-shaped tracts of tissue known as the **medullary pyramids**. At the tips of these medullary pyramids, papillary tissue projects into minor **calyces**, which are hollow projections of the renal pelvis. Urine draining from tiny ducts in the papillary tissue passes from the minor into the major calyces, finally reaching the renal pelvis where the lining of transitional epithelium distends to accommodate it. Contractions of the smooth muscle in the walls of the calyces and pelvis propel urine downwards into the ureter.

The nephron

Each kidney contains approximately 1–1.5 million nephrons. Each nephron consists of a tuft of blood vessel, the **glomerulus**, within the invagination of a tubule 6 cm long, lined throughout by a columnar epithelium (Fig. 5.4.2). Five anatomically distinct regions can be distinguished in the tubule of each nephron: Bowman's capsule, the proximal convoluted tubule, the loop of Henlé, the distal convoluted tubule, and collecting ducts.

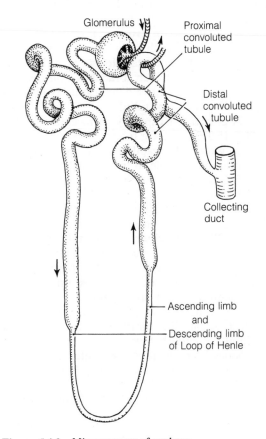

Figure 5.4.2 Microanatomy of nephron

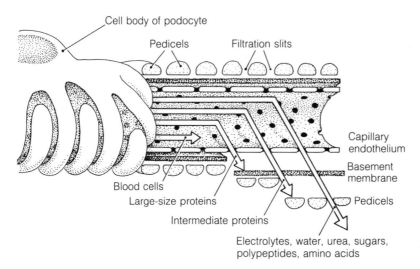

Figure 5.4.3 Schematic representation of the filtration barrier (redrawn from Creager, J. G. (1983) *Human Anatomy and Physiology*. Belmont, California: Wadsworth)

THE GLOMERULUS AND BOWMAN'S CAPSULE

Bowman's capsule forms the spherical, dilated upper end of the tubule which surrounds the glomerulus; the entire structure is only 150 nm in diameter. All glomeruli lie in the cortex and originate from an afferent arteriole and give rise to an efferent arteriole which branches (on leaving Bowman's capsule) into a dense capillary network surrounding the renal tubule.

Total glomerular capillary surface area is vast, at 5000–15 000 cm^2/100 g tissue. Research suggests that the glomerular capillaries are one hundred times more permeable to water and solutes than extrarenal capillaries and that they contain 'pores' of 75–100 nm diameter. The capillary endothelium rests on a basement membrane, on the other side of which rests the epithelium lining Bowman's capsule. This epithelium is probably podocytic (Greek *podus* = foot), that is to say, the cells are elongated and divide towards the base to form pedicels (foot processes) which rest on the basement membrane and are separated by filtration slits 25 nm wide. Fine diaphragms appear to bridge the filtration slits. The entire structure, comprising the capillary endothelium, basement membrane and podocytic epothelium, constitutes the selective filtration barrier, as shown in Fig. 5.4.3.

THE PROXIMAL CONVOLUTED TUBULE

Following Bowman's capsule, the proximal convoluted tubule extends for a length of 12–24 mm through the cortex. This region is 50–65 nm in diameter and lined by large columnar epithelial cells which are modified on the internal surface to form a brush border of microvilli, a device for increasing the surface area inside the proximal tubule where most of the solute reabsorption takes place.

THE LOOP OF HENLÉ

Extending from the proximal convoluted tubule, the thin-walled descending limb of the Loop of Henlé moves down into the medulla and then makes a U-turn, moving back into the cortex via a thicker walled ascending limb. In both limbs of Henlé's Loop, the columnar cells are flatter and they contain fewer microvilli on the internal (luminal) surfaces.

THE DISTAL NEPHRON

The ascending limb of the loop of Henlé leads into the distal convoluted tubule, the first part of which folds back in proximity to the afferent arteriole, where it forms a region known as the **macula densa** (Fig. 5.4.4). Specialized epithelial cells of the macula densa monitor the sodium chloride concentration of fluid flowing through this area and comprise part of the juxtaglomerular apparatus.

The distal convoluted tubule is comparatively short (4–8 mm), and leads into the collecting ducts which join together as they move through the medulla, finally opening at the tips of medullary papillae into the calyces of the renal pelvis.

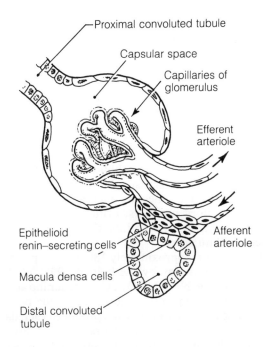

- Proximal convoluted tubule
- Capsular space
- Capillaries of glomerulus
- Efferent arteriole
- Afferent arteriole
- Epithelioid renin–secreting cells
- Macula densa cells
- Distal convoluted tubule

Figure 5.4.4 The juxtaglomerular apparatus showing the macula densa (redrawn from Creager, J. G. (1983) *Human Anatomy and Physiology*. Belmont, California: Wadsworth)

CORTICAL AND JUXTAMEDULLARY NEPHRONS

Two types of nephron are present: superficial cortical nephrons and juxtamedullary nephrons. Comprising seven-eighths of the total, cortical nephrons lie in the superficial areas of the cortex.

In contrast, juxtamedullary nephrons, comprising only one-eighth of the total, lie in deeper areas of the cortex. They have larger glomeruli, longer loops of Henlé with a thin and thick ascending limb, and their efferent arterioles give rise not only to the peritubular network previously described, but also to looped capillaries which lie closely alongside the loop of Henlé, known as the **vasa recta**.

Alterations in the blood flow to these two groups of nephrons are seen in conditions of sodium imbalance. For example, in the presence of an excessive salt intake, the renal blood flow to the cortical nephrons is increased. Because they have short loops of Henlé, it is likely that these nephrons are 'salt losing', allowing more sodium ions to be excreted in the urine. This is one factor which allows the kidney to regulate sodium balance.

Renal blood supply

Twenty-five per cent of the cardiac output is delivered to the kidneys each minute. In terms of blood flow rate, this is $400\,\mathrm{ml.\,min^{-1}.\,100\,g^{-1}}$ at rest, higher than that of any other tissue.

The two renal arteries arise from either side of the abdominal aorta, enter at the hilum and divide in the renal tissue to form interlobar arteries between the pyramids (Fig. 5.4.5). Arcuate arteries arising from these give rise to cortical interlobar arteries which branch and ultimately give rise to afferent arterioles supplying each glomerulus.

Efferent arterioles emerging from the glomerulus form a dense peritubular capillary network. Venous plexi drain this capillary bed around the tubule, delivering venous blood via the inter-

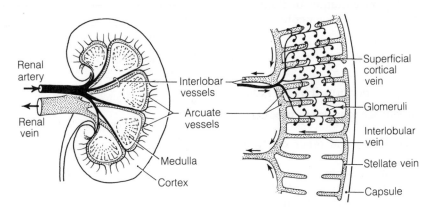

- Renal artery
- Renal vein
- Interlobar vessels
- Arcuate vessels
- Medulla
- Cortex
- Superficial cortical vein
- Glomeruli
- Interlobular vein
- Stellate vein
- Capsule

Figure 5.4.5 The renal blood supply

lobular, arcuate and interlobar veins to the renal vein. A renal vein leaving the hilum returns venous blood to the inferior vena cava.

Renal nerve supply

The kidneys are richly innervated by sympathetic nerve fibres (see Chap. 2.4). A few parasympathetic fibres are also present. Sympathetic nerve fibres supply the smooth muscle of the arterioles and the juxtaglomerular apparatus. Stimulation of these nerves brings about vasoconstriction, reduced renal blood flow, a reduced glomerular filtration rate and the release of renin from the juxtaglomerular apparatus.

In addition, the kidney is supplied with some afferent nerve fibres which mediate the sensation of pain. Stimulation of these fibres occurs when the renal capsule is distended, such as may be caused by bleeding, inflammation associated with glomerulonephritis or pyelonephritis, and obstruction by calculi. Ischaemia caused by renal artery occlusion may also stimulate the pain fibres.

Juxtaglomerular apparatus

Two major cell types are found here (see Fig. 5.4.4).

(a) Renin-secreting cells which are located in the wall of the afferent arteriole, and are derived from smooth muscle cells.
(b) Macula densa cells, which are found in the epithelium lining the first part of the distal convoluted tubule. These cells monitor the sodium chloride concentration of fluid as it flows through this area of the nephron.

RENAL FUNCTION

Four main functions are performed by the kidney.

1 Removal of the nitrogenous waste products of metabolism.
2 Regulation of acid–base and other electrolyte balances.
3 Maintenance of water balance.
4 Production of erythrogenin – renal erythropoietic factor (see Chap. 4.1) – and the biologically active form of vitamin D – 1,25-dihydroxycholecalciferol (see Chap. 2.5).

The first three of these functions are achieved by processes of ultrafiltration, reabsorption and secretion in the nephron. An ultrafiltrate of plasma is produced at the glomerulus. As this flows through regionally specialized areas of the nephron, water and vital solutes, including glucose, electrolytes, amino acids and vitamins, are conserved by reabsorption. Waste products of metabolism, including urea, creatinine, uric acid, sulphates and nitrates, remain in the filtrate. These, together with secreted metabolites, and water and electrolytes which are surplus to requirements reach the terminal collecting ducts and renal pelvis as the excretory products, urine. This is finally conveyed via the ureter into the lower urinary tract for elimination.

In the adult, approximately 180 litres of plasma are filtered each day by the kidneys; 99% of the water, glucose, Na^+, HCO_3^-, Cl^- and amino acids filtered is subsequently reabsorbed in the nephrons. The final volume of urine produced each day by an adult in health is in the range 1–1.5 litres, although its volume and composition vary depending on fluid intake, diet and extrarenal losses of water and electrolytes, for example in sweat and expired air. Over a 24-hour period, total fluid intake balances fluid output, as can be seen in Table 5.4.1.

Some of the properties and the composition of urine in a healthy adult are summarized in Table 5.4.2. The levels of protein and glucose listed are not detectable by routine methods of ward investigation.

Table 5.4.1   Water balance in an adult over a 24-hour period (from Valtin, 1979)

Intake (ml/day)		Output (ml/day)	
In oral fluids	1500	In urine	1500
In food	1000	In faeces	200
Water of metabolic oxidation	500	Insensible	
		skin	900
		lungs	400
Total	3000		3000
Recycling of gastrointestinal fluid			
Secreted		Reabsorbed	
Saliva	1500		
Gastric juice	2500		
Bile	500		
Pancreatic juice	700		
Intestinal juice	3000		
Total	8200		8000
Therefore, 200 ml lost in faeces			

Table 5.4.2　The composition of urine in an adult (from Valtin, 1979)

pH	5.0–6.0
Osmolality	500–800 mosmol
Specific gravity	1.003–1.030
Urea	200–500 mmol/l
Creatinine	9–17 mmol/l
Na^+	50–30 mmol/l
K^+	20–70 mmol/l
Organic acids	10–25 mmol/l
Protein	0–50 mg/24 h
Urochrome	Traces
Glucose	0–11 mmol/l
Cellular components (epithelial cells, leucocytes)	< 20 000/l

A number of disorders are characterized by the presence of abnormal urinary constituents. Glomerulonephritis and its sequel the nephrotic syndrome feature pathological changes in the glomeruli. Alterations in the permeability properties of the filtration barrier then result in proteinuria, which may be so severe that hypoproteinaemic oedema is precipitated. Glycosuria is a characteristic feature of diabetes mellitus, and bile pigments may appear in the urine in hepatic disorders when jaundice is present, imparting a brown coloration to the urine.

Polyuria

Polyuria is the term used to describe a persistent output of urine which is greater than 2 litres per day. In healthy individuals it can occur when a high fluid intake is in excess of body requirements. It may also be a feature of disorders such as diabetes mellitus, diabetes insipidus and early chronic renal failure. In all of these, the ability to form a concentrated urine in the nephron is impaired.

Oliguria

This is the term used to describe a urine output which is in the range 100–400 ml/24 hours; anuria a range of 0–100 ml/24 hours. Both of these symptoms are caused by the reduced glomerular filtration rate which characterizes hypovolaemia, and different stages of acute and chronic renal failure.

Glomerular filtration

Glomerular filtration is the first stage in urine production. Ultrafiltration of plasma takes place across the barrier of the glomerular endothelium, basement membrane and podocytic epithelium of Bowman's capsule into the intracapsular space. The selective permeability of the barrier prevents the filtration of blood cells and protein macromolecules with molecular weights greater than 70 000. In other respects, the filtrate has approximately the same solute concentrations, pH and osmolality as plasma. Very small amounts of albumin of mol. wt 69 000 are filtered each day. If the filtration barrier is damaged by diseases such as the nephrotic syndrome, or by trauma, then the effective 'pore' size may be increased and large molecules above mol. wt 70 000 may then appear in the urine, e.g. proteins (proteinuria, albuminuria) and red or white blood cells. Red cells may then haemolyse in the urine, releasing detectable amounts of haemoglobin (haemoglobinuria). Alternatively, haemoglobin may be released from red blood cells if haemolysis occurs in the bloodstream as a result of haemolytic diseases or blood transfusion reactions. As haemoglobin has a molecular weight of 68 000, which is near the limits of pore size, it can cross the healthy filtration barrier in small quantities and appear in the urine.

Fluid filters out of the glomeruli into Bowman's capsule because a pressure gradient exists between the two areas. If a hydrostatic pressure difference exists across any porous membrane, water will flow across from the side of higher pressure to that of lower pressure, dragging with it molecules which are smaller than any pores in the membrane.

In effect, the forces involved in glomerular filtration are the forces which govern fluid exchanges between other systemic capillaries and tissues (see Chap. 4.2). Hydrostatic pressure in the glomerular capillary is the major force moving fluid out of the capillary. In lower mammals this has a value of approximately 45 mmHg, but it may be higher in humans. Two forces oppose movement of fluid from the capillary – the colloid osmotic pressure exerted by plasma proteins (25 mmHg), and the hydrostatic pressure in Bowman's space (10 mmHg). The net ultrafiltration pressure is therefore low

$$45 - (25 + 10\,\text{mmHg}) = 10\,\text{mmHg}$$

In addition to their unique permeability characteristics, other important differences exist between glomerular and systemic capillaries.

Due to the movement of fluid exclusively *out* of glomerular capillaries, the oncotic pressure rises along the length of the capillary until it equals the hydrostatic pressure. At this point no further filtration can take place, i.e. filtration equilibrium is reached. This factor limits the fraction of plasma filtered to one-fifth (120 ml) of the total 600 ml entering the glomerular capillaries per minute.

Hydrostatic pressure decreases only very slightly along the glomerular capillary length. It falls markedly in systemic capillaries.

As a consequence of these differences, note that in systemic capillaries the balance of forces favours a movement of fluid out at the arteriolar end and back at the venous end. In glomerular capillaries, fluid moves exclusively out and across the filtration barrier.

The **glomerular filtration rate (GFR)** is defined as the volume of plasma filtered through the glomeruli in 1 minute. In health, the balance of forces in the glomerular capillaries limits the volume of plasma filtered to about 120 ml/min. However, as the glomerular filtration rate is decreased by acute or chronic renal and cardiovascular disorders, its measurement is used to assist in diagnosis, and to monitor the progression of such illnesses.

MEASUREMENT OF THE GLOMERULAR FILTRATION RATE

In order to measure the GFR, it is necessary to estimate the renal clearance of a harmless chemical marker which is freely filtered by the glomerulus and is neither secreted nor reabsorbed in the nephron, i.e. the amount of marker that finally appears in the urine depends only on the rate at which it is filtered at the glomerulus. Furthermore, whilst the GFR is measured, plasma concentrations of the marker must remain constant.

Renal clearance is defined as the volume of plasma from which the kidneys remove the marker in 1 minute. For such a marker, which is freely filtered and unmodified during its transit through the nephron, clearance is equal to the glomerular filtration rate.

The GFR is most conveniently and routinely estimated in humans by measuring the clearance of creatinine, an end-product of muscle metabolism. Creatinine has the advantages of continuous internal production, and its plasma levels are virtually constant over a 24-hour period. The method for estimating creatinine clearance is relatively straightforward too: urine is collected from

the patient over a 24-hour period, during which time a blood sample is taken. Creatinine concentrations are then measured in both samples by routine biochemistry and the glomerular filtration rate is calculated from the urine volume excreted per minute (V), concentration of creatinine in mg/ml urine (U), and the plasma concentration of creatinine in mg/ml (P).

Calculation of the GFR
For solutes such as creatinine which are freely filtered at the glomerulus and neither secreted nor reabsorbed in the nephron, the amount filtered from the plasma must equal the amount that is excreted in the urine.

Filtered creatinine. The amount of creatinine filtered at the glomerulus is dependent on the glomerular filtration rate in ml/min and the plasma concentration of creatinine (P) in mg/ml, i.e.

$$\text{creatinine filtered} \ = \ \text{GFR} \times P$$

Excreted creatinine. The amount of creatinine excreted is dependent on the urine flow rate (V) in ml/min and the urine-concentration of creatinine (U) in mg/ml, i.e.

$$\text{creatinine excreted} \ = \ U \times V$$

The amount of creatinine filtered from the plasma equals the amount excreted in the urine. Therefore,

$$\text{GFR} = P \ = \ U \times V$$

and

$$\text{GFR (ml/min)} \ = \ \frac{U \times V}{P}$$

As it varies with body size, the normal range of GFR in humans is expressed per $1.73\,\text{m}^2$ surface area.

REGULATION OF RENAL BLOOD FLOW AND GLOMERULAR FILTRATION RATE

As described earlier, the hydrostatic blood pressure in the glomerulus is the main force which brings about the ultrafiltration of plasma, moving fluid out of the glomerular capillary across the filtration barrier and into Bowman's space. It would be logical to expect the hydrostatic pressure, and therefore the glomerular filtration rate, to vary directly according to changes in the arterial

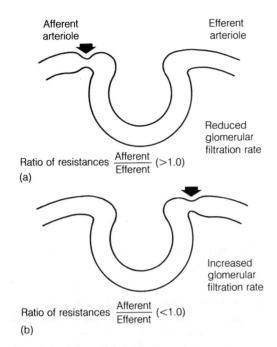

Afferent arteriole

Efferent arteriole

Reduced glomerular filtration rate

Ratio of resistances $\dfrac{\text{Afferent}}{\text{Efferent}}$ (>1.0)

(a)

Increased glomerular filtration rate

Ratio of resistances $\dfrac{\text{Afferent}}{\text{Efferent}}$ (<1.0)

(b)

Figure 5.4.6 Changes in glomerular filtration brought about by altered resistance in arterioles

blood pressure. However, both remain fairly constant over a range of arterial systolic blood pressures, extending from 80 mmHg to 180 mmHg. How then is the glomerular hydrostatic pressure controlled and why does it, and the GFR, show such small changes when the arterial blood pressure varies over such a wide range?

Consider first the structure of the afferent and efferent arterioles supplying each glomerulus. A thin layer of smooth muscle, richly supplied with sympathetic nerve fibres, is present in the walls of these blood vessels. Acting like a sphincter, the degree of muscle contraction controls the diameter of the arterioles and thereby the resistance to blood flow. If the afferent arteriole is constricted, glomerular hydrostatic pressure and GFR will decrease; whereas constriction of the efferent arteriole raises the glomerular hydrostatic pressure and GFR (Fig. 5.4.6).

A number of factors, including the following, are known to influence changes in the diameter of the afferent and efferent arterioles.

Sympathetic nervous stimulation
Sympathetic nerve fibres supply the renal arterioles. Stimulation of these nerves in situations of emotional stress, pain, shock, haemorrhage and vigorous exercise causes constriction of the afferent arterioles, reducing the renal blood flow and glomerular filtration rate. In life-threatening situations, reduction in blood flow to the kidney makes additional blood available to perfuse such vital structures as the brain and heart.

In a healthy individual, resting in a supine position, the sympathetic nerves transmit few impulses to the renal blood vessels. Changes in posture, such as standing up, cause a transient pooling of blood in the veins, and as a result the venous return and cardiac output decrease. In order to maintain the systemic blood pressure, this triggers a compensatory reflex response by the sympathetic nervous system, producing constriction of blood vessels supplying the skin and visceral organs including the kidney. Due to autoregulation, no changes in GFR accompany this alteration in renal blood flow. Autoregulatory mechanisms adjust the degree of constriction in the afferent/efferent arterioles such that glomerular hydrostatic pressure and GFR remain stable.

Angiotensin II
Angiotensin II is a powerful vasoconstrictor produced following the release of the hormone renin from the juxtaglomerular apparatus in the kidney (see Chap. 2.5). Renal blood vessels are highly sensitive to the effects of angiotensin II, which causes constriction of the efferent arteriole and an increase in the GFR.

Autoregulation
The kidney is able to regulate its own blood flow over a wide range of arterial blood pressures from 80 mmHg to 180 mmHg. This phenomenon is termed autoregulation and it has been shown to persist in isolated, perfused, denervated kidneys. It suggests that the kidney has some internal mechanisms which maintain the GFR constant by altering the diameter of the afferent and efferent arterioles whenever the systemic blood pressure changes. One factor which appears to play an important role in autoregulation is prostaglandin E_2 (PGE_2). PGE_2 is synthesized by interstitial cells in the renal medulla, and has been shown to cause renal vasodilation, increasing renal blood flow and GFR. These effects oppose those of sympathetic nerve stimulation described earlier. It seems likely that prostaglandins and angiotensin II are only two of a

number of factors probably involved in the phenomenon of autoregulation.

Autoregulation has its limitations and it does not sustain the renal blood flow and GFR when the sympathetic nerves are stimulated. Neither will it maintain the GFR during shock when the systolic blood pressure is less than 80 mmHg, which is outside the autoregulatory range.

AGE AND GLOMERULAR FILTRATION

At birth the glomerular filtration rate is only 30 ml/min per m², and the volume of urine passed each day is only 20–50 ml. Although the GFR rises steadily afterwards as the infant matures, up to the age of 3 months the low GFR restricts the rate at which water can be excreted. This is one reason why babies are so vulnerable to circulatory overload and oedema when undergoing fluid replacement therapy.

At the other end of the spectrum, the GFR markedly declines in old age due to the decrease in functioning nephrons. In an 80 year old, the GFR has fallen by 50% to 60–70 ml/min per 1.73 m². Obliteration of the afferent arterioles by degenerative vascular disease is partly responsible for these changes. Reduction of GFR in the elderly must be taken into consideration when drugs which are excreted unchanged by the kidney are prescribed, e.g. digoxin, sulphonamides. Reduced excretion of a drug can lead to an elevation of its concentration in the plasma, increasing the likelihood of toxic effects. The same problem applies in the oliguric/anuric stages of renal failure. Therefore, for elderly patients, it may be necessary to reduce the dosage of these drugs to minimize the risk of toxicity. However, at both ends of the age spectrum, the changes in GFR usually only cause problems during illness which demand that the kidney responds rapidly in the excretion of water, electrolyte and acid or base loads.

GLOMERULAR FILTRATION IN RENAL FAILURE

In order for most excretory products to be eliminated by the kidney, they must first be filtered. If the filtering mechanism is impaired by renal, ureteric or cardiovascular disorders, the ability of the kidney to excrete nitrogenous waste products and regulate water, electrolyte and acid–base balance may also be impaired. As a consequence of severely impaired filtration, the concentrations of excretory products in plasma, such as urea and creatinine, increase, precipitating uraemia, and the urine output falls. If during this time the individual is maintained on a normal dietary intake, then the failure to excrete urea and fluid and electrolytes which are surplus to requirements produces serious problems. The severity of these problems depends on a number of factors.

(a) The extent to which the GFR is reduced.
(b) The total number of functioning nephrons which remain.
(c) Whether or not it is possible to control the uraemia, fluid and electrolyte problems by limiting protein, fluid and electrolyte intake.
(d) The temporary or permanent nature of the renal damage.

Unfortunately, overt symptoms or renal failure may not appear until the GFR has fallen below 30 ml/min, by which time the blood urea levels have quadrupled (normal 3.0–7.0 mmol/l) and the individual is uraemic and oliguric. One major consideration, then, is whether the uraemia can be controlled by limiting the protein intake or if alternative routes for the excretion of nitrogenous wastes must be provided by dialysis.

Acute renal failure

In acute renal failure there is an abrupt cessation of normal renal function which in most cases is apparent as a drastic fall in the glomerular filtration rate. As a result, oliguria or anuria is present in the early stages, together with uraemia. Fortunately, the loss of renal function is frequently reversible. The reduction in GFR and resulting oliguria are brought about by the following.

PRERENAL CAUSES

(a) Reduced renal perfusion due to low cardiac output in shock, heart failure, severe dehydration and haemorrhage.
(b) Constriction of the afferent arteriole which occurs in these emergency situations as an autonomic reflex response in order to divert blood to other vital organs.

Both factors reduce the glomerular hydrostatic pressure and thus the major force for glomerular filtration.

RENAL CAUSES

(a) Tubular obstruction, which may be brought about by sloughing necrosis (acute tubular necrosis – ATN) of the epithelium lining the nephron due to ischaemic or toxic damage. This causes the hydrostatic pressure in Bowman's capsule to rise, reducing the GFR.
(b) Back leakage of filtered fluid through the damaged tubular epithelium may occur, causing oliguria.
(c) Constriction of the afferent arteriole by sympathetic nerve stimulation (see Prerenal causes).

POST-RENAL CAUSES

In this case the reduced GFR and oliguria are caused by an obstruction in the lower urinary tract. Prostatic hypertrophy, bilateral ureteric strictures, stones in the renal pelvis or ureter, and compression of the ureter and bladder by external neoplasms are the precipitating causes. Obstruction to the flow or urine generates a back pressure which eventually raises the hydrostatic pressure in Bowman's capsule, reducing the GFR.

Chronic renal failure

In the chronic form of renal failure the number of functioning nephrons is irreversibly reduced due to the progression of diseases such as pyelonephritis, glomerulonephritis, hypertensive disease and diabetic nephropathy. Glomerular damage, with distortion of the cortex and medulla due to fibrosis and scarring, are common features of these disorders. The progression from health to illness, in contrast to acute renal failure, is slow, taking place over months to years. The GFR falls slowly. In the early stages of chronic failure, loss of the ability to concentrate urine causes polyuria, but in the terminal stages of the GFR is reduced to such a level that oliguria and severe uraemia occur.

PROBLEMS IN OLIGURIC/ANURIC RENAL FAILURE

It is beyond the scope of this text to give a detailed consideration of the problems arising in the oliguric/anuric stages of acute or chronic renal failure, when the glomerular filtration rate is severely reduced. Three major problems will be briefly discussed here.

Failure to excrete a fluid load
Oliguric patients rapidly expand their body fluid compartments, precipitating dyspnoea and peripheral oedema due to heart failure, unless their water intake is restricted. Fluids should be given to replace extrarenal losses and the urinary losses over a 24-hour period. Assuming an insensible loss of 1 litre per day, and a gain of 600 ml water from oxidative metabolism, the fluid intake in a totally anuric patient should not exceed 1000 − 600 ml = 400 ml.

Expansion of the extracellular fluid compartment is exacerbated in the presence of a positive sodium balance, therefore in anuria or very severe oliguria, restricting the sodium intake may be necessary. As a general rule, positive Na^+ balances are avoided by restricting the dietary intake where necessary.

Failure to maintain electrolyte balance
In the severely oliguric or anuric patient, failure to excrete a potassium load may result in fatal hyperkalaemia, and failure to excrete the hydrogen ions of 'fixed' acids may result in metabolic acidosis. Restricting the dietary K^+ intake, administration of Na^+/K^+ exchange resins, and the provision of a high-energy carbohydrate diet which minimizes K^+ produced from tissue breakdown, may be adequate measures to control the patient's serum K^+ level. A high-energy carbohydrate diet with restricted protein content may also reduce the production of 'fixed' acids from protein and lipid metabolism.

Uraemia
A complex array of problems is associated with uraemia, including anorexia, nausea, vomiting due to gastrointestinal irritation, and bleeding from the gut wall. Uraemic irritation of the skin may cause pruritus, and muscle cramps and stomatitis may be present. The effects of uraemia on the central nervous system include irritability, confusion, drowsiness and coma, and may lead to death.

Uraemia may be conservatively controlled by the following.

(a) Decreasing the protein intake in order to

curtail the production of urea and other nitrogenous wastes.

(b) Preventing the catabolism of protein stores in muscle by supplying a high-energy carbohydrate diet. Any sources of infection should be eliminated to prevent catabolism.

In the presence of severe uraemic symptoms, uncontrollable fluid overload, hyperkalaemia and metabolic acidosis, the only options available are to institute haemodialysis or peritoneal dialysis. In those patients with chronic renal failure, where the return of normal renal function cannot occur, a renal transplant offers the only hope of escape from lifelong intermittent dialysis. Ideally, dialysis should be instituted before severe uraemia sets in; it is mandatory once the GFR falls below 3 ml/min.

Tubular reabsorption and secretion: the modification of the glomerular filtrate

In the first stage of urine production, during the process of glomerular filtration, an ultrafiltrate is formed of approximately the same solute concentrations, osmolality and pH as plasma. Blood cells and protein macromolecules of a molecular weight greater than 70 000 are not filtered.

During the second stages in the production of urine – tubular reabsorption and secretion – the filtrate is greatly modified as it moves along the nephron. Vital solutes which must be conserved, such as glucose, amino acids and electrolytes, are reabsorbed, together with water, passing from the lumen of the nephron across the epithelial layer, to be transported away by peritubular blood capillaries. A few substances are actually secreted into the filtrate in the reverse direction, including hydrogen ions, ammonia and drug metabolites. By the time the final excretory product – urine – drains from the collecting duct into the renal pelvis it is greatly reduced in volume, and contains nitrogenous waste products together with electrolytes which are surplus to body requirements.

An idea of the vast extent of solute reabsorption accomplished by nephrons is realized by considering the volumes and quantities involved. During

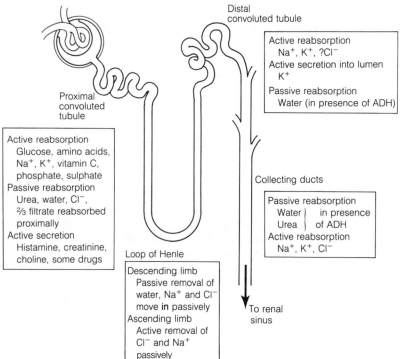

Figure 5.4.7 Regional specialization in reabsorption and secretion in the nephron. N.B. Throughout the nephron, exchange of Na^+ for H^+, HCO_3^- reabsorption and NH_3 secretion occur. (See Acid–base balance p. 534.)

the course of one day, 180 litres of plasma are filtered through the kidneys, containing about 500 g sodium bicarbonate, 100 g amino acids, more than 1 kg sodium chloride, and 270 g glucose. Only negligible amounts of these solutes appear in the 1–1.5 litres of urine eliminated each day.

Some variation exists in the volume of fluid and type of solutes reabsorbed in different regions of the nephron. By far the greater degree of reabsorption occurs in the proximal convoluted tubule, where two-thirds of the filtrate is removed. Glucose and amino acids are reabsorbed almost exclusively here, whilst water and electrolytes are also reabsorbed in more distal areas. A summary of regional specialization in reabsorption in the nephron is shown in Fig. 5.4.7.

In this section, some of the transport mechanisms by which reabsorption and secretion take place are described, together with some of the major disorders which impair them. Reabsorption of HCO_3^- and excretion of H^+ are considered separately in the section on acid–base balance. Similarly, factors which regulate Na^+ and K^+ balance are also described at a later stage.

TRANSPORT MECHANISMS IN THE NEPHRON

In order for any substance to be reabsorbed from the nephron, it must be transferred across the plasma membranes and cytoplasm of the epithelial cell into interstitial fluid, and from there into the peritubular blood capillaries. In the case of secretion, the reverse applies. The tubular transport mechanisms which permit reabsorption and secretion are either passive or active.

Passive transfer

Passive transfer is defined as the movement of non-electrolytes and ions across cell membranes according to the chemical and electrical gradients which prevail. In effect, these solutes move 'downhill', from an area of high to low chemical concentration or electrical potential. In addition to passive movement down a concentration gradient, charged solutes such as ions must also negotiate electrical gradients and move across a 'polarized' membrane to the side which is oppositely charged. In a polarized membrane there is a distribution of electrical charge such that one side of the membrane is positively charged and the other side negative; this is called a **potential difference**.

Ions with a negative charge will move passively across the membrane to the side which is positively charged, and vice versa. No energy is directly used in the process of passive transfer, but it is used in setting up the gradients in the first place. All living cells have an intracellular ionic composition which is different from their environment, and a negative electrical potential operates across the plasma membrane. (This is discussed further in Chap. 2.1.)

Active transfer

In contrast to passive movements, active transfer is defined as the 'uphill' movement of solutes against an unfavourable chemical or electrical gradient. In this case, solutes are moved from an area of low to high concentration, and ions are transported against an unfavourable electrical potential. A direct use of energy in the form of ATP is consumed in powering the active transport of solutes against unfavourable gradients. However, *net* movement of an ion may take place down a favourable concentration gradient but still require energy if a greater unfavourable electrical gradient is also present, and vice versa.

Both active and passive mechanisms may be involved in the reabsorption of solutes across the epithelium of the nephron, but the overall process is called 'active' as long as one of the steps involves active transport.

A number of models have been put forward to explain active transport across cell membranes, but such are the limitations of experimental techniques at present that they are difficult to prove. Nevertheless, an active 'pump' mechanism must have some sort of 'carrier' mechanism for binding a specific solute on one side of the cell membrane and discharging it on the other, with energy supplied in the form of ATP to move the solute plus carrier. In fact, the enzyme Na^+/K^+ ATPase mediates the transport of Na^+ and K^+ across the plasma membrane in all cells, not just the kidney epithelium. This is a protein (as are all enzymes) with specific binding sites for Na^+ and K^+, which extrudes Na^+ from the cell and moves K^+ inside.

Little is known about the nature of other membrane carriers, but it seems likely that they too are proteins and are part of the plasma membrane. Carriers mediate not only active transport but some passive transfer systems too where the favourable concentration gradient is the driving force which moves the carrier across the membrane. It is an interesting feature of carrier-mediated active transfer systems that they may be interdependent, and that more than one solute can use each carrier. This is certainly the case where glucose is reabsorbed in the nephron in conjunction with other sugars; and glucose transport is linked to the active transport of sodium ions.

Glucose

Glucose is reabsorbed in the proximal convoluted tubule against a concentration gradient by active transport. The carrier system for glucose also transports other sugars such as galactose, mannose, xylose and fructose. An interesting feature is that glucose transfer is linked to the active transport of sodium ions, a phenomenon known as **cotransfer**. Glucose is preferentially bound by sugar carrier mechanism, so that in the presence of an increased glucose concentration, reabsorption of the other sugars is decreased.

Active transport systems such as that involved in glucose reabsorption can only operate up to a certain limit, known as the **transport maximum** (Tm value). In healthy individuals with plasma glucose concentrations in the range 4.2–6.7 mmol/l, virtually all the filtered glucose is reabsorbed. However, the maximum rate at which the carrier systems can reabsorb glucose is 375 mg/min. At any plasma glucose level which exceeds the critical renal threshold of 10 mmol/l, so much glucose is filtered that the carrier system is saturated and unable to reabsorb the load. As a result, glucose appears in the urine. In health, the plasma glucose concentration is maintained in the normal range, mainly by the hormone insulin, so that glycosuria does not normally occur. Glycosuria is therefore a feature of diabetes mellitus in which plasma glucose levels are elevated due to insulin deficiency.

Phosphate

Phosphate is also actively reabsorbed in the nephron, mainly in the proximal tubule, by a system with a transport maximum. Plasma phosphate levels are low, at 1.0 mmol/l, so a large part of the phosphate which is absorbed from the gut each day is excreted in the urine to maintain balance. In effect, this means that alterations in renal phosphate reabsorption play a vital role in controlling the total body phosphate pool. Phosphate reabsorption is linked in some way to glucose transport for, if the carrier system for glucose is saturated,

an increased amount of phosphate appears in the urine.

A number of hormones are known to exert a regulatory role on phosphate reabsorption in the nephron. Parathyroid hormone, vitamin D and calcitonin all promote the excretion of phosphate in urine. In hyperparathyroidism, the excessive secretion of parathyroid hormone promotes calcium loss from bone and reabsorption from the gut, leading to hypercalcaemia. Increased quantities of calcium are then filtered by the kidney and eventually calcium deposits appear in the epithelium lining the nephron and also inside the tubules. Together with the reduction in renal phosphate reabsorption, this predisposes to the formation of calcium phosphate calculi. In contrast, severe hypoparathyroidism is marked by increases in the plasma phosphate concentration (phosphataemia) due to the increase in phosphate reabsorption in the nephron.

Sulphate and vitamin C

Sulphate and vitamin C are both actively reabsorbed in the proximal convoluted tubule by systems which are also linked in some way to glucose transfer. Saturation of the carrier system for glucose decreases both sulphate and vitamin C reabsorption.

Amino acids

Amino acids are actively reabsorbed, almost entirely in the proximal convoluted tubules, by systems which show a transport maximum for each amino acid which is transferred. Several carrier mechanisms have been identified.

Amino acid transport is independent of glucose transfer but, as yet, the exact mechanisms involved are not fully understood.

A number of rare inherited disorders, for example, cystinuria and Fanconi's syndrome, are caused by defective amino acid reabsorption and feature aminoaciduria – the presence of amino acids in the urine.

Active mechanisms for ion reabsorption

Sodium and potassium ions are actively reabsorbed throughout the nephron, as shown in Fig. 5.4.8. Chloride ions are passively reabsorbed in the proximal convoluted tubule, but must be actively transported elsewhere. Potassium ions are normally secreted in the distal convoluted tubule. Mechanisms concerned with the reabsorption of

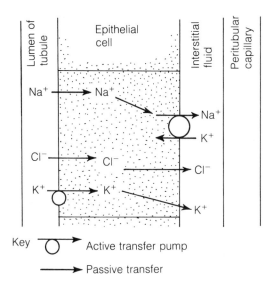

Key ⟳ Active transfer pump

──▶ Passive transfer

Figure 5.4.8 Ion transport in the proximal convoluted tubule

bicarbonate ions are described in the section on acid–base balance.

Passive transport

Urea. The transfer of urea across the tubular epithelium is unusual in that it is passively reabsorbed in the proximal and distal tubules and collecting ducts but secreted into the loop of Henlé. In effect, there is some recycling of urea between the nephron and medullary tissue, a process which is discussed later in the section on the formation of concentrated and dilute urine.

Water. Water is passively reabsorbed throughout the nephron along the osmotic gradients created by the transport of other solutes, notably sodiun ions. In the distal regions of the nephron, water reabsorption is controlled by antidiuretic hormone (ADH).

TUBULAR SECRETION

Tubular secretion is the reverse of reabsorption, entailing the movement of solutes from the peritubular capillary, across the epithelial cell, into the lumen of the nephron. Both active and passive transport mechanisms facilitate this process. Actively secreted solutes include drugs such as penicillins, sulphonamides and chlorothiazides.

In addition, active secretion of thiamine, choline, creatinine and histamine also takes place. However, active secretion is not the main route of removal for any of these substances, for all are primarily eliminated by glomerular filtration.

The circumstances under which potassium ions are passively secreted into the distal convoluted tubule are described in the section on K^+ balance.

Countercurrent multiplication and exchange: the formation of concentrated and dilute urine

In the proximal convoluted tubule, water is passively reabsorbed along the osmotic gradient largely set up by the active transport of sodium ions. In contrast to this, water reabsorption in the distal convoluted tubule and collecting ducts is controlled by antidiuretic hormone (ADH). When the osmolality of plasma rises, signalling a physiological requirement for water, ADH is secreted into the circulation from the posterior pituitary and renders the epithelium lining the distal tubule and collecting duct permeable to water. As a result, water moves passively out of these areas, drawn by the osmotic effects of trapped solutes in the medullary interstitial fluid (interstitium) outside. It is subsequently transported away by peritubular blood capillaries, into the circulation. Under these circumstances, enough water is reabsorbed to return the osmolality of plasma to normal (285 mosmol/kg H_2O) and a concentrated urine is formed. In the absence of ADH, the distal tubule and collecting duct are almost impermeable to water, so that virtually no reabsorption takes place and a dilute urine is eliminated. Before considering the release mechanisms and actions of antidiuretic hormone in greater detail, we must first consider how solutes such as urea, sodium and chloride ions are trapped in the medullary interstitium around the nephron, creating a gradient of increasing osmolality, 300–1200 mosmol/kg H_2O, through the renal medulla (Fig. 5.4.9).

The trapping of solutes in the medullary interstitial fluid forms the basis of the **countercurrent theory**, advanced to explain how concentrated urine is formed by the passive reabsorption of water. According to this theory, the U-shaped countercurrent arrangement of the loop of Henlé and vasa recta operates to cycle or trap solutes in such a way that the osmolality of the medullary

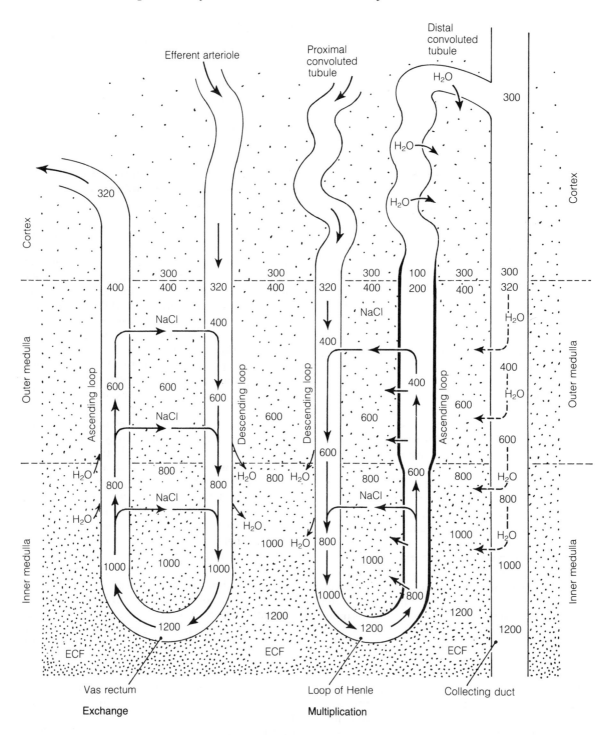

Figure 5.4.9   Countercurrent multiplication and exchange (units = osmolality in mosmol/kg H_2O). (From Creager, J. G. (1983) *Human Anatomy and Physiology*. Belmont, California: Wadsworth)

interstitium is increased. Furthermore, due to countercurrent multiplication in the loop of Henlé, the fluid reaching the distal convoluted tubule is dilute. This is subsequently concentrated by passive reabsorption of water when tubular fluid is exposed to the medullary osmotic gradient exclusively under the influence of ADH.

THE LOOP OF HENLÉ: COUNTERCURRENT EXCHANGE AND THE FORMATION OF CONCENTRATED URINE

Fluid leaving the proximal convoluted tubule is approximately isosmolal with plasma, at 300 mosmol/kg H_2O. In the descending limb of Henlé's loop, the epithelium is relatively permeable to water which moves out passively under the influence of the hyperosmotic medullary interstitium and is removed by peritubular capillaries. It is not known exactly how much water is lost from the descending limb, but the contents become increasingly concentrated towards the tip of the loop, and attain the same osmolality as the interstitium by the time the U-bend is reached.

In the thick ascending limb of Henlé's loop, the epithelium is impermeable to water. However, chloride ions are actively transported out of this area into interstitial fluid accompanied passively by sodium ions. As ions are removed from the tubular fluid and water is not, the fluid delivered to the distal convoluted tubule is dilute and has an osmolality of approximately 100 mosmol/kg H_2O.

The chloride and sodium ions move through the interstitium and diffuse passively back into the descending limb. Urea also diffuses passively into the descending limb; both these mechanisms contribute to the increased concentration of tubular fluid at the bend of the loop.

This cyclic movement of ions from the ascending limb across to the descending limb without the simultaneous transport of water multiplies the concentration of solutes in the descending limb, hence the term countercurrent multiplication. The longer the loop of Henlé, the greater will be the concentration at the tip of the loop. In essence, the countercurrent operation of the loop of Henlé continuously moves ions into the interstitial fluid of the medulla and raises its osmolality, such that the gradient shown in Fig. 5.4.9 exists. Countercurrent multiplication also ensures that a dilute fluid reaches the distal convoluted tubule.

In the presence of ADH, the epithelium lining

the distal convoluted tubule and collecting duct becomes water permeable. Water then moves out passively by osmosis into the medullary interstitium until the osmolality of the interstitium equilibrates with that of the tubular fluid. Water is eventually reabsorbed into the peritubular capillaries and transported away into the circulation. Urine which is concentrated in this way then flows out of the collecting duct into the renal pelvis and from there down the ureter to the bladder.

FORMATION OF DILUTE URINE

The stages are the same as those described above until the dilute fluid reaches the distal convoluted tubule. In these circumstances, ADH is absent and the epithelium lining the distal tubule and collecting duct is virtually impermeable to water, so that very little water is reabsorbed despite the huge osmotic gradient operating between the tubule and interstitium outside. As a result, a dilute fluid reaches the collecting ducts and a dilute urine is excreted.

THE VASA RECTA

Juxtamedullary nephrons possess a U-shaped capillary network closely surrounding the loop of Henlé. By acting as countercurrent exchangers, these blood vessels also help to trap solutes in the medullary interstitium which, combined with the actions of the loop of Henlé, raises the osmolality of interstitial fluid.

Blood moving down the descending limb of the vasa recta (Fig. 5.4.9) gains solutes which move passively into the limb from their higher concentration in the medulla outside. At the same time the descending limb loses water passively, due to the osmotic gradient existing between the interstitium and capillary blood.

In the ascending limb of the vasa recta the reverse process takes place. As blood flows through this region towards the cortex, the external gradient diminishes. As blood has gained solutes in the descending limb, such that its osmolality approaches 1200 mosmol/kg H_2O, these solutes now pass out from their high concentration in blood into the region of lower concentration in the medulla outside. At the same time, water moves back into the ascending limb by osmosis.

The *net* result is that the vasa recta remove water from the inner regions of the kidney, but solutes are recycled in the medulla, helping to raise the osmolality of the interstitial fluid. Blood leaving the vasa recta has an osmolality of 325 mosmol/kg H_2O, having gained a small amount of solutes.

THE ROLE OF UREA

In addition to increasing the water permeability of the distal tubules and collecting ducts, ADH also increases the permeability of the terminal collecting ducts to urea. By the time fluid reaches this area of the nephron, water has been reabsorbed in the distal tubule, therefore it contains a high concentration of urea. Urea moves out of the collecting duct, down its concentration gradient into the medullary and papillary interstitium until the concentrations are the same. The effect of this movement of urea into the interstitium under the influence of ADH is that it too raises the osmolality of this area, aiding the urine-concentrating mechanism. Eventually a large part of the urea is secreted back into the loop of Henlé as part of a recycling process, and some is removed by the vasa recta.

WATER BALANCE: CONTROL OF ADH SECRETION

Water balance is controlled by the actions of ADH which are exerted on the distal convoluted and collecting ducts. In the presence of ADH, a concentrated urine is formed and water is conserved to meet body requirements. In the absence of ADH, a dilute urine is formed to unload water which is in excess of body needs.

Antidiuretic hormone is synthesized by cells of the supraoptic nucleus in the hypothalamus, following which it is transported down axons for storage in the posterior pituitary (see also Chap. 2.5). A number of mechanisms are responsible for triggering the secretion of ADH into the circulation (Fig. 5.4.10).

Primary factors
Osmoreceptors
Pressure receptors (baroreceptors)

Secondary factors
Pain
Emotional stress
Hypoxia
Severe exercise
Surgery, anaesthetics such as cyclopropane and nitrous oxide
Angiotensin II

Osmoreceptors. Osmoreceptors trigger ADH secretion in response to the following.

(a) Dehydration produced by water loss which increases the osmolality of plasma.
(b) 'Relative' dehydration: no net loss of water but a gain of sodium ions which raises plasma osmolality.

In an individual deprived of water, the plasma osmolality rises and ADH is secreted from the posterior pituitary, resulting in water reabsorption in the distal nephron. A reduced volume of concentrated urine is produced and the plasma osmolality is returned to normal (285 mosmol/kg

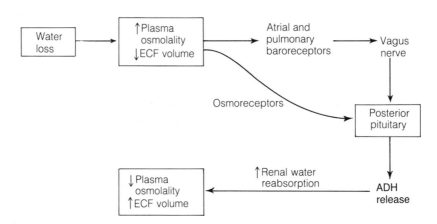

Figure 5.4.10 Stimuli to ADH secretion

H_2O). In contrast, in an individual who consumes a large volume of water, the dilution of plasma decreases its osmolality and ADH secretion is inhibited. As the distal nephron remains impermeable to water in this situation, a large volume of dilute urine is eliminated.

Precisely where the osmoreceptors are located remains unclear, but the available evidence suggests that they are near the supraoptic nucleus or adjacent areas of the hypothalamus and third ventricle.

Pressure receptors (baroreceptors). Antidiuretic hormone secretion is also stimulated by changes in the circulating volume. A reduction of 8–10% in blood volume due to haemorrhage evokes ADH secretion. Conversely, there is some evidence to suggest that expansion of the plasma volume inhibits ADH secretion. The pressure receptors are located in the walls of the atria and large pulmonary blood vessels, and the nerve impulses which trigger ADH release are relayed by the vagus nerve.

Thirst
Alterations in fluid balance that trigger the sensation of thirst also increase ADH secretion. Thus, thirst is experienced when the plasma osmolality rises and when the extra- and intracellular fluid volumes are decreased. It is not known whether the osmoreceptors involved are the same as those which stimulate ADH secretion. Angiotensin II is a hormone which is produced in response to a decrease in blood volume and this appears to bring about the sensation of thirst by constricting blood vessels in an area of the third ventricle known as the subfornical organ. The role of the renin–angiotensin–aldosterone system, which complements the actions of ADH in controlling the extracellular fluid volume, is discussed later.

IMPAIRED CONCENTRATION AND DILUTION OF URINE

Impaired concentration
Damage to the medullary countercurrent system, inadequate secretion of ADH and osmotic diuresis all result in a reduced ability to concentrate urine. Polyuria is the consequence of this, as long as the GFR is not critically reduced at the same time.

Damage to the medullary countercurrent system. Progression of chronic renal disorders such as pyelonephritis results in distortion of the medullary tissue and damage to the loop of Henlé and its associated blood capillaries. As a result, hyperosmolality of the medulla cannot be maintained and so urine cannot be concentrated.

Inadequate ADH secretion. This may be caused by deficient secretion of ADH as is found in the disorder known as **diabetes insipidus**. Five to twenty litres of urine may be eliminated each day in individuals afflicted with this disorder.

A psychological condition known as **polydipsia** (excessive water drinking) results in a dilution of the plasma, lowering its osmolality, and causing inhibition of ADH secretion.

Solute (osmotic) diuresis. In the *early* stages of chronic renal failure there is a reduction in the number of functioning nephrons. As a result, an increased amount of solute is delivered to those nephrons which are still functioning normally. The osmotic pressure exerted by this increase in solute concentration in remaining nephrons prevents the reabsorption of water, and as a result large volumes of urine are eliminated.

In the *late*, 'diuretic', phase of acute renal failure, which follows recovery from oliguria, the ability to reabsorb solutes, particularly sodium and potassium ions, is impaired. Again, this results in the elimination of large amounts of urine.

In diabetes mellitus the elevated blood glucose concentration brings about a glucose load in the nephron which exceeds the reabsorptive capacity, causing an osmotic diuresis.

In the majority of these conditions, continued water and electrolyte losses due to polyuria must be balanced by adequate replacement therapy, or dehydration, hyponatraemia and even hypovolaemic shock may supervene.

Impaired diluting ability
The ability to form a dilute urine is impaired in the following conditions.

Renal failure. The diluting ability is impaired in early acute renal failure and late chronic renal failure once the glomerular filtration rate has fallen below 30 ml/min.

Liver failure. A failing liver may no longer efficiently metabolize hormones such as aldosterone and therefore the plasma concentrations of the hormone rise, resulting in an increase in the reabsorption of sodium ions and water from the nephron. This impairs normal diluting mechanisms.

Heart failure. In heart failure, aldosterone secretion is enhanced due to the low cardiac output reducing renal perfusion, and triggering renin release from the juxtaglomerular apparatus. The retention of sodium ions and water due to the actions of aldosterone on the nephron impairs the normal diluting mechanism.

Excessive ADH secretion. Rarely, excessive ADH secretion by tumours of the lung, brain or pancreas may result in increased water reabsorption from the nephron.

A major problem for the patient who cannot dilute urine is that failure to excrete a fluid load results in expansion of the extracellular fluid volume and may lead to oedema. If the fluid intake is not restricted, signs of water intoxication may become apparent, including anorexia, muscle weakness and confusion. Heart failure may be worsened or precipitated by prolonged circulatory overload.

ACTIONS OF DIURETIC DRUGS ON THE NEPHRON

As their name suggests, diuretic drugs are used to promote diuresis, augmenting the excretion of solutes and water and increasing the volume of urine eliminated by the kidneys. Conventionally, diuretics are widely used for the treatment of fluid retention and oedema in heart failure, liver disorders and renal disorders such as the nephrotic syndrome.

Osmotic diuretics, such as mannitol, are inert substances which are freely filtered at the glomerulus and are not reabsorbed in the tubule of the nephron. The osmotic effects exerted by filtered mannitol prevent the reabsorption of water, increasing urine flow rate and volume. Mannitol can promote a diuresis even when the GFR is reduced. For this reason it is used to prevent acute renal failure occurring in patients who are hypovolaemic and potentially oliguric, for example following haemorrhage.

In contrast, **thiazide diuretics** such as hydrochlorothiazide inhibit the reabsorption of sodium and chloride ions in the ascending loop of Henlé and proximal and distal nephron, thereby promoting water loss. The **loop diuretic**, frusemide (Lasix), which is related to the thiazides, acts in a similar way, enhancing the excretion of chloride ions. However, both drugs produce an increase in the excretion of potassium ions so that it is necessary to provide K^+ supplements during the therapy. Frusemide is a very potent diuretic and the diuresis it evokes may reach 10 litres/day in a grossly oedematous patient. Its action must therefore be very closely monitored by observing strict fluid balance recordings, and any impending signs of dehydration should be immediately reported.

Potassium-sparing diuretics, such as spironolactone, act as antagonists to the hormone aldosterone, resulting in decreased reabsorption of sodium in the distal convoluted hibule, and hence promoting sodium and water loss. Decreased Na^+ reabsorption is balanced by K^+ retention at this site and so K^+ supplements are not necessary with these drugs. Potassium retention is one potentially toxic side-effect of drugs which act as aldosterone antagonists.

The renin–angiotensin–aldosterone system

The operation of the renin–angiotensin–aldosterone system plays an important part in the maintenance of the extracellular fluid volume, sodium and potassium balance and the regulation of blood pressure.

Renin is a proteolytic enzyme which is secreted into the circulation by the juxtaglomerular apparatus of the kidney; its source appears to be the granular epithelioid cells in the wall of the afferent arteriole. Renin cleaves an α_2 globulin in plasma – angiotensinogen – to form the decapeptide (i.e. composed of 10 amino acids) angiotensin I. As blood circulates through the lungs and kidneys, this is converted by an enzyme in these two sites to the octapeptide, angiotensin II (Fig. 5.4.11). Angiotensin II brings about the following physiological effects.

(a) Increased peripheral resistance in arterioles and small arteries, increasing systemic blood pressure.

(b) Aldosterone release from the adrenal cortex,

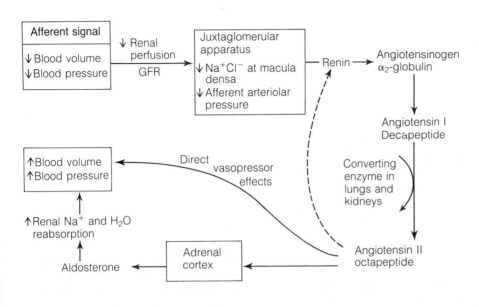

Figure 5.4.11 The renin–aldosterone system

- - - - Negative feedback

which increases the reabsorption of sodium ions and water in the nephron.

(c) Thirst.

The operation of the entire system is summarized in Fig. 5.4.11. Angiotensin II exerts a negative feedback effect on renin release.

RENIN RELEASE

Renin is released from the juxtaglomerular apparatus in response to the following.

(a) A fall in perfusion pressure in the afferent arteriole. This stimulus is probably mediated by receptors (baroreceptors) in the wall of the afferent arteriole as the blood pressure falls.
(b) A decrease in the delivery of Na^+ and Cl^- to the distal convoluted tubule, which is detected by macula densa cells. This takes place whenever blood pressure and volume are decreased and the GFR is reduced. A low flow rate of fluid through the nephron also allows more time for Na^+ and Cl^- reabsorption, reducing their concentration by the time fluid reaches the distal convoluted tubule.
(c) An increase in the activity of the renal sympathetic nerve supply in shock and haemorrhage, which brings about renin release via a direct action on the juxtaglomerular cells and also constriction of the afferent arteriole.

(d) A fall in the plasma concentrations of Na^+ and K^+ increases renin release, and vice versa.

A reduction in plasma Na^+ concentration brought about by excessive losses due to vomiting, sweating, and severe diarrhoea may be compensated in part by increased aldosterone secretion which promotes renal Na^+ reabsorption.

The role of renin in the maintenance of normal blood pressure is not fully understood, and even more difficult to assess is its role in the aetiology of hypertension.

Hypertension and the kidney

Hypertension is secondary to a definable cause in only 10% of cases and, of these, renal disorders account for the majority. Renal causes of hypertension may be considered in three categories.

RENOVASCULAR HYPERTENSION

Stenosis of the renal artery, which is most commonly brought about by arteriosclerosis, causes this form of hypertension. Plasma renin activity is elevated in 65% of the patients suffering from this disorder. The reduction in renal perfusion brought about by renal artery stenosis leads to renin release, culminating in the formation of angiotensin II which elevates blood pressure via direct effects on arterioles and via aldosterone, increasing Na^+ and water retention and hence the circulating volume.

RENAL PARENCHYMAL HYPERTENSION

This form of hypertension is commonly found in individuals suffering from chronic renal failure. Many of these patients have normal plasma renin levels, although in a few cases slight elevations have been detected. It is possible that this form of hypertension may be associated with failure to excrete a salt load or with an increased sensitivity to angiotensin II.

RENOPRIVAL HYPERTENSION

Renoprival hypertension occurs in those patients whose kidneys are totally non-functioning or have been removed and who are undergoing permanent haemodialysis. In this situation, renin cannot be involved in the cause of hypertension, but it is possible that following nephrectomy or chronic renal damage, a renal vasodilator, which lowers the systemic blood pressure, is lost. This rasodilator may be prostaglandin in E_2 (PGE_2) which is produced in healthy individuals by interstitial cells of the renal medulla. The observed effects when PGE_2 is administered to normotensive and hypertensive individuals are a lowering of the blood pressure and promotion of the renal excretion of Na^+ and water. It may be involved in the autoregulation of renal blood flow and GFR, and may fulfil a role as a **natriuretic** hormone.

As yet, a role for renin in the cause of essential hypertension is speculative, as plasma renin activity is elevated in very few individuals suffering from this disorder. However, the malignant phase of hypertension is associated with increased renin secretion.

Strategies for reducing the hypertension associated with increases renin secretion include

(a) restriction of dietary Na^+ intake
(b) administration of diuretics in the presence of oedema
(c) reduction of plasma renin activity by administration of drugs which block the effects of the sympathetic nervous system, e.g. propranolol, oxprenolol, acebutalol
(d) inhibition of the enzyme which converts angiotensin I to angiotensin II by drugs such as captopril.

Sodium balance

In health, sodium balance is regulated exclusively by the kidneys, the plasma Na^+ concentration remaining within the narrow limits of 135–146 mmol/l. In the face of any challenge to Na^+ balance, rapid adjustments are made by the kidney in order to prevent Na^+ depletion or overload.

Of the filtered load of Na^+ (approximately 1 kg/24 hours), 67% is reabsorbed in the proximal convoluted tubule and a further 33% is reabsorbed in the remainder of the nephron, mainly in the loop of Henlé.

If the GFR rises, Na^+ excretion will be increased, and if it falls, Na^+ retention will result. Compensatory mechanisms must then be brought into operation to prevent excessive Na^+ losses or gains whenever the GFR alters. These include

(a) autoregulation, which stabilizes the GFR, preventing Na^+ depletion or overload
(b) reabsorption of a constant percentage of the filtered Na (67%) in the proximal convoluted tubule, despite changes in filtration rate.

CHALLENGES TO SODIUM BALANCE

In response to an increased intake of Na^+ in the diet or intravenously, the following mechanisms act to readjust the balance; the *reverse* takes place if the Na^+ intake is reduced or excessive Na^+ losses occur in sweating, prolonged vomiting or diarrhoea.

(a) In the presence of an Na^+ load, the GFR increases in order to increase the amount of Na^+ excreted in the urine. The increase in GFR is probably brought about by expansion of the extracellular fluid volume which increases the blood pressure.
(b) Tubular reabsorption of Na^+ decreases in the nephron if the intake of Na^+ is increased. Two hormonal factors are responsible here: Na^+ loading decreases aldosterone secretion and increases natriuretic hormone secretion (probably PGE_2), both of which restore Na^+ balance by reducing tubular Na^+ reabsorption and increasing the urinary Na^+ excretion.

Also involved in alterations in tubular Na^+ reabsorption are the two different types of nephrons – juxtamedullary and cortical nephrons. Cortical nephrons have shorter proximal convoluted tubules and loops of Henlé, and hence less Na^+ reabsorptive capacity than the juxtamedullary nephrons. In the presence of an Na^+ load, it is probable that renal blood flow and GFR are increased in the salt-losing cortical nephrons, in an attempt to readjust the balance. The reverse takes place in situations where Na^+ depletion is a problem: renal blood flow is directed to the salt-saving juxtamedullary nephrons which are better equipped to reabsorb Na^+.

Potassium

Net reabsorption of K^+ takes place in the nephron; only 10–20% of the filtered load of K^+ is excreted in health. Most of the filtered K^+ is reabsorbed in the proximal convoluted tubule and loop of Henlé. In a healthy person taking a normal diet, K^+ is secreted into the distal convoluted tubule, whilst further reabsorption of K^+ takes place in the collecting ducts. Alterations in distal reabsorption or secretion of K^+ are seen where the dietary intake of K^+ varies. In the presence of a low intake, net secretion of K^+ is reversed in favour of reabsorption in the distal tubule. If the K^+ intake is high, enhanced secretion into the distal tubule takes place.

Other factors which increase K^+ excretion in the distal nephron include the hormones aldosterone and cortisol. Both act to promote the reabsorption of Na^+ particularly in the distal nephron, at the expense of K^+, which are excreted. Thus hypokalaemia is a feature of Cushing's disease and primary aldosteronism (where hypersecretion of ACTH and aldosterone, respectively, occurs) and is responsible for the severe muscle weakness, flaccid paralysis, cardiac arrhythmias and alkalosis which feature in these disorders.

Acid–base disorders also affect K^+ excretion in the nephron. In acidotic states, K^+ excretion is decreased, whereas in alkalosis it is enhanced.

Acid–base balance

It is essential that the pH of blood is maintained within the narrow limits of 7.35–7.45 as the activities of enzymes which govern intracellular chemical reactions are dependent on an optimum pH range, as are the structures of other proteins. Deviation outside the normal pH range may produce fatal changes in metabolism; the range compatible with life is 7.0–7.8.

Each day, the pH of blood is maintained in the face of a massive acid onslaught. Over a 24-hour period, 10 000–20 000 mmol CO_2 are produced by oxidative metabolism and converted to the weak acid, carbonic acid

$$H_2O + CO_2 \rightleftharpoons H_2CO_3 \rightleftharpoons H^+ + HCO_3^-$$
Carbonic anhydrase

In addition to this daily source of hydrogen ions are the fixed acids such as sulphuric and phosphoric acid which are derived from protein and lipid metabolism respectively. A fixed acid production of 40–70 mmol each day is a feature of the high meat protein diet typical of Western civilizations.

How then, is the threat of acidaemia controlled? Three systems of body defence are employed.

(a) Blood buffer systems.
(b) The lungs.
(c) The kidneys.

BUFFERS (see also Chap. 5.3)

Buffers are chemicals which resist changes in pH on dilution or on addition of acid or alkali. They usually consist of either of the following pairs in solution.

A weak acid with one of its salts.
A weak base with one of its salts.

In order for blood pH to be maintained in the range 7.35–7.45, the ratios of weak acid to salt concentration must be preserved in all the buffer pairs in body fluids. Major buffer pairs in humans include the following.

(a) Hydrogen carbonate (bicarbonate)/carbonic acid pair, predominately located in extracellular fluid $[HCO_3^-]/[H_2CO_3]$.
(b) Hydrogen phosphate/dihydrogen phosphate pair, predominately located in intracellular fluid $[HPO_4^{2-}]/[H_2PO_4]$.
(c) Protein/acid protein pair, located in plasma and intracellular sites including haemoglobin [protein]/[acid protein].

All buffers in extra- and intracellular fluid are in equilibrium with each other, with the result that addition of hydrogen ions to any individual system is eventually reflected in changes throughout all the buffer systems. Addition of hydrogen ions to the hydrogen carbonate system, the major extracellular buffer, will ultimately result in alterations in all the other buffer pairs. It follows, therefore, that regulating the hydrogen carbonate buffer pair $[HCO_3^-]/[H_2CO_3]$ ratio will ultimately control the ratio of [salt]/[acid] in all other buffer systems and maintain the hydrogen carbonate/carbonic acid ratio at 20:1 and the pH of the blood at 7.4.

The capacity of the buffering systems in humans is vast, and 1000 mmol of H^+ can be buffered before the pH in the body fluid compartments falls to fatal levels.

THE LUNGS (see Chap. 5.3)

The lungs play a vital role in controlling the threat to H^+ homeostasis from carbonic acid accumulation, by eliminating 10 000–20 000 mmol CO_2 each day.

Carbon dioxide formed as a result of metabolism diffuses from the tissues into the erythrocytes and, in the presence of the enzyme carbonic anhydrase, is converted to carbonic acid. Hydrogen ions resulting from this are extensively buffered by haemoglobin.

In the lungs the reverse process takes place, and CO_2 is unloaded in expired air. If CO_2 unloading were suppressed for as little as half an hour, this would produce a fatal lowering of blood pH. However, adjustments in ventilation can adjust the pH in minutes following the introduction of an acid load.

As soon as the pH declines below normal limits, the respiratory centre in the brain-stem is stimulated, and the ventilation rate is increased, unloading carbonic acid as H_2O and CO_2 in expired air. This restores the $[HCO_3^-]/[H_2CO_3]$ ratio to 20:1.

THE KIDNEYS

The kidneys provide the third line of defence in maintaining H^+ homeostasis. Unlike the other two defence systems which become effective almost immediately, the renal mechanisms involved in acid–base balance take hours to days to complete.

The renal mechanisms which operate against acidaemia include the following.

(a) Reabsorption of virtually all filtered HCO_3^- to restore the buffering capacity in extracellular fluid.
(b) Excretion of the H^+ of fixed acids.
(c) Reclamation of the HCO_3^- originally consumed in buffering fixed acids.
(d) Excretion of the anions of fixed acids which cannot be converted to CO_2, and therefore cannot be removed by respiration.
(e) Compensation for acidosis by increasing tubular H^+ secretion, and vice versa in alkalotic states.

Reabsorption of hydrogen carbonate
Eighty to ninety per cent of the HCO_3^- which is filtered is reabsorbed in the proximal convoluted tubule and the remainder is reabsorbed throughout the other areas of the nephron. The mechanism which is concerned with HCO_3^- reabsorption is shown in Fig. 5.4.12. The process for conserving

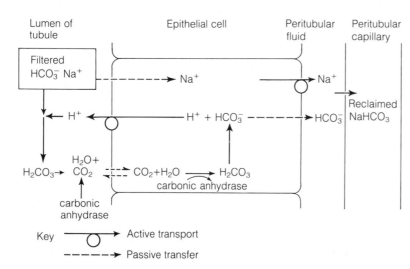

Figure 5.4.12 Acid–base balance: hydrogen carbonate reabsorption in the nephron

HCO$_3^-$ is an indirect one, as filtered HCO$_3^-$ is used to generate CO$_2$ in the tubular fluid in the presence of the enzyme carbonic anhydrase. In turn, the CO$_2$ formed in the tubular fluid is used to generate HCO$_3^-$ inside the tubular epithelial cell and it is this ion which is reabsorbed, not the hydrogen carbonate ion originally filtered. One hydrogen carbonate ion is reabsorbed together with a sodium ion in exchange for each hydrogen ion which is actively transported out into the tubular fluid, as shown in Fig. 5.4.12. Several factors affect the renal conservation of HCO$_3^-$.

Arterial PCO$_2$. If the arterial PCO$_2$ rises due to lung disorders which cause hypoventilation, the blood pH will fall and the patient will develop a respiratory acidosis. In this situation, renal HCO$_3^-$ reabsorption is increased to return the blood pH to normal limits. Exactly the reverse takes place when an individual hyperventilates; the increased unloading of CO$_2$ precipitates a respiratory alkalosis. Here, the kidney compensates by reducing HCO$_3^-$ reabsorption

Serum potassium concentration. An increase in the serum K$^+$ concentration reduces the renal reabsorption of HCO$_3^-$ by disturbing the mechanism shown in Fig. 5.4.12. The reason is that less H$^+$ can be secreted if, due to an increased serum K$^+$, more K$^+$ enter the tubular cells from the blood. If H$^+$ cannot be secreted, then NaHCO$_3$ cannot be reabsorbed. The reverse occurs if the serum K$^+$ concentration falls.

Replacement of hydrogen carbonate used in buffering fixed acids; excretion of H$^+$ of fixed acids

As previously described, fixed acids, including phosphoric and sulphuric acids, are derived from the metabolism of lipids and proteins, respectively. As soon as these acids are formed, they are buffered by the hydrogen carbonate buffer system in plasma as follows.

sulphuric + sodium → disodium + water + carbon
acid hydrogen sulphate dioxide
carbonate

phosphoric + sodium → disodium + water + carbon
acid hydrogen hydrogen dioxide
carbonate phosphate

The carbon dioxide formed is unloaded in expired air through the lungs and the two salts disodium hydrogen phosphate and disodium sulphate are filtered into the nephron. The role of the kidney is to recover the HCO$_3^-$ originally used in buffering these fixed acids, to unload H$^+$ and to excrete the anions of the fixed acids (sulphate and phosphate) which cannot be converted to CO$_2$ and eliminated via respiration. There are two ways in which this can be achieved by the kidney.

Renal excretion of ammonium ions. Disodium sulphate formed by the buffering of sulphuric acid is filtered into the nephron. Figure 5.4.13 summarizes the reactions which subsequently take place in all areas of the nephron. Inside the epithelial cell, H$^+$ and HCO$_3^-$ are produced by the

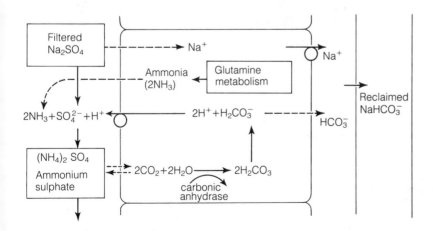

Figure 5.4.13 Acid–base balance: excretion of ammonium ions

same intracellular reactions previously described. Ammonia (NH_3) is also produced in the epithelial cells by the metabolism of glutamine, an amino acid. Ammonia is fat soluble and easily diffuses across the cell membrane into tubular fluid, where it accepts actively secreted H^+, forming ammonium sulphate which is excreted in urine. The reabsorption of sodium and internally produced HCO_3^- into peritubular fluid and blood is dependent on the active secretion of H^+ into tubular fluid. In fact, a large proportion of the ammonium ions (NH_4^+) excreted in urine are in the form of ammonium chloride. The *net* results of the process outlined in Fig. 5.4.13 are as follows.

(a) Indirect replenishment of HCO_3^- stores originally used in buffering fixed acids by intracellular generation in the epithelial cell.
(b) Excretion of H^+.
(c) Reabsorption of Na^+.
(d) Excretion of the anion SO_4^{2-} of sulphuric acid.

In acidotic states, more ammonia is produced by the tubular cells and this enables more H^+ to be excreted in the form of ammonium ions (NH_4^+). In chronic renal failure, where renal tissue is damaged and the cells can no longer form ammonia, less H^+ can be excreted and a metabolic acidosis may ensue.

Formation of titratable acid. Disodium hydrogen phosphate formed from the hydrogen carbonate buffering of phosphoric acid is filtered into the nephron. Hydrogen carbonate ions are generated by the intracellular mechanisms shown in Fig. 5.4.14 and are reabsorbed with Na^+ in exchange for H^+ which are actively secreted. The H^+ are then combined with sodium hydrogen phosphate in tubular fluid to form sodium dihydrogen phos-

phate – **titratable acid** – which is excreted. Again, the ultimate result of this process is that

(a) HCO_3^- are reabsorbed, replenishing the HCO_3^- originally consumed in buffering phosphoric acid
(b) Na^+ are reabsorbed
(c) H^+ are excreted
(d) the fixed acid anion, PO_4^{2-} is excreted.

Thus, in an indirect way, the HCO_3^- originally used in buffering fixed acids have been reclaimed and H^+ excreted, restoring the acid–base balance.

More titratable acid is excreted in urine in acidotic states. As the H^+ concentration in blood rises, eventually the intracellular H^+ concentration also increases, enhancing the active secretion of H^+ into urine.

Approximately 75% of the daily fixed acid load is excreted as NH_4^+, the rest as titratable acid.

Acid–base disturbances

In health, the lungs, kidneys and blood buffers regulate the HCO_3^-/H_2CO_3 ratio at 20:1, so maintaining blood pH within normal limits. The HCO_3^- component of this buffer pair is regulated by the kidney and the H_2CO_3 by the lungs. Acid–base imbalances occur if either the HCO_3^- component is altered by metabolic disorder, or the H_2CO_3 component is affected by respiratory disturbances.

Any decrease in arterial pH to below 7.35 (i.e. an increase in H^+ to above 45 nmol/l) is termed **acidosis**, whereas a rise in pH to above 7.5. (i.e. a decrease in H^+ to below 32 nmol/l) is called **alkalosis**. It is not always possible to classify a

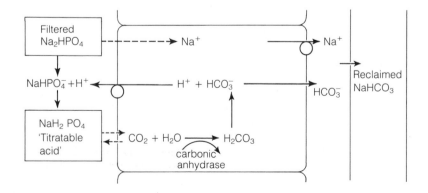

Figure 5.4.14 Acid–base balance: excretion of titratable acid

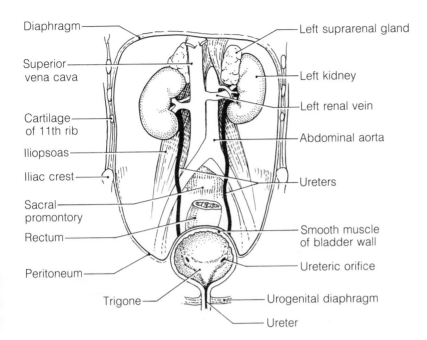

Diaphragm

Superior vena cava

Cartilage of 11th rib

Iliopsoas

Iliac crest

Sacral promontory

Rectum

Peritoneum

Trigone

Ureter

Left suprarenal gland

Left kidney

Left renal vein

Abdominal aorta

Ureters

Smooth muscle of bladder wall

Ureteric orifice

Urogenital diaphragm

Figure 5.4.15 Anatomy of the lower urinary tract

disturbance as purely metabolic or respiratory since mixed disorders frequently occur. For example, an overdose of aspirin (which is acidic) will cause metabolic acidosis, but this stimulates the respiratory centre, and the hyperventilation that follows frequently results in profound respiratory alkalosis.

THE LOWER URINARY TRACT (Fig. 5.4.15)

The ureters

The two ureters are hollow tubes which extend from the renal pelvis to the posterior wall of the bladder. In an adult, each ureter is approximately 30 cm long and lies behind the peritoneum. The wall of the ureter is formed by layers of smooth muscle, lined inside by mucous membrane. The renal pelvis is the funnel-shaped upper end of the ureter.

Urine formed in the nephrons drains from tiny collecting ducts into hollow projections (calyces) of the renal pelvis. Urine passes from here, down through the ureters into the bladder, where it is stored before micturition takes place. Peristaltic contractions in the muscle walls of the ureters help to propel urine down into the bladder. Back-

flow of urine from the bladder is prevented by the oblique angle taken by the ureters as they pass through the bladder wall.

Renal stones (calculi) may pass into the ureters and obstruct the flow of urine. If a stone blocks the ureter, a sudden onset of severe loin pain (colic) occurs which is usually accompanied by nausea, vomiting and haematuria. The upper end of the ureter may dilute to accommodate the urine which accumulates behind the stone; this is known as hydronephrosis. If prolonged, it may lead to acute renal failure.

The bladder

The bladder is a hollow, highly muscular organ, which when empty lies low in the pelvis and expands upwards and forwards in the abdomen as it fills. It is lined by a transitional epithelium which allows stretching as the bladder fills and is able to withstand the hypertonicity of urine. Beneath this lies a sheet of smooth muscle known as the **detrusor**.

At the base of the bladder, the smooth muscle fibres form an internal sphincter which surrounds the urethra. As the detrusor contracts, the internal sphincter also contracts, widening the neck of the urethra and allowing urine to enter. At the same

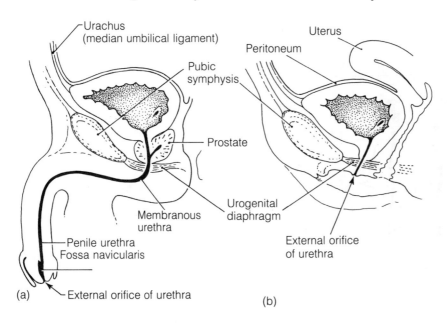

Figure 5.4.16 Anatomy of (a) the male, and (b) the female urethra

time, the levator ani muscles of the pelvic floor relax, 'funnelling' the bladder base and helping urine to flow into the urethra.

The **trigone** is a triangular area formed by the openings of the ureters and bladder neck. Smooth muscle fibres at the apex of this area help to form the internal sphincter.

The urethra

The urethra extends from the neck of the bladder to the external urethral opening (meatus). Due to the tone of surrounding smooth muscle and the activity of the external sphincter, no urine enters the urethra except when micturition takes place. A stratified squamous epithelium lines the internal surface of the urethra, and its lower part is surrounded by the external sphincter, which is formed from striated muscle (Fig. 5.4.16).

In the female the urethra is approximately 4 cm long, running down and along the anterior wall of the vagina; the external meatus opens in front of the vaginal orifice. In the male (Fig. 5.4.16) it is 20 cm long, penetrating through the prostate; the prostatic glands and ejaculatory duct open into it. Three distinct regions are present: the prostatic, membranous and spongy areas. The external sphincter is situated in the membranous region.

Micturition

Micturition requires the coordinated activity of parasympathetic, sympathetic and somatic nerves. It is controlled by higher brain centres in the cerebral cortex, basal ganglia and reticular formation. A centre for the reflex control of micturition is situated in the second, third and fourth sacral segments of the spinal cord. It connects with nerve tracts which descend from, and ascend to, the higher centres, and also with parasympathetic and somatic nerves which supply the bladder and sphincters.

PARASYMPATHETIC NERVE SUPPLY

Motor (efferent) and sensory (afferent) parasympathetic nerves pass in the pelvic splanchnic nerve to and from the bladder. The sensory fibres supply the detrusor and, as the bladder distends, impulses are relayed by these nerves to the sacral segments of the spinal cord. From here they are transmitted in the ascending ventral columns of the cord to the higher centres concerned with the control of micturition. As a result of this passage of impulses, the individual becomes aware of the need to pass urine.

Parasympathetic motor nerve fibres relay to the bladder from the sacral segments of the spinal cord. These nerves supply the detrusor muscle and the internal sphincter and when stimulated during micturition bring about contraction of the detrusor muscle and relaxation of the internal sphincter.

Acetylcholine is the chemical transmitter which relays the signal from the parasympathetic nerve fibres to the smooth muscle cells of the detrusor; contraction of the detrusor follows this. If the drug atropine is administered, for example as a premedication, this can prevent acetylcholine combining with receptors in the muscle cell membrane. As a result, the nerve impulses which cause contraction of the detrusor are blocked, and retention of urine may occur.

SYMPATHETIC NERVE SUPPLY

The sympathetic nerve supply consists of sensory (afferent) and motor (efferent) fibres but their role in micturition in humans appears to be minor.

Sympathetic motor fibres originate from the lower thoracic and upper regions of the spinal cord. It is thought that stimulation of these nerves inhibits contraction of the detrusor and closes the internal sphincter.

Sympathetic sensory fibres relay impulses from the bladder to the thoracic region of the spinal cord and these are transmitted upwards in the spinothalamic tracts to higher centres. The sensory nerve fibres relay impulses in response to painful stimuli; they are stimulated by overdistension, spasm and inflammation caused by bladder infections, stones or carcinoma.

SOMATIC NERVES

Somatic sensory and motor nerve fibres supply the bladder in the pudendal nerves. More specifically, the sensory fibres supply the wall of the urethra and the motor nerves the external spincter.

THE MECHANISM OF MICTURITION

As urine accumulates, it causes the bladder to distend increasing the wall tension. As a result, stretch receptors are stimulated which generate impulses transmitted in the parasympathetic sensory nerves to the sacral centre. When small volumes of urine are present in the bladder, an individual is not aware of any distension. This is because the impulses reaching the sacral centre are prevented from travelling upwards to the cortex by inhibitory impulses descending from these higher centres. However, the presence of approximately 250 ml urine in the bladder sufficiently increases the intensity of impulses to the sacral centre so that the descending inhibition is overcome. Nerve impulses are then relayed upwards to the sensory cortex, and the individual becomes aware of the need to pass urine. Micturition is postponed until the time and place are socially acceptable. When these are satisfactory, inhibition exerted by the motor cortex on the spinal centre is lifted and micturition takes place. Contraction of the detrusor and relaxation of the internal sphincter are brought about by impulses relayed in parasympathetic motor nerves. As soon as urine enters the urethra, stretch receptors are stimulated which generate impulses in sensory afferent nerves to the sacral centre. By reflex, this produces inhibition of the somatic motor nerves which maintain the external sphincter closed while the bladder is filling. As the external sphincter relaxes, urine passes out through the meatus. The detrusor contracts until the bladder is empty, helped by a volley of excitatory impulses relayed down from the higher centres.

Expulsion of urine is also helped by contractions of the abdominal muscles and the performance of Valsalva's manoeuvre, that is, forced expiration against a closed glottis. Control of micturition is learned in infancy and is usually attained at about 2 years of age, when the infant acquires the power to inhibit spinal reflexes and contract the external sphincter at will.

DISORDERS WHICH IMPAIR MICTURITION

For normal micturition to take place, the following criteria must be met.

(a) An intact nerve supply to the urinary tract at all levels, including the higher centres.
(b) Normal muscle tone in the detrusor, sphincters, periurethral and pelvic floor muscles, together with maintenance of an acute angle between the urethra and posterior bladder wall.
(c) Absence of any abnormal obstruction to the outflow of urine in any part of the upper or lower urinary tract.
(d) Normal bladder capacity.
(e) Absence of any psychological or environmental factors which may impair micturition.

Abnormal micturition: some definitions

Frequency is defined as the passage of urine on seven or more occasions during the day and more than twice each night. It is a characteristic feature of the uninhibited neurological bladder, cystitis, lower urinary tract obstructions and conditions in which bladder capacity is reduced due to intrinsic or extrinsic tumours, or pregnancy.

Urgency is most commonly described as a strong desire to micturate which can become irrepressible and lead to uncontrolled voiding of urine – urge incontinence (see below).

Retention is an inability to pass urine despite experiencing the desire to do so. It may be associated with frequency, overflow or incontinence and be preceded by a reduced or interrupted flow of urine. The bladder becomes distended, tense and often painful. Obstruction of the lower urinary tract, for example by prostatic hypertrophy, fibroids and strictures, or compression due to external tumours may all cause retention of urine.

Postmicturition dribbling after the bladder has apparently emptied is a common feature of obstructive disorders due to prostatic hypertrophy in males and the presence of bladder tumours or calculi. It may also occur as a complication following prostatectomy or, in young men, may be caused by failure of the bulbospongiosus muscle to evacuate the urethral bulb during micturition.

Incontinence is defined as the passage of urine at unsuitable times in unacceptable places, due to loss of the voluntary control of micturition. It may occur in neurological disorders, resulting in an uninhibited, atonic, automatic or autonomous bladder. In addition, it can feature in bladder infection (**urge incontinence**), obstructive and compressive disorders (**overflow incontinence**) and in conditions where muscles at the urethrovesicular junction are weakened (**stress incontinence**).

Neurological disturbances

An essential feature of most of the neurological disorders which can impair micturition is that the nerve supply to the bladder and/or sphincters is disrupted by trauma or disease occurring at different levels, i.e. between the bladder and sacral centre, ascending and/or descending tracts in the spinal cord or the higher centres in the cerebral cortex, reticular formation and basal ganglia. Such lesions result in varying degrees of loss of control of bladder function.

Incontinence in neurological disorders presents some unique problems in management. For many patients the incontinence may be intractable, and the sequelae to an increased residual urine in the bladder are formation of calculi, infection and occasionally renal failure. In addition the immobility which results from many neurological disorders predisposes to pressure sores, a problem which is exacerbated in the presence of incontinence. Immobility and recumbency can also hinder the flow of urine from the renal pelvis down through the urinary tract, enhancing the risk of infection and calculus formation.

The major aim in management is to utilize to the maximum whatever bladder function remains to the individual. Some patients may be helped by bladder-retraining regimes, particularly if urge incontinence is the problem, and the drugs described earlier also have a valuable role to play. Patients who have automatic and autonomous bladders should be encouraged to empty their bladders at regular intervals using a moderate degree of manual compression (which is inadequate to cause ureteric reflux), or tapping trigger spots on the inside of the thighs or on the abdomen. In suitable cases a urinary diversion may be performed. Other aids to management include intermittent self-catheterization and, in males, the use of a sheath connected to a urinary drainage bag.

Infective disorders

The term 'significant bacteruria' is used to denote the presence of at least 10^5 colony-forming bacterial units per millilitre of voided urine and indicates that a urinary infection has occurred, since it is unlikely that contaminants could account for this level of colonization. However, Asscher (1980) has maintained that this definition is limiting when applied to urinary infections and suggests instead that: 'if in a symptomatic subject, pus cells are present and pure growth of an organism is obtained, then urinary tract infection is present, regardless of bacterial numbers.' The Medical Research Council Bacteruria Committee defined three states of urinary tract infection in 1979.

Cystitis: an inflammation caused by the presence of bacteria in the bladder or urethra, associated with frequency, dysuria, loin pain and fever.

Covert bacteruria: bacteria are present in the bladder urine, but no symptoms of cystitis are present.

Pyelonephritis: a localized inflammation of the kidneys resulting in scarring, one cause of which is bacteria ascending from the lower urinary tract. This is accompanied by loin pain, fever, dysuria.

The organisms which commonly cause urinary infections are *Escherichia coli* and *Streptococcus faecalis*; the former has been established as the cause of 25% of all hospital-acquired infections. Both of these bacteria are common in the normal gut flora and colonize the perineum. *Staphylococcus albus*, found in the skin flora, is another frequent cause of infection. Other organisms which cause urinary infections, less frequently, include *B. proteus* and *Pseudomonas aeruginosa*.

SYMPTOMS OF INFECTION

These include frequency and urgency, as irritation of the pressure receptors in the bladder wall causes the detrusor to contract when only small volumes of urine have collected. A debilitating dysuria or abdominal pain may be present together with pyrexia, shivering, a raised erythrocyte sedimentation rate and leucocyte count. The urine looks cloudy, smells offensive and contains pus cells and leucocytes. Serious complications can arise from a urinary tract infection, including pyelonephritis which can cause chronic renal failure, and bacteraemia which can lead to septicaemia.

DEFENCE MECHANISMS

Normally, bladder urine is a sterile fluid, but some microorganisms may enter the urethra in both sexes. As the female urethra is short (4 cm), it is relatively easy for microorganisms to enter from the perineal skin surface.

A number of protective mechanisms in the bladder and urethra are known to prevent the entry and growth of bacteria. Immunoglobulins IgA and IgG, which are locally synthesized and secreted by the mucosa, are present in human urine and increased in concentration if an infection occurs. Lactic acid secreted by the mucosal cells also helps to inhibit bacterial growth and, in males, prostatic fluid contains an antibacterial agent. Another important defence mechanism is

that the bladder empties completely during micturition and so no residual urine remains to provide a reservoir for bacterial growth, although approximately 1 ml of urine may remain in the compressed folds of the mucosa. If a residual urine accumulates, for example in an autonomous or automatic bladder or obstructive disorder, the risk of infection is significantly increased, as is that of stone formation. The optimal pH for bacterial growth is at pH 6–7; inhibition of growth occurs at pH values below 5.5 and above 7.55 (normal urine pH 5–6).

SOURCES OF INFECTION

An increased risk of bladder infection exists if a residual urine is present, providing a potential reservoir for bacterial growth, and in any situation where microorganisms may be pushed through the urethral meatus and ascend to the bladder. An infection may then be a sequel to any of the following.

(a) Bladder catheterization and cystoscopy.
(b) Urethral trauma during childbirth or intercourse.
(c) A vaginal discharge.
(d) Obstruction of the lower urinary tract.
(e) Any condition requiring prolonged bedrest when urinary stasis may occur.

In addition, any individual suffering from immune depression is at risk of developing a urinary tract infection, for example while taking a course of cytotoxic or corticosteroid drugs, or in the presence of uraemia. Other vulnerable groups include neonates who have an immature immune system, the elderly, and diabetic patients.

DIAGNOSIS

The presence of urinary tract infection is confirmed by taking a **midstream specimen of urine (MSU)** for culture and sensitivity. An MSU is a specimen of urine which is not contaminated by organisms outside the urinary tract. Before taking the sample, the perineum and urethral meatus in females and the penis in males should be cleaned. To obtain an MSU, the patient first micturates into either a toilet or non-sterile container to clear contaminants from the lower urethra and its meatus and then stops urination in midstream. On recommencing micturition, 30–50 ml urine is collected in a sterile container which is removed, following which the patient can then complete passing urine into the non-sterile receptacle. The MSU is sent to the laboratory immediately in a sterile container, labelled correctly, for culture and sensitivity tests.

MANAGEMENT

After culture and sensitivity of the infecting organism, an appropriate antibiotic is prescribed, usually for a period of 7–10 days. However, in the case of chronic infections, relatively longer term antibacterial drugs may be given, for example co-trimoxazole (Septrin) for 3 weeks duration. Other important aspects of management are to increase the fluid intake to 2.5–3 litres daily to aid expulsion of bacteria from the urinary tract, unless fluid restrictions are essential. Scrupulous attention to perineal hygiene is required, particularly so in patients who are suffering from faecal incontinence. Analgesia is prescribed for abdominal or loin pain and the patient should be advised to avoid intercourse until the symptoms have disappeared. As a prophylactic measure it is vital that an adequate fluid intake of at least 2 litres daily is attained by immobile, bedridden patients in whom urinary stasis is likely to occur, and that remobilization is attempted as soon as the patient's condition allows. A repeat MSU is taken when treatment is complete, to confirm that the infection has been eradicated.

CATHETERIZATION

A positive correlation has been shown to exist between the procedure of urethral catheterization and bladder infection. It is a sobering thought that 19.5% of catheterized males and 22.7% of females had a demonstrable infection in a study by Meers in 1981.

It is vital that this aseptic nursing procedure is only undertaken when absolutely necessary (Table 5.4.3) and that the catheter is removed as soon as possible. Furthermore, a sterile, closed drainage system must be maintained throughout the period of catheterization to minimize the likelihood of an ascending infection. Guidelines which can reduce the risk of infection in catheterized patients have emerged from studies by Killion (1982), Jenner (1982), Meers and Stronge (1980), Seal (1982) and the study of the Southampton Infection Control Team of 1982.

Table 5.4.3   Indications for bladder catheterization

To relieve retention associated with lower urinary tract obstruction or neurological disorders

To provide preoperative bladder decompression prior to abdominal surgery

To obtain an accurate estimate of urine output in critically ill and/or unconscious patients

To irrigate the bladder in infections and clot retention and to install antimicrobial drugs in the former

To alleviate incontinence of urine when other methods have failed

Obstructive disorders

Urinary tract obstruction is most likely to occur at the following sites.

(a) The junction of the renal pelvis and ureter.
(b) Any level within the ureter or urethra due either to blockage of the lumen or external compression.
(c) The junction of the ureter and bladder neck.

At any level, an obstruction impedes the outflow of urine, which gradually accumulates behind it, and can lead to dilation of the bladder, ureter or renal pelvis. For example, an obstruction in the ureter can lead to a dilation known as megaureter and ultimately to a dilation of the renal pelvis, described as hydronephrosis. As urine accumulates behind the obstruction, it generates a back pressure which can be so high that the glomerular filtration rate is reduced, leading to renal failure. Other sequelae to obstruction include stasis of urine in the bladder, which predisposes to infection and calculus formation. Therefore, prompt recognition and treatment of urinary tract obstruction are essential to prevent any serious complications setting in.

Altered muscle tone

Mandelstam (1986) has defined stress incontinence as a small leakage of urine which occurs when the intra-abdominal pressure is raised due to sudden exertion, sneezing, coughing, bending or even hitting a tennis ball! Normally, urinary continence is maintained when intra-abdominal pressure rises, by the preservation of an acute angle between the urethra and posterior bladder wall, and by support from periurethral muscles and those in the perineal floor. In stress incontinence, the acute angle is obliterated by small increases in intra-abdominal pressure, which in turn cause the pressure inside the bladder to rise.

This situation is likely to occur when the outlet musculature is no longer competent, due to laxity and weakness of muscles at the bladder neck, around the urethra and in the pelvic floor. In males this can occur following prostatectomy and in females following muscular stretching during childbirth, or at the menopause when mucosal turgor is diminished due to decreased oestrogen secretion. Loss of muscle tone as part of the ageing process can also lead to stress incontinence.

Pelvic floor excercises can make a most valuable contribution to the prevention and treatment of stress incontinence. The exercises are simple to perform and the rapid improvement which they bring is encouraging to the affected individual (Shepherd, 1980).

Incontinence: aids to assessment

Incontinence is a problem which can cause discomfort, embarrassment and guilt. It undermines human dignity, depresses morale and can lead to social ostracism. Early, effective, sensitive management is vital, based on a nursing assessment which is carried out on an individual basis. Some forms of urinary incontinence may well prove intractable if severe neurological damage has occurred. However, for many individuals incontinence is a temporary problem associated with infection, faecal impaction and the prescription of high-dose diuretic or hypnotic drugs, and in these cases to remedy the underlying cause is relatively straightforward. A number of contributing factors must also be borne in mind by the nurse when making an assessment of the problem, particularly if the patient is physically handicapped, elderly and/or immobile to any degree. These include poor access to toilet facilities and lack of handgrips or rails. In addition, lack of privacy, unclean toilets, and an icy environmental temperature are not designed to promote continence. Fortunately, many of these environmental factors can be easily remedied.

A number of life events may also precipitate incontinence; these include grief, apathy, giving up one's home and any admission for institutionalized care.

Frequency–volume charts are useful aids to assessment and can identify the pattern of micturition in those afflicted with frequency, urgency and incontinence. For a period of several days, the times of micturition and volumes of urine passed are charted, together with any instances of incontinence. The information obtained from such a chart will identify the extent of frequency, number of incontinent episodes, functional bladder capacity and the 24-hour total as well as the day–night differential in volumes of urine passed. Depending on the pattern which emerges, either a bladder or toilet training regime may be introduced.

A number of other strategies and practical aids are available to help alleviate incontinence and its associated problems of pressure sores, odour control, and the logistics of coping with wet beds and soiled garments. For an excellent review of these aspects of nursing management, the reader is directed to the publications by Mandelstam and Blannin cited at the end of this chapter.

Review questions

The answers to all these questions can be found in the text. In each case there is at least one correct and at least one incorrect answer.

1 The renal blood flow comprises

   (a) 25% of the cardiac output each minute
   (b) 400 ml. min^{-1}. 100 g^{-1} tissue.
   (c) 55 ml. min^{-1}. 100 g^{-1} tissue.
   (d) 15% of the cardiac output each minute.

2 Renin-secreting cells are found in

   (a) the epithelium lining the loop of Henlé
   (b) the wall of the afferent arteriole
   (c) the juxtamedullary apparatus
   (d) the epithelium lining the collecting ducts.

3 Solutes filtered into Bowman's space

   (a) include glucose, amino acids and electrolytes
   (b) any blood component of molecular weight > 70 000 daltons
   (c) large amounts of albumin
   (d) all the components of plasma of molecular weight < 70 000 daltons.

4 Active transport of any solute involves

(a) 'uphill' movement of the solute against an unfavourable chemical or electrical gradient
(b) 'uphill' movement of the solute against a favourable chemical or electrical gradient
(c) a requirement for energy in the form of ATP to power movement of the solute
(d) 'downhill' movement from an area of high to an area of low chemical concentration or electrical potential.

5 Water reabsorption in the nephron is

(a) passive, along the osmotic gradients created by the reabsorption of other solutes
(b) performed by active transport
(c) controlled by antidiuretic hormone in the distal convoluted tubule and collecting duct
(d) controlled by antidiuretic hormone in the proximal tubule and loop of Henlé.

6 A reduced ability to concentrate urine, leading to polyuria, can arise if there is

(a) loss of hyperosmolality in the medullary interstitium associated with chronic renal diseases
(b) excessive secretion of antidiuretic hormone
(c) deficient secretion of antidiuretic hormone in diabetes insipidus
(d) excessive aldosterone secretion.

7 Renal reabsorption of hydrogen carbonate ions is increased in the presence of

(a) respiratory acidosis
(b) an increased serum potassium concentration
(c) a decreased serum potassium concentration
(d) a decreased arterial $p\mathrm{CO}_2$.

8 In health, the specific gravity of urine is in the range

(a) 1.305–1.406
(b) 1.503–1.603
(c) 1.003–1.030
(d) > 2.0.

Answers to review questions

1 a and b
2 b
3 a and d
4 a and c
5 a and c
6 a and c
7 a and c
8 c

Short answer and essay topics

1 List the functions of the kidneys.
2 Briefly describe the morphological features of juxtamedullary and cortical nephrons.
3 Describe the structure of the filtration barrier in the nephron and discuss the forces which bring about glomerular filtration.
4 Discuss the mechanisms involved during the formation of concentrated and dilute urine.
5 Discuss regional specialization in the nephron with regard to reabsorption and secretion.
6 Review the role of the kidney in the regulation of acid–base balance.
7 'The kidney is the ultimate regulator of homeostasis.' Discuss.
8 Outline the events which take place during micturition.

References

Asscher, A. W. (1980) *The Challenge of Urinary Tract Infections*. London: Academic Press.
Jenner, E. A. (1982) Education related to urinary tract infection and catheter care. *Conference Proceedings, Northwick Park Hospital*, May 1982, p. 52.
Killion, A. (1982) Reducing the risk of infection from indwelling urethral catheters. *Nursing, 15*, No. 2: 84–88.
Mandelstam, D. (1986) *Incontinence and its Management*, 2nd edn. London: Croom Helm.
Medical Research Council Bacteruria Committee (1979) Recommended terminology for urinary tract infection. *British Medical Journal 2*: 717–719.
Meers, P. (1981) Report on the National Survey of Infection in Hospitals 1980. *Journal of Hospital Infection 2* (Suppl.) 53.
Meers, P. & Stronge, J. L. (1980) Hospitals should do the sick no harm. *Nursing Times Supplement 76*.

Seal, D. V. (1982) Chlorhexidine and urinary drainage bags. *Lancet i*: 965.

Shepherd, A. M. (1980) Re-education of the muscles of the pelvic floor. In *Incontinence and its Management*, Mandelstam, D. (ed.). London: Croom Helm.

Southampton Infection Control Team (1982) Evaluation of aseptic techniques and chlorhexidine in the rate of catheter-associated urinary tract infection. *Lancet i*: 89–91.

Valtin, H. (1979) *Renal Dysfunction: Mechanisms Involved in Fluid and Solute Imbalance*. Boston: Little Brown.

Suggestions for further reading

Andersson, B. (1977) Regulation of body fluids. *Annual Review of Physiology 39*: 185.

Arruda, J. A. L. & Kurtzman, N. A. (1978) Relationship of renal sodium and water transport to hydrogen ion secretion. *Annual Review of Physiology 40*: 43.

Blannin, J. (1984) Assessment of incontinence. *Nursing 2(29)*: 863.

Brundage, D. (1980) *Nursing Management of Renal Problems*. St Louis: C. V. Mosby.

Chapman, A. (1979) *Acute Renal Failure*. New York: Churchill Livingstone.

Cogan, M. G. & Garovoy, M. R. (1985) *Introduction to Dialysis*, 1st edn. Edinburgh: Churchill Livingstone.

Cox, C. E. (1966) The urethra and its relationship to urinary tract infection: the flora of the normal female urethra. *Southern Medical Journal (USA) 59*: 621.

De Wardener, H. E. (1985) *The Kidney: An Outline of Normal and Abnormal Function*, 5th edn. Edinburgh: Churchill Livingstone.

Katz, A. I. & Lindheimerr, N. D. (1977) Actions of hormones on the kidney. *Annual Review of Physiology 39*: 97

Mandelstam, D. (1977) *Incontinence*. London: Heinemann.

Norton, C. (1986) *Nursing for Continence*. Beaconsfield: Beaconsfield Publishers.

Pitts, R. F. (1974) *Physiology of the Kidney and Body Fluids*, 3rd edn. Chicago: Year Book Medical.

Ullrich, K. L. (1979) Sugar and sodium cotransport in the proximal tubule. *Annual Review of Physiology 41*: 181.

Valtin, H. (1983) *Renal Function: Mechanisms Preserving Fluid and Solute Balance in Health*, 2nd edn. Boston: Little Brown.

Wright, F. S. (1979) Feedback control of glomerular blood flow pressure and filtration rate. *Physiology Review 59*: 958.

Section 6
Protection and Survival

Chapter 6.1
Innate Defences

Susan M. Hinchliff

Learning objectives

After studying this chapter, the reader should be able to

1 Describe the structure of the epidermis and relate this to its protective function.
2 Relate the structure of the dermis to its water-binding capacity, and describe the part the sebaceous glands play in water-proofing the skin.
3 State the function of melanin and give examples of abnormal skin pigmentation.
4 Discuss cutaneous blood flow and its alteration in relation to local factors and temperature, demonstrating an understanding of how this is controlled.
5 Demonstrate links between skin sensory receptors and the sensory nerve pathways described in Chapters 2.2 and 2.3.
6 Describe the functions of hair and nails and relate these to their structure and (briefly) to disease states.
7 Explain the role of sweat glands in the regulation of body temperature.
8 Carry out a full assessment of a person's skin and make an intelligent attempt to interpret the results.
9 Explain the factors which predispose to pressure sores, assess a patient's susceptibility and plan care to prevent their occurrence.
10 Discuss the role of inflammation as a defence mechanism.
11 State how wound healing occurs after surgery.
12 Write an essay entitled 'The control of body temperature'.
13 Explain changes that occur in temperature regulation during the onset and resolution of a fever.
14 Teach a junior nurse how to assess a patient's temperature.
15 Describe local defence mechanisms throughout the body that prevent micro-organisms gaining entry to body tissues.
16 Summarize the body's innate defences against mechanical, physical, chemical and biological influences.

Introduction

This chapter is concerned with those body defences with which we are born.

The body, if it is to survive, must be protected against environmental influences which may harm it. Specifically, the cells which are contained within the outer coverings of the body must be protected from trauma – this may be mechanical, chemical, physical or microbiological. To put this idea another, and by now more familiar, way, the defence mechanisms which are discussed in this chapter function to maintain homeostasis.

THE SKIN AS AN ORGAN OF DEFENCE

The skin, in terms of surface area covered, is the

largest organ of the body. It is the organ we see first (and indeed one of the few organs we can actually see when looking at a naked human body) and is of great importance in the defence of the body and thus in the delivery of nursing care. Further, since it reflects physiological and pathological changes in other areas of the body, skin changes can be used to aid both nursing and medical diagnosis.

Although we consider the skin as a single organ, there are great variations in it over the total body surface. For example, it varies in thickness: on the eyelids, the skin is only approximately 1 mm thick, whereas on the palms of the hands and the soles of the feet it is often over 3 mm thick. In those who go barefoot, or who subject the sole of the foot to a great deal of friction (e.g ballet dancers), the skin may be as thick as 1 cm.

The skin also varies in its degree of pigmentation. Throughout the body, exposed areas are generally darker than areas which are normally kept covered. The skin varies, too, in its degree of hairiness (*hirsutism*), over different areas of the body and between individuals.

The reasons for these variations in structure become clear when the functions of the skin are examined. Subsequent sections will link these functions with the structures that facilitate them.

Summary of the functions of the skin

PROTECTION

Another name for the skin is the integument (Latin *integere*, to cover over, protect). The *Concise Oxford Dictionary* defines integument as 'skin' rind or husk'. A further name, sometimes used for the superficial skin, is the cuticle, hence the use of the word cutaneous, meaning pertaining to the skin.

Just as the rind or husk on a fruit or berry protects it from drying up in drought or swelling up in rain, so too does the outer covering protect the body from the undue entry or loss of water. The skin also protects from all the minor mechanical blows that the environment deals, for example pressure and friction. When intact, the skin is virtually impermeable to micro-organisms, and also protects from chemicals (weak acids, alkalis etc.) and most gases – although some gases developed for use in chemical warfare can be absorbed

through the skin. The integument gives some protection from physical trauma, for instance from some forms of particulate radiation such as alpha-rays and, to a lesser extent, from beta-rays. The former cannot penetrate skin at all, and the latter can penetrate a few millimetres of skin and are therefore unable to reach the underlying organs. The melanin produced within the skin gives protection from ultraviolet radiation.

SENSATION

The skin forms the largest sensory organ of the body. Its total surface area (which can be calculated from nomograms) varies with the height and weight of the individual and ranges in adults from $1.5\,m^2$ to $2.0\,m^2$. It weighs about 9 kg, or 14% of the body weight. Nerve endings are found in the dermis throughout the whole organ, although they are more concentrated in certain areas (for example the fingertips and lips). They are sensitive to heat, cold, pain or touch, and give information about the environment that can be acted upon both at the reflex level and consciously. Similarly, skin hairs are supplied by nerve endings sensitive to touch. Action can thus be taken to avoid injury to the body.

TEMPERATURE REGULATION

The maintenance of a constant core temperature is a major function of the skin, and is controlled by the hypothalamus. In the British temperate and in hotter climates, the problem is mainly one of losing the heat produced by metabolism. Homeostasis is achieved via conduction, convection and radiation of heat from the skin surface. This heat loss can be varied by dilation or constriction of the blood vessels which supply the skin, so increasing or decreasing its blood flow. Heat is also lost by the evaporation of sweat.

The skin protects the body from excessive heat loss in cold weather, by a decrease in both the blood supply to the skin surface and the production of sweat. The organs lying deeper in the body are further insulated from the environment by subcutaneous connective tissue and fat.

EXCRETION

There is a very small amount of gas exchange through the skin, a negligible amount of carbon dioxide being lost in this way – about 0.5% of that lost through the lungs. The skin also serves as a minor excretory route for urea and sodium chloride, via sweat.

SYNTHESIS OF VITAMIN D

Ultraviolet light falling on the skin brings about the synthesis within it of vitamin D from 7-dehydrocholecalciferol, and this indirectly promotes calcium absorption from the gut. However, in the UK, dietary intake usually provides an adequate supply of vitamin D and, in this situation, the production of vitamin D by the skin is not essential for health. A full discussion of calcium metabolism can be found in Chapter 2.5. This is a further example of a mechanism which maintains a stable internal environment; in this case, vitamin D helps to maintain an adequate amount of calcium in body fluids.

ENERGY AND WATER RESERVE

The skin forms a reserve of energy and water for use in emergencies. In order to restore a fall in blood volume (e.g. in haemorrhage), fluid can be withdrawn from the dermis. Subcutaneous fat forms an energy reserve which can be called upon in starvation.

The above is a brief summary of the functions of the skin, the structural basis of which will now be examined in some detail.

THE STRUCTURE OF THE SKIN RELATED TO ITS FUNCTIONS
(Fig. 6.1.1a)

Epidermis

Although the epidermis and dermis are described as though they were two discrete layers, in fact they function as a single layer and both develop from the embryonic ectoderm.

The epidermis forms a protein-based barrier between the internal and external environments. Its surface is covered by the micro-organisms which form the normal flora of the skin and which will be discussed later in this section. The epidermis contains no blood vessels and no nerve endings. This latter fact becomes clear when we

consider a blister formed, for example, on the heel by a too-tight shoe. The top white skin of the blister can be removed without blood loss or pain, although to remove this top layer would not be sensible since it protects the lower layers.

Histologically, the epidermis is formed of squamous epithelial tissue and this is usually made up of five layers of cells (Fig. 6.1.1b). It is therefore a stratified epithelium, with cells being produced in the basal layer and moving through the other layers over a period of approximately 40–56 days.

BASAL CELLS (STRATUM BASALE)

The basal cells comprise the layer nearest to the dermis, where cell division occurs. The cells in the basal layer dip down into the dermis to surround sweat glands and hair follicles, a fact which is relevant in the care of patients suffering from burns or undergoing plastic surgery. As long as that part of the dermis is retained which contains the hair roots or sweat glands, then some epidermal regeneration can occur from the remaining basal cells. Cells called keratinocytes in this layer manufacture the protein keratin.

PRICKLE CELLS (STRATUM SPINOSUM)

These cells lie distal to the basal cells. They are so named because of their intercelluar bridges or desmosomes, which prevent cell separation (something which could otherwise occur as a result of surface stresses).

GRANULAR CELLS (STRATUM GRANULOSUM)

These cells have cytoplasmic granules which contain the precursor of keratin.

CLEAR CELLS (STRATUM LUCIDUM)

These cells form a layer which is only present in thick skin such as that on the palms of the hands and soles of the feet. The cells here are starting to undergo nuclear degeneration and contain a great deal of keratin.

HORNY CELLS (STRATUM CORNEUM)

These form the surface layer of cells (often called squames), all of which are dead and therefore

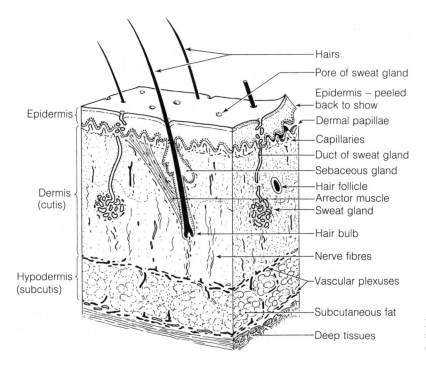

Epidermis

Dermis
(cutis)

Hypodermis
(subcutis)

Hairs
Pore of sweat gland
Epidermis – peeled back to show
Dermal papillae
Capillaries
Duct of sweat gland
Sebaceous gland
Hair follicle
Arrector muscle
Sweat gland
Hair bulb
Nerve fibres
Vascular plexuses
Subcutaneous fat
Deep tissues

Figure 6.1.1a Generalized structure of skin, showing epidermis, dermis and appendages

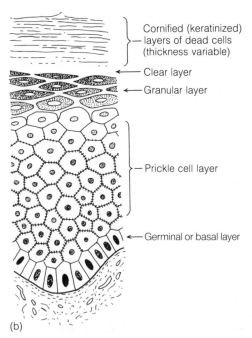

Cornified (keratinized) layers of dead cells (thickness variable)
Clear layer
Granular layer
Prickle cell layer
Germinal or basal layer

(b)

Figure 6.1.1b Cell layers of epidermis.

constantly shed from the body's surface with movement. Up to one million of these cell flakes, formed of keratin, are shed every 40 minutes, amounting to approximately 1 g lost per day. This process is called **desquamation** or **exfoliation**. Cell flakes are shed into the atmosphere, particularly during dressing and undressing – about 80–90% of the dust found in bedrooms and in bed linen is made up of such dead skin cells. Although this may initially seem of little relevance, it provides the major basis for the way in which hospital staff gown up for the operating theatre; it also provides the reason why nurses should not make undue movements or disturb clothing when dressing open wounds, or carry out wound-dressing just after bed-making.

In general, it takes 40–56 days for one cell to progress through all the cell stages until it is shed from the body. About 20% of an adult's dietary protein intake is used by the body in health for skin growth and repair. It therefore follows that it is necessary to increase the dietary protein intake for patients who have sustained burns or serious skin lesions, whether due to accidental or operative trauma.

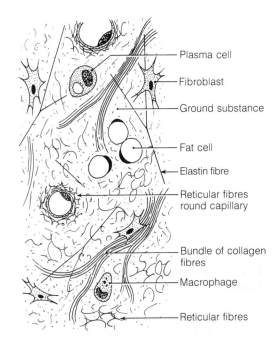

Plasma cell
Fibroblast
Ground substance
Fat cell
Elastin fibre
Reticular fibres round capillary
Bundle of collagen fibres
Macrophage
Reticular fibres

Figure 6.1.2 Diagram to show loose (areolar) connective tissue

The substance that makes the epidermis tough as a barrier is keratin, a small, insoluble fibrous protein molecule made up of 18 amino acids. The process of **keratinization** starts in keratinocytes in the basal layer and progresses throughout the layers until the horny cells are filled with this protein. Keratinization is most evident in those areas subjected to the greatest friction, for example, the palms and soles. It is absent from areas that are not exposed to external stresses, e.g. inside the lips. When keratinization occurs in excess, callosities are formed. Students may notice that after years of note taking and essay writing they may have a hard patch of skin on their middle finger which supports their pen. This is an example of a hyperkeratotic area. **Psoriasis** is a skin condition characterized by rapid and excessive production of keratinized cells which form silvery flakes on the skin surface and which result in excessive desquamation. **Dandruff** (seborrhoeic dermatitis) is a minor but unsightly hyperplasic condition of the scalp which results in flakes of keratin being exfoliated.

Dermis

The dermis is formed of loose **areolar connective tissue** (Fig. 6.1.2) of mesodermal origin. It is, in essence, a fibrous elastic bed, supporting and providing nourishment for the epidermis and its appendages (the hairs, sweat glands, blood vessels, lymphatic vessels and nerve endings; see Fig. 6.1.1a).

Histologically, the dermis is formed of two distinct layers, the papillary layer and the reticular layer. The papillary layer lies next to the basal layer of the epidermis and forms a series of undulations called **dermal papillae** or **rete pegs**. In fact, the two layers slot into one another rather like two pieces of corrugated paper (see Fig. 6.1.1a). This cellular arrangement prevents the epidermis shearing off the dermis when shearing forces are applied to the skin. Excessive force (e.g. dragging a helpless patient up the bed) can cause the two skin layers to shear apart. This is what happens when, for example, a too-tight shoe is worn and a blister forms in response to shearing forces generated during walking. The conclusion is that the skin should always be treated gently, especially in areas subjected to pressure. The deeper reticular layer contains fewer blood vessels and is less reactive.

Like all true connective tissues, the dermis is formed of a ground substance or matrix, fibres and cells.

THE GROUND SUBSTANCE

This looks rather like aspic jelly. It is fairly firm but by no means solid. It supports fibres and cells and fills the spaces between them. Substances in solution travelling between blood vessels, and cells must cross through this matrix. The gel is made up of hyaluronic acid, mucopolysaccharides and chondroitin sulphate, and is synthesized by the cells suspended in it. Certain microorganisms, for example *Streptococcus pyogenes* and *Clostridium perfringens*, produce an enzyme called hyaluronidase, which can lyse (dissolve) the hyaluronic acid in the ground substance and thus open up tracks in the dermis for dissemination of infections. It is thought that the susceptibility of the skin of diabetic people to infections may be due to an abnormality of their ground substance.

THE FIBRES

There are three main fibres, all produced by mesodermal fibroblasts.

Collagen. This is a fibrous protein with the tensile strength of steel wire of the same diameter. The fibres lie parallel to one another in the dermis in bundles. Apart from its strength and protective function, the major characteristic of collagen is that it binds water avidly, so much so that the dermis contains 18–40% of the total body water. The mucopolysaccharides in the ground substance aid in this function. This water can be mobilized in dehydration or haemorrhage.

The water-binding properties of collagen decrease with age, and it is thought that the wrinkled appearance of the skin in the elderly is due not so much to degeneration of elastin fibres (as was previously thought) as to decreased water-holding power of the collagen and mucopolysaccharides. The skin therefore appears to be less 'plumped out' and so more wrinkled.

Ascorbic acid seems to be necessary for collagen formation and hence an adequate intake is essential for efficient wound healing.

Clostridium perfringens produces a collagenase, an enzyme which can lyse collagen fibres, in addition to hyaluronidase.

Reticular fibres. These form a loose framework in the dermis and envelop the collagen bundles. They help to disperse forces applied to the dermis.

Elastin fibres. These are branching yellow fibres, which are elastic but may rupture when stretched by pregnancy, obesity or prolonged ascites. When this occurs, silvery linear scars appear, called **striae** or stretch marks.

THE CELLS

These may either be resident (synthesizing ground substance or fibres) or blood-borne and thus transient.

Fibroblasts. These lie between the bundles of collagen and are concerned with collagen and elastin synthesis. Their numbers increase considerably during wound healing.

Tissue macrophages or histiocytes. These are wandering phagocytic cells which engulf particulate matter and are protective in function.

Tissue mast cells. These cells (analogous to the basophils in the bloodstream) produce both histamine and heparin, and are usually found in the vicinity of blood vessels and hair follicles.

Transient cells. These include neutrophils, lymphocytes and monocytes. They move constantly between the blood vessels of the dermis, moving out of the blood in large numbers as part of the inflammatory reaction which occurs in response to trauma. Normally, the dermis is free from bacteria.

Melanocytes and the production of melanin

These are pigment-producing cells which are present in the dermo-epidermal junction over the entire skin. Their function is to protect the body from ultraviolet light, and in some animals they have a further function of affording camouflage.

Their numbers do not vary with age, race or sex, but do vary, though, over the individual's body, being relatively few on the trunk and numerous on the penis, scrotum and areola of the nipple. Melanocytes form the pigment melanin from the amino acid tyrosine, in melanosomes (melanin-producing organelles in their cytoplasm). Melanocytes then transfer the melanin to the epidermal cells.

Melanin, which colours hair, originates from melanocytes near the hair papillae, and there is a biochemical difference in the melanin produced by blondes, brunettes and redheads which results in different tones of hair pigment colour. The melanin in the iris of the eye, too, differs structurally between individuals. Melanin production is influenced by sunlight. After exposure of skin to sunlight, new melanin is not formed immediately. Erythema (reddening of the skin) due to inflammation occurs initially, and this lasts approximately 2 days. Then melanin production is evident as the skin colour changes from pink to brown. Apart from being a protective mechanism, skin tanning frequently induces a feeling of psychological well-being. However, excessive prolonged exposure to the sun can lead to early ageing and wrinkling of the skin. In extreme cases, carcinoma of the skin may result, probably due to excessive exposure to ultraviolet radiation. Melanin production is under genetic control, and is regulated by **melanocyte-stimulating hormone (MSH)** secreted from the anterior lobe of the pituitary. This hormone is of particular interest in frogs, chameleons and certain other animals in which it allows a rapid camouflage response to changes in background colour. The MSH is very similar in structure to adrenocorticotrophic hormone (ACTH) and, in Addison's disease (in which there is an increase in ACTH secretion), abnormal skin pigmentation occurs in exposed areas and in areas where the skin is subjected to mild, repeated pressure or trauma (e.g. the belt, cuff or collar areas).

Pigmentation is affected by oestrogen production, which causes the marked darkening of the nipple, the surrounding areola and linea nigra in pregnancy. It occasionally causes patchy brown pigmentation of the face in pregnancy which can be upsetting for those affected; this phenomenon is sometimes referred to as the 'mask of pregnancy'.

Albinism is an autosomal recessive hereditary condition characterized by a lack of ability to synthesize melanin in the skin, hair and eyes. Consequently, sufferers have no protection from ultraviolet light and, unless precautions are taken such as constant wearing of dark glasses, retinal problems may develop.

Vitamin D synthesis

The term vitamin D actually refers to a group of sterols. The vitamin which we obtain in our diet from fish-liver oils and dairy produce should more correctly be called vitamin D_3, or **cholecalciferol**. It is synthesized when ultraviolet light falls on uncovered skin and acts on the 7-dehydrocholecalciferol present in the dermis, converting it to previtamin D_3. This is then slowly converted to vitamin D_3.

Problems of vitamin D_3 deficiency (rickets in children and osteomalacia in adults) can occur when dietary intake is insufficient (i.e. below $2.5 \mu g$ per day for adults and $10 \mu g$ per day for children and pregnant women). In Britain, our dietary intake is usually adequate for our needs, but with the increasing emphasis on the need to cut our intake of the saturated fats found in dairy produce, it is possible that deficiencies could occur.

Deficiencies can also occur when the body is deprived of sunlight, for example if sunlight is prevented from reaching the body by atmospheric pollution or excessive clothing. Dark-skinned immigrants to temperate countries may be particularly at risk, as their dietary intake of vitamin D_3 may be inadequate and they are no longer able to supplement it by ultraviolet-stimulated synthesis of vitamin D_3, as would have been possible in their sunny countries of origin.

Blood vessels and lymphatics

The cutaneous blood vessels lie entirely within the dermis and have an essential role in transporting nutrients to, and waste substances from, the dermal tissues. They are also of major importance in the regulation of body temperature. (This will be discussed in greater detail later in the chapter.) The number of vessels is much greater than would be necessary for purely nutritional purposes, and so the skin is capable of accommodating varying amounts of blood. The number of vessels carrying blood at any one time varies with the environmental temperature. At very low temperatures, skin blood flow may fall to 20 ml/min; in very high temperatures, it can rise to 3000 ml/min. In temperatures of 25–30°C a naked adult at rest might have a skin blood flow of about 200 ml/min. This variation in flow is made possible by the existence of **precapillary sphincters** which direct the flow, often in response to local metabolites (Fig. 6.1.3).

In some exposed areas, for example the hands, feet, ears or nose, large numbers of **arteriovenous anastomoses** exist (Fig. 6.1.3b). At normal body temperatures, these anastomoses are kept almost closed by sympathetic vasoconstrictor nerves. When body temperature rises, the sympathetic vasoconstrictor tone is reduced and the anastomoses open, allowing blood to short-circuit the deeper capillary beds, so that a large volume of warm blood enters the superficial vessels, promoting heat loss.

Cutaneous blood vessels have a rich sympathetic nerve supply. When the environmental temperature drops, the sympathetic nerves are stimulated and so vasoconstriction occurs; vasodilation results from sympathetic inhibition in warm conditions.

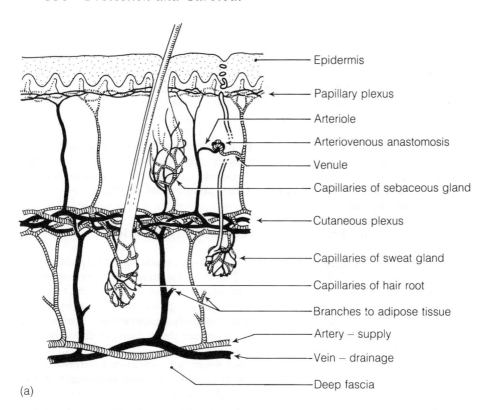

Epidermis

Papillary plexus

Arteriole

Arteriovenous anastomosis

Venule

Capillaries of sebaceous gland

Cutaneous plexus

Capillaries of sweat gland

Capillaries of hair root

Branches to adipose tissue

Artery – supply

Vein – drainage

Deep fascia

(a)

Figure 6.1.3 Diagram to show cutaneous blood flow

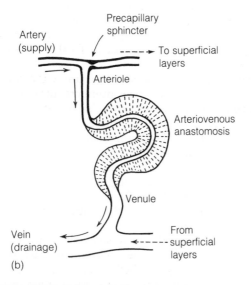

(b)

Figure 6.1.3b   Diagram to illustrate arteriovenous anastomosis

Under normal conditions, there is always a certain amount of background vasoconstrictor tone. Circulating adrenaline and noradrenaline enhance the vasoconstrictor effect of the sympathetic nerves.

LOCAL ALTERATIONS IN SKIN BLOOD FLOW

Reactive hyperaemia

If the blood supply to a limb is occluded, for example by inflating the cuff of a sphygmomanometer on the upper arm to above systolic pressure and holding blood flow arrested for several minutes, release of the cuff will re-establish the blood flow, and the arm below the cuff will flush red. The arteriolar dilation which brings about the reddening is caused by hypoxia and local collection of metabolites which are unable to be carried away in the venous blood. It is called reactive hyperaemia. The increase in blood flow in the arm if the circulation is occluded for 10 minutes can be as much as twentyfold. Nurses will be familar with this phenomenon when treating pressure areas. Areas of skin subjected to pressure are rendered bloodless and accumulate a blood-flow debt. Once the pressure is relieved, and blood flow re-established, reactive hyperaemia occurs.

White reaction
If a fingernail is drawn lightly over the skin, the stroke line becomes pale. This is due initially to displacement of blood and subsequently to contraction of the precapillary sphincters and draining of blood from the distal capillaries.

Triple response
If the skin is scratched (and therefore injured), after about 10 seconds a red area appears at the site of the injury. This is due to vasodilation of the arteriole and venules, and is called the **red reaction**.

Next, a **wheal** appears when the capillaries become more permeable, and fluid diffuses out of the capillaries and into the tissue spaces. This is **inflammatory oedema**. The increase in capillary permeability is due to the release of chemicals collectively called **H-substance**. Histamine is probably one of its constituents.

Finally, a diffuse **flare** appears around the injured area as the surrounding arterioles dilate. The flare is almost certainly the result of a local **axon reflex**. This is thought to be due to the fact that the sensory nerves in the skin that are stimulated by the injury are connected not only with the skin but also with the surrounding arterioles (Fig. 6.1.4).

Impulses generated pass back along the axon of the sensory nerve to bring about vasodilation of the arterioles. **Substance P** is the transmitter,

released at the arteriolar axon terminations, which brings about the vasodilation and consequent flare. This response to injury is independent of the central nervous system and is present after a total sympathectomy.

LYMPHATIC VESSELS

These are found throughout the dermis and they have a major role in draining excess tissue fluid and any plasma proteins that may have leaked into the tissues. It is possible that up to 15% of all tissue fluid formed is reabsorbed by this route. The lymphatics are therefore important in maintaining interstitial fluid volume and composition.

Nerve supply

The skin is the largest sensory organ in the body. Its dermis contains sensory nerves whose cell bodies lie in the dorsal root ganglia. These afferent nerves may then either link with a reflex arc and/or travel up the spinal cord in specific spinal pathways for each sensory modality to the sensory cortex, where the information they carry is interpreted.

Sensory receptors, in general, are specialized so that they each respond to a different form of energy – chemical, mechanical or thermal – thus the modalities of touch-pressure, cold, warmth and pain can be consciously appreciated. The receptors act as transducers, and convert the energy into action potentials. The skin contains three main types of sensory nerve endings (Fig. 6.1.5): naked nerve endings, encapsulated endings (Pacinian corpuscles, Meissner's corpuscles and Krause's end-bulbs), and expanded

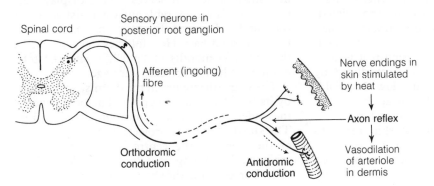

Figure 6.1.4 Diagram to show axon reflex

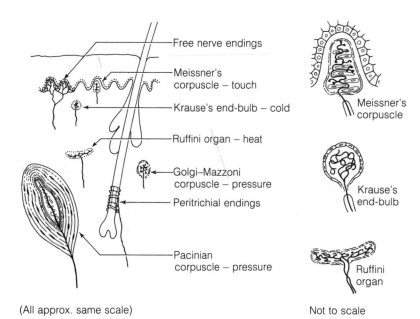

(All approx. same scale)

Not to scale

Figure 6.1.5   Diagram to illustrate skin sensory receptors

endings (Ruffini's organs). These are in addition to the sensory endings supplying hair follicles (see also Chap. 2.2).

Ruffini's endings respond to heat; Meissner's, Pacinian and Golgi-Mazzoni corpuscles to touch and pressure; Krause's end-bulbs to cold; and naked nerve endings to mechanical and chemical stimuli of varying intensity. The position, though, is not as clearcut as this may seem. In practice, a receptor may react to more than one stimulus. A factor that seems to be of importance in the appreciation of skin sensations is changed stimulus production. In general, alterations produce sensation, and continuous stimulation after a while fails to produce an effect. One is aware, for instance, of clothes touching the body when first dressing in the morning. After a while, though, the sensation seems no longer to be appreciated. Changes in stimuli are thus more important than their intensity, except where pain is concerned. Appreciation of pain is not lessened if it is constant. The sensation of itching is produced by stimulation of pain receptors at a low threshold, and thus it can be abolished by the application of a more intense stimulus such as scratching.

The sensory receptors in the skin serve to protect from injurious elements in the environment by coding information which can be acted upon rapidly at reflex level and subsequently interpreted by the sensory cortex. As a result, the body can respond to the environment in both a rapid automatic way (as when the hand is reflexly drawn away from a hot plate) and in a slower, more deliberate way (as when deciding to put on an extra sweater in cold weather).

Hair

Hair is formed of fibrous protein (largely keratin) and is exclusive to mammals. It functions to protect the skin from ultraviolet light, extremes of temperature and also from trauma. Owing to the presence of nerve endings around hair follicles, hair also acts as a tactile organ. To a certain extent, in human society, it serves sexual display purposes. Hair is very durable and can resist decay for thousands of years: Egyptian mummies show traces of head hair. Hair follicles develop during the first few months of fetal life, and the total number of hairs decreases with age.

The hair **follicle** lies in the dermis, surrounded by its blood and nerve supply. The basal layer of the epidermis dips down to surround it (Fig. 6.1.6). The **shaft** of the hair projects beyond the surface of the skin. In the centre of the shaft is the medulla, containing loosely packed keratinized cells. Surrounding the shaft is the **cortex**, formed

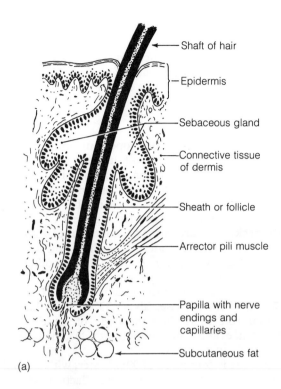

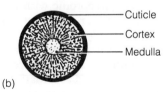

(a)

—— Shaft of hair

—— Epidermis

—— Sebaceous gland

—— Connective tissue of dermis

—— Sheath or follicle

—— Arrector pili muscle

—— Papilla with nerve endings and capillaries

—— Subcutaneous fat

—— Cuticle
—— Cortex
—— Medulla

(b)

Figure 6.1.6 Diagram of hair in (a) longitudinal section, and (b) cross-section

of keratinized cells which are cemented together and which contain granules of melanin. The melanin is secreted by melanocytes which are concentrated around the germinal base of the hair. If air spaces occur in the cortex, the hair will appear grey or silver.

The **cuticle** surrounds the cortex and is formed of overlapping pigment-free cells which, under the microscope, look rather like fish-scales. These scales can be prised off the cortex and hence the hair may be damaged by the technique of 'back-combing' it.

The lower end of the shaft in the dermis is distended to form the **bulb**. If a hair is sharply pulled out of the scalp, the bulb at the end can be seen quite clearly. Cell division occurs in the bulb

and hence this area is affected by cytotoxic drugs. A small bundle of muscle fibres is attached to the portion of the shaft near the bulb called the **arrector pili muscle**. When this contracts, the hair is pulled more vertically in the dermis. This action is more effective with short body hair than with heavy head hair, and it results in the non-functional appearance of goose-pimples. In animals, contraction of the arrector pili muscles is an efficient response to cold: when the hairs of the pelt are held perpendicular to the body surface, a deeper layer of insulating air is held next to the skin and this prevents heat loss. In humans, this function is non-existent as they do not have sufficient hair for it to be an effective mechanism.

Erection of the hair in animals, referred to as horripilation, is also effective in frightening off predators since the animal looks larger and more fearsome when the hairs are erect. All that remains of this in humans is the production of goose-pimples when one is cold or frightened.

TYPES OF HAIR

Human hair is of three types:

Lanugo
This is fine, long, silky non-pigmented hair present in the first 7–8 months of fetal life, after which time it is shed in utero. Premature babies may be born still covered in lanugo, and this may be a cause of some distress to their parents, who need to be reassured that the presence of lanugo at this stage is normal and that it will soon disappear. In malignant disease, it occasionally happens that lanugo may replace terminal hair (see below), following chemotherapy.

Vellus
This is short hair similar to lanugo. It is colourless, forms the child's body hair and remains on the adult female's face.

Terminal hair
This refers to the hair described at the start of this section, as found on the adult head and pubis. When, as a result of disease such as Cushing's syndrome, a person becomes hirsute, it is terminal hair not vellus which appears on the face. This is much more noticeable and may cause distress.

Hair production is influenced by several factors.

Racial. Negro hair is flat in cross-section, and these ribbon-shaped hairs tend to spiral, producing the characteristic crinkly hair. In Caucasians, straight hair tends to be round, and curly hair oval in cross-section.

Hormonal. The level of androgens is a critical factor in hair growth. Testosterone stimulates beard growth but not that of head hair. It was Aristotle who pointed out that neither boys, women, nor castrated men go bald. It is testosterone, in conjunction with an inherited predisposition, that produces male pattern baldness. Thyroxine deficiency affects the quality of hair produced (cf. the coarse hair seen in myxoedema).

Genetic. Hair colour and texture are inherited characteristics. The tendency to premature balding or greying is familial.

Hair tends to grow in cycles of activity, and at any one time about 85% of scalp hairs are in an actively growing stage which lasts for about 2 years. This active period is followed by a period of rest, then regression. About 50 scalp hairs are lost daily out of the total of about 100 000 scalp follicles.

Permanent waving techniques employ solutions which alter the bonds in the keratin molecule and thus soften the hair, which is then set in the required position and a 'neutralizer' employed to reset the bonds in the keratin. If the solutions used are too strong, the hair can be damaged. Hydrogen peroxide used in some hair lighteners may damage the keratin molecule so that the hair fractures. People who habitually bleach their hair are likely to make it brittle, dull and damaged.

Nails

These are protective keratinized plates resting on the stratified squamous epithelium that forms the highly sensitive and vascular nailbed. The nails protect the tips of the fingers and toes from injury. In humans, their importance as a clawing device is minimal, although it is quite difficult to carry out some fine manipulation, for example separating paper, without using the nails.

The nail root extends deeply into the dermis towards the interphalangeal joint. Nails can give an indication of certain diseases to the observer. This will be discussed under assessment of the skin.

Sweat glands

There are two distinct types of sweat gland (sometimes known as sudoriferous glands) in the skin, each of which has a separate function.

ECCRINE GLANDS

These glands are well developed only in the higher apes and humans. In the adult, there are between 3 and 4 million sweat glands, not all of which are active at the same time. The distribution of glands throughout the body is uneven; there are over $400/cm^2$ on the hands and feet and only $70/cm^2$ on the back. They are formed as coiled tubular downgrowths from the epidermis, each having its own blood supply and innervation from cholinergic sympathetic fibres (see Fig. 6.1.1a).

Sweat glands produce sweat when the skin surface temperature rises above 35°C. They have a minor role in ridding the body of some waste substances and a more important role in helping to bring about heat loss.

Sweat is composed of 99% water, with a small amount of sodium chloride, urea, lactic acid and potassium in solution. It is hypotonic with reference to plasma, containing fewer electrolytes and very little glucose. It is usually acid in reaction. The specific gravity of sweat is normally about 1.004, but this varies with the rate of secretion, aldosterone levels in the blood and acclimatization to extreme heat. Sweat forms a minor excretory route for some drugs and allergens, for instance garlic.

Sweat glands have a cholinergic sympathetic innervation. There is always some slight activity, with a resultant insensible loss of some 400–500 ml per day. This loss is not all via the sweat glands; although fairly waterproof, the skin does allow some loss through diffusion and osmosis.

In a temperate climate, approximately 500 ml of sweat each day is produced in addition to the insensible loss. This amount varies with the degree of activity and the environmental temperature. Sweat production can rise to 12 l/day in very hot climates. Secretion increases in response to spicy foods and, for example, those who eat very hot curries may sweat profusely over their faces. This is called gustatory sweating. Sweat production also increases in stressful and emotional states. This forms the basis for lie-detector machines (polygraphs) which measure galvanic skin response.

The evaporation of sweat, from the skin surface, requires latent heat from the surface of the skin and is thus a method of heat loss. The areas which are most active in this thermoregulatory sweating are the face, trunk, axillae, palms and soles. The evaporation of 1 litre of sweat from the skin requires 2400 kJ (580 kcal) of heat, and it is the evaporation of sweat that causes heat to be lost (not simply the production of sweat). Sweat will not evaporate if the environment is already laden with water vapour.

Tepid sponging of a pyrexial person is based on this principle that evaporation of water from the body surface causes heat loss. The water should not be allowed to drip off the skin since this will not result in heat loss. Tepid water is used to avoid unnecessary discomfort to the patient, and also to avoid the vasoconstriction of surface blood vessels that would occur if cold was used.

APOCRINE GLANDS

These are glands which are also found in the lower animals, in whom sweat secretion, with its distinctive odour, is used for the recognition of mates and territory. In humans there are racial differences in their structure. They appear to be well developed in Negroes and less so in Asians. Structurally, they are related to the hair follicles, being situated near the sebaceous glands.

Apocrine glands are mainly present in the areas of the pubis, genitalia, areola, axillae, umbilicus and external auditory canals. The apocrine glands are ten times larger than the eccrine glands, and the openings of their coiled ducts are just visible in a dissected cadaver.

The secretion from these glands is scanty and sticky. It is odourless when first secreted, but is quickly acted upon by bacteria and then gives rise to a distinctive, rather unpleasant, smell. Deodorants work on a bactericidal principle and also attempt, with perfumes, to mask any odour produced. Antiperspirants act by plugging the openings of the ducts, usually with metal salts such as those of aluminium.

In childhood, the apocrine glands are non-functional, but at puberty they begin actively to secrete and thus appear to be affected by the sex hormones. Their secretion does not seem to be under the control of the nervous system, though secretion increases in stress and there may be some sympathetic nervous influence.

Apocrine secretions may contain **pheromones** – sexually attractant substances which were first described in insects. Human females are thought to produce fatty acid pheromones in their vaginal secretions in the middle of the menstrual cycle, and it is possible that the apocrine glands may secrete a similar substance. The individual secreting the pheromone is unable to detect its presence.

Sebaceous glands

These are formed as outgrowths of the developing hair follicles (see Fig. 6.1.6). They start to produce their secretions at about 15 weeks of fetal life. Their activity, which is low in childhood, increases during puberty. These glands are particularly plentiful over the scalp and face where there may be up to $900/cm^2$, and also over the middle of the back, the auditory canal and genitalia. There are very few sebaceous glands on the hands and feet. Over the forehead of a person with particularly active sebaceous glands, more than 2 g of slightly acid sebum may be produced each week. **Sebum** is a mixture of triglycerides, waxes, paraffins and cholesterol acid. The sebaceous glands are holocrine glands, their secretion being formed by the disintegration of the glandular cells. New lobes of the gland are therefore continually reforming. Most, but not all, of the glands open onto a hair shaft. The function of sebum is to waterproof the skin; in addition, it is thought possible that it protects from fungal and bacterial infections: athlete's foot, for instance, occurs in an area between the toes where there are no or very few sebaceous glands and a damp environment. The skin forms a dry barrier, and pathogenic bacteria on its surface often die simply through desiccation. Sebum production is stimulated by a rise in temperature and by androgens; it is inhibited by oestrogens.

A child has very small sebaceous glands which are few in number, and hence babies and children are prone to chapping of the skin on exposure to damp conditions. This predisposes to nappy rash. In old age, the number of sebaceous glands diminishes, with resulting potential skin care problems. **Acne** may be caused by streptococci, staphylococci, or *Corynebacterium acnes* within the hair follicle. These bacteria act on sebum and liberate free fatty acids from it, which cause the wall of the hair follicle to burst thus allowing the bacteria to come into contact with epidermal tissue. In the epidermis, they produce inflammation: leucocytes come to the defence of the tissue, pus formed of living and dead leucocytes and bacteria is produced, and pustules develop. The condition can be treated with tetracycline or can sometimes be alleviated by ultraviolet light, which has a bactericidal effect, or by the administration of oestrogen in females.

Subcutaneous fat

Adipose tissue, or fat, forms a valuable store of triglycerides for the body, and this is a potential source of energy. Sixty percent of the total body fat stores are subcutaneous, and this fat insulates the body and prevents heat loss from the core. Fat also protects from trauma, in that it acts as a shock absorber: for example, an examination of the palm of the hand will show how the fat is distributed in pockets on the base of each finger, at the fingertips and between the finger joints (where the fat best fulfils a protective function). Blows to the body surface cause lateral displacement of the skin due to the presence of subcutaneous fat, and hence the full force of the blow is not transmitted to the deeper structures.

Fat distribution, under the skin, differs between the sexes: females have greater deposits on the upper limbs, breasts and buttocks than do males. Generally, the adult female has greater fat stores than the male – some 15 kg as opposed to the male's 7.5 kg in humans. Fat stores are determined by diet, genetic factors and sex hormones.

Adipose tissue has very little ground substance, the tissue being divided into lobes by septa which carry the blood vessels and the nerves. The cells that make up adipose tissue are large – up to 50 μm in diameter – and consist of a flat nucleus surrounded by a large single fat globule. When fat is catabolized, considerable metabolic water is produced – a factor of importance to humans in starvation.

BROWN FAT

Some fat appears darker in colour. Microscopically, each brown fat cell contains many droplets of fat instead of a single globule, many mitochondria and some pigment. Brown fat tissue is also more vascular. The fact that the cells contain many small droplets of fat means that brown fat forms a more readily and rapidly available energy supply. The presence of many mitochondria suggests that the tissue is capable of an increased rate of metabolism.

Brown fat has an important role in the neonate as it provides an easily mobilized energy source. The neonate has a large body surface area relative to its weight and it is also unable to shiver. The newborn baby therefore has the potential problem of inability to conserve body heat and a propensity to lose too much heat. Thus, the neonate uses the brown fat stored around the back of its neck and kidneys, free fatty acids being liberated from this in response to cold, noradrenaline and glucagon. Brown fat breakdown increases oxygen consumption and brings about localized energy release. It loses its importance in thermal regulation as the child develops muscular control and hence uses muscular activity as a method of heat production. It is possible that some individuals may retain their stores of brown fat into adulthood, within the para-aortic, renal and peritoneal fat depots, and that such individuals may be less prone to obesity. They may also be able to lose weight more easily.

NURSING ASSESSMENT OF THE SKIN

The skin is an organ from which a great deal of information about a patient's nutritional status, fluid balance, circulation, emotional state and age can be obtained. Furthermore, the skin can provide clues leading to the diagnosis of a patient's health problems and to an evaluation of the effectiveness of the patient's care, both nursing and medical.

In assessing a patient generally, and in particular the patient's skin (whether this assessment occurs in hospital or in the community), all the senses – sight, touch, smell and hearing – may be used.

Observation of the skin

Before attempting to observe a patient's skin, it is important to ensure that the light is adequate; for example, it is not possible to assess whether a patient is cyanosed if he is in a very dark corner of the ward. Also, the nurse may need patience. On first acquaintance, she may see only those areas of skin exposed in normal social contacts. She may be able to assess more of the patient's skin later, when delivering personal nursing care, for example when helping the patient to bath himself. Even then, she may not be able to assess the skin fully, because the patient may not wish her to do so; some patients go to great lengths to conceal features of their body that they feel are disfiguring, ugly or embarrassing, and those with actual skin lesions may attempt to cover them with make-up, clothing or jewellery.

Age

From observation, it is possible to gain a fairly accurate impression of a person's age. Skin tends to become drier or more wrinkled with age; it is worth remembering that the skin on the neck may belie an apparently youthful face.

Racial origin

This may be obvious; certainly, it will be possible to attempt a broad categorization such as Asian, negroid or Caucasian.

General state of grooming. This may give a clue to the patient's mental state. The depressed patient may not have the energy or incentive to care for himself, while the manic patient may not have time.

Skin colour

The colour of the skin is of great importance in assessment.

Pallor. The degree of pallor or redness of the skin depends largely on the blood flow through the surface vessels. Hence, the person who has collapsed peripheral blood vessels, which can occur in shock, with the severe pain experienced after myocardial infarction, or in cold conditions, may appear very pale. Adrenaline causes vasoconstriction, so the frightened or anxious patient may also appear pale. If assessment is carried out very shortly after admission, the patient may appear pale if he is anxious about his hospitalization, and so later reassessment is often advisable.

In anaemia, surface vessel blood flow is adequate, but the haemoglobin concentration of the blood is low. Anaemia can be most accurately assessed by looking at the mucous membranes,

e.g. inside the lips or lower eyelid. Here the blood vessels lie nearer the surface and colour can therefore be observed.

In myxoedema (hypopituitarism), the patient's skin may appear pale and puffy due to the excess oedematous and fatty tissue between the blood vessels and the surface.

Flushing. The skin will appear red when it has an increased blood flow of normal haemoglobin content. For example, in hot weather or when the patient has a raised body temperature, cutaneous vessels will be dilated to facilitate heat loss from the skin surface. This is also the reason for the flushing that may follow after exercise.

In inflammation (which will be discussed later), vasodilation occurs over the affected area, and redness or rubor is a characteristic feature of the process.

Patients who are used to a high alcohol intake may appear permanently flushed. Flushing may also occur in anger or agitation, or during the climacteric (see Chap. 6.3)

Cyanosis. This occurs when more than 5 g/dl (0.74 mmol/l) of haemoglobin is in the reduced state. Thus, cyanosis, a blue coloration, occurs relatively easily in patients who are polycythaemic but is rarely seen in those who are anaemic. Cyanosis occurs in individuals suffering from diseases which result in a reduced amount of oxygen being carried by the blood (hypoxaemia – see Chap. 5.3).

Cyanosis may be *central*, e.g. over the face or lips, or *peripheral*, where the extremities are affected. The latter usually indicates inadequate or sluggish blood flow in the peripheral tissues. Cyanosis is difficult to assess in coloured patients whose skin pigments may obscure the condition. The inside of the lips, palms and soles may, however, give some indication of the problem.

Jaundice. Yellow discoloration of the skin is most easily assessed from the conjunctiva. In the assessment of jaundice, it is helpful to devise some sort of scale so that deterioration or improvement can be determined. Other conditions, such as a fading suntan or the ingestion of certain drugs such as mepacrine hydrochloride (Quinacrine), should be excluded.

Jaundice is evident when plasma bilirubin levels rise above 34 μmol/l. (The normal level is less than 19 μmol/l.) A slightly yellow appearance may be apparent in the skin in the later stages of malignant disease when cachexia exists.

Pigmentation. This was discussed earlier under the heading of melanocytes.

Skin type and texture
It is helpful to make an assessment of the patient's skin type and texture. Oily skins are less likely to get chapped, but they are more prone to acne and minor infections – a point which may be of relevance if surgery to the face or back is proposed, since these areas contain numerous sebaceous glands.

Patients who come from the tropics or who are very tanned may have a rather leathery skin. Myxoedematous patients tend to have skin which has a coarse texture.

Scars
The presence of scars, striae and bruising may be noted. Injection marks may give a clue to drug abuse or to conditions requiring prophylactic medication by injection, such as diabetes or haemophilia.

Abnormal skin conditions
Abnormal conditions of the skin, such as rashes, areas of erythema, desquamation or pustules, may be observed.

The hair
Evidence about the patient's general status can be obtained by observing whether the hair looks cared for and clean. It can also be useful to note the patient's preferred hair style, especially for patients who are undergoing treatment with cytotoxic drugs and are thus likely to experience some hair loss, which may make it necessary for them to be fitted with a wig. The amount of hair is important in conditions such as ringworm, where loss (alopecia) occurs. The hair may also be examined for evidence of infestation; areas of hair at the nape of the neck and behind the ears may provide evidence of existing pediculosis. Abnormal patterns of hair distribution may be observed. These may be due to hormonal conditions or to drug therapy, e.g. long-term steroid administration.

Nails
These can give further information about the patient's general grooming and self-care. In the elderly, the toenails may be neglected because of arthritis, which makes it difficult for the patient to reach them. In anaemia, the nails become concave and spoon shaped; this is called *koilonychia*. *Paronychia* is the name given to infection of the margin of the nail. The nail may show numerous small vertical haemorrhages (splinter haemorrhages) with subacute bacterial endocarditis or injury to the nail. *Clubbing* describes a condition in which the fingertip and nail appear expanded, rather like a drumstick. The nail curves over the tip of the finger. Clubbing is due to an increase in vascularity of the chronically hypoxic peripheral tissue. The phenomenon is associated with congenital cyanotic heart disease and cyanotic respiratory conditions.

Palpation of the skin

The feel of the skin can give information about the patient's water balance, state of nutrition and health.

Moderate and severe dehydration
This can be assessed by gently but firmly pinching up a fold of skin on the back of the hand or on the inner forearm. In the well-hydrated person, it will immediately return to its normal position. In the patient who is in an advanced state of dehydration, the fold of skin may stay pinched for up to 30 seconds. This tends to happen normally (but to a lesser extent) with advancing years.

Oedema
This can be identified by pressing firmly over a bony prominence for about 5 seconds. Waterlogged tissue retains the imprint of the finger (so-called pitting oedema).

Obesity
This can be assessed by palpation. Skinfold calipers can be used to assess superfluous subcutaneous fat. Obese skin feels flabby and may wobble when pushed. An obese patient who has experienced rapid weight loss may have folds of skin on the abdomen and buttocks.

Temperature
A reasonable estimate of relative temperature can be obtained by feeling the skin, which will feel

warm over an inflamed area – a characteristic sign of inflammation – or over an area of increased blood flow. A suspected deep vein thrombosis in a leg may be provisionally diagnosed from the extra warmth and colour of the skin and the pain in the calf that results when the foot is dorsiflexed (Homans' sign).

Areas of chapped or scaly skin can be assessed by touch.

Using the sense of smell

Normal skin does not smell. People who do not pay sufficient attention to their personal hygiene may develop a characteristic odour that results from bacterial decomposition of their eccrine and, in particular, their apocrine secretions.

Skin infections may lead to quite characteristic smells, for example the mousey smell that occurs with gangrene infections.

Using hearing

Apart from the use of this sense in communication, there is not a great deal that one can directly assess about the skin through hearing. However, it is vital in assessing one particular condition – **surgical emphysema**. When air enters the tissue spaces in the skin (this may sometimes occur around the entry point of an intercostal drainage tube if there is a small leak), a distinctive crackling noise can be heard when pressure is applied to the area.

A full assessment of the skin is most important in the nursing assessment of the patient, as a great deal can be learned about the person's actual, possible and potential problems. Care can therefore be planned rationally to meet the patient's perceived needs.

PRESSURE AREA CARE: A RATIONAL BASIS FOR THE PREVENTION OF PRESSURE SORES

Much has been written about the prevention of pressure sores, sometimes called bed sores, chair sores or decubitis ulcers (from the Latin *decumbo*, meaning I lie down).

The problem is not a recent one. Walker (1971) reported evidence of pressure sores on the mummified body of an Egyptian princess. Today care can be planned rationally in order to prevent this painful and potentially dangerous breech in the defensive barrier provided by the skin. The basic problem is one of cessation of blood flow to an area of superficial tissue, due to pressure occluding or collapsing the capillaries and small blood vessels. Consequently, insufficient oxygen and nutrients reach the cells, and the resultant ischaemia leads to cell death and tissue necrosis. An ulcer forms and is surrounded by a hyperaemic inflammatory area. In severe cases, the subcutaneous fat also necroses.

The phenomenon of capillary collapse as a result of pressure can be seen quite clearly if one holds a clear glass tumbler. White, bloodless pressure areas are clearly visible on looking through the glass at one's fingers. Pressure sores develop over the bony prominences of the body, because here body weight is concentrated on a smaller area of skin and so the degree of tissue compression is greater. An analogous situation would be that of standing on a soft cork floor in (a) a pair of flat walking shoes, and (b) a pair of shoes with high thin heels. In the walking shoes, one's body weight is distributed over a wide area and so the floor is undamaged. In the high, thin-heeled shoes the body weight is concentrated on a very small area and consequently the floor is likely to be pitted and damaged.

Sustained pressure, sufficient to cause capillary collapse, over an area tends to be more damaging than a high pressure over a short period. If a patient is very debilitated, a sore can, however, develop after the first half-hour of unrelieved pressure.

Application of shearing forces can also compress blood vessels and lead to cessation of blood flow to an area of skin. When this occurs, the two skin layers are torn apart and deeper blood vessels are damaged. As a result necrosis may occur in the deeper tissues while the surface skin initially appears normal. Eventually, a surface ulcer forms; this results in a deep pressure sore. Shearing forces occur when a patient is pulled up the bed roughly or when he slips down the bed, especially if the sacral skin is actually adherent to (for example) a drawsheet. Similarly, shearing forces may occur when a bedpan is pulled roughly from under the patient.

Patients at risk of pressure sore formation

Those who do not move enough
It is a useful and informative exercise when on night duty to watch sleeping patients and see how often they change position, to relieve pressure, even while asleep. Some patients, though, are not able to move because they may be paralysed, unconscious, in pain, too weak, too obese, heavily sedated or have musculoskeletal mobility problems. Such patients are more likely to experience tissue damage as a result of sustained pressure unless they are helped to move and so redistribute their weight.

Poorly nourished or dehydrated patients
These patients are often thin and in negative nitrogen balance, with skin that is already in a poor condition. They may well be vitamin deficient (especially of vitamin C) and their skin thus has poor healing properties.

Patients subjected to mechanical injury
Mechanical injury may include, for example, a too-tight plaster cast, crumbs in the bed, wrinkles in the sheet, hard lavatory paper, damage from the nurse's sharp finger-nails, watch or rings. Shearing forces may occur when strapping or dressing tape is removed roughly.

Incontinent or oedematous patients
Skin that is habitually moist at the surface or waterlogged in the dermis is prone to pressure sores because it becomes soft and fragile. Similarly, problems may occur when two moist body surfaces are juxtaposed and so subjected to friction, e.g. under heavy breasts and between obese thighs.

Patients who are infected or very ill
Skin infections or generalized sepsis increase the risk of tissue damage as a result of pressure. Febrile conditions increase the body's metabolic rate and therefore oxygen demand in an area that is already potentially hypoxic (Williams, 1972). Patients who already have circulatory problems, for example anaemia, atherosclerosis or hypoxaemic conditions, and those with malignant disease are similarly susceptible to pressure area problems. Shock, with its associated peripheral circulatory failure, can predispose to rapid pressure sore formation (Barton, 1983).

Prevention of pressure sores

Pressure sores are a very serious problem. Once developed, they may be very difficult to cure. They can become infected, and may eventually need skin grafting. They increase the length of the patient's hospital stay and impose extra strain on both the patient and nursing staff (David, 1983). Prevention starts with the assessment of the patient's susceptibility on a four- or five-point scale, such as that derived by Norton *et al.* (1975).

Thermography and radiometry have been employed with effect to aid the identification of areas of poor skin blood flow and thus allow early preventive measures to be taken (Barton, 1983).

When the potentiating problems have been elicited, steps can be taken to correct or alleviate them by avoiding prolonged circulatory blockage. This may be achieved through shifting the patient's weight regularly and by

(a) spreading the patient's weight over a wide area, which may involve the use of mechanical aids to relieve pressure
(b) avoiding shearing forces and mechanical injury
(c) skin care
(d) adequate nutrition and hydration.

References and suggestions for further reading on pressure sores and their prevention and treatment are given at the end of this chapter.

TEMPERATURE REGULATION

In order for body temperature to be regulated at an optimal level, a balance must be achieved between heat gained and heat lost by the body. Overall control is exerted via the hypothalamic heat-regulating centre in the brain. Humans are homeothermal, that is, they maintain a constant core temperature of about 37°C independent of the environment. They are capable of living at environmental temperatures ranging from −52°C to +49°C.

The term core temperature refers to that inside the skull and the abdominal organs. The skin over the trunk when the body feels comfortably warm may be 33–34°C and the core temperature of 37°C is reached about 2 cm below the body surface. When feeling comfortably warm, the toes may be at 27°C, arms and legs at 31°C, and the forehead at about 34°C. There is thus a temperature gradient between the deep tissues and the skin or, in other words, between the core and the periphery.

Diurnal variations

Generally, body temperature is lower during the night than during the day. It starts to rise from about 5 a.m. to 11 a.m., the early morning oral temperature ranging from 36.3°C to 37.1°C. After 11 a.m., body temperature tends to level out, until about 5 p.m., by which time it may have risen by 0.5–0.7°C from the early morning level. After this time, the body temperature starts to fall, until 1 a.m. A stable period then follows until the next cycle starts at 5 a.m. This pattern varies slightly between individuals, but it is interesting to note that it persists for a time even when night shifts are worked. Thus, the normal day-time pattern is retained in nurses who work on night-duty for, say, one week in every four. Those who go on night-duty for, say 3 months at a time will experience a reversal of their temperature biorhythms.

It is vital that body temperature is retained at a more or less constant level in order to maintain a stable internal environment for the optimal functioning of cellular enzymes. The respiratory and other cellular proteins are not only pH specific but are also temperature specific, and so fail to function efficiently in cell metabolism once the body temperature differs from its normal level.

As already mentioned, balance must be achieved between heat gained by the body and heat lost. The various factors that may influence both sides of this equation are shown in Table 6.1.1.

Table 6.1.1 Factors affecting heat gain and loss

Heat gain	Heat loss
Metabolism	Conduction
External environment	Convection
Hot food and drinks	Radiation
Specific dynamic action of food	Evaporation
Shivering	Loss via lungs, urine, faeces
Hormones	Behavioural responses
Brown fat and adipose tissue	
Behavioural responses	

Heat gain

Metabolism
This is the major way in which heat is gained in the body. Organs such as the liver produce a more or less constant amount of heat through their oxidative processes. The amount produced by skeletal and cardiac muscle varies with their activity; after a prolonged bout of exercise, for example, body temperature may rise to 39°C.

If no heat were to be lost by the body, metabolic processes produce an amount sufficient to raise the body temperature by 1°C per hour, at rest. Activity might raise that figure to 2–3°C per hour. Thus, it can be seen that the problem for humans is, in general, one of losing heat rather than gaining it.

The resting metabolic rate varies throughout life: it is higher in babies and children, and tends to drop slightly in old age. It is usually up to 10% lower in females than in males, and is affected in both sexes by thyroxine production.

External environment
Heat is gained from the environment when the ambient temperature is higher than that of the body. Thus, heat is gained from the sun, a coal fire or radiator, mainly by radiation and conduction, and a hot bath can raise body temperature.

Hot food and drinks
These will also raise body temperature. The effect is, however, minimal, but can be of some psychological benefit in cold weather.

Specific dynamic action (SDA) of food
The ingestion of any food, hot or cold, will raise body temperature as a result of the energy expenditure that occurs during the assimilation of the food into the body. Carbohydrates and fats can cause a rise in the basal metabolic rate of about 3%; proteins bring about a more sustained rise of 20–30% for several hours. The high SDA of proteins is probably related to the oxidative deamination of the constituent amino acids in the liver.

Shivering
This is uncoordinated muscle activity, triggered off by cooling the body surface to below 28°C. Both this critical temperature and the ability to shiver vary with age. Neonates and the elderly are less able to control their body temperature

effectively. Shivering, that is, out-of-phase contraction and relaxation of muscles, tends to occur in bursts, and is not sustained continuously. It is possible, by shivering, to raise the basal metabolic rate by three fold, for short periods. Shivering is controlled by a shivering centre in the posterior hypothalamus, which is influenced by the anterior heat-regulating centre.

It should be remembered that the administration of muscle relaxants during anaesthesia or to the ventilated patient abolishes the ability to shiver, and thermoregulation may therefore be problematic in these patients.

Hormones
Thyroxine output is increased slightly in response to cold, and this raises the metabolic rate. Similarly, adrenaline and noradrenaline both increase the metabolic rate, and so their release leads to an increase in heat production. They also both cause constriction of cutaneous blood vessels, which decreases heat loss; this action, though, is of relatively short duration. In females, a rise in body temperature of $0.3-0.5°C$ occurs at ovulation; this is thought to be due to the increased endometrial activity brought about by progesterone. The rise in temperature continues until progesterone levels fall before the onset of menstruation.

Brown fat
In infants this is a considerable source of heat production and has been discussed earlier in this chapter. Adipose tissue generally serves as insulation for the body core and prevents undue loss of heat.

Behavioural responses to cold
In cold weather, people tend to undertake activities that generate heat, for example, exercising and clapping their hands together. They also try to prevent heat loss, for example by increasing insulating layers in the form of extra clothing. As little of the body surface area as possible is exposed to the cold environment and huddled postures may be adopted in order to reduce the surface area exposed.

Heat loss

Conduction
Heat is conducted to any solid in direct contact with the skin, as long as the surface temperature of the object is lower than that of the skin. For example, heat is lost to cold tile floors from the soles of the feet, and from the buttocks to a lavatory seat. Conduction is, however, not a major method of heat loss.

Convection
Air coming into contact with naked skin is warmed by it; the warm air rises and is replaced by cooler air. The efficiency of this as a method of heat loss depends on the skin blood flow, the temperature difference between the skin and the air, the air flow and the total area of exposed skin. Fans used to cool pyrexial patients increase the efficiency of the convection currents.

Radiation
Heat rays leave the warm body, potentially warming objects in their path. If the objects encountered by the rays are light in colour or shiny, then the heat is reflected back. This is the principle on which aluminium foil 'space-blankets' are based when these are used to warm neonates and those suffering from hypothermia. If black, matt objects are encountered by the radiated heat, these objects tend to absorb the heat. The human body behaves like a black body as it absorbs radiated heat. At an environmental temperature of $35°C$, heat loss by radiation ceases.

About 85% of the heat lost by the body is lost by conduction, convection and radiation via the skin.

Evaporation
Water is lost continually and insensibly from the surface of the body. It evaporates when the environmental temperature is higher than that of the body surface, and in so doing takes heat from the body. The evaporation of 1 litre of water requires 2400 kJ (580 kcal) of heat. Sweat loss can vary between 500 ml and 12 l (1–24 pints) per day; evaporation of 12 l of sweat would require 30,000 kJ (approx. 7000 kcal) of heat.

Evaporation of sweat can no longer occur when the environmental humidity is high.

Via urine, faeces and the respiratory tract
Heat loss via evaporation and by the above routes normally accounts for the remaining 15% of heat loss that has to occur to achieve balance. Antidiuretic hormone production tends to fall in cold

weather, so that a 'cold diuresis' occurs. The opposite occurs in hot weather.

Behavioural responses to warmth

In extremely warm temperatures, clothing is removed to expose more of the body surface and thus maximize heat loss; intake of hot food and drinks is restricted; and less exercise is undertaken. Furthermore, postures that encourage maximum heat loss are adopted, for example arms and legs spread wide when sunbathing.

Control of body temperature

The heat-regulating centres in the anterior hypothalamus receive impulses via afferent nerves from peripheral thermoreceptors in the skin and also from central thermoreceptors in the hypothalamus which monitor the temperature of the blood flowing through them. The centres initiate responses leading to conservation or production of heat, or to heat dissipation, when they receive signals from the thermoreceptors indicating that the core temperature is below or above the 'set' level of 37°C. The centres thus act as a thermostat (Table 6.1.2).

The posterior heat production centre is stimulated by relative cold and initiates measures to prevent heat loss – shivering and vasoconstriction of the cutaneous blood vessels. The anterior heat loss centre is stimulated by relative heat and initiates measures designed to bring about heat loss – vasodilation in the skin blood vessels and increased sweat production. The effects of both centres are mediated via the sympathetic nervous system.

Temperature control and skin blood flow

In cold weather, when the need is to conserve heat, the hypothalamic heat-production centre responds to impulses from the thermoreceptors by causing peripheral vasoconstriction via the sympathetic nervous system (Fig. 6.1.7b). When the superficial vessels are constricted, blood flows via the deeper vessels and so heat loss via the skin is minimized. The skin may appear white, or cyanosed; the latter occurs when blood is trapped in the vasoconstricted vessels and loses oxygen to the tissues.

If vasoconstriction of peripheral vessels is prolonged (e.g. at temperatures below 10°C), damage can occur to the peripheral tissues. This is avoided by the phenomenon of **cold vasodilation**, which is a transient local protective mechanism in which vasodilation occurs in one area, for example fingers or nose, and is compensated for by vasoconstriction in another area, for example the nasal mucosa. As soon as the tissue hypoxia is relieved, vasoconstriction occurs once more.

In warm weather, when the need is to lose heat, surface blood vessels in the skin dilate (Fig. 6.1.7a). This is due to inhibition of sympathetic impulses, initiated by the hypothalamic heat loss centre. Blood is therefore brought nearer to the surface of the body and heat loss occurs via conduction, convection and radiation. A **countercurrent heat exchange system** has developed in order to conserve heat within the body core when the environmental temperature drops. Under cold conditions the temperature of the skin and cutaneous blood in the feet, for example, may be only 19°C. Cold blood returning from the skin to the heart flows in a network of veins called the **venae comitantes**. These pass round the arteries carrying warm blood from the heart to the periphery. Heat is exchanged from the arteries to the veins, and hence warmed blood is returned to the core (Fig. 6.1.8).

Disturbance of temperature control

FEVER

This results from infection by bacteria, viruses or protozoa or from the presence of necrotic tissue, all of which produce pyrogens that affect the temperature-regulating centre. In addition, brain damage, cerebral tumour or head injury may directly affect the temperature-regulating centre.

In fever, the thermostat of the hypothalamus (which is normally set at 37°C) functions as if it

Table 6.1.2 Hypothalamic temperature-regulating centre set at 37°C

When temperature falls below 37°C	When temperature rises above 37°C
Posterior heat production centre initiates	Anterior heat loss centre initiates
shivering	sweating
peripheral vasoconstriction	peripheral vasodilation

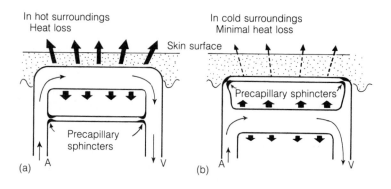

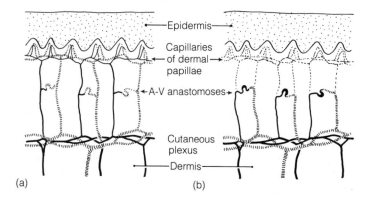

Figure 6.1.7 Diagrams to show heat loss and heat conservation from skin surface: (a) diagrammatic representation, (b) anatomical representation

has been reset to a new, higher level. As a result, heat production is increased and heat loss inhibited in order to raise the core temperature to the new level. With each rise in temperature of 0.5°C, tissue oxygen requirements increase by 7% and so heart rate and respiratory rate both increase. At this stage, the patient feels cold until his core temperature reaches the new, higher level. The cutaneous blood vessels are constricted, with resultant pallor; the patient may shiver, experience rigors, and generally feels cold and unwell. These responses are independent of the ambient temperature, and their net result is that the patient's core temperature is raised, and the patient becomes pyrexial.

When recovery occurs, either naturally, or due to the administration of antibiotics, or when antipyretics such as aspirin are given, the hypothalamic thermostat is reset to its original level of 37°C. If the core temperature was at 39°C, measures will be initiated to bring about heat loss; cutaneous vasodilation and consequent flushing occurs. The patient sweats and complains of feeling warm. Eventually body temperature returns to its normal level. A slight fever is normal for one or two days after surgery and is the result of tissue damage.

HEAT SYNCOPE

With a rapid rise in the environmental temperature in the unacclimatized person, fainting may occur, as a result of a fall in blood pressure, due to inadequately compensated peripheral vasodilation. Adaptation to tropical temperatures (acclimatization) occurs with gradual exposure, for example 2 hours per day. Before this occurs, the person may become very uncomfortable, unable to work or concentrate, and may collapse.

HEAT STROKE

When the body temperature rises to 42°C, the ability to sweat is lost, core temperature rises further and disturbances in brain function occur. Coma and cardiovascular collapse ensue and death may follow rapidly. When the body temperature rises above 43°C, coagulation or precipitation of body proteins occurs. Heat stroke is likely to occur in the elderly or the unacclimatized in the tropics,

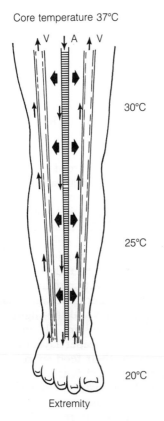

Core temperature 37°C

30°C

25°C

20°C

Extremity

Figure 6.1.8 Diagram to show countercurrent mechanism in leg

when such people exercise in extreme heat. It usually happens after prolonged exposure to high environmental temperatures.

HYPOTHERMIA

Normally, when the body temperature falls, shivering occurs. This is a protective mechanism which functions to raise core temperature. The ability to shiver, however, gradually decreases as the core temperature falls below 34°C. Muscle weakness then occurs, as the core temperature drops further. Activity becomes difficult as muscular movements become uncoordinated, and dulling of mental faculties become evident. Consciousness is lost at a body temperature of 30–32°C. At 22–28°C cardiac arrhythmias occur, and at 18–22°C the heart stops. The ability spontaneously to bring about cessation of heat loss and to initiate heat production is lost at 28°C. At this

temperature, however, gradual rewarming of the patient in hospital may still permit survival. Hypothermia of 26–28°C may be induced surgically to enable the circulation to be stopped for relatively long periods. The oxygen needs of the tissues are substantially reduced at low temperatures, for example at 28–30°C, the metabolic rate is reduced by half. Accidental hypothermia is a potential problem with four main groups of people

(a) those who experience accidents in circumstances likely to produce a fall in body temperature, such as divers or mountaineers
(b) neonates
(c) the elderly
(d) vagrants: in this group, susceptibility may be further increased by misuse of alcohol and other drugs.

The survival time of those immersed in water at 0°C is about 30 minutes; at 15°C, survival time is about 7 hours. In neonates, problems occur with a temperature-regulating centre that is not fully operational and an inability to shiver. A baby exposed to cold may kick to try to increase heat production, but this may lead to kicking off the bedclothes, so accelerating heat loss. In spite of the hypothermia, the baby may appear deceptively pink and healthy. This is because dissociation of oxygen from haemoglobin does not occur so rapidly at low temperatures. The baby may also exhibit cold vasodilation, which was referred to earlier.

The ability of people to respond appropriately to changes in the ambient temperature decreases with age: old people may not realize how cold they are, and therefore fail to act to increase their body temperature. The peripheral blood vessels of the elderly are not so responsive to sympathetic stimulation in many cases, due to atheromatous deposits. Socioeconomic problems, excessive sedation, paralysis, muscle weakness, arthritis, myxoedema or alcoholism are further factors which may make individuals unable to respond appropriately to a fall in environmental temperature.

Assessment of body temperature

ORALLY

The normal oral temperature range is 36.5–37.5°C.

To record this temperature for clinical purposes, it is necessary to leave a clinical mercury-in-glass thermometer in place beneath the tongue for at least four minutes. It is important not to take the temperature within 15 minutes of eating, drinking or smoking, or shortly after any exertion has occurred, and talking should not be allowed while the thermometer is in situ, as all these can affect the oral temperature and so the reading will be an inaccurate reflection of core temperature. Placing a cold glass thermometer in the mouth serves to lower the temperature of the mouth tissues. Frequently, insufficient time is allowed for an accurate reading to be established (Nichols *et al.*, 1966; Nichols and Verhonick, 1967, 1968; Baker *et al.*, 1984). Early morning estimations will usually be up to 0.5°C lower than early evening readings due to diurnal temperature variations, and in females postovulatory readings will be slightly higher than preovulatory ones.

AXILLARY METHOD

In many cases, this is a safer area from which to record body temperature as it avoids the possibility of trauma to the patient who may accidentally bite the thermometer. Excessive sweating or axillary air pockets may, however, lead to inaccuracies. In order to obtain an accurate reading, the thermometer should be left in place for at least 10 minutes. The recorded temperature is approximately 0.5°C lower than the oral temperature (Nichols *et al.*, 1966).

RECTAL METHOD

This gives the closest approximation to the core temperature. It will, on average, register a temperature of up to 0.5°C higher than the oral temperature; one reason for this is thought to be the extra metabolic heat that is produced in the rectum by the bacterial colonization of the area. The lubricated thermometer is inserted approximately 2–3 cm into the rectum and left in situ for at least 3 minutes (Nichols *et al.*, 1966). It must be adequately cleaned after use, and should be used solely for temperature recording by the rectal route, to eliminate the risk of introducing micro-organisms from the rectum into a patient's mouth.

The hands may be used to give an approximation of body temperature. The backs of the fingers and not the tips should be used, as nerve endings on the backs of the hands are more sensitive to changes in temperature. Whichever method is used to record a patient's temperature, the site of measurement should be recorded and used consistently.

THE NORMAL FLORA OF THE SKIN

On each square centimetre of skin there may be up to 3 million micro-organisms, most of which are commensals. A harmless association exists between humans and their commensals (literally, 'table companions', from the Latin), to the sole benefit of the latter.

The skin does not provide a very hospitable environment for bacteria unless they have become adapted through evolution to live there, and commensals have, in general, become adapted to live off human skin scales and the slightly acid secretions produced by the skin. The micro-organisms tend to live in the deeper layers of the stratum corneum, near to their food source. Hence, they are not normally shed with desquamation. The fine balance which exists between us and our parasites is upset by the application of strong deodorants, whether applied to the axillae or vulva, and the use of strong soaps which alter the skin pH from acid to alkaline. These tend to kill or inhibit the normal flora, leaving the area open to potential invasion by pathogens.

Babies are born with no resident microbial flora. During a normal delivery, a baby starts to pick up organisms during the passage down the vagina. A resident population of commensals makes it harder for pathogens to survive, and so a neonate is particularly prone to skin infections until his skin becomes colonized. If skin is poorly cared for, the normal microbial commensals may be overcome and pathogens may colonize the area.

Micro-organisms thrive in moist conditions, and so the axillae and groins provide favourable areas for their growth. A waterproof plaster applied to a cut on the forearm for one or two days causes an increase in the resident population from a few thousand/cm^2 to a few million. Washing and bathing both increase the numbers of bacteria released from the skin for up to 10 hours. Heat and moisture cause the break-up of large colonies, so more organisms tend to be shed.

The skin can never be sterilized. The topical

application of alcohol and iodine-based lotions may cause the death of a large percentage of resident organisms, but such applications do not remove those bacteria which colonize the hair follicles (and these account for at least 20% of the total numbers). In 1972, Selwyn found strains of skin bacteria present in about one in every five individuals which actively inhibited the growth of resistant *Staphylococcus aureus* by the production of their own antibodies. He isolated 32 strains of micrococci that had this ability. The fortunate people who possess these bacteria are less likely to develop wound infections (Andrews, 1978).

Man and his commensals can live in harmony as long as the skin remains an intact barrier. However, if these micro-organisms are given the opportunity to invade the dermis, or normally sterile body cavities, for example the bladder, then the association of man and host might no longer be harmless as many commensals may then act as opportunistic pathogens.

GENERAL INNATE BODY DEFENCES AGAINST INFECTION

In order for infection to occur, micro-organisms must gain entry to the body. For each portal of entry, there are local defence mechanisms.

Mouth (see also Chap. 5.1)

This cavity is lined with a fairly tough mucous membrane which is irrigated by a constant backward flow of saliva (i.e. directed towards the throat). The flow of saliva also prevents organisms from entering the salivary glands, and traps organisms which can then be swallowed.

Saliva contains **lysozyme**, an antibacterial enzyme, and mucus (which contains the immunoglobulin IgA). Patients who become dehydrated have a reduced flow of saliva and a dry mouth. This may result in mouth infections.

The resident bacteria in the mouth are in general harmless; indeed, α-haemolytic *Streptococcus* is of positive benefit as it produces hydrogen peroxide which helps to clean the mouth. If these resident flora are destroyed by, for example, the prolonged administration of antibiotics, then pathogens (commonly *Candida albicans* which causes thrush) may colonize the mouth. The tonsils, formed of lymphoid tissue, further defend the

oral cavity; however, the tonsillar epithelial covering is very thin and so is easily traumatized. Tonsillar infections are not uncommon, especially in childhood.

Stomach (see also Chap. 5.1)

The hydrochloric acid produced by the stomach kills most organisms entering in food, drink or swallowed sputum. Some organisms can, however, resist this strong acid, for example tubercle bacilli, enteroviruses and salmonella. Milk and proteins are both effective buffers against gastric acid, and so some organisms (if ingested with these foodstuffs) may be protected from the gastric acid. This explains how, in spite of gastric defences, typhoid and dysentery can still occur.

Vomiting can be regarded as a reflex defence mechanism, ridding the body of mechanical and chemical irritants, such as alcohol and bacterial toxins.

Intestines (see also Chap. 5.1)

To a certain extent, the small and large intestines rely on the stomach's bactericidal activity. The resident flora of the large intestine – *Escherichia coli*, non-haemolytic streptococci, anaerobic *Bacteroides* – all contribute to the normal function of the area. Their role becomes more evident when they are removed by the administration of broad-spectrum antibiotics. The area is then open to colonization by pathogens (e.g. *Staphylococcus pyogenes*), which may be resistant to antibiotics. Such superinfections can be fatal in the debilitated patient.

The small and large bowels are both liberally supplied with patches of lymphoid tissue throughout their length. Plasma cells from these areas produce immunoglobins (IgA) and these provide local defence.

Diarrhoea, like vomiting, can be regarded as a defensive response to an established infection. It occurs too late to be protective, however.

Respiratory tract (see also Chap. 5.3)

UPPER RESPIRATORY TRACT

Insects and large particles are prevented from

entering the tract by the presence of hairs, or vibrissae, in the nose. The ciliated nasal mucosa secretes a backward-flowing stream of mucus which traps smaller particles and has both bactericidal and virucidal properties. Lysozyme is also present in nasal secretions.

The epithelium of the upper respiratory tract is thin and is unfortunately prone to infections by rhinoviruses and adenoviruses which are not affected by the nasal secretions. Sneezing is a protective reflex which expels irritants.

TRACHEA AND LUNGS

The trachea and bronchi are lined with a ciliated mucous membrane which serves to trap any organisms in debris that may have escaped through the upper tract. Here the cilia beat upwards, and hence shift a stream of mucus away from the lungs and towards the pharynx to be swallowed. Should any organisms reach the alveoli, alveolar macrophages phagocytose them. The hilum of the lung is well supplied with lymph nodes which act as a further filter.

Coughing is a defensive reflex which removes particulate matter or excess mucus in the lower tract.

Genitourinary tract (see also Chap. 5.4)

The constant downward flow of urine through the ureter and bladder tends to militate against ascending infections. Micturition itself irrigates the urethra. This is an effective response in the long male urethra, but less so in the female. The adult female urethra is only 2 or 3 cm long and hence forms a relatively short and readily available portal of entry for organisms into the bladder. Sexual activity may predispose to the occurrence of cystitis (inflammation of the urethra), the most common offending organism usually being coliform bacillus from the perineal area. Faulty aseptic technique during catheterization can, therefore, result in urinary tract infection, which is the commonest cause of hospital-acquired infection accounting for 25% of all such infections (Meers and Stronge, 1980).

Reproductive tract (see also Chap. 6.3)

The acid medium of the vagina (pH 4.0–4.5) is maintained by the resident flora, notably *Lactobacillus*; this acts upon the glycogen in the vagina to produce lactic acid, which forms an inhospitable environment for pathogens. If non-prescribed douches or vaginal deodorants are used, the acid environment of the vagina is disturbed and pathogens may colonize the area. *Candida albicans* may supervene and produce vaginal thrush. In postmenopausal females, the production of glycogen in the vagina decreases, less acid is produced and a senile vaginitis may occur; this can be treated with oestrogens.

Eye (see also Chap. 2.2)

The conjunctival sac is constantly irrigated with tears produced by the lachrymal glands. Tears contain more lysozyme than any other body fluid. The eye infections that occur with vitamin A deficiency do so as a result of the decrease in lysozyme secretion in this condition. Blinking is a defensive reflex which helps to rid the eye of irritants and to distribute the tears. In conditions such as facial nerve paralysis, this reflex is lost, and it becomes necessary to prevent the conjunctiva drying or becoming ulcerated, for example by keeping the eye closed and irrigating it regularly.

INFLAMMATION

This is a local defensive response to tissue damage. It functions to eliminate the cause of the tissue damage, removing the consequent dead cells and restoring the constancy of the internal environment.

Inflammation occurs immediately after any physical, chemical or microbiological injury. As such, it is a local non-specific response. It may occur just as much in response to a sliver of glass entering the skin as to a burn or local staphylococcal infection. The last-mentioned, which results in a boil, is often considered as a classic example of inflammation, exhibiting the five characteristic features of the process.

(a) Redness over the area (rubor).
(b) Swelling (tumor).
(c) Heat (calor).
(d) Pain (dolor).
(e) Loss of function (laesio functi).

The stages of inflammation that result in these features are as follows.

(a) As soon as injury occurs, the blood vessels at the edge of the wound constrict momentarily – perhaps for 2–3 minutes. The length of time this lasts depends on the suddenness of the injury, being longer with rapid injury. This constriction is the result of a myogenic reflex.

(b) All blood vessels in the area then dilate, leading to hyperaemia (redness) over the area. Blood flow to the area is increased, and chemical mediators such as histamine, serotonin, bradykinin and leucotaxin are released by the injured tissue cells. These increase the capillary permeability in the area, leading to the next stage.

(c) Fluid leaves the blood vessels and enters the tissue spaces, resulting in oedema, pain and eventual loss of function. The enlarged pores in the capillary are such that plasma proteins escape in the fluid exudate and hence exert an osmotic effect in the tissues. The net result of this is that blood flow slows in the dilated vessels.

(d) White blood cells, especially neutrophils, marginate (move to the sides of the blood vessels) and leave the vessels to enter the inflamed area. This process is called **diapedesis**. They are attracted there by chemicals released by the injured tissue cells and by bacteria, if any are present (**chemotaxis**). Neutrophils and macrophages phagocytose the dead tissue and the micro-organisms, if the latter are the causative agents. Micro-organisms are rendered more susceptible to phagocytosis by antibodies present in the exudate. Antibodies enhance the adherence of phagocytic leucocytes to the micro-organisms, a process which is initiated by chemotaxis and determined by the surface charges of the cell membranes.

(e) Finally, the inflamed area becomes enclosed, or walled off, by fibrin. The framework of fibrin threads so formed prevents the damage or infection extending to neighbouring tissues.

White blood cells engulf bacteria by putting out pseudopodia around them and enclosing them in a vacuole within their cytoplasm. Lytic enzymes such as lysozyme and phagocytin are then poured into the vacuole by the white cell. It may take up to 15 minutes for the micro-organisms to be destroyed in this way; alternatively, the white cell may be killed by the engulfed micro-organism. Occasionally, the two may live together for much longer until one or the other supervenes. This occurs, for example, with *Mycobacterium tuberculosis*. Pus is formed from the collection of dead white cells, tissue cells, exudate and micro-organisms, and may need to be drained from the wound before healing can occur.

WOUND HEALING

After injury to the body has occurred, healing of the wound takes place in order to restore the intact barrier provided by the skin. The healing process and the inflammatory process, although described separately here, overlap to a considerable degree.

When the edges of a wound are opposed, for example after a surgical incision, healing is likely to be rapid, and is sometimes said to occur by **primary** or **first intention**. If this is not possible, for example when there is a deep and wide ulcer or a gaping wound, the healing process takes longer and occurs by the formation of granulation tissue from the bottom of the wound; this is sometimes referred to as healing by **secondary intention**.

Healing of a surgical wound (Fig. 6.1.9)

Eight hours after surgery, a blood clot with its fibrin framework has filled the incision track. Necrosis occurs on either side of the incision and extends for 100–200 μm. This is due to the disturbance in the blood supply to the surrounding tissues brought about by the incision. By 16 hours after surgery, epidermal cells have started to invade the boundary between the living and the necrotic tissue. Eventually, after about 40–80 hours, these two arms of epithelial cells unite in the dermis below the level of the incision. This serves to sever the connection between the necrotic area and the living area. The necrotic area forms the scab and, when the underlying epithelium becomes keratinized, the scab will be shed.

Epithelial invasion also occurs down the suture tracks, so that in effect the sutures are contained within tubes of epithelium. These epithelial cells are often removed with the sutures, and can be seen as debris on the suture material. This leaves a raw track, a potential point of entry for micro-organisms. During this initial healing period, there is a rapid proliferation of cells in the

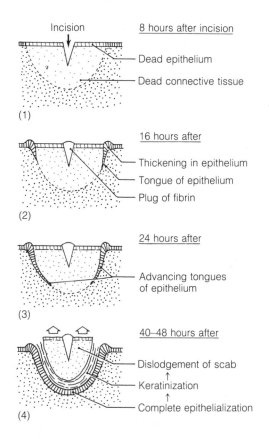

Incision | 8 hours after incision
— Dead epithelium
— Dead connective tissue
(1)

16 hours after
— Thickening in epithelium
— Tongue of epithelium
— Plug of fibrin
(2)

24 hours after
— Advancing tongues of epithelium
(3)

40–48 hours after
— Dislodgement of scab
— Keratinization
— Complete epithelialization
(4)

Figure 6.1.9 Diagrammatic representation of stages of wound healing

surrounding dermis. Fibroblasts in the area are stimulated to produce collagen fibres and capillaries grow into the area from surrounding vessels; as a result a soft, pink, delicate tissue called **granulation tissue** fills the wound area.

As the fibroblasts mature and form more collagen, both the cells and the fibres start to shrink, causing some contraction of the area and eventual obliteration of some of the capillaries. The end-result is a firm fibrous epithelial scar.

For the first 3–5 days postoperatively, the tensile strength of the wound is low. It can be measured experimentally by estimating the force required to distract the edges of a wound.

The rate of increase in strength is maximal between days 5 and 12 postoperatively, when collagen formation is occurring rapidly. The remaining gain in strength occurs gradually over subsequent weeks, as collagen continues to be deposited for up to 80 days postoperatively.

The role of sutures in securing wound adherence is thus complete by 12–14 days after surgery, as by this time sufficient collagen has been formed to ensure skin healing.

Factors which affect healing

Vitamin C deficiency. This results in an inability to form hydroxyproline, an amino acid necessary for collagen manufacture (Dowding, 1986).

Protein depletion. This causes delay in healing and in the development of tensile strength in the scar tissue, since protein is necessary for all reproduction and repair.

Raised corticosteroid levels. Raised plasma levels of corticosteroids (as occur in Cushing's syndrome, steroid therapy and in physiological and emotional stress) suppress the inflammatory and immune responses and hence delay wound healing and reduce resistance to infection.

Infection. This results in increased inflammation, further necrosis and delayed healing. The wound may need to be kept open while there is deep infection, to prevent re-epithelialization over pockets of infection and subsequent abscess formation (Serter and Pringle, 1985). Wound healing would therefore take place by secondary intention and the formation of granulation tissue. Wound infections form 24% of all hospital-acquired infections (Meers and Stronge, 1980) and the majority of these could be avoided by more carefully planned care.

Many factors play a part in the success or failure of wound healing, and it may be difficult to ascertain the cause of failure in any one patient. However, it is undoubtedly one of the major responsibilities of the nurse in a surgical unit to ensure optimal conditions for healing to occur and thus to protect the patient.

The aim of this chapter has been to indicate the body's innate defences against environmental trauma, whether this is of a mechanical, physical, chemical or biological nature. For the most part, the innate defences are very effective. When they are breeched, however, actively or passively acquired defences come into play, and these are described in the next chapter.

Review questions

The answers to all these questions can be found in the text. In each case, there is at least once correct answer and least one incorrect answer.

1 Which of the following statements about melanin production is/are correct?

  (a) Melanin is produced in response to infra-red radiation.
  (b) Caucasian races have fewer melanocytes than other races.
  (c) Ability to produce vitamin D_3 depends on the total number of melanocytes.
  (d) Melanin production is absent in albinism.

2 Cutaneous blood flow

  (a) is greater in the epidermis than in the dermis
  (b) can rise in 3000 ml/min at very high temperatures
  (c) increases with the ingestion of alcohol
  (d) increases with sympathetic inhibition.

3 Which of the following statements about hair is/are true?

  (a) Terminal hair is produced in response to cytotoxic drugs.
  (b) Body hair on humans helps to conserve heat.
  (c) Testosterone stimulates head hair growth.
  (d) Cytotoxic drugs affect the actively dividing bulb of the hair.

4 Which of the following statements concerning sweat is/are correct?

  (a) Sweat is hypotonic with reference to plasma.
  (b) Its production can rise above 20 l per day.
  (c) Its production is influenced by circulating aldosterone.
  (d) Sweating occurs only in hot weather.

5 Which of the following conditions may affect the colour of the skin?

  (a) Myxoedema.
  (b) Alcoholism.
  (c) Pulmonary oedema.
  (d) Diabetes mellitus.

6 Which of the following would be the most effective method of raising core temperature?
  (a) Holding a hot water bottle.
  (b) Ingesting hot food.
  (c) Lying under a duvet.
  (d) Jogging for half an hour.

Answers to review questions

1 d
2 b and c
3 d
4 a and c
5 a, b and c
6 d

Essay topics

1 Discuss how the structure of the skin is related to its functions.
2 Explain in detail how body temperature is controlled.
3 Describe how wound healing occurs after surgery and relate this to the nurse's role in the management of wound care.

Suggestions for practical work

1 Elicit a white line reaction using the blunt end of a key, and explain to a junior nurse why this occurs.
2 Scrape the forearm firmly with the sharp edge of a key, and elicit the triple response. Account for the changes you see.
3 Record your oral temperature at rest, then have a hot bath lasting at least 10 minutes. Retake your temperature. Why does a change occur?
4 Females only: measure your oral temperature every morning immediately on waking for two menstrual cycles. Record and assess when ovulation occurs.
5 Record your oral temperature at rest. Step on and off a safely positioned sturdy stool for 5–10 minutes (do this only if you are healthy). Measure your oral temperature at the end of the exercise and account for the result.

6 Using a clinical thermometer, record your oral temperature every 30 seconds until a stable reading is established. Note the total time taken and compare this with ward practice.

7 Measure your oral temperature. Drink a cup of hot tea or a cup of cold water and note how long it takes for the temperature to return to the original level.

References

Andrews, M. (1978) *The Life that Lives on Man.* London: Arrow Books.

Baker, N. et al. (1984) Effect of type of thermometer and length of time inserted on oral temperature measurements. *Nursing Research, 33*(2), pp. 109–111.

Barton, A. (1983) Pressure sores. In *Pressure Sores*, Barbenel, J. C., Forbes, C. F. and Lowe, G. D. O. (eds.). London: Macmillan.

Forbes, C. F. & Lowe, G. D. O. (eds), pp. 53–57. London: Macmillan.

David, J. (1983) *An Investigation of the Current Methods used in Nursing for the Care of Patients with Established Pressure Sores.* Nursing Practice Research Unit. London: DHSS.

Dowding, C. (1986) Nutrition in wound healing. *Nursing*, 5, No. 3 (May).

Meers, P. D. & Stronge, J. L. (1980) Urinary tract infection. *Nursing Times 76* (July 24): Centre pages

Nichols, G. A., Ruskin, M. M., Glor, B. A. K. & Kelly, W. H. (1966) Oral, axillary and rectal temperature determinations and relationships. *Nursing Research 15* (4): 307–310.

Nichols, G. A. & Verhonick, P. J. (1967) Time and temperature. *American Journal of Nursing 67* (11): 2304–2306.

Nichols, G. A. & Verhonick, P. J. (1968) Placement times for oral thermometers: a nursing replication. *Nursing Research 17*(2): 159–161.

Norton, D., McLaren, R. & Exton-Smith, A. (1975) *An Investigation of Geriatric Nursing Problems in Hospital.* Edinburgh: Churchill Livingstone.

Serter, H. & Pringle, A. (1985) *How Wounds Heal: A Practical Guide for Nurses.* London: Wellcome Foundation.

Walker, K. (1971) *Pressure Sores – Prevention and Treatment.* Nursing in Depth Series. London: Butterworth.

Williams, A. (1972) A study of the factors contributing to skin breakdown. *Nursing Research 21*: 238–243.

Suggestions for further reading

THE SKIN

Burton, J. L. (1985) *Essentials of Dermatology*, 2nd edn. Edinburgh: Churchill Livingstone.

Fry, L. & Cornell, M. N. P. (1984) *Dermatology*. Lancaster: MTP Press.

Leigh, I. & Wojnarowska, F. (1985) *Coping with Skin and Hair Problems*. Edinburgh: Chambers.

Nursing (1983) The skin. *Nursing*, 2 No. 9 (January), No. 10 (February).

Solomons, B. (1983) *Lecture Notes on Dermatology*, 5th edn. Oxford: Blackwell.

Williams, S. & McVan, B. (1983) *Assessing Vital Functions Accurately*, 2nd edn. New York: Springhouse.

PRESSURE SORES

Barbenel, J. C., Forbes, C. F. & Lowe, G. D. O. (eds.) (1983) *Pressure Sores*. London: Macmillan.

Conduct and Utilization of Research in Nursing Project (1981) *Preventing Decubitus Ulcers: CURN project.* London: Grune & Stratton.

Lee, B. Y. (ed) (1985) *Chronic Ulcers of the Skin*. New York: McGraw-Hill.

Torrance, C. (1983) *Pressure Sores: Aetiology, Treatment and Prevention.* London: Croom Helm.

WOUND CARE

David, J. (1986) Wound Management. London: Martin Dunitz.

Lawrence, J. C. (ed.) (1983) *Wound Healing Symposium: Proceedings of a Symposium held at the Queen Elizabeth Postgraduate Medical Centre, Birmingham, England, October, 1982.* London: Medicine Publishing Foundation.

Nursing (1986) Wound care. *Nursing*, 3, No. 5 (May), No. 6 (June).

Westaby, S. (1985) *Wound Care*. London: Heinemann Medical.

BODY TEMPERATURE

Age Concern (1985) *Hypothermia: The Facts*. London: Age Concern Information Department.

Collins, K. J. (1983) *Hypothermia: The Facts*. Oxford: Oxford University Press.

Chapter 6.2
Acquired Defences

Diana G. Langelaan

Learning objectives

After studying this chapter, the reader should be able to

1 State the characteristics of an acquired (specific) defence mechanism.
2 Define the term antigen.
3 Compare and contrast the development and method of function of the cell-mediated and humoral immune systems.
4 Describe the characteristics of a typical antibody.
5 Name the classes of immunoglobulin and state where each may be found in the body.
6 Explain the ways in which interaction between antigens and antibodies conveys protection.
7 Outline the ways in which interaction between cell antigens and immunologically competent T lymphocytes conveys protection.
8 Compare and contrast the features of artificially induced active and passive immunity.
9 Demonstrate an understanding of the acquired defence mechanisms by stating the criteria to be considered for an effective immunization programme.
10 Explain the relevance of the humoral immune system to resistance to infection and indicate some ways in which this may be disrupted by drugs or disease.
11 Explain briefly the mechanisms underlying

each type of hypersenstivity reaction, giving examples of clinical significance.
12 Describe the clinical manifestations of anaphylactic shock, outlining the underlying cause of each.
13 Outline some possible ways in which autoimmunity may occur and give examples of disorders of known autoimmune aetiology.
14 Outline the possible role of the immune system in the normal prevention of cancer and potential treatments based on immunotherapy.
15 Explain the role of the cell-mediated immune system in the rejection of transplanted tissue.
16 Define the term HLA system in the context of tissue typing.
17 Describe the anatomical features of the lymphatic system.
18 Discuss the importance of the lymphatic system in the formation and reabsorption of interstitial fluid.
19 Outline the clinical implications of the filtering function of lymph glands.
20 Describe the functions of the lymph glands in relation to the immune system.

Introduction

The body, if it is to survive, must be protected, both on its surface and internally, from invasion by any potentially harmful substance. As well as a variety of non-specific defence mechanisms such

579

as the phagocytic activity of certain leucocytes discussed in Chapter 6.1, there are also mechanisms which protect against specific noxious substances and agents. These mechanisms work most efficiently when the body has previously come into contact with the foreign substance and thus it seems that the mechanisms are acquired following exposure, and that the substance is in some way remembered.

Having a particular infectious disease imparts immunity (resistance) to that disease but not to others – this was observed before even the existence of micro-organisms was known. For instance, people who had recovered from smallpox frequently nursed those suffering from the disease. It can thus be concluded that following exposure to a foreign substance, whether it is a bacterium, virus or pollen, the body develops something which neutralizes or destroys the substance and which acts specifically with only that substance or one very similar to it. As an example of this, **vaccination** (*vacca* from the Latin for cow) involves injection of the cowpox virus which confers immunity against smallpox, the virus of which is similar. It can, however, occasionally result in vaccinia, a human form of cowpox. Edward Jenner first used fluid extracted from the cowpox blisters of a milkmaid to vaccinate a child against smallpox in 1796. His research which culminated in the development of this vaccination arose from the observations of farmers that milkmaids who had suffered from cowpox were immune to smallpox. The term vaccination has come to be used loosely for any immunization.

The immune system is primarily a defensive system which protects the individual against a variety of foreign substances including micro-organisms, transplanted cells and irritants. The mechanisms involved are relevant to a wide variety of clinical situations, including immunity against infectious disease, the rejection of transplanted organs, compatibility of blood for transfusion, allergic disorders, autoimmune diseases and malignant conditions.

Although the agents conferring immunity, which are antibodies (specialized blood proteins) and components of certain lymphocytes, have been recognized since the end of the nineteenth century, immunity remains a subject of intensive research, and many questions remain unanswered.

THE NATURE OF IMMUNITY

An **antigen** is an agent or substance which can be recognized by the body as 'foreign'. Often it is only one relatively small chemical group of a larger foreign, substance which acts as the antigen, for example a component of the cell wall of a bacterium. Most antigens are proteins, though carbohydrates may act as weak antigens.

The body reacts to antigens by making **antibodies**, themselves proteins, or special lymphocytes carrying an antibody-like component on their cell surface. These antibodies or lymphocyte components interact chemically with the antigen in a highly specific manner, rather like a key which is made for only one lock. However, occasionally antibodies or lymphocyte components may 'fit' other similar antigens, in much the same way as a key will sometimes work in other locks (Fig. 6.2.1).

When the antigen and antibody interact, they bind together to form an antigen/antibody complex, or **immune complex**, and this binding neutralizes or brings about the destruction of the antigen. A bacterial toxin may, for example, cause disease by binding to a particular type of cell in the body. When an antibody (in this case called an **antitoxin**) interacts with the toxin it may neutralize its toxicity by, perhaps, binding to the active site of the toxin, that is, the part which binds to the target cell. Alternatively, interaction between antitoxin and toxin may alter the shape of the toxin so that it no longer has the capacity to bind to cells.

Other ways in which such interaction destroys the antigen include an increased susceptibility of the immune complex to phagocytosis by leucocytes. More specific details of the ways in which antigen/antibody interaction protects the body will be given later, but already it can be seen that the acquired and innate defence mechanisms require mutual cooperation.

Immunity due to antibody-like components of lymphocytes is referred to as **cell-mediated immunity** since the lymphocyte cells themselves interact with the antigen. Immunity conveyed by antibodies is referred to as **humoral immunity** since in most cases the antibodies reach their destination via the blood.

THE DEVELOPMENT OF THE CELL-MEDIATED IMMUNE SYSTEM

Lymphocytes originally derive from stem cells of the bone marrow. At around the time of birth, lymphocytes derived in this way leave the marrow and pass to the thymus gland in the chest, where they multiply. The lymphocytes are processed in some way (the mechanisms of which are unknown at present) by the thymus gland, so that between them they carry the genetic information necessary to react with a multitude of possible antigens. Each lymphocyte (or possibly a few) is thus able to react with one antigen. They are called **T lymphocytes** since they have been processed by the thymus gland. Antigens of the body's own cells are excluded so that the population of T lymphocytes processed in this way is potentially capable of recognizing 'self' and 'non-self' and only reacting against the latter. Autoimmune diseases are those in which this recognition is defective, and certain of the body's cells may be recognized erroneously as 'non-self' and subsequently destroyed.

Processing of T lymphocytes occurs largely during fetal life and early childhood, after which the thymus gland shrinks in size. Mice experimentally deprived of the thymus gland immediately after birth, before they have begun to process T lymphocytes, are incapable of cell-mediated immune responses, and so, for example, they fail to reject transplanted tissue. In addition, the humoral immune system works less efficiently, indicating some cooperation between the two systems (Miller, 1962).

The T lymphocytes, each processed to 'recognize' and interact with a specific antigen, circulate permanently between the blood and lymphatic systems. On recognition of the antigen, one or a group of lymphocytes takes up residence in secondary lymphoid tissue (e.g. lymph glands, Peyer's patches, spleen, bone marrow) and divides to form two types of cells, **memory cells**, which are lymphocytes processed in the same way as themselves, and **killer cells**, which interact with the antigen.

Recognition of the antigen by the appropriate lymphocytes is achieved by receptors on the surface of the T lymphocytes which fit the anti-

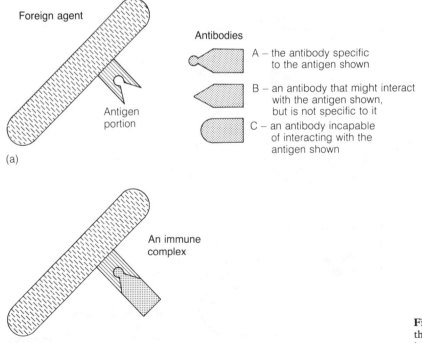

Foreign agent

Antibodies

Antigen portion

A – the antibody specific to the antigen shown

B – an antibody that might interact with the antigen shown, but is not specific to it

C – an antibody incapable of interacting with the antigen shown

(a)

An immune complex

(b)

Figure 6.2.1 Diagram to illustrate the specificity of an antibody/antigen interaction

gen in a similar way to that previously described for antigen/antibody interaction. It is thus a chemical interaction which is highly specific (Fig. 6.2.2a).

THE DEVELOPMENT OF THE HUMORAL IMMUNE SYSTEM

A separate population of lymphocytes derived from the stem cells of the bone marrow undergoes multiplication and processing in lymphoid tissue elsewhere than in the thymus gland. In birds, the lymphoid tissue concerned has been located in the gut and called the bursa of Fabricius. In humans, the site is unknown, though there is some evi-

dence to suggest that such processing occurs in the bone marrow itself. Lymphocytes processed in this way are called **B lymphocytes** because they are processed by the bursa or its equivalent in humans (Fig. 6.2.2b).

B lymphocytes, like T lymphocytes, have surface receptors which enable them to recognize the appropriate antigen, but do not themselves interact to neutralize or destroy the antigen. On recognition of the antigen they take up residence in secondary lymphoid tissue and proliferate to form daughter lymphocytes, processed in the same way as themselves (again these can be referred to as memory cells). They also form short-lived **plasma cells** which make and secrete the appropriate antibody. The plasma cells them-

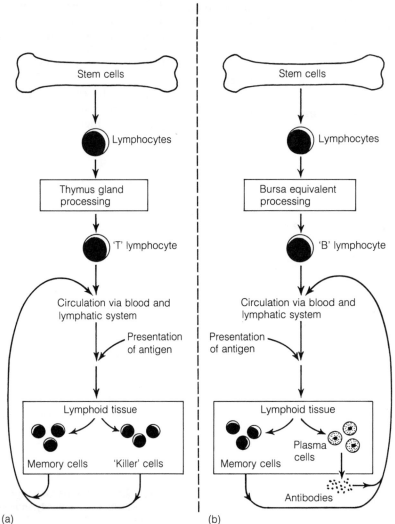

Figure 6.2.2 Summary of the development of (a) the cell-mediated immune system, and (b) the humoral immune system

selves remain in the lymphoid tissue for their short lives.

The population of circulating lymphocytes of the humoral immune system is made up largely of so-called memory cells derived from B lymphocytes, as well as some B lymphocytes themselves.

THE CLONAL THEORY

Burnet (1959) proposed this theory, which accounts, in particular, for the more rapid formation and increased levels of antibodies or killer cells in response to the introduction of an antigen on a second or subsequent occasion – the **booster response**. Burnet postulated that one or a group of processed T or B lymphocytes can recognize one or a group of closely allied antigens. On 'meeting' the antigen for the first time, the processed lymphocyte (or lymphocytes) proliferates and produces a group, or clone, of identical daughter lymphocytes, all capable of recognizing the antigen and responding to it. A **clone** is a group of genetically identical cells derived from a common ancestor. There are thus more immunologically processed cells to react on a subsequent invasion by the antigen, and antibodies and killer cells are therefore produced in much larger numbers.

ANTIBODIES AND THEIR MODE OF ACTION

Antibodies are protein molecules produced by plasma cells within lymphoid tissue. They circulate in the blood, along with many other proteins. They are found mostly in the gamma-globulin fraction of the plasma proteins, and may be referred to as **immunoglobulins**. Their structure is such that they may be able to clump antigens together, since they may have two or more antigen-binding sites (Fig. 6.2.3). The formation of large aggregates of antigens bound together by antibodies (referred to as **precipitation**) may, for instance, limit the dissemination of antigen-bearing bacteria. A special case is the agglutination (clumping) of red blood cells which occurs when incompatible blood is transfused into a patient (see Chap. 4.1).

There are five classes of immunoglobulin (Ig), separated by variations in structure and function:

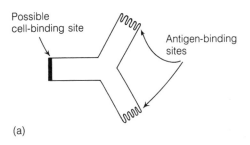

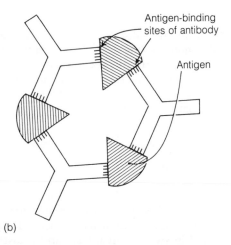

Figure 6.2.3 Schematic diagrams (a) to represent the structure of a simple (IgG type) antibody, and (b) to illustrate the way in which this bivalent structure enables antigens to be clumped together

IgG, IgA, IgM, IgD and IgE. (Details of molecular structure can be found in Weir, 1977.)

IgG antibodies

These are the commonest immunoglobulins and are found not only in the blood but in all fluid compartments of the body, as well as in the cerebrospinal fluid and urine. Their molecules are small enough to be able to cross the placenta from mother to fetus.

The structure of IgG is that depicted schematically in Fig. 6.2.3, and not only are they bivalent and thus able to bind to two antigens, but their third limb is able to bind to special receptor sites on certain cells, for example phagocytes. This enables phagocytes to be brought into close proximity with antigens and thus enhances phagocytosis.

When the foreign substance bearing antigens is a cell, antibodies can coat its surface by binding with the cell's surface antigens. Phagocytes and other cells which carry appropriate surface receptors then bind to the cell-binding sites of the antibodies and the foreign cell is destroyed. Such interaction may be involved when tissue transplants are rejected, when autoimmune diseases cause damage to the body's own cells, and when cancer cells which develop spontaneously in the body are eradicated.

IgG antibodies also activate the complement system (a non-specific defence system which will be discussed later in this section). It is important to realize that it is through the inflammatory response (discussed in Chap. 6.1), which is entirely innate, that antibodies exert their protective function.

IgA antibodies

These are found in the blood as well as in mucous secretions such as those of the respiratory, gastrointestinal and genitourinary tracts. They are also found in sweat, saliva, tears and colostrum.

IgA antibodies present in mucous secretions are produced by plasma cells located in the mucous membranes themselves and pass into the lumen of the tract with mucus. They are thought to protect the body from invasion by pathogens by coating the surface of the pathogens and thus preventing them from adhering to the mucous lining.

IgM antibodies

These are the largest immunoglobulins and are confined to the bloodstream. They are the first antibodies to be produced when the body is confronted by a new antigen, but are short lived. The antibodies responsible for agglutination of the red blood cells are of this class and are well equipped for this role since they have ten antigen-binding sites.

IgM antibodies are unable to cross the placenta from mother to fetus and thus high levels of IgM antibodies in a newborn infant suggest intrauterine infection such as syphilis.

IgD antibodies

These are large and are therefore only found in the blood. Little is known of their function at present.

IgE antibodies

These are present only in small amounts in the blood but are also found attached to mast cells and basophils. Mast cells are mostly located around capillaries. Mast cells and basophils both contain granules of histamine, the release of which, in large quantities, results in allergic states such as hay fever and asthma. IgE antibodies have a cell-binding site (as illustrated in Fig. 6.2.3) which allows them to bind to basophils and mast cells, enabling the release of histamine from these cells (see also Chap. 4.1).

Antibody levels in the newborn

IgG antibodies are able to cross the placenta freely from mother to fetus so that a newborn infant has a high level of these antibodies, which gradually falls during the first few months of life. This conveys some passive immunity on the infant – passive because the antibodies have been produced by the mother and the baby's own immune system has not been involved in their production.

IgM antibodies are the first to be produced by the infant, IgG antibodies not being produced in appreciable amounts until about 3 months after birth. For this reason, immunization programmes are usually begun after 3 months of age, when the infant's immune system can respond to administered vaccine.

IgA, IgD and IgE antibody levels begin to rise 1 month after birth and have reached half the adult level by 3 years of age.

The complement system

This is an innate defence system which can be activated in a variety of ways, the most significant of which is by antibodies. It consists of a number of enzymes found in the beta-globulin fraction of the plasma proteins. Each is activated in turn in a cascade, similar to that described in Chapter 4.1

for blood clotting. The first enzyme in the series is activated by antigen/antibody (IgG or IgA) complexes; this activated enzyme in turn activates the second enzyme, which activates the third, and so on, each enzyme of the complement system being activated in sequence.

Activated complement enzymes produce a variety of effects which protect against antigens. One brings about the release of histamine and other substances called **kinins** from mast cells. These substances cause vasodilation and increased permeability of capillary walls – part of the normal protective inflammatory response. They also cause contraction of the smooth muscle in bron-

chioles which gives rise to dyspnoea in anaphylactic reactions (which will be described later). The same enzyme causes neutrophils to be attracted to the site of the inflammation, this chemical attraction being called **chemotaxis**.

A combination of several activated enzymes causes enhanced phagocytosis. The activated enzymes adhere to antigens and also to phagocytes, bringing the two together. One activated enzyme breaks down phospholipids in the cell membrane of foreign cells such as bacteria, incompatible transfused red blood cells or transplanted cells, and thus causes their disintegration.

The complement system is non-specific and

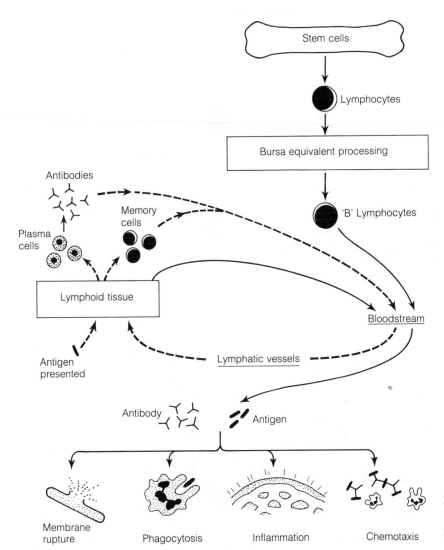

Figure 6.2.4 Diagram to summarize the formation and protective function of the humoral immune system

can, for instance, be activated merely by the presence of bacterial products. However, a more vigorous response occurs in the presence of immune complexes.

To summarize, one can take the example of invading bacteria: antibodies adhere to antigen sites on the bacterial wall and the immune complex thus formed activates the complement system. Various activated enzymes of this system, singly or in combination, cause blood flow to the affected area to increase and the capillaries to become more permeable. As a result, phagocytes, lymphocytes and other protective cells can reach the bacteria, which may then be ingested and destroyed, or may disintegrate due to membrane damage (Fig. 6.2.4).

MODE OF ACTION OF THE CELL-MEDIATED IMMUNE SYSTEM

T lymphocytes are processed in the thymus gland. They recognize the appropriate antigen by means of receptors on their cell surface, and, having done so, take up resistance in secondary lymphoid tissue. Here, they proliferate to form both killer cells, which have a limited length of life and do not divide further, and memory cells, which can further subdivide, like T lymphocytes, if they come into contact with the appropriate antigen. Killer cells are so named because they become attached to foreign cells carrying the antigen for which they have been processed, and cause damage to the

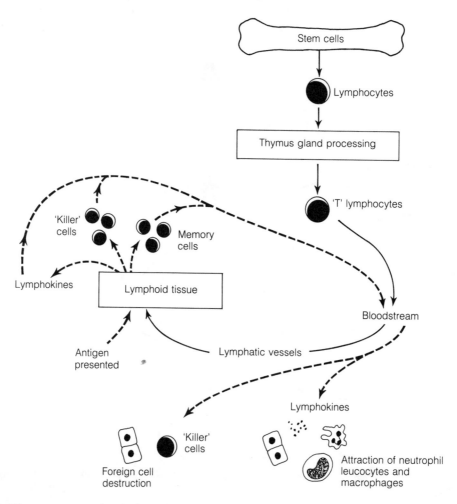

Figure 6.2.5   Diagram to summarize the formation and protective function of the cell-mediated immune system

membrane of the cell. Foreign cells may be, for example, transplanted cells, neoplastic cells, or cells infected by a virus (Fig. 6.2.5).

Viruses are composed of genetic material. They have no energy-producing or protein-building organelles of their own but rely on those of the host cells which they invade. They take over the energy- and protein-making resources of the host cell to enable their own replication, after which the host cell ruptures, liberating daughter viruses which promptly invade further host cells. If not destroyed by antibodies during their viraemic phase (the phase during which the viruses are in the bloodstream), they and their host cells may be destroyed by killer cells because the viral antigens appear on the cell membrane of the host cells.

T lymphocytes activated by contact with the appropriate antigen also produce a variety of protein secretions, collectively called **lymphokines**, which have protective functions. These include the attraction of leucocytes and macrophages to the antigen-containing cell. One of these factors, called **interferon**, is the subject of much current research as a possible therapeutic agent (Scott and Tyrell, 1980). Interferon is made both by activated T lymphocytes and by most cells of the body once they have been infected by viruses. It prevents virally infected cells from synthesizing vital proteins and thus prevents virus multiplication. Lymphokines are non-specific in their function and thus interferon produced in response to invasion by one virus prevents successful invasion by other viruses. Such protection lasts for about 3 weeks.

T lymphocytes are also thought to have a role in the regulation of both the cell-mediated and humoral immune systems. They can stimulate (helper T cells) or suppress (suppressor T cells) the immune responses to a given antigen. For this reason they are thought to be significant in the lack of immune response to the body's own cells. Lack of suppression may be a factor in auto- the cells of the individual as if they were foreign. Similarly, suppression may be responsible for immunological tolerance. This phenomenon forms the basis of desensitization programmes for patients with allergic disorders: repeated administration of increasing doses of an allergen may result in reduction of abolition of the allergic response.

IMMUNIZATION

When a pathogen invades for the first time, it normally takes a minimum of 4 or 5 days for the individual to produce measurable amounts of antibody, and in most cases it takes much longer, nearer to 3 weeks, before the antibodies have reached a protective level. The length of this lag phase depends largely on the pathogen concerned and its mode of entry. During this interval the pathogen can multiply very rapidly (for example, many can divide every 20 minutes under optimum conditions) and reach disease-producing numbers before the immune system has had time to respond. If the individual survives this first attack, a second or subsequent exposure to the pathogen will usually be met by an almost immediate and larger (booster) immune response (with a lag phase of perhaps 24 hours) and the individual will either fail to develop any clinical manifestations of the disease, or will suffer it much more mildly. Such immunity to infectious diseases can also be induced artificially by immunization.

Passive immunization

This involves the injection of antibodies against the organism. Its effect is therefore immediate but short lived, lasting only until the antibodies are destroyed in the recipient's body. This immunity is not sustained until a second attack, since the recipient's B and T lymphocytes have not come into contact with the antigen. Passive immunization is therefore reserved for situations in which a patient with no previous immunity to a potentially dangerous disease comes into contact with it, for example hepatitis, poliomyelitis and rabies.

It should be remembered when nursing patients who have received passive immunization that antibody-containing serum obtained from any species other than humans can itself induce an immune reaction, which may manifest itself as a dangerous anaphylactic response. This has led to the abandonment of the use of horse serum containing tetanus antitoxin for the treatment of patients potentially infected with the tetanus organism.

Another important use of passive immunization is in the prevention of Rhesus (Rh) incompatibility in pregnancy (see Chap. 4.1).

Active immunization

This involves the administration of antigen-containing organisms or their toxins. The individual is induced to mount his own immune response to the antigens, which includes the production of large numbers of memory cells, so that if the organism invades in the future there will be only a very short lag phase before large numbers of protective antibodies are produced.

Preparation of a vaccine from organisms or toxins involves altering them in such a way that they can no longer produce disease yet their antigen components remain intact. It is perhaps not surprising that vaccines are occasionally produced which fail in one or other of these respects. Sometimes **adjuvants** are included in vaccines, such as aluminium hydroxide or phosphate. These are substances which increase the antigenic properties of a vaccine which would normally produce a rather weak immune response, for example in triple vaccine which contains diphtheria and tetanus toxoids and pertussis (whooping cough) vaccine.

A **toxoid** is a preparation of a bacterial exotoxin which no longer produces disease; one such preparation involves treatment of the toxin with formalin. The organisms which cause diphtheria and tetanus both produce exotoxins, and protection is therefore achieved by the administration of toxoids.

Killed vaccines are those in which dead organisms are used. Examples of diseases which can be prevented by this method include pertussis (whooping cough), typhoid and paratyphoid. Both toxoids and killed vaccines require administration on two or three occasions, often with booster doses after the initial programme of vaccination, because only small numbers of antigens are introduced on each occasion.

Attenuated vaccines are those in which the organisms are live, but have been cultured artificially to produce a strain which is no longer pathogenic. Since the organisms are live, they are able to multiply within the body after administration and give rise to a full immune response closely mimicking that following a natural infection. For this reason, often only one administration is required to give protection which may last a lifetime. Diseases which can be prevented by this type of vaccine include smallpox, poliomyelitis, measles, rubella (German measles) and tuber-

culosis. The strain of organism used for the last-mentioned has been called after the workers who developed it – the bacillus Calmette–Guérin (BCG).

Variation of response to immunization

As mentioned earlier, the infant's immune responses are not as efficient as the adult's. IgG levels are rising by 3 months of age, but ideally immunization programmes should be delayed until about 6 months of age, when a better immune response can be expected. Balanced against this is the need to protect infants from infectious diseases such as pertussis which could kill them when very young. For this reason the immunization programme starts at 3 months of age, usually with triple vaccine against pertussis, diphtheria and tetanus. Vaccination for the latter two conditions induces a strong immune response and so it will be effective in spite of the immaturity of the infant's immune system.

Live poliomyelitis vaccine, administered orally, is usually given at the same time as triple vaccine. Although it is a live vaccine, and therefore should need to be given only once to achieve complete immunity, polio vaccine contains three strains of the virus and often all of these do not 'take' at the first administration. As a result it is given in three doses (along with triple vaccine which consists of toxoids and killed pertussis organisms). The first dose is given at 3 months, the second 6–8 weeks later and the third 4–6 months later.

Measles immunization is usually delayed until after 1 year of age, because maternal IgG antibodies present until 6–9 months of age tend to destroy the injected organisms before the infant's immune system has been able to recognize and respond to them. Immunization against measles is therefore likely to be ineffective unless delayed beyond this time. Since the vaccine consists of live attenuated organisms, protection is achieved by one dose.

Rubella (German measles) is not, of itself, a dangerous infectious disease, but can cause serious developmental abnormalities during fetal life if the mother contracts the disease during the first trimester of pregnancy. In consequence, all girls are routinely offered immunization in Britain at the age of 11 and 12. Rubella vaccine is also offered to women of childbearing age following a negative result from a blood test for antibodies to rubella. Women having received rubella vaccine must avoid pregnancy for at least 4 months after vaccination as the attenuated virus may damage the fetus. For this reason the vaccine is often given to non-immune women immediately after childbirth.

There has been no attempt, in this section, to survey systematically all the vaccines in current routine use. Further information can be found in Dick (1986). An attempt has been made to show some of the principles underlying immunization as one method of preventing infectious disease, principles which depend largely on the type of vaccine used and the age at which protection is most desirable.

In addition to these factors, other variables may need to be taken into account when deciding whether, or when, to immunize an individual with a particular vaccine. If the vaccine to be given is a virus, immunization will not be effective if interferon is present in the body and should therefore be postponed if the individual is currently suffering from, or has recently had, any viral infection such as the common cold. Similarly, since live poliomyelitis vaccine is given orally, it is likely to be ineffective if the individual has diarrhoea or is vomiting.

High levels of glucocorticoid steroids suppress the immune response. For this reason patients receiving steroid therapy or with overactivity of the adrenal glands may not respond effectively to immunization. In addition, due to the failure of the immune response, they may suffer ill-effects when given live vaccine, and in consequence a vaccine such as that against poliomyelitis may be contraindicated in such patients.

Allergic reactions can occur to some of the vaccines used, particularly to the virus vaccines which are prepared in tissue cultures, and to vaccines containing whole bacterial cells. Such hypersensitivity may involve the nervous system. This means that children with a history (or possibly family history) of severe allergy such as eczema or hay fever or of convulsions may be considered unfit to receive immunization against, for example, measles. Individuals with known allergy to certain antibiotics or to hen's eggs may not be able to receive live viral vaccines. This is because these vaccines are often grown in egg-based tissue culture, or in tissue cultures to which antibiotics have been added in order to suppress bacterial contamination. Similarly, severe reac-

tion to a preceding dose of vaccine will preclude its further use. Live vaccine will not be administered during pregnancy. Manifestations of allergy will be discussed in more depth in the next section.

HYPERSENSITIVITY REACTIONS

When an immune reaction gives rise to severe or even fatal ill-effects, this is referred to as a hypersensitivity reaction. Hypersensitivity arises mostly from the effects of substances such as histamine on the body tissues. These are released in the course of an antigen–antibody interaction. Hypersensitivity reactions may occur immediately (**immediate hypersensitivity reaction**), in which case they are brought about by agents such as histamine which are activated by antigen–antibody interactions (called antibody-mediated reactions). Alternatively, reactions may be delayed for about 24 hours (**delayed hypersensitivity reaction**), in which case they are caused largely by lymphokines released from T lymphocytes which have reacted with antigen (called cell-mediated reactions).

There are three types of antibody-mediated reactions – types I, II & III – one type of cell-mediated reaction – type IV – and a reaction type which is both antibody and cell mediated – type V.

Antibody-mediated (type I) reaction

This is an anaphylactic reaction (anaphylaxis, from the Greek meaning guarding) occurring in, for example, eczema, asthma, hay fever and drug reactions. Many people, described as atopic, have a familial tendency to anaphylactic allergy and react to a multiplicity of antigens such as pollen, house dust mites, animal fur and dandruff, and cow's milk. Antigens which give rise to an allergic reaction are referred to as allergens.

The mechanism of the reaction is that IgE antibodies formed in response to the allergen bind to mast cells and basophils. Mast cells are found in abundance in the respiratory tract, gastrointestinal tract and skin and thus reactions are often localized in any or all of these sites. Histamine, 5-hydroxytryptamine (serotonin), slow reaction substance, prostaglandins and kinins are released from the granules of basophils and mast cells and bring about vasodilation, increased permeability of capillaries, contraction of smooth muscle and attraction of eosinophils to the site affected. These responses bring about the typical symptoms and signs of anaphylactic-type reactions, including swelling and redness of the skin or mucous membranes, bronchospasm or gastro-intestinal upset. Eosinophils phagocytose the immune complexes and their granules release an antihistamine substance.

Skin testing to identify the offending allergens prior to desensitization therapy involves injecting purified allergens under the skin and observing for a local flare and wheal occurring within half an hour. Antihistamine drugs should be stopped 3 days before skin testing.

Desensitization therapy involves giving subcutaneous injections of the offending allergen in gradually increasing doses. It is believed to promote the formation of IgG antibodies against the allergen, which subsequently bind to the allergen and thus prevent its binding to IgE antibodies.

Since anaphylactic reaction is an immune response, it does not normally occur on first meeting the allergen (called the sensitizing dose), but may occur on second or subsequent exposures, which is of particular relevance in drug allergy. **Anaphylactic shock** is a very rare, but extremely serious and potentially fatal condition which occurs in very sensitive individuals. It consists of a widespread allergic response with massive release of chemicals from mast cells and basophils, occurring within minutes of secondary contact with the allergen. The major effects are profound hypotension, bronchoconstriction and the development of oedema of the face, tongue and larynx which may obstruct the airway. Immediate treatment with intramuscular 1 in 1000 adrenaline, to raise blood pressure and dilate the airways, is essential.

Antibody-mediated (type II) reaction

This is a cytotoxic reaction involving IgG, IgM or IgA antibodies which bind to antigens attached to various cells of the body, notably blood cells. The complement system is activated and phagocytosis of the affected cells promoted.

Transfused red cells of an incompatible ABO group promote such a reaction. The affected donor cells first agglutinate (clump) and are then lysed by the recipient's phagocytes. Rhesus incompatibility may also result in haemolysis of a Rh-positive fetus' blood, brought about in the same way, but initiated by maternal antibodies which cross the placenta (see also Chap. 4.1). In addition, it is becoming apparent that drugs, for example methyldopa, which as a side-effect cause haemolytic anaemia, do so by coating the red cells and provoking an antibody response which results in haemolysis of the coated red cells. Leucocytes and platelets may be destroyed in the same way. The endotoxins of certain bacteria, for example *Salmonella*, may also coat red cells and result in haemolysis.

Transplanted organs, if they survive other immune reactions, may become necrotic due to thrombosis in the vasculature of the donated organ. A cytotoxic reaction caused by antibodies to the endothelium of blood vessels in the donated organ results in damage to the endothelium, adherence of platelets and thrombus formation.

Antibody-mediated (type III) reaction

This is a reaction in which immune complexes (antigens and antibodies) are deposited in various tissues and bring about varying degrees of tissue damage. The complement system may be activated if IgG or IgM antibodies are involved.

Serum sickness (reaction to serum injected from a foreign source) results in a type III reaction. Immune complexes may be deposited in joints, causing arthritis; in glomeruli, causing glomerulonephritis; in the heart, causing myocarditis and valvulitis; and in the skin, causing urticaria. Drugs such as penicillin and the sulphonamides have the same effect in susceptible individuals, as do certain bacteria and viruses, notably a strain of the *Streptococcus* bacterium which may cause glomerulonephritis.

The **Arthus reaction** is a localized type III reaction. When antigens are injected, IgG antibodies may form complexes with them, causing local vasculitis and an inflammatory response. This is sometimes seen in patients with diabetes mellitus who develop IgG antibodies against some antigenic component of their insulin preparation.

Erythema nodosum is due to the deposition of immune complexes in the blood vessels of the legs, with resultant vasculitis. This may occur in response to an infection elsewhere in the body, such as tuberculosis.

Cell-mediated (type IV) reaction

This reaction results from sensitized T lymphocytes rather than antibodies and is delayed for 24–48 hours following exposure to the antigen.

The **positive tuberculin test** (Mantoux, Heaf or Tine test) is a typical example of this reaction. A suspension of tuberculin, a protein derived from tubercle bacilli, is injected intradermally. If the individual has been previously sensitized to tuberculin, an inflammatory reaction occurs at the site of the injection after 24–48 hours and lasts for some weeks. It is thought that sensitized T lymphocytes interact with the antigen and subsequently attract macrophages and other lymphocytes to the area. Lymphocytes release a variety of lymphokines and the proteolytic enzymes of macrophage lysosomes cause tissue damage.

A variety of bacteria, e.g. salmonellae and streptococci, and many viruses which multiply intracellularly also produce this type of reaction. Various metals (e.g. nickel and chromium), dyestuffs and plants (e.g. primula and poison ivy) cause the same response when in contact with the skin, and the reaction is then referred to as **contact dermatitis**. These substances only become antigenic when they combine with proteins in the skin. Topical application of penicillin may cause contact dermatitis in the same way.

This type of reaction is currently being investigated as a potential method of treatment for skin cancers since it has been found that, if painted with a substance which has antigenic properties, the cells of the affected area undergo a type IV hypersensitivity reaction and the malignant cells are destroyed. This type of reaction is also involved in transplant rejection.

Antibody- and cell-mediated (type V) reaction

This reaction involves antibody directed at a target cell, to which lymphocytes may subsequently attach at the cell-binding site of the immunoglobulin. Phagocytic cells may also be attracted, resulting in lysis of the target cells. Examples of this type of reaction include herpes simplex (where the cells are infected with the herpes simplex virus), and antibody-coated tumour cells, e.g. melanoma. In addition, some cases of thyrotoxicosis may occur by a similar reaction. IgG antibodies have been discovered which react with the thyroid-stimulating hormone receptors of the thyroid gland and cause excessive amounts of thyroid hormones to be produced.

All these hypersensitivity reactions are, it must be emphasized, normal protective responses which in excess bring about ill-effects. Exposure to an antigen may lead to one or several types of reaction.

AUTOIMMUNE DISEASES

These diseases result from an inability of the body to recognize certain tissues or cell components as 'self'. In consequence, an immune response occurs to the body's own tissues. Ways in which such non-recognition can arise are, to a large extent, speculative, but many diseases can now be regarded as being the result of an immune reaction to the body's own tissues.

One explanation of the failure to recognize the body's own tissues is that during the development and processing of the lymphocytes to recognize 'self' and 'non-self', certain antigens may be inaccessible to the immune system. If these become accessible later, they will then induce an immune response. The lens of the eye is such a tissue since it has no blood supply. Sometimes surgical removal of one lens for cataract results in an inflammatory response in the other lens, as surgical intervention has allowed lymphocytes to come into contact with the antigens of lens tissue.

Another, better documented, example of a likely cause of failure of self-recognition is that antibodies may be formed to a foreign antigen, e.g. a bacterium. These antibodies are then able to interact to some extent with a 'self' antigen. Group A streptococci have an antigen similar to one found in the heart, and the heart lesion occurring in some cases of rheumatic fever is probably due to antibody formed during a Group A streptococcal infection. Infection with virus may actually alter cell antigens. Drugs such as methyldopa may bind to tissue proteins and the combination may then act as an antigen, resulting in destruction of cells, in this case erythrocytes.

Another possible explanation for the failure to recognize 'self' may lie in the lymphocytes themselves. Drugs, bacteria or viruses may induce changes in the lymphocytes. Suppressor T lymphocytes, mentioned in the previous discussion of the mode of action of the cell-mediated immune system, may be concerned in the normal recognition and non-response to 'self'. An inherited or acquired defect of these lymphocytes may be responsible for autoimmunity. There is some experimental evidence to suggest, at least in animals, that certain viruses may alter the structure of the cell surface of lymphocytes, interfering with their ability to recognize antigens. In the case of autoimmune diseases, this allows the lymphocytes to respond to antigens to which they would not normally respond.

Autoantibodies (antibodies against 'self') may be organ specific, that is, directed against an antigen peculiar to the organ concerned. This is so in some cases of thyrotoxicosis, Addison's disease and pernicious anaemia. In the last-mentioned case, the cells are the parietal cells of the gastric mucosa, which manufacture intrinsic factor. Treatment consists of correcting whatever dysfunction results from the structural damage produced. Alternatively, autoantibodies may be directed at a tissue type which is not confined to one organ, at, for example, the nuclear material of cells or at connective tissue, and appropriate management may be immunosuppressive therapy.

The reaction of the tissues involved in autoimmunity is like that of hypersensitivity reactions. Cytotoxic (type II) reaction occurs in autoimmune haemolytic anaemia. IgG and IgM antibodies coat the subject's erythrocytes and, although the cells are not agglutinated or lysed, they are destroyed by the spleen and liver, resulting in anaemia. Type III reaction, in which the antigen/antibody complex causes tissue damage, is responsible for the glomerulonephritis seen, for instance, in systemic lupus erythematosus (SLE). In this instance the antigen is in the nuclear material of cells, which, in a complex with antinuclear antibodies and complement, collects in the glomeruli and brings about damage to the basement membrane. It may also have similar effects in the skin and central nervous system. Cell-mediated (Type IV) reaction may be involved in certain autoimmune diseases, although at present this has only been demonstrated in animal experiments.

Although a number of diseases are of known autoimmune aetiology, there are also a number in which specific antibodies can be demonstrated in the patient's serum, but it is not known how, or even if, these antibodies are responsible for the

disease. (These antibodies are collectively called the rheumatoid factor.) The joint damage found in rheumatoid arthritis may result from the deposition of immune complexes and complement in the joint. The rheumatoid factor is also frequently found in patients with other conditions, such as SLE.

Autoimmunity is the subject of a great deal of research at present, as, indeed, is the whole matter of immunity.

CANCER SURVEILLANCE AND IMMUNOTHERAPY

It has been postulated that abnormal cells are detected and subsequently destroyed by the immune system because they carry abnormal antigens. Supportive evidence for this comes from many sources, notably from the increased incidence of malignancy in individuals with immunodeficiency states, those receiving immunosuppressive drugs or with autoimmune diseases. In addition, the function of the immune system becomes less efficient with advancing age and this may explain the increased incidence of malignancy in the elderly.

Malignant cells may evade the immune system if, for instance, they multiply more rapidly than the immune system can respond to them. Some cancers may be produced by viruses which cause a specific antigen to adhere to the malignant cells. The antigen is the same for any one type of virus, regardless of the tissue or host involved. This makes it theoretically possible for the host to be treated by immunization. Certain human cancers, e.g. bladder carcinoma, have common antigens in all patients and thus may be virally induced and possibly amenable to treatment by immunization. By contrast, chemical carcinogens produce different antigens in different tissues and different hosts.

For immunotherapy to be successful in the treatment of cancer, the tumour must be small, and immunotherapy is probably only likely to be effective in conjunction with other forms of treatment such as cytotoxic chemotherapy. Passive immunization, in this case, involves the transfer of sensitized lymphocytes from a patient who has had a complete regression of a similar tumour. These lymphocytes would then attack the recipient's tumour. Active immunization involves vaccination of the patient with his own tumour cells or those from another patient with a similar tumour. The cells are first inactivated by irradiation, but are still able to produce an immune response.

Certain vaccines, e.g. BCG and pertussis vaccine, are thought to have a non-specific stimulating effect on the immune system and have been injected topically into skin cancers with some success in their treatment, as well as systemically into patients with leukaemia (Mathé, 1967).

ORGAN TRANSPLANT

All nucleated cells in humans carry antigens on their surface membrane, the nature of these antigens being determined by the genetic make-up of the individual.

Blood groups are determined by the erythrocytes, which carry a series of antigens, some causing a strong immune response, such as the ABO and Rhesus systems, and some of only minor clinical significance since they cause only a weak immune response (see Chap. 4.1). Other body cells carry different antigens, some being strong and therefore important in transplant rejection, and some being weak and therefore of only minor significance in organ transplant. The leucocytes carry all the 'strong' antigens of the **HLA system** (human leucocyte locus A system) and these antigens are widely distributed throughout the tissues. Tissue compatibility can thus be determined using leucocytes. Cell-mediated immunity, through sensitized T lymphocytes, is the main agent of rejection, and most immunosuppression therapy is aimed against T lymphocytes.

The more similar the genetic make-up of the donor and recipient, the less likely is the transplanted tissue to be rejected. An **autograft** is one in which the person's own tissue is used for grafting and thus no foreign antigens are involved. Similarly, an **isograft** involves transplanting tissue from a donor of identical genetic make-up, e.g. a monozygotic twin.

An **allograft** (formerly called a homograft) is one in which the donor and recipient are of the same species but not genetically identical, and a **xenograft** (formerly called a heterograft) is one in which the donor and recipient are of different species, e.g. pig and human. The more closely related the donor and recipient, the closer the genetic match, but there are thousands of possible genetic, and thus cell antigen, combinations and a complete match is extremely unlikely except in related individuals. Since the HLA genes are linked on one chromosome, they will be inherited in a block and so there is a 1 in 4 chance of a match between siblings. Even where there is complete HLA matching however, late rejection can occur and it is likely therefore that there are other cell antigens not yet discovered. In consequence, however good the match, immunosuppressive therapy is required.

The incidence of some diseases, for example coeliac disease, multiple sclerosis, acute lymphatic leukaemia and juvenile onset diabetes, has been found to show a correlation with the presence of certain HLA antigens in the patient's tissues. The significance of this remains to be clarified.

Immunosuppressive therapy

Drugs which interefere with and suppress the immune response make a crucial contribution to the prevention of rejection in tissue and organ transplantation. They are also of value in treating people with severe hypersensitivity states and autoimmune conditions. The aim of treatment is to suppress T and/or B lymphocyte production and activity, without interfering with other cells.

Agents which inhibit cell division – cytotoxic drugs, for example methotrexate and 6-mercaptopurine – are used as immunosuppressants and also in treating neoplasms. These drugs particularly affect rapidly dividing cells and so, in addition to their effect on lymphocytes and malignant cells, they suppress other cells, for example in bone marrow, skin and those lining the gastrointestinal tract. This can produce undesirable side-effects such as anaemia, thrombocytopenia, leucopenia and increased susceptibility to infection, hair loss, skin disorders and gastrointestinal upsets.

Antilymphocytic serum (ALS) depletes T cells, but also has the effect of damaging other lymphocytes, making the recipient more susceptible to infection. This serum is produced by immunizing horses or rabbits with human lymphocytes. It has a limited use in preventing the rejection of transplanted organs.

Corticosteroids, for example hydrocortisone and prednisone, are used extensively for their immunosuppressive action in preventing rejection of transplanted tissues, and also in the treatment of hypersensitivity and autoimmune diseases. They cause a gradual destruction of lymphoid tissue and may directly deplete T cells. However, probably their major action is to decrease the phagocytic activity of polymorphonuclear leucocytes and macrophages. High doses of corticosteroids are associated with undesirable side-effects, which are described in Chapter 2.5.

IMMUNOLOGICAL DEFICIENCY STATES

There is a variety of conditions which result from some failure of the innate or acquired defence mechanisms. A brief categorization of these will be given here, and further details can be found in the immunology texts in the bibliography.

Immunological deficiency states can be present at birth (congenital) and due to genetic defects, or acquired later in life. They can be due either to defects of the phagocytes or complement system (the innate defence mechanisms) or to defects of the lymphoid system (the acquired immune defence mechanisms).

Congenital defects affecting B lymphocytes may result either from a reduced number of these cells or from a failure to synthesize immunoglobulins in the presence of normal numbers of B lymphocytes. The result is that as the infant's maternally derived IgG antibodies decline in number at about 3–6 months of age, he is unable to make his own antibodies. This condition is referred to as **agammaglobulinaemia** (since most antibodies are plasma proteins of the gamma-globulin class) and affected infants suffer recurrent infections. In some conditions, only the synthesis of one class of immunoglobulins is affected.

Congenital T lymphocyte deficiency may result from failure of development of the thymus gland, often associated with other defects affecting organs such as the heart or thyroid gland. Failure of the cell-mediated immune response usually becomes apparent with infection with organisms which multiply intracellularly, e.g. viruses, fungi or protozoa.

Congenital stem cell defects result in the combined deficiency of both B and T lymphocytes and may thus be called **combined immunodeficiency**.

Acquired lymphocyte deficiency can be second-

ary to a number of other conditions, for example gross malnutrition (in which there is a deficiency of raw materials, notably protein, for cell synthesis), cytotoxic drug treatment and radiation therapy (in which rapidly dividing cells, among these lymphocytes, are destroyed), corticosteroid therapy (which suppresses the immune response), certain infections, diseases of the lymphoid tissue such as Hodgkin's disease, lymphoid leukaemia and multiple myeloma. In the last-mentioned, a malignant proliferation of one clone of plasma cells results in the production of one class of immunoglobulin which is associated with a reduced production of immunoglobulins of the other classes. Hypogammaglobulinaemia may also occur in protein-losing conditions such as nephrotic syndrome or enteropathy (for example ulcerative colitis in which not only is there reduced absorption of amino acids from the intestine, but also a loss of protein from the ulcerated mucosa).

Phagocytosis, an innate defence mechanism dependent on polymorphonuclear leucocytes, monocytes and macrophages, is an integral part of the acquired mechanisms, both humoral and cell-mediated. Frequently, complement is also involved. Congenital defects of the phagocytes may affect their numbers or the processes of chemotaxis, phagocytosis or degradation of ingested antigens. Acquired phagocyte deficiency may be secondary to any condition which suppresses the bone marrow and therefore affects their production, for example chemical damage or myeloid leukaemia. In diabetes mellitus, chemotaxis and ingestion by phagocytes are reduced as long as insulin is deficient. Deficiency of complement can be congenital or can result from any condition which reduces plasma protein synthesis, for example liver cell failure.

The clinical effect of all immunodeficiency states is an increased susceptibility to infection, relatively minor infections frequently proving fatal. Upper respiratory tract infection is the most common, though immunodeficiency may manifest itself as diarrhoea, enlargement of the liver or spleen, or skin rashes. Treatment will be aimed at replacing the deficient component, for example by transfusion of gamma-globulins extracted from whole blood, or by bone marrow transplant, as well as antibiotic treatment of infections. An important component of care is the measures taken to minimize the pathogens with which the patient is in contact. This may involve isolation and barrier nursing within a sterile unit, with contacts such as hospital staff, relatives and friends being kept to a minimum. Recent technological developments have done much to improve the effectiveness of such measures and to reduce the social isolation felt by patients. One such development is the Trexler tent in which there is a physical barrier between the patient and those handling him, everything else with which the patient has contact being capable of sterilization. The use of laminar flow units in suitable cases enables the air within the bed area or cubicle to be purified and effectively isolated from that in the ward without the need for a physical barrier. This does not overcome the problems associated with the lack of physical contact between the patient and those caring for him, but does much to reduce social isolation.

Acquired immune deficiency syndrome (AIDS)

This almost invariably fatal syndrome was first recognized in the United States, in 1979. It occurs principally, but not exclusively, in homosexual/bisexual men and to a lesser extent in haemophiliacs and intravenous drug users. Human immunodeficiency virus (HIV) has been identified as the cause of AIDS. The virus is spread parenterally (i.e. via body fluids), for example in semen or blood. It produces a severe cell-mediated immune deficiency.

In AIDS the helper–suppressor T cell ratio is reversed and killer cell activity is reduced. The major clinical feature is the occurrence of opportunistic infections which indicate severely suppressed T cell activity, for example pneumonia caused by the protozoan *Pneumocystis carinii*, (Waterson, 1983). Such infections are usually only seen in patients undergoing immunosuppressive therapy. The usually uncommon cancer, Kaposi's sarcoma, occurs frequently in AIDS and in the wake of immunosuppressive therapy. Concern has been expressed about the possible transmission of AIDS through donated blood and particularly blood products prepared from the blood of many donors, for example factor VIII. For this reason haemophiliacs form an at-risk group, numbering some 4500 males in the UK. British haemophilia centres are now issuing heat-treated replacement clotting factors to their patients to eliminate the risk of infection.

Screening tests for antibodies to the AIDS virus are currently being used to screen potential blood donors and all haemophiliacs.

For full details of the condition, its treatment, nursing care and prevention, the reader should consult the text by Wells (1986) suggested at the end of this chapter. Since research is advancing so rapidly in this field, the reader is urged to consult the most up to date sources for details of current knowledge and its application.

The basic mechanisms of the acquired immune system have been described above and their clinical relevance indicated. Immunology is a fascinating and relatively new area of scientific study. It is a subject in which many questions still remain unanswered and one in which the new insights and developments, produced by research, are often of significance in the aetiology or treatment of current major problems, for example, controlling viral infection and cancer.

THE LYMPHATIC SYSTEM

This circulatory system has received scant attention in the past when compared with, for instance, the cardiovascular system which has been researched for centuries. The lymphatic system is only found in higher vertebrates which have developed an extensive vascular system. A high intravascular pressure is necessary within such a vascular system if it is to be adequately perfused with blood. Mayerson (1963) suggests that the lymphatic system evolved to cope with the fluid seepage from capillaries which is the result of this high intravascular pressure. In addition to this role, the lymphatic system includes lymph glands which filter debris from the lymph, facilitate recognition of antigens by lymphocytes of the immune system, and are the main centres of lymphocyte proliferation.

Hydrostatic pressure at the arterial end of blood capillaries forces water and solutes with molecular weights of less than 68 000 into the interstitial spaces. Most of this returns to the blood capillaries towards the venous end, where hydrostatic pressure is less (see Chap. 4.2), but a little remains to be drained away via the lymphatic capillaries. It is estimated that some 50% of the plasma protein enters the interstitial fluid over 24 hours. Without the constant clearance of this protein, the osmotic pressure of the interstitial fluid would increase, resulting in fluid accumulation in the tissues (oedema). Patients with breast cancer involving the lymph nodes in the axilla, for example, may develop gross oedema of the affected arm, due to failure of lymphatic drainage.

Within the interstitial spaces lies a network of lymphatic capillaries. These start as blind-ended vessels, fluid and solutes entering along the length of the capillaries. Like blood capillaries, they consist of a layer of endothelial cells, but unlike blood capillaries, they have no basement membrane. Large molecular weight substances such as plasma proteins are able to enter between the endothelial cells. Lacteals, the equivalent of lymphatic capillaries in the villi of the small intestine, permit the uptake of triglycerides with long-chain fatty acids. It is thought (Mayerson, 1963) that some hormones which have large molecular weights probably reach the bloodstream by way of the lymphatics. The rate of lymph formation and flow depends on the rate of accumulation of interstitial fluid. Normally, 2–4 litres of interstitial fluid accumulate and return to the blood in 24 hours. This results in a constant turnover of the interstitial fluid.

The network of lymphatic capillaries joins to form larger lymphatic vessels. These contain some smooth muscle in their walls and have one-way valves. Flow of lymph is maintained, like the flow of blood in veins, by skeletal muscle contraction in the limbs, pulsation of adjacent arteries and negative intrathoracic pressure, but in addition lymphatic vessels themselves contract rhythmically, their rate of contraction being proportional to the volume of lymph in the vessels.

At certain points (Fig. 6.2.6) these vessels feed, via afferent lymphatic vessels, into lymph glands (nodes). Efferent lymphatic vessels drain from the lymph glands and ultimately unite to form the thoracic duct (from the lower limbs, digestive tract, left arm and left side of thorax, head and neck), or the right lymphatic duct (from the right arm and right side of head, neck and thorax) and drain into the great veins in the neck (Fig. 6.2.6).

Lymph glands

Each lymph gland is approximately the size and shape of a broad bean. Other lymphoid tissue is

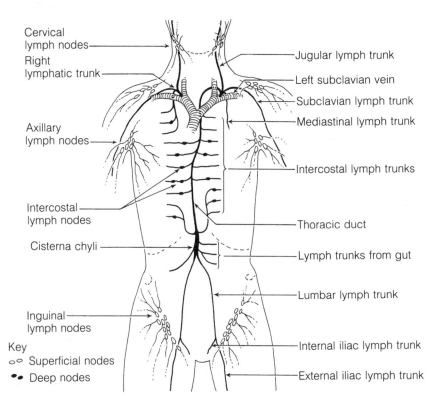

Cervical lymph nodes
Right lymphatic trunk
Axillary lymph nodes
Intercostal lymph nodes
Cisterna chyli
Inguinal lymph nodes

Key
oo Superficial nodes
•• Deep nodes

Jugular lymph trunk
Left subclavian vein
Subclavian lymph trunk
Mediastinal lymph trunk
Intercostal lymph trunks
Thoracic duct
Lymph trunks from gut
Lumbar lymph trunk
Internal iliac lymph trunk
External iliac lymph trunk

Figure 6.2.6 General arrangement of the lymphatic system

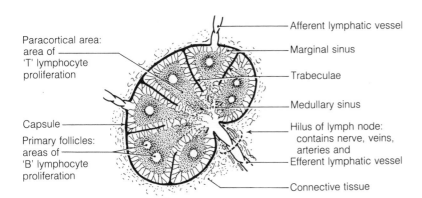

Figure 6.2.7 Section through a lymph gland

found in specific organs, notably the spleen, bone marrow, lung and liver, and functions in a similar way to the lymph glands. Lymph enters the capsule of the gland via afferent lymphatic vessels (Fig. 6.2.7) and passes via the marginal sinuses to trabeculae which extend radially into the gland. The substance of the gland consists of a meshwork of reticular cells in which lymphocytes are embedded. Large numbers of macrophages are found throughout the gland but are concentrated in the medulla. Having ingested an antigen, a macrophage partially breaks it down, exposing antigen sites to B lymphocytes. In addition to this role of aiding recognition of antigens by suitably processed lymphocytes, macrophages probably facilitate the concentration of antigens in lymphoid tissue as well as engulfing and destroying foreign material, such as dead cells and bacteria.

Primary follicles in the cortex of the gland are centres for the proliferation of B lymphocytes. Cells at the centre of the follicle are actively dividing whilst those at the periphery are antibody forming. T lymphocyte centres of proliferation are found in the paracortical area.

An efferent lymphatic vessel leaves the gland at the hilum, the point at which blood vessels enter and leave. Lymphocytes migrate from the blood vascular system within the gland. Most of the cells of the gland, for example macrophages and lymphocytes, are part of a mobile population which circulates between the blood and lymphatic tissue.

Lymph glands are the major centres for lymphocyte proliferation and antibody production as well as for filtering the lymph. Carbon particles introduced artificially have been shown to accumulate in the lymph glands distal to the point of introduction. Malignant cells, if they invade lymphatic vessels, can be carried away in the lymph until their progress is impeded by the filtering effect of the lymph glands, where they become deposited and can give rise to a secondary tumour.

Normally, lymphatic vessels cannot be seen with the naked eye and lymph glands cannot be palpated. However, when bacteria are carried away from a focus of infection by the lymphatic vessels, this can give rise to inflammation in the lymphatic vessels themselves and subsequently in the lymph glands into which they drain. For example, when the focus of infection is superficial, such as in the hand, the inflamed lymphatics can be seen as red streaks extending up the arm, and the lymph glands in the axilla may be tender, hard and enlarged.

Review questions

The answers to all these questions can be found in the text. In each case, there is at least one correct and at least once incorrect answer.

1 Which of the following statements is/are true of antibodies?

   (a) They are immunoglobulins.
   (b) They are produced by T lymphocytes.
   (c) They are complex carbohydrate molecules.
   (d) They interact chemically with antigens.

2 Which of the following may occur in response to bacterial infection?

   (a) Release of histamine from mast cells.
   (b) Ingestion of bacteria by complement.
   (c) Migration of polymorphonuclear cells to the affected area.
   (d) Production of a clone of B lymphocytes in secondary lymphoid tissue.

3 T lymphocytes

   (a) are processed by the thyroid gland during fetal life
   (b) can differentiate into plasma cells when sensitized
   (c) have surface receptors which interact with antigens on the surface of cells
   (d) are formed exclusively in the liver.

4 Which of the following might be expected to disrupt the humoral immune response?

   (a) Severe proteinuria.
   (b) Neonatal thymectomy.
   (c) Severe anaemia.
   (d) Overactivity of the adrenal cortex.

5 Would you expect one administration of a live, attenuated vaccine to convey immunity?

   (a) No.
   (b) Probably not.
   (c) Probably.
   (d) Definitely.

Answers to review questions

1 a and d
2 a, c and d
3 c
4 a, b and d
5 c

Short answer and essay topics

1 Compare and contrast the development and method of function of the cell-mediated and humoral immune systems.
2 Outline the mechanisms by which auto-immunity occurs and give examples of diseases of established autoimmune aetiology.
3 Describe the role of the cell-mediated immune system in the rejection of transplanted tissue.
4 Describe the functions of lymph glands and outline the clinical implications of these functions.

Suggestions for practical work

1 Take note of the current, routine immunization programme used for the protection of infants and children against infectious diseases and account for its rationale.
2 Collect and read
   (a) The leaflets provided with any vaccines you have observed and administered. Note associated precautions, contra-indications and side-effects as well as how the preparation achieves its objectives of producing an adequate antibody response, without ill-effects.
   (b) Health Education Council and DHSS leaflets on immunization.

References

Burnet, F. M. (1959) *The Clonal Selection Theory of Acquired Immunity*. London: Cambridge University Press.
Dick, G. (1986) *Practical Immunization*. Lancaster: MTP Press
Mathé, G. (1967) The immunological approach to the treatment of cancer. *Annals of the Royal College of Surgeons of England 41*: 93.
Mayerson, H. S. (1963) The lymphatic system. In: *Readings from the Scientific American. Immunology* (1976). San Francisco: W. H. Freeman.
Miller, J.F.A.P. (1962) Immunological function of the thymus. *Lancet ii*: 748.
Scott, G. M. & Tyrell, D. A. (1980) Interferon: therapeutic fact or fiction for the 80's. *British Medical Journal 283*: 1558–1562.
Waterson, P. (1983) Acquired immune deficiency syndrome. *British Medical Journal 286*: 743–746.
Weir, D. M. (1983) *Immunology – An Outline for Students of Medicine and Biology*, 5th edn. Edinburgh: Churchill Livingstone.
Wells, N. (1986) *The AIDS Virus – Forecasting its Impact*. London: Office of Health Economics.

Suggestions for further reading

Bowry, T. R. (1984) *Immunology Simplified*, 2nd edn. Oxford: Oxford University Press.
Joint Committee on Vaccination and Immunisation (1984). *Immunisation Against Infectious Disease*. London: DHSS.
Jones, P. (1985) *AIDS & the Blood – A Practical Guide* London: Haemophilia Society.
Nursing (1982) Body defences. *Nursing 2*, No 6 (October).
Playfair, J. H. L. (1984) *Immunology at a Glance*, 3rd edn. Oxford: Blackwell Scientific.
Riedman, S. R. (1974) *The Story of Vaccination*. Folkestone: Bailey Brothers & Swinfen.

Chapter 6.3
Reproduction

Rosamund A. Herbert

Learning objectives

After studying this chapter the reader should be able to

1 Describe the mechanisms involved in the determination of the sex of an individual.
2 Describe the normal physiology of the male and female reproductive systems.
3 Demonstrate how disruption of normal function can lead to abnormal states or disease.
4 Relate any altered physiology to the treatment of common health problems, particularly in the fields of obstetrics, gynaecology, urogenital medicine and surgery, family planning, well-woman and infertility clinics.
5 Base the planning, delivery and evaluation of nursing care, in the above fields, on sound physiological principles.
6 Give adequate and comprehensive explanations of normal and abnormal conditions to patients.
7 Discuss the complex anatomical and physiological integration necessary for the optimum functioning of the male and female reproductive systems to ensure continuation of the species.

Introduction

Reproductive biology is the science of the transmission of life. It is essential that animals reproduce in order to ensure the survival of their species.

The efficient and complex reproductive system in humans has enabled them to evolve and survive for at least a million years. The mechanisms involved in reproduction will be discussed in this chapter.

The function of the reproductive system is to produce gametes (germ cells) and to provide the optimum conditions for the fusion of two gametes, one from the male and one from the female. The human female also has to provide a life-support system for the first 9 months of a new individual's life; thus her body must adapt and provide a suitable environment during that time.

Research into some aspects of human reproductive physiology is still in its infancy. For ethical reasons, most of this research has been conducted on animals, and great care must be taken in extrapolating conclusions from animal experiments to humans. Man is very much an 'animal with culture' and, in many instances, social and cultural influences greatly modify his basic instincts and behaviour. This is especially true in some aspects of reproductive biology.

Nevertheless, pure research into reproductive physiology is expanding, particularly in areas applicable to medicine. For example, research by Steptoe and Edwards in the 1970s enabled an infertile woman to have a baby by the development of a method to extract a mature ovum, fertilize it in vitro (in a laboratory) and return it to the mother. In vitro fertilization and embryo transfer are common now. Also we owe much to the work over the past 30 years of Kinsey and his associates,

Masters and Johnson, and Schofield, who conducted the first objective studies, free from moral or social interpretations, of human reproductive physiology and sexual behaviour.

It seems appropriate to consider meiosis, the process of cell division by which gametes are formed, and the factors that determine the genetic sex and phenotype (i.e. physical characteristics) of an individual, before discussing the male and female reproductive systems in detail.

THE FORMATION OF GAMETES

The formation of gametes (germ cells) is known as **gametogenesis** and the type of cell division that results in the formation of gametes is **meiosis**. Meiotic cell division is unique to reproductive cells and gamete production. Meiosis does, however, have some features in common with mitosis (see Section 1). Meiosis begins in immature reproductive cells and involves two successive nuclear divisions. The end-result is four haploid daughter cells, each containing 23 chromosomes.

Meiosis, like mitosis, is divided into stages, but as prophase in the first division is considerably longer than that in mitosis, it is further subdivided into recognizable phases. Prior to the commencement of meiosis, the chromatin material in the nucleus doubles in quantity (again like mitosis). Chromosomes are not clearly visible between cell divisions (Fig. 6.3.1i).

Prophase I

Leptotene. The chromosomes are in the form of very fine single threads, hence the derivation of the name of this phase, meaning 'slender ribbon'. The chromosomes become clearly visible (Fig. 6.3.1ii).

Zygotene. The homologous, or matched, chromosomes (one from each parent) come together. This is called synapsis or pairing. An homologous pair is also known as a bivalent (Fig. 6.3.1iii).

Pachytene. The two chromosomes become coiled around each other, and also become shorter and thicken (pachytene means 'thick ribbon') (Fig. 6.3.1iv).

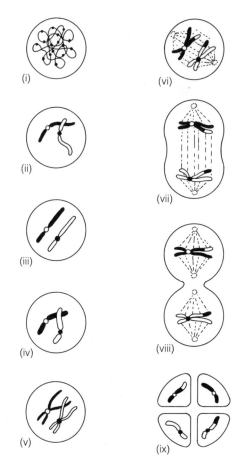

Figure 6.3.1   The stages of meiosis (for explanation, see text). (Only one chromosome pair is shown for clarity.)
   (i) Interphase
  (ii) Prophase I: leptotene
 (iii)   zygotene
  (iv)   pachytene
   (v)   diplotene
  (vi) Metaphase I
 (vii) Anaphase I
(viii) Telophase I
  (ix) Second meiotic division

Diplotene. The two chromosomes in each pair begin to part. With microscopes it is possible to see that each chromosome has itself divided into two daughter chromatids and so each unit now has four units, that is, it is quadrivalent. The separation of the homologous pairs is incomplete: some of the chromatids remain attached at one or more points along their length. These 'crossing-overs', or **chiasmata** (singular, chiasma), allow exchange of chromatin between chromatids (Fig. 6.3.1v).

This is an important stage because it is one of the mechanisms by which genetic variation between parents and offspring is produced. The resultant chromatid is not identical to either of the two parent chromosomes, but a mixture of the two. It is during this stage that the female gametes go into a prolonged resting phase (see later section on oogenesis).

Diakenesis. The two bivalents continue to contract and move away from each other. The nucleolus also disappears at this point.

Metaphase I

The nuclear membranes break down and the spindle apparatus, composed of contractile protein, appears. The bivalents arrange themselves on the spindle so that the centromeres of the two homologous chromosomes lie on either side of the equator of the spindle (Fig. 6.3.1vi). This arrangement is important because it ensures that each daughter cell only receives *one* of the chromosome pair (this contrasts to the arrangement of the chromosomes during mitosis where each chromosome lies independently on the equator of the spindle).

Anaphase I

The chromosomes move to opposite poles of the spindle but the centromeres do not divide as they do during mitosis. The remaining chiasmata slip apart and free the homologous chromosomes from each other (Fig. 6.3.1vii). Thus at each pole there are 23 chromosomes, that is, half the total complement of 46 chromosomes. This is referred to as the diploid (2n) number of chromosomes.

Telophase I

The nuclear membrane forms around the haploid (n) set of chromosomes at each pole (Fig. 6.3.1viii). At this stage there are two daughter cells, each containing 23 chromosomes.

Then there is a short interphase period during which time *no* DNA synthesis occurs. The second meiotic division (starting with prophase II) closely resembles mitotic division as the units involved are chromosomes, not bivalents, and the centro-

mere takes up a position on the equatorial plane at metaphase II. It differs from mitosis in that only half the normal number of chromosomes are present. During anaphase II the centromeres divide and the daughter chromatids go to opposite poles. During telophase II the nuclear membranes reform.

Thus, at the end of meiosis, four haploid nuclei have been produced, each containing 23 chromosomes, from the original diploid parent nucleus. One of the most important aspects of meiosis is the halving of the chromosome number: each gamete must contain the haploid number of chromosomes so that when the male and female gametes fuse the resulting zygote will have the full complements of 46 chromosomes – the same number as either parent.

Errors can occur during meiosis; occasionally, for example, the homologous chromosomes fail to separate and migrate together into the same gamete. This is called **non-disjunction**. The resulting individual is abnormal to a greater or lesser extent, depending upon which chromosome

Figure 6.3.2 A Down's syndrome child

pair has failed to split. Errors can occur during chiasmata formation too.

Ninety-five per cent of all cases of **Down's syndrome** (mongolism) are caused by non-disjunction. In this case, the two chromosomes (bivalents) of pair 21 do not part during meiosis. Thus, if this 'imperfect' gamete is fertilized, the zygote will have three chromosome 21s and a total complement of 47 chromosomes. This example is known as **trisomy 21**. The incidence of Down's syndrome is approximately 1 in 600 live births, but the occurrence varies with the age of the mother: it occurs in mothers of 25 years of age once in every 2000 births, but the risk rises fortyfold for mothers aged 45 years. The maternal age factor is thought to be associated with the increased length of time that an oocyte remains dormant (an oocyte remains dormant from before birth until just prior to ovulation). In oocytes which are dormant for 30–40 years there is a greater risk of exposure to events that may bring about non-disjunction, e.g. radiation, toxins, viruses.

The additional chromosome in Down's syndrome results in an individual who is mentally retarded (a mean IQ of 50) with a characteristic appearance (mongoloid eyes), an increased susceptibility to infection and very often a happy personality (Fig. 6.3.2).

THE DETERMINATION OF THE SEX OF AN INDIVIDUAL

In each human somatic cell there are 46 chromosomes, that is, 23 pairs. These are divided up into 22 pairs of autosomes and 1 pair of sex chromosomes; the latter are given the names X and Y due to the shape of the chromosomes. The human female has the complement XX in all her cells and the male XY in all his cells (Fig. 6.3.3).

During the first division of meiosis, when the homologous pairs split, each resulting gamete receives one sex chromosome. The gametes from the female will all contain one X chromosome, whereas those from the male will contain *either* an

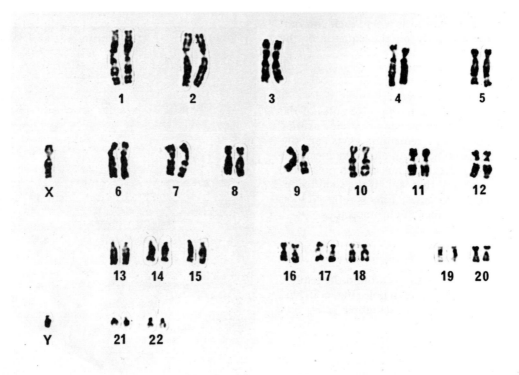

Figure 6.3.3 The chromosome complement of a normal human male (46, XY) (courtesy of the Paediatric Research Unit, Guy's Hospital Medical School, London)

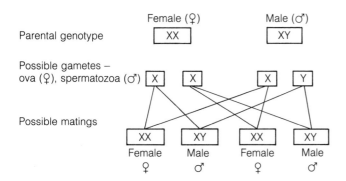

Parental genotype

Female (♀) XX   Male (♂) XY

Possible gametes – ova (♀), spermatozoa (♂)   X X   X Y

Possible matings

XX
Female
♀

XY
Male
♂

XX
Female
♀

XY
Male
♂

Figure 6.3.4 Diagram showing the mechanism by which the chromosomal sex of an individual is determined

X chromosome *or* a Y chromosome. The chromosomal sex of an individual is determined by the nature of the two gametes that unite at the time of fertilization, that is, by the sex chromosomes carried in the gametes from the parents (Fig. 6.3.4). So the chromosomal or genetic sex of the offspring is determined by the spermatozoon (the male gamete).

There is general agreement now that, in females, one of the two X chromosomes is inactivated in all body cells soon after fertilization and it is precipitated out near the nuclear membrane, where it is known as a **Barr body** (Fig. 6.3.5). If a Barr body is present, the individual is said to be 'chromatin positive'. This is thought to be a device to compensate for the extra genetic information, that is, the second X chromosome, carried by the female, as the small Y chromosome in the male carries very little genetic information (Lyon, 1974).

For each mating there is, in theory, a 50%

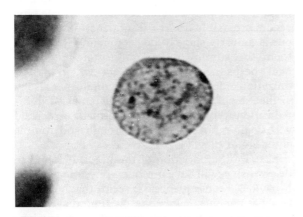

Figure 6.3.5 A cell nucleus from a female, with a Barr body precipitated out near the nuclear membrane (on the right-hand side) (courtesy of the Paediatric Research Unit, Guy's Hospital Medical School, London)

chance (two out of four) that a female will result and a 50% chance that a male will result. However, the observed sex ratio at birth shows that usually more males are born than females. The exact ratio varies from place to place and from time to time but, at present, in England and Wales the ratio at birth is approximately 106 males to every 100 females. It is not known what factor or factors favour conception by Y spermatozoa (for further reading see Parkes, 1976).

However, more than the sex chromosome complement is involved in the determination of the gender of an individal. For example, for an individual to be truly male he needs an XY chromosome complement, normal male gonads (testes) and genitalia and the presence of normal quantities of male hormones. True females need an XX chromosome complement, normal female gonads (ovaries) and genitalia and the presence of normal quantities of female hormones.

The male and female gonads are derived embryologically from the same site in the body and the primitive gonads are initially identical in both sexes. During the seventh week of embryonic life in genetic males, testes begin to develop. Testicular differentiation depends upon some influence exerted by the Y chromosome on the undifferentiated gonad. The exact nature of this influence is as yet unknown. In the absence of such direction the undifferentiated gonad becomes an ovary, but this does not occur until about the thirteenth week of embryonic development.

The primitive genital tracts, composed of the Müllerian (paramesonephric) and Wolffian (mesonephric) ducts, are also identical initially in XX and XY embryos (Fig. 6.3.6). After 7–8 weeks the normal male tract begins to develop from the Wolffian ducts in a male XY embryo, and the Müllerian ducts regress. Hormones produced by

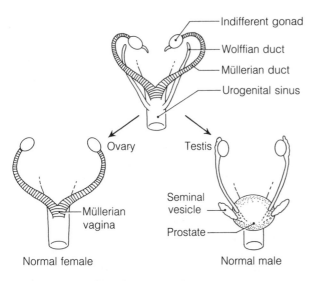

Indifferent gonad
Wolffian duct
Müllerian duct
Urogenital sinus

Ovary   Testis

Seminal
vesicle
Prostate

Müllerian
vagina

Normal female   Normal male

Figure 6.3.6 Differentiation of Müllerian and Wolffian ducts in male and female embryos

the fetal testes are responsible for these changes. The fetal testes are capable of secreting two distinct classes of hormones which act directly on adjacent structures before being secreted into the systemic circulation to produce more generalized effects. One of these secretions, from the Leydig cells, is an androgen (male hormone) and it causes development of the Wolffian duct into the vas deferens, seminal vesicles and other male sexual organs. However, it does not inhibit the Müllerian duct; the other testicular hormone, produced by the Sertoli cells, does this. This is a large molecular weight protein (*not* an androgen) known as the Müllerian-inhibitory substance.

The female genital tract develops later from the Müllerian ducts, and the presence or absence of ovaries seems to have little or no obvious effect on its development. However, it is possible that the presence of high levels of oestrogens within the fetal ovaries may have some, as yet undefined, role to play in the development of the female reproductive system.

The development of the external genitalia in the male depends upon androgenic stimulation before the twelfth week. In the absence of androgen the external genitalia are those of a female, irrespective both of the chromosomal sex and even of gonadal sex. Thus the sexual differentiation of the gonads appears to be primarily dependent on chromosomal make-up, but the later differentiation

of the genital tract and external genitalia largely depends upon the presence or absence of fetal testicular hormones.

Gradually, during intrauterine life, the ovaries or testes descend from their original position high in the abdomen. In the female the ovaries and their associated tubes come to lie within the pelvis; in the male the testes normally descend into the scrotum before birth, by the eighth fetal month. The production of male gametes (**spermatogenesis**) occurs optimally just below body temperature, and the temperature within the descended testes is approximately 35°C.

The descent of the testes is under hormonal control, although the true biological stimulus for descent is still unidentified. It may be partly under the control of androgens and/or the anterior pituitary. Occasionally the testes may be retained within the abdomen (**cryptorchidism**) or their descent may be arrested at the abdominal end of the inguinal canal. If the condition is bilateral and the testes are not lowered during childhood (this can be done by hormone therapy, human chorionic gonadotrophin, or surgery), the male will be sterile. An undescended testis left in the abdomen may undergo malignant changes, when it is thought that the higher body temperature causes degeneration of the tubular epithelium in the testes.

The sex hormones secreted even during embryonic life have effects on other aspects of the embryo's development. One of the most important organs to become sexually differentiated is the brain. In experimental animals there is clear evidence that the early embryonic brain is similar in both sexes and that conversion to the male type occurs ostensibly under the influence of testosterone at an early stage of development. Goy (1968) found that administration of testosterone to pregnant rhesus monkeys resulted not only in masculinization of female offspring's genitalia but also in virilization of their behaviour: the females exhibited various types of behaviour normally associated with male monkeys – threat, rough and tumble and pursuit play.

Sexual differentiation of the brain, if found to be extensive, will have profound physiological and behavioural consequences. It seems that hormones can permanently alter the 'wiring diagram' if present at the appropriate stage of brain development, leading to differences between male and female brains in later life. For example, in humans there are marked physiological differences between male and female in the hypothalamic and/or pituitary response to injected oestrogen: the female shows a positive feedback discharge of luteinizing hormone that is completely lacking in the male (van Look *et al.*, 1977).

The complex process of sexual differentiation, as just described, helps to explain the many variations in male and female differentiation that are observed clinically and also explains manifestations of intersex states. Aberrations of sexual development can arise either from changes in sex chromosomes, as already discussed, or from abnormalities in sex differentiation due to hormonal or environmental causes. The changes that occur at puberty to differentiate males from females are also hormone dependent and will be discussed later in the chapter.

THE FUNCTIONING OF THE MALE REPRODUCTIVE SYSTEM

A discussion of male reproductive physiology can be divided into three sections: the production of spermatozoa (the male gametes), the endocrine function of the testis, and the endocrine control of these processes. A brief description of the male reproductive system will be given first.

The male genital system consists of two testes and their ducts, several accessory glands and the penis (Fig. 6.3.7). The **testes** are situated in the scrotum. The tubules and ducts from each testis unite to form the **epididymis** (Fig. 6.3.8) and from here the **vas deferens** (ductus deferens) travels up into the pelvis, passes anterior to the pubic symphysis and then loops around the ureter.

At this point the vas deferens enlarges to become the ampulla. A **seminal vesicle**, one of the accessory sex glands, joins each vas at the lower end of the ampulla. The duct is then known as the **ejaculatory duct**. The two ejaculatory ducts (one from each ampulla and testis) fuse with the urethra in the middle of the prostate gland. The resultant duct, known as the **prostatic urethra**, is a common duct for both urination and the carriage of semen. The ducts from two additional accessory glands, the **bulbo-urethral glands** (Cowper's glands), join the urethra which enters the penis.

The **penis** is an elongated organ composed of mainly vascular spaces making up the erectile tissue and consists of three cylindrical bodies: two dorsal corpora cavernosa and ventrally one corpus spongiosum (Fig. 6.3.9). The penile urethra, which is lined with mucus-secreting glands, traverses the corpus spongiosum to the external urethral meatus. The head, or **glans penis**, is usually covered by the **prepuce**, or foreskin. Circumcision is the removal of the prepuce. If the prepuce is retracted for a long period of time, a paraphimosis may develop which involves constriction and swelling of the glans and an inability to restore it to the natural position. This can occur after catheter care, if the prepuce is not returned to its original position covering the glans.

Spermatogenesis

The production of spermatozoa, or spermatogenesis, occurs in the seminiferous tubules of the testis (see Fig. 6.3.8). Each testis is divided into many compartments, each containing one or more

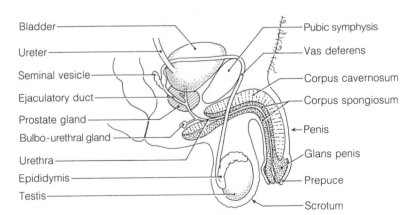

Figure 6.3.7 The male reproductive system

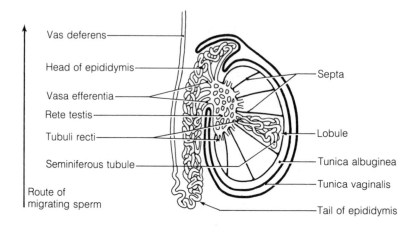

Vas deferens
Head of epididymis
Vasa efferentia
Rete testis
Tubuli recti
Seminiferous tubule
Route of migrating sperm

Septa
Lobule
Tunica albuginea
Tunica vaginalis
Tail of epididymis

Figure 6.3.8 The testis

minute **seminiferous tubules**. Within each testicular compartment, connective tissue containing the **Leydig cells** (or **interstitial cells**) surrounds the seminiferous tubules. The Leydig cells are concerned with the synthesis and release of androgenic (male) hormones.

There are two types of cell in the seminiferous tubule of an active testis: the germ cells and the Sertoli cells (Fig. 6.3.10). At any one time the germ cells are at various stages of development, but they all originate from **spermatogonia**. The spermatogonia undergo continuous mitotic divisions to ensure a constant supply of germ cells. Some of these cells mature and increase in size to become primary **spermatocytes**. These primary spermatocytes undergo the first meiotic division to become secondary spermatocytes, which then contain only the haploid number of chromosomes. The secondary spermatocytes undergo the second division of meiosis to become **spermatids** (Fig. 6.3.11).

The spermatids are in close association with the Sertoli cells. The **Sertoli cells** are polymorphic

cells (occurring in many shapes) that are attached to the basement membrane but extend into the lumen of the seminiferous tubule (see Fig. 6.3.10). The exact role of the Sertoli cells is unclear, but they are probably involved in the nutrition and

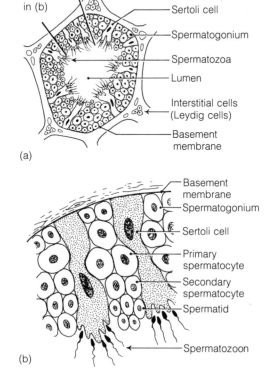

Part shown in (b)

Sertoli cell
Spermatogonium
Spermatozoa
Lumen
Interstitial cells (Leydig cells)
Basement membrane

(a)

Basement membrane
Spermatogonium
Sertoli cell
Primary spermatocyte
Secondary spermatocyte
Spermatid
Spermatozoon

(b)

Figure 6.3.10 Cross-section of a seminiferous tubule (a), with a small section enlarged to show the microscopic structure of the testis (b)

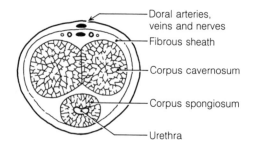

Doral arteries, veins and nerves
Fibrous sheath
Corpus cavernosum
Corpus spongiosum
Urethra

Figure 6.3.9 Cross-section of the penis

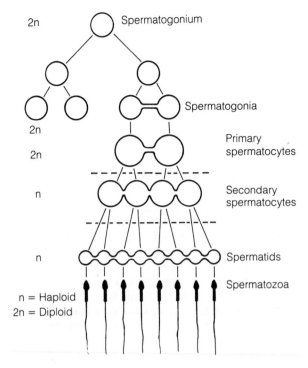

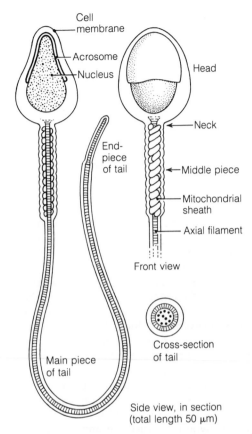

2n Spermatogonium

Spermatogonia

2n
2n Primary
spermatocytes

n Secondary
spermatocytes

n Spermatids

Spermatozoa

n = Haploid
2n = Diploid

Figure 6.3.11 Diagram to show spermatogenesis

Cell membrane
Acrosome
Nucleus
Head
Neck
End-piece of tail
Middle piece
Mitochondrial sheath
Axial filament
Front view
Main piece of tail
Cross-section of tail
Side view, in section
(total length 50 μm)

Figure 6.3.12 Structure of the mature spermatozoon, with a small cross-section of the tail enlarged to show the 9 + 2 arrangement of the axial filaments

support of the developing germ cells and may have some phagocytic action and endocrine influences.

The spermatids, still in close association with the Sertoli cells, undergo transformation from relatively simple cells into the highly specialized **spermatozoa**. Changes occur in both the nucleus and cytoplasm of the cell. The chromatin of the nucleus condenses to become the head of the spermatozoon; the Golgi apparatus contributes towards the formation of the **acrosome** (which contains hyaluronidases and proteinases that help the spermatozoon penetrate both the mucus plug of the cervix and ovum); one of the centrioles in the cell lengthens to form the tail, and the mitochondria aggregate in the middle section of the spermatozoon. Any superfluous cytoplasm is lost (Fig. 6.3.12).

Once the spermatozoa have been produced, they are released from the Sertoli cells into the lumen of the seminiferous tubule. However, they are not functionally mature at this stage; completion of the maturation process occurs while the spermatozoa are stored in the epididymis. The total time taken for the production of mature spermatozoa from spermatogonia is approximately 70

days. After puberty and throughout adult life spermatogenesis occurs continuously. The normal human male may manufacture several hundred million spermatozoa per day. With advancing age, the seminiferous tubules undergo gradual involution and in the testes of a 70 year old, many tubules show extensive atrophy and are depleted of germ cells but still contain Sertoli cells.

Once fully formed, the spermatozoa are pushed out of the seminiferous tubules along the tubuli recti and into the rete testis and then via the vasa efferentia into the head of the epididymis (see Fig. 6.3.8). The **epididymis** is a long, coiled tube (if unravelled it would be approximately 6 m long) and is divided into sections – the head, body and tail – all closely applied to the posterior surface of the testis. The cilia in the tubuli recti beat and the smooth muscle around the tubules contracts, moving the spermatozoa towards the epididymis.

The time spent by the spermatozoa in the epididymis is an organized delay: it allows them to mature before ejaculation. It is thought that the secretory columnar epithelium of the epididymis secretes hormones, enzymes and nutrients that may be important for sperm maturation. The growth of the epithelium of the epididymis is dependent on adequate levels of male sex hormones.

The tail of the epididymis is the main storehouse for the spermatozoa, although some may be stored in the vas. Storage is necessary because spermatogenesis is a continuous process, whereas ejaculation occurs at irregular intervals. If no ejaculation occurs, the spermatozoa in the epididymis ultimately degenerate and undergo liquefaction. It is also thought that the epithelial cells of the ducts may have phagocytic properties and may be able to remove abnormal spermatozoa. The spermatozoa mature in their abilities both to move and to fertilize ova during the passage through the excretory ducts, but the exact nature of this maturation process is not known. Spermatozoa can be stored in the genital ducts for as long as 42 days.

The vas deferens (or ductus deferens) is a muscular tube that begins in the scrotum as a continuation of the tail of the epididymis and ends in the pelvis by joining with the duct of the seminal vesicle in the formation of the ejaculatory duct. It ascends from the scrotum in the spermatic cord. The **spermatic cord** also contains blood vessels and nerves that supply the testes and also muscular and connective tissue extensions from the anterior abdominal wall. In cases of torsion of the testis, the blood and nerve supply to the testis can be cut off and, if not corrected quickly, may necessitate an orchidectomy (removal of the testis).

The wall of the vas deferens is composed of an outer layer of loose connective tissue and three layers of smooth muscle with an abundant autonomic nerve supply. This arrangement accounts for the ability of the vas to contract quickly and efficiently during ejaculation. In addition to these peristaltic contractions, the spermatozoa are able to 'swim' by means of two-dimensional bending waves which pass from the front to the back end of their tails. The tails of the spermatozoa are composed of a central filament containing a pair of microfibrils surrounded by a circle of nine fibrils.

The vas deferens is ligated and cut during the operation of **vasectomy** for sterilization of the male. Small sections of both vas are usually removed from the epididymal end. Men having undergone a vasectomy may remain fertile for 6–8 weeks after the operation since spermatozoa may remain viable within the genital tract for this period (the duration of fertility after vasectomy also depends on the frequency of ejaculation). Therefore, additional contraceptive methods are needed for at least a couple of months after surgery. The men are usually required to go back to the clinic twice with a specimen of their ejaculate to check that there are no spermatozoa in it before sterility is assured. With the vas occluded, the man is sterile, but there should be no changes in either somatic or psychological sexual characteristics. These characteristics are androgen dependent and the secretion of these hormones from the interstitial cells of the testis into the bloodstream is not altered by the procedure. The postvasectomy changes that occur in the seminiferous tubules are variable: sometimes complete degeneration of the cells occurs, but more often some spermatogenesis continues, but the spermatozoa are destroyed by liquefaction and replaced by new cells. Antibodies to the spermatozoa may also be produced. Normal ejaculation still occurs, with the emission of fluids from the accessory sex glands, minus the spermatozoa, of course. Reversal of vasectomy is possible in some instances.

The role of the accessory sex glands

The accessory sex glands provide a transport medium, with the necessary nutrients, for the spermatozoa to leave the male and enter the female.

THE SEMINAL VESICLES

These two glands are so called because they were originally thought to store the spermatozoa, but that is not their function; they are secretory glands. They are located behind the prostate gland and lateral to each ampulla of vas. Each vesicle is a muscular convoluted tube lined by secretory epithelium with a maximum capacity of $3 \, cm^3$.

The secretion of the seminal vesicles is an alkaline, viscid, yellowish fluid containing, amongst other compounds, fructose, globulin, ascorbic acid and prostaglandins. It forms the fluid vehicle

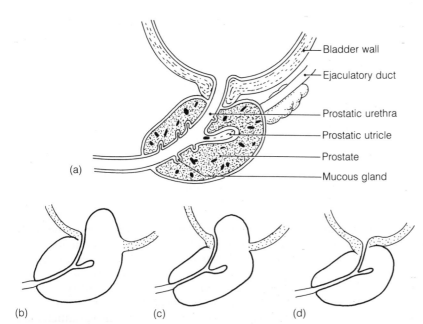

(a)

- Bladder wall
- Ejaculatory duct
- Prostatic urethra
- Prostatic utricle
- Prostate
- Mucous gland

(b)　　　　　(c)　　　　　(d)

Figure 6.3.13 The normal prostate gland and diagrams showing the effects of hypertrophy: (a) the normal gland, (b) hypertrophy of the lateral and middle lobes, (c) hypertrophy of the middle lobe, (d) hypertrophy of the lateral lobes

for the spermatozoa. Its secretory activity is under the control of the testicular hormones and in old age the vesicles diminish in size because of decreased hormone stimulation.

THE PROSTATE GLAND

The prostate gland is situated around the bladder neck and the first part of the urethra, into which its secretions pass (Fig. 6.3.13). Its actual size varies considerably. At puberty the prostate gland increases rapidly in size and in the normal adult its shape is likened to a chestnut with an approximate 3 cm diameter in all directions; it remains fairly constant in size until middle age when it may involute or, more often, undergo benign hypertrophy which frequently results in urological problems. The main glandular tissue is situated in the lateral and posterior portions (known as the outer zone) of the prostate, whereas the inner zone (the middle of the gland) consists mainly of mucosal glands.

The **prostatic secretion** consists of a thin, slightly acidic, milky fluid containing enzymes, for example acid hydrolase, acid phosphatase, protease and fibrinolysin; it is also rich in calcium and citrates. It is responsible for the characteristic odour of semen. Prostatic secretion is thought to stimulate the motility of the spermatozoa, coagulate the fluid from the seminal vesicles and go

some way towards neutralizing the prevailing vaginal acidity. As with the secretions of the seminal vesicles, prostatic secretory activity is dependent upon adequate stimulation by the testicular hormones. Any reduction of this stimulus results in involution of the gland and its secretory elements.

Virtually every man over the age of 40 has some degree of benign enlargement of the prostate. The inner zone of the gland hypertrophies in particular and projects into the bladder, impeding the passage of urine by elongating and distorting the prostatic urethra (Fig. 6.3.13). Benign prostatic enlargement is thought to be due to a disturbance of the ratio and quantity of the circulating androgens and oestrogens in the more advanced years of life.

The term **prostatism** refers to the urinary problems associated with lesions of the prostate. The common symptoms that manifest, usually between the ages of 60 and 70 years, include difficulty in initiating micturition, a poor urinary stream and urinary frequency with nocturia. Prostatism can result in incomplete voiding with stasis, infection and back pressure, resulting in renal damage. Approximately 1 in 10 men in the West require surgical treatment for prostatic enlargement (the condition has a racial distribution, being common in Western white races, rarer in Indian races and extremely rare in Negroid males). Treatment for benign hypertrophy usually involves removal of the enlarged lobe of the prostate. Particularly in the case of inner lobe enlargement, the removal can be performed transurethrally, i.e. via the urethra. After transurethral resection of the prostate, ejaculation may be impaired but erection should be possible.

Hormone dependency also occurs in carcinoma of the prostate, which more often involves enlargement of the outer or posterior part of the gland. It appears to be androgen dependent, thus treatment of this carcinoma consists of giving large doses of oestrogens and/or orchidectomy to reduce the relative levels of circulating androgens. Radical surgery is only appropriate for a few patients because the carcinoma has often spread by the time of diagnosis; urinary symptoms do not occur until the disease is well advanced as it is associated with enlargement of the outer zones.

THE BULBO-URETHRAL GLANDS

The bulbo-urethral, or Cowper's, glands are small globular glands and are roughly pea sized. They lie between the lower prostate and the penis, and their ducts open into the urethra. They secrete mucus which serves as a lubricant prior to ejaculation.

The composition of semen

Semen, or seminal fluid, consists of spermatozoa, the secretions of the genital tract, especially the epididymis, and the secretions of the associated accessory glands. The bulk of the semen (approximately 60%) originates from the seminal vesicles. The pH of the combined secretions is slightly alkaline (pH 7.2–7.4), the acid prostatic secretions being neutralized by the other components. The pH of semen is important because spermatozoa are rapidly immobilized in an acid medium. The pH of the vaginal secretions is acid (approximately pH 4.5) and the alkaline semen neutralizes the inhibitory effect of the acid vagina. Semen is also rich in hyaluronidase, an enzyme which causes breakdown of mucopolysaccharides and which facilitates passage through the cervical mucus and the chemicals surrounding the ovum.

The average volume of ejaculate is $3 \, cm^3$ (range of $2–6 \, cm^3$) and contains approximately 300 million spermatozoa. The number, morphology and motility of the spermatozoa give an indication of the fertility of the male and are used clinically in the assessment of infertility. The normal is considered to be

Volume of ejaculate	$2–6 \, cm^3$
Density of spermatozoa	$60–150 \, million/cm^3$
Morphology	60–80% normal shape
Motility	50% should be motile after incubation for 1 hour at 37°C

A specimen of semen for investigation is obtained by masturbation. The volume and density will depend on the previous period of abstinence from sexual intercourse: frequent ejaculations lead to a progressive reduction in the spermatozoa count in the semen. Values outside these normal ranges may be an indication of infertility.

Spermatozoa may be stored at $-70°C$ for weeks or even months and their motility and fertilizing ability reappear when unfrozen. This property is sometimes used when specimens are used for artificial insemination either by husband or donor.

The endocrine function of the testis

The testes are responsible for spermatogenesis, as already discussed, but they also function as an endocrine gland. The interstitial cells of Leydig (see Fig. 6.3.10) synthesize, store and secrete androgens, principally testosterone. Very small quantities of oestrogens are also produced but their role, in the male, is obscure. Approximately 95% of the androgen is produced in the testis and the remainder in the adrenal glands.

TESTOSTERONE

Testosterone is a steroid molecule synthesized from cholesterol (Fig. 6.3.14). Total plasma levels of testosterone in the adult male are $12–30 \, nmol/l$ (in the adult female the testosterone level is $0.5–2.0 \, nmol/l$, of adrenal origin). Most of the testosterone is loosely bound with plasma proteins once it is released into the blood; it circulates in the plasma before becoming fixed to the target tissues and then it is finally metabolized and excreted by the liver.

Testosterone has widespread effects on the body, both on the reproductive organs and on the somatic tissues. Most of the effects of testosterone

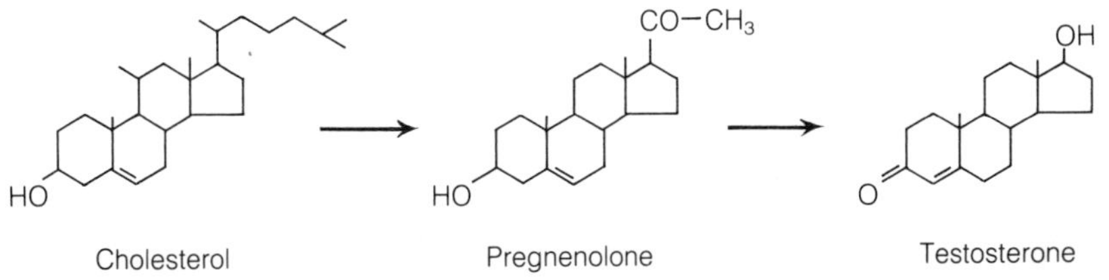

Figure 6.3.14 Pathway for the synthesis of testosterone

(and other androgens) can be directly related to the fact that it is an important anabolic agent, synthesizing complex molecules from simpler ones. Once in the cells, testosterone stimulates an increase in the synthesis of proteins; it probably directly influences the DNA and RNA in the cell. Its major functions are as follows.

1 It stimulates the growth of the seminiferous tubules and is therefore an important regulator of spermatogenesis.
2 It is necessary for the development and maintenance of the accessory sex organs, including the penis and prostate gland.
3 It is responsible for the changes that occur at puberty. These include

   (a) the development of a typical male physique, for example muscle development
   (b) the stimulation of epiphyseal growth in the long bones and the subsequent fusion of the epiphyses – there is thus an initial growth spurt at puberty when testosterone levels increase, followed by cessation of growth in the late teens
   (c) the growth of facial and body (chest, axillary and pubic) hair and recession of the scalp line
   (d) the lowering of the voice pitch due to a thickening of the vocal cords and enlargement of the larynx
   (e) increased sebum secretion in the skin, which may predispose towards the development of acne
   (f) the mild retention of sodium, potassium, calcium, phosphate, sulphate and water by the kidneys
   (g) the development and maintenance of libido (female libido is probably also androgen dependent).

4 Testosterone may also be partly responsible for the changes in male behaviour after puberty, for instance increased aggression, but social conditioning plays an important part too.

The endocrine control of the male reproductive system

There are three levels of hormones involved in the control of the male reproductive system: namely, hypothalamic hormone, anterior pituitary hormones, and testicular hormones. Both spermatogenesis and androgen production are controlled by the anterior pituitary hormones, the gonadotrophins, namely **follicle-stimulating hormone (FSH)** and **luteinizing hormone (LH)**. The latter is also known as **interstitial cell-stimulating hormone (ICSH)** in the male. FSH acts primarily on the seminiferous tubules as a stimulus for spermatogenesis. Small quantities of LH are also required, however, for the completion of spermatogenesis. LH, or ICSH, as this alternative name indicates, stimulates the production of androgen from the interstitial cells of Leydig (Fig. 6.3.15).

The release of these gonadotrophins is in response to a stimulus from the hypothalamus, and the release of both FSH and LH appears to be controlled by one hypothalamic releasing hormone, known as **gonadotrophin-releasing hormone (GnRH)**. It is also known as follicle-stimulating hormone/luteinizing hormone-releasing hormone (FSH/LH-RH) or simply LHRH).

The exact nature of the feedback control system involved is still uncertain but it involves a negative feedback acting at either or both hypothalamic and pituitary levels. If a male is castrated, that is, has his testes removed, there is a marked increase in the levels of plasma gonadotrophins. This suggests that there is normally some negative feedback, and it is thought that testosterone levels are one of the stimuli involved; for example, relatively high levels of testosterone exert an inhibitory effect on the hypothalamus (and possibly the anterior pituitary), reducing the release of GnRH (and FSH and LH).

Inhibin, a non-steroidal factor, has been isolated from testicular extracts and this appears to

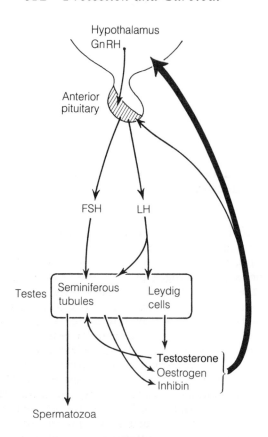

Figure 6.3.15 Possible pathways controlling the release of male reproductive hormones (GnRH, gonadotrophin-releasing hormone; FSH, follicle-stimulating hormone; LH, luteinizing hormone)

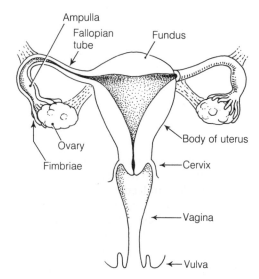

Figure 6.3.16 The female reproductive organs (with one uterine tube dissected out)

inhibit FSH secretion. It is probable that inhibin is produced from the Sertoli cells in response to stimulation by FSH, and it then exerts a negative feedback on FSH release. Oestrogens secreted by the testes could also be involved in the negative feedback process (Lincoln, 1979). Figure 6.3.15 suggests, diagrammatically, the possible pathways involved.

THE FUNCTIONING OF THE FEMALE REPRODUCTIVE SYSTEM

The discussion of female reproductive physiology will be divided into the following sections: the production of ova (the female gametes) and ovulation; the menstrual cycle and its relationship

with ovulation; and finally, the events leading to fertilization and a brief review of pregnancy. A brief description of the anatomy of the female reproductive system will be given first.

The female genital system consists of two ovaries, two uterine tubes, the uterus, vagina and the external genitalia or vulva (Fig. 6.3.16 and 6.3.17). It is completely separate from the urinary system, unlike that of the male. The two ovaries, lying close to the lateral wall of the pelvis, are suspended from the posterior layer of the broad ligament by mesentery, through which blood vessels, lymphatics and nerves pass to and from the ovaries. The broad ligament is a fold of peritoneum passing from the uterus to the side wall of the pelvis.

The **uterine tubes** (also known as oviducts or Fallopian tubes) lie in the upper margin of the broad ligament. Each uterine tube is approximately 10 cm long and has an outer smooth muscle coat and an inner mucous membrane. Both tubes have an outer funnel-shaped part, known as the **infundibulum**, the lumen of which communicates with the peritoneal cavity. The finger-like projections, or fimbriae, 'catch' the released ovum from the ovary. When fertilization occurs it usually takes place within the ampulla. Peristaltic waves occurring in the uterine tubes, together with the movements of the cilia on the epithelium, are responsible for moving the ovum into the uterus. Some non-

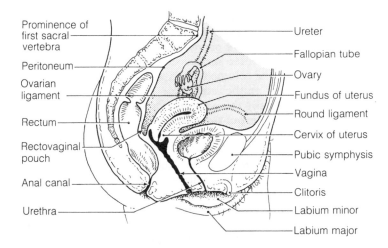

Prominence of first sacral vertebra
Peritoneum
Ovarian ligament
Rectum
Rectovaginal pouch
Anal canal
Urethra

Ureter
Fallopian tube
Ovary
Fundus of uterus
Round ligament
Cervix of uterus
Pubic symphysis
Vagina
Clitoris
Labium minor
Labium major

Figure 6.3.17 Median sagittal section of the female pelvis

ciliated cells in the epithelium secrete nutrients for the ovum.

The uterine tubes lead into the uterus or womb. In the adult the **uterus** is a pear-shaped structure with an upper expanded section known as the **body of the uterus** and a lower cylindrical section known as the **cervix** or neck of the uterus. In the non-pregnant state, it measures about 7.5 cm in length, 5 cm in breadth at its upper border and 2.5 cm in thickness. The part of the uterus that projects above the level of entry of the uterine tubes is known as the **fundus** (this part is used in the palpation of the uterus during pregnancy). The uterus is normally flattened dorsoventrally (Fig. 6.3.17) and is supported in position by the broad, round and uterosacral ligaments.

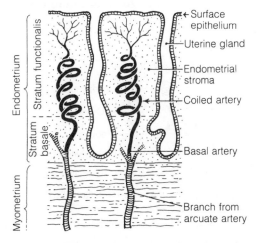

Endometrium
Stratum functionalis
Stratum basale
Myometrium

Surface epithelium
Uterine gland
Endometrial stroma
Coiled artery
Basal artery
Branch from arcuate artery

Figure 6.3.18 The vascular supply to the endometrium

The greater part of the wall of the uterus is composed of a mass of smooth muscle, the **myometrium**, and the cavity, which is normally collapsed, is lined by a mucous membrane, the **endometrium**. From the time of puberty until the menopause the endometrium undergoes monthly cyclical changes in response to hormonal secretions from the ovary (described later). The endometrium is a very vascular tissue, the nature of which is important in understanding menstruation (Fig. 6.3.18). The surface epithelium of the endometrium contains numerous tubular uterine glands.

The cervix of the uterus does not exhibit the same cyclical activity as the body of the uterus, but the nature of the mucous secretions from the glands in this region varies during the cycle. Normal columnar epithelial cells of the cervix may exhibit metaplasia (a change in form) and, under the influence of some other stimulus, possibly as yet undefined but viral in nature, the changes can become neoplastic (that is, cancerous).

Precancerous or cancerous changes can be detected easily by examination of cervical cells taken in a smear test. The cervix can also be directly examined by means of a colposcope which is inserted vaginally (this procedure is called a colposcopy). It is often possible to remove abnormal areas by laser treatment performed on an out-patient basis.

The cervix leads into the **vagina** which is a tubular organ, about 8 cm long, with an outer muscle and elastic coat. It is thus capable of expanding or stretching, as occurs during sexual intercourse or childbirth. During reproductive

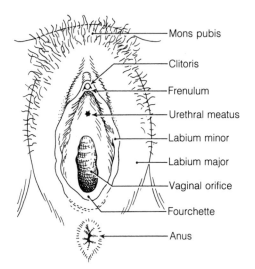

Figure 6.3.19 The vulva

life the vagina is colonized by Döderlein's bacilli (*Lactobacillus* species). These bacilli ferment glycogen to produce lactic acid which renders the vaginal environment acid, with a pH in the region of 4.5; this serves to inhibit the growth of many pathogenic bacteria which otherwise invade the genital tract. There are no glands in the vagina. The vagina opens into the vulva, with the vaginal orifice lying between the urethra anteriorly and the anus posteriorly (Fig. 6.3.19). Two small glands, **Bartholin's glands**, are situated at each side of the external orifice of the vagina; occasionally cysts occur in these glands.

The **vulva** is composed of the mons pubis, the inner labia minora, the outer labia majora and the clitoris. The **clitoris** is a small erectile organ, the female homologue of the male penis, situated at the anterior junction of the labia minora. The **hymen** is a membranous partition partially blocking the orifice of the vagina. Its extent, even in virgins, is very variable, and if present it is almost always ruptured with the use of internal tampons or at first coitus.

Oogenesis and the ovarian cycle

The production of ova, or oogenesis, occurs in the ovaries. The **ovaries** vary in size and appearance according to the age of the female and the stage of the reproductive cycle. In the adult they are approximately 4 cm long, 2 cm wide and 1 cm

thick. They have an irregular outer appearance resulting from deposition of scar tissue where follicles have previously ruptured.

The outer surface of the ovary is formed of a layer of columnar cells and is known as the germinal epithelium. This is a misnomer; it was originally thought that the ova were produced in this layer but this is not so. Next there is a poorly defined layer of fibrous connective tissue – the tunica albuginea – and then the cortex where the female germ cells (oocytes) are located and develop. The innermost layer is the vascular medulla (Fig. 6.3.20).

The adult ovarian cortex contains two types of structure – the follicles and the corpora lutea. The follicles contain the **oocytes** which are all at different stages of development at any one time. During fetal life several million germ cells develop; however, by the time of birth only 200 000 or so remain, and during a female's reproductive life (that is, 30–40 years) only about 400 ova will be released. The number of germ cells is continually being reduced by cell degeneration or atresia. A **follicle** consists of the developing oocyte and its surrounding follicular cells, and changes occur in both during the ovulatory cycle. The changes are easier to understand if considered separately.

The follicular cell changes will be considered first. The primordial follicle (Fig. 6.3.21a) consists of flat cuboidal cells and these divide to form several layers of granulosa cells in the primary follicle (Fig. 6.3.21b). Amorphous material begins to accumulate between the granulosa cells and the oocyte, known as the **zona pellucida**. Outside

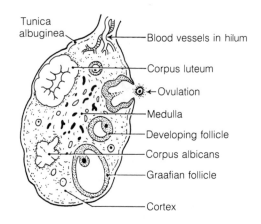

Figure 6.3.20 General plan of the ovary

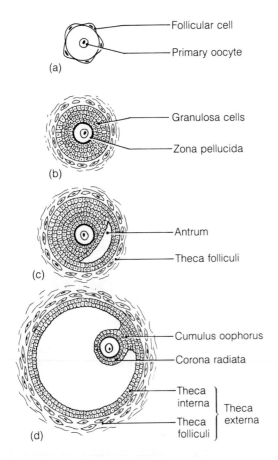

Follicular cell
Primary oocyte
(a)

Granulosa cells
Zona pellucida
(b)

Antrum
Theca folliculi
(c)

Cumulus oophorus
Corona radiata
Theca interna ⎫
Theca folliculi ⎬ Theca externa
⎭
(d)

Figure 6.3.21 Diagrams showing the stages of development in the follicle: (a) primordial follicle, (b) primary follicle, (c) secondary follicle, (d) Graafian follicle

the follicle the interstitial cells change to become the theca folliculi which is then invaded by capillaries. The inner layer of the theca produces oestrogens. The follicle continues to increase in size and an antrum, or cleft, appears which fills with follicular fluid. At this stage the structure is known as the **secondary follicle** (Fig. 6.3.21c). After a further period of maturation the granulosa cells split: one layer forms the corona radiata around the oocyte and the other outer layer, the membrana granulosa. The two layers are continuous at the cumulus oophorus. At this stage the whole structure is called a **Graafian follicle**, after the Dutch physiologist, de Graaf, who first described it in the seventeenth century

(Fig. 6.3.21d). Gradually the Graafian follicle moves towards the surface of the ovary.

Simultaneously, a mature **ovum** is developing within the follicle. The primordial germ cells differentiate into **oogonia** and by the third month of the intrauterine life of the fetus they begin to undergo mitotic division to form primary **oocytes**. The primary oocytes are located in the primordial follicle. The first meiotic division occurs at this stage. Meiosis in the primary oocyte begins in utero and division up to the diplotene stage of prophase I is completed shortly before birth. Then there is a long resting phase; in the case of the human oocyte it may be anywhere between 10 and 50 years. Thus the first meiotic division is not completed until around the time of ovulation.

In each ovarian cycle a few oocytes become selectively reactivated and proceed through meiotic division. However, usually only one continues through to the Graafian follicle stage, probably from alternate ovaries. The first meiotic division is completed before ovulation, giving a **secondary oocyte** (with only 23 chromosomes) and the **first polar body** (a polar body is a minute cell containing one of the nuclei formed during meiotic cell division, but virtually no cytoplasm; the secondary oocyte retains the major portion of cytoplasm). The second meiotic division begins almost immediately, but stops again at metaphase II and there is another, comparatively short, resting phase until fertilization. So the secondary oocyte (with the first polar body still in the zona pellucida) is released from the Graafian follicle at ovulation. Completion of the second meiotic division is dependent upon fertilization, that is, penetration of the ovum by a spermatozoon (Fig. 6.3.22).

As is apparent, the production of mature female gametes differs from the production of spermatozoa. In the male, meiosis does not begin until around puberty and then continues without interruption, whereas in the female it commences before birth and has two resting phases. Also, the secondary spermatocytes share equal amounts of chromatin and cytoplasmic material, whereas in the female there is loss of genetic material via the polar bodies.

The exact mechanism of **ovulation** – the release of the ovum from the Graafian follicle (in fact as a secondary oocyte) – is not fully understood but it probably results from increasing quantities of follicular fluid which raise the press-

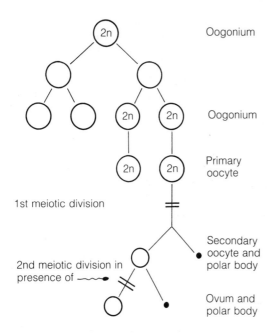

Figure 6.3.22   Diagram showing oogenesis

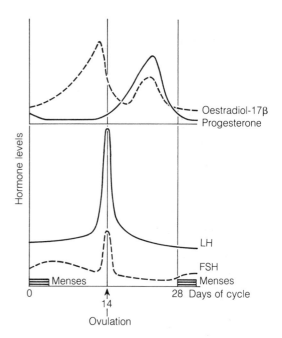

Figure 6.3.23   Hormone changes during the menstrual cycle (LH, luteinizing hormone; FSH, follicle-stimulating hormone)

ure and cause the follicle to burst. Prostaglandins may well be involved in the process of follicular rupture. The secondary oocyte, together with its follicular cells of the cumulus, is released into the abdominal cavity after which it is usually trapped by the fimbriae of the uterine tube. After ovulation the follicle collapses and the membrana granulosa becomes folded.

It is difficult to be certain on clinical grounds that ovulation has occurred: up to 10% of ovarian cycles can be anovulatory, that is, no follicle ruptures (Wilson and Rennie, 1976). The only absolute proof is pregnancy, but a regular cycle and dysmenorrhoea (see section on menstrual cycle) are indications that cycles are ovulatory. Some women notice lower abdominal pain for a brief period at ovulation, known as **mittelschmerz** (German for middle pain). The pain may be bilateral or unilateral. It is usually cramp-like and lasts a day or so and is often replaced by a dull ache. The cause of the pain is uncertain: it may be due to some local irritation within the pelvis caused by the presence of follicular fluid and blood, or from the ovary itself. Most women have some microscopic bleeding into the vagina at that time and a few experience overt bleeding; this is probably due to a temporary fall in sex hormone production

between the time of the follicle rupturing and before the establishment of a corpus luteum (Fig. 6.3.23).

The collapsed follicle becomes an endocrine gland and is termed the **corpus luteum** or yellow body (the yellowish carotenoid pigment, lutein, is formed within the cells). Both the granulosa and theca cells proliferate. The corpus luteum secretes the steroids oestrogen and progesterone and both play an important role in the reproductive cycle and in the maintenance of pregnancy should fertilization occur (androgens are also secreted from the stroma of the ovary). The corpus luteum, however, has a finite life, the length of which depends on whether pregnancy occurs or not. If no pregnancy ensues, the lutein cells degenerate after approximately 12–14 days and hormone production is reduced. Accompanying this degeneration, the corpus luteum becomes infiltrated by fibroblasts that produce scar tissue and it becomes known as the **corpus albicans** (the white body). If pregnancy occurs, the corpus luteum persists and continues to secrete oestrogen and progesterone until about the third month of gestation, when the fetoplacental unit assumes this function (see later).

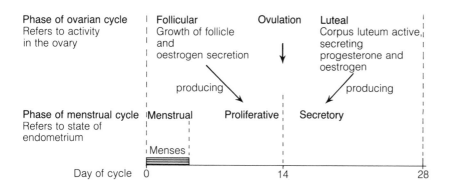

Figure 6.3.24 Diagram showing the relationship between the ovarian and menstrual cycles

The cycle of events previously described is known as the **ovarian** or **ovulatory cycle**. The cycle lasts approximately 28 days in most women, although it can vary between 21 and 35 days. It is usually divided into two phases.

(a) The **follicular phase**, when the ovarian follicles grow, mature and finally rupture; during this time the theca interna cells secrete oestrogen.
(b) The **luteal phase**, when the formation, development and degeneration of the corpus luteum occur; the granulosa lutein cells secrete progesterone and the theca lutein cells secrete oestrogens (Fig. 6.3.24).

The luteal phase of the cycle lasts approximately 14 days, although slight variations do occur amongst different individuals and from cycle to cycle in one individual; the follicular phase is more variable and accounts for the wide range in cycle length. To determine the timing of ovulation it is usual to count *back* 14 days from the first day of the next menstruation.

Successful release of a secondary oocyte from a mature Graafian follicle is dependent upon the appropriate level of circulating gonadotrophins from the anterior pituitary and this is first achieved around the time of puberty (during the years from birth until puberty some follicular growth and activity occur in the ovary but follicles degenerate before completing their development).

The gonadotrophins involved are follicle stimulating hormone (FSH) and luteinizing hormone (LH). FSH stimulates the initial development of the follicles, but the process is *not* completed to the Graafian follicle stage. Subsequent release of FSH leads to completion of follicular growth, ovulation, and development of a corpus luteum. LH then maintains the corpus luteum, stimulating the release of progesterone and oestrogens. It is now generally accepted that it is a surge in the level of LH that causes ovulation. There is a smaller peak in FSH release but this is probably due to the secretion of only one releasing hormone from the hypothalamus that raises both LH and FSH levels at the same time (see Fig. 6.3.23). (A fuller description of the hormonal control is given after discussion of the menstrual cycle.)

Ovulation is not an event in isolation; other parts of the female reproductive system are undergoing changes in preparation for fertilization. These changes will be discussed later.

The ovarian hormones

Oestrogens, progesterone and small quantities of androgens are produced in the ovaries.

OESTROGENS

The oestrogens are hormones produced by the theca interna of the ovary during the follicular phase and from the theca lutein cells in the luteal phase (they are also produced in the placenta during pregnancy). The three main oestrogens produced are all steroid molecules, each with 19 carbon atoms, and are synthesized from cholesterol (Fig. 6.3.25).

Oestradiol is the most potent of the oestrogens, followed by **oestrone** and finally **oestriol**. Seventy per cent of the oestrogens are bound to plasma proteins, principally to sex-hormone-binding globulin (SHBG). The plasma oestrogen levels vary during the monthly ovarian cycle, as shown in Fig. 6.3.23. There are two peaks in plasma levels of oestrogens: the larger peak occurs just before ovulation and the smaller one during the luteal phase.

Oestrogens produce many changes in the body

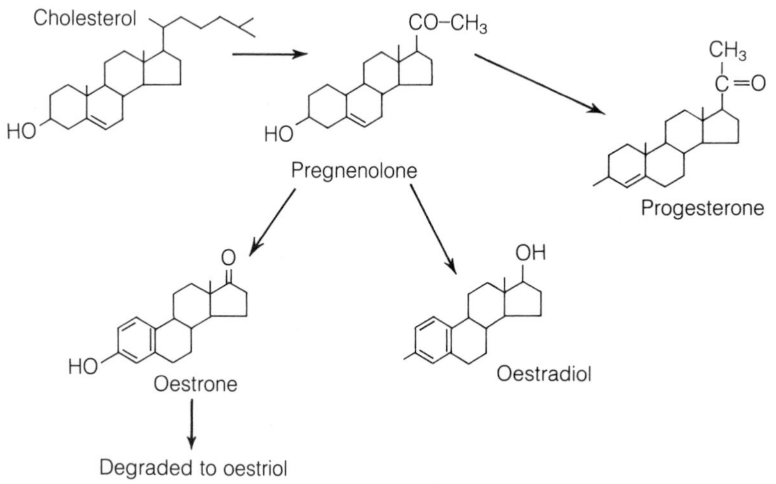

Figure 6.3.25 Pathway for the synthesis of ovarian hormones

which combine together to facilitate fertilization. For example, oestrogens are necessary for the development and maintenance of the uterine tubes, uterus, cervix, vagina, labia and for priming the duct tissue in the breast. Oestrogens also increase the motility of the uterine tubes and the excitability of the uterine muscle (these changes will be dealt with in greater detail in the next section). In animals, oestrogens heighten the female's awareness, particularly to male-associated smells, and increase visual acuity and touch sensitivity. Sensitivity to pain is decreased with raised oestrogen levels, a factor that may facilitate copulation and may be significant at the end of pregnancy when oestrogen levels are high (it has also been suggested that raised endorphin levels at the end of pregnancy may account for this decrease in sensitivity to pain (Fletcher *et al.*, 1980).

Females of many species, including *Homo sapiens*, produce odoriferous substances, **pheromones**, attractive to males. Some pheromones are produced in sweat (see Chap. 6.1) and some in or around the vagina in response to oestrogenic stimulation. It has been suggested (McClintock, 1971) that a relatively common phenomenon in women can be explained by the existence of pheromones; namely, women living together, particularly room-mates and very close friends, show highly significant synchrony in their menstrual cycles. Russell *et al.* (1981) believe some substance secreted in sweat is responsible. In general, visual cues are probably of greater significance in attracting a mate in our species. However, oestro-

gen is also responsible for all the obvious attributes of femininity, for example breast growth and the female pattern of body fat deposition in the buttocks giving the female curved outline. Oestrogen is probably involved in mood changes. Human females are generally sexually receptive throughout the cycle, unlike most animals that exhibit an oestrous cycle.

High oestrogen levels are also probably one of the factors responsible for the lower incidence of atherosclerosis recorded in women. Ischaemic heart disease is approximately six times more frequent in white men than in white women aged 40–50 years in the USA and UK but this ratio slowly diminishes with increasing age. Oestrogen, until the time of the menopause, may decrease circulating plasma cholesterol levels, by a mechanism which has not been completely elucidated.

PROGESTERONE

Progesterone is a hormone produced by the granulosa lutein cells of the corpus luteum during the luteal phase of the ovarian cycle (see Fig. 6.3.23). During pregnancy it is produced by the placenta. It is a steroid molecule with 21 carbon atoms and is an important intermediate in the synthesis of many steroids (see Fig. 6.3.25).

All the changes produced by progesterone can be thought of as facilitating gestation and it often acts on oestrogen-primed tissue; for example, progesterone increases the thickness of the endometrium, depresses myometrial activity, decreases cornification of the vagina and increases the

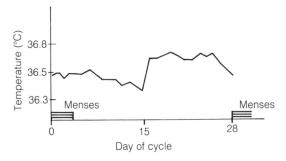

Figure 6.3.26 Temperature chart showing a typical rise in basal body temperature during the luteal phase of the menstrual cycle

secretory gland tissue in the breast (this effect is minimal during a normal cycle). Premenstrual water retention, which may be associated with clinically detectable oedema, is attributed to the increased level of progesterone in the second half of the cycle.

Progesterone is also responsible for the slight rise in body temperature during the luteal phase. There is a 0.2–0.6°C rise in *basal* body temperature during the luteal phase that is sustained until just before menstruation. The rise in temperature usually occurs over a period of 24 hours, but in some individuals it can take 3–4 days to reach its peak (Fig. 6.3.26). The cause of the increased body temperature is thought to be directly related to the increased secretion of progesterone by the corpus luteum.

If the basal body temperature is recorded daily (that is, taken each morning on waking before any activity or refreshments are taken), it can be used as an indication of ovarian progesterone production and thus, by observing the timing of the rise, gives an *approximate* guide to the timing of ovulation. This method is used to time ovulation retrospectively by women who want to conceive and by those who do not (as it can be incorporated with the rhythm method of birth control, enabling sexual intercourse to be avoided until a few days after ovulation).

ANDROGENS

Small quantities of androgens are produced from the stroma of the ovary, but have little significance. Androgens of adrenal origin have a greater influence on the body.

The menstrual cycle

The hormones released by the ovaries have functional and structural repercussions throughout the body, but particularly on the reproductive system. The changes occurring in the endometrium of the uterus constitute the menstrual cycle, which terminates in the loss of blood per vagina, i.e. menstruation. The length of the cycle is said to be approximately 28 days but its actual length may vary considerably.

The menstrual cycle is usually divided into three phases: the proliferative, secretory and menstrual phases (see Fig. 6.3.24). The **proliferative phase**, which lasts 10–11 days, coincides with the growth of the ovarian follicles and the secretion of oestrogenic hormones. The endometrium is gradually built up from the stratum basale, the epithelium regenerates from the stumps of the uterine glands left from the previous cycle (see Fig. 6.3.18), and the vascularity of the stroma increases. All these changes are brought about by the influence of oestrogens. By the time the Graafian follicle is fully mature, the regenerative changes in the uterus are complete.

The next stage is the **secretory (or progestation) phase** and it coincides with the period when the corpus luteum is functionally active and secreting progesterone and oestrogens and lasts for approximately 14 days. Under the influence of these hormones, particularly progesterone, the cells of the endometrial stroma become oedematous, the glands dilate and secrete a glycogen-rich watery mucus and the spiral arteries become increasingly prominent and tightly coiled. These spiral arteries undergo rhythmic dilations and contractions which are under the control of the ovarian hormones. The endometrium is approximately 5 mm thick at this stage.

After approximately 12–14 days, if fertilization has not occurred, the corpus luteum begins to degenerate and the secretion of ovarian hormones wanes. Thus the hormonal support to the endometrial tissue is withdrawn; there is a loss of water and a decreased blood flow to the endometrium due to spasm of the arteries, which ultimately leads to endometrial necrosis (death of the tissue). However, when the endometrial arteries dilate again bleeding occurs into the stroma of the necrotic endometrium. Thus blood enters the lumen of the uterus and menstruation, or the menstrual phase, commences. The endometrium produces prostaglandins in increasing amounts during the secretory phase and these reach a peak at the time of menstruation. It is possible that prostaglandins are involved in the initiation of menstruation and the shedding of the endometrium.

The **menstrual loss** (or **menses**) is composed of blood and epithelial and stromal cells of the endometrium discharged per vagina (the most common cause of a positive blood urinalysis in women is contamination of urine with menstrual loss). By the end of menstruation the endometrium is only 0.5 mm in thickness. There are usually between 3 and 7 days of external bleeding, which is referred to as the **menstrual phase** of the cycle. For convenience, day one of the menstrual cycle is taken from the first day of menstrual bleeding, although this is really the end of the previous cycle. Endometrial regeneration, and hence the next cycle, can begin as early as the third day of menstrual bleeding. The endometrium regenerates from the remaining glandular, stromal and vascular elements in the stratum basale, hence the reason for the separate blood supply.

The mean **menstrual blood loss** is approximately 50 cm³, although there is a wide individual variation of 10–80 cm³. Sometimes some small, darkly coloured 'clots' of blood are observed in the menstrual blood, but these are usually aggregations of red blood cells in a mass of mucoid material or even glycogen. Menstrual blood collected from within the uterine cavity does not contain any fibrinogen and therefore is incapable of clotting. The 'clots' are not harmful and are more common when bleeding is excessive. The incoagulability of menstrual blood has long been a puzzle, but it may be due to proteolysis and immediate digestion of fibrinogen by enzymes (released from the damaged endometrial cells) within the uterus and cervix, but the precise nature remains obscure.

Excessive menstrual flow, **menorrhagia**, can lead to iron deficiency anaemia due to depletion of iron stores. The iron status of women seems to be very delicately balanced. A woman taking a normal diet with a haemoglobin concentration of 12 g/dl and a roughly regular cycle, will remain in iron balance only if her blood loss does not exceed 65 cm³ (Wilson and Rennie, 1976). Thus, women that experience heavy menstrual bleeding may need to be advised to take intermittent courses of iron therapy. Women with an intrauterine contraceptive device in situ often experience an

Table 6.3.1   Signs and symptoms of the premenstrual syndrome

Lower abdominal discomfort and distension
Nausea
Breast discomfort
General 'bloated' feeling
Weight gain of up to 3 kg
Frequency of micturition
Change of bowel habit
Increase or decrease in acne
Swelling of ankles and hands
Darkening of the skin under the eyes
Headache
Increased emotional lability, especially increased irritability, depression
Decreased libido

increased menstrual loss and thus may become iron deficient.

The absence of menstruation, or **amenorrhoea**, is usually a sign of failure to ovulate and there are many possible causes. If a woman never establishes menstruation (known as primary amenorrhoea), the cause may be chromosomal, for example Turner's syndrome, or due to endocrine imbalances. The most common cause of secondary amenorrhoea (or the absence of menses once they have commenced) is pregnancy; some diseases may cause the cessation of menstruation too, for example tumours of the hypothalamus or anterior pituitary, anorexia nervosa, endocrine disturbances (thyrotoxicosis or post-pill amenorrhoea).

Some women experience mood changes and unpleasant physical symptoms during the 10 days prior to the onset of menstruation. This has been given the name **premenstrual tension (PMT)** or the **premenstrual syndrome (PMS)**. The clinical features can include a varying and complex range of symptoms (Table 6.3.1).

In the past it was thought that there was a considerable emotional component to PMS but now it is generally accepted that there is a physiological basis, although the exact nature of this is unknown. The cause cannot be ascribed to increasing levels of either oestrogen or progesterone as levels of both are usually declining at this stage of the cycle. It has been postulated that it is the rate at which the hormone levels change rather than the absolute levels that is significant in its causation. In one study carried out at St Thomas' Hospital, London (Trimmer, 1978), approximately 30% of women sufferers with PMS were shown to have a deficiency of circulating progesterone in the second half of the cycle. Higher prolactin levels have also been implicated in PMS, as has lack of vitamin B_6 (pyridoxine).

Treatment varies, but includes the prescription of oral contraceptives or additional progestagens (synthetic progesterone) in the second half of the cycle. Recently, pyridoxine has been given; pyridoxine has the effect of lowering prolactin levels and this may be its mechanism of action. (Lack of vitamin B_6 leads to low dopamine levels in the brain which in turn results in increased prolactin levels. Pyridoxine also has an effect on tryptophan metabolism and a deficiency of this can lead to mood disturbances). Oil of Evening Primrose, a rich source of essential fatty acids, especially gamma-linoleic acid, has been used with success in some PMS clinics (Brush, 1984). As fluid retention appears to be a major part of this syndrome, treatment with diuretics often alleviates much of the discomfort. The symptoms reach their maximum intensity 1–2 days preceding the onset of menstruation and are invariably relieved by menstruation.

Whatever the physiological basis for PMS, it is undoubtedly the cause of much suffering in women. It has been estimated that several million working days are lost each year in Britain because of it. There are other problems beside absenteeism: there is an increased incidence of road traffic, home and industrial accidents during the premenstrual period and menstruation, ascribed to a decreased ability to concentrate; work efficiency and judgement are also impaired. Exam performance during this period can be impaired and a handicap of as much as 5% has been found. There is an increased admission rate to psychiatric hospitals during this period and suicide rates in women of reproductive age are seven times higher in the second half of the menstrual cycle. The incidence of all types of crime committed by women increases in the premenstrual and menstrual phases. England and France have recognized the significance of PMS to such an extent that during a few criminal trials allowance has been made for the fact that the women were premenstrual (for a more detailed discussion of PMS see Dalton (1983) and further reading).

The other common problem is **dysmenorrhoea**, or painful menstruation. Some pain at the time of menstruation is almost universal, but a few women have severe and disabling pain. About 45% of menstruating women report moderate or severe dysmenorrhoea. The pain is lower abdominal, either suprapubic or lateralized. Dysmenorrhoea is described as being cramp-like or as a dull ache. It is most severe on the first day of bleeding and in some young women may be associated with faintness, nausea and vomiting. The incidence decreases with age and after childbearing. Sufferers can be reassured that dysmenorrhoea is usually only experienced in cycles in which ovulation occurs, although the reason for this remains obscure.

The origin of the pain is almost certainly within the uterus, but again the exact mechanism is unknown. It may be an ischaemic type of pain due to uncoordinated uterine contractions caused by hypersensitivity of nerve endings within the uterus. It is now thought likely that the release of prostaglandins from the disintegrating endometrium may play a major part in the causation of dysmenorrhoea. Some cases of dysmenorrhoea have been successfully treated with flufenamic acid, a non-steroidal anti-inflammatory drug which acts by suppressing prostaglandin formation.

OTHER CHANGES ASSOCIATED WITH THE OVARIAN AND MENSTRUAL CYCLES

Changes occur in other regions of the reproductive system, also under the influence of the ovarian hormones. During the follicular phase the epithelium of the uterine tubes proliferates and during the luteal phase the secretory cells become more active, presumably to suppy nutrients for the ovum as it moves to the uterus. There are changes in the motility of the muscular elements of the uterine tubes and uterus: tubal movements and myometrial contractions predominate under oestrogenic influence, whilst progesterone decreases the motility. Again it has been suggested that prostaglandins may be involved in tubal contractility and ovum transport.

Cyclical changes are seen in the composition of the mucus secreted by the cervix (the cervix itself dos not exhibit marked cyclical activity). As a result of the oestrogen secretion at the time of ovulation, the water and electrolyte content of the mucus increases. This thinner mucus, produced around the time of ovulation, is thought to allow easier penetration of the cervix by spermatozoa. During the luteal phase, under the influence of progesterone, the volume of mucus decreases and it becomes thicker.

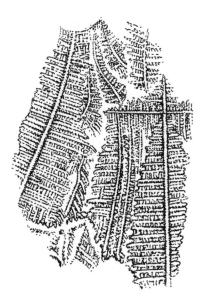

Figure 6.3.27 'Ferning' pattern of cervical mucus at the time of ovulation

These changes in the cervical mucus form the basis of some tests to ascertain whether ovulation has occurred, and also its timing. If a cervical smear of mucus is taken around the time of ovulation, allowed to dry on a glass slide and examined under a microscope, a characteristic pattern of crystallization occurs, known as ferning or a fern-leaf pattern (Fig. 6.3.27). This is indicative of a well-oestrogenized cervical mucus. This phenomenon disappears after ovulation due to the effects of progesterone and it is absent during pregnancy. Also at the time of ovulation, the mucus develops the property of 'spinnbarkheit', which means that it can be drawn out into long threads.

There are marked changes in the squamous epithelial cells of the vagina during the menstrual cycle. Under oestrogenic stimulation, there is an increased tendency to cornification (an increase in 'horny' tissue) of the cells which decreases under the influence of progesterone. This change may increase the vagina's resistance to trauma.

The breasts may increase in size and tenderness in the premenstrual week, due to oedema and hyperaemia (increased blood flow) in the intra-lobular connective tissue. There is often an increase in skin pigmentation premenstrually, especially around the eyes, but also in the areola of the nipple. These changes may be due to an increase in the level of melanocyte-stimulating hormone from the anterior pituitary, and are similar to those seen in pregnancy, but less marked.

The integration of the hormonal systems

Several components of the hormonal systems involved in female reproductive physiology have been discussed in isolation, but it is crucial to understand the integration and interrelationships of these systems. Gonadotrophins from the anterior pituitary induce both ovulation and the secretion of the ovarian hormones, and these in turn have widespread effects on the body. But how are all these events coordinated?

The hypothalamus is the vital integrating centre. The release of gonadotrophins FSH and LH from the pituitary is controlled by a releasing hormone produced in the median eminence of the hypothalamus. There appears to be only one hormone produced and this one hormone stimulates the release of both FSH and LH at the same time. Hence it is given the name gonadotrophin-releasing hormone (GnRH) or sometimes follicle-stimulating hormone/luteinizing hormone-releasing hormone (FSH/LH-RH).

The release of GnRH from the hypothalamus causes the release of FSH and LH from the anterior pituitary, which in turn causes development of the ovarian follicles, release of an ovum, maintenance of the corpus luteum and, as a result, the secretion of oestrogens and progesterone. The regulation and integration of this system are complicated, involving fine balances between the levels of gonadotrophins and ovarian hormones, incorporating both negative and positive feedback pathways and influences from other parts of the brain. The important sensor in the system is the part of the hypothalamus that is sensitive to circulating levels of oestrogens. It is logical for oestrogen to be the important factor, as oestrogen levels give a direct indication of the stage of follicular development and are also responsible for producing most of the preparatory changes necessary to ensure fertilization.

When there are low levels of circulating oestrogens, during and following menstruation, a negative feedback system operates, i.e. the hypothalamus detects the low oestrogen levels and release of GnRH is increased. This is turn increases secretion of FSH, which stimulates follicular

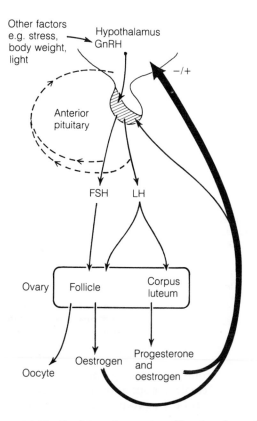

Figure 6.3.28 Possible pathways controlling the release of female reproductive hormones (GnRH, gonadotrophin-releasing hormone; FSH, follicle-stimulating hormone; LH, luteinizing hormone)

development (in the presence of basal levels of LH). Several follicles begin to develop during each cycle under the influence of FSH and they all contribute initially to the increasing oestrogen levels. The higher level of oestrogen is then detected by the hypothalamus and levels of FSH are subsequently reduced; only the most mature follicle will be able to complete its development without the FSH stimulus, the other follicles degenerate (Fig. 6.3.28).

Oestrogen probably affects both the anterior pituitary and the hypothalamus, although the latter is thought to be the main site of action. There may also be a short feedback loop, with levels of LH and FSH directly influencing release of GnRH.

However, a simple negative feedback loop is not an adequate explanation. The LH surge that produces ovulation (see Fig. 6.3.23) is due to a positive feedback mechanism, that is, *high* oestro-

gen levels, acting on the hypothalamus, produce the surge in LH secretion (the second surge of FSH is presumably a result of the surge in releasing hormone and is of secondary importance to the LH surge). The precise mechanism for this paradoxical negative/positive feedback system is uncertain and many aspects require confirmation by further research. The proposed mechanism stated here is no doubt a gross simplification of the complex monitoring system involved.

External stimuli are also known to affect the occurrence and timing of ovulation and menstruation. Ovulation is commonly delayed in females subject to mild stress, for example when taking examinations, and it may cease altogether under conditions of severe stress, for example some women in prisoner-of-war camps ceased to menstruate.

Even though our knowledge of the control mechanisms in the human reproductive system is far from complete, the use of fertility drugs involves the hypothalamic–pituitary–ovarian axis just described. One such drug is clomiphene, an active non-steroidal agent that acts at the hypothalamic level. Sometimes this is given together with human chorionic gonadotrophin later in the cycle to induce ovulation. When clomiphene is given, the hypothalamus responds as if the oestrogen level is lower than it actually is, and thus levels of FSH and LH are raised. It has been postulated that clomiphene blocks the oestrogen receptor complexes in the hypothalamus. The intended result is that, with raised levels of gonadotrophins, ovulation will occur. In fact, often multiple ovulation, or superovulation as it is sometimes called, occurs. This may result in multiple pregnancies.

Gonadotrophins are sometimes directly administered, with good results, in the treatment of infertility, but administration of GnRH has not yet been effective.

As much as 20% of cases of anovulation have been associated with high prolactin levels. Bromocriptine, a dopamine antagonist, has been found to inhibit the release of excessive prolactin from the anterior pituitary and to correct anovulatory cycles (this mechanism is also implicated in the lactational postpartum amenorrhoea – see later).

The female reproductive system is thus governed by a complex control system, the main points of which can be summarized as follows (Fig. 6.3.29).

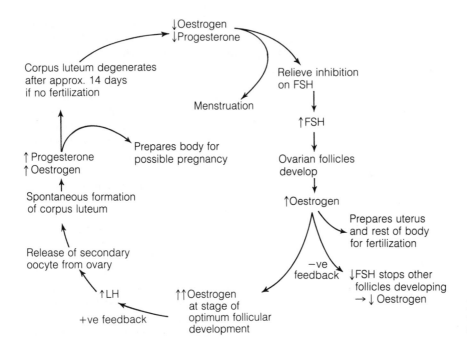

Figure 6.3.29 Summary of events occurring in the female reproductive system

1 Falling oestrogen and progesterone levels initiate menstruation and also relieve the inhibition on FSH secretion; thus FSH levels rise and follicles begin to develop.

2 As the follicles mature, increasing levels of oestrogen are secreted which in turn inhibits the secretion of FSH (the negative feedback in operation); oestrogen also prepares the uterus and the rest of the body for ovulation.

3 As a result of optimum follicular maturation, oestrogen levels rise high enough at midcycle to stimulate the release of sufficient LH to result in ovulation. This forms the positive feedback mechanism of oestrogen on the hypothalamus.

4 A corpus luteum forms spontaneously from the collapsed follicle after ovulation and secretes progesterone and oestrogen, which maintain the suppression of the gonadotrophins and prepare the body for possible pregnancy.

5 In the absence of fertilization and the implantation of the embryo, the corpus luteum degenerates and the levels of ovarian hormones fall, inducing endometrial breakdown.

6 The cycle is repeated.

PUBERTY

Puberty is the time at which the reproductive organs become functionally active, and in both the male and female it is accompanied by development of the **secondary sexual characteristics**.

In the male

In the male, the testes increase rapidly in size and spermatogenesis commences. The interstitial cells of Leydig secrete more androgens and the accessory organs of reproduction, e.g. penis and prostate gland, begin to grow. The higher androgen levels are responsible for the development of secondary sexual characteristics, for example hair grows on the face, trunk, axillae and pubic region; the larynx enlarges and, as a consequence, the voice 'breaks' or deepens; considerable muscle development starts and the male experiences occasional ejaculations. In boys some growth of the testes takes place between 6 and 10 years of age, but marked changes start to occur at a mean age of approximately $11\frac{1}{2}$ years (in the UK and USA), followed a year later by penile growth. During puberty there is a general growth spurt, with a peak at about 14 years. The onset of pubertal changes and the rate of progress vary from one individual to another and puberty is insidious in nature; indeed, hair growth on the chest continues into the mid-twenties.

In the female

In the female, breast development and growth of sexual hair occur first, the former commonly beginning just after 11 years of age. As in the male, there is a growth spurt and differentiation of the skeleton to form the typical female pelvis, and fatty tissue is deposited both in the breasts and buttocks. Higher levels of oestrogen are responsible for most of the female's pubertal changes, but raised androgen levels lead to the growth of pubic and axillary hair even in the female. The most dramatic event during puberty is the **menarche**, the first menstrual period, but it is only one event in a complex series of changes which occur at puberty. The mean age of menarche in most developed countries lies between 12 and 13 years, although it can occur between the ages of 9 and 17 years. In Western Europe and the USA the age of menarche has declined steadily since the middle of the last century by as much as 4 months per decade.

Ovulation is normally an infrequent occurrence during the first 2 or 3 years after the menarche. Menstruation itself may be irregular during adolescence and intervals of 4–6 months between periods are not uncommon, especially during the first five cycles or so.

However, some mature gametes are produced from the time of puberty onwards; simultaneously, there are changes in the levels of reproductive hormones secreted by the gonads. The actual onset of puberty, in both sexes, is thought to be due to a significant fall in the sensitivity of the hypothalamus to the sex hormones. In the pre-puberty period, *small* amounts of the sex hormones of adrenal origin are sufficient to suppress the activity of those parts of the central nervous system responsible for the synthesis and/or release of GnRH. However, at puberty there is a decreased sensitivity to the negative feedback effect of the sex hormones in some hypothalamic neural 'centres'; thus the pituitary gonadotrophin levels increase, leading to an increase in the secretion of the gonadal sex hormones. Changes in pituitary sensitivity to GnRH and adrenal androgens may also be involved.

After puberty, male and female gonadotrophin secretion differs: the sexually mature female exhibits marked fluctuations in an approximately monthly cycle (see Fig. 6.3.23), whereas in the sexually mature male the gonadotrophins are secreted at a fairly constant rate. This difference is governed by the sexual differentiation of the hypothalamus, established during intrauterine life as a result of male and female fetuses being exposed to different hormones (Holmes and Fox, 1979).

Other factors influence the timing of puberty. It seems that a critical body weight (or possibly a critical percentage of body fat) is essential for sexual maturation. This 'body weight' concept may account for the decline in the mean age of menarche seen over the past 100 years, which is now flattening out. This reduction in the age at which menarche occurs coincides with improved nutrition and better standards of hygiene in developed countries, leading to improved growth and development and this, in turn, could well influence sexual maturity. This idea also explains the amenorrhoea experienced by females with anorexia nervosa who stop menstruating once their weight falls below a critical point, about 42–47 kg in the UK. Menstruation recommences during treatment once the 'critical weight' of these patients is regained. This cannot explain, however, how women near starvation in famine areas manage to maintain their high fecundity (birth rates).

The **pineal gland** (a small reddish-grey structure on the dorsal surface of the midbrain) may be involved in changes occuring at puberty. Animal experiments suggest that the pineal gland produces antigonadotrophic hormones, and this is supported by some clinical examples, although precise evidence is limited; for example, some children with destructive pineal lesions exhibit precocious puberty, whereas those with actively secreting pineal tumours show delayed puberty and hypogonadism (Mullen and Smith, 1981). It has been suggested that melatonin, one of the main secretions of the pineal, modifies the secretion of the gonadotrophins, that is, it has some antigonadotrophic effect.

The knowledge concerning puberty is extensive but far from complete, and the relationships between the hypothalamus and other regions, such as the pineal gland, are largely unexplained.

THE MENOPAUSE

The menopause is a single event occurring during the **climacteric**, a period which extends for some years either side of the menopause. The menopause is defined as the stage in a woman's life when menstruation ceases. The systemic changes associated with it may occur months before or after the cessation of menstruation; therefore the term menopause is often used to embrace both the last menses and the changes that occur.

During the climacteric, ovulation ceases and there is a deficiency of oestrogen and progesterone. The hypothalamus and anterior pituitary respond to this deficiency by increasing secretions of GnRH, and FSH and LH, respectively. This increased output of gonadotrophins, which can reach ten times the level in a normal menstruating woman, remains raised for some 20 years or so until senescence occurs. The menopausal symptoms are caused by the deficiency of ovarian hormones and the increase in FSH and LH.

The human is the only animal which has a significant period of life beyond the cessation of reproductive ability. In the majority of women, the menopause occurs between the ages of 46 and 52 years, but it can occur anywhere between 40 and 55 years. In the UK, the mean age of menopause is around 50 years (Parkes, 1976). Pregnancy resulting in a spontaneous abortion has been reliably reported as late as the age of 56 years.

The years preceding the menopause show increasing menstrual irregularity, and the incidence of ovulatory cycles decreases. The usual pattern of events is a decrease in either the frequency of menstruation, the amount of blood lost, or both. It is *not* normal for a woman to experience an increased loss or irregularity resulting in more frequent menstruation. It is also usual to regard episodes of bleeding occurring more than 1 year after the menopause as abnormal. Women who present with either of these two sets of symptoms need careful investigation, as there is an increased incidence of genital cancer in women with postmenopausal bleeding.

The low climacteric oestrogen levels are responsible for the atrophy of the breasts, labia, uterus and vaginal epithelium that occurs. Atrophy of the vaginal epithelium results in dryness and there is an increase in vaginal pH (it becomes less acidic) which together render the vagina more susceptible to infection. Other commonly associated symptoms are pruritus (itching), dyspareunia (painful or difficult coitus) and the urethral syndrome (symptoms of urinary infection with sterile urine).

Most women also experience vasomotor hot flushes, sweating and emotional instability. The precise aetiology for these changes is unknown but a rise in the gonadotrophin levels (rather than a reduction in ovarian hormones) and endorphins have been implicated. The skin may become thinner, causing wrinkles; there is thinning of the distribution of hair too, both on the head and in the pubic region. Senile osteoporosis is closely associated with long periods of low oestrogen levels. Also, as oestrogen levels fall, plasma cholesterol levels rise and the 'protective' influence of oestrogens against ischaemic heart disease is thus removed.

Many non-specific symptoms are also experienced, for example fatigue, insomnia, depression, headaches and palpitations. These may have a hormonal basis, but the menopause is also a time when some women suffer from psychological or social stresses; for example, many women feel that they have lost their physical attraction to the opposite sex; their children have grown up and left home and in consequence their motherhood role is lost; a feeling of unwantedness predominates, which may account, in part, for the above symptoms.

In some instances **hormone replacement therapy (HRT)** is prescribed for a woman during or after the menopause. It is debatable whether HRT should be used to maintain 'youthfulness', but it is certainly justified if the menopausal symptoms cause distress. Oestrogen is the hormone prescribed (sometimes in combination with progestagens); oestrogen can be replaced orally, by injection or by the implantation of an oestrogen pellet. The long-term effects of HRT have not yet been evaluated, but there is a known risk of developing endometrial carcinoma if prolonged therapy is prescribed. Individuals contemplating long-term HRT should be followed-up at regular intervals for cervical smears, blood pressure and weight recordings. If the major menopausal problems are vaginal, application of a topical oestrogen cream is often helpful.

Many women are advised to undergo a **hysterectomy** (removal of the uterus) at some stage for problems such as fibroids or uterine prolapse. Premenopausal women who undergo a hysterectomy do not experience an 'artificial' menopause straightaway as their ovaries are not removed, unless they are abnormal. Many women naturally feel anxiety about this preoperatively. If an

oophorectomy (removal of the ovaries) is performed at the same time, the woman will experience an artificial menopause within days of the operation.

After a simple hysterectomy, pregnancy is impossible and monthly menstruation will stop, but there should be no other significant changes. The uterus itself does not produce any sex hormones, therefore a woman's femininity or enjoyment of sex should not be altered. In most instances the vagina is not made smaller by the operation and so normal intercourse is possible. Advice as to the resumption of sexual intercourse should be given postoperatively: it is advisable to wait 4–6 weeks before having gentle intercourse, particularly if there are internal sutures. Similarly, energetic sports should not be recommenced until the same time. Intercourse may be uncomfortable at first and use of a lubricant such as KY jelly may be helpful. The vagina may temporarily 'shrink' in size, but intercourse will actually help the tissues to stretch and become supple again.

If the patient has had a hysterectomy for carcinoma of the uterus, some of the vagina may have been removed. However, intercourse will stretch the vagina again, but in this case it is advisable to wait 3 months before resuming sexual activities. Radiotherapy to the genital region will probably decrease the lubricant properties and decrease the possible expansion and lengthening capacity of the vagina, but coitus is still usually feasible.

Patients should be encouraged to voice their anxieties at all stages of treatment. Practical points to help the patient include seeing and talking with both partners together and providing privacy for them to talk to each other. Junior nurses should be given the opportunity where possible to listen to the ward sister giving advice. Amias (1975) and Steele and Goodwin (1975) give detailed accounts of advice that can be given to patients who have had hysterectomies. A study by Webb and Wilson-Barnett (1983) emphasizes the importance of giving information to patients in aiding recovery after hysterectomy.

There is no equivalent climacteric in males to that occurring in females – males retain their fertility for a longer period, and there are well-authenticated cases of men in their eighties becoming fathers. Male hormonal output shows a gradual decline between the ages of 20 and 70 years, although they continue to produce fertile sperm well past 70 years. The age at which an individual male loses his sexual drive has come to be called, quite wrongly, the male menopause. Loss of sexual drive, in either sex, can result from many causes and may occur at any age.

THE PHYSIOLOGY OF SEXUAL INTERCOURSE

The male and female gametes are brought together by the act of sexual intercourse (coitus). A brief description of the events of coitus will now be given (for a more detailed description see further reading).

The male

The normal state of the penis is flaccid, but under conditions of sexual excitement it becomes erect. *Erection* is purely a vascular phenomenon: during sexual excitement, the arterioles in the erectile tissue of the penis (corpus spongiosum and corpora cavernosa, see Fig. 6.3.9) dilate and become engorged with blood. As the erectile bodies are surrounded by a strong fibrous coat the penis becomes rigid, elongated and increases in girth. As the erectile tissue expands, the veins emptying the corpora are compressed and thus the outflow of blood is minimal. The process occurs rapidly, in 5–10 seconds. Erection is controlled by a spinal reflex.

The erection reflex (Fig. 6.3.30) can start from direct stimulation of the genitals – the glans penis or the skin around the genitals. There are highly sensitive mechanoreceptors located in the tip of the penis. The afferent synapse in the lower spinal cord and the efferent flow, via the nervi erigentes, produce relaxation of the arterioles in the penis. The higher brain centres, via descending pathways, can have profound facilitative or inhibitory effects. Thoughts, visual cues or emotions can cause erection in the complete absence of any mechanical stimulation.

The parasympathetic nerves simultaneously stimulate the urethral glands to secrete a mucoid-like material which aids lubrication. Erection allows entry of the penis into the female vagina, and the angle which the erect penis makes with the male's trunk closely follows the angle of the vagina in the female's pelvis.

After intromission (the insertion of the erect

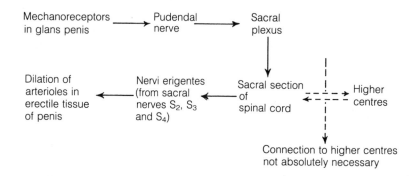

Figure 6.3.30 The nervous pathways (simplified) involved in the erection reflex

penis into the vagina), **ejaculation of semen** into the female vagina may occur. This is again basically a spinal reflex and the afferent pathway is the same as for erection. When the level of stimulation reaches a critical peak a patterned automatic sequence of efferent discharge is elicited to the smooth muscle of the genital ducts and to the skeletal muscle at the base of the penis. The exact nature of the nervous pathways involved is complex but includes sympathetic stimulation to the ducts, via L_1 and L_2 nerve roots. The first stage is known as emission, and the genital ducts and accessory glands empty their contents into the posterior urethra. During the second stage, ejaculation proper, the semen is expelled from the penis, by a series of rapid muscle contractions, into the female genital tract.

During ejaculation the sphincter at the base of the bladder is closed, therefore no spermatozoa can enter the bladder nor can urine be voided. This again is under the control of the sympathetic nervous system. A feeling of intense pleasure arises with ejaculation and the event is referred to as an **orgasm**. There is simultaneously a noticeable skeletal muscle contraction throughout the body which is rapidly followed by muscular and psychological relaxation. After ejaculation there is a **latent period** during which time a second erection is not possible. The latent period varies from individual to individual but can range from a few minutes to several hours in 'normal' men. Loss of erection occurs due to vasoconstriction of the arterioles in the penis, hence venous compression is reduced.

Any interference with the spinal reflexes may result in **impotence** or other sexual dysfunction (although libido is unaffected). For example, ejaculation is usually not possible after a bilateral lumbar sympathectomy below L_2. Erection may not be possible after an abdominoperineal resection due to damage to the nervi erigentes. Administration of drugs that inhibit the release of noradrenaline from postganglionic sympathetic nerve endings, e.g. methyldopa and reserpine, can lead to ejaculatory failure, although erection and sensation would be normal. Ganglion-blocking drugs, e.g. hexamethonium, inhibit both parasympathetic and sympathetic nerve pathways and reduce both erection and ejaculation; these drugs are rarely used now because of their widespread anticholinergic side-effects.

It is worth noting that hypertensive individuals report few problems in sexual performance whilst undiagnosed, and thus untreated, but once hypotensive therapy has been initiated, the incidence of impotence and erection failure increases considerably due to the nature of the drugs prescribed. Diabetics also suffer from problems with impotence, although the cause is probably a result of metabolic, neuropathological and vascular disturbance. Some patients who have spinal cord injuries have problems too. The extent of return to normal sexual function in these patients varies considerably according to the nature and position of the injury: Trimmer (1978) estimates that erections return eventually in approximately 75% of cases, but there is a low incidence of orgasm. Sex education and counselling for patients with all types of handicap are important. There are several organizations that can help patients (see further reading).

The female

In the female, sexual excitement is characterized by erection of the clitoris and labia minora, both of which are largely composed of erectile tissue.

The neural control of erection is the same as for the male. The breasts may enlarge during sexual excitement and the nipples become erect. As the sexual tension increases, there may be a flushing of the skin which begins on the chest and spreads upwards over the breasts, neck and up to the face.

The female provides most of the lubrication for coitus by the transudation of fluid through the vaginal walls. The exact source of the mucus is unclear as there are no glands in the vagina. Additional secretions may come from the glands in the vulva and from the Bartholin's glands.

The movement of the penis in and out of the vagina causes pleasure in both the male and female. The female may experience a climax (or orgasm) similar to that of the male. Stimulation of the clitoris may heighten the state of excitement and contributes to the orgasm of the female. The female is potentially capable of several orgasms within a short period of time, unlike the male. During coitus and orgasm the uterus may contract rhythmically and this may serve to aspirate the semen into the uterine lumen. Females do not always experience orgasm, and it is not necessary for successful fertilization, for example, orgasm does not occur in artificial insemination. Orgasm may, however, contribute to fertilization in some cases of subfertility where uterine aspirations hasten the movement of spermatozoa towards the ovum.

Cardiovascular and respiratory changes during coitus

During orgasm in both the male and female there is a marked increase in the heart rate, blood pressure and respiratory rate. The respiratory rate may rise to 40 respirations/minute, the heart rate to between 100 and 170 beats/minute and the systolic blood pressure may be increased by 30–80 mmHg and the diastolic by 20–40 mmHg.

Individuals who suffer from cardiovascular disease are often anxious about resuming normal sexual activities after, for example, a myocardial infarction, because of the extra strain it might cause.

Fox and Fox (1969) showed that peak coital heart rates of middle-aged married men were similar to those obtained during light exercise, i.e.

heart rate during light exercise
107–130 beats/min

heart rate during actual orgasm
90–144 beats/min
(mean value 117 beats/min)
(Rates before and after orgasm were considerably lower.)

Thus, in individuals who can tolerate light exercise, sexual activities are perfectly feasible. In a few patients the effort tolerance is more limited and may be exceeded during intercourse, and coital or postcoital angina or arrhythmias may be precipitated, but this does not rule out coitus, as glycerol trinitrate or some similar drug could be taken beforehand. Fox (1978) suggested that there may be a greater demand on the cardiovascular system, i.e. showing increased signs of physiological stress, if intercourse is being performed with a new partner.

Both the patient and his or her spouse should be given advice on discharge from hospital, whether or not they verbally express anxieties. The vast majority of patients with ischaemic heart disease can enjoy normal sexual relations without risk. Extremely few people die during sexual intercourse! It is generally safe to tell the individual that he can start sexual activities as soon as he has returned to mild or moderate levels of physical activity and can probably return to the same level of sexual activity as before the attack.

CONCEPTION

The egg is released from the ovary at the second metaphase stage of meiosis and it enters the uterine tube still surrounded by follicular cells. If **fertilization** occurs, it does so in the ampulla of the uterine tube. Damaged or blocked uterine tubes, possibly as a result of salpingitis (inflammation of the uterine tubes), are an important cause of infertility. In November 1977, Patrick Steptoe, an obstetrician, and Robert Edwards, a physiologist from Cambridge, made history with the first successful **human fertilization in vitro** (i.e. in artificial conditions). An oocyte that had undergone its initial maturation in vivo (i.e. within the ovary) was removed by laparoscopy, fertilized in vitro, and then the resultant embryo was transferred back to the uterus of the mother. Thus the mother's blocked uterine tubes were bypassed. In July 1978, a perfectly healthy female infant weighing 2.7 kg was born. Many successful operations of this nature have been performed since then and it is likely to become more common as the technical problems are overcome.

Normally, however, the spermatozoa reach the oocyte by traversing the lumen of the uterus and moving along the uterine tubes. Estimates vary, but spermatozoa on their own can probably move only a few millimetres per hour, by propelled movement of their tails. The fact that after coitus spermatozoa can reach the ampulla of the uterine tube within 30 minutes or so implies that their movement is assisted. As already discussed, coitus provides the initial impetus to spermatozoa in their journey. After coitus, the primary transport mechanism is contraction of the musculature in the uterus and uterine tubes. Prostaglandins present in the semen may cause the smooth muscle to contract. The wastage rate of spermatozoa is huge: of the several hundred million spermatozoa deposited in the vagina, only a few thousand actually reach the uterine tubes. Fructose present in the semen probably provides an important energy source for the spermatozoa.

As the spermatozoa are transported to the site of fertilization, they undergo their final maturation which enables them to pass through the follicular cells and zona pellucida and penetrate the oocyte. The maturation processes are known as **capacitation** and the **acrosome reaction**. These processes involve changes in the acrosome and the release of hyaluronidase and other proteolytic enzymes which assist the passage of the spermatozoa through the layers of cells around the oocyte. It has been postulated that a high density of spermatozoa is required to produce sufficient hyaluronidase to remove most of the follicular cells. However, only one spermatozoon is able to enter one egg. The entry of additional spermatozoa appears to be blocked in some way.

The time available for fertilization of the oocyte in the female is approximately 24 hours after ovulation, after which time the egg begins to degenerate. Spermatozoa maintain their fertilizing ability for up to 48 hours (some suggest even as long as 72 hours) inside the female genital tract, so there is only a limited period when fertilization is at all possible.

The penetration of the oocyte by the head of the spermatozoon is followed by completion of the second meiotic division of the ovum and the formation of the second polar body. The nuclei of the spermatozoon and ovum, containing the maternal and paternal haploid sets of chromosomes, come

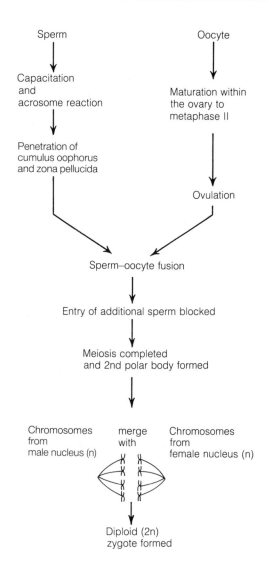

Figure 6.3.31 Events in the female reproductive tract leading up to fertilization

together on the mitotic spindle (Fig. 6.3.31). Fertilization is completed with the restoration of the diploid complement of chromosomes.

The fertilized ovum, usually referred to as the **embryo** (or **zygote**), continues a series of mitotic divisions as it passes along the uterine tube, although it does not increase in size. A round mass of cells is formed, still surrounded by the zona pellucida, known as a **morula** (Fig. 6.3.32). A central cavity develops in the morula so that a

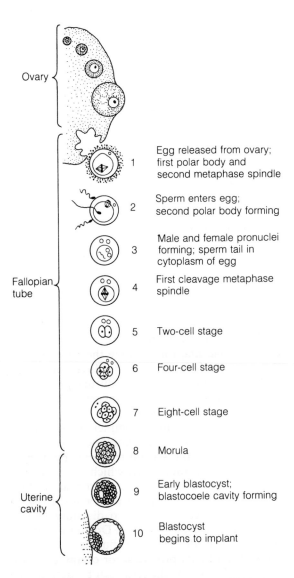

Ovary		
	1	Egg released from ovary; first polar body and second metaphase spindle
	2	Sperm enters egg; second polar body forming
	3	Male and female pronuclei forming; sperm tail in cytoplasm of egg
Fallopian tube	4	First cleavage metaphase spindle
	5	Two-cell stage
	6	Four-cell stage
	7	Eight-cell stage
	8	Morula
Uterine cavity	9	Early blastocyst; blastocoele cavity forming
	10	Blastocyst begins to implant

Figure 6.3.32 Diagrammatic representation of follicular growth, ovulation, fertilization and preimplantation

fluid-filled cyst is formed called the **blastocyst**. The cells of the blastocyst become arranged into an outer layer, the **trophoblast**, and an inner cell mass bulging into the central fluid-filled cavity.

Normally only one oocyte is released during each cycle. Sometimes, however, two or more oocytes are released almost simultaneously. When this is the case and two oocytes are fertilized, non-identical (dizygotic) twins will develop. Identical (monozygotic) twins result when a single fertilized ovum, at a very early stage of development, becomes completely divided into two independently growing cell masses.

The developing zygote normally reaches the uterus approximately 3–5 days after fertilization, at the morula stage, and lies free within the lumen of the uterus for a short time. Chemical substances are capable of passing from the embryo to the mother and vice versa whilst the embryo is free living in the uterus, that is, before attachment or implantation.

By the sixth or seventh day after fertilization the process of **implantation**, or embedding, into the oestrogen and progesterone primed endometrium occurs. The trophoblast lies next to the uterine epithelium and the two layers of cells become intimately associated as the cell boundaries between the trophoblast and uterine epithelial cells disappear. This probably occurs as a result of cellular phagocytosis by the trophoblast. The cells of the trophoblast actively invade the endometrium, which responds by undergoing a hypertrophic reaction which converts the endometrium into the **decidua**. The trophoblast plays an active part in the nutrition of the inner cell mass from which the embryo and associated structures, such as the amnion, develop. The blastocyst becomes completely embedded in the decidua.

Occasionally the embryo becomes implanted in the uterine tube and this is known as an **ectopic pregnancy**. It is not known why ectopic pregnancies occur, but there is a higher incidence in women who use intrauterine contraceptive devices (IUCD). It can be a dangerous condition as rupture of the uterine tube inevitably follows if the pregnancy is not terminated.

The **placenta** is derived from maternal (decidual) and embryonic (trophoblastic) components (Fig. 6.3.33). After embedding, the trophoblast proliferates and divides into two layers, an inner one where the cell structure persists, called the **cytotrophoblast**, and an outer one where the cell boundaries largely break down, called the **syncytiotrophoblast**. In the embedding process endometrial capillaries are broken down and so maternal blood oozes around the trophoblast. Initial nutrition of the embryo is provided from the cell debris produced by the trophoblastic phagocytosis, with the placenta and fetal circulation taking over this role after a few weeks. During the first month of life the embryo develops a comprehensive blood supply, including

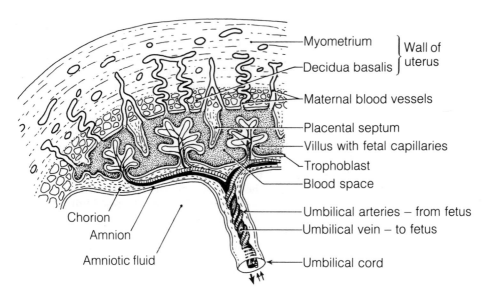

Figure 6.3.33 The structure of the placenta

arteries and veins in the umbilical cord that connects the embryo and mother. Five weeks after implantation this system has become well established; the fetal heart has begun to pump blood and the entire mechanism for nutrition of the fetus is functional. Nutrients move from the maternal blood, across the placental membranes into the fetal blood; waste products move in the opposite direction. Oxygen and carbon dioxide move by simple passive diffusion whereas other substances are carried mainly by active transport mechanisms in the placental membranes.

Some drugs taken by the mother can reach the fetus via the placenta; thalidomide is an example of a drug which was transmitted to the embryo with disastrous effects, as affected children were born with grossly deformed or absent limbs. Physiologically, some chemicals may cause in the fetus exactly what they would do in the mother, for instance narcotics not only induce sleep in the mother, but also make the fetus sleepy. Nicotine and lead can cross the placenta too; the former has been associated with 'small for dates' babies and a reduction in placental size. Even salicylates, taken as aspirin, may cause neonatal bleeding at term, alter the prothrombin time of the fetus and delay mechanisms of haemostasis. Pregnant women should be warned against taking drugs unless prescribed by the doctor, and this information should be included in the health education given to women during pregnancy. Ideally, alcohol

should be avoided. Certain live viruses can also cross the placenta, for example the rubella virus (German measles). There is no direct mixing of the fetal and maternal blood during pregnancy; thus it is feasible for a rhesus-negative mother to carry to term a rhesus-positive fetus.

The fetus floats in a completely fluid-filled cavity. Specimens of this amniotic fluid can be removed during the second trimester by a technique called **amniocentesis** which is usually carried out around the 15th week of pregnancy. This process, which is associated with a small risk of complications, can be used to diagnose certain congenital defects, for example Down's syndrome and neural tube defects, as the fluid contains some fetal cells and fetal products, for example α-feto-proteins. The sex of the fetus can be determined from chromosomal examination. As the fetal cells have to be cultured, the complete results from amniocentesis are often not available for 3 or 4 weeks. A method for first trimester prenatal diagnosis has been developed involving chorion villus sampling (CVS) – the chorion is part of the embryonic sac (see Fig. 6.3.33). CVS can be performed between 8 and 11 weeks of pregnancy, using either the transcervical or transabdominal route, and would allow a much earlier termination of pregnancy if necessary.

The fetus remains connected to the mother via the umbilical cord for the whole period of **gestation**. The average duration of pregnancy is 266 days (38 weeks) from conception and 280 days (40 weeks) from the date of the last menstrual period in a woman with a 26–32-day cycle. A convenient method of calculating the expected date of delivery (EDD) is to add 1 year and 10 days to the date of the last menstrual period and to take 3 months from the total.

A detailed description of embryological changes in pregnancy can be found in any textbook of human embryology (see further reading). The maintenance of pregnancy and changes in the mother are under hormonal influences and these will be considered shortly, but first there will be a brief discussion of methods of preventing conception, that is, contraception.

Contraception

Conception refers to both the fertilization of the oocyte by a spermatozoon and the successful implantation (nidation) of the embryo into the endometrium. Thus any contraceptive method aims to prevent fertilization and/or implantation from occurring.

First, there are contraceptives that prevent fertilization, that is, prevent the male and female gametes uniting. Examples of these include simple mechanical barriers, e.g. the diaphragm or cap, the sheath and spermicides that interfere with spermatozoa viability. Sterilization by vasectomy and cutting of the uterine tubes are further examples.

Second, intrauterine contraceptive devices (IUCDs), i.e. the coil, act by preventing implantation of the embryo into the endometrium once fertilization has occurred. The exact mechanism by which the IUCD acts is not certain, but it is known that foreign bodies within the uterus will prevent implantation.

The combined oral contraceptive pill acts by a combination of both previously described methods. The various types of pill contain oestrogen and progestogen (a synthetic progesterone). Tablets are usually taken for 21 days, with a break of 7 days during which time there is a withdrawal bleed, i.e. simply due to the withdrawal of the hormonal support. The raised levels of these hormones disrupt the normal hypothalamic–pituitary–ovarian axis and its feedback system, with the consequence that no ovulation occurs. Higher levels of oestrogen and progestogen have other effects on the woman too: the cervical mucus remains viscid due to the progestogen, and penetration of it by the spermatozoa is difficult; the endometrium is altered and is not in a state capable of accepting any embryo, and the tubal transport mechanism is altered, interfering with the transport of the oocyte from the ovary to the uterus. These combined effects account for the almost complete effectiveness of combined oral contraceptives (providing that they are taken as prescribed). Thus, if ovulation were to take place, pregnancy would be unlikely to occur since conditions in the reproductive tract are unfavourable for spermatozoa and ovum transport and also for implantation. The pill thus seems to be an ideal contraceptive, but it does have some serious side-effects, for example in some individuals there is an increased risk of thromboembolus formation. Some high-risk factors have been identified and these include increasing age, obesity and cigarette smoking in the pill user.

Other steroid contraceptives can be prescribed, such as the progestogen-only 'mini-pill'. This is not as effective as the combined pill because ovulation still occurs in many cycles. It acts primarily by inducing unfavourable conditions for fertilization and implantation.

Slow-release depot injections of steroids can also be used. An intramuscular injection of the drug, e.g. Depo-Provera, can be given to a women, giving her protection from pregnancy for 2–3 months whilst the steroids are released. The use of these depot injections remains controversial due to a lack of precise information on the effects of prolonged high levels of the steroids. They are sometimes prescribed for women who need contraceptive protection either after a rubella vaccination or after their partners have undergone a vasectomy and before sterility is certain.

Under some circumstances, it is possible to take a combined oral contraceptive as a postcoital contraceptive, within 72 hours of unprotected intercourse. The exact mode of action of the postcoital 'pill' is not fully understood, but it has an effect between fertilization and implantation. (Insertion of an IUCD up to 5 days after intercourse can also be effective postcoitally, by interfering with implantation of the fertilized ovum.)

HORMONE CHANGES DURING PREGNANCY

Progesterone and oestrogen are essential for the

initiation and continuance of pregnancy. Fertilization results in the persistence of the corpus luteum which continues to develop and increases its secretion of these hormones. Progesterone maintains the endometrium in its 'progestational' state essential for pregnancy, and is necessary to depress the contractile activity of the uterus, thus allowing the blastocyst to implant and preventing its expulsion. Oestrogens are necessary for uterine growth, which involves both general protein synthesis and the production of specific enzymes necessary for muscular contraction and energy mechanisms – these are important during parturition.

The non-pregnant corpus luteum is maintained by the gonadotrophin LH from the anterior pituitary. However, LH levels fall 12–14 days after ovulation. The corpus luteum is maintained in a pregnant woman by a hormone called **human chorionic gonadotrophin (HCG)**, produced by the trophoblast of the developing blastocyst. HCG is a glycoprotein very similar in structure to LH and is found only in the presence of a trophoblast. It can be detected in maternal blood and urine about 10 days after ovulation, i.e. 5 days before the next menstrual period would have occurred. Thus the blastocyst must begin to produce HCG very soon after fertilization and before implantation is complete. HCG maintains the secretion of progesterone and oestrogen from the corpus luteum in early pregnancy.

Secretion of HCG reaches a peak 8–9 weeks after the last menstrual period and then the level drops dramatically to a lower one that is maintained until the end of pregnancy (Fig. 6.3.34). Thus the function of the corpus luteum also begins to decline after 8 weeks of pregnancy. Oophorectomy (excision of an ovary) before the sixth week leads to abortion, but after that time has no effect on pregnancy.

The presence of HCG in the maternal urine forms the basis of the immunological **pregnancy test**. This will give a reliable positive result approximately 28 days after conception but HCG can be detected as early as 14 days after conception with some tests. Recent development of a radioimmunoassay technique to detect the presence of a subunit of HCG in maternal serum now allows diagnosis of pregnancy even before the first missed menstrual period.

In humans the placenta takes over steroid production from the corpus luteum by the twelfth

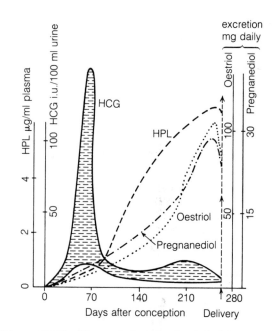

Figure 6.3.34 Changes in hormone levels during pregnancy (HCG, human chorionic gonadotrophin; HPL, human placental lactogen)

week of pregnancy. Oestrogen, particularly oestriol, and progesterone are synthesized in the placenta from precursors originating mainly from the fetal adrenal glands and liver and from the maternal adrenal glands. It is likely that the placental hormones are synthesized and released by the syncytiotrophoblast, the outer layer of the trophoblast. The interdependence of fetal and placental tissues for the production of oestrogen in particular has given rise to the concept of the **fetoplacental unit**, the term implying that both are necessary and neither can function in isolation.

Progesterone and oestrogen levels rise throughout pregnancy until just prior to delivery and these elevated levels inhibit the release of FSH and LH. The oestrogen and progesterone levels can be detected in the form of breakdown products – oestriol and pregnanediol – in the maternal urine. Measurement of oestriol in 24-hour urine collections can be used to assess the function of the fetoplacental unit in pregnancy, reflecting both placental function and fetal well-being. Usually, several successive 24-hour urine determinations are necessary. These indicate trends of secretion rather than absolute amounts as hormone levels vary considerably at different stages of pregnancy.

Another hormone produced by the feto-placental unit is **human placental lactogen (HPL)**, also known as human chorionic somato-mammotrophin (HCS). Levels of HPL rise steadily throughout pregnancy, with the curve flattening off towards term (see Fig. 6.3.34). HPL is structurally very similar to growth hormone. The exact role of HPL is unknown, but it exhibits growth-promoting and lactogenic (able to induce lactation) properties. It may also act with HCG to help maintain the corpus luteum. HPL has, in addition, an anti-insulin effect and may be responsible for the diabetogenic changes of pregnancy. It has been suggested that it may also in some way protect the fetus from rejection by the mother. The control of HPL secretion is poorly understood, but since its production appears to be related to placental and fetal weight, its measurement serves as a valuable indicator of the condition of the fetus. Falling levels in early pregnancy often indicate inevitable abortion.

The role of the placenta in producing other hormones is uncertain; it has been proposed that relaxin, renin and possibly a substance similar to thyroid-stimulating hormone, known as human chorionic thyrotrophin (HCT), are also produced by the placenta. **Relaxin** has also been isolated from the corpora lutea and so may be synthesized in the ovaries and simply stored in the placenta. Relaxin is thought to cause the softening of the elastic ligaments of the symphysis pubis and pelvic girdle in order that the fetal head may descend through the bony arch at delivery without damage.

Maternal oxytocin levels rise throughout pregnancy. **Oxytocin**, secreted from the posterior pituitary gland, stimulates the contraction of the smooth muscle of the pregnant uterus and lactating mammary glands, but this effect is held in check by the high levels of progesterone that inhibit uterine motility. Only at parturition is oxytocin left unopposed to act on the smooth muscle of the uterus.

Many other physiological changes occur in the mother during pregnancy. A few of these changes are given below.

1 There is increased secretion of hormones from the anterior pituitary, adrenal cortex and thyroid gland (the thyroid gland can increase in size by as much as 50% during pregnancy).
2 There are cardiovascular changes, for example
   (a) the resting heart rate increases to an average of 85 beats/min by the end of pregnancy
   (b) the venous pressure rises in the legs due to the pressure of the growing fetus which impedes venous return, and this can lead to the development of dependent oedema.
3 The pregnant woman retains water and sodium as a consequence of increased production of steroid hormones by the placenta and adrenal cortex.
4 The nausea and constipation frequently experienced are probably due to a lowered tone in the stomach and gut musculature which occurs as a result of rising oestrogen and progesterone levels. The phenomenon of 'morning sickness' is possibly also associated with the rapidly increasing and then declining HCG levels.
5 Metabolic changes, probably mediated via the hypothalamus, account for the changes in appetite.
6 There is an increase in pigmentation on the nipples and face and also the appearance of the linea nigra (a pigmented line from the umbilicus to the pubis).
7 Increased corticosteroid output alters the tensile strength of the dermal fibres in the skin which, when stretched, often result in striae gravidarum, 'stretch marks'.

Pregnancy is a highly emotional time: there are psychological changes and wide mood swings that probably have, at least in part, a hormonal causation. These psychological changes are evident too in the puerperium, most obviously in post-natal or puerperal depression.

Thus widespread changes can be seen to occur in the general maternal metabolism.

HORMONES AND PARTURITION

The precise factor or factors responsible for the initiation of labour are still unknown. **Labour** is usually spontaneously initiated about 38 weeks after conception. Just before term the oestrogen and progesterone levels fall and the ratio of these two hormones becomes such that the uterine smooth muscle becomes particularly sensitive to the posterior pituitary hormone oxytocin, which stimulates uterine contractions. Oxytocin is probably released in spurts via a neurogenic reflex when the cervix is stretched as a result of uterine contractions. A few hours before labour contractions start, under the influence of relaxin and oestrogen, the cervix rapidly becomes compliant, enabling it to dilate.

It is now widely accepted that prostaglandins, particularly prostaglandin $F_{2\alpha}$, which is produced by the endometrium, initiate parturition. The factors that actually determine the timing of the prostaglandin release have not been elucidated, but an increase in the oestrogen : progesterone ratio may be involved. Thus, several factors are known to be involved, but the whole mechanism underlying the initiation of labour is still unclear.

The role of prostaglandins is confirmed by the ability of prostaglandin inhibitors and antagonists to prolong pregnancy. Prostaglandins can induce labour at all stages of pregnancy. Both oxytocin and prostaglandins are used therapeutically to induce parturition both at term and in abortion.

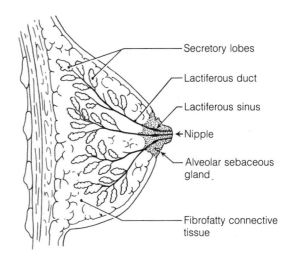

Figure 6.3.35   The female breast (prepubertal)

THE BREASTS AND LACTATION

The paired mammary glands (breasts) are specialized skin derivations which produce milk during the lactation period following childbirth. They exist in the male too, but only in the rudimentary state. The female breasts have a composite structure in which a radiating compound alveolar gland is embedded in a mound of fat and connective tissue in the pectoral region. The nipple of each gland is surrounded by an area of pigmented skin, known as the **areola**, which changes from pink to brown during the first pregnancy and remains so thereafter.

Each breast consists of between 15 and 25 independent glandular units called breast lobes. The lobes are arranged radially at different depths around the nipple. A single large duct, the **lactiferous duct**, drains each lobe via a separate opening on the surface of the nipple. Just before each duct opens onto the surface, it forms a dilation called the **lactiferous sinus** (Fig. 6.3.35). Each breast lobe is divided into a variable number of lobules; the lobules consist of a system of alveolar ducts from which large numbers of secretory alveoli develop during pregnancy.

The rudimentary breast in the female enlarges under hormonal influences during puberty and undergoes slight cyclical changes throughout the menstrual cycle. Some abnormal growths in the breast, for example an adenocarcinoma, are often female hormone dependent; this is confirmed by

the regression of some tumours if treated with testosterone.

The true transformation from an inactive to an active gland occurs during pregnancy and the breasts only become fully functional at parturition. Many hormones are involved in the transformation; oestrogen, in the presence of adrenal steroids and growth hormone, stimulates the development of the duct system, and progesterone stimulates growth of the secretory alveoli. Insulin, thyroid hormones, HPL (human placental lactogen) and prolactin from the anterior pituitary are also required for complete development.

From the fifth week of pregnancy onwards the breasts enlarge, there is an increased blood flow to them, the nipples increase in size, the areolae extend and become darker and small cutaneous glands, the glands of Montgomery, which open onto the areola, become more prominent and appear as small tubercles. In the last 3 months of pregnancy growth of the breasts slows down and **colostrum**, a fluid precursor of milk, is secreted. Colostrum is also secreted in the immediate post-partum days until replaced by milk. Colostrum is deep yellow in colour and rich in protein and salts, with less lactose than milk; it also carries some immunoglobulins.

Prolactin, also known as lactotrophic hormone (LTH), exerts a major influence on the initiation and maintenance of lactation. During pregnancy the level of prolactin begins to rise after about the eighth week and continues to rise until delivery.

The release of prolactin is due partly to the inhibition of prolactin inhibitory factor from the hypothalamus; it may also be due to the stimulation of some, as yet unknown, hypothalamic releasing factor. If suckling does not follow parturition, prolactin levels fall to pre-pregnancy values within a few weeks. Prolactin primarily stimulates milk production in the alveoli. The milk produced consists of water, lactose, fat, protein (casein and lactalbumin) and electrolytes (Na^+, K^+, Ca^{2+}, Cl^- and HCO_3^-). A number of hormones with general metabolic effects, for instance insulin and corticosteroids, also influence lactation. Human milk contains considerably less protein, fewer salts and more carbohydrate than cow's milk.

Milk does not flow out of the nipple unless suckling takes place. The tactile stimulation of the nipple sets up a neuroendocrine reflex releasing oxytocin; this in turn stimulates the contraction of the myoepithelial cells surrounding the alveoli and the smooth muscle of the ductile system. Milk is forced into the main ducts of the breast which ultimately connect to the nipple, and the milk is ejected. This is sometimes known as the 'let-down' reflex. Suckling also stimulates prolactin release, thus neuroendocrine reflexes are responsible for the secretion of both oxytocin and prolactin. Secretion and expulsion of milk from lactating breasts occur simultaneously. Breast-feeding can cause uterine contractions to occur at the same time, due to the effect of oxytocin on the uterine musculature (see section on parturition). This helps to involute (reduce the size of) the uterus after the pregnancy.

If suckling does not occur, milk production diminishes. Oestrogen given in large doses postpartum effectively blocks lactation, indicating that oestrogen itself may act directly on mammary tissue. At the end of the lactation period the breasts return to their resting condition – as quiescent glands – although they often remain larger than they were before the start of the pregnancy.

The level of prolactin remains relatively high for the duration of lactation. If full lactation with frequent demand feeding occurs, prolactin levels remain higher than if 4-hourly feeding takes place, especially if there is a complete break overnight. The number of suckling episodes seems to be important in maintaining high prolactin levels. This has practical relevance as lactation (more specifically, a high prolactin level) delays the return of normal ovulatory cycles and can result in amenorrhoea. Lactation therefore can seem to act as a natural contraceptive, but it cannot be considered to be a reliable form of contraception postnatally. Lactation probably increases the intervals between births in some developing countries where full lactation is practised. Short (1976) concluded that, throughout the world as a whole, more births are prevented as a result of lactation than by all other forms of contraception put together.

High blood levels of prolactin are closely associated with both lactational and some pathological forms of amenorrhoea, although the exact linking mechanism is uncertain. There are adequate levels of LH and FSH in the pituitary during lactation, but oestrogen fails to induce a positive feedback release of LH and the hypothalamic–pituitary axis appears to be more sensitive to the negative feedback effects of oestrogen.

After this discussion of reproductive physiology in humans, the reader may now appreciate that there are many aspects of the subject that remain to be more fully researched and explained. Progress in understanding all aspects of reproductive physiology will not only lead to more effective treatment and care on an individual basis, but may also offer societies alternative ways of controlling population growth.

Review questions

The answers to all these questions can be found in the text. Except for questions 3, 8 and 11, there is at least one correct and at least one incorrect answer in each case.

1 Which of the following contain(s) a single set of chromosomes?

   (a) A spermatid.
   (b) A fertilized ovum.
   (c) An ordinary body cell.
   (d) An ovum just after ovulation.

2 Males differ from females in that

   (a) their cells have Barr bodies
   (b) their pituitary glands secrete different gonadotrophic hormones
   (c) they continue to produce gametes until later in life.

3 Place the following in order to indicate the pathway followed by spermatozoa in ejaculation.

(a) Urethra.
(b) Epididymis.
(c) Seminiferous tubules.
(d) Vas deferens.
(e) Ejaculatory duct.

4 Human spermatozoa

(a) are produced optimally at 37°C
(b) are motile in the seminiferous tubules
(c) contain enzymes which facilitate penetration of the ovum.

5 The normal seminal ejaculation

(a) is about 2–5 cm^3 in volume
(b) comes mainly from the epididymis
(c) contains prostaglandins
(d) contains buffers which make the pH of the vaginal fluids more suitable for sperm viability.

6 Testosterone causes

(a) increased production of LH
(b) increased possibility of baldness
(c) the epiphyses of long bones to unite
(d) a positive nitrogen balance.

7 Androgens

(a) are formed in the prostate
(b) are steroids
(c) are secreted in small amounts in adult females.

8 Which of the phrases on the right apply to which of the hormones listed on the left?

(i) LH
(ii) oestrogen
(iii) progesterone
(iv) FSH

(a) produced by corpus luteum
(b) causes development of follicles
(c) causes ovulation
(d) causes growth of endometrium

9 Compared with the 7th day of the menstrual cycle, on the 21st day there is a greater

(a) progesterone level in the blood
(b) thickness of uterine muscle
(c) body temperature.

10 Temporary cessation of menstruation (secondary amenorrhoea)

(a) may occur for psychological reasons
(b) occurs if the body weight rises above a critical level
(c) may occur after taking oral contraceptives.

11 Which hormone

(a) maintains the corpus luteum during the early stages of pregnancy?
(b) causes the development of secretory acini in the breasts during pregnancy?
(c) is released from the posterior pituitary to reinforce and maintain uterine contractions?
(d) stimulates the secretion of milk by the acini?

12 Are the following statements true or false?

(a) Birth usually occurs 38 weeks after the last menstruation.
(b) Milk acini only develop during puberty.
(c) Breast-feeding can cause uterine contractions.

13 Fertilization of the ovum

(a) normally occurs in the cervix
(b) by one spermatozoon prevents other spermatozoa from entering the ovum
(c) may occur up to 4 days after ovulation
(d) usually occurs about 4 days before implantation.

14 Oral contraceptive treatment

(a) acts mainly by preventing implantation of the fertilized ovum
(b) may cause an increase in body weight in some patients
(c) probably depresses anterior pituitary secretion of gonadotrophic hormones.

Answers to review questions

1 a and d
2 c
3 c, b, d, e and a
4 c
5 a, c and d

6 b, c and d

7 b and c

8 (i) and (c), (ii) and (d), (iii) and (a), (iv) and (b)

9 a and c

10 a and c

11 (a) HCG, (b) progesterone, (c) oxytocin, (d) prolactin

12 (a) false, (b) false, (c) true

13 b and d

14 b and c

References

Amias, A. G. (1975) Sexual life after gynaecological operations I and II. *British Medical Journal 2*: 608–609, 680–681.

Brush, M. G. (1984) *Understanding Premenstrual Tension*. London: Pan.

Dalton, K. (1983) *Once a Month*. Glasgow: Fontana.

Fletcher, J. E., Thomas, T. A. & Hill, R. G. (1980) β-Endorphin and parturition. *Lancet i*: 310.

Fox, C. A. (1978) Recent research in human coital physiology. *British Journal of Sexual Medicine 5* (41): 13–19.

Fox, C. A. & Fox, B. (1969) Blood pressure and respiratory patterns during human coitus. *Journal of Reproductive Fertility 19*: 405–415.

Goy, R. W. (1968) Organizing effects of androgen on the behaviour of Rhesus monkeys. In *Endocrinology and Human Behaviour* (Michael, R. P. ed.) Oxford: Oxford University Press.

Holmes, R. L. & Fox, C. A. (1979). *Control of Human Reproduction*. London: Academic Press.

Lincoln, G. A. (1979) Pituitary control of the testis. *British Medical Bulletin 35* (2): 167–172.

van Look, P. F. A., Hunter, W. M., Corker, C. S. & Baird, D. T. (1977) Failure of positive feedback in normal men and subjects with testicular feminization. *Clinical Endocrinology 7*: 353–366.

Lyon, M. F. (1974) Mechanisms and evolutionary origins of variable X-chromosome activity in mammals. *Proceedings of the Royal Society of London, B 187*: 243–268.

McClintock, M. K. (1971) Menstrual synchrony and suppression. *Nature 229*: 244.

Masters, W. H. & Johnson, V. E. (1966) *Human Sexual Response*. Boston: Little Brown.

Mullen, P. E. & Smith, I. (1981) The endocrinology of the human pineal. *British Journal of Hospital Medicine 25* (3): 248–256.

Parkes, A. S. (1976) *Patterns of Sexuality and Reproduction*. Oxford: Oxford University Press.

Russell, M., Switz, G. & Thompson, K. (1981) Sweat synchronizes menstrual cycles. *New Scientist 89* (1235): 71.

Short, R. V. (1976) Breastfeeding and the mother. In *Ciba Foundation Symposium (New Series) No. 45*, Amsterdam: Elsevier Excerpta Medica.

Short, R. V. (1979) Sex determination and differentiation. *British Medical Bulletin 35* (2): 121–127.

Steele, S. J. & Goodwin, M. F. (1975) A pamphlet to answer the patient's questions before hysterectomy. *Lancet ii*: 492–493.

Trimmer, E. (1978) *Basic Sexual Medicine* London: Heinemann Medical.

Webb, C. & Wilson-Barnett, J. (1983) Hysterectomy: a study in coping with recovery. *Journal of Advanced Nursing 8* (4): 311–319.

Wilson, E. W. & Rennie, P. I. C. (1976) *The Menstrual Cycle*. London: Lloyd-Luke.

Suggestions for further reading

Dalton, K. (1984) *The Premenstrual Syndrome and Progesterone Therapy*. London: Heinemann Medical.

Edlund, B. (1982) The needs of women with gynaecologic malignancies. *Nursing Clinics of North America 17* (1): 165–177.

Edwards, R. & Steptoe, P. (1980) *A Matter of Life: The Story of a Medical Breakthrough*. London: Hutchinson.

Emery, A. E. H. (1983) *Elements of Medical Genetics*, 6th edn. Edinburgh: Churchill Livingstone.

Fraser Roberts, J. A. & Pembrey, M. E. (1985) *An Introduction to Medical Genetics*, 8th edn. Oxford: Oxford University Press.

Hutt, C. (1975) *Males and Females*. Harmondsworth: Penguin Education.

Johnson, M. & Everitt, B. (1984) *Essential Reproduction*, 2nd edn. London: Blackwell Scientific.

Kinsey, A. C., Pomeroy, W. B. & Martin, C. E. (1948) *Sexual Behaviour in the Human Male*. Philadelphia: Saunders.

Kinsey, A. C., Pomeroy, W. B., Martin, C. E. & Gebhard, P. H. (1953) *Sexual Behaviour in the Human Female*. Philadelphia: Saunders.

Laycock, J. & Wise, P. (1983) *Essential Endocrinology*, 2nd edn. Oxford: Oxford University Press.

Nazzaro, A. & Lombard, D. (with Horrobin, D.) (1985) *The PMT Solution*. London: Adamantine Press.

Potts, M. & Diggory, P. (1983) *Textbook of Contraceptive Practice*, 2nd edn. Cambridge: Cambridge University Press.

Royal College of General Practitioners' Oral Contraception Study (1981) Further analyses of mortality in oral contraceptive users. *Lancet i*: 541–546.

Sadow, J. I. D., Gulamhusein, A. P., Morgan, M. J., Naftalin, N. J. & Peterson, S. A. (1980) *Human Reproduction: An Integrated View*. London: Croom Helm.

Schofield, M. (1968) *The Sexual Behaviour of Young People*, 2nd edn. Harmondsworth: Penguin.

Webb, C. & Wilson-Barnett, J. (1983) Self-concept, social support and hysterectomy. *International Journal of Nursing Studies 20* (2): 97–107.

Williams, P. L., Wendell-Smith, C. P. & Treadgold, S. (1984) *Basic Human Embryology*, 3rd edn. London: Pitman Medical.

Useful address

Committee on Sexual and Personal Relationships of the Disabled (SPOD)
Brook House
2–16 Torrington Place
London WC1E 7HN

Telephone: 01 637 4712

Appendix 1
SI Units and Their Relevance to Clinical Practice

Système Internationale d'Unités, or the international system of units (SI for short), based on metric measurements, is now widely used in science and medicine throughout the world. In the UK the metric system replaced the Imperial system in the 1960s and 1970s; however, there was a wide variety of metric units in use at that time and so the SI system was adopted in an attempt to standardize these units. The units used in medicine are sometimes inconsistent, mainly because of the impracticality of some SI units and also in order to avoid having to replace equipment using Imperial units, e.g. sphygmomanometers for measuring blood pressure.

SI uses seven basic, so-called fundamental, units and many derived units (made from two or more of the fundamental units). Before describing the SI units and their relationship to Imperial or non-SI units, a brief description will be given of the accepted way of writing numbers.

Metric systems use decimal numbers, that is, numbers based on powers of 10. Very large and very small numbers are often encountered when describing chemical and biological measurements. The scientific notation is to write such numbers in terms of a power of ten, e.g. 1000 can be written as 10^3. This is because $10 \times 10 \times 10 = 1000 = 10^3$ (i.e. ten to the power three). Similarly

$$1000 \times 1000 = 1\,000\,000 = 10^6$$

$$10^3 \times 10^3 = 10^6$$

When a number contains other numerals, the number is written as follows:

$$5 \text{ million} = 5\,000\,000 = 5 \times 10^6$$

$$5.5 \text{ million} = 5\,500\,000 = 5.5 \times 10^6$$

The accepted form of scientific notation is to place only one numeral to the left of the decimal point.

A similar principle operates for numbers smaller than 1. For example;

$$0.001 = 1 \times 10^{-3}$$

This is equivalent to

$$0.1 \times 0.1 \times 0.1 = 1 \times 10^{-3}$$

$$\tfrac{1}{10} \times \tfrac{1}{10} \times \tfrac{1}{10} = \tfrac{1}{1000} = 1 \times 10^{-3}$$

$$\text{or} \quad 10^{-1} \times 10^{-1} \times 10^{-1} = 1 \times 10^{-3}$$

The negative or minus sign in front of the raised numeral indicates that the powers of 10 are being divided into the digit.

With numbers other than 1:

$$0.0004 = 4 \times 10^{-4}$$

$$0.000\,045 = 4.5 \times 10^{-5} \quad \text{i.e.} \quad \frac{4.5}{100\,000}$$

Numbers with many figures are divided into groups of three digits by gaps and *not* commas as previously (e.g. earlier printing: 5,613,100; modern printing: 5 613 100). The exceptions to this are 4-digit numbers, which are usually printed without a gap, e.g. 1629.

Commas are no longer used because, although the decimal point is indicated by a full stop on the line in English-speaking countries (e.g. 1.96), in some other countries the decimal point is indicated by a comma (e.g. 1,96). Thus to avoid any possible confusion commas are no longer inserted to space large numbers (see earlier). If a number is less than 1, a zero must always be inserted before the decimal point, e.g. 0.88 and not .88.

Decimal multiples and submultiples of SI units are commonly required in everyday use of units. Powers of 1000 are preferred e.g. kilo-, milli-, micro-, but any of the prefixes in the following table can be used.

Prefix	Symbol	Value	Factor by which the unit is multiplied
tera-	T	1 000 000 000 000	10^{12}
giga-	G	1 000 000 000	10^{9}
mega-	M	1 000 000	10^{6}
*kilo-	k	1000	10^{3}
hecto-	h	100	10^{2}
deca-	da	10	10^{1}
deci-	d	0.1	10^{-1}
*centi-	c	0.01	10^{-2}
*milli-	m	0.001	10^{-3}
*micro-	$\mu^{\dagger}$	0.000001	10^{-6}
*nano-	n	0.000 000 001	10^{-9}
pico-	p	0.000 000 000 001	10^{-12}
femto-	f	0.000 000 000 000 001	10^{-15}
atto-	a	0.000 000 000 000 000 001	10^{-18}

*Prefixes in common clinical use
†This symbol is the Greek letter 'mu' – however, it is read as 'micro-'.

Basic SI units

The seven basic units are:

Physical quantity	Name of SI base unit	Symbol of SI unit
length	metre	m
mass	kilogram	kg
time	second*	s
temperature	kelvin	K
amount of substance	mole	mol
electrical current	ampere	A
luminous intensity	candela	cd

*Although the second is the basic unit for time, for general use the minute (symbol min), hour (symbol h) and day (symbol d) may be used.

No 's' should be added to the symbols if plural and the symbol should *not* be followed by a full stop (unless it occurs at the end of a sentence).

e.g. 60 kg and *not* 60 kgs.

Thus: 'Mrs Smith, who is only 1.5 m tall, weighs 80 kg and so is severely overweight.'

The prefixes listed previously are frequently combined with these basic units to give workable and "sensible sized" units. For example:

kilometre(km) = 1000 m

centimetre(cm) = 0.01 m

nanometre(nm) = 10^{-9} m

millimole(mmol) = 0.001 mol

or 1×10^{-3} mol

Where necessary the fundamental units can be combined. For example a fluid infusion rate of 100 millilitres per hour can be written as 100 ml/h or $100 \, \text{ml} \, \text{h}^{-1}$. More specific combinations are possible, for example a fluid infusion rate may be described as 10 millilitres per kilogram bodyweight per hour. This may be written as 10 ml/kg/h, or $10 \, \text{ml} \, \text{kg}^{-1} \, \text{h}^{-1}$.

ADDITIONAL NOTES AND EXPLANATIONS ON THE USE OF THE BASIC UNITS

Length

The SI unit for length is the metre: multiples and sub-multiples of the metre replace the units of miles, yards, feet and inches (1 metre is equal to 39.37 inches).

Two metric units used in the past for very small lengths were the micron (10^{-6} m) and the Ångstrom (10^{-10} m). These terms are now obsolete: the micron is referred to as the micrometer (i.e. $1 \, \mu\text{m}$ or 10^{-6} m), while the Ångstrom unit is equivalent to 0.1 nanometres.

Volume

The volume of an object is found by multiplying length, breadth and depth together.

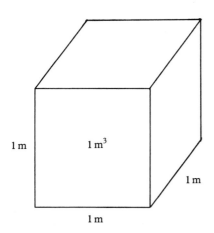

Volume of this cube = 1 m × 1 m × 1 m
= 1 m³
= 1 metre cubed or 1 cubic metre.

This is a large volume and not practical for many clinical purposes. Thus the litre (symbol 'l') is commonly used as the unit for measuring volumes of liquids and gases. The litre is not a separate fundamental unit of the SI system, but is related to both length and mass; 1 litre is equivalent to 1 cubic decimetre (1 dm³).

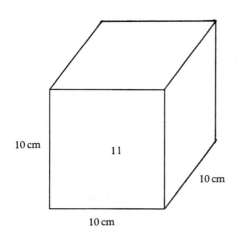

Volume of this cube = 10 cm × 10 cm × 10 cm
= 1000 cm³
= 1 dm × 1 dm × 1 dm
= 1 dm³
= 1 l

1 m³ is thus equivalent to 1000 l.
An even smaller unit of volume is the millilitre (ml) i.e. one thousandth of a litre

$$1\,\text{ml} = 1 \times 10^{-3}\,\text{l}$$

The millilitre is familiar to most people – for instance the standard spoon for medicines is 5 ml. One millilitre is equivalent to one cubic centimetre.

The litre is also related to mass, in that one litre of water weighs approximately 1 kg.

Mass
Although the kilogram is the basic unit for mass, the smaller unit gram (g) is used very commonly (1000 g = 1 kg). These units replace stones, pounds and ounces. The prefixes described previously are used in combination with gram, for example

milligram (mg) = 0.001 g

*microgram (μg) = 0.001 mg or 1×10^{-6} g

(*sometimes microgram is written mcg – this is incorrect and should not be used).

Strictly speaking, the weight of an object is a measure of force, i.e. the mass multiplied by the acceleration due to gravity. However, as in most measurements the acceleration due to gravity is constant, the weight of an object is usually expressed in kilograms, indicating its mass. This point is demonstrated by considering a person in outer space who is 'weightless' but still has a mass of, say, 75 kg.

Temperature
The SI unit for temperature is the kelvin (symbol K), but this is not used clinically: for ordinary use the *Celsius* scale is internationally recognized. On this scale water freezes at zero degrees Celsius (0°C) and boils at one hundred degrees Celsius (100°C). These units are frequently called degrees centigrade: strictly speaking, the term centigrade should not be used in SI units since in some countries it means something quite different – a measure of angle. The Celsius scale replaces the Fahrenheit scale, where water freezes at 32°F and boils at 212°F.

Absolute zero or 0 K is equal to − 273°C. A change of 1 K is identical to a change of 1°C. Thus 37°C is equal to 310 K as 37 + 273 = 310 K. Note that the symbol K is not preceded by a degree (°) symbol.

The amount and concentration of a substance

The SI unit for the amount of a substance is the mole. For most practical purposes one mole is the molecular weight of a substance in grams. There is one mole of molecules present for every 6.023×10^{23} molecules (an analogy to this is a ream of paper representing 480 sheets of paper).

Concentration refers to the quantity of substance present in a given volume, and is usually expressed in terms of moles and litres; for example a blood glucose concentration of 4.2 mmol/l, or 4.2 mmol l^{-1} or intravenous saline containing 150 mmol/l, or 150 mmol l^{-1}.

Proteins are of variable molecular weight and there are problems in calculating the molar concentrations, particularly when mixtures of proteins are present. Therefore it is usual to report protein concentration in grams per litre (g/l or g l^{-1}) rather than in moles per litre, e.g. a plasma albumin concentration of 40 g/l or 40 g l^{-1}. Values for enzymes are given in international units (abbreviated as U, IU, or i.u.) per litre and are a measure of enzyme activity. The values may vary from hospital to hospital if the assay procedures differ.

By convention, haemoglobin concentration is given in grams per decilitre, e.g. 14 g/dl. This is an internationally accepted inconsistency! Grams per decilitre is equivalent to grams per hundred millilitres; for example, 14 g/dl may be written as 14 g/100 ml.

Concentrations were often previously expressed in milliequivalents per litre (mEq/l), but these units have been replaced by the millimole.

$$\text{Number of equivalents (Eq)} = \frac{\text{weight in grams} \times \text{valency}}{\text{molecular weight}}$$

$$\text{cf. number of moles (mol)} = \frac{\text{weight in grams}}{\text{molecular weight}}$$

In the case of univalent ions (eg Na^+, K^+) the values will be numerically the same, i.e. a sodium concentration of 140 mEq/l becomes 140 mmol/l. For polyvalent ions (e.g. Ca^{2+}, Mg^{2+}) the old units are numerically divided by the valency i.e. a magnesium ion concentration of 2.0 mEq/l becomes 1.0 mmol/l.

Some values were previously expressed as mg/100 ml; therefore the method of conversion to mmol/l is to divide by the molecular weight (to convert from mg to mmol) and to multiply by 10 (to convert from 100 ml to a litre).

In some clinical areas now, pH is being expressed in nanomoles of hydrogen ions per litre

$$\text{i.e. pH } 7.4 = 40 \text{ nmol/l}$$

DERIVED UNITS

The seven basic units do not cover all the parameters that need units; therefore, other units, all derived from the basic units, have been given special names and symbols of their own.

Quantity/parameter	Name of derived unit	Symbol	In terms of other SI derived units
Work Energy Quantity of heat	joule	J	$N\,m$
Force	newton	N	$kg\,m\,s^{-2}$ $(kg\,m/s^2)$
Power	watt	W	$J\,s^{-1}$ (J/s)
Pressure	pascal	Pa	$N\,m^{-2}$ (N/m^2)
Frequency (for periodic phenomena – replaces cycles per second)	hertz	Hz	s^{-1} $(\ /s)$
Electrical Potential Potential difference Electromotive force	volt	V	$W\,A^{-1}$ (W/A)
Absorbed dose (ionizing radiation)	gray	Gy	$m^2\,s^{-2}$ (m^2/s^2)
Radionuclide activity (for measurements of numbers of nuclear transformations per second)	becquerel	Bq	s^{-1} $(\ /s)$

ADDITIONAL EXPLANATIONS FOR DERIVED UNITS

Pressure units

The SI unit for pressure is the pascal (Pa) and this replaces the Imperial unit of millimetres of mercury (mmHg). As the pascal is a very small unit of pressure, measurements are frequently given in kilopascals (kPa), i.e. 1000 Pa.

Medical science is inconsistent in its usage of pressure units. Blood gas pressure estimations are now often given in kilopascals; for example, an arterial pO_2 of 12.7 kPa rather than 95 mmHg. However, blood pressure recordings are still expressed in mmHg; for example, a blood pressure of 120/75 mmHg rather than 16/10 kPa.

The pressure in kilopascals = pressure in mmHg ÷ 7.5.

Energy units
The SI unit for all forms of energy is the joule (J) or commonly kilojoule (kJ). Thus the energy of food is usually measured in kJ and this replaces the Calorie (or kilocalorie), which was a measure of units of heat.

One Calorie is approximately 4.2 kJ. Thus a 2000 Calorie diet is equivalent to an 8400 kJ diet.

Approximation conversions

Parameter	Accepted clinical SI unit in common use	Imperial units	Approximate Conversions	
			From SI units	To SI units
Temperature	°C	°F	$(9/5°C) + 32 = °F$	$5/9(°F - 32) = °C$
Pressure	kPa	mmHg	kPa × 7.5 = mmHg	mmHg ÷ 7.5 = kPa
Volume	l (or dm^3)	pint	l ÷ 0.57 = pint	pint × 0.57 = l
	ml	fluid ounce	ml ÷ 28.4 = fl. oz	fl. oz × 28.4 = ml
Length	m	yard	m ÷ 0.91 = yard	yard × 0.91 = m
	mm	inch	mm ÷ 25.4 = inch	inch × 25.4 = mm
Mass	kg	pound	kg × 2.2 = lb	lb ÷ 2.2 = kg
	g	pound	g ÷ 454 = lb	lb × 454 = g
Concentration	mmol/l	mEq/l	see text	
Energy	kJ	Calorie or kilocalorie	kJ ÷ 4.2 = Cal	Cal × 4.2 = kJ

Appendix 2
Anatomical Terminology

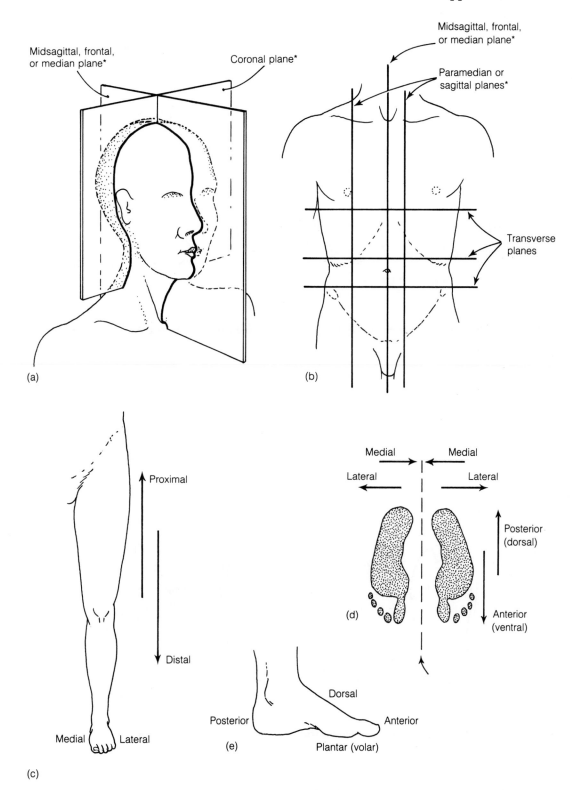

(a)

(b)

(c)

(d)

(e)

Figure App. 2.1 Diagrams illustrating anatomical directional terms.

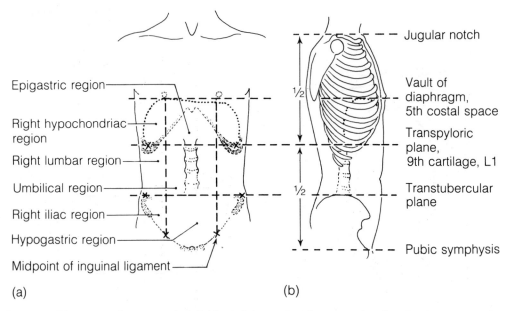

Figure App. 2.2 Diagrams to show anatomical divisions of the trunk and terms used to describe them. (a) anterior view; (b) lateral view

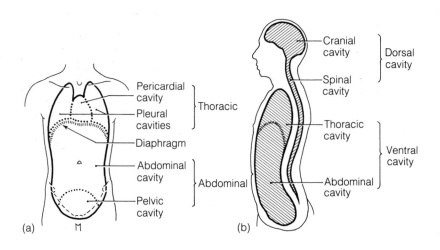

Figure App. 2.3 Diagrams to show cavities within the trunk. (a) anterior view; (b) lateral view

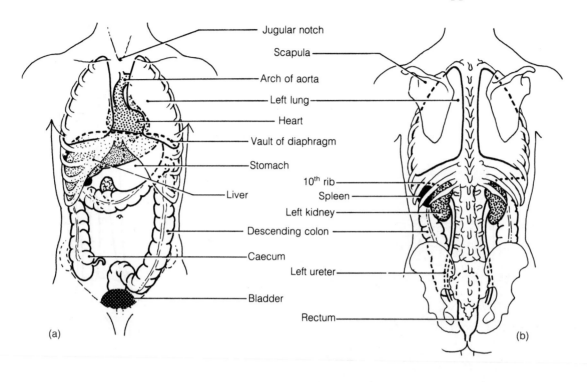

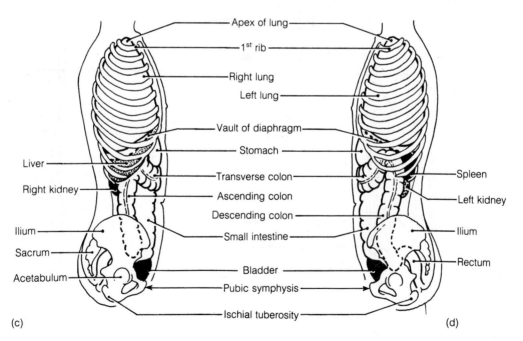

Figure App. 2.4 Diagrams to show anatomical relationship of organs within the trunk. (a) anterior view; (b) posterior view; (c) right lateral view; (d) left lateral view

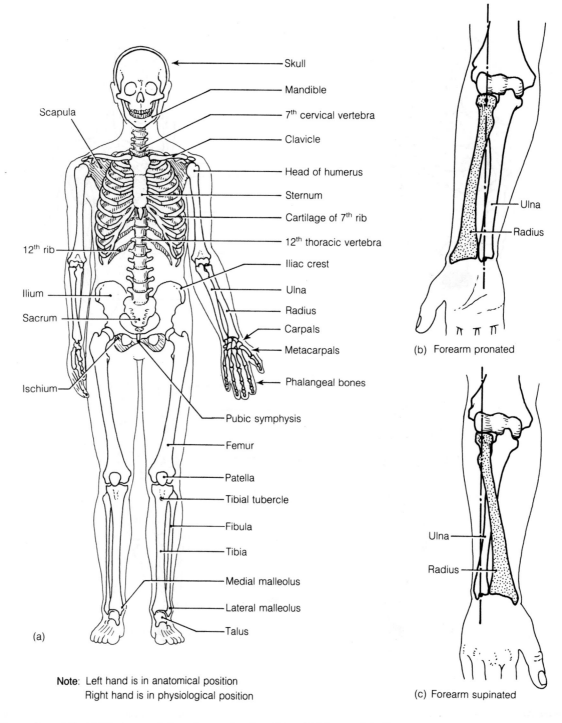

Skull
Mandible
7th cervical vertebra
Clavicle
Head of humerus
Sternum
Cartilage of 7th rib
12th thoracic vertebra
Iliac crest
Ulna
Radius
Carpals
Metacarpals
Phalangeal bones
Pubic symphysis
Femur
Patella
Tibial tubercle
Fibula
Tibia
Medial malleolus
Lateral malleolus
Talus

Scapula
12th rib
Ilium
Sacrum
Ischium

(a)

Note: Left hand is in anatomical position
Right hand is in physiological position

(b) Forearm pronated
Ulna
Radius

(c) Forearm supinated
Ulna
Radius

Figure App. 2.5   (a) The human skeleton in anterior view; (b) and (c) bones of the forearm

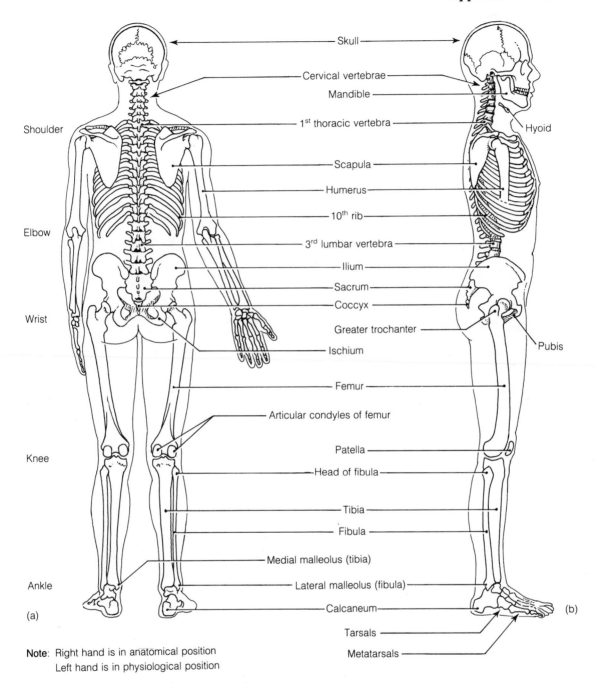

Note: Right hand is in anatomical position
      Left hand is in physiological position

Figure App. 2.6 The human skeleton. (a) posterior view; (b) right lateral view

Index

Page numbers in **bold type** refer to main sections in the text. Page numbers in *italics* refer to illustrations.

Motor cortex, *98*, *99*, 117–118, 120, **121**
  lesions, 122
  topographical organization, *121*
Motor end-plates, *see* Neuromuscular junctions
Motor homunculus, *121*
Motor nerve fibres, 26
  initiation of muscle contraction by, 211, 212
  myelinated alpha efferents, 207–208
  myelinated gamma efferents, 208
  non-myelinated autonomic efferents, 208
Motor neurones, 27, 47
  lower, 118, 212
    lesions of, 212
  upper, 118
    lesions of, 212, 221
Motor system
  central nervous system control of, **117–130**
  cerebellum, 117, 126–130
  extrapyramidal system, 117, 123–126
  pyramidal system, 117, 120–123
  spinal cord descending pathways, 118–120
  upper and lower motor neurones, 118
Motor units, 118, 208
Mountain sickness, 481
Mouth, **396–401**, *401*
  breathing through, 482–483
  defence mechanisms of, 573
  nursing assessment of, 401
  odour detected in, 399
  pain in, 437
Movements
  disorders of, 125–126, 129–130
  lever systems, 243–247
  at synovial joints, types of, 252–254
MSU (midstream specimen of urine), 541–542
Mucosa, gastrointestinal, *402*, 403
Mucus
  cervical, 621–622
  gastric secretion of, 407–408
  intestinal secretion of, 415, 419, 431
  respiratory tract, 474–476, *475*
Mucus-secreting cells, 403, 415, 419, 421, 474
Müllerian ducts, 603–604, *604*
Multinucleate cells, 10
Multiple sclerosis (MS), 39, 212
Mumps, 399
Murmurs, heart, 323
Muscarinic receptors, 43, 145
Muscle fibres
  intrafusal, 49
  necrosis of, 220
  skeletal, 206, *209*, 209–211
  smooth, *403*, 403
Muscle relaxant drugs, 214, 568

Muscle spindles, 49, *50*, 55, 60, 208
Muscles, 17, 26, **205–226**
  antagonists, 207
  atrophy of, 118, 206, 220, 223
  attachments (origin and insertion), 207
  biopsy, 221
  blood supply, 208
  cardiac, 26, **323–324**
  connective tissue, *207*, 207
  contraction, 211–219, 324
    electrical events, 213
    energy supply, 219
    excitation-contraction coupling, 211, 216–217, 324
    isometric, 218
    isotonic, 218
    recruitment, 219
    sliding filament hypothesis, 211
    stimulation of motor nerves, 211, 212
    tonic, 218
    twitch, 218–219
    types of, 217–219
    see also Neuromuscular transmission
  contracture deformities, 206, 208, 223–224, 239, 259, 261
  cramps, 224
  development at puberty, 624
  disorders of, *see* Myopathies
  fasciculations, 118
  fasciculi, *206*, 206, 403
  function, 205
  glycogen in, 170, 172, 219
  gross anatomy, 205–208
  hypertrophy, 206, 220, 223
  microanatomy, 209–211
  necrosis, 220
  nerve supply, 207–208, *208*
  pain, 224
  prime movers, 206–207
  pseudohypertrophy, 220
  regeneration of, 220–221
  rigidity, 125, 126, 218
  sensory receptors in, 56, 59, 60, 208
  smooth, *403*, 403–404
  spasms, 224, 260
  stiffness, 223
  synergistic, 207
  tone, 218
  transverse tubular system (T-system), 210, *210*, 216, 324
  venous return and, 360
  wasting of, 118
  weakness and fatigue, 222–223
Muscular dystrophies
  degenerative changes in, 220
  Duchenne type, 216–217, 220, 222, 223
Muscularis, gastrointestinal, *402*, 403–404

Muscularis mucosae, 403
Mutations, 15
Myasthenia gravis, **214–215**, 222, 223, 224
Myelencephalon, 89
Myelin sheath, 27
Myelinated neurones, 27
  propagation of action potentials along, 37, *38*
Myelination, 90
Myelitis, 103
Myeloma, multiple, 594
Myenteric plexus, 402, 404
Myenteric reflex, 402
Myocardial contractility, 349
Myocardial infarction, 154, 300, **327–329**, 369
  cardiogenic shock in, 329
  diagnosis, 328
  muscle damage caused by, 327–328, *328*
  serum enzymes after, 272, 328, 456
  sexual intercourse after, 629
  treatment, 328–329
Myocardium, 320
Myoclonic jerks, 134
Myofibrils, 9, 209, *209*
  abnormalities in, 220
Myogenic reflex, 290
Myoglobin, 210–211, 429
Myometrium, 613
Myopathies, **219–225**
  degenerative changes in, 220
  effects on activities of daily living, 222
  investigations in, 221
  medical assessment, 52, 221
  planning nursing care, 222–225
  primary, 219
  secondary, 219
  signs of, 221
Myopia (short-sightedness), 65
Myosin, 209, *209*, 211, 216
Myositis, 219
Myotonia, 216, 223
Myotonia congenita, 220, 223
Myotonia dystrophica, 216, 222, 223
Myotubular myopathy, 222
Myxoedema, *see* Hypothyroidism

Nails, 560, 564
Narcotic analgesics, 433, 435
Nasal cavity, 473, 477
  defence mechanisms, 574
Natriuretic hormone, 532
Nausea, **412**, 635
Near-point, visual, 65
Negative feedback, 162–163
Neocerebellum, 127, *128*
Neonates, *see* Newborn infants
Neostigmine, 153, 214, 215, 401